Lecture Notes in Computer Scie

T0238495

Commenced Publication in 1973
Founding and Former Series Editors:
Gerhard Goos, Juris Hartmanis, and Jan van Leeuwen

Juan A. Garay   Rosario Gennaro (Eds.)

Advances in Cryptology – CRYPTO 2014

34th Annual Cryptology Conference
Santa Barbara, CA, USA, August 17-21, 2014
Proceedings, Part I

 Springer

Volume Editors

Juan A. Garay
Yahoo Labs
701 First Avenue
Sunnyvale, CA 94089, USA
E-mail: garay@yahoo-inc.com

Rosario Gennaro
The City College of New York
160 Convent Avenue
New York, NY 10031, USA
E-mail: rosario@cs.ccny.cuny.edu

ISSN 0302-9743 e-ISSN 1611-3349
ISBN 978-3-662-44370-5 e-ISBN 978-3-662-44371-2
DOI 10.1007/978-3-662-44371-2
Springer Heidelberg New York Dordrecht London

Library of Congress Control Number: 2014944726

LNCS Sublibrary: SL 4 – Security and Cryptology

Typesetting: Camera-ready by author, data conversion by Scientific Publishing Services, Chennai, India

Printed on acid-free paper

Springer is part of Springer Science+Business Media (www.springer.com)

Preface

CRYPTO 2014, the 34rd Annual International Cryptology Conference, was held August 17–21, 2014, on the campus of the University of California, Santa Barbara. The event was sponsored by the International Association for Cryptologic Research (IACR) in cooperation with the UCSB Computer Science Department.

The program represents the recent significant advances and trends in all areas of cryptology. Out of 227 submissions, 60 were included in the program; these two-volume proceedings contains the revised versions of all the papers. Two of the papers shared a single presentation slot in the program. The program also included two invited talks. On Monday, Mihir Bellare from UCSD delivered the IACR Distinguished Lecture, entitled "Caught in Between Theory and Practice." On Wednesday, Yael Tauman Kalai from Microsoft Research New England spoke about "How to Delegate Computations: The Power of No-Signalling Proofs." As usual, the rump session took place on Tuesday evening, and was chaired by Dan Bernstein and Tanja Lange.

This year's program continued the trend started last year of trying to accommodate as many high-quality submissions as possible, yielding a high number of accepted papers. As a result, sessions were also held on Tuesday and Thursday afternoons, and presentations were kept short (20 minutes per paper, including questions and answers). The option of having parallel sessions, which would allow for longer presentations and an early adjournment on Thursday, was also discussed and decided against, since we assessed that our research field is still sufficiently homogeneous and the community would benefit from the option of attending all the talks. However, we believe that future Program Committees should continue to explore possible options to implement some form of parallel sessions.

The submissions were reviewed by a Program Committee (PC) consisting of 38 leading researchers in the field, in addition to the two co-chairs. Each PC member was allowed to submit one paper, plus an additional one if co-authored with a junior researcher (a student or a postdoc). PC-authored submissions were held to higher standards during the review process. Papers were reviewed in a double-blind fashion. Initially, each paper was assigned to three reviewers (four for PC-authored papers); during the discussion phase, when necessary, extra reviews were solicited. The process also included a rebuttal phase after preliminary reviews were finalized, where authors received them and were given the option to comment on the reviews within a window of several days. The authors' comments were then taken into account in the discussions within the PC and the final reviews. Despite being labor-intensive, we feel the rebuttal phase was a worthwhile process as it resulted in the significantly better understanding of many submissions. As part of the discussion phase, the PC held a 1.5-day in-person meeting on May 15 and 16 in Copenhagen, Denmark, right after Eurocrypt.

We would like to sincerely thank the authors of all submissions—those whose papers made it into the program and those whose papers did not. Our deep appreciation also goes out to the PC members, who invested an extraordinaty amount of time in reviewing papers, interacting with the authors via the rebuttal mechanism, and participating in so many discussions on papers, their contribution, and the state of the art in their areas of expertise. We also sympathize with the occasional frustration from seeing decisions go against personal recommendations and preferences, in spite of all the hard work.

We are also indebted to the many external reviewers who significantly contributed to the comprehensive evaluation of the submissions. A list of PC members and external reviewers appears after this note. Despite all our efforts, the list of external reviewers may contain errors or omissions; we apologize for that in advance.

We would like to thank Sasha Boldyreva, the general chair, for working closely with us throughout the whole process and providing the much needed support at every step, including artfully creating and maintaining the website and taking care of all aspects of the conference's logistics—especially the in-person PC meeting arrangements.

As always, special thanks are due to Shai Halevi for his tireless support regarding the *websubrev* software, which we used for the whole conference planning and operation, including paper submission and evaluation and interaction among PC members and with the authors. Alfred Hofmann and his colleagues at Springer provided a meticulous service for the timely production of these proceedings.

Finally, we would like to thank Google, Microsoft Research, and the National Science Foundation for their generous support.

August 2014

Juan A. Garay
Rosario Gennaro

CRYPTO 2014

The 34rd International Cryptology Conference

Sponsored by the *International Association for Cryptologic Research*

General Chair

Alexandra Boldyreva   Georgia Institute of Technology, USA

Program Co-Chairs

Juan A. Garay   Yahoo Labs, USA
Rosario Gennaro   The City College of New York – CUNY, USA

Program Committee

Yevgeniy Dodis   New York University, USA
Orr Dunkelman   University of Haifa, Israel
Serge Fehr   CWI, The Netherlands
Pierre-Alain Fouque   Université Rennes I, France
Craig Gentry   IBM Research, USA
Vipul Goyal   MSR India
Nadia Heninger   University of Pennsylvania, USA
Thomas Holenstein   ETH, Switzerland
Yuval Ishai   Technion, Israel
Dimitar Jetchev   EPFL, Switzerland
Aggelos Kiayias   University of Athens, Greece
Kaoru Kurosawa   Ibaraki University, Japan
Alexander May   Ruhr-Universität Bochum, Germany
Ilya Mironov   MSR, USA
Payman Mohassel   University of Calgary, Canada
Jörn Müller-Quade   Karlruhe Institute of Technology, Germany
María Naya-Plasencia   Inria Paris-Rocquencourt, France
Claudio Orlandi   Aarhus University, Denmark
Rafael Pass   Cornell University, USA
Christopher Peikert   Georgia Institute of Technology, USA
Krzysztof Pietrzak   Institute of Science and Technology, Austria
Leonid Reyzin   Boston University, USA
Ron Rivest   MIT, USA

Amit Sahai	UCLA, USA
Gil Segev	Hebrew University, USA
Elaine Shi	University of Maryland, USA
Tom Shrimpton	Portland State University, USA
Alice Silverberg	UC Irvine, USA
Marc Stevens	CWI, The Netherlands
Katsuyuki Takashima	Mitsubishi Electric, Japan
Stefano Tessaro	UC Santa Barbara, USA
Vinod Vaikuntanathan	MIT, USA
Gilles Van Assche	STMicroelectronics, Belgium
Muthu Venkitasubramanian	University of Rochester, USA
Ivan Visconti	University of Salerno, Italy
Bogdan Warinschi	University of Bristol, UK
Brent Waters	UT Austin, USA
Vassilis Zikas	ETH, Switzerland

External Reviewers

Michel Abdalla
Masayuki Abe
Arash Afshar
Divesh Aggarwal
Martin Albrecht
Joel Alwen
Scott Ames
Prabhanjan Ananth
Daniel Apon
George Argyros
Gilad Asharov
Nuttapong Attrapadung
Christian Badertscher
Abhishek Banerjee
Carsten Baum
Amos Beimel
Mihir Bellare
David Bernhard
Dan Bernstein
Guido Bertoni
Raghav Bhaskar
Joppe Bos
Elette Boyle
Brandon Broadnax
Christina Brzuska
Ran Canetti

Anne Canteaut
Ignacio Cascudo
David Cash
Dario Catalano
Andr Chailloux
Nishanth Chandran
Jie Chen
Cheng Chen
Céline Chevalier
Kai-Min Chung
Aloni Cohen
Henry Cohn
Sandro Coretti
Jean-Sebastien Coron
Craig Costello
Dana Dachman-Soled
Joan Daemen
Ivan Damgård
Bernardo David
Gregory Demay
Yi Deng
Itai Dinur
Nico Doettling
Rafael Dowsley
Chandan Dubey
Alexandre Duc

Leo Ducas
Alina Dudeanu
Markus Duermuth
Frédéric Dupuis
Aner Ben Efraim
Xiong Fan
Antonio Faonio
Sebastian Faust
Dario Fiore
Marc Fischlin
Georg Fuchsbauer
Benjamin Fuller
Jun Furukawa
Steven Galbraith
Nicolas Gama
Chaya Ganesh
Peter Gaži
Ran Gelles
Essam Ghadafi
Sasha Golovnev
Sergey Gorbunov
Dov Gordon
Robert Granger
Jens Groth
Divya Gupta
Tim Gneysu

Shai Halevi
Sean Hallgren
Moritz Hardt
Brett Hemenway
Yan Huang
Jan Hazla
William Skeith III
Vincenzo Iovino
Takashi Ito
Ioana Ivan
Tibor Jager
Abhishek Jain
David Jao
Stanislaw Jarecki
Mahavir Jhawar
Antoine Joux
Marc Joye
Yael Kalai
Seny Kamara
Jean-Gabriel Kammerer
Pierre Karpman
Jonathan Katz
Yutaka Kawai
Nathan Keller
Dakshita Khurana
Eike Kiltz
Thorsten Kleinjung
Vlad Kolesnikov
Venkata Koppula
Daniel Kraschewski
Hugo Krawczyk
Sara Krehbiel
Abishek
   Kumarasubramaniam
Ranjit Kumaresan
Robin Künzler
Tanja Lange
Gregor Leander
Nikos Leonardos
Anthony Leverrier
Kevin Lewi
Allison Bishop Lewko
Benoit Libert
Huijia (Rachel) Lin
Yehuda Lindell

Feng-Hao Liu
Adriana Lopez-Alt
Steve Lu
Stefan Lucks
Atul Luykx
Vadim Lyubashevsky
Mohammad Mahmoody
Hemanta Maji
Alex Malozemoff
Mohammad Mammody
Christian Matt
Daniele Micciancio
Andrea Miele
Eric Miles
Andrew Miller
Brice Minaud
Toru Nakanishi
Jesper Buus Nielsen
Valeria Nikolaenko
Tobias Nilges
Ryo Nishimaki
Adam O'Neill
Wakaha Ogata
Cristina Onete
Pascal Paillier
Omkant Pandey
Omer Paneth
Dimitris Papadopoulos
Charalampos
   Papamanthou
Sunoo Park
Anat
   Paskin-Cherniavsky
Valerio Pastro
Kenny Paterson
Michal Peeters
Ludovic Perret
Christophe Petit
Le Trieu Phong
Stefano Pironio
Manoj Prabhakaran
Ananth Raghunathan
Kim Ramchen
Vanishree Rao
Pavel Raykov

Mariana Raykova
Christian Rechberger
Oded Regev
Thomas Ristenpart
Ben Riva
Mike Rosulek
Aaron Roth
Yannis Rouselakis
saeed Sadeghian
Yusuke Sakai
Katerina Samari
Alessandra Scafuro
Christian Schaffner
Thomas Schneider
Lior Seeman
Nicolas Sendrier
Karn Seth
Yannick Seurin
Barak Shani
Nigel Smart
Ben Smith
Florian Speelman
François-Xavier
   Standaert
Damien Stehlé
John Steinberger
Noah
   Stephens-Davidowitz
Mario Strefler
Takeshi Sugawara
Koutarou Suzuki
Björn Tackmann
Qiang Tang
Sidharth Telang
Aris Tentes
Isamu Teranishi
R. Seth Terashima
Abhradeep Guha
   Thakurta
Justin Thaler
Emmanuel Thom
Mehdi Tibouchi
Jean-Pierre Tillich
Joana Treger
Roberto Trifiletti

Eran Tromer
Yiannis Tselekounis
Hoang Viet Tung
Dominique Unruh
Berkant Ustaoglu
Prashant Vasudevan
Thomas Vidick

Dhinakaran
   Vinayagamurthy
Akshay Wadia
Gaven Watson
Hoeteck Wee
Daniel Wichs
Shota Yamada

Kazuki Yoneyama
Thomas Zacharias
Hila Zarosim
Mark Zhandry
Bingsheng Zhang
Hong-Sheng Zhou
Jens Zumbrägel

Table of Contents – Part I

Symmetric Encryption and PRFs

Formal Methods

Hash Functions

Side Channels and Leakage Resilience I

Obfuscation I

FHE

Table of Contents – Part II

Composable Security

Secure Computation – Foundations

Secure Computation – Implementations

Security of Symmetric Encryption against Mass Surveillance

Mihir Bellare[1], Kenneth G. Paterson[2], and Phillip Rogaway[3]

[1] Dept. of Computer Science and Engineering,
University of California San Diego, USA
cseweb.ucsd.edu/~mihir
[2] Information Security Group, Royal Holloway, University of London, UK
www.isg.rhul.ac.uk/~kp
[3] Dept. of Computer Science, University of California Davis, USA
www.cs.ucdavis.edu/~rogaway

Abstract. Motivated by revelations concerning population-wide surveillance of encrypted communications, we formalize and investigate the resistance of symmetric encryption schemes to mass surveillance. The focus is on algorithm-substitution attacks (ASAs), where a subverted encryption algorithm replaces the real one. We assume that the goal of "big brother" is undetectable subversion, meaning that ciphertexts produced by the subverted encryption algorithm should reveal plaintexts to big brother yet be indistinguishable to users from those produced by the real encryption scheme. We formalize security notions to capture this goal and then offer both attacks and defenses. In the first category we show that successful (from the point of view of big brother) ASAs may be mounted on a large class of common symmetric encryption schemes. In the second category we show how to design symmetric encryption schemes that avoid such attacks and meet our notion of security. The lesson that emerges is the danger of choice: randomized, stateless schemes are subject to attack while deterministic, stateful ones are not.

1 Introduction

OVERVIEW. This paper is about the troubling possibility of mass surveillance by *algorithm-substitution attack* (ASA). Suppose that encryption scheme $\Pi = (\mathcal{K}, \mathcal{E}, \mathcal{D})$ is to be implemented in closed-source software—think, for example, of implementing the CBC-AES encryption underlying the TLS record layer within Microsoft's Internet Explorer or Apple's Safari browsers, or in corresponding server-side code. An ASA replaces the executable code for the desired encryption algorithm $\mathcal{E}$ with, for example, the code of an NSA-authored alternative $\tilde{\mathcal{E}}$.

ASAs have been discussed before, under various names, in particular falling under the banner of *kleptography*. This prescient idea was developed by Young and Yung starting in the 1990s [27,28]. While some cryptographers seem to have dismissed kleptography as far-fetched, recent revelations suggest this attitude to

J.A. Garay and R. Gennaro (Eds.): CRYPTO 2014, Part I, LNCS 8616, pp. 1–19, 2014.

be naïve [1]. ASAs may well be going on today, possibly on a massive scale. In this light we aim to provide a *formal* and *practical* treatment of ASAs, with a focus on *symmetric encryption*, an attractive target for real-world attacks. Building on, yet going further than, prior work, we fully and formally define security goals. We then come at ASAs from both ends, showing on the one hand how successful (from the point of view of big brother) ASAs may be mounted on standard schemes, and showing on the other hand how to design schemes that provably resist them. Our findings surface what we call *the danger of choice*: the trend towards flexibility and open-ended choices in protocols, often present for vendor flexibility or political compromise, works against us with regard to protection against ASAs, which are best defeated by stateful, deterministic encryption that curtails randomness and choice.

MODEL AND DEFINITIONS.   The real encryption algorithm $\mathcal{E}$ takes, as usual, user key K, message M, and associated data A. It returns a ciphertext C. The subverted algorithm $\widetilde{\mathcal{E}}$ that substitutes for $\mathcal{E}$ takes the same inputs but also an additional, big-brother key, $\widetilde{K}$. It also returns a ciphertext.

With no restrictions on $\widetilde{\mathcal{E}}$, there would appear to be no hope of security, for $\widetilde{\mathcal{E}}$ can fold K into the ciphertext, say encrypted under $\widetilde{K}$, and big brother can use $\widetilde{K}$ to recover K. However, such an attack would be detected by users, who would see that ciphertexts fail to decrypt normally. Big brother aims to achieve compromise without detection: subverted ciphertexts should look like real ones, yet enable recovery of K or M. ASAs, in this view, live in a tension between detectability and success, the former working to curtail the latter. We will formally define metrics of both detectability and success.

We will require that ciphertexts produced by $\widetilde{\mathcal{E}}$ decrypt normally under the decryption algorithm $\mathcal{D}$ of the base scheme. This decryptability condition is the most basic form of undetectability. But we expect that big brother will aim to evade more sophisticated forms of detection. We formalize detection security as requiring that real and subverted ciphertexts are indistinguishable even to a test that knows some users' keys but does not know $\widetilde{K}$.

Success refers to big brother's ability to obtain knowledge about user data from subverted ciphertexts. Certainly an ASA allowing big brother to recover the user key K from any ciphertext is successful, but for positive results (defeating big brother) we want more. We formalize surveillance security as the requirement that big brother, even with its key $\widetilde{K}$, cannot differentiate real ciphertexts from subverted ones.

The duality between detection and surveillance security is reflected in our formalizations. Both require indistinguishability of real and subverted ciphertexts to an adversary, the difference being that in detection the adversary knows the user keys but not the big-brother key, and in surveillance it's the other way around. We remark that, in both cases, our formalizations are multi-user, meaning there are many users (but a single subverter).

MOUNTING ASAs.   We show that most symmetric encryption schemes succumb to damaging ASAs. Our attacks recover the user key K from subverted ciphertexts while remaining undetectable. These attacks apply to base schemes that are randomized and stateless. Building on [9], we first describe what we call IV-replacement attacks, where the initial vector in a blockcipher mode of operation is used to communicate to big brother an encryption under $\widetilde{K}$ of the user key K. Then we describe a more general ASA that we call the biased-ciphertext attack. This makes few assumptions on the structure of the base scheme and succeeds by creating ciphertexts that are not distributed quite like real ones. They are biased in a way that reveals bits of the user key to a holder of $\widetilde{K}$, but we show that the bias is undetectable without knowledge of $\widetilde{K}$. The difficulty here is showing undetectability even for tests that know the user key K, and for the analysis we prove an information-theoretic lemma about biased functions. Beyond presenting generic attacks [4], we discuss how encryption in SSL/TLS, IPsec, and SSH can be subverted by these means. The conclusion is that randomized, stateless schemes, including deployed ones, invariably fall to even generic ASAs.

DEFEATING ASAs. We aim to build symmetric encryption schemes that resist ASAs, meaning achieve surveillance security in the formal sense we define. Given the above, such schemes need to be stateful and deterministic. But not every such scheme works. The difficulty with provably achieving surveillance security is that standard security properties of the base scheme, such as its privacy or authenticity, are of no particular use towards the new goal. The reason is that these properties rely on the adversary not knowing the key K. But in the surveillance setting, the subverted ciphertexts are being created by an algorithm, $\widetilde{\mathcal{E}}$, that knows K, and can thus compromise privacy or authenticity to make subverted ciphertexts look different from real ones, and in a way useful to big brother. Nonetheless, we show that security is achievable by relying on combinatorial properties of the scheme. We define what it means for a base symmetric encryption scheme to have *unique ciphertexts* and then show that every unique-ciphertext scheme meeting the decryptability condition is secure against ASAs. This provides a strong anti-surveillance guarantee: no ASA will succeed in differentiating real from subverted ciphertexts, let alone recovering the message or a user's key. We show this assuming only minimal undetectability—decryptability, meaning that subverted ciphertexts must remain decryptable by the decryption algorithm of the base scheme.

To realize concrete benefits from this general result, we need to find unique-ciphertext symmetric encryption schemes. Here we give a simple construction based on a variable-input-length PRP. In [4], we present a more practical result, showing how any nonce-based symmetric encryption scheme [22,23] may be transformed into a unique ciphertext stateful deterministic scheme while preserving efficiency. Using existing nonce-based encryption schemes like CCM, GCM, or OCB, this yields practical designs of surveillance-resistant symmetric encryption.

ASYMMETRIC ASAs. For simplicity, our main definitions only capture the case in which big brother embeds a symmetric key $\widetilde{K}$ into subverted software. It is obviously useful to replace this with a public key, the corresponding secret key being held by big brother, so that reverse engineering of a subverted encryption algorithm will not confer the capabilities that big brother aims to keep to itself. The necessary definitional extensions, which are small, are described in [4].

SCOPE. Our paper is deliberately of restricted scope: we consider ASAs only for symmetric encryption schemes. In reality, encryption schemes are deployed as part of larger cryptographic protocols and these protocols will afford additional opportunities for algorithmic subversion. To pick one example, a protocol might involve the transmission of a nonce for authentication purposes during a key-exchange phase. This nonce could be chosen so as to directly leak an ensuing session key. Or it could be chosen to leak the internal state of a back-doored PRNG, indirectly revealing future session keys. This technique has been posited as a subversion method for SSL/TLS [7].

Our scope also means that we exclude subversion attempts that exploit side-channels in implementations. For example, our model does not capture timing information, so attacks in which the encryption key is leaked through fine-grained timing behaviour of the encryption algorithm fall outside our notions. Big brother's subverted $\tilde{\mathcal{E}}$ could stutter the times at which ciphertexts or their blocks are produced; this might be sufficient to build a covert channel with adequate bandwidth to convey the session key. Such timing approaches have been used to infer information about user keystrokes over SSH connections [25].

The limitations on scope imply that our positive security results are certainly not definitive in terms of eliminating all subversion possibilities for a symmetric encryption scheme deployed within a real-world system. Still, a limited scope has merit. First, symmetric encryption is fundamental to secure communications, so it's important to study this primitive's susceptibility to subversion. Second, our model fits well within the scenario where an agency subverts encryption software, like a crypto library, rather than a particular protocol built on that library. Third, the positive results we provide, showing that ASAs on certain schemes are impossible, confine big brother to other avenues of attack, which may be less attractive. Finally, we aim to lay foundational results, in the modern, provable-security style, that can be built upon by succeeding researchers to broaden the scope of surveillance-resistant protocols to include tasks such as authenticated key exchange. It should eventually be possible to have a corpus of protocols, and even system-level code analysis, to provide strong guarantees on the ineffectiveness ASAs.

THE DANGER OF CHOICE. The characteristic of modern encryption schemes that makes ASAs possible is the freedom-of-choice routinely provided by protocols, as well as the unverifiability of mandated randomness. Consider a symmetric encryption scheme that requires a user to select a 128-bit IV. The specification might say that the IV *should* be chosen uniformly at random, or it might even say that it *must* be so chosen. But, either way, the black-box behavior of the encryption scheme will never reveal if uniform random bits were used. Because of this, there is no way to ensure that the IV is not selected in a manner that will covertly communicate a session key to an agency engaged in mass surveillance—which we exploit in our IV-replacement attack. Similarly, if a scheme permits variable-length padding there will be no way to ensure that the amount of padding is not used as a covert channel to transmit a user's key.

The ultimate conclusion of this paper is that unverifiable algorithmic choice can be a significant liability. We have in some sense come full-circle. In their classical paper on probabilistic encryption [10], Goldwasser and Micali explained the danger of deterministic public-key encryption: leaking that one ciphertext is the repetition of another, or allowing a ciphertext to be decrypted by trial-encryption. But these threats can be eliminated without the use of probabilism—namely, through the use of state. For the most conventional setting in symmetric encryption—realizing a reliable, encrypted channel—ASAs provide one motivation for deterministic, stateful schemes, for sender and receiver both. We believe that there are further benefits to such schemes, including improved utility for software testing and the elimination of any need, post key-generation, to harvest unpredictable random bits.

RELATED WORK. Young and Yung have developed an extensive body of work on what they call *kleptography*, beginning with [27,28]. This concerns the deliberate subversion of cryptosystems to provide backdoor capabilities; our work is a special case. While much of their work has focused on the public-key setting, Young and Yung have also considered attacks on protocols like Kerberos, and developed blockciphers containing backdoors for the black-box setting (ie, where the code of the blockcipher is not made available for inspection) [29,31,30]. In the light of recent revelations, we contend that kleptography deserves to play a larger role in the future development of our field. Additional work on back-doored blockciphers can be found in [21,19,20]. This entire line of work has focused on building schemes with deliberately-inserted and hard-to-detect backdoors. By contrast, we also provide positive results, constructing schemes that are provably hard to subvert.

Goh, Boneh, Pinkas and Golle [9] consider the problem of adding key recovery to the SSL/TLS and SSH protocols. Some passages of this 2003 paper now sound prophetic: *The government can convince major software vendors to distribute SSL/TLS or SSH2 implementations with hidden and unfilterable key recovery. ... Users will not notice the key recovery mechanism because the scheme is hidden.* [9, Section 2.2]. Goh et al. suggest that when the server needs a random nonce, it can use in its place an encryption of the session key computed under the escrow key. We build on this idea to consider more general classes of attack on symmetric encryption schemes.

The problem of inserting backdoors and key-recovery defects into cryptographic schemes is closely related to the topic of *subliminal channels*, whose extensive literature begins with [24] and the study of covert channels [17]. There is a similarly extensive body of work on the exploitation, measurement, and elimination of timing side channels, both in cryptographic and non-cryptographic settings, with representative examples including [6,15].

FURTHER REMARKS. We posed our initial question in the context of closed-source software. However the sheer complexity of cryptographic libraries like OpenSSL, and the small number of experts who review such code, makes it plausible that ASAs might be carried out against open-source software. Note too that even when code appears to be "clean," there's always the possibility of code being subverted at compilation or run time, by subverting the compiler or interpreter [26]. And there's certainly the possibility of performing ASAs on hardware-based cryptography, a prospect rendered all the easier by the widespread use of countermeasures *intended* to shield algorithmic internals from inspection.

We do not know if ASAs are among the techniques used to make TLS-encrypted traffic available under warrantless surveillance [1]. We offer no empirical evidence in this direction. We hope that other researchers are seeking it out, which is necessary for understanding the actual nature of our communication infrastructure.

2   Preliminaries

NOTATION. A string means a member of $\{0,1\}^*$, and $\perp \notin \{0,1\}^*$ denotes a special symbol standing for "invalid" or "reject." If S is a set then $x \leftarrow S$ denotes sampling x uniformly at random from S.

SYNTAX. Our syntax for symmetric encryption encompasses encryption that is probabilistic, deterministic, or stateful; and decryption that is deterministic or stateful. We allow associated data (AD), in order that our basic syntax encompass this practically-important component of authenticated encryption.

A scheme for *symmetric encryption* is a triple $\Pi = (\mathcal{K}, \mathcal{E}, \mathcal{D})$. The *key space* $\mathcal{K}$ is a finite nonempty set. The *encryption algorithm* $\mathcal{E}$ is a possibly randomized algorithm that maps a four-tuple of strings K, M, A, σ to a pair of strings $(C, \sigma') \leftarrow \mathcal{E}(K, M, A, \sigma)$. The arguments to $\mathcal{E}$ represent the key, message (plaintext), associated data and current state. The output consists of the ciphertext C and revised state σ'. The *decryption algorithm* $\mathcal{D}$ is a deterministic algorithm that maps a four-tuple of strings (K, C, A, σ) to a pair of strings $(M, \sigma') \leftarrow \mathcal{D}(K, C, A, \sigma)$.

Algorithms $\mathcal{E}$ and $\mathcal{D}$ are said to *reject* if they return a pair with first component of $\perp$, and to *accept* otherwise. We may write $\mathcal{E}_K(M, A, \sigma)$ and $\mathcal{D}_K(C, A, \sigma)$ for $\mathcal{E}(K, M, A, \sigma)$ and $\mathcal{D}(K, C, A, \sigma)$, respectively. We adopt the convention that $\mathcal{E}$ and $\mathcal{D}$ return $(\perp, \perp)$ if any argument is $\perp$. In addition, whether or not $C_i = \perp$ is allowed to depend only on $|M_1|, |A_1|, \ldots, |M_{i-1}|$, and $|A_{i-1}|$. This eliminates pointless degeneracies.

We say that $\mathcal{E}$ is stateless if the second component of any output of $\mathcal{E}$ on any inputs is ε, and likewise for $\mathcal{D}$. We say that Π is stateless if both $\mathcal{E}$ and $\mathcal{D}$ are stateless. In this case, we drop the second component of the output of both algorithms, so that $\mathcal{E}$ now returns just a ciphertext and $\mathcal{D}$ just a message. We also drop the last (state) input to $\mathcal{D}$ and, for $\mathcal{E}$, think of it as the coins of the algorithm, dropping which is regarded as having the coins being chosen at random. In this way, when Π is stateless, we recover the conventional syntax.

It is well understood that encryption must be stateful or probabilistic to achieve IND-CPA privacy and decryption must be stateful to avoid replay attacks. Our work will show that decryption must be stateful to avoid algorithm-substitution attacks.

CORRECTNESS. We say that $\Pi = (\mathcal{K}, \mathcal{E}, \mathcal{D})$ is *correct*, or meets the correctness condition, if, when the sender encrypts a sequence of messages and the receiver decrypts the resulting sequence of ciphertexts in order, the receiver will get back what the sender started with. To be clear what this means in our current stateful context, we now proceed more formally. Saying that encryption scheme $\Pi = (\mathcal{K}, \mathcal{E}, \mathcal{D})$ is correct means that for all q, all $M_1, \ldots, M_q \in \{0,1\}^*$ and all $A_1, \ldots, A_q \in \{0,1\}^*$, the following game returns true with probability zero:

$\sigma_0, \tau_0 \leftarrow \varepsilon$
For $i = 1, \ldots, q$ do $(C_i, \sigma_i) \leftarrow \mathcal{E}(K, M_i, A_i, \sigma_{i-1})$; $(M_i', \tau_i) \leftarrow \mathcal{D}(K, C_i, A_i, \tau_{i-1})$
Return $((\forall i : C_i \neq \bot)$ and $(\exists i : M_i \neq M_i'))$

We will only consider schemes that are correct in this sense.

SECURITY NOTIONS. We recall a standard notion of privacy for symmetric encryption [2,3,22]. Let $\Pi = (\mathcal{K}, \mathcal{E}, \mathcal{D})$ be a symmetric encryption scheme and let $\mathscr{A}$ be an adversary. Consider the following game:

Game $\mathrm{PRIV}_{\Pi}^{\mathscr{A}}$	$\mathrm{ENC}(M, A)$
$K \leftarrow \mathcal{K}$; $\sigma \leftarrow \varepsilon$; $b \leftarrow \{0,1\}$	If $b = 1$ then $(C, \sigma) \leftarrow \mathcal{E}(K, M, A, \sigma)$
$b' \leftarrow \mathscr{A}^{\mathrm{ENC}}$; Return $(b = b')$	Else $(C, \sigma) \leftarrow \mathcal{E}(K, 0^{|M|}, A, \sigma)$
	Return C

Let $\mathbf{Adv}_{\Pi}^{\mathrm{priv}}(\mathscr{A}) = 2\Pr[\mathrm{PRIV}_{\Pi}^{\mathscr{A}} \Rightarrow \mathrm{true}] - 1$ be the privacy advantage of adversary $\mathscr{A}$. Positive results will provide schemes secure in this sense and also resistant to surveillance as we will define in Section 3.

3 Subverting Encryption

We now ask what it would mean for a symmetric encryption scheme $\Pi = (\mathcal{K}, \mathcal{E}, \mathcal{D})$ to fall to an algorithm substitution attack (ASA). An attacker $\mathscr{B}$ (for "big brother") wants to subvert an encryption scheme *en masse*. We assume it is able to arrange that subverted encryption code $\widetilde{\mathcal{E}}_{\widetilde{K}}$ is used in place of $\mathcal{E}$. (The subscript indicates that a key $\widetilde{K}$ chosen by $\mathscr{B}$ may be embedded in the code.) $\mathscr{B}$ wants its subversion to be successful and yet undetected. The former means that from observing only ciphertexts computed under the subverted algorithm, $\mathscr{B}$ can compromise privacy. (For example, it can, using $\widetilde{K}$, efficiently recover the plaintexts underlying the ciphertexts.) This captures the relevant attack scenario where $\mathscr{B}$ is able, through mass surveillance of network traffic, to intercept bulk ciphertexts at will. The latter means that the subverted encryption algorithm should produce ciphertexts that look alright. The most basic form of the latter requirement is that they correctly decrypt under the decryption algorithm $\mathcal{D}$ of the base scheme, but we expect that big brother would prefer to evade even more sophisticated attempts at detection.

One can consider subverting an encryption scheme's privacy, authenticity, or both. One can also consider subversion for public-key schemes or for other cryptographic goals, like key exchange. There are possibilities for algorithm-substitution attacks (ASAs) in all these settings. Here we limit the scope to subversion aimed at compromising the privacy of a symmetric encryption scheme. The extensions to cover additional schemes is an obvious and important target for future research.

SUBVERSIONS. Let $\Pi = (\mathcal{K}, \mathcal{E}, \mathcal{D})$ be a symmetric encryption scheme. A *subversion* of Π is a triple $\widetilde{\Pi} = (\widetilde{\mathcal{K}}, \widetilde{\mathcal{E}}, \widetilde{\mathcal{D}})$. The *master-key space* $\widetilde{\mathcal{K}}$ is a finite nonempty set. The *subverted encryption algorithm* $\widetilde{\mathcal{E}}$ is a (possibly randomized) algorithm that maps a six-tuple of strings $(\widetilde{K}, K, M, A, \sigma, i)$ to a pair of strings (C, σ'). Here σ and σ' are the current and updated states, respectively, indicating that $\widetilde{\mathcal{E}}$ may be stateful. The input i represents some public information identifying a user encrypting under K and is assumed different for all keys. Such information is usually available in a system, perhaps a MAC address or an IP address, and we allow $\widetilde{\mathcal{E}}$ to take it as input because we cannot realistically disallow a subverter from having or using such information.

The *plaintext-recovery algorithm* $\widetilde{\mathcal{D}}$ takes $\widetilde{K}, \boldsymbol{C}, \boldsymbol{A}, i$ where $\boldsymbol{C}$ is a vector of ciphertexts, $\boldsymbol{A}$ is a vector of associated data and i is again the identity associated to the key K whose usage is being subverted. The algorithm attempts to produce a vector of corresponding plaintexts $\boldsymbol{M}$. How effectively it does this will vary. For example, the plaintext-recovery algorithm $\widetilde{\mathcal{D}}$ may *always* find the plaintext, for every ciphertext in the list, regardless of the length of the list. Or it may effectively perform a key recovery attack first, then simply decrypt the ciphertexts, but require many ciphertexts. In describing the severity of a practical ASA, we will explicitly specify $\widetilde{\mathcal{D}}$ and quantify how good a job it does—a break that always finds the plaintext, or something else. For defining our security notion, however, we will ignore $\widetilde{\mathcal{D}}$, for the very strong notion we shall give implies the inexistence of any practical plaintext-recovery algorithm $\widetilde{\mathcal{D}}$.

DECRYPTABILITY. We say that $\widetilde{\Pi} = (\widetilde{\mathcal{K}}, \widetilde{\mathcal{E}}, \widetilde{\mathcal{D}})$ satisfies the *decryptability condition* relative to $\Pi = (\mathcal{K}, \mathcal{E}, \mathcal{D})$ if $(\widetilde{\mathcal{K}} \times \mathcal{K}, \widetilde{\mathcal{E}}, \mathcal{D}')$ is a correct encryption scheme where $\mathcal{D}'$ is defined by $\mathcal{D}'((\widetilde{K}, K), C, A, \sigma) = \mathcal{D}(K, C, A, \sigma)$. Thus, although algorithm $\widetilde{\mathcal{E}}$ operates on a key $(\widetilde{K}, K)$ different from the key K of the base scheme Π, a party possessing only K can decrypt $\widetilde{\mathcal{E}}$-encrypted plaintexts using the legitimate decryption algorithm $\mathcal{D}$. This represents the most basic form of resistance to detection, and we will assume any subversion must meet it.

DETECTION ADVANTAGE. By detectability, we refer to the ability of ordinary users—they know their secret keys, but not the master key—to tell, from the ciphertexts, if encryption is happening by the real or subverted algorithm. In the absence of any detectability condition, subversion is always possible. The decryptability condition we gave above embodies a particularly basic form of detection, in that failure to meet this condition is likely to lead to detection. However, we expect that big brother wants to evade not just this, but more

Game $\mathrm{DETECT}^{\mathcal{U}}_{\Pi,\widetilde{\Pi}}$	Game $\mathrm{SURV}^{\mathcal{B}}_{\Pi,\widetilde{\Pi}}$
$b \leftarrow \{0,1\};\ \widetilde{K} \leftarrow \widetilde{\mathcal{K}};\ b' \leftarrow \mathcal{U}^{\mathrm{KEY,ENC}}$	$b \leftarrow \{0,1\};\ \widetilde{K} \leftarrow \widetilde{\mathcal{K}};\ b' \leftarrow \mathcal{B}^{\mathrm{KEY,ENC}}(\widetilde{K})$
Return $(b = b')$	Return $(b = b')$
$\underline{\mathrm{KEY}(i)}$	$\underline{\mathrm{KEY}(i)}$
If $(K_i = \bot)$ then $K_i \leftarrow \mathcal{K};\ \sigma_i \leftarrow \varepsilon$	If $(K_i = \bot)$ then $K_i \leftarrow \mathcal{K};\ \sigma_i \leftarrow \varepsilon$
Return K_i	Return ε
$\underline{\mathrm{ENC}\,(M, A, i)}$	$\underline{\mathrm{ENC}\,(M, A, i)}$
If $(K_i = \bot)$ then return $\bot$	If $(K_i = \bot)$ then return $\bot$
If $(b = 1)$ then $(C, \sigma_i) \leftarrow \mathcal{E}(K_i, M, A, \sigma_i)$	If $(b = 1)$ then $(C, \sigma_i) \leftarrow \mathcal{E}(K_i, M, A, \sigma_i)$
Else $(C, \sigma_i) \leftarrow \widetilde{\mathcal{E}}(\widetilde{K}, K_i, M, A, \sigma_i, i)$	Else $(C, \sigma_i) \leftarrow \widetilde{\mathcal{E}}(\widetilde{K}, K_i, M, A, \sigma_i, i)$
Return C	Return C

Fig. 1. Games used to define detection and surveillance security of subversion $\widetilde{\Pi} = (\widetilde{\mathcal{K}}, \widetilde{\mathcal{E}}, \widetilde{\mathcal{D}})$ of encryption scheme $\Pi = (\mathcal{K}, \mathcal{E}, \mathcal{D})$

sophisticated forms of detection. We now define what it means to do so. Let $\Pi = (\mathcal{K}, \mathcal{E}, \mathcal{D})$ be an encryption scheme and let $\widetilde{\Pi} = (\widetilde{\mathcal{K}}, \widetilde{\mathcal{E}}, \widetilde{\mathcal{D}})$ be a subversion of it. Let $\mathcal{U}$ be an algorithm representing a detection test being run by users. Let

$$\mathbf{Adv}^{\mathrm{det}}_{\Pi,\widetilde{\Pi}}(\mathcal{U}) \;=\; 2\Pr[\mathrm{DETECT}^{\mathcal{U}}_{\Pi,\widetilde{\Pi}} \Rightarrow \mathsf{true}] - 1$$

where game DETECT is shown on the left of Fig. 1. This measures the ability of test $\mathcal{U}$ to detect an ASA. In this game, $\mathcal{U}$ must detect whether it receives ciphertexts produced by $\mathcal{E}$ or by $\widetilde{\mathcal{E}}$. Via oracle KEY the test $\mathcal{U}$ can obtain keys, reflecting that users may use their own keys in detection. The test of course does not have access to the subversion key $\widetilde{K}$. A subversion $\widetilde{\Pi}$ in which this advantage is negligible for all practical tests $\mathcal{U}$ is said to be *undetectable* and would be one that evades detection in a powerful way. If such a subversion permitted plaintext recovery, big brother would consider it a very successful one. Attacks we will present in Section 4 show that such subversion is possible for a broad class of schemes Π.

We emphasize that the above definition captures the *users'* inability to know which encryption scheme is being used, the real one or the subverted one, even if it knows the private underlying keys. The adversary $\mathcal{U}$ in this setting might be regarded as the good guys—the population of users intent on seeing if they are all being surveilled based on the input/output behavior of the encryption code. We note that even if the detection advantage above is large, it is not clear that users would actually be able to detect subversion: for one thing, they probably wouldn't know what to look for. Thus detection advantage is only interesting when, for a scheme, it is demonstrably small. In that case big-brother has effectively forced detection to work by way of reverse-engineering the subverted code, not by looking at its black-box behavior.

SURVEILLANCE ADVANTAGE. Now we want to define what it means for a scheme Π to resist, meaning be secure against, ASAs. The first thought is to ask that big brother, even given its subversion key $\widetilde{K}$, cannot recover the plaintexts underlying subverted ciphertexts. We ask for something stronger, namely that big brother, even given $\widetilde{K}$, cannot tell whether ciphertexts are being produced by the real encryption algorithm $\mathcal{E}$ or by the subverted algorithm $\widetilde{\mathcal{E}}$. Formally let $\Pi = (\mathcal{K}, \mathcal{E}, \mathcal{D})$ be an encryption scheme and let $\widetilde{\Pi} = (\widetilde{\mathcal{K}}, \widetilde{\mathcal{E}}, \widetilde{\mathcal{D}})$ be a subversion of it. Let $\mathscr{B}$ be an adversary representing big brother. Let

$$\mathbf{Adv}^{\mathrm{srv}}_{\Pi, \widetilde{\Pi}}(\mathscr{B}) \ = \ 2\Pr[\mathrm{SURV}^{\mathscr{B}}_{\Pi, \widetilde{\Pi}} \Rightarrow \mathsf{true}] - 1$$

where game SURV is shown on the right of Fig. 1. In the game, adversary $\mathscr{B}$ is given the subversion key $\widetilde{K}$, but is not given user keys $K_1, K_2, \ldots$. (We remark that the SURV and DETECT games are very similar, effectively duals of each other, the ENC oracle in particular being the same. The difference is that in the former the adversary gets $\widetilde{K}$ but not $K_1, K_2, \ldots$ while in the latter it is the other way around.) For Π to be secure against surveillance requires that this advantage is small for all subversions $\widetilde{\Pi}$ of Π and all $\mathscr{B}$. This is the desired notion for *positive* results, and we will present schemes secure in this sense in Section 5. (We will assume minimal detection security in the form of the decryptability condition. Without some resistance to detection, surveillance security is not possible.) In offering a scheme secure in this sense we are asserting that big-brother can't come close to achieving surveillance *en masse*.

We have formulated surveillance security with multiple users, but a hybrid argument shows that the advantage relative to the one-user game can grow by at most a factor of the number of users. We will use this result to simplify proofs, which will restrict attention to the game with a single user. We remark that a similar claim is not true for detection security.

4   Mounting ASAs

This section shows that typical randomized, stateless encryption schemes are subvertible. We first describe an attack on modes of operation that surface their IV. Then we describe what we call a universal attack, so named because it applies regardless of the specifics of the scheme being attacked. In [4] we explain to what extent such attacks are applicable to the most important secure communications protocols for the Internet, namely SSL/TLS, IPsec, and SSH.

4.1   IV-Replacement Attacks

Following Young and Yung [28], Goh, Boneh, Pinkas and Golle [9] consider the problem of adding a hidden key recovery to protocols. They suggest that when the server needs a random nonce, it can use in its place an encryption of the session key computed under the escrow key. We expand on this idea, letting the escrow key be the subversion key. We show how to subvert stateless encryption schemes that put a random nonce into the ciphertext.

We consider randomized, stateless schemes $\Pi = (\mathcal{K}, \mathcal{E}, \mathcal{D})$, writing $C \leftarrow \mathcal{E}(K, M, A; IV)$, where we now surface the randomness input IV (for initial vector, IV) to the encryption algorithm and suppress the state input. Such a scheme is said to surface its IV if there is an efficient algorithm $\mathcal{X}$ such that $\mathcal{X}(\mathcal{E}(K, M, A; IV)) = IV$ for all K, M, A, IV. The condition says that $\mathcal{X}$ can recover the IV from the ciphertext. A simple example of a scheme that surfaces its IV is CBC\$, namely CBC mode with random IV. Another example is CTR\$, counter mode with random starting point.

The first requirement of a subversion attack is undetectability, but other attributes are relevant too. We will describe two attacks.

STATEFUL ATTACK. This is the simplest attack, in which the IV is simply replaced by an encipherment, under the subversion key $\widetilde{\mathcal{K}}$, of the encryption key K. For simplicity of presentation, we assume that the IV length and key length are the same. (The attack extends easily to accommodate cases where the key length is greater than the IV length.) In order to prevent repeated IVs being seen across ciphertexts, we must limit the IV substitution to one ciphertext. This necessitates the use of a stateful subversion scheme. To avoid this repetition, one might consider replacing the IV by the encryption of K under a randomized symmetric encryption scheme that is IND\$-CPA secure, but, since this encryption will usually be longer than the IV and thus cannot replace the IV in a single ciphertext, we would need to adopt a stateful approach to implement it too.

In more detail, let the bit length of the IV and key be n and assume we have a blockcipher $E \colon \widetilde{\mathcal{K}} \times \{0,1\}^n \to \{0,1\}^n$ with block length n. The subversion of Π is the triple $\widetilde{\Pi} = (\widetilde{\mathcal{K}}, \widetilde{\mathcal{E}}, \widetilde{\mathcal{D}})$ where:

$\underline{\widetilde{\mathcal{E}}(\widetilde{K}, K, M, A, \sigma, i)}$	$\underline{\widetilde{\mathcal{D}}(\widetilde{K}, \boldsymbol{C}, \boldsymbol{A}, i)}$
If $\sigma = 0$ then $IV \leftarrow E(\widetilde{K}, K)$	$IV \leftarrow \mathcal{X}(\boldsymbol{C}[1])$
Else $IV \twoheadleftarrow \{0,1\}^n$	$K \leftarrow E^{-1}(\widetilde{K}, IV)$
$C \leftarrow \mathcal{E}(K, M, A, IV)$	$\boldsymbol{M}[1] \leftarrow \mathcal{D}(K, \boldsymbol{C}[1], \boldsymbol{A}[1])$
$\sigma \leftarrow \sigma + 1$; Return C	Return $\boldsymbol{M}$

The state σ maintained by $\widetilde{\mathcal{E}}$ is an integer initialized at 0. When the state has this initial value, $\widetilde{\mathcal{E}}$ sets the IV to an encryption of the key K, and otherwise performs no subversion, picking the IV at random. Now assume user i has requested an encryption of a message $\boldsymbol{M}[1]$ under associated data $\boldsymbol{A}[1]$ with $\sigma = 0$, resulting in ciphertext $\boldsymbol{C}[1] = \widetilde{\mathcal{E}}(\widetilde{K}, K, \boldsymbol{M}[1], \boldsymbol{A}[1], 0, i)$. The subverter's decryption algorithm gets input $\widetilde{K}$ together with i and the length-one vectors $\boldsymbol{C}, \boldsymbol{A}$, and recovers the key K as shown. Once obtained, the key can be used to decrypt not only the current but any future ciphertexts.

This subversion $\widetilde{\Pi}$ meets the decryptability condition. Furthermore, as long as E is a PRP/PRF, the subverted IV is indistinguishable from a random one, even to an observer that knows K (the observer does not know $\widetilde{K}$), making the subversion undetectable. Formally:

Theorem 1. *Let $\Pi = (\{0,1\}^n, \mathcal{E}, \mathcal{D})$ be a randomized, stateless symmetric encryption scheme that surfaces an IV of length n. Let $E\colon \widetilde{\mathcal{K}} \times \{0,1\}^n \to \{0,1\}^n$ be a blockcipher. Let the subversion $\widetilde{\Pi} = (\widetilde{\mathcal{K}}, \widetilde{\mathcal{E}}, \widetilde{\mathcal{D}})$ of Π be defined as above. Let $\mathcal{U}$ be a test that makes q queries to its KEY oracle. Then we can construct an adversary $\mathcal{A}$ such that $\mathbf{Adv}^{\mathrm{det}}_{\Pi,\widetilde{\Pi}}(\mathcal{U}) \leq q^2/2^n + \mathbf{Adv}^{\mathrm{prf}}_E(\mathcal{A})$. Adversary $\mathcal{A}$ makes q oracle queries and its running time is that of $\mathcal{U}$.*

The $q^2/2^n$ term corresponds to the chance that two users have the same key, in which case their subverted IVs will be the same while the real ones would be random and independent.

Suppose, however, that a user system, and hence the state of $\widetilde{\mathcal{E}}$, is reset. Then the subverted IV will be recreated and the observer detects a repeated IV, something not likely to happen in the absence of the subversion (though plausibly explainable as a randomness failure). This reduces the effectiveness of this simple attack. One solution to this problem is to adopt the above-mentioned idea of replacing the IV by the encryption of K under a randomized symmetric encryption scheme. This would result in a subversion $(\widetilde{\mathcal{K}}, \widetilde{\mathcal{E}}, \widetilde{\mathcal{D}})$ that is both randomized *and* stateful. This subversion would have the practical advantage of being able to continuously leak the key K, rather than relying on big brother to intercept ciphertext $\boldsymbol{C}[1]$. In our next attack, we present a subversion that preserves this property and only requires randomisation.

STATELESS ATTACK. We present an attack where $\widetilde{\mathcal{E}}$ is stateless. In this attack the subversion is undetectable even under resets of the encryptor system, making the attack harder to detect in practice. Let k be the key length of Π and let $v = \lceil \log_2(k) \rceil$. (For example if $k = 128$ as for AES then $v = 7$.) Let $E\colon \widetilde{\mathcal{K}} \times \{0,1\}^n \to \{0,1\}^n$ be a blockcipher where n is the length of the IV of Π as before. The subversion of Π is the triple $\widetilde{\Pi} = (\widetilde{\mathcal{K}}, \widetilde{\mathcal{E}}, \widetilde{\mathcal{D}})$ where:

$\underline{\widetilde{\mathcal{E}}(\widetilde{K}, K, M, A, i)}$	$\underline{\widetilde{\mathcal{D}}(\widetilde{K}, \boldsymbol{C}, \boldsymbol{A}, i)}$
$\ell \leftarrow [1..k]$	For $j = 1, \ldots, \lvert \boldsymbol{C} \rvert$ do
$R \leftarrow \{0,1\}^{n-v-1}$	$\quad b \lVert \ell \rVert R \leftarrow E^{-1}(\widetilde{K}, \mathcal{X}(\boldsymbol{C}[j])); \; K'[\ell] \leftarrow b$
$IV \leftarrow E(\widetilde{K}, K[\ell] \lVert \ell \rVert R)$	For $j = 1, \ldots, \lvert \boldsymbol{C} \rvert$ do
$C \leftarrow \mathcal{E}(K, M, A, IV)$	$\quad \boldsymbol{M}[j] \leftarrow \mathcal{D}(K', \boldsymbol{C}[j], \boldsymbol{A}[j])$
Return C	Return $\boldsymbol{M}$

In computing $E(\widetilde{K}, K[\ell] \lVert \ell \rVert R)$ the integer ℓ is encoded as a v-bit string. After around $k \ln(k)$ encryptions, we expect that every $\ell \in [1..k]$ has been chosen at least once, so that if a vector of this many ciphertexts is passed to $\widetilde{\mathcal{D}}$, the latter will succeed. Undetectability again follows if E is a PRP/PRF, exploiting the fact that the observer does not know $\widetilde{K}$:

Theorem 2. *Let $\Pi = (\{0,1\}^k, \mathcal{E}, \mathcal{D})$ be a randomized, stateless symmetric encryption scheme that surfaces an IV of length n. Let $E\colon \widetilde{\mathcal{K}} \times \{0,1\}^n \to \{0,1\}^n$ be a blockcipher. Let $v = \lceil \log_2(k) \rceil$. Let the subversion $\widetilde{\Pi} = (\widetilde{\mathcal{K}}, \widetilde{\mathcal{E}}, \widetilde{\mathcal{D}})$ of Π be defined as above. Let $\mathcal{U}$ be a test that makes q queries to its ENC oracle.*

Then we can construct an adversary $\mathscr{A}$ such that $\mathbf{Adv}^{\mathrm{det}}_{\Pi,\widetilde{\Pi}}(\mathscr{U}) \leq q^2/2^{n-v-1} +$ $\mathbf{Adv}^{\mathrm{prf}}_{E}(\mathscr{A})$. It makes q oracle queries and its running time is that of $\mathscr{U}$.

This subversion achieves an even stronger form of undetectability than Theorem 2 captures. Since the subversion is stateless, reset of the system does not lead to detection. (It is assumed that the subvertor has access to fresh coins at every invocation. If a reset results in re-use of coins, our claim would no longer be true.) The subversion obviously extends to one leaking more than bit of K per ciphertext, at the cost of a weaker bound on detection advantage.

4.2   The Biased-Ciphertext Attack

The above IV-replacement attacks apply to several common modes in their "textbook" form and to some of their deployments in Internet protocols, but there are many encryption schemes to which they do not apply. These include schemes that do not surface the IV, for example encrypted-IV schemes like CBC2 [23], IACBC [14] and XCBC$ [8].

In this section we present a more general attack that we call the biased ciphertext attack. This attack is "universal" in that it applies to any randomized and stateless encryption scheme $\Pi = (\mathcal{K}, \mathcal{E}, \mathcal{D})$ that uses a minimal amount of randomness, say 7 bits. Undetectability holds in a strong form, namely even under reset of the state of the subverter.

Suppose the user asks its system to use this scheme to encrypt a message M with key K and associated data A, which means that the system is expected to pick coins δ at random from the space D of coins for $\mathcal{E}$ and return ciphertext $C \leftarrow \mathcal{E}(K, M, A; \delta)$ (where we now replace IV by δ to emphasise the fact that δ may not be surfaced). Our subverted encryption algorithm will compute C the same way, except that δ will not be chosen quite at random. Instead, it will be chosen to ensure that $F(\widetilde{K}, C) = K[j]$ is the j-bit of the key, where F is a PRF. The subverter decryption algorithm, on receiving C, will recompute $K[j]$ as $F(\widetilde{K}, C)$. The counter j will be maintained by the subverter algorithms in their state, so that over $|K|$ encryptions, the entire key is leaked. The challenge here is showing that the bias created in the distribution of C is not detectable, *even given the key K*. Exploiting PRF security, we can move to a setting where $F(\widetilde{K}, \cdot)$ is replaced by a random function. Then we use an information-theoretic argument to show that the statistical distance between the real and subverted ciphertexts is small even given K. In terms of our formal definitions, big brother is undetectable.

We highlight the following features of the attack. First, big brother does not pick, or care, what messages or associated data is encrypted – this is no chosen-message attack. Big brother will succeed no matter what the user chooses to encrypt, as long as it encrypts $|K|$ or more messages. Second, the attack does not merely distinguish between real and subverted ciphertexts; rather, it recovers the encryption key. Although presented as a key recovery attack, it is not hard to see that, in terms of our formal definitions, big brother has surveillance advantage close to 1.

Let us say that Π is *coin injective* if the mapping of coins to ciphertext, for each fixed key, message and associated data, is injective. The analysis in our current proof of undetectability requires that Π have this property. The assumption is not particularly restrictive. Schemes that surface their IV are coin injective, not just the ones to which the IV-replacement attack applies, but also ones like OCB with random nonce that, as we indicated, were harder to handle. Schemes that encrypt the IV are also coin injective and thus covered. More generally, our analysis applies when the mapping is not injective but is regular.

Proceeding, suppose $g \colon D \to R$ where $D \subseteq \{0,1\}^*$, and $f \colon \{0,1\}^* \to \{0,1\}$. For $b \in \{0,1\}$ we let $S^{f,g}(b,D) = \{\delta \in D \colon f(g(\delta)) = b\}$. Here think of g as taking coins δ and returning an encryption under them, the key, message, and associated data being fixed as part of g. Let $F \colon \widetilde{\mathcal{K}} \times \{0,1\}^* \to \{0,1\}$ be a PRF that returns a bit. The subversion of Π is the triple $\widetilde{\Pi} = (\widetilde{\mathcal{K}}, \widetilde{\mathcal{E}}, \widetilde{\mathcal{D}})$ where:

$\widetilde{\mathcal{E}}(\widetilde{K}, K, M, A, \sigma, i)$	$\widetilde{\mathcal{D}}(\widetilde{K}, \boldsymbol{C}, \boldsymbol{A}, i)$
$j \leftarrow \sigma \bmod \lvert K \rvert; \; j \leftarrow j+1$	For $j = 1, \ldots, \lvert \boldsymbol{C} \rvert$ do
$g(\cdot) \leftarrow \mathcal{E}(K, M, A; \cdot) \lVert \sigma \rVert i$	$\quad K'[j] \leftarrow F(\widetilde{K}, \boldsymbol{C}[j] \lVert j-1 \rVert i)$
$\delta \leftarrow S^{F(\widetilde{K}, \cdot), g(\cdot)}(K[j], D)$	For $j = 1, \ldots, \lvert \boldsymbol{C} \rvert$ do
$C \leftarrow \mathcal{E}(K, M, A; \delta)$	$\quad \boldsymbol{M}[j] \leftarrow \mathcal{D}(K', \boldsymbol{C}[j], \boldsymbol{A}[j])$
$\sigma \leftarrow \sigma + 1; \; \text{Return } C$	Return $\boldsymbol{M}$

The state σ maintained by $\widetilde{\mathcal{E}}$ is an integer, initially zero. Encryption lets g be the function that has K, M, A, j, σ, i hardwired and on input coins δ in the space D of coins of $\mathcal{E}$, returns $\mathcal{E}(K, M, A; \delta) \lVert \sigma \rVert i$, the last two components ensuring no collisions in output values of the function across different users and states. Picking δ at random from the indicated set means that the ciphertext $C = \mathcal{E}(K, M, A; \delta)$ will satisfy $F(\widetilde{K}, C \lVert j-1 \rVert i) = K[j]$, except with some probability of error when the set is empty.

Let $k = \lvert K \rvert$. Now assume that user i has requested encryptions of messages $\boldsymbol{M}[1], \ldots, \boldsymbol{M}[k]$ under associated data $\boldsymbol{A}[1], \ldots, \boldsymbol{A}[k]$, respectively, to result in ciphertexts $\boldsymbol{C}[1], \ldots, \boldsymbol{C}[k]$, created via $\boldsymbol{C}[j] = \widetilde{\mathcal{E}}(\widetilde{K}, K, \boldsymbol{M}[j], \boldsymbol{A}[j], j-1, i)$ for $j = 1, \ldots, k$. The big-brother decryption algorithm gets input $\widetilde{K}, \boldsymbol{C}, \boldsymbol{A}, i$ and recovers the key K' as shown. It then decrypts under the true decryption algorithm to return the corresponding vector of messages. Except in the case of an error, the event $K \neq K'$ whose probability we will bound below, not only does decryption succeed, but the process does more, recovering the key, and once this is done the key can be stored and further ciphertexts decrypted directly.

The error probability of the key recovery attack is at most $e_1 + \cdots + e_k$ where $e_j = \Pr[K'[j] \neq K[j]] = \Pr[S^{F(\widetilde{K}, \cdot), g(\cdot)}(K[j], D) = \emptyset]$. Assuming F is a good PRF, our estimate can be made with a random function f in its place. Due to the inclusion of $\sigma \lVert i$ in the argument to f, the applications of f are independent. Assuming g is injective, each time, the set has chance 2^{-d} to be empty where $d = \lvert D \rvert$, so the error probability is at most $k 2^{-d}$. This is small as long as the scheme uses a minimal amount of randomness, for example 7 bits, resulting in $d = 2^7 = 128$. (A randomized mode will typically use 96–128 bits of randomness, in which case the error probability is entirely negligible.) A similar analysis can be carried out for the formal surveillance attack.

We claim that the subversion is undetectable. Our analysis first uses the PRF security of F to replace $F(\widetilde{K}, \cdot)$ with a random function f. The key claim is then the following information theoretic lemma. The proof is in [4].

Lemma 1. *Suppose* $g\colon D \to R$. *Let* $b \in \{0,1\}$ *and* $\overline{\delta} \in D$. *Let* $d = |D|$. *Let* $p = \Pr[\delta = \overline{\delta}]$ *where we first draw* $f\colon g(D) \to \{0,1\}$ *at random and then draw* δ *at random from* $S^{f,g}(b, D) = \{\delta \in D\colon f(g(\delta)) = b\}$.
(1)   *If* g *is injective then* $p = (1 - 2^{-d})/d$.
(2)   *More generally, if* g *is* k-*regular, then* $p = (1 - 2^{-d/k})/d$.

We use this lemma to estimate the undetectability of the subversion:

Theorem 3. *Let* $\Pi = (\mathcal{K}, \mathcal{E}, \mathcal{D})$ *be a randomized, stateless, coin-injective symmetric encryption scheme with randomness-length* r, *and let* $d = 2^r$. *Let* $F\colon \widetilde{\mathcal{K}} \times \{0,1\}^* \to \{0,1\}$ *be a PRF. Let the subversion* $\widetilde{\Pi} = (\widetilde{\mathcal{K}}, \widetilde{\mathcal{E}}, \widetilde{\mathcal{D}})$ *of* Π *be defined as above. Let* $\mathcal{U}$ *be a test that makes* q *queries to its* ENC *oracle. Then we can construct an adversary* $\mathcal{A}$ *such that* $\mathbf{Adv}^{\mathrm{det}}_{\Pi, \widetilde{\Pi}}(\mathcal{U}) \leq q/2^d + \mathbf{Adv}^{\mathrm{prf}}_F(\mathcal{A})$. *Adversary* $\mathcal{A}$ *makes* q *oracle queries and its running time is that of* $\mathcal{U}$.

So again as long as the scheme uses a non-trivial amount of randomness, for example $r \geq 7$ bits resulting in $d \geq 128$, Theorem 3 implies that the subversion is undetectable. The proof makes crucial use of Lemma 1, which, letting $D = \{0,1\}^r$ be the space of coins of $\mathcal{E}$, implies that the statistical distance between the real and subverted ciphertexts is 2^{-d}. A reset of the state will lead to increased detection ability for an observer, but if Π draws its coins from a reasonably large space, this increase does not appear to be enough to lead to actual detection. However the attack continues to be randomized, so if a system reset results in re-use of entropy, detection becomes possible.

5   Defeating ASAs

We turn to finding schemes that resist ASAs. Given the results of Section 4, such schemes must be deterministic and stateful. But not any such scheme works. The challenge here is that security properties of a scheme, such as privacy and authenticity, are of no evident use in showing resistance to ASAs, for these properties hold relative to adversaries that do not know the key K, while in the surveillance game, the subverted encryption algorithm has the key K. Thus surveillance security will rely on combinatorial properties of the scheme. We pinpoint one such property, defining what it means for a symmetric encryption scheme to have unique ciphertexts. We then show that any such scheme is surveillance-resistant. We then present some designs of unique-ciphertext, and thus surveillance-secure, schemes.

UNIQUE CIPHERTEXTS. Let $\Pi = (\mathcal{K}, \mathcal{E}, \mathcal{D})$ be a symmetric encryption scheme. For any possible state τ of $\mathcal{D}$ with respect to key K, any message $M \in \{0,1\}^*$ and any associated data $A \in \{0,1\}^*$, let $\mathcal{C}_\Pi(K, M, A, \tau)$ be the set of all ciphertexts C such that $\mathcal{D}(K, C, A, \tau)$ accepts with message M, meaning its output is (M, τ') for some τ'. We say that Π has unique ciphertexts if the set $\mathcal{C}_\Pi(K, M, A, \tau)$ has size at most one for all K, M, A, τ. This means that, for any given key, message, associated data and state, there exists at most one ciphertext that the decryptor will decrypt to the message in question.

Due to the correctness condition, any unique-ciphertext scheme is deterministic. The converse is not true, meaning Π being deterministic does not necessarily mean it has unique ciphertexts. If Π is deterministic there is only one ciphertext an honest encryptor will produce given a particular key, message, associated data and state, but determinism does not ensure that there is not some other ciphertext that the decryptor will decrypt to the same message. As an analogy, the difference is the same as between deterministic and unique signature schemes [11,16].

SURVEILLANCE-SECURITY. The following says that a unique-ciphertext scheme cannot be subverted without violating the decryptability condition. The proof is in [4].

Theorem 4. *Let* $\Pi = (\mathcal{K}, \mathcal{E}, \mathcal{D})$ *be a unique ciphertext symmetric encryption scheme. Let* $\widetilde{\Pi} = (\widetilde{\mathcal{K}}, \widetilde{\mathcal{E}}, \widetilde{\mathcal{D}})$ *be a subversion of* Π *that obeys the decryptability condition relative to* Π. *Let* $\mathcal{B}$ *be an adversary. Then* $\mathbf{Adv}^{\mathrm{srv}}_{\Pi, \widetilde{\Pi}}(\mathcal{B}) = 0$.

A UNIQUE-CIPHERTEXT SCHEME. We give an example of a symmetric encryption scheme that has unique ciphertexts and hence, by Theorem 4, is not subvertible. Our scheme is based on the encode-then-encipher paradigm of [5] which we extend to allow associated data. Let $P\colon \{0,1\}^k \times \{0,1\}^* \to \{0,1\}^*$ be a family of permutations. By P^{-1} we denote the inverse of P, satisfying $P_K^{-1}(P_K(x)) = x$ for all $x \in \{0,1\}^*$. We also let $F\colon \{0,1\}^k \times \{0,1\}^* \to \{0,1\}^t$ be a family of functions. (It will be used as a MAC.) The state σ in our scheme will be a counter, and we denote by $\langle\sigma\rangle$ its representation as a ℓ-bit string. Our symmetric encryption scheme $\Pi = (\mathcal{K}, \mathcal{E}, \mathcal{D})$ has key space $\mathcal{K} = \{0,1\}^{2k}$ and encryption and decryption algorithms defined as follows:

$\mathcal{E}(K, M, A, \sigma)$	$\mathcal{D}(K, C, A, \tau)$
If $(\sigma = 2^\ell)$ then return $(\perp, \sigma)$	If $(\tau = 2^\ell)$ then return $(\perp, \tau)$
$K_1 \| K_2 \leftarrow K$	$K_1 \| K_2 \leftarrow K$; $(W, T) \leftarrow C$; $x \leftarrow P^{-1}(K_1, W)$
$W \leftarrow P(K_1, \langle\sigma\rangle \| M)$	If $(|x| < \ell)$ then return $(\perp, \tau)$
$T \leftarrow F(K_2, W \| A)$	$\langle\sigma\rangle \| M \leftarrow x$
$C \leftarrow (W, T)$	If $(T \neq F(K_2, W \| A))$ then return $(\perp, \tau)$
$\sigma \leftarrow \sigma + 1$	If $(\sigma \neq \tau)$ then return $(\perp, \tau)$
Return (C, σ)	$\tau \leftarrow \tau + 1$; Return (M, τ)

In the 4th line of the code of $\mathcal{D}$, we are interpreting the first ℓ bits of x as the binary encoding of an integer denoted σ, and letting M be the rest of the bits of x. If P is a PRP and F is a PRF then Π is a secure authenticated encryption scheme. This is a standard claim that can be proved following [5]. Of interest in our context is instead the following, which says that Π has unique ciphertexts. This makes no security assumptions on P or F. The proof is in [4].

Theorem 5. *Let* $P\colon \{0,1\}^k \times \{0,1\}^* \to \{0,1\}^*$ *be a family of permutations and* $F\colon \{0,1\}^k \times \{0,1\}^* \to \{0,1\}^t$ *a family of functions. Let* $\Pi = (\mathcal{K}, \mathcal{E}, \mathcal{D})$ *be the symmetric encryption scheme associated to them as above. Then* Π *satisfies the correctness condition and has unique ciphertexts.*

SURVEILLANCE-RESISTANCE FROM NONCE-BASED SCHEMES. Above we gave a simple scheme to illustrate that surveillance-resistance is possible. However, likely candidates to instantiate the PRP are two pass [12,13], making the scheme potentially slower than standard, deployed ones. In [4] we describe a better solution. We show that any nonce-based scheme meeting a natural non-degeneracy condition, called "tidiness" in [18], can be turned into a stateful symmetric encryption scheme (by using the nonce as a counter) that has unique ciphertexts. Most existing and practical nonce-based schemes meet our condition, so this results in a number of surveillance-secure schemes that may be easily deployed.

Acknowledgments. Bellare was supported in part by NSF grants CNS-1228890 and CNS-1116800, Paterson by EPSRC Leadership Fellowship EP/H005455/1, and Rogaway by NSF grants CNS-1228828 and CNS-1314885.

References

1. Ball, J., Borger, J., Greenwald, G.: Revealed: How US and UK Spy Agencies Defeat Internet Security and Privacy. The Guardian (September 5, 2013)

2. Bellare, M., Desai, A., Jokipii, E., Rogaway, P.: A Concrete Security Treatment of Symmetric Encryption. In: 38th FOCS. IEEE (1997)

3. Bellare, M., Kohno, T., Namprempre, C.: Authenticated Encryption in SSH: Provably Fixing the SSH Binary Packet Protocol. In: ACM CCS 2002. ACM (2002)

4. Bellare, M., Paterson, K., Rogaway, P.: Security of Symmetric Encryption against Mass Surveillance. Full version of this paper. Cryptology ePrint Archive, Report 2014/438 (2014)

5. Bellare, M., Rogaway, P.: Encode-then-Encipher Encryption: How to Exploit Nonces or Redundancy in Plaintexts for Efficient Cryptography. In: Okamoto, T. (ed.) ASIACRYPT 2000. LNCS, vol. 1976, pp. 317–330. Springer, Heidelberg (2000)

6. Cabuk, S., Brodley, C., Shields, C.: IP Covert Channel Detection. ACM Trans. Inf. Syst. Secur. 12(4) (2009)

7. Checkoway, S., Fredrikson, M., Niederhagen, R., Everspaugh, A., Green, M., Lange, T., Ristenpart, T., Bernstein, D.J., Maskiewicz, J., Shacham, H.: On the Practical Exploitability of Dual EC in TLS Implementations. In: USENIX Security Symposium (2014)

8. Gligor, V.D., Donescu, P.: Fast Encryption and Authentication: XCBC Encryption and XECB Authentication Modes. In: Matsui, M. (ed.) FSE 2001. LNCS, vol. 2355, pp. 92–108. Springer, Heidelberg (2002)

9. Goh, E.-J., Boneh, D., Pinkas, B., Golle, P.: The Design and Implementation of Protocol-Based Hidden Key Recovery. In: Boyd, C., Mao, W. (eds.) ISC 2003. LNCS, vol. 2851, pp. 165–179. Springer, Heidelberg (2003)

10. Goldwasser, S., Micali, S.: Probabilistic Encryption. Journal of Computer and System Sciences 28(2), 270–299 (1984)

11. Goldwasser, S., Ostrovsky, R.: Invariant Signatures and Non-Interactive Zero-Knowledge Proofs are Equivalent (Extended Abstract). In: Brickell, E.F. (ed.) CRYPTO 1992. LNCS, vol. 740, pp. 228–245. Springer, Heidelberg (1993)

12. Halevi, S., Rogaway, P.: A Tweakable Enciphering Mode. In: Boneh, D. (ed.) CRYPTO 2003. LNCS, vol. 2729, pp. 482–499. Springer, Heidelberg (2003)

13. Halevi, S., Rogaway, P.: A Parallelizable Enciphering Mode. In: Okamoto, T. (ed.) CT-RSA 2004. LNCS, vol. 2964, pp. 292–304. Springer, Heidelberg (2004)

14. Jutla, C.: Encryption Modes with Almost Free Message Integrity. Journal of Cryptology 21(4), 547–578 (2008)

15. Kocher, P.C.: Timing Attacks on Implementations of Diffie-Hellman, RSA, DSS, and Other Systems. In: Koblitz, N. (ed.) CRYPTO 1996. LNCS, vol. 1109, pp. 104–113. Springer, Heidelberg (1996)

16. Lysyanskaya, A.: Unique Signatures and Verifiable Random Functions from the DH-DDH Separation. In: Yung, M. (ed.) CRYPTO 2002. LNCS, vol. 2442, pp. 597–612. Springer, Heidelberg (2002)

17. Millen, J.: 20 years of Covert Channel Modeling and Analysis. In: IEEE Symposium on Security and Privacy (1999)

18. Namprempre, C., Rogaway, P., Shrimpton, T.: Reconsidering Generic Composition. In: Nguyen, P.Q., Oswald, E. (eds.) EUROCRYPT 2014. LNCS, vol. 8441, pp. 257–274. Springer, Heidelberg (2014)

19. Patarin, J., Goubin, L.: Asymmetric Cryptography with S-Boxes. In: Han, Y., Quing, S. (eds.) ICICS 1997. LNCS, vol. 1334, pp. 369–380. Springer, Heidelberg (1997)

20. Paterson, K.G.: Imprimitive Permutation Groups and Trapdoors in Iterated Block Ciphers. In: Knudsen, L.R. (ed.) FSE 1999. LNCS, vol. 1636, pp. 201–214. Springer, Heidelberg (1999)

21. Rijmen, V., Preneel, B.: A Family of Trapdoor Ciphers. In: Biham, E. (ed.) FSE 1997. LNCS, vol. 1267, pp. 139–148. Springer, Heidelberg (1997)

22. Rogaway, P.: Authenticated-Encryption with Associated-Data. In: ACM CCS 2002. ACM (2002)

23. Rogaway, P.: Nonce-Based Symmetric Encryption. In: Roy, B., Meier, W. (eds.) FSE 2004. LNCS, vol. 3017, pp. 348–359. Springer, Heidelberg (2004)

24. Simmons, G.: The Prisoners' Problem and the Subliminal Channel. In: CRYPTO 1983. Springer (1983)

25. Song, D., Wagner, D., Tian, X.: Timing Analysis of Keystrokes and Timing Attacks on SSH. In: USENIX Security Symposium (2001)

26. Thompson, K.: Reflections on Trusting Trust. Commun. ACM 27(8), 761–763 (1984)

27. Young, A., Yung, M.: The Dark Side of "Black-Box" Cryptography, or: Should We Trust Capstone? In: Koblitz, N. (ed.) CRYPTO 1996. LNCS, vol. 1109, pp. 89–103. Springer, Heidelberg (1996)

28. Young, A., Yung, M.: Kleptography: Using cryptography against Cryptography. In: Fumy, W. (ed.) EUROCRYPT 1997. LNCS, vol. 1233, pp. 62–74. Springer, Heidelberg (1997)

29. Young, A., Yung, M.: Monkey: Black-Box Symmetric Ciphers Designed for MON*opolizing* KEY*s*. In: Vaudenay, S. (ed.) FSE 1998. LNCS, vol. 1372, p. 122. Springer, Heidelberg (1998)

30. Young, A., Yung, M.: A Subliminal Channel in Secret Block Ciphers. In: Handschuh, H., Hasan, M.A. (eds.) SAC 2004. LNCS, vol. 3357, pp. 198–211. Springer, Heidelberg (2004)

31. Young, A., Yung, M.: Backdoor Attacks on Black-Box Ciphers Exploiting Low-Entropy Plaintexts. In: Safavi-Naini, R., Seberry, J. (eds.) ACISP 2003. LNCS, vol. 2727, pp. 297–311. Springer, Heidelberg (2003)

The Security of Multiple Encryption
in the Ideal Cipher Model

Yuanxi Dai[1], Jooyoung Lee[2], Bart Mennink[3], and John Steinberger[1]

[1] Institute for Interdisciplinary Information Sciences,
Tsinghua University, Beijing, P.R. China
{shustdc,jpsteinb}@gmail.com
[2] Faculty of Mathematics and Statistics, Sejong University, Seoul, Korea
jlee05@sejong.ac.kr
[3] Dept. Electrical Engineering, ESAT/COSIC, KU Leuven, and iMinds, Belgium
bart.mennink@esat.kuleuven.be

Abstract. Multiple encryption—the practice of composing a blockcipher several times with itself under independent keys—has received considerable attention of late from the standpoint of provable security. Despite these efforts proving definitive security bounds (i.e., with matching attacks) has remained elusive even for the special case of triple encryption. In this paper we close the gap by improving both the best known attacks and best known provable security, so that both bounds match. Our results apply for arbitrary number of rounds and show that the security of ℓ-round multiple encryption is precisely $\exp(\kappa+\min\{\kappa(\ell'-2)/2), n(\ell'-2)/\ell'\})$ where $\exp(t) = 2^t$ and where $\ell' = 2\lceil\ell/2\rceil$ is the smallest even integer greater than or equal to ℓ, for all $\ell \geq 1$. Our technique is based on Patarin's H-coefficient method and relies on a combinatorial result of Chen and Steinberger originally required in the context of key-alternating ciphers.[1]

1 Introduction

Let $E : \{0,1\}^\kappa \times \{0,1\}^n \to \{0,1\}^n$ be a blockcipher with key space $\{0,1\}^\kappa$ and message/ciphertext space $\{0,1\}^n$. The ℓ-*cascade of* E, denoted $E^{(\ell)}$, is the blockcipher of key space $\{0,1\}^{\ell\kappa}$ and of message space $\{0,1\}^n$ obtained by composing E ℓ times with itself under independent keys. Thus

$$E_k^{(\ell)}(x) = E_{k_\ell}(E_{k_{\ell-1}}(\ldots(E_{k_1}(x))\ldots)) \tag{1}$$

where $k = k_1\|\ldots\|k_\ell \in \{0,1\}^{\ell\kappa}$. (The inverse of $E^{(\ell)}$ is computed the obvious way.) In particular $E^{(1)} = E$.

Since $E^{(\ell)}$ has longer keys than E for $\ell \geq 2$, the ℓ-cascade can be viewed as a natural mechanism for increasing the key space of a blockcipher and, hence, potentially, enhancing the security level. Security does not necessarily increase linearly with the key length, however. For example there exist meet-in-the-middle

---

[1] This paper is an independently initiated merge of preprints [9, 23, 30], that were separately submitted to CRYPTO 2014.

J.A. Garay and R. Gennaro (Eds.): CRYPTO 2014, Part I, LNCS 8616, pp. 20–38, 2014.
© International Association for Cryptologic Research 2014

(key-recovery) attacks against cascades of length 2 that cost no more[2] than generic (key-recovery) attacks against cascades of length 1 [11]. Indeed, when a variant of DES with longer keys was needed, designers eschewed double encryption (cascades of length 2) in favor of triple encryption [11, 31]. The standard which eventually resulted, so-called Triple DES [2,15,35], is still widely deployed.

Even while generic attacks have guided the considerations of designers since the beginning, finding nontrivial provable security results for multiple encryption in idealized models remained an open problem for a very long time. In the ideal model which we and most previous authors envisage [1,4,16,17,22] the security of the ℓ-cascade is quantified by the information-theoretic indistinguishability of two worlds, "real" and "ideal". In the "real" world the adversary A is given oracle access to an ideal[3] cipher E, to its inverse E^{-1}, and to a randomly keyed ℓ-cascade instance $E_k^{(\ell)}$ of E (for hidden k) as well as to the inverse $(E_k^{(\ell)})^{-1}$ of the ℓ-cascade; in the "ideal" world the ℓ-cascade instance $E_k^{(\ell)}$ is replaced by a random independent permutation π and its inverse π^{-1}. The adversary knows the value ℓ in question.

The case $\ell = 1$, while quite simple, is already instructive to analyze. In that case the adversary must distinguish between $E_k^{(1)} = E_k$ and a random permutation π, while being given oracle access to E. Since E is ideal, it is easy to argue that the adversary has no advantage as long as it has not queried its oracle E on key k. With k being uniform at random, and with other queries to $E/\pi/E_k$ giving no clue as to the value of k, the adversary's distinguishing advantage is thus upper bounded by—and in fact basically equal to—$q/2^\kappa$, where q is the number of queries made. (We note this bound holds even if n is very small compared to κ, e.g., $n = 1, 2$. For the sake of completeness, we formalize the argument just sketched in Appendix C of our full version [10].) An easy reduction[4] argument, moreover, shows that $E^{(\ell)}$ is at least as secure as $E^{(r)}$ for all $r \leq \ell$. Hence $E^{(\ell)}$ achieves *at least* κ bits of security for all $\ell \geq 1$, and the basic question is to determine how security grows with ℓ.

The first nontrivial results obtained pertaining to this question were by Aiello et al. [1] who show that $E_k^{(2)}$ is *slightly* harder to distinguish from a random π than $E_k^{(1)} = E_k$. More precisely, Aiello et al. show that A's distinguishing advantage for $E^{(2)}$ is upper bounded by an expression of the form $q^2/2^{2\kappa}$, as opposed to $q/2^\kappa$ for $E^{(1)}$, where q is the number of queries made by A. In either event, thus, $E^{(1)}$ and $E^{(2)}$ both essentially offer κ bits of security, given the meet-in-the-middle attack for length two cascades of cost $q = 2^\kappa$ [11]. (See also the full version of this paper [10], which revisits Aiello et al.'s result.)

Subsequently we will write $\exp(\kappa)$ for 2^κ, somewhat in line with the computer science convention of writing $\log(t)$ for $\log_2(t)$. We thus say, e.g., that $E^{(1)}$ and $E^{(2)}$ "achieve security $\exp(\kappa)$", in the sense that it requires about $\exp(\kappa) = 2^\kappa$

---

[2] This should be qualified: the memory costs are much larger and the query complexity is *slightly* greater [1].

[3] I.e., $E(k, \cdot) : \{0, 1\}^n \to \{0, 1\}^n$ is a random permutation for each key $k \in \{0, 1\}^\kappa$.

[4] Since the adversaries considered are information-theoretic, we note that we don't even have to consider the reduction's running time lossiness.

Table 1. Security lower and upper bounds for cascaded encryption (in log). Here, $\ell' = 2\lceil \ell/2 \rceil$. All results in **bold** are derived in this work.

$E^{(\ell)}$	security	tight
$\ell = 1, 2$	κ [1,11]	✓
$\ell = 3, 4$	$\kappa + \min\{\kappa/2, n/2\}$ [4,17]	✗
	$\boldsymbol{\kappa + \min\{\kappa, n/2\}}$	✓
$\ell \geq 5$	$\kappa + \min\{\kappa(\ell' - 2)/\ell', n/2\}$ [17]*	✗
	$\boldsymbol{\kappa + \min\{\kappa(\ell' - 2)/2, n(\ell' - 2)/\ell'\}}$	✓

*Starting from $\ell \geq 16$, Lee [22] proved an improved security bound of $\exp(\kappa + \min\{\kappa, n\} - 8n/\ell)$.

queries to achieve constant distinguishing advantage between the real and ideal worlds for those cascade lengths.

After Aiello et al. a complicated history of improved security bounds ensues, including work by Bellare and Rogaway [4] for length 3 cascades, by Gaži and Maurer [17] (who corrected some errors in Bellare-Rogaway and who generalized their approach to larger numbers of rounds), and by Lee [22]. For reasons of space, however, we eschew a detailed discussion of these prior results in this proceedings version, and refer the reader to the synopsis in Table 1.

On the attack side Lucks [26] found an attack of cost $\kappa + n/2$ for length 3 cascades (thus matching the Bellare-Rogaway security bound for length 3 cascades in the regime $\kappa \geq n$). Gaži found an attack of cost $\kappa + n(\ell - 2)/\ell$ for arbitrary ℓ generalizing Lucks's attack. (Moreover Gaži was the first to give a mathematically rigorous analysis of Lucks's attack.)

Despite this series of results obtaining matching upper and lower bounds on security has remained elusive for all $\ell \geq 3$. In the case $\ell = 3$, for example, all we know is that the security of $E^{(3)}$ lies somewhere in the interval

$$[\exp(\kappa + \min\{\kappa/2, n/2\}), \exp(\kappa + n/2)]$$

which leaves open the question of the true security for $\kappa < n$. For $\ell \geq 5$, moreover, exact security remained open regardless of the ratio between ℓ and κ.

OUR RESULTS. In this paper we close the remaining gaps between upper and lower bounds for all ℓ, up to customary lower-order terms. More precisely, we show that $E^{(\ell)}$ has security

$$\exp(\kappa + \min\{\kappa(\ell' - 2)/2, n(\ell' - 2)/\ell'\}) \tag{2}$$

by exhibiting matching attacks and security proofs, for all $\ell \geq 1$. (Note by the form of (2) that new attacks are only needed when $\kappa(\ell' - 2)/2 < n(\ell' - 2)/\ell'$; otherwise the attacks of Gaži suffice.) One can observe from (2) that $\ell = 2r$ rounds buy the same amount of security as $\ell = 2r - 1$ rounds. In fact, we expect the curve describing the adversary's advantage to be slightly more advantageous for $2r - 1$ rounds than for $2r$ rounds, as observed by Aiello et al. for $r = 1$, but our analysis is not fine-grained enough to verify this.

TECHNIQUES. Tightening the security bounds for triple encryption is already an interesting problem in itself. Besides devising a new rather easy attack of cost $\exp(2\kappa)$, it turns out that the bound directly follows from tightening a key combinatorial lemma in Bellare and Rogaway's original proof (Lemma 10 in [5]).

We found the case of larger number of rounds (in particular, $\ell \geq 5$) to be more challenging. While we copied the basic approach of Bellare and Rogaway [4] and of Gaži and Maurer [17] some significant structural changes were required in order to achieve tightness. In particular, we had to rebundle a key two-step game transition from [17] into a single-step transition. Moreover we found that the best way to handle this (now rather delicate) single-step transition was by Patarin's H-coefficient technique [37]. Here we drew inspiration from Chen and Steinberger [8] and, indeed, reused the key combinatorial lemma of that paper. Roughly speaking, this lemma gives an explicit expression for the probability that

$$(P_\ell \circ \cdots \circ P_1)(a) = b$$

where each P_i is a *partially defined* random permutation of $\{0,1\}^n$, where $\circ$ denotes function composition, where $a, b \in \{0,1\}^n$ are two values such that $P_1(a)$ and $P_\ell^{-1}(b)$ are undefined. Here the probability is expressed (in particular, lower-bounded) as a function of the number of edges[5] already defined in the P_i's as well as of the number of "chains" of various lengths[6] formed by those edges in the composition $P_1 \circ \cdots \circ P_\ell$. (In our case $P_i = E_{k_i}$ where $k = k_1 \| \ldots \| k_\ell$ is the secret key.) It is noteworthy that the security proofs for three different classes of composed ciphers (key-alternating ciphers [8], cascade ciphers (this paper), and XOR-cascade ciphers [8,16,18]) now rely on this lemma.

In order to successfully apply the H-coefficient technique and Chen and Steinberger's lemma a crucial step is to upper bound the probability of the adversary obtaining (too many) long chains in $P_\ell \circ \cdots \circ P_1 = E_{k_\ell} \circ \cdots \circ E_{k_1}$. Like Bellare and Rogaway [4] and like Gaži and Maurer [17] before us, we do this by upper bounding the *total* number of query chains of a given length formed by *all* of the adversary's queries to E, regardless of the underlying key, and then by applying a Markov inequality—but in our case we strive for tight bounds on the total number of query chains. At first glance the combinatorial question is nonobvious (especially given the presence of an adaptive adversary) but we observe that on any path of queries at least half the queries are "backwards" (meaning contrary to the path's direction, in this instance) *for at least one of the two possible ways of orienting the path* (as a given path can be traversed right-to-left or left-to-

---

[5] If $x \in \{0,1\}^n$ is a value such that $y = P_i(x)$ is defined, then the pair (x, y) is also called an *edge* of P_i, equating P_i with a bipartite graph (more precisely, a partial matching) from $\{0,1\}^n$ to $\{0,1\}^n$. The composition $P_\ell \circ \cdots \circ P_1$ is visualized by "gluing" these bipartite graphs sequentially next to one another.

[6] See the previous footnote.

right). Together with some classical balls-in-bins occupancy results, this simple symmetry-breaking observation gives an easy means of upper bounding the total number of query chains formed, and the bounds obtained are also tight. We refer to Proposition 1 for more details.

OTHER RELATED WORK. We have already briefly mentioned related work on key-alternating ciphers [7,8,14,21,38] as well as on XOR cascades [16,18,22], to which the beautiful work of Rogaway and Kilian on DESX (a special case of an XOR-cascade) should be added [19].

Coming back to cascade ciphers, Merkle and Hellman [31] show an attack on two-key triple encryption, which attack is revisited by Oorschot and Wiener [34]. (See also [33].) Even and Goldreich [13] present a medley of observations on multiple encryption in various models, including some conclusions which are disputed by Maurer and Massey [27]. The best paper award at CRYPTO 2012, by Dinur et al. [12], concerns, in large part, non-information-theoretic key-recovery attacks on cascade ciphers.

We finally point that similar questions (though using very different techniques) have been pursued in the computational setting, in which one seeks to amplify the *computational* indistinguishability of a PRP by composing it with itself [25, 28,29,32]. See in particular [39] which culminates this line of work.

OPEN QUESTIONS. As will be seen, our results actually hold even if the adversary is always allowed to make 2^n queries to its permutation oracle (which is $E_k^{(\ell)}$ or π) for free, i.e., to entirely learn its permutation oracle for free. It would be interesting to know if better bounds can be achieved by restricting the number of permutation queries. This is all the more relevant given that many applications will impose limitations on the number of encryptions/decryptions available to the adversary.

2   Definitions

BLOCKCIPHERS AND CASCADES. A blockcipher is a function $E : \{0,1\}^\kappa \times \{0,1\}^n \to \{0,1\}^n$ such that $E(k, \cdot) : \{0,1\}^n \to \{0,1\}^n$ is a permutation for each key $k \in \{0,1\}^\kappa$. We also write $E_k(x)$ for $E(k,x)$. By the "inverse" E^{-1} of E we mean the blockcipher $E^{-1} : \{0,1\}^\kappa \times \{0,1\}^n \to \{0,1\}^n$ such that E_k^{-1} is the inverse permutation of E_k for each $k \in \{0,1\}^\kappa$.

For a blockcipher E and an integer $\ell \geq 1$ we define the ℓ-*cascade* of E, written $E^{(\ell)}$, by equation (1). We note that $E^{(\ell)}$ is a blockcipher of key space $\{0,1\}^{\ell\kappa}$ and of message space $\{0,1\}^n$.

IDEAL CIPHERS. A blockcipher $E : \{0,1\}^\kappa \times \{0,1\}^n \to \{0,1\}^n$ which is sampled uniformly at random from the space of all blockciphers of key space $\{0,1\}^\kappa$ and of message space $\{0,1\}^n$ is called an *ideal cipher*. In this case E_k is a random independent permutation of $\{0,1\}^n$ for each $k \in \{0,1\}^\kappa$.

SECURITY GAME. Let ℓ, κ and n be given. Let A be an information-theoretic adversary (or "distinguisher") with oracle access to, among others, an ideal cipher

$E : \{0,1\}^\kappa \times \{0,1\}^n \to \{0,1\}^n$, which we write A^E but by which we mean that A can query *both* E and E^{-1}. (Along the same lines writing A^π indicates that A has access to both π and π^{-1} when π is a permutation.) Then A's *distinguishing advantage* against ℓ-cascades, written $\mathbf{Adv}^{\mathsf{casc}}_{\ell,\kappa,n}(A)$ is defined as

$$\mathbf{Adv}^{\mathsf{casc}}_{\ell,\kappa,n}(A) = \Pr[k^* \xleftarrow{\$} \{0,1\}^{\ell\kappa}; A^{E,E^{(\ell)}_{k^*}} = 1] - \Pr[\pi \xleftarrow{\$} \mathcal{P}; A^{E,\pi} = 1]$$

where the notation

$$k^* \xleftarrow{\$} \{0,1\}^{\ell\kappa}; A^{E,E^{(\ell)}_{k^*}} = 1$$

indicates the event that A outputs 1 after interacting with oracles E/E^{-1} and $E^{(\ell)}_{k^*}/(E^{(\ell)}_{k^*})^{-1}$ where k^* is sampled uniformly at random from the key space of $E^{(\ell)}$, and hidden from A; whereas the notation

$$\pi \xleftarrow{\$} \mathcal{P}; A^{E,\pi} = 1$$

indicates the event that A outputs 1 after interacting with oracles E/E^{-1} and π/π^{-1} where π is a permutation of $\{0,1\}^n$ sampled uniformly at random from the set of all permutations of $\{0,1\}^n$, here denoted $\mathcal{P}$; and where in either case the sampling of the ideal cipher E at the start of the experiment is kept implicit for the sake of succinctness.

We write

$$\mathbf{Adv}^{\mathsf{casc}}_{\ell,\kappa,n}(q)$$

for the supremum of $\mathbf{Adv}^{\mathsf{casc}}_{\ell,\kappa,n}(A)$ taken over all q-query information-theoretic adversaries A. (The notation $\mathbf{Adv}^{\mathsf{casc}}_{\ell,\kappa,n}$ is thus overloaded.)

3   Statement of Results

LOWER BOUNDS. Our paper's main result is the following theorem (as always, $\ell' = 2\lceil \ell/2 \rceil$; we also write $(\ell+1)'$ for $2\lceil (\ell+1)/2 \rceil$, etc):

Theorem 1. *(a) If $q \geq 2^n$ then, for every real number $C \geq 1$,*

$$\mathbf{Adv}^{\mathsf{casc}}_{\ell,\kappa,n}(q) \leq \frac{\ell^2}{2^{\kappa+1}} + \frac{4}{2^n} + \frac{\alpha}{C} + 2^n \ell C^{(\ell+1)'/2} \left(\frac{8q}{2^{\kappa+n}} \right)^{\ell'/2}$$

where $\alpha = \ell^2 2^\ell (7n)^{\ell'/2}$. Furthermore if $q \geq n2^n$ we can improve α to $\alpha' = \ell^2 2^\ell 14^{\ell'/2} \leq \ell^2 8^{\ell'}$.
(b) If $q \leq 2^n$ then, for every $C \geq 1$ such that $Cq < 2^{\kappa+n-2}$,

$$\mathbf{Adv}^{\mathsf{casc}}_{\ell,\kappa,n}(q) \leq \frac{\ell^2}{2^{\kappa+1}} + \frac{4}{2^n} + \frac{\beta}{C} + \frac{q^2\ell}{2^n} C^{(\ell+1)'/2} \left(\frac{8}{2^\kappa} \right)^{\ell'/2} + \frac{q\beta}{\ell 2^{\kappa\ell'/2}}$$

where $\beta = \ell^2 2^\ell (3 \log q + 2)^{\ell'/2}$.
Moreover (a) and (b) also hold if the adversary is allowed to ask, for free, all possible 2^n queries to its second oracle.

The presence of the adjustable constant C is typical of security proofs involving a threshold-based "bad event". For given parameters q, n, κ and ℓ there some optimal C that minimizes the bound.

Theorem 1 is, unfortunately and evidently, hard to parse. By analytically optimizing C and making a few other simplifications, however, Theorem 1 yields the following, slightly more digestible corollary:

Corollary 1. (a) If $q \geq 2^n$ then

$$\mathbf{Adv}^{\mathsf{casc}}_{\ell,\kappa,n}(q) \leq \frac{\ell^2}{2^{\kappa+1}} + \frac{4}{2^n} + \alpha(\ell/2 + 2)\ell^{1/2}\left(\frac{8q}{2^{\kappa+n(\ell'-2)/\ell'}}\right)^{\ell'/(\ell+3)'}$$

where $\alpha = \ell^2 2^{\ell}(7n)^{\ell'/2}$. Furthermore if $q \geq n2^n$ we can improve α to $\alpha' = \ell^2 2^{\ell} 14^{\ell'/2} \leq \ell^2 8^{\ell}$.

(b) If $q \leq 2^n$ and $2^{\ell}(3n + 2)^{\ell'/2} \leq 2^n$ then

$$\mathbf{Adv}^{\mathsf{casc}}_{\ell,\kappa,n}(q) \leq \frac{\ell^2}{2^{\kappa+1}} + \frac{4}{2^n} + \beta(\ell/2 + 2)\left(\frac{\ell 3^{\ell'} q^2}{2^{\kappa\ell'/2+n}}\right)^{2/(\ell+3)'} + \frac{q\beta}{\ell 2^{\kappa\ell'/2}}$$

where $\beta = \ell^2 2^{\ell}(3\log q + 2)^{\ell'/2}$.

Moreover (a) and (b) also hold if the adversary is allowed to ask, for free, all possible 2^n queries to its second oracle.

The proof of Corollary 1 from Theorem 1 can be found in the full version [10].

We note the constraint $2^{\ell}(3n + 2)^{\ell'/2} \leq 2^n$ that appears in the second part of Corollary 1 is almost always satisfied by practical parameters and is always asymptotically verified as $n \to \infty$. (Indeed, we imagine ℓ as fixed whereas $n, \kappa \to \infty$ according to some fixed ratio.)

It directly follows from Corollary 1 that $\mathbf{Adv}^{\mathsf{casc}}_{\ell,\kappa,n}(q)$ is small if

$$q \ll \exp(\kappa + \min\{\kappa(\ell' - 2)/2, n(\ell' - 2)/\ell'\})$$

(note $\kappa + \kappa(\ell' - 2)/2 = \kappa\ell'/2$ and $q^2/2^{\kappa\ell'/2+n} \leq q/2^{\kappa\ell'/2}$ when $q \leq 2^n$) or, a little more precisely, if

$$q \ll (2^{-\ell/2}(7n)^{-\ell'/4}\ell^{-2})^{(\ell+3)'}$$
$$\cdot \exp(\kappa + \min\{\kappa(\ell' - 2)/2 - 2\ell, n(\ell' - 2)/\ell' - 3\}). \tag{3}$$

We emphasize that the above threshold is a coarse estimate, which takes into account the factors of all three non-negligible expressions in Corollary 1. (Note that $\log q \leq n$ in the second part of Corollary 1, so $\beta \leq \alpha$.) Indeed, if q is a factor r smaller than the expression on the right of (3), then it is easy to see from Corollary 1 that the adversary's advantage is upper bounded by either $r^{\ell'/(\ell+3)'}$ or $r^{4/(\ell+3)'} + r$, disregarding the negligible terms $\ell^2/2^{\kappa+1}$ and $4/2^n$.

UPPER BOUNDS. In Section 4 we present a simple attack of query complexity

$$\ell \cdot \exp(\kappa\ell'/2)$$

that succeeds in distinguishing $(E, E_k^{(\ell)})$ from (E, π) with overwhelming advantage. This complements the previously quoted attack by Gaži, of query complexity

$$\ell \cdot \exp(\kappa + n(\ell' - 2)/\ell')$$

and which also succeeds with overwhelming advantage. Hence the gap left between lower and upper bounds is essentially the gap left between

$$\min\{\ell \cdot \exp(\kappa\ell'/2), \ell \cdot \exp(\kappa + n(\ell' - 2)/\ell')\}$$

and the right-hand side of (3).

4   An Attack of Cost $\exp(\kappa\ell'/2)$

In this section we describe a new "meet-in-the-middle-attack" on $E^{(\ell)}$ of complexity $\exp(\kappa\ell'/2)$, which complements Gaži's attack of query complexity $\exp(\kappa + n(\ell' - 2)/\ell')$. A precise statement is given by the following theorem.

Theorem 2. *For any integer ρ, $1 \leq \rho \leq 2^{n-1}$, there exists an adversary A making at most $\rho\ell 2^{\kappa\ell'/2}$ queries to E and at most ρ queries to $E_k^{(\ell)}/\pi$, such that*

$$\mathbf{Adv}_{\ell,\kappa,n}^{\mathsf{casc}}(A) \geq 1 - 2^{\kappa\ell - \rho(n-1)}.$$

Proof. The adversary A, which implements a meet-in-the-middle attack, is given by the pseudocode of Fig. 1. A starts by querying ρ messages $m_1, \ldots, m_\rho$ to $E_k^{(\ell)}/\pi$, thus obtaining their corresponding ciphertexts $c_1, \ldots, c_\rho$. Then for each of these message/ciphertext pairs (m_i, c_i) it evaluates the first $\lceil \ell/2 \rceil$ block ciphers for all possible keys starting from m_i and the last $\lfloor \ell/2 \rfloor$ block ciphers in inverse direction starting from c_i. One possible key $k = (k_1 \| \ldots \| k_\ell)$ must "stand out" unless A is in the ideal world. Thus A returns 1 if and only if there is a key compatible with all ρ message-ciphertext pairs (m_i, c_i). It is easy to see that A makes ρ queries to $E_k^{(\ell)}/\pi$ and

$$\rho(\lceil \ell/2 \rceil 2^{\kappa\lceil \ell/2 \rceil} + \lfloor \ell/2 \rfloor 2^{\kappa\lfloor \ell/2 \rfloor}) \leq \rho\ell 2^{\kappa\ell'/2}$$

queries to E, as claimed.

Clearly, in the real world $(E_k^{(\ell)}, E)$, for $(k_L^*, k_R^*) = k$ we have $a_{i,k_L^*} = b_{i,k_R^*}$ for all $i = 1, \ldots, \rho$, so A returns 1. We consider the probability that A returns 1 in the ideal world (π, E). For each key $k = (k_1 \| \ldots \| k_\ell)$, $E_{k_\ell} \circ \cdots \circ E_{k_1}$ becomes a truly random permutation, independent of π. For this key, the probability that $E_{k_\ell} \circ \cdots \circ E_{k_1}(m_i) = c_i$ for every $i = 1, \ldots, \rho$ is upper bounded by

$$\frac{(2^n - \rho)!}{(2^n)!} \leq \left(\frac{1}{2^n - \rho + 1}\right)^\rho \leq \frac{1}{2^{\rho(n-1)}}.$$

The theorem follows by a union bound over all possible keys.   □

$$\begin{array}{l}
\textbf{fix distinct } m_1, \ldots, m_\rho \in \{0,1\}^n \\
\textbf{for } r = 1 \textbf{ to } \rho \textbf{ do} \\
\quad \textbf{query } c_i \leftarrow \mathcal{R}(m_i) \\
\quad \textbf{forall } k_L^* = (k_1^* \| \ldots \| k_{\lceil \ell/2 \rceil}^*) \in \{0,1\}^{\kappa \lceil \ell/2 \rceil} \\
\qquad \textbf{query } a_{i,k_L^*} \leftarrow E_{k_{\lceil \ell/2 \rceil}^*} \circ \cdots \circ E_{k_1^*}(m_i) \\
\quad \textbf{forall } k_R^* = (k_{\lceil \ell/2 \rceil+1}^* \| \ldots \| k_\ell^*) \in \{0,1\}^{\kappa \lfloor \ell/2 \rfloor} \\
\qquad \textbf{query } b_{i,k_R^*} \leftarrow E_{k_{\lceil \ell/2 \rceil+1}^*}^{-1} \circ \cdots \circ E_{k_\ell^*}^{-1}(c_i) \\
\quad \textbf{forall } (k_L^*, k_R^*) \in \{0,1\}^{\kappa \lceil \ell/2 \rceil} \times \{0,1\}^{\kappa \lfloor \ell/2 \rfloor} \\
\qquad \textbf{if } a_{i,k_L^*} = b_{i,k_R^*} \textbf{ for all } i = 1, \ldots, \rho \\
\qquad\quad \textbf{return } 1 \\
\textbf{return } 0
\end{array}$$

Fig. 1. The adversary A for Theorem 2. The oracle to $E_k^{(\ell)}/\pi$ is denoted $\mathcal{R}$.

5   Preliminary Reductions and Proof Overview

MODIFICATIONS OF BELLARE AND ROGAWAY [4]. In view of proving Theorem 1, we start by modifying the distinguishability game in the following way. At the very start of the experiment we send a symbol $\star \in \{\bot, \top\}$ to the adversary. In the ideal world we send $\star = \top$, and in the real world we also send $\star = \top$ unless $k_\ell^* = k_i^*$ for some $i < \ell$, where $k^* = k_1^* \| \ldots \| k_\ell^*$ is the secret key, in which case we send $\star = \bot$. Since the adversary is free to disregard $\star$, this modification is without loss of generality.

Next, we make a second modification, namely that if $\star = \bot$ then we forbid the adversary from making any queries. Since $\star$ can only be $\bot$ in the real world this is without loss of generality either (as the adversary already knows which world it is in anyway).

Now we make yet another modification to the real world, by generating a random permutation π like in the ideal world at the beginning of the experiment. If $\star = \top$ we answer queries to $E_{k^*}^{(\ell)}$ by π instead and, to compensate, we define $E_{k_\ell^*} = \pi \circ E_{k_1^*}^{-1} \circ \cdots \circ E_{k_{\ell-1}^*}^{-1}$ (thus "overwriting" $E_{k_\ell^*}$). Since this simply trades the randomness of $E_{k_\ell^*}$ for the randomness in π, it is easy to see that this is an equivalent way of defining the real world.

Note that both worlds now involve an independent[7] random permutation π. For each fixed permutation S one can also consider the distinguishing experiment where π is set to S in each world. A simple averaging argument over π shows, moreover, that there must exist some S for which the adversary's distinguishing advantage is at least as great when π is fixed to S as when π is random. We can thus assume without loss of generality that π is not sampled at random, but set

---

[7] The real world now has three "random tapes": one for k^*, one for π, and one for the ideal cipher E. Every query made by the adversary is deterministically answered as a function of these three random tapes, and these random tapes are independently sampled. This is the sense in which π is "independent" from other randomness in the real world.

to the same fixed permutation S in both worlds. Since S is fixed, now, and since we are quantifying over all information-theoretic adversaries A, we can assume that A knows S and, hence, makes no queries to its second oracle.

To summarize, modifications so far amount to this: in the real world, we abort the experiment if $k_\ell^* = k_i^*$ for some $i < \ell$, whereas in the contrary (generic) case there is some fixed permutation S, known to the adversary, such that $E_{k_\ell^*} = S \circ E_{k_1^*}^{-1} \circ \cdots \circ E_{k_{\ell-1}^*}^{-1}$. The ideal world never aborts.

FURTHER NORMALIZATIONS. Since A is information-theoretic we can assume without loss of generality that A is *deterministic*.

As in [8] we will also modify the experiment by *giving the secret key to A after it has finished making all its queries.* More precisely, in the real world we give the "real" key k^* used to key the second oracle $E_{k^*}^{(\ell)}$ whereas in the ideal world (where no such key exists) we sample a "dummy" key $k^* \in \{0,1\}^{\kappa\ell}$ uniformly at random and give this dummy key to A. Since A is free to disregard this extra information this is also without loss of generality.

TRANSCRIPTS. The interaction of A with its oracles is encoded by a *transcript* which, basically, is a list of questions asked and answers received, together also with the key value received at the end of the experiment.

More precisely, a transcript can be encoded by a triple of the form $(\star, Q_E, k^*)$ where $\star \in \{\bot, \top\}$, where $k^* \in \{0,1\}^{\kappa\ell}$ is the final key value received, and where Q_E is an *unordered* set of triples of the form $(k, x, y) \in \{0,1\}^\kappa \times \{0,1\}^n \times \{0,1\}^n$ with each such tuple indicating that either $E(k,x)$ was queried with answer y or that $E^{-1}(k,y)$ was queried with answer x. Indeed, A's interaction with its oracles can be unambiguously reconstructed from such an "unordered and undirected" set Q_E by using the fact that A is deterministic, cf. [8].

We write $\mathcal{T}$ for the set of all possible transcripts.

PROBABILITY SPACE OF ORACLES. Let $\mathcal{P}$ be the set of all permutations from $\{0,1\}^n$ to $\{0,1\}^n$. Then a blockcipher of key space $\{0,1\}^\kappa$ and message space $\{0,1\}^n$ can be viewed as an element of $\mathcal{P}^{\exp(\kappa)}$ (2^κ-fold direct product). Thus, an ordered pair

$$(E', k^*) \in \mathcal{P}^{\exp(\kappa)} \times \{0,1\}^{\kappa\ell}$$

uniquely determines a real-world environment for A. More precisely, unless $\star = \bot$ in which case A receives no further information except for k^*, A's ideal cipher oracle E is defined by

$$E_k = \begin{cases} E_k' & \text{if } k \neq k_\ell^* \\ S \circ E_{k_1^*}'^{-1} \circ \cdots \circ E_{k_{\ell-1}^*}'^{-1} & \text{if } k = k_\ell^* \end{cases}$$

where $k^* = k_1^* \| \ldots \| k_\ell^*$. We thus identify elements of

$$\Omega_X := \mathcal{P}^{\exp(\kappa)} \times \{0,1\}^{\kappa\ell}$$

with real-world oracles. We view Ω_X as a probability space with uniform measure (indeed, the definition of the real-world experiment induces uniform measure on Ω_X).

We similarly define

$$\Omega_Y := \mathcal{P}^{\exp(\kappa)} \times \{0,1\}^{\kappa\ell}$$

to be identified with the set of all ideal-world oracles, and which we also view as a probability space with uniform measure. Here the last coordinate corresponds to the "dummy key" given to the adversary at the end of the experiment. We emphasize that, for $(E, k^*) \in \Omega_Y$, the ideal cipher oracle to which A has access is precisely E, i.e., with no key being overwritten as a function of k^* and S; this is precisely the difference between the real and ideal worlds in the (generic) case when $k_\ell^* \notin \{k_1^*, \ldots, k_{\ell-1}^*\}$.

We can view the transcript produced by A in the real world as a random variable defined over Ω_X. Formally, let $X : \Omega_X \to \mathcal{T}$ be the function defined by letting $X(\omega)$ be the transcript obtained by running A on oracle ω. Thus X is a random variable of range $\mathcal{T}$, and the distribution of X is exactly the distribution of transcripts in the real world. We similarly define $Y : \Omega_Y \to \mathcal{T}$, so that Y is the transcript distribution in the ideal world.

The H-coefficient technique [36,37], in its simplest form, states that if we can divide $\mathcal{T}$ into a set of (so-called) "good" transcripts $\mathcal{T}_1$ and (so-called) "bad" transcripts $\mathcal{T}_2$, such that[8]

$$\frac{\Pr[X = \tau]}{\Pr[Y = \tau]} \geq 1 - \varepsilon_1 \qquad (4)$$

for some $\varepsilon_1 > 0$ and for all $\tau \in \mathcal{T}_1$, then the adversary's distinguishing advantage is upper bounded by

$$\Pr[Y \in \mathcal{T}_2] + \varepsilon_1.$$

We refer to [8] for more details.

COMPUTING TRANSCRIPT PROBABILITIES. Another key insight of the H-coefficient technique is that the probability of obtaining a transcript in either world can be computed via the formulas

$$\Pr[X = \tau] = \frac{|\mathsf{comp}_X(\tau)|}{|\Omega_X|}, \qquad \Pr[Y = \tau] = \frac{|\mathsf{comp}_Y(\tau)|}{|\Omega_Y|} \qquad (5)$$

as long as $\Pr[Y = \tau] > 0$, and where $\mathsf{comp}_X(\tau) \subseteq \Omega_X$ (resp. $\mathsf{comp}_Y(\tau) \subseteq \Omega_Y$) is the set of real-world (resp. ideal-world) oracles that are compatible with a transcript τ, where "compatibility" is defined the obvious[9] way: an oracle ω is compatible with a transcript τ if each individual query in τ is compatible with ω (in particular, τ's key value should match ω's key value). See [8] and Appendix D of our full version [10] for further discussion of these identities.

---

[8] By convention, the ratio $\Pr[X = \tau]/\Pr[Y = \tau]$ is considered to be ∞ if $\Pr[Y = \tau] = 0$.

[9] Slightly more formally—but less intuitively—an oracle (or "environment") ω is compatible with a transcript τ if there exists *some* (wlog, deterministic) adversary A' that produces τ as transcript when given ω as oracle.

TERMINOLOGY: CHAINS. Let $\tau = (\star, Q_E, k^*)$ be a transcript, where $k^* = k_1^* \| \ldots \|$ k_ℓ^*. Loosely following [17], a tuple $(h, x_h, k_{h+1}, x_{h+1}, k_{h+2}, \ldots, k_{h+r}, x_{h+r})$ where $0 \leq h \leq \ell - 1$ is an integer, where $1 \leq r \leq \ell$, and where

$$\begin{cases} (k_i, x_{i-1}, x_i) \in Q_E & \text{if } i - 1 \neq \ell \\ (k_i, S^{-1}(x_{i-1}), x_i) \in Q_E & \text{if } i - 1 = \ell \end{cases}$$

for $h + 1 \leq i \leq h + r$ (in particular, $x_i \in \{0,1\}^n$ and $k_i \in \{0,1\}^\kappa$ for each x_i, k_i) is called an r-chain of τ starting at index h or simply an r-chain of τ. Moreover, an r-chain is said to fit τ if $k_{h+i} = k_{h+i}^*$ for $1 \leq i \leq r$, indices taken mod ℓ and in the range $\{1, \ldots, \ell\}$. We sometimes commit a slight abuse of language by saying that a chain "fits k^*" instead of "fits τ" when it is clear which transcript τ is intended.

By means of emphasis, a chain which doesn't (necessarily) fit the key of τ is said to be *generic*; thus all r-chains of τ are by definition generic.

THE REST OF THE PROOF IN A NUTSHELL. Broadly, our "bad transcripts" are transcripts that either have a bad key (i.e., $k_i^* = k_j^*$ for some $i \neq j$) or transcripts with too many (long) fitting chains, where "too many" depends geometrically on the chain length r, as might be expected. When there are not too many long chains that fit the transcript's key, indeed, we are in a position to apply the lemma of Chen and Steinberger [8] to show that the probability of obtaining the given transcript in the real world is not far off from the probability of obtaining the same transcript in the ideal world, as required by (4).

The main technical challenge that arises is that of upper bounding the probability of obtaining too many length r chains that fit the key. Here one must emphasize that this probability (which is the probability of obtaining a "bad" transcript) is being computed in the ideal world. In the ideal world, the key value $k^* \in \{0,1\}^{\kappa\ell}$ is chosen at random *after* all queries are completed. Hence, by a Markov bound, it suffices to show that, with high probability, not too many *generic* r-chains are created by the adversary's queries. We deliver a tight bound on the number of generic chains by using a fairly simple argument, as already discussed in the paper's introduction (see in particular Proposition 1 in Section 6). See further details in Section 6.

6 Proof of Theorem 1

For the remainder of the proof of Theorem 1 we will assume that $n \geq 2$ and also, if $q \geq 2^n$, that

$$4Cq \leq 2^{\kappa+n} \quad \text{and} \quad C2^n \left(\frac{q}{2^{\kappa+n}} \right)^{\lceil \ell/2 \rceil} < 1. \tag{6}$$

These assumptions are without loss of generality because the first part of Theorem 1 is void otherwise, as can easily be checked. We also let $N = 2^n$.

We start by making a few more definitions that will be useful for the definition of bad transcripts and thereafter. Firstly, for a transcript $\tau = (\star, Q_E, k^*)$ we let

Q_E^+, Q_E^- be the sets of queries in Q_E obtained respectively by *forward* and *backward* queries to E by the adversary. (To wit, a query to E is forward, a query to E^{-1} is backward.) We note that while Q_E does not explicitly encode forward/backward information by design, such information can be uniquely reconstructed from Q_E given the fact that A is deterministic; hence, this information is implicitly contained in Q_E.

The *maximum forward query occupancy* of τ, denoted $\mathsf{fwd}(\tau)$, is given by

$$\mathsf{fwd}(\tau) := \max_{y_0 \in \{0,1\}^n} |\{(k, x, y) \in Q_E^+ : y = y_0\}| \tag{7}$$

and $\mathsf{bwd}(\tau)$, the *maximum backward query occupancy*, is similarly given by

$$\mathsf{bwd}(\tau) = \max_{x_0 \in \{0,1\}^n} |\{(k, x, y) \in Q_E^- : x = x_0\}|.$$

We also define

$$\mathsf{fitkey}(\tau, r, h)$$

as the number of r-chains in τ that fit k^* and that start at position h.

Note that back-of-the-envelope computations suggest that $\mathsf{fwd}(\tau)$ and $\mathsf{bwd}(\tau)$ should be around q/N for $q \geq N = 2^n$ and should be around $\log(q) \leq n$ for $q \leq N$. This motivates the definition of the following threshold $\zeta(q)$:

$$\zeta(q) := \begin{cases} 3\log(q) + 2 & \text{if } q \leq N, \\ 7nq/N & \text{if } N \leq q \leq nN, \\ 14q/N & \text{if } nN \leq q. \end{cases}$$

For now, the factors $3\log(q) + 2$, $7n$ and 14 that appear in the definition of $\zeta(q)$ should be more or less ignored; these coefficients are necessary to make bad transcripts, as defined next, unlikely. (We distinguish between the cases $N \leq q \leq nN$ and $nN \leq q$ only so that we can give a slightly sharper bound in the latter case. Also we allow cases to overlap for the sake of typographical and conceptual convenience.) In fact, we find it convenient to factor $\zeta(q)$ into "essential" an "non-essential" parts $\zeta'(q)$ and $\zeta''(q)$:

$$\zeta''(q) = \begin{cases} 3\log(q) + 2 & \text{if } q \leq N, \\ 7n & \text{if } N \leq q \leq nN, \\ 14 & \text{if } nN \leq q. \end{cases} \qquad \zeta'(q) = \begin{cases} 1 & \text{if } q \leq N, \\ q/N & \text{if } q \geq N. \end{cases} \tag{8}$$

Thus $\zeta(q) = \zeta''(q)\zeta'(q)$. Note also that $\zeta(q) \leq 2^\kappa$ by the wlog assumptions made in (6).

BAD TRANSCRIPTS. We say that a transcript $\tau = (\star, Q_E, k^*)$ is *bad* if either (i) $k_i^* = k_j^*$ for some $i \neq j$, or (ii) $\mathsf{fwd}(\tau) \geq \zeta(q)$ or $\mathsf{bwd}(\tau) \geq \zeta(q)$, or (iii) there exists some h, $0 \leq h \leq \ell - 1$ such that

$$\mathsf{fitkey}(\tau, \ell, h) \geq 1,$$

or (iv) there exists some r, $1 \leq r \leq \ell$ and some h, $0 \leq h \leq \ell - 1$ such that

$$\mathsf{fitkey}(\tau, r, h) \geq Cz_r.$$

where

$$z_r := \min\{q, N\} \cdot \left(\frac{\zeta'(q)}{2^\kappa} \right)^{\lceil r/2 \rceil}. \tag{9}$$

We let $\mathcal{T}_2$ be the set of bad transcripts, and let $\mathcal{T}_1 = \mathcal{T} \backslash \mathcal{T}_2$. One can note that every transcript with $\star = \bot$ is a bad transcript, since in that case $k_\ell^* = k_i^*$ for some $i \neq \ell$.

BOUNDING THE PROBABILITY OF BAD TRANSCRIPTS. Here we attach ourselves to upper bounding $\Pr[Y \in \mathcal{T}_2]$, as required by the H-coefficient technique. This is the probability of obtaining a bad transcript in the ideal world.

The probability that two subkeys of k^* are equal is obviously at most $\binom{\ell}{2} 2^{-\kappa} \leq \ell^2 / 2^{\kappa+1}$. For the other two events we need the help of the following lemmas:

Lemma 1. *One has*

$$\Pr_{\tau \sim Y}[\mathsf{fwd}(\tau) \geq \zeta(q)] \leq \frac{2}{N} \qquad and \qquad \Pr_{\tau \sim Y}[\mathsf{bwd}(\tau) \geq \zeta(q)] \leq \frac{2}{N}$$

for all q, n.

(Here $\Pr_{\tau \sim Y}$ indicates that τ is sampled according to the ideal world distribution on transcripts. The same probabilities could equivalently be written $\Pr[\mathsf{fwd}(Y) \geq \zeta(q)]$, $\Pr[\mathsf{bwd}(Y) \geq \zeta(q)]$.)

Lemma 2. *One has*

$$\Pr_{\tau \sim Y}[\mathsf{fitkey}(\tau, \ell, h) \geq 1 \wedge \mathsf{fwd}(\tau) \leq \zeta(q) \wedge \mathsf{bwd}(\tau) \leq \zeta(q)] \leq 2^\ell \zeta''(q)^{\lceil \ell/2 \rceil} z_\ell$$

for each $0 \leq h \leq \ell - 1$, and

$$\Pr_{\tau \sim Y}[\mathsf{fitkey}(\tau, r, h) \geq Cz_r \wedge \mathsf{fwd}(\tau) \leq \zeta(q) \wedge \mathsf{bwd}(\tau) \leq \zeta(q)] \leq \frac{2^r \zeta''(q)^{\lceil r/2 \rceil}}{C}$$

for each $1 \leq r \leq \ell$, $0 \leq h \leq \ell - 1$ with z_r as defined in (9).

We can combine Lemmas 1 and 2 by a union bound. When $q \geq N$ condition (iii) is implied by condition (iv) since

$$Cz_\ell = CN \cdot \left(\frac{q}{2^{\kappa+n}} \right)^{\lceil \ell/2 \rceil}$$

is less than 1 by (6). In this case, therefore, we don't need to incorporate the first part of Lemma 2 into the union bound. Using the fact that $r \leq \ell$ and that $\zeta''(q) \geq 1$ we can upper bound $\zeta''(q)^{\lceil r/2 \rceil}$ by $\zeta''(q)^{\lceil \ell/2 \rceil}$, thus obtaining

$$\Pr[Y \in \mathcal{T}_2] \leq \begin{cases} \frac{\ell^2}{2^{\kappa+1}} + \frac{4}{N} + 2^\ell \zeta''(q)^{\lceil \ell/2 \rceil} \cdot (\ell^2/C) & \text{if } q \geq N, \\ \frac{\ell^2}{2^{\kappa+1}} + \frac{4}{N} + 2^\ell \zeta''(q)^{\lceil \ell/2 \rceil} \cdot (\ell z_\ell + \ell^2/C) & \text{if } q \leq N \end{cases} \tag{10}$$

since there are ℓ choices for h and ℓ^2 choices for the pair (r, h).

The proof of Lemma 1 (which involves a few subtleties because permutations "lose randomness" after $\approx 2^n$ queries) can be found in the paper's full version [10].

For Lemma 2, a key component is given by the following proposition, which happens to be a key part of our proof and which sharpens similar bounds found in [4,17]:

Proposition 1. *Assume* $\tau = (\star, Q_E, k^*)$ *is a q-query transcript such that* $\mathsf{fwd}(\tau)$ $\leq \zeta(q)$, $\mathsf{bwd}(\tau) \leq \zeta(q)$. *Then the total number of r-chains of* τ *starting at position* h *is at most*

$$2^r \cdot \min\{q, N\} \cdot \zeta(q)^{\lceil r/2 \rceil} 2^{\kappa \lfloor r/2 \rfloor}.$$

Proof. Let $\nu = (h, x_h, k_{h+1}, x_{h+1}, \ldots, k_{h+r}, x_{h+r})$ be an r-chain of τ. Thus either $(k_i, x_{i-1}, x_i) \in Q_E^+$ or $(k_i, x_{i-1}, x_i) \in Q_E^-$ for $h+1 \leq i \leq h+r$. Let ν's *signature* be the string $sig^\nu \in \{+, -\}^r$ such that $(k_i, x_{i-1}, x_i) \in Q_E^{sig_i^\nu}$ for $h+1 \leq i \leq h+r$.

We start by fixing a signature $sig^0 \in \{+, -\}^r$ and by upper bounding the number of r-chains ν of τ starting at position h such that $sig^\nu = sig^0$. We can assume without loss of generality that sig^0 contains at least as many $-$'s as $+$'s, i.e., that the number of $-$'s is at least $\lceil r/2 \rceil$.

If $\nu = (h, x_h, k_{h+1}, x_{h+1}, \ldots, k_{h+r}, x_{h+r})$ is a ν-chain with signature sig^0 then there are, firstly, at most

$$\min\{q, N\}$$

choices for x_h given that $(k_{h+1}, x_h, x_{h+1}) \in Q_E$. Then, presuming x_h fixed, there are at most 2^κ choices for x_{h+1} if $sig_1^0 = +$ and at most $\zeta(q)$ choices for x_{h+1} if $sig_1^0 = -$, given that τ is a transcript such that $\mathsf{bwd}(\tau) \leq \zeta(q)$. Similarly, each subsequent step introduces a factor of either 2^κ or $\zeta(q)$ depending on the sign of that step in sig^0. Hence (and since $2^\kappa \geq \zeta(q)$) the total number of choices for $x_h, k_{h+1}, \ldots, x_{h+r}$ is at most

$$\min\{q, N\} \cdot \zeta(q)^{\lceil r/2 \rceil} 2^{\kappa \lfloor r/2 \rfloor}.$$

Multiplying by 2^r to account for all possible signatures concludes the proof. $\square$

Proof of Lemma 2. Since $\Pr[A \wedge B] \leq \Pr[A|B]$ we have

$$\Pr_{\tau \sim Y}[\mathsf{fitkey}(\tau, r, h) \geq T \wedge \mathsf{fwd}(\tau) \leq \zeta(q) \wedge \mathsf{bwd}(\tau) \leq \zeta(q)]$$
$$\leq \Pr_{\tau \sim Y}[\mathsf{fitkey}(\tau, r, h) \geq T \mid \mathsf{fwd}(\tau) \leq \zeta(q) \wedge \mathsf{bwd}(\tau) \leq \zeta(q)]$$

where $T \in \{Cz_r, 1\}$ is the bound we want to prove. When we condition on $\mathsf{fwd}(\tau) \leq \zeta(q) \wedge \mathsf{bwd}(\tau) \leq \zeta(q)$, however, k^* is still independent uniformly at random (being entirely independent from Q_E in the ideal world), and so the expected number of r-chains that fit τ at position h is upper bounded by

$$2^r \cdot \min\{q, N\} \cdot \zeta(q)^{\lceil r/2 \rceil} 2^{\kappa \lfloor r/2 \rfloor} \frac{1}{2^{\kappa r}} \tag{11}$$

by Proposition 1. (Each r-chain of Q_E, indeed, has probability of exactly $1/2^{\kappa r}$ of being "hit" by k^*.) Since $r - \lfloor r/2 \rfloor = \lceil r/2 \rceil$, (11) can be written

$$2^r \zeta''(q)^{\lceil r/2 \rceil} \min\{q, N\} \left(\frac{\zeta'(q)}{2^\kappa} \right)^{\lceil r/2 \rceil} = 2^r \zeta''(q)^{\lceil r/2 \rceil} z_r$$

with z_r as defined in (9). It thus follows by Markov's inequality that

$$\Pr_{\tau \sim Y}[\text{fitkey}(\tau, \ell, h) \geq 1 \wedge \text{fwd}(\tau) \leq \zeta(q) \wedge \text{bwd}(\tau) \leq \zeta(q)] \leq 2^\ell \zeta''(q)^{\lceil \ell/2 \rceil} z_\ell$$

and

$$\Pr_{\tau \sim Y}[\text{fitkey}(\tau, r, h) \geq C z_r \mid \text{fwd}(\tau) \leq \zeta(q) \wedge \text{bwd}(\tau) \leq \zeta(q)] \leq \frac{2^r \zeta''(q)^{\lceil r/2 \rceil}}{C}$$

which proves Lemma 2 and inequality (10). $\qquad\square$

REMAINING STEPS. Having upper bounded the probability of bad transcripts, the rest of the proof concerns itself with lower bounding the ratio

$$\frac{\Pr[X = \tau]}{\Pr[Y = \tau]}$$

for good transcripts τ, and more precisely of showing this ratio is at least $1 - \varepsilon$ for

$$\varepsilon = \begin{cases} \ell C^{\lceil (\ell+1)/2 \rceil} N \left(\frac{8q}{2^{\kappa+n}} \right)^{\lceil \ell/2 \rceil} & \text{if } q \leq N, \\ \frac{q^2 \ell}{N} C^{\lceil (\ell+1)/2 \rceil} \left(\frac{8}{2^\kappa} \right)^{\lceil \ell/2 \rceil} & \text{if } q \geq N. \end{cases}$$

For reasons of space we leave this part of the proof to the full version [10]. The overall approach, however, is quite similar to that espoused by Chen and Steinberger [8], and the main technical tool required for this part of the proof is indeed their own "path-completion lemma".

Acknowledgments. Yuanxi Dai was supported by the National Basic Research Program of China Grant 2011CBA00300, 2011CBA00301 and the National Natural Science Foundation of China Grant 61033001, 61361136003. Jooyoung Lee was supported by Basic Science Research Program through the National Research Foundation of Korea (NRF) funded by the Ministry of Education (NRF-2013R1A1A2007488). Bart Mennink was supported by the Research Fund KU Leuven, OT/13/071 and the Research Council KU Leuven: GOA TENSE (GOA/11/007). John Steinberger was supported by the National Basic Research Program of China Grant 2011CBA00300, 2011CBA00301, the National Natural Science Foundation of China Grant 61033001, 61361136003, and by the China Ministry of Education grant number 20121088050.

References

1. Aiello, W., Bellare, M., Di Crescenzo, G., Venkatesan, R.: Security amplification by composition: the case of doubly-iterated, ideal ciphers. In: Krawczyk, H. (ed.) CRYPTO 1998. LNCS, vol. 1462, pp. 390–407. Springer, Heidelberg (1998)
2. ANSI X9.52: Triple Data Encryption Algorithm Modes of Operation, withdrawn (1998)
3. Armknecht, F., Fleischmann, E., Krause, M., Lee, J., Stam, M., Steinberger, J.: The preimage security of double-block length compression functions. In: Lee, D.H., Wang, X. (eds.) ASIACRYPT 2011. LNCS, vol. 7073, pp. 233–251. Springer, Heidelberg (2011)
4. Bellare, M., Rogaway, P.: The security of triple encryption and a framework for code-based game-playing proofs. In: Vaudenay, S. (ed.) EUROCRYPT 2006. LNCS, vol. 4004, pp. 409–426. Springer, Heidelberg (2006)
5. Bellare, M., Rogaway, P.: Code-based game-playing proofs and the security of triple encryption. IACR eprint report, http://eprint.iacr.org/2004/331
6. Black, J.A., Rogaway, P., Shrimpton, T.: Black-Box Analysis of the Block-Cipher-Based Hash-Function Constructions from PGV. In: Yung, M. (ed.) CRYPTO 2002. LNCS, vol. 2442, pp. 320–335. Springer, Heidelberg (2002)
7. Bogdanov, A., Knudsen, L.R., Leander, G., Standaert, F.-X., Steinberger, J., Tischhauser, E.: Key-Alternating Ciphers in a Provable Setting: Encryption Using a Small Number of Public Permutations. In: Pointcheval, D., Johansson, T. (eds.) EUROCRYPT 2012. LNCS, vol. 7237, pp. 45–62. Springer, Heidelberg (2012)
8. Chen, S., Steinberger, J.: Tight security bounds for key-alternating ciphers. In: Nguyen, P.Q., Oswald, E. (eds.) EUROCRYPT 2014. LNCS, vol. 8441, pp. 327–350. Springer, Heidelberg (2014)
9. Dai, Y., Steinberger, J.: Tight security bounds for multiple encryption. IACR Cryptology ePrint Archive, 2014/096, http://eprint.iacr.org/2014/096.pdf
10. Dai, Y., Lee, J., Mennink, B., Steinberger, J.: The security of multiple encryption in the ideal cipher model (Full version of this paper.) IACR Cryptology ePrint Archive
11. Diffie, W., Hellman, M.: Exhaustive cryptanalysis of the NBS data encryption standard. Computer 10(6), 74–84 (1997)
12. Dinur, I., Dunkelman, O., Keller, N., Shamir, A.: Efficient Dissection of Composite Problems, with Applications to Cryptanalysis, Knapsacks, and Combinatorial Search Problems. In: Safavi-Naini, R., Canetti, R. (eds.) CRYPTO 2012. LNCS, vol. 7417, pp. 719–740. Springer, Heidelberg (2012)
13. Even, S., Goldreich, O.: On the power of cascade ciphers. ACM Transactions on Computer Systems 3(2), 108–116 (1985)
14. Even, S., Mansour, Y.: A Construction of a Cipher From a Single Pseudorandom Permutation. In: Matsumoto, T., Imai, H., Rivest, R.L. (eds.) ASIACRYPT 1991. LNCS, vol. 739, pp. 210–224. Springer, Heidelberg (1993)
15. FIPS46-3: Data Encryption Standard. National Institute of Standards and Technology, withdrawn (1999)
16. Gaži, P.: Plain versus Randomized Cascading-Based Key-Length Extension for Block Ciphers. In: Canetti, R., Garay, J.A. (eds.) CRYPTO 2013, Part I. LNCS, vol. 8042, pp. 551–570. Springer, Heidelberg (2013)
17. Gaži, P., Maurer, U.: Cascade encryption revisited. In: Matsui, M. (ed.) ASIACRYPT 2009. LNCS, vol. 5912, pp. 37–51. Springer, Heidelberg (2009)

18. Gaži, P., Tessaro, S.: Efficient and Optimally Secure Key-Length Extension for Block Ciphers via Randomized Cascading. In: Pointcheval, D., Johansson, T. (eds.) EUROCRYPT 2012. LNCS, vol. 7237, pp. 63–80. Springer, Heidelberg (2012)

19. Kilian, J., Rogaway, P.: How to protect DES against exhaustive key search (an analysis of DESX). Journal of Cryptology 14(1), 17–35 (2001)

20. Krause, M., Armknecht, F., Fleischmann, E.: Preimage resistance beyond the birthday bound: Double-length hashing revisited. IACR eprint report, http://eprint.iacr.org/2010/519.pdf

21. Lampe, R., Patarin, J., Seurin, Y.: An Asymptotically Tight Security Analysis of the Iterated Even-Mansour Cipher. In: Wang, X., Sako, K. (eds.) ASIACRYPT 2012. LNCS, vol. 7658, pp. 278–295. Springer, Heidelberg (2012)

22. Lee, J.: Towards Key-Length Extension with Optimal Security: Cascade Encryption and Xor-cascade Encryption. In: Johansson, T., Nguyen, P.Q. (eds.) EUROCRYPT 2013. LNCS, vol. 7881, pp. 405–425. Springer, Heidelberg (2013)

23. Lee, J.: Tight Security for Triple Encryption. IACR Cryptology ePrint Archive, 2014/015, http://eprint.iacr.org/2014/015.pdf

24. Lee, J., Steinberger, J., Stam, M.: The preimage security of double-block-length compression functions. IACR eprint report, http://eprint.iacr.org/2011/210.pdf

25. Luby, M., Rackoff, C.: Pseudo-random permutation generators and cryptographic composition. In: STOC 1986: Proceedings of the 18th Annual ACM Symposium on Theory of Computing, pp. 356–363 (1986)

26. Lucks, S.: Attacking triple encryption. In: Vaudenay, S. (ed.) FSE 1998. LNCS, vol. 1372, pp. 239–253. Springer, Heidelberg (1998)

27. Maurer, U., Massey, J.L.: Cascade ciphers: The importance of being first. Journal of Cryptology 6(1), 55–61 (1993)

28. Maurer, U., Pietrzak, K., Renner, R.: Indistinguishability amplification. In: Menezes, A. (ed.) CRYPTO 2007. LNCS, vol. 4622, pp. 130–149. Springer, Heidelberg (2007)

29. Maurer, U., Tessaro, S.: Computational indistinguishability amplification: Tight product theorems for system composition. In: Halevi, S. (ed.) CRYPTO 2009. LNCS, vol. 5677, pp. 355–373. Springer, Heidelberg (2009)

30. Mennink, B., Preneel, B.: Triple and Quadruple Encryption: Bridging the Gap. IACR Cryptology ePrint Archive, 2014/016, http://eprint.iacr.org/2014/016.pdf

31. Merkle, R., Hellman, M.: On the Security of Multiple Encryption. Communications of the ACM 24(7), 465–467 (1981); See also: Communications of the ACM 24(11), 776 (1981)

32. Myers, S.: On the development of block-ciphers and pseudo-random function generators using the composition and XOR operators. Master's thesis, University of Toronto (1999)

33. van Oorschot, P.C., Wiener, M.: Improving implementable meet-in-the-middle attacks by orders of magnitude. In: Koblitz, N. (ed.) CRYPTO 1996. LNCS, vol. 1109, pp. 229–236. Springer, Heidelberg (1996)

34. van Oorschot, P.C., Wiener, M.: A Known-Plaintext Attack on Two-Key Triple Encryption. In: Damgård, I.B. (ed.) EUROCRYPT 1990. LNCS, vol. 473, pp. 318–325. Springer, Heidelberg (1991)

35. NIST SP 800-67, Revision 1: Recommendation for the Triple Data Encryption Algorithm (TDEA) Block Cipher. National Institute of Standards and Technology (2012)

36. Patarin, J.: Etude de Génerateurs de Permutations Bases sur les Schemas du DES. In Ph.D. Thesis. Inria, Domaine de Voluceau, France (1991)
37. Patarin, J.: The "Coefficients H" Technique. In: Avanzi, R.M., Keliher, L., Sica, F. (eds.) SAC 2008. LNCS, vol. 5381, pp. 328–345. Springer, Heidelberg (2009)
38. Steinberger, J.: Improved Security Bounds for Key-Alternating Ciphers via Hellinger Distance, http://eprint.iacr.org/2012/481.pdf
39. Tessaro, S.: Security Amplification for the Cascade of Arbitrarily Weak PRPs: Tight Bounds via the Interactive Hardcore Lemma. In: Ishai, Y. (ed.) TCC 2011. LNCS, vol. 6597, pp. 37–54. Springer, Heidelberg (2011)

Minimizing the Two-Round
Even-Mansour Cipher

Shan Chen[1], Rodolphe Lampe[2], Jooyoung Lee[3],
Yannick Seurin[4], and John Steinberger[5]

[1] Tsinghua University, P.R. China
dragoncs16@gmail.com
[2] University of Versailles, France
rodolphe.lampe@gmail.com
[3] Sejong University, Seoul, Korea
jlee05@sejong.ac.kr
[4] ANSSI, Paris, France
yannick.seurin@m4x.org
[5] Tsinghua University, P.R. China
jpsteinb@gmail.com

Abstract. The r-round (iterated) *Even-Mansour cipher* (also known as *key-alternating cipher*) defines a block cipher from r fixed public n-bit permutations $P_1, \ldots, P_r$ as follows: given a sequence of n-bit round keys $k_0, \ldots, k_r$, an n-bit plaintext x is encrypted by xoring round key k_0, applying permutation P_1, xoring round key k_1, etc. The (strong) pseudo-randomness of this construction in the random permutation model (i.e., when the permutations $P_1, \ldots, P_r$ are public random permutation oracles that the adversary can query in a black-box way) was studied in a number of recent papers, culminating with the work of Chen and Steinberger (EUROCRYPT 2014), who proved that the r-round Even-Mansour cipher is indistinguishable from a truly random permutation up to $\mathcal{O}(2^{\frac{rn}{r+1}})$ queries of any adaptive adversary (which is an optimal security bound since it matches a simple distinguishing attack). All results in this entire line of work share the common restriction that they only hold under the assumption that *the round keys $k_0, \ldots, k_r$ and the permutations $P_1, \ldots, P_r$ are independent*. In particular, for two rounds, the current state of knowledge is that the block cipher $E(x) = k_2 \oplus P_2(k_1 \oplus P_1(k_0 \oplus x))$ is provably secure up to $\mathcal{O}(2^{2n/3})$ queries of the adversary, when k_0, k_1, and k_2 are three independent n-bit keys, and P_1 and P_2 are two independent random n-bit permutations. In this paper, we ask whether one can obtain a similar bound for the two-round Even-Mansour cipher *from just one n-bit key and one n-bit permutation*. Our answer is positive: when the three n-bit round keys k_0, k_1, and k_2 are adequately derived from an n-bit master key k, and the same permutation P is used in place of P_1 and P_2, we prove a qualitatively similar $\widetilde{\mathcal{O}}(2^{2n/3})$ security bound (in the random permutation model). To the best of our knowledge, this is the first "beyond the birthday bound" security result for AES-like ciphers that does not assume independent round keys.

J.A. Garay and R. Gennaro (Eds.): CRYPTO 2014, Part I, LNCS 8616, pp. 39–56, 2014.

Keywords: generalized Even-Mansour cipher, key-alternating cipher, indistinguishability, pseudorandom permutation, random permutation model, sum-capture problem.

1   Introduction

BACKGROUND. An elementary way to construct a block cipher with message space $\{0,1\}^n$ from r fixed and public n-bit permutations $P_1, \ldots P_r$ is to encrypt a plaintext x by computing

$$y = k_r \oplus P_r(k_{r-1} \oplus P_{r-1}(\cdots P_2(k_1 \oplus P_1(k_0 \oplus x)) \cdots)),$$

where $(k_0, \ldots, k_r)$ is a sequence of n-bit round keys which are usually derived from some master key K. This construction, which captures the high-level structure of (most) block cipher designs known as Substitution-Permutation Networks (SPNs), such as AES [12], PRESENT [7], or LED [20] to name a few, was coined a *key-alternating cipher* by Daemen and Rijmen [13].

In the random permutation model (i.e., when permutations $P_1, \ldots, P_r$ are modeled as public random permutation oracles), provable security results for this construction were first obtained for $r = 1$ round by Even and Mansour [16], who showed that the block cipher encrypting x into $k_1 \oplus P_1(k_0 \oplus x)$, where k_0 and k_1 are independent n-bit keys, and P_1 is a random permutation oracle, is secure up to $\mathcal{O}(2^{n/2})$ queries of the adversary.[1] For this reason, this construction is often referred to as the *Even-Mansour cipher*. Curiously, the general construction with $r > 1$ remained unstudied for a long while until a paper by Bogdanov *et al.* [8], who showed that for $r \geq 2$, security is guaranteed up to $\mathcal{O}(2^{2n/3})$ queries of the adversary. They also conjectured that the security should be $\mathcal{O}(2^{\frac{rn}{r+1}})$ for general r, which matches a simple distinguishing attack. Progress towards solving this conjecture was rather quick: Steinberger [32] proved security up to $\mathcal{O}(2^{3n/4})$ queries for $r \geq 3$, Lampe *et al.* [26] proved security up to $\mathcal{O}(2^{\frac{rn}{r+2}})$ queries for any even r, and finally Chen and Steinberger [9] resolved the conjecture and proved the $\mathcal{O}(2^{\frac{rn}{r+1}})$-security bound for any r. We stress that *all these results* only hold assuming that the $r + 1$ round keys and the r permutations are independent.[2]

OUR PROBLEM. Let us quickly recapitulate existing provable security results on the Even-Mansour cipher for a low number of rounds. For $r = 1$, we know that the single-key Even-Mansour cipher $x \mapsto k \oplus P(k \oplus x)$ ensures security up to $\mathcal{O}(2^{n/2})$ queries of the adversary. As pointed out by Dunkelman *et al.* [15], this construction is "minimal" in the sense that if one removes any component (either

---

[1] Actually it is not very hard to prove that a similar result holds when using $k_0 = k_1$.

[2] Actually, this is not perfectly accurate: one only needs the $r + 1$ round keys $(k_0, \ldots, k_r)$ to be r-wise independent [9], which can be obtained from only an rn-bit long master key, the most simple example being round keys of the form $(k'_1, k'_1 \oplus k'_2, k'_2 \oplus k'_3, \ldots, k'_{r-1} \oplus k'_r, k'_r)$, in which case the resulting iterated Even-Mansour cipher is exactly the cascade of r single-key one-round Even-Mansour ciphers $x \mapsto k'_i \oplus P_i(k'_i \oplus x)$.

Fig. 1. Two constructions of "minimal" two-round Even-Mansour ciphers provably secure up to $\widetilde{\mathcal{O}}(2^{\frac{2n}{3}})$ queries of any (adaptive) adversary. Left: π is a (fixed) linear orthomorphism of $\mathbb{F}_2^n$, and P is a public random permutation oracle. Right: P_1 and P_2 are two independent public random permutation oracles.

the addition of one of the keys, or the permutation P), the construction becomes trivially breakable. For the two-round Even-Mansour cipher, the best provable security result we have so far requires two independent n-bit permutations P_1 and P_2, and two independent n-bit keys (k, k') to construct three pairwise independent round keys, for example $(k, k' \oplus k, k')$. Concretely, the block cipher $x \mapsto k' \oplus P_2((k' \oplus k) \oplus P_1(k \oplus x))$ ensures security up to $\mathcal{O}(2^{2n/3})$ queries of the adversary. In this paper, we tackle the following question:

Can we obtain a $\mathcal{O}(2^{2n/3})$-security bound similar to the one proven for the two-round Even-Mansour cipher with (pairwise) independent round keys and independent permutations, from just one n-bit key k and one n-bit random permutation P?

This question is natural since in most (if not all) SPN block ciphers, round keys are derived from an n-bit master key (or more generally an ℓ-bit master key, where $\ell \in [n, 2n]$ is small compared with the total length of the round keys), and the same permutation, or very similar ones, are used at each round. It is therefore fundamental to determine whether security can actually benefit from the iterative structure and increase beyond the birthday bound, even though one does not use more key material nor more permutations than in the single-key one-round Even-Mansour cipher.

OUR RESULTS. We answer positively to the question above. Our main theorem states sufficient conditions on the way to derive three n-bit round keys (k_0, k_1, k_2) from one n-bit master key k so that the two-round Even-Mansour cipher defined from a single permutation $x \mapsto k_2 \oplus P(k_1 \oplus P(k_0 \oplus x))$ is secure up to $\widetilde{\mathcal{O}}(2^{2n/3})$ queries of the adversary, where the $\widetilde{\mathcal{O}}(\cdot)$ notation hides logarithmic (in $N = 2^n$) factors. In particular, such a good key-schedule $k \mapsto (k_0, k_1, k_2)$ can be constructed from any (fixed) linear orthomorphism of $\mathbb{F}_2^n$. A permutation π of $\{0, 1\}^n$ is called an orthomorphism if $x \mapsto x \oplus \pi(x)$ is also a permutation. The good cryptographic properties of orthomorphisms have already been noticed in a number of papers [29, 19], and are in particular used in Lai-Massey schemes [25, 34] such as the block ciphers IDEA [25] and FOX [22]. Our main theorem is as follows.

Theorem (Informal). *Let π be any (fixed) linear orthomorphism of $\mathbb{F}_2^n$, and let P be a public random n-bit permutation oracle. Then the block cipher with message space and key space $\{0,1\}^n$ defined as (see Figure 1, left)*

$$\mathsf{EM}_k^P(x) = k \oplus P(\pi(k) \oplus P(k \oplus x)) \qquad (\star)$$

is secure against any adversary making up to $\widetilde{\mathcal{O}}(2^{\frac{2n}{3}})$ queries to EM_k^P and P. (Queries can be adaptive and are allowed in both directions for EM_k^P and P).

We remark that if one omits π in construction $(\star)$, i.e., if one adds the same round key k each time, security drops back to $\mathcal{O}(2^{n/2})$ queries. More generally, if round keys are all equal and the same permutation P is used at each round of the iterated Even-Mansour cipher, security caps at $\mathcal{O}(2^{n/2})$ queries of the adversary, independently of the number r of rounds. This seems to be known as a folklore result about slide attacks [5, 6], but since we could not find a detailed exposition in the literature, we precisely describe and analyze this attack (as well as a simple extension for two rounds when the key-schedule simply consists in xoring constants to the master key) in this paper. Hence, construction $(\star)$ can be regarded as a "minimal" two-round Even-Mansour cipher delivering security beyond the birthday bound, since removing any component causes security to drop back to $\mathcal{O}(2^{n/2})$ queries at best (for π this follows from the slide attack just mentioned, while removing any instance of permutation P brings us back to a one-round Even-Mansour cipher). Additionally, we show that when using two independent public random permutations P_1 and P_2, the trivial key-schedule is sufficient: adding the same round key k at each round (see Figure 1, right) also yields a $\widetilde{\mathcal{O}}(2^{2n/3})$-security bound.

To the best of our knowledge, these are the first results proving "beyond the birthday bound" security for key-alternating ciphers such as AES that do not rely on the assumption that round keys are independent. This sheds some light on which exact properties are required from the key-schedule in order to lift the round keys independence assumption in provable security results. In particular, this seems to point out that a *pseudorandom* key-schedule is not needed (we remind the reader that our results come with the usual caveat that they are only proved in the very strong Random Permutation Model, and hence can only be taken as a heuristic security insurance once the inner permutation(s) are instantiated).

OVERVIEW OF OUR TECHNIQUES. In order to prove our results, we use the indistinguishability framework, namely we consider a distinguisher which must tell apart two worlds: the "real" world where it interacts with (EM_k^P, P), where EM_k^P is the Even-Mansour cipher instantiated with permutation P and a random key k, and the "ideal" world where it interacts with (E, P) where E is a random permutation independent from P. The distinguisher can make at most q_e queries to EM_k^P/E and at most q_p queries to P (all queries are adaptive and can be forward or backward, and we work in the information-theoretic setting, i.e., the adversary is computationally unbounded). In order to upper bound the distinguishing advantage of this attacker, we use, as already done in [9], the

H-coefficient method of Patarin [31]. In a nutshell, this technique consists in partitioning the set of all possible transcripts of the interaction between the distinguisher and the tuple of permutations into a set $\mathcal{T}_1$ of "good" transcripts and a set $\mathcal{T}_2$ of "bad" transcripts. Good transcripts $\tau \in \mathcal{T}_1$ have the property that the ratio of the probabilities to obtain τ in the real and in the ideal world is greater that $1 - \varepsilon_1$ for some small $\varepsilon_1 > 0$, while the probability to obtain any bad transcript $\tau \in \mathcal{T}_2$ (in the ideal world) is less than some small $\varepsilon_2 > 0$. Then the advantage of the distinguisher can be upper bounded by $\varepsilon_1 + \varepsilon_2$.

In order to get intuition about what hides behind good and bad transcripts, it helps to first look at an example of how an adversary might "get lucky" during an attack. Specifically, we focus on the following attack scenario (we assume that $q_e = q_p = q$ for simplicity). The distinguisher (adversary) $\mathcal{D}$ starts by making q arbitrary queries to EM_k^P/E, resulting in a set of q pairs $\mathcal{Q}_E = \{(x_1, y_1), \ldots, (x_q, y_q)\}$; then $\mathcal{D}$ determines the pair of sets (U, V) with $|U| = |V| = q$ and $U, V \subseteq \{0,1\}^n$, that maximizes the size of the set

$$\mathcal{K}(\mathcal{Q}_E, U, V) \stackrel{\text{def}}{=} \{k' \in \{0,1\}^n : \exists (x_i, y_i) \in \mathcal{Q}_E \text{ s.t. } x_i \oplus k' \in U, y_i \oplus k' \in V\}, \quad (1)$$

and $\mathcal{D}$ queries $P(u)$, $P^{-1}(v)$ for all $u \in U$, $v \in V$. (This makes $2q$ queries to P instead of q, but this small constant factor is unimportant for the sake of intuition.) Note that if $\mathcal{D}$ is in the real world and if the real key k is in the set $\mathcal{K}(\mathcal{Q}_E, U, V)$ defined in (1), then $\mathcal{D}$ can see that one of its EM_k^P/E-queries is compatible with two of its P-queries with respect to k (in more detail, there exists a value i and queries (u, v), (u', v') to P such that $x_i \oplus k = u$, $v \oplus \pi(k) = u'$, and $v' \oplus k = y_i$). Elementary probabilistic considerations show that such a "complete cycle" will occur for at most a handful of keys in $\mathcal{K}(\mathcal{Q}_E, U, V)$, so that "false alerts" can be quickly weeded out and the correct key k validated in a few extra queries, all assuming $k \in \mathcal{K}(\mathcal{Q}_E, U, V)$. Moreover, heuristic considerations indicate that k will be in $\mathcal{K}(\mathcal{Q}_E, U, V)$ with probability $|\mathcal{K}(\mathcal{Q}_E, U, V)|/2^n$. In particular, thus, it becomes necessary to show that $|\mathcal{K}(\mathcal{Q}_E, U, V)|$ is significantly smaller than 2^n with high probability over $\mathcal{Q}_E$, i.e., that

$$\max_{\substack{U,V \subseteq \{0,1\}^n \\ |U|=|V|=q}} |\{k' \in \{0,1\}^n : \exists (x_i, y_i) \in \mathcal{Q}_E \text{ s.t. } x_i \oplus k' \in U, y_i \oplus k' \in V\}| \quad (2)$$

is significantly smaller than 2^n with high probability over $\mathcal{Q}_E$, in order to show that $\mathcal{D}$ has small advantage at q queries. One of the criteria that can make a transcript "bad" in our proof happens to be, precisely, if the set of queries $\mathcal{Q}_E$ to EM_k^P/E contained within the transcript is such that (2) is larger than desirable. (Jumping ahead, $\mathcal{K}(\mathcal{Q}_E, U, V)$ will be re-baptized BadK_1 in Definition 1 of a bad transcript.)

To elaborate a little more on this, note that

$$|\mathcal{K}(\mathcal{Q}_E, U, V)| \leq |\{(k', u, v) \in \{0,1\}^n \times U \times V :$$
$$k' \oplus u = x_i, k' \oplus v = y_i \text{ for some } 1 \leq i \leq q\}|$$
$$= |\{(i, u, v) \in \{1, \ldots, q\} \times U \times V : x_i \oplus y_i = u \oplus v\}|.$$

Also note that the set of values $\{x_i \oplus y_i : (x_i, y_i) \in \mathcal{Q}_E\}$ is essentially a random set since if the i-th query to EM_k^P/E is forward then y_i comes at random from a large set, whereas otherwise x_i comes at random from a large set. Moreover, as a matter of fact, the problem of upper bounding

$$\mu(A) \stackrel{\text{def}}{=} \max_{\substack{U,V \subseteq \{0,1\}^n \\ |U|=|V|=q}} |\{(a, u, v) \in A \times U \times V : a = u \oplus v\}$$

for a *truly random* set $A \subseteq \{0,1\}^n$ of size q has already been studied before [3, 21, 1, 24, 33], being dubbed[3] the *sum-capture problem* in [33]. One of the main known results [3, 33] on the sum-capture problem is that $\mu(A)$ is upper bounded by roughly $q^{3/2}$ for $q \leq 2^{2n/3}$. Surprisingly enough, this bound is exactly sufficient for our application, since $q^{3/2} \ll 2^n$ for $q \ll 2^{2n/3}$. (Implying, thus, that (2) is far from 2^n as long as q remains beneath $2^{2n/3}$, as desired.) Our own setting is, of course, slightly different, since the set $\{x_i \oplus y_i : (x_i, y_i) \in \mathcal{Q}_E\}$ isn't, unlike A, a purely random set of size q. Other complications also arise: in the general case where the three round keys (k_0, k_1, k_2) are derived from the n-bit master key k using non-trivial (bijective) key derivation functions $\gamma_i : k \mapsto k_i$, $\mathcal{K}(\mathcal{Q}_E, U, V)$ takes the more complicated form

$$\{k' \in \{0,1\}^n : \exists (x_i, y_i) \in \mathcal{Q}_E \text{ s.t. } x_i \oplus \gamma_0(k') \in U, y_i \oplus \gamma_2(k') \in V\},$$

so that we have to upper bound

$$|\{(i, u, v) \in \{1, \dots, q\} \times U \times V : x_i \oplus u = \gamma_0 \circ \gamma_2^{-1}(y_i \oplus v)\}|.$$

All this means that we have to carefully adapt (and to some degree significantly extend) the Fourier-analytic techniques used in [3, 33].

Once the probability to obtain a bad transcript has been upper bounded, the second part of the proof is to show that the ratio between the probabilities to obtain any good transcript in the real and the ideal world is close to 1. This part is in essence a permutation counting argument. When the two permutations are independent (Figure 1, right), the counting argument is not overly complicated. While we could, in principle, re-use the general results of [9], we expose it in the full version of this paper [10] since it constitutes a good warm-up for the reader before the more complicated counting in the subsequent section. For the single-permutation case, things become much more involved: first, we need to consider more conditions defining bad transcripts; and second, the permutation counting itself becomes much more intricate. Interestingly, this part is related to the following simple to state (yet to the best of our knowledge unexplored) problem: how many queries are needed to distinguish a random squared permutation $P \circ P$ (where P is uniformly random) from a uniformly random permutation E?

RELATED WORK. Two recent papers analyzed a stronger security property of the iterated Even-Mansour cipher than mere pseudorandomness, namely indifferentiability from an ideal cipher [2, 27]. Aside with provable security results

---

[3] The terminology is attributed to Mario Szegedy.

already mentioned, a number of papers explored attacks on the (iterated) Even-Mansour cipher for one round [11, 6, 15], two rounds [30], three rounds [14], and four rounds [4].

Gazi and Tessaro [18] considered a construction they named 2XOR, which is a variant of the DESX [23] and "Xor-Cascade" [17, 28] key-length extension methods. Given a block cipher E with message space $\{0,1\}^n$ and key space $\{0,1\}^\kappa$, the 2XOR construction defines a new block cipher with message space $\{0,1\}^n$ and key space $\{0,1\}^{\kappa+n}$ as

$$2\mathsf{XOR}^E_{z,k}(x) = E_{z_2}(k \oplus E_{z_1}(k \oplus x)),$$

where (z_1, z_2) are pairwise distinct sub-keys derived from $z \in \{0,1\}^\kappa$. They showed that, when the underlying block cipher E is modeled as an ideal cipher, this construction is secure up to $\mathcal{O}(2^{\kappa+n/2})$ queries to E, even when the adversary can make all possible 2^n queries to the permutation oracle (which, in the indistinguishability experiment, is either $2\mathsf{XOR}^E_{z,k}$ or an independent random permutation). Considering a block cipher E with key-length $\kappa = 1$, one obtains a construction which is similar to the two-round Even-Mansour cipher of Figure 1, right, where the last key addition would be omitted.[4] Hence, the Gazi-Tessaro result says that this construction is secure for $q_e = 2^n$ and $q_p = \mathcal{O}(2^{n/2})$.[5] Our own results are incomparable with the one of [18]. First, the third key addition is omitted in the 2XOR construction. Second, our bounds are more general: they hold for any value of q_e and q_p as long as $q_e < 2^{2n/3}$ and $q_p < 2^{2n/3}$. Though our bounds become meaningless for $q_e = 2^n$, they show that when $q_e < 2^{2n/3}$ (an interesting case in practice since an attacker will not always have access to the entire codebook), security is ensured up to $\widetilde{\mathcal{O}}(2^{2n/3})$ queries to the internal permutations (something that cannot be derived from the result of [18]).

OPEN QUESTIONS. Currently, our results only apply when the key derivation functions mapping the master key to the round keys are *linear* bijective functions of $\mathbb{F}_2^n$. This is due to the fact that the proof of our sum-capture theorem in Section 3 requires linear mappings. It is an open question whether this theorem can be extended to nonlinear (bijective) mappings as well. A second tantalizing yet challenging open problem is of course to generalize our results to larger

---

[4] There is a slight subtlety here: in the 2XOR construction used with a block cipher with key-length $\kappa = 1$, i.e., a pair of permutations (P_1, P_2), there is an additional key bit z (hidden to the distinguisher) which tells in which order the two permutations are called.

[5] This is in fact very closely related to the security result for the single-key one-round Even-Mansour cipher up to $\mathcal{O}(2^{n/2})$ queries to the inner and outer permutations [15]. In the Gazi-Tessaro case with $\kappa = 1$, the adversary is given an arbitrary permutation E, and must distinguish, given access to (P_1, P_2), whether P_1 and P_2 are independent, or whether $P_2(k \oplus P_1(k \oplus x)) = E(x)$ for some random key k. In the single-key one-round Even-Mansour case, the adversary must distinguish, given access to (P_1, P_2), whether P_1 and P_2 are independent, or whether $k \oplus P_1(k \oplus x) = P_2(x)$, i.e., $P_2^{-1}(k \oplus P_1(k \oplus x)) = x$. These are very similar problems, the latter being (up to changing P_2 into P_2^{-1}) a special case of the former with E the identity.

numbers of rounds. Namely, for $r > 2$, can we find sufficient conditions on the key-schedule such that the r-round single-permutation Even-Mansour cipher ensures security up to $\widetilde{\mathcal{O}}(2^{\frac{rn}{r+1}})$ queries of the adversary? We stress that even the simpler case where permutations are independent and round keys are identical seems hard to tackle for $r > 2$: we currently have no idea of how to extend our sum-capture result in order to upper bound the probability of bad transcripts even in the case $r = 3$.

It would also be interesting to reduce the *time* complexity of attacks against the two-round Even-Mansour cipher (potentially down to $\mathcal{O}(2^{2n/3})$). Currently, the best known attack (for the case of independent permutations and identical round keys) has time complexity $\mathcal{O}(2^{n-\log_2 n})$ [15]. Since our focus in this paper is on query complexity, we have not investigated whether this attack applies to the single-permutation variant $(\star)$ as well.

ORGANIZATION. We start in Section 2 by setting the notation, giving the necessary background on the H-coefficient technique, and proving some helpful lemmas. In Section 3, which is self-contained, we prove our new sum-capture result, which might be of independent interest. Section 4 contains our main provable security result for the "minimized" variant of the single-permutation two-round Even-Mansour cipher (Figure 1, left). The case where the two permutations are independent and the three round keys are identical (Figure 1, right) is treated in the full version of the paper [10]. The permutation counting argument in that case serves as a good exercise before reading the corresponding one for the single-permutation case (Lemma 3). In the full version of the paper [10], we also detail slide attacks against the iterated Even-Mansour cipher.

2   Preliminaries

NOTATION. In all the following, we fix an integer $n \geq 1$, and we write $N = 2^n$. The set of all permutations on $\{0,1\}^n$ will be denoted $\mathcal{P}_n$. For integers $1 \leq s \leq t$, we will write $(t)_s = t(t-1)\cdots(t-s+1)$ and $(t)_0 = 1$ by convention. Given $\mathcal{Q} = ((x_1, y_1), \ldots, (x_q, y_q))$, where the x_i's are pairwise distinct n-bit strings and the y_i's are pairwise distinct n-bit strings, and a permutation $P \in \mathcal{P}_n$, we say that P extends $\mathcal{Q}$, denoted $P \vdash \mathcal{Q}$, if $P(x_i) = y_i$ for $i = 1, \ldots, q$. When two sets A and B are disjoint, we denote $A \sqcup B$ their (disjoint) union. We denote $\mathbb{F}_2 \simeq \{0,1\}$ the field with two elements, and $\mathbb{F}_2^n$ the vector space of dimension n over $\mathbb{F}_2$. The general linear group of degree n over $\mathbb{F}_2$, i.e., the set of all automorphisms (linear bijective mappings) of $\mathbb{F}_2^n$, will be denoted $\mathsf{GL}(n)$.

THE GENERALIZED EVEN-MANSOUR CIPHER. Fix integers $n, r, m, \ell \geq 1$. Let $\phi : \{1, \ldots, r\} \to \{1, \ldots, m\}$ be an arbitrary function, and $\gamma = (\gamma_0, \ldots, \gamma_r)$ be a $(r+1)$-tuple of functions from $\{0,1\}^\ell$ to $\{0,1\}^n$. The r-round Generalized Even-Mansour construction $\mathsf{EM}[n, r, m, \ell, \phi, \gamma]$ specifies, from any m-tuple $\boldsymbol{P} = (P_1, \ldots, P_m)$ of permutations on $\{0,1\}^n$, a block cipher with message space $\{0,1\}^n$ and key space

$\{0, 1\}^\ell$, simply denoted EM^P in the following (parameters $[n, r, m, \ell, \phi, \gamma]$ are implicit and will always be clear from the context), which maps a plaintext $x \in \{0, 1\}^n$ and a key $K \in \{0, 1\}^\ell$ to the ciphertext defined by (see Figure 2):

$$\mathsf{EM}^P(K, x) = \gamma_r(K) \oplus P_{\phi(r)}(\gamma_{r-1}(K) \oplus P_{\phi(r-1)}(\cdots P_{\phi(1)}(\gamma_0(K) \oplus x) \cdots)).$$

We denote $\mathsf{EM}_K^P : x \mapsto \mathsf{EM}^P(K, x)$ the Even-Mansour cipher instantiated with key K (hence, syntactically, EM_K^P is a permutation on $\{0, 1\}^n$).

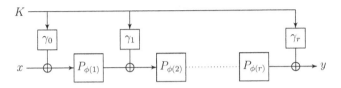

Fig. 2. The r-round Generalized Even-Mansour cipher

All previous work about the indistinguishability of the Even-Mansour cipher [8, 26, 32, 9] considered the case where all permutations and all round keys are independent, namely $m = r$, ϕ is the identity function, $\ell = (r + 1)n$, and γ_i simply selects the i-th n-bit string of $K = (k_0, \dots, k_r)$.

In the following, we will focus in particular on two special cases:

- the case where permutations are independent and the same n-bit key k is used at each round, namely $m = r$, ϕ is the identity function, $\ell = n$, and all γ_i's are the identity function, in which case we will simply denote $\mathsf{EMIP}[n, r]$ the resulting construction. Hence, for an r-tuple of permutations $P = (P_1, \dots, P_r)$, the block cipher EMIP^P maps a plaintext $x \in \{0, 1\}^n$ and a key $k \in \{0, 1\}^n$ to the ciphertext defined by:

$$\mathsf{EMIP}^P(k, x) = k \oplus P_r(k \oplus P_{r-1}(\cdots P_2(k \oplus P_1(k \oplus x)) \cdots)).$$

- the case where a single permutation P is used at each round, namely $m = 1$ and $\phi(i) = 1$ for $i = 1, \dots, r$, in which case the resulting construction will simply be denoted $\mathsf{EMSP}[n, r, \ell, \gamma]$. Hence, for a permutation P, the block cipher EMSP^P maps a plaintext $x \in \{0, 1\}^n$ and a key $K \in \{0, 1\}^\ell$ to the ciphertext defined by:

$$\mathsf{EMSP}^P(K, x) = \gamma_r(K) \oplus P(\gamma_{r-1}(K) \oplus P(\cdots P(\gamma_1(K) \oplus P(\gamma_0(K) \oplus x)) \cdots)).$$

When additionally $\ell = n$ (namely the master key length is equal to the block length), we overload the notation and simply denote $\mathsf{EMSP}[n, r, \gamma]$ the resulting construction.

SECURITY DEFINITION. To study the indistinguishability of the Generalized Even-Mansour cipher (in the Random Permutation Model), we consider a distinguisher $\mathcal{D}$ which interacts with a set of $m + 1$ permutation oracles on n bits that we denote generically $(P_0, P_1 \ldots, P_m) = (P_0, \boldsymbol{P})$. The goal of $\mathcal{D}$ is to distinguish whether it is interacting with $(\mathsf{EM}_K^{\boldsymbol{P}}, \boldsymbol{P})$, where $\boldsymbol{P} = (P_1, \ldots, P_m)$ are random and independent permutations and K is randomly chosen from $\{0, 1\}^\ell$ (we will informally refer to this case as the "real" world), or with $(E, \boldsymbol{P})$, where E is a random n-bit permutation independent from $\boldsymbol{P}$ (the "ideal" world). Note that in the latter case the distinguisher is simply interacting with $m + 1$ independent random permutations. We sometimes refer to the first permutation P_0 as the *outer* permutation, and to permutations $P_1, \ldots, P_m$ as the *inner* permutations. The distinguisher is adaptive, and can make both forward and backward queries to each permutation oracle, which corresponds to the notion of adaptive chosen-plaintext and ciphertext security (CCA). We consider computationally unbounded distinguishers, and we assume *wlog* that the distinguisher is deterministic and never makes useless queries (which means that it never repeats a query, nor makes a query $P_i^{-1}(y)$ if it received y as the answer to a previous query $P_i(x)$, or vice-versa).

The distinguishing advantage of $\mathcal{D}$ is defined as

$$\mathbf{Adv}(\mathcal{D}) = \left| \Pr\left[\mathcal{D}^{\mathsf{EM}_K^{\boldsymbol{P}}, \boldsymbol{P}} = 1\right] - \Pr\left[\mathcal{D}^{E, \boldsymbol{P}} = 1\right] \right|,$$

where the first probability is taken over the random choice of K and $\boldsymbol{P}$, and the second probability is taken over the random choice of E and $\boldsymbol{P}$. We recall that, even though this is not apparent from the notation, the distinguisher can make both forward and backward queries to each permutation oracle.

For q_e, q_p non-negative integers, we define the insecurity of the ideal[6] Generalized Even-Mansour cipher with parameters $(n, r, m, \ell, \phi, \boldsymbol{\gamma})$ as:

$$\mathbf{Adv}_{\mathsf{EM}[n, r, m, \ell, \phi, \boldsymbol{\gamma}]}^{\mathrm{cca}}(q_e, q_p) = \max_{\mathcal{D}} \mathbf{Adv}(\mathcal{D}),$$

where the maximum is taken over all distinguishers $\mathcal{D}$ making exactly q_e queries to the outer permutation and exactly q_p queries to each inner permutation. The notation is adapted naturally for the two special cases EMIP and EMSP defined above.

THE H-COEFFICIENT TECHNIQUE. We give here all the necessary background on the H-coefficient technique [31, 9] that we will use throughout this paper. All the information gathered by the distinguisher when interacting with the system of $m + 1$ permutations can be summarized in what we call the *transcript* of the interaction, which is the ordered list of queries and answers received from the system (i, b, z, z'), where $i \in \{0, \ldots, m\}$ names the permutation being queried, b is a bit indicating whether this is a forward or backward query, $z \in \{0, 1\}^n$ is the actual value queried and z' the answer. We say that a transcript is *attainable* (with respect to some fixed distinguisher $\mathcal{D}$) if there exists a tuple of permutations $(P_0, \ldots, P_m) \in (\mathcal{P}_n)^{m+1}$ such that the interaction of $\mathcal{D}$ with

---

[6] By ideal, we mean that this insecurity measure is defined in the Random Permutation Model for $P_1, \ldots, P_m$.

$(P_0, \ldots, P_m)$ yields this transcript (said otherwise, the probability to obtain this transcript in the "ideal" world is non-zero). In fact, an attainable transcript can be represented in a more convenient way that we will use in all the following. Namely, from the transcript we can build $m + 1$ lists of directionless queries $\mathcal{Q}_E = ((x_1, y_1), \ldots, (x_{q_e}, y_{q_e}))$, $\mathcal{Q}_{P_1} = ((u_{1,1}, v_{1,1}), \ldots, (u_{1,q_p}, v_{1,q_p}))$, $\ldots, \mathcal{Q}_{P_m} = ((u_{m,1}, v_{m,1}), \ldots, (u_{m,q_p}, v_{m,q_p}))$ as follows. For $j = 1, \ldots, q_e$, let $(0, b, z, z')$ be the j-th query to P_0 in the transcript: if this was a forward query then we set $x_j = z$ and $y_j = z'$, otherwise we set $x_j = z'$ and $y_j = z$. Similarly, for each $i = 1, \ldots, m$, and $j = 1, \ldots, q_p$, let (i, b, z, z') be the j-th query to P_i in the transcript: if this was a forward query then we set $u_{i,j} = z$ and $v_{i,j} = z'$, otherwise we set $u_{i,j} = z'$ and $v_{i,j} = z$. A moment of thinking should make it clear that for attainable transcripts there is a one-to-one mapping between these two representations. (Essentially this follows from the fact that the distinguisher is deterministic). Moreover, though we defined $\mathcal{Q}_E, \mathcal{Q}_{P_1}, \ldots, \mathcal{Q}_{P_m}$ as ordered lists, the order is unimportant (our formalization keeps the natural order induced by the distinguisher).

For convenience, and following [9], we will be generous with the distinguisher by providing it, at the end of its interaction, with the actual key K when it is interacting with (EM_K^P, P), or with a dummy key K selected uniformly at random when it is interacting with (E, P). This is without loss of generality since the distinguisher is free to ignore this additional information. Hence, all in all a transcript τ is a tuple $(\mathcal{Q}_E, \mathcal{Q}_{P_1}, \ldots, \mathcal{Q}_{P_m}, K)$. We refer to $(\mathcal{Q}_E, \mathcal{Q}_{P_1}, \ldots, \mathcal{Q}_{P_m})$ (without the key) as the *permutation transcript*, and we say that a transcript τ is attainable if the corresponding permutation transcript is attainable. We denote $\mathcal{T}$ the set of attainable transcripts. (Thus $\mathcal{T}$ depends on $\mathcal{D}$, as the notion of attainability depends on $\mathcal{D}$.) In all the following, we denote T_{re}, resp. T_{id}, the probability distribution of the transcript τ induced by the real world, resp. the ideal world (note that these two probability distributions depend on the distinguisher). By extension, we use the same notation to denote a random variable distributed according to each distribution.

In order to upper bound the advantage of the distinguisher, we will repeatedly use the following strategy: we will partition the set of attainable transcripts $\mathcal{T}$ into a set of "good" transcripts $\mathcal{T}_1$ such that the probabilities to obtain some transcript $\tau \in \mathcal{T}_1$ are close in the real and in the ideal world, and a set $\mathcal{T}_2$ of "bad" transcripts such that the probability to obtain any $\tau \in \mathcal{T}_2$ is small in the ideal world. More precisely, we will use the following result, which is proved in the full version of the paper [10].

Lemma 1. *Fix a distinguisher $\mathcal{D}$. Let $\mathcal{T} = \mathcal{T}_1 \sqcup \mathcal{T}_2$ be a partition of the set of attainable transcripts. Assume that there exists ε_1 such that for any $\tau \in \mathcal{T}_1$, one has[7]*

$$\frac{\Pr[T_{\mathrm{re}} = \tau]}{\Pr[T_{\mathrm{id}} = \tau]} \geq 1 - \varepsilon_1,$$

and that there exists ε_2 such that $\Pr[T_{\mathrm{id}} \in \mathcal{T}_2] \leq \varepsilon_2$. Then $\mathbf{Adv}(\mathcal{D}) \leq \varepsilon_1 + \varepsilon_2$.

---

[7] Recall that for an attainable transcript, one has $\Pr[T_{\mathrm{id}} = \tau] > 0$.

3   A Sum-Capture Theorem

In this section, we prove a variant of previous "sum-capture" results [3, 24, 33]. Informally, such results typically state that when choosing a random subset A of $\mathbb{Z}_2^n$ (or more generally any abelian group) of size q, the value

$$\mu(A) = \max_{\substack{U,V \subseteq \mathbb{Z}_2^n \\ |U|=|V|=q}} |\{(a,u,v) \in A \times U \times V : a = u \oplus v\}|$$

is close to its expected value q^3/N (if A, U, V were chosen at random), except with negligible probability. Here, we prove a result of this type for the setting where A arises from the interaction of an adversary with a random permutation P, namely $A = \{x \oplus y : (x,y) \in \mathcal{Q}\}$, where $\mathcal{Q}$ is the transcript of the interaction between the adversary and P. In fact our result is even more general, the special case just mentioned corresponding to Γ being the identity in the theorem below.

Theorem 1. *Fix an automorphism $\Gamma \in \mathsf{GL}(n)$. Let P be a uniformly random permutation of $\{0,1\}^n$, and let $\mathcal{A}$ be some probabilistic algorithm making exactly q (two-sided) adaptive queries to P. Let $\mathcal{Q} = ((x_1, y_1), \ldots, (x_q, y_q))$ denote the transcript of the interaction of $\mathcal{A}$ with P. For any two subsets U and V of $\{0,1\}^n$, let*

$$\mu(\mathcal{Q}, U, V) = |\{((x,y), u, v) \in \mathcal{Q} \times U \times V : x \oplus u = \Gamma(y \oplus v)\}|.$$

Then, assuming $9n \leq q \leq N/2$, one has

$$\Pr_{P,\omega}\left[\exists U, V \subseteq \{0,1\}^n : \mu(\mathcal{Q}, U, V) \geq \frac{q|U||V|}{N} + \frac{2q^2\sqrt{|U||V|}}{N} + 3\sqrt{nq|U||V|}\right]$$
$$\leq \frac{2}{N},$$

where the probability is taken over the random choice of P and the random coins ω of $\mathcal{A}$.

Proof. Deferred to the full version [10] for reasons of space. □

4   Security Proof for the Single Permutation Case

In this section, we study the security of the two-round Even-Mansour construction where a single permutation P is used instead of two independent permutations, namely $\mathsf{EMSP}[n, r, \ell, \gamma]$ (depicted on Figure 3). Because of the slide attack described in the full version of the paper [10], we know that we cannot simply use the same n-bit key k at each round if we aim at proving security beyond the birthday bound, so that some non-trivial key-schedule $\gamma = (\gamma_0, \gamma_1, \gamma_2)$, with $\gamma_i : \{0,1\}^\ell \to \{0,1\}^n$, is needed (we remain as general as possible in a first phase, and will only specify the key-schedule later on). Given a key $K \in \{0,1\}^\ell$, we denote $k_0 = \gamma_0(K)$, $k_1 = \gamma_1(K)$, and $k_2 = \gamma_2(K)$, so that:

$$\mathsf{EMSP}_K^P(x) = P(P(x \oplus k_0) \oplus k_1) \oplus k_2.$$

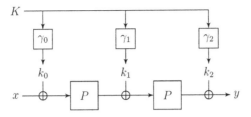

Fig. 3. The two-round Even-Mansour cipher with a single permutation and an arbitrary key-schedule

Let $\tau = (\mathcal{Q}_E, \mathcal{Q}_P, K)$, with $|\mathcal{Q}_E| = q_e$, $|\mathcal{Q}_P| = q_p$, and $K \in \{0,1\}^\ell$ be an attainable transcript. As previously, we start by defining the set of bad transcripts. In all the following, we let

$$M = \frac{q_e}{N^{\frac{1}{3}}}.$$

Definition 1 (Bad transcript, single-permutation case). *We say that a transcript* $\tau = (\mathcal{Q}_E, \mathcal{Q}_P, K) \in \mathcal{T}$ *is* bad *if*

$$K \in \mathsf{BadK} = \bigcup_{1 \leq i \leq 10} \mathsf{BadK}_i$$

where

$K \in \mathsf{BadK}_1 \Leftrightarrow \exists(x,y) \in \mathcal{Q}_E, \exists(u,v),(u',v') \in \mathcal{Q}_P : k_0 = x \oplus u \text{ and } k_2 = v' \oplus y$

$K \in \mathsf{BadK}_2 \Leftrightarrow \exists(x,y) \in \mathcal{Q}_E, \exists(u,v),(u',v') \in \mathcal{Q}_P : k_0 = x \oplus u \text{ and } k_1 = v \oplus u'$

$K \in \mathsf{BadK}_3 \Leftrightarrow \exists(x,y) \in \mathcal{Q}_E, \exists(u,v),(u',v') \in \mathcal{Q}_P : k_1 = v \oplus u' \text{ and } k_2 = v' \oplus y$

$K \in \mathsf{BadK}_4 \Leftrightarrow \exists(x,y),(x',y') \in \mathcal{Q}_E, \exists(u,v) \in \mathcal{Q}_P :$
$$k_0 = x \oplus u \text{ and } k_0 \oplus k_1 = v \oplus x'$$

$K \in \mathsf{BadK}_5 \Leftrightarrow \exists(x,y),(x',y') \in \mathcal{Q}_E, \exists(u,v) \in \mathcal{Q}_P :$
$$k_1 \oplus k_2 = y' \oplus u \text{ and } k_2 = v \oplus y$$

$K \in \mathsf{BadK}_6 \Leftrightarrow |\{((x,y),(u,v)) \in \mathcal{Q}_E \times \mathcal{Q}_P : x \oplus u = k_0\}| > \dfrac{M}{3}$

$K \in \mathsf{BadK}_7 \Leftrightarrow |\{((x,y),(u,v)) \in \mathcal{Q}_E \times \mathcal{Q}_P : v \oplus y = k_2\}| > \dfrac{M}{3}$

$K \in \mathsf{BadK}_8 \Leftrightarrow |\{((x,y),(u,v)) \in \mathcal{Q}_E \times \mathcal{Q}_P : x \oplus v = k_0 \oplus k_1\}| > \dfrac{M}{3}$

$K \in \mathsf{BadK}_9 \Leftrightarrow |\{((x,y),(u,v)) \in \mathcal{Q}_E \times \mathcal{Q}_P : u \oplus y = k_1 \oplus k_2\}| > \dfrac{M}{3}$

$K \in \mathsf{BadK}_{10} \Leftrightarrow |\{((x,y),(x',y')) \in \mathcal{Q}_E \times \mathcal{Q}_E : x \oplus y' = k_0 \oplus k_1 \oplus k_2\}| > M.$

Otherwise τ *is said* good. *We denote* $\mathcal{T}_2$ *the set of bad transcripts, and* $\mathcal{T}_1 = \mathcal{T} \backslash \mathcal{T}_2$ *the set of good transcripts.*

In this section, we focus on the case where $\ell = n$, namely the master key length is equal to the block length (and hence to the round keys length). We treat the (simpler) cases where the three round keys are independent, or derived from two independent n-bit keys, in the full version of the paper [10]. First, we specify conditions on the key-schedule that will allow us to upper bound the probability to obtain a bad transcript (in the ideal world).

Definition 2 (Good key-schedule). *We say that a key-schedule* $\gamma = (\gamma_0, \gamma_1, \gamma_2)$, *where* $\gamma_i : \{0,1\}^n \to \{0,1\}^n$, *is* good *if it satisfies the following conditions:*

(i) $\gamma_0, \gamma_1, \gamma_2 \in \mathsf{GL}(n)$ *(i.e., each* γ_i *is a linear bijective map of* $\mathbb{F}_2^n$*)*;
(ii) $\gamma_0 \oplus \gamma_1 \in \mathsf{GL}(n)$ *and* $\gamma_1 \oplus \gamma_2 \in \mathsf{GL}(n)$;
(iii) $\gamma_0 \oplus \gamma_1 \oplus \gamma_2$ *is a permutation over* $\{0,1\}^n$ *(non-necessarily linear over* $\mathbb{F}_2^n$*)*.

A simple way to build a good key-schedule is to take for γ_0 and γ_2 the identity, and $\gamma_1 = \pi$, where π is a linear orthomorphism of $\mathbb{F}_2^n$ (recall that a permutation π of $\{0,1\}^n$ is an orthomorphism if $x \mapsto x \oplus \pi(x)$ is also a permutation), so that the sequence of round keys is $(k, \pi(k), k)$. We give two simple examples of linear orthomorphisms which are attractive from an implementation point of view:

- When n is even, and $k = (k_L, k_R)$ where k_L and k_R are respectively the left and right halves of k, then

$$\pi : (k_L, k_R) \mapsto (k_R, k_L \oplus k_R)$$

  is a linear orthomorphism.
- Fix an irreducible polynomial p of degree n over $\mathbb{F}_2$ and identify $\mathbb{F}_2^n$ and the extension field $\mathbb{F}_{2^n}$ defined by p in the canonical way. Then, for any $c \in \mathbb{F}_{2^n} \setminus \{0, 1\}, k \mapsto c \odot k$ (where $\odot$ denotes the extension field multiplication) is a linear orthomorphism.

Lemma 2. *Let* $\gamma = (\gamma_0, \gamma_1, \gamma_2)$ *be a good key-schedule. Assume that* $9n \leq q_e, q_p \leq N/2$. *Then*

$$\Pr[\mathcal{T}_{id} \in \mathcal{T}_2] \leq \frac{10}{N} + \frac{4q_e^2 q_p + 7q_e q_p^2 + 4q_p^2 \sqrt{q_e q_p}}{N^2}$$

$$+ \frac{9q_p \sqrt{nq_e} + 6q_e \sqrt{nq_p}}{N} + \frac{q_e + 12q_p}{N^{\frac{2}{3}}}.$$

Proof. In the ideal world, sets BadK_i only depend on the random permutations E and P, and not on the key k, which is drawn uniformly at random at the end of the interaction of the distinguisher with (E, P). Moreover, the size of BadK_i for $i = 6$ to 10 can be upper bounded independently of E, P. Indeed, since γ_0, γ_2, $\gamma_0 \oplus \gamma_1$, $\gamma_1 \oplus \gamma_2$, and $\gamma_0 \oplus \gamma_1 \oplus \gamma_2$ are all permutations of $\{0,1\}^n$, one has, for any permutation transcript $(\mathcal{Q}_E, \mathcal{Q}_P)$,

$$|\mathsf{BadK}_6|, |\mathsf{BadK}_7|, |\mathsf{BadK}_8|, |\mathsf{BadK}_9| \leq \frac{3q_e q_p}{M} \quad \text{and} \quad |\mathsf{BadK}_{10}| \leq \frac{q_e^2}{M},$$

so that

$$\Pr\left[k \leftarrow_\$ \{0,1\}^n : k \in \bigcup_{i=6}^{10} \mathsf{BadK}_i\right] \leq \frac{12q_e q_p}{NM} + \frac{q_e^2}{NM} \leq \frac{q_e + 12q_p}{N^{\frac{2}{3}}}.$$

On the other hand, in order to upper bound $|\mathsf{BadK}_i|$ for $i = 1$ to 5, we need to appeal to the sum-capture theorem of Section 3. For a permutation transcript $(\mathcal{Q}_E, \mathcal{Q}_P)$, let

$$X = \{x \in \{0,1\}^n : (x,y) \in \mathcal{Q}_E\}, \qquad Y = \{y \in \{0,1\}^n : (x,y) \in \mathcal{Q}_E\},$$
$$U = \{u \in \{0,1\}^n : (u,v) \in \mathcal{Q}_P\}, \qquad V = \{v \in \{0,1\}^n : (u,v) \in \mathcal{Q}_P\}$$

denote the domains and the ranges of $\mathcal{Q}_E$ and $\mathcal{Q}_P$, respectively. Then one has

$$|\mathsf{BadK}_1| \leq \mu(\mathcal{Q}_E, U, V)$$
$$\overset{\mathrm{def}}{=} |\{((x,y),u,v) \in \mathcal{Q}_E \times U \times V : x \oplus u = \gamma_0 \circ \gamma_2^{-1}(y \oplus v)\}|$$
$$|\mathsf{BadK}_2| \leq \mu(\mathcal{Q}_P, X, U)$$
$$\overset{\mathrm{def}}{=} |\{((u,v),x,u') \in \mathcal{Q}_P \times X \times U : x \oplus u = \gamma_0 \circ \gamma_1^{-1}(v \oplus u')\}|$$
$$|\mathsf{BadK}_3| \leq \mu(\mathcal{Q}_P, V, Y)$$
$$\overset{\mathrm{def}}{=} |\{((u',v'),v,y) \in \mathcal{Q}_P \times V \times Y : v \oplus u' = \gamma_1 \circ \gamma_2^{-1}(v' \oplus y)\}|$$
$$|\mathsf{BadK}_4| \leq \mu(\mathcal{Q}_P, X, X)$$
$$\overset{\mathrm{def}}{=} |\{((u,v),x,x') \in \mathcal{Q}_P \times X \times X : x \oplus u = \gamma_0 \circ (\gamma_0 \oplus \gamma_1)^{-1}(v \oplus x')\}|$$
$$|\mathsf{BadK}_5| \leq \mu(\mathcal{Q}_P, Y, Y)$$
$$\overset{\mathrm{def}}{=} |\{((u,v),y,y') \in \mathcal{Q}_P \times Y \times Y : y' \oplus u = (\gamma_1 \oplus \gamma_2) \circ \gamma_2^{-1}(v \oplus y)\}|.$$

By our assumption that the key-schedule is good, we have that $\gamma_0 \circ \gamma_2^{-1}$, $\gamma_0 \circ \gamma_1^{-1}$, $\gamma_1 \circ \gamma_2^{-1}$, $\gamma_0 \circ (\gamma_0 \oplus \gamma_1)^{-1}$, and $\gamma_0 \circ (\gamma_0 \oplus \gamma_1)^{-1}$ are all automorphisms of $\mathbb{F}_2^n$. Hence, we can apply Theorem 1 (note that in order to apply this theorem to upper bound, say, $|\mathsf{BadK}_1|$, we consider the combination of the distinguisher $\mathcal{D}$ and permutation P as a probabilistic adversary $\mathcal{A}$ interacting with permutation E, resulting in transcript $\mathcal{Q}_E$). Thus, if we set

$$C_1 = \frac{q_e q_p^2}{N} + \frac{2q_e^2 q_p}{N} + 3q_p\sqrt{nq_e}$$
$$C_2 = C_3 = \frac{q_e q_p^2}{N} + \frac{2q_p^2\sqrt{q_e q_p}}{N} + 3q_p\sqrt{nq_e}$$
$$C_4 = C_5 = \frac{q_e^2 q_p}{N} + \frac{2q_e q_p^2}{N} + 3q_e\sqrt{nq_p},$$

one has $\Pr[E, P \leftarrow_\$ \mathcal{P}_n : |\mathsf{BadK}_i| \geq C_i] \leq 2/N$ for each $i = 1$ to 5. Since

$$\Pr[T_{\mathrm{id}} \in \mathcal{T}_2] \leq \sum_{i=1}^{5} \Pr[E, P \leftarrow_\$ \mathcal{P}_n : |\mathsf{BadK}_i| \geq C_i] + \frac{\sum_{i=1}^5 C_i}{N} + \frac{q_e + 12q_p}{N^{\frac{2}{3}}},$$

we get the final result. $\qquad\square$

In the second stage of the proof, it remains to show that for any good transcript τ, the ratio between the probabilities to obtain τ in the ideal world and the real world is close to 1. We have the following lemma, proved in the full version of the paper [10].

Lemma 3. *Assume that $N \geq 7^3$ and $4q_e + 2q_p \leq N$. Let $\tau = (\mathcal{Q}_E, \mathcal{Q}_P, K) \in \mathcal{T}_1$ be a good transcript. Then*

$$\frac{\Pr[T_{\mathrm{re}} = \tau]}{\Pr[T_{\mathrm{id}} = \tau]} \geq 1 - \varepsilon_1,$$

where

$$\varepsilon_1 = \frac{4q_e(q_e + q_p)^2}{N^2} + \frac{2q_e^2}{N^{\frac{4}{3}}} + \frac{20q_e}{N^{\frac{2}{3}}}.$$

Combining Lemmas 1, 2, and 3, we obtain the main theorem of this paper.

Theorem 2 (Single permutation and non-independent round keys).
Consider the single-permutation two-round Even-Mansour cipher $\mathsf{EMSP}[n, 2, \gamma]$ *with a good key-schedule γ (see Definition 2). Assume that $N \geq 7^3$, $9n \leq q_e, q_p \leq N/2$, and $4q_e + 2q_p \leq N$. Then*

$$\mathbf{Adv}^{\mathrm{cca}}_{\mathsf{EMSP}[n,2,\gamma]}(q_e, q_p) \leq \frac{10}{N} + \frac{4q_e^3 + 12q_e^2 q_p + 11 q_e q_p^2 + 4q_p^2 \sqrt{q_e q_p}}{N^2} + \frac{2q_e^2}{N^{\frac{4}{3}}}$$
$$+ \frac{9q_p\sqrt{nq_e} + 6q_e\sqrt{nq_p}}{N} + \frac{21q_e + 12q_p}{N^{\frac{2}{3}}}.$$

Letting $q = \max(q_e, q_p)$, and assuming $q \leq N^{\frac{2}{3}}$, the upper bound of Theorem 2 simplifies into

$$\frac{10}{N} + \frac{31q^3}{N^2} + \frac{2q^2}{N^{\frac{4}{3}}} + \frac{15\sqrt{n}q^{\frac{3}{2}}}{N} + \frac{33q}{N^{\frac{2}{3}}} \leq \frac{10}{N} + \frac{81\sqrt{n}q}{N^{\frac{2}{3}}} = \frac{10}{2^n} + \frac{81q}{2^{\frac{2n}{3} - \frac{1}{2}\log_2 n}}.$$

Hence, security is ensured up to $\mathcal{O}(2^{\frac{2n}{3} - \frac{1}{2}\log_2 n}) = \widetilde{\mathcal{O}}(2^{\frac{2n}{3}})$ queries of the adversary.

Acknowledgments. Shan and John were supported by National Basic Research Program of China Grant 2011CBA00300, 2011CBA00301, the National Natural Science Foundation of China Grant 61033001, 61361136003, and by the China Ministry of Education grant number 20121088050. Rodolphe was supported by the French Direction Générale de l'Armement and the French National Agency of Research through the PRINCE project (contract ANR-10-SEGI-015). Jooyoung was supported by Basic Science Research Program through the National Research Foundation of Korea (NRF) funded by the Ministry of Education (NRF-2013R1A1A2007488). Yannick was partially supported by the French National Agency of Research through the BLOC project (contract ANR-11-INS-011).

References

[1] Alon, N., Kaufman, T., Krivelevich, M., Ron, D.: Testing Triangle-Freeness in General Graphs. SIAM J. Discrete Math. 22(2), 786–819 (2008)

[2] Andreeva, E., Bogdanov, A., Dodis, Y., Mennink, B., Steinberger, J.P.: On the Indifferentiability of Key-Alternating Ciphers. In: Canetti, R., Garay, J.A. (eds.) CRYPTO 2013, Part I. LNCS, vol. 8042, pp. 531–550. Springer, Heidelberg (2013), http://eprint.iacr.org/2013/061

[3] Babai, L.: The Fourier Transform and Equations over Finite Abelian Groups: An introduction to the method of trigonometric sums. Lecture notes (December 1989), http://people.cs.uchicago.edu/~laci/reu02/fourier.pdf

[4] Biham, E., Carmeli, Y., Dinur, I., Dunkelman, O., Keller, N., Shamir, A.: Cryptanalysis of Iterated Even-Mansour Schemes with Two Keys. IACR Cryptology ePrint Archive, Report 2013/674 (2013), http://eprint.iacr.org/2013/674

[5] Biryukov, A., Wagner, D.: Slide Attacks. In: Knudsen, L.R. (ed.) FSE 1999. LNCS, vol. 1636, pp. 245–259. Springer, Heidelberg (1999)

[6] Biryukov, A., Wagner, D.: Advanced Slide Attacks. In: Preneel, B. (ed.) EUROCRYPT 2000. LNCS, vol. 1807, pp. 589–606. Springer, Heidelberg (2000)

[7] Bogdanov, A.A., Knudsen, L.R., Leander, G., Paar, C., Poschmann, A., Robshaw, M., Seurin, Y., Vikkelsoe, C.: PRESENT: An Ultra-Lightweight Block Cipher. In: Paillier, P., Verbauwhede, I. (eds.) CHES 2007. LNCS, vol. 4727, pp. 450–466. Springer, Heidelberg (2007)

[8] Bogdanov, A., Knudsen, L.R., Leander, G., Standaert, F.-X., Steinberger, J., Tischhauser, E.: Key-Alternating Ciphers in a Provable Setting: Encryption Using a Small Number of Public Permutations - (Extended Abstract). In: Pointcheval, D., Johansson, T. (eds.) EUROCRYPT 2012. LNCS, vol. 7237, pp. 45–62. Springer, Heidelberg (2012)

[9] Chen, S., Steinberger, J.: Tight Security Bounds for Key-Alternating Ciphers. In: Nguyen, P.Q., Oswald, E. (eds.) EUROCRYPT 2014. LNCS, vol. 8441, pp. 327–350. Springer, Heidelberg (2014)

[10] Chen, S., Lampe, R., Lee, J., Seurin, Y., Steinberger, J.: Minimizing the Two-Round Even-Mansour Cipher. Full version of this paper, http://eprint.iacr.org/2014/443

[11] Daemen, J.: Limitations of the Even-Mansour Construction. In: Matsumoto, T., Imai, H., Rivest, R.L. (eds.) ASIACRYPT 1991. LNCS, vol. 739, pp. 495–498. Springer, Heidelberg (1993)

[12] Daemen, J., Rijmen, V.: The Design of Rijndael: AES - The Advanced Encryption Standard. Springer (2002)

[13] Daemen, J., Rijmen, V.: Probability Distributions of Correlations and Differentials in Block Ciphers. ePrint Archive, Report 2005/212 (2005), http://eprint.iacr.org/2005/212.pdf

[14] Dinur, I., Dunkelman, O., Keller, N., Shamir, A.: Key Recovery Attacks on 3-round Even-Mansour, 8-step LED-128, and Full AES^2. In: Sako, K., Sarkar, P. (eds.) ASIACRYPT 2013, Part I. LNCS, vol. 8269, pp. 337–356. Springer, Heidelberg (2013), http://eprint.iacr.org/2013/391

[15] Dunkelman, O., Keller, N., Shamir, A.: Minimalism in Cryptography: The Even-Mansour Scheme Revisited. In: Pointcheval, D., Johansson, T. (eds.) EUROCRYPT 2012. LNCS, vol. 7237, pp. 336–354. Springer, Heidelberg (2012)

[16] Even, S., Mansour, Y.: A Construction of a Cipher from a Single Pseudorandom Permutation. Journal of Cryptology 10(3), 151–162 (1997)

[17] Gaži, P.: Plain versus Randomized Cascading-Based Key-Length Extension for Block Ciphers. In: Canetti, R., Garay, J.A. (eds.) CRYPTO 2013, Part I. LNCS, vol. 8042, pp. 551–570. Springer, Heidelberg (2013)

[18] Gaži, P., Tessaro, S.: Efficient and Optimally Secure Key-Length Extension for Block Ciphers via Randomized Cascading. In: Pointcheval, D., Johansson, T. (eds.) EUROCRYPT 2012. LNCS, vol. 7237, pp. 63–80. Springer, Heidelberg (2012)

[19] Golomb, S.W., Gong, G., Mittenthal, L.: Constructions of Orthomorphisms of $\mathbb{Z}_n^2$. In: Proceedings of The Fifth International Conference on Finite Fields and Applications, pp. 178–195. Springer (1999)

[20] Guo, J., Peyrin, T., Poschmann, A., Robshaw, M.: The LED Block Cipher. In: Preneel, B., Takagi, T. (eds.) CHES 2011. LNCS, vol. 6917, pp. 326–341. Springer, Heidelberg (2011)

[21] Hayes, T.P.: A Large-Deviation Inequality for Vector-Valued Martingales. Manuscript (2005), http://www.cs.unm.edu/~hayes/papers/VectorAzuma

[22] Junod, P., Vaudenay, S.: FOX: A New Family of Block Ciphers. In: Handschuh, H., Hasan, M.A. (eds.) SAC 2004. LNCS, vol. 3357, pp. 114–129. Springer, Heidelberg (2004)

[23] Kilian, J., Rogaway, P.: How to Protect DES Against Exhaustive Key Search (an Analysis of DESX). Journal of Cryptology 14(1), 17–35 (2001)

[24] Kiltz, E., Pietrzak, K., Szegedy, M.: Digital Signatures with Minimal Overhead from Indifferentiable Random Invertible Functions. In: Canetti, R., Garay, J.A. (eds.) CRYPTO 2013, Part I. LNCS, vol. 8042, pp. 571–588. Springer, Heidelberg (2013)

[25] Lai, X., Massey, J.L.: A Proposal for a New Block Encryption Standard. In: Damgård, I.B. (ed.) EUROCRYPT 1990. LNCS, vol. 473, pp. 389–404. Springer, Heidelberg (1991)

[26] Lampe, R., Patarin, J., Seurin, Y.: An Asymptotically Tight Security Analysis of the Iterated Even-Mansour Cipher. In: Wang, X., Sako, K. (eds.) ASIACRYPT 2012. LNCS, vol. 7658, pp. 278–295. Springer, Heidelberg (2012)

[27] Lampe, R., Seurin, Y.: How to Construct an Ideal Cipher from a Small Set of Public Permutations. In: Sako, K., Sarkar, P. (eds.) ASIACRYPT 2013, Part I. LNCS, vol. 8269, pp. 444–463. Springer, Heidelberg (2013), http://eprint.iacr.org/2013/255

[28] Lee, J.: Towards Key-Length Extension with Optimal Security: Cascade Encryption and Xor-cascade Encryption. In: Johansson, T., Nguyen, P.Q. (eds.) EUROCRYPT 2013. LNCS, vol. 7881, pp. 405–425. Springer, Heidelberg (2013)

[29] Mittenthal, L.: Block Substitutions Using Orthomorphic Mappings. Advances in Applied Mathematics 16(1), 59–71 (1995)

[30] Nikolica, I., Wang, L., Wu, S.: Cryptanalysis of Round-Reduced LED. In: Fast Software Encryption, FSE 2013 (2013) (to appear)

[31] Patarin, J.: The "Coefficients H" technique. In: Avanzi, R.M., Keliher, L., Sica, F. (eds.) SAC 2008. LNCS, vol. 5381, pp. 328–345. Springer, Heidelberg (2009)

[32] Steinberger, J.: Improved Security Bounds for Key-Alternating Ciphers via Hellinger Distance. IACR Cryptology ePrint Archive, Report 2012/481 (2012), http://eprint.iacr.org/2012/481

[33] Steinberger, J.: Counting solutions to additive equations in random sets. arXiv Report 1309.5582 (2013), http://arxiv.org/abs/1309.5582

[34] Vaudenay, S.: On the Lai-Massey Scheme. In: Lam, K.-Y., Okamoto, E., Xing, C. (eds.) ASIACRYPT 1999. LNCS, vol. 1716, pp. 8–19. Springer, Heidelberg (1999)

Block Ciphers – Focus on the Linear Layer (feat. PRIDE)*

Martin R. Albrecht[1,**], Benedikt Driessen[2,***], Elif Bilge Kavun[3,†],
Gregor Leander[3,‡], Christof Paar[3], and Tolga Yalçın[4,***]

[1] Information Security Group, Royal Holloway, University of London, UK
[2] Infineon AG, Neubiberg, Germany
[3] Horst Görtz Institute for IT Security, Ruhr-Universität Bochum, Germany
[4] University of Information Science and Technology, Ohrid, Macedonia

Abstract. The linear layer is a core component in any substitution-permutation network block cipher. Its design significantly influences both the security and the efficiency of the resulting block cipher. Surprisingly, not many general constructions are known that allow to choose trade-offs between security and efficiency. Especially, when compared to Sboxes, it seems that the linear layer is crucially understudied. In this paper, we propose a general methodology to construct good, sometimes optimal, linear layers allowing for a large variety of trade-offs. We give several instances of our construction and on top underline its value by presenting a new block cipher. PRIDE is optimized for 8-bit micro-controllers and significantly outperforms all academic solutions both in terms of code size and cycle count.

Keywords: block cipher, linear layer, wide-trail, embedded processors.

1 Introduction

Block ciphers are one of the most prominently used cryptographic primitives and probably account for the largest portion of data encrypted today. This was facilitated by the introduction of Rijndael as the Advanced Encryption Standard (AES) [1], which was a major step forward in the field of block cipher design. Not only does AES offer strong security, but its structure also inspired many cipher designs ever since. One of the merits of AES (and its predecessor SQUARE [20]) was demonstrating that a well-chosen linear layer is not only crucial for the

---

* Due to page limitations, several details are omitted in this proceedings version. A full version is available at [2].
** Most of this work was done while the author was at the Technical University of Denmark
*** Most of this work was done while the authors were at Ruhr-Universität Bochum.
† The research was supported in part by the DFG Research Training Group GRK 1817/1.
‡ The research was supported in part by the BMBF Project UNIKOPS (01BY1040).

J.A. Garay and R. Gennaro (Eds.): CRYPTO 2014, Part I, LNCS 8616, pp. 57–76, 2014.

security (and efficiency) of a block cipher, but also allows to argue in a simple and thereby convincing way about its security.

There are two main design strategies that can be identified for block ciphers: Sbox-based constructions and constructions without Sboxes, most prominently those using addition, rotation, and XORs (ARX designs). Furthermore, Sbox-based designs can be split into *Feistel-ciphers and substitution-permutation networks (SPN)*. Both concepts have been successfully used in practice, the most prominent example of an SPN cipher being AES and the most prominent Feistel-cipher being the former Data Encryption Standard (DES) [22].

It is also worth mentioning that the concept of SPN has not only been used in the design of block ciphers but also for designing cryptographic permutations, most prominently for the design of several sponge-based hash functions including SHA-3 [11]. In SP networks, the round function consists of a non-linear layer composed of small Sboxes working in parallel on small chunks of the state and a linear layer that mixes those chunks. Thus, designing an SPN block cipher essentially reduces to choosing one (or several) Sboxes and a linear layer.

A lot of research has been devoted to the study of Sboxes. All Sboxes of size up to 4 bits have been classified (indeed, more than once – cf. [14,36,46]). Moreover, Sboxes with optimal resistance against differential and linear attacks have been classified up to dimension 5 [17]. In general, several constructions are known for good and optimal Sboxes in arbitrary dimensions. Starting with the work of Nyberg [43], this has evolved into its own field of research in which those functions are studied in great detail. A nice survey of the main results of this line of study is provided by Carlet [18].

The situation for the other main design part, the linear layer, is less clear.

1.1 The Linear Layer

For the design of the linear layer, two general approaches can be identified. One still widely-used method is to design the linear layer in a rather ad-hoc fashion, without following general design guidelines. While this might lead to very secure and efficient algorithms (cf. Serpent [3] and SHA-3 as prominent examples), it is not very satisfactory from a scientific point-of-view. The second general design strategy is the wide-trail strategy introduced by Daemen in [19] (see also [21]). Especially for the security against linear [41] and differential [12] attacks, the wide-trail strategy usually results in simple and strong security arguments. It is therefore not surprising that this concept has found its way in many recent designs (e.g. Khazad [9], Anubis [8], Grøstl [25], PHOTON [29], LED [30], PRINCE [16], mCrypton [39] to name but a few). In a nutshell, the main idea of the wide-trail strategy is to link the number of active Sboxes for linear and differential cryptanalysis to the minimal distance of a certain linear code associated with the linear layer of the cipher. In turn, choosing a good code (with some additional constraints) results in a large number of active Sboxes.

While the wide-trail strategy does provide a powerful tool for arguing about the security of a cipher, it does not help in actually designing an efficient linear layer (or the corresponding linear code) with a suitable number of active Sboxes.

Here, with the exception of early designs in [19] and later PRINCE and mCrypton, most ciphers following the wide-trail strategy simply choose an MDS matrix as the core component. This might guarantee an (partially) optimal number of active Sboxes, but usually comes at the price of a less efficient implementation. The only exception here is that, in the case of MDS matrices, the authors of PHOTON and LED made the observation that implementing such matrices in a serialized fashion improves hardware-efficiency. This idea was further generalized in [47,53], and more recently in [5].

It is our belief that, in many cases, it is advantageous to use a near-MDS matrix (or in general a matrix with a sub-optimal branch number) for the overall design. Furthermore, it is, in our opinion, utmost surprising that there are virtually no general constructions or guidelines that would allow an SPN design to benefit from security vs. efficiency trade-offs. This is in particular important when it comes to ciphers where specific performance characteristics are crucial, e.g. in lightweight cryptography.

1.2 The Current State of Lightweight Cryptography

In recent years, the field of lightweight cryptography has attracted a lot of attention from the cryptographic community. In particular, designing lightweight block ciphers has been a very active field for several years now. The dominant metric according to which the vast majority of lightweight ciphers have been optimized was and still is the chip area. While this is certainly a valid optimization objective, its relevance to real-world applications is limited. Nowadays, there are several interesting and strong proposals available that feature a very small area but simultaneously neglect other, important real-world constraints. Moreover, recent proposals achieve the goal of a small chip area by sacrificing execution speed to such an extent that even in applications where speed is supposedly uncritical, the ciphers are getting too slow[1].

Note that software solutions, i.e. low-end embedded processors, actually dominate the world of embedded systems and dedicated hardware is a comparably small fraction. Considering this fact, it is quite puzzling that efficiency on low-cost processors was disregarded for so long. Certainly, there were a few exceptions: Several theoretical and practical studies have already been done in this field. Practical examples include several proposals for instruction set extensions [38,42,48,37]. Among these, the Intel AES instruction set [31] is the most well-known and practically relevant one. There have also been attempts to come up with ciphers that are (partially) tailored for low-cost processors [51,50,54,26,10,32]. Of these, execution times of both SEA and ITUbee are rather high, mostly due to the high number of rounds. Furthermore, ITUbee uses 8-bit Sboxes, which occupy a vast amount of program memory storage. SPECK, on the other hand, seems to be an excellent *lightweight software cipher* in terms of both speed and program memory.

---

[1] See also [35] asking "Is lightweight = light + wait?".

It is obvious that there are quite some challenges to be overcome in this relatively untouched area of lightweight software cryptography. The software cipher for embedded devices of the future should not only be compact in terms of program memory, but also be relatively fast in execution time. It should clearly be secure and, preferably, its security should be easily analysed and verified. The latter can possibly be achieved by building on conservative structures, which are conventionally costly in software implementation, thereby posing even harder challenges.

One major component influencing all or at least most of those criteria outlined above is the linear layer. Thus, it is important to have general constructions for linear layers that allow to explore and make optimal use of the possible trade-offs.

1.3   Our Contribution

In this paper, we take steps towards a better understanding of possible trade-offs for linear layers. After introducing necessary concepts and notation in Section 2, we give a general construction that allows to combine several strong linear mappings on a few number of bits into a strong linear layer for a larger number of bits (cf. Section 3). From a coding theory perspective, this construction corresponds to a construction known as block-interleaving (see [40], pages 131-132). While this idea is rather simple, its applicability is powerful. Implicitly, a specific instance of our construction is already implemented in AES. Furthermore, special instances of this construction are recently used in [7] and [28].

We illustrate our approach by providing several linear layers with an optimal or almost optimal trade-off between hardware-efficiency and number of active Sboxes in Section 4. Along similar lines, we present a classification of all linear layers fulfilling the criteria of the block cipher PRINCE in [2], Appendix C. Those examples show in particular that the construction given in Section 3 allows the construction of non-equivalent codes even when starting from equivalent ones. Secondly, we show that our construction also leads to very strong linear layers with respect to efficiency on embedded 8-bit micro-controllers. For this, we adopt a search strategy from [52] to find the *most efficient linear layer possible* within our constraints. We implemented this search on an FPGA platform to overcome the big computational effort involved and to have the advantage of reconfigurability. Details are described in Section 5.1.

With this, and as a second main contribution of our paper, we make use of our construction to design a new block cipher named PRIDE that significantly outperforms all existing block ciphers of similar key-sizes, with the exception of SIMON and SPECK [10]. One of the key-points here is that our construction of strong linear layers is nicely in line with a bit-sliced implementation of the Sbox layer. Our cipher is comparable, both in speed and memory size, to the new NSA block ciphers SIMON and SPECK, dedicated for the same platform. We conclude the paper in Section 6 with some open problems and pressing topics for further investigation. Finally, we note that while in this paper we focus on SPN ciphers, most of the results translate to the design of Feistel ciphers as well.

2 Notation and Preliminaries

In this section, we fix the basic notation and furthermore recall the ideas of the wide-trail strategy.

We deal with SPN block ciphers where the Sbox layer consist of n Sboxes of size b each. Thus the block size of the cipher is $n \times b$. The linear layer will be implemented by applying k binary matrices in parallel.

We denote by $\mathbb{F}_2$ the field with two elements and by $\mathbb{F}_2^n$ the n-dimensional vector space over $\mathbb{F}_2$. Note that any finite extension field $\mathbb{F}_{2^b}$ over $\mathbb{F}_2$ can be viewed as the vector space $\mathbb{F}_2^b$ of dimension b. Along these lines, the vector space $(\mathbb{F}_{2^b})^n$ can be viewed as the (nested) vector space $\left(\mathbb{F}_2^b\right)^n$.

Given a vector $x = (x_1, \dots, x_n) \in \left(\mathbb{F}_2^b\right)^n$ where each $x_i \in \mathbb{F}_2^b$ we define its weight[2] as

$$\mathrm{wt}_b(x) = |\{1 \leq i \leq n \mid x_i \neq 0\}|.$$

Following [21], given a linear mapping $L : (\mathbb{F}_2^b)^n \to (\mathbb{F}_2^b)^n$ its differential branch number is defined as

$$\mathcal{B}_d(L) := \min\{\mathrm{wt}_b(x) + \mathrm{wt}_b(L(x)) \mid x \in \left(\mathbb{F}_2^b\right)^n , \; x \neq 0\}.$$

The cryptographic significance of the branch number is that the branch number corresponds to the minimal number of active Sboxes in any two consecutive rounds. Here an Sbox is called active if it gets a non-zero input difference in its input.

Given an upper bound p on the differential probability for a single Sbox along with a lower bound of active Sboxes immediately allows to deduce an upper bound for any differential characteristic[3] using

$$\text{average probability for any non-trivial characteristic } \leq p^{\#\text{active Sboxes}}.$$

For linear cryptanalysis, the linear branch number is defined as

$$\mathcal{B}_l(L) := \min\{\mathrm{wt}_b(x) + \mathrm{wt}_b(L^*(x)) \mid x \in \left(\mathbb{F}_2^b\right)^n, x \neq 0\}$$

where L^* is the adjoint linear mapping. That is, with respect to the standard inner product, L^* corresponds to the transposed matrix of L.

In terms of correlation (cf., for example, [19]), an upper bound c on the absolute value of the correlation for a single Sbox results in a bound for any linear trail (or linear characteristic, linear path) via

$$\text{absolute correlation for a trail } \leq c^{\#\text{active Sboxes}}.$$

The differential branch number corresponds to the minimal distance of the $\mathbb{F}_2$-linear code C over $\mathbb{F}_2^b$ with generator matrix

$$G = [I \mid L^T]$$

---

[2] Of course $\left(\mathbb{F}_2^b\right)^n$ is isomorphic to $\mathbb{F}_2^{nb}$, but the weight is defined differently on each.
[3] Averaging over all keys, assuming independent round keys.

where I is the $n \times n$ identity matrix. The length of the code is $2n$ and its dimension is n (here dimension corresponds to $\log_{2^b}(|C|)$ as it is not necessarily a linear code). Thus, C is a $(2n, 2^n)$ additive code over $\mathbb{F}_2^b$ with minimal distance $d = \mathcal{B}_d(L)$.

The linear branch number corresponds in the same way to the minimal distance of the $\mathbb{F}_2$-linear code $C^\perp$ with generator matrix

$$G^* = [L \mid I].$$

Note that $C^\perp$ is the dual code of C and in general the minimal distances of $C^\perp$ and C do not need to be identical.

Finally, given linear maps L_1 and L_2, we denote by $L_1 \times L_2$ the direct sum of the mappings, i.e.

$$(L_1 \times L_2)(x, y) := (L_1(x), L_2(y)).$$

3   The Interleaving Construction

Following the wide-trail strategy, we construct linear layers by constructing a $(2n, 2^n)$ additive codes with minimal distance d over $\mathbb{F}_2^b$. The code needs to have a generator matrix G in standard form, i.e.

$$G = [I \mid L^T]$$

where the submatrix L is invertible, and corresponds to the linear layer we are using.

Hence, the main question is how to construct "efficient" matrices L with a given branch number. Our construction allows to combine small matrices into bigger ones. We hereby drastically reduce the search-space of possible linear layers. This in turn makes it possible to construct efficient linear layers for various trade-offs, as demonstrated in the following sections.

As mentioned above, the construction described in [21] can be seen as a special case of our construction. The main difference (except the generalization) is that we shift the focus of the construction in [21] from the 4 round super-box view to a 2 round-view. While Daemen and Rijmen focused on the bounds for 4 rounds, we make use of their ideas to actually construct linear layers. Moreover, a particular instance of the general construction we elaborate on here, was already used in the linear layer of the hash function Whirlwind [7]. There, several small MDS matrices are used to construct a larger one.

We give a simple illustrative example of our approach in [2], Appendix A.

3.1   The General Construction

We are now ready to give a formal description of our approach. First define the following isomorphism

$$P_{b_1,\dots b_k}^n : \left(\mathbb{F}_2^{b_1} \times \mathbb{F}_2^{b_2} \times \cdots \times \mathbb{F}_2^{b_k}\right)^n \to \left(\mathbb{F}_2^{b_1}\right)^n \times \left(\mathbb{F}_2^{b_2}\right)^n \times \cdots \times \left(\mathbb{F}_2^{b_k}\right)^n$$

$$(x_1, \dots, x_n) \mapsto \left(\left(x_1^{(1)}, \dots, x_n^{(1)}\right), \dots, \left(x_1^{(k)}, \dots, x_n^{(k)}\right)\right)$$

$$\text{where } x_i = \left(x_i^{(1)}, \dots, x_i^{(k)}\right) \text{ with } x_i^{(j)} \in \mathbb{F}_2^{b_j}.$$

This isomorphism performs the transformation of mapping Sbox outputs to our small linear layers L_i. For example, in Appendix A of [2], we considered individual bits (i.e. $b_1, \ldots, b_k = 1$) from 4 (i.e., $k = 4$) 4-bit Sboxes (i.e $n = 4$).

Note that, for our purpose, there are in fact many possible choices for P. In particular, we may permute the entries within $(\mathbb{F}_2^b)^n$. Given this isomorphism we can now state our main theorem. The construction of P follows the idea of a diffusion-optimal mapping as defined in [21, Definition 5].

Theorem 1. *Let $G_i = [I \mid L_i^T]$ be the generator matrix for an $\mathbb{F}_2$-linear $(2n, 2^n)$ code with minimal distance d_i over $\mathbb{F}_2^{b_i}$ for $0 \leq i < k$. Then the matrix $G = [I \mid L^T]$ with*

$$L = \left(P_{b_1, \ldots b_k}^n\right)^{-1} \circ (L_0 \times L_1 \times \cdots \times L_{k-1}) \circ P_{b_1, \ldots b_k}^n$$

is the generator matrix of an $\mathbb{F}_2$-linear $(2n, 2^n)$ code with minimal distance d over $\mathbb{F}_2^b$ where

$$d = \min_i d_i \quad and \quad b = \sum_i b_i.$$

Proof. Since $P_{b_1, \ldots b_k}^n$ and $\left(P_{b_1, \ldots b_k}^n\right)^{-1}$ are permutation matrices, by construction L has full rank. To see that $\mathrm{wt}_b(w) + \mathrm{wt}_b(v) \geq \min_i d_i$ for any $v \in \mathbb{F}_2^b \setminus \{0\}$ and $w = L \cdot v$, observe that $\mathrm{wt}_b(w) + \mathrm{wt}_b(v)$ is minimal when all entries in v are zero except those mapped to the positions acted on by L_j where L_j is the matrix with the minimal branch number. □

Remark 1. The interleaving construction allows to construct non-equivalent codes even when starting with equivalent L_i's. This is shown in a particular case in Appendix C of [2], where different choices of (equivalent) L_i's lead to different numbers of minimum-weight codewords.

A special case of the construction above is implicitly already used in AES. In the case of AES, it is used to construct a $[8, 4, 5]$ code over $\mathbb{F}_2^{32}$ from 4 copies of the $[8, 4, 5]$ code over $\mathbb{F}_2^8$ given by the *MixColumn* operation. In the Superbox view on AES, the *ShiftRows* operation plays the role of the mapping P (and its inverse) and *MixColumns* corresponds to the mappings L_i.[4]

In the following, we use this construction to design efficient linear layers. Besides the differential and linear branch number, we hereby focus mainly on three criteria:

– Maximize the diffusion (cf. Section 3.3)
– Minimize the density of the matrix (cf. Section 4)
– Software-efficiency (cf. Section 5)

---

[4] Note that the cipher PRINCE implicitly uses the construction twice. Once for generating the matrix M as in Appendix A of [2] and second for the improved bound on 4 rounds, just like in AES.

The strategy we employ is as follows. We first find candidates for L_0, i.e., $(2n, 2^n)$ additive codes with minimal distance d_0 over $\mathbb{F}_{2^{b_0}}$. In this stage, we ensure that the branch number is d_0 and our efficiency constraints are satisfied. We then apply permutations to L_0 to produce L_i for $i > 0$. This stage maximizes diffusion.

3.2   Searching for L_0

The following lemma (which is a rather straightforward generalization of Theorem 4 in [53]) gives a necessary and sufficient condition that a given matrix L has branch number d over $\mathbb{F}_2^b$.

Lemma 1. *Let L be a $bn \times bn$ binary matrix, decomposed into $b \times b$ submatrices $L_{i,j}$.*

$$L = \begin{pmatrix} L_{0,0} & L_{0,1} & \cdots & L_{0,n-1} \\ L_{1,0} & L_{1,1} & \cdots & L_{1,n-1} \\ \vdots & \vdots & \ddots & \vdots \\ L_{n-1,0} & L_{n-1,1} & \cdots & L_{n-1,n-1} \end{pmatrix} \tag{1}$$

Then, L has differential branch number d over $\mathbb{F}_2^b$ if and only if all $i \times (n-d+i+1)$ block submatrices of L have full rank for $1 \leq i < d - 1$. Moreover, L has linear branch number d if and only if all $(n - d + i + 1) \times i$ block submatrices of L have full rank for $1 \leq i < d - 1$.

Based on Lemma 1 we may instantiate various search algorithms which we will describe in Section 4 and Section 5. In our search we focus on cyclic matrices, i.e. matrices where row $i > 0$ is constructed by cyclic shifting row 0 by i indices. These matrices have the advantage of being efficient both in software and in hardware. Furthermore, since these matrices are symmetric, considering the dual code $C^\perp$ to $C = [I \mid L^T]$ is straightforward.

3.3   Ensuring High Dependency

In this section, we assume we are given a matrix L_0 and wish to construct $L_1, \ldots, L_{k-1}$ that maximize the diffusion of the map $L = \left(P_{b_1, \ldots b_k}^n \right)^{-1} \circ (L_0 \times L_1 \times \cdots \times L_{k-1}) \circ P_{b_1, \ldots b_k}^n$.

Given an $bn \times bn$ binary matrix L decomposed as in Eq. (1), we define its support as the $n \times n$ binary matrix $\mathrm{Supp}(L)$ where

$$\mathrm{Supp}(L)_{i,j} = \begin{cases} 1 & \text{if } L_{i,j} \neq 0 \\ 0 & \text{else} \end{cases}$$

Now assume that $\mathrm{Supp}(L_0)$ has a zero entry at index i', j'. If we apply the same L_i in all k positions this means that the outputs from the i'th Sbox have no impact on the inputs of the j'th Sbox after the linear layer. In other words, a linear-layer following the construction of Theorem 1 ensure full dependency if and only if

$$\left(\bigvee_{0 \leq i < k} \mathrm{Supp}(L_i) \right)_{i',j'} = 1 \quad \forall \, 0 \leq i',j' < n.$$

Hence, we want to apply different matrices L_i in each of the k positions, such that in at least one $\mathrm{Supp}(L_i)$ has a non-zero entry at index i',j' for all $0 \leq i',j' < n$. In order to construct matrices L_i for $i > 0$ from a matrix L_0 we may apply block-permutation matrices from the left and right to L_0 as these clearly neither impact the density nor the branch number. Hence, we focus on finding permutation matrices P_i, Q_i such that the density of $\bigvee_{0 \leq i < b} \mathrm{Supp}(P_i \cdot L_0 \cdot Q_i)$ is maximized. In Appendix F of [2], we give two strategies for finding such P_i, Q_i, one is heuristic but computationally cheap, the other is guaranteed to return an optimal solution – based on Constraint Integer Programming – but can be computationally intensive.

We note that the difficulty of the problem depends on the size of the Sbox and the density of L_i. As MDS matrices always have density 1, the problem of full dependency does not occur when combining such matrices. Finally, if the construction ensures full dependency for a given k, it is always possible to achieve full dependency for any $k' \geq k$.

In contrast with the branch number, if a linear layer ensures high dependency, its inverse does not necessarily achieve the same dependency. Thus, it is in general necessary to check the dependency of the inverse separately.

4 Optimizing for Hardware

In this section, we give examples of $[2n, n, d]$ codes over $\mathbb{F}_2^b$ and give algorithms for finding such instances. First, the following lemma gives a lower bound on the density of a matrix with branch number d. Our aim here is to find linear layers that are efficiently implementable in hardware. More precisely, we aim for an implementation in one cycle. PHOTON and LED demonstrated that there is a trade-off between clock cycles and number of gate equivalence for the linear layer. The trade-off we consider here is, complementary to PHOTON and LED, between efficient implementation in one clock cycle and the (linear and differential) branch number. Note that in our setting, the cost of implementation is directly connected to the number of ones in the matrix representation of the linear layer.

Lemma 2. *Let matrix* $G = [I \mid L^T]$ *be the generator matrix for an* $\mathbb{F}_2$-*linear* $(2n, 2^n)$ *code with minimal distance* d *such that the dual code has minimum distance* d *as well. Then* L *has at least* $d - 1$ *ones per row and per column.*

Proof. Computing $w = L \cdot v$ where v is a vector with one non-zero entry 1, we have that w must be a vector with $d-1$ non-zero entries if the minimum distance of $[I \mid L^T]$ is d. Hence, there must be at least $d-1$ ones per row. Applying the same argument to $w = L^T \cdot v = v \cdot L$ shows that at least $d-1$ entries per column must be non-zero. □

The main merit of the above lemma is that it allows to determine the optimal solutions in terms of efficiency. This is in contrast to the case for software implementation, where the optimal solution is unknown.

Lemmas 1 and 2 give rise to various search strategies for finding $(2n, 2^n)$ additive codes with minimal distance d over $\mathbb{F}_2^b$. We discuss those strategies in Appendix B of [2] and present results of those strategies next.

4.1   Hardware-Optimal Examples

Below we give some examples for our construction. We hereby focus on $[2n, n, d]$ codes over $\mathbb{F}_2$, i.e. we use $b_i = 1$.[5] Note that this naturally limits the achievable branch number. For binary linear codes the optimal minimal distance is known for small length (cf. [27] for more information). We give a small abridgement of the known bounds on the minimal distance for linear $[2n, n]$ codes over $\mathbb{F}_2, \mathbb{F}_4$, and $\mathbb{F}_8$ in Appendix E of [2]. As can be seen in this table, in order to achieve a high branch number, it might be necessary to consider linear codes over $\mathbb{F}_{2^m}$, or (more general) additive codes over $\mathbb{F}_2^m$ for some small $m > 1$.

The examples in Figure 1 are optimal in the sense that they achieve the best possible branch number (both linear and differential) for the given length (with the exception of $n = 11, 13$, and 14) with the least possible number of ones in the matrix (cf. Lemma 2). The number D corresponds to the average number of ones per row/column and D_{inv} to the average number of ones per row/column of the inverse matrix. The only candidate which does not satisfy $D = d - 1$ is $n = 8$. This candidate was found using the approach from Appendix B.3 of [2], which guarantees to return the optimal solution. Hence, we conclude that $4\frac{1}{8}$ is indeed the lowest density possible. That is, there is no 8×8 binary matrix with branch number 5 with only 32 ones, but the best we can do is 33 ones.

For each example we list the dimension (i.e the number of Sboxes), the achieved branch number and the minimal k such that it is possible to achieve full dependency with two Sbox layers interleaved with one linear layer. These values were found using the CIP approach in Section 3.3. Note that in this case (i.e. $b_i = 1$) the value k actually corresponds to the minimal Sbox size that allows full dependency. Finally, k_{inv} is the minimum Sbox size to achieve full diffusion for the inverse matrix. Note that for all these examples, the corresponding code is actually equivalent to its dual. In particular this implies that the linear and differential branch number are equal.

---

[5] We refer to Appendix C of [2] for an exemplary comparison of the set of linear layers constructed by Theorem 1 and the entire space with the same criteria for $[8, 4, 4]$ codes over $\mathbb{F}_2^4$.

n	max(d)	d	D	D_{inv}	k	k_{inv}	Technique	Matrix
2	2	2	1	1	2	2	[2], App. B.1	cyclic shift (10) to the left.
3	2	2	1	1	3	3	[2], App. B.1	cyclic shift (100) to the left.
4	4	4	3	3	2	2	[2], App. B.1	cyclic shift (1110) to the left.
5	4	4	3	3	2	2	[2], App. B.1	cyclic shift (11100) to the left.
6	4	4	3	3	2	2	[2], App. B.1	cyclic shift (110100) to the left.
7	4	4	3	$4\frac{3}{7}$	3	2	[2], App. B.3	in Figure 2
8	5	5	$4\frac{1}{8}$	$4\frac{7}{8}$	3	2	[2], App. B.3	in [2], Appendix F.3
9	6	6	5	$5\frac{6}{9}$	2	2	[2], App. B.3	in Figure 2
10	6	6	5	5	2	2	[2], App. B.1	cyclic shift (1111010000) to the left.
11	7	6	5	5	3	3	[2], App. B.1	cyclic shift (11110100000) to the left.
12	8	8	7	7	2	2	[2], App. B.1	cyclic shift (110111101000) to the left.
13	7	6	5	5	≤ 4	≤ 4	[2], App. B.1	cyclic shift (1110110000000) to the left.
14	8	6	5	5	≤ 4	≤ 4	[2], App. B.1	cyclic shift (11101010000000) to the left.
15	8	8	7	7	3	3	[2], App. B.1	cyclic shift (101101110010000) to the left.
16	8	8	7	7	3	3	[2], App. B.1	cyclic shift (1111011010000000) to the left.

Fig. 1. Examples of hardware efficient linear layers over $\mathbb{F}_2$

$$\begin{pmatrix} 0110001 \\ 1000011 \\ 0100011 \\ 0001110 \\ 1001100 \\ 0110100 \\ 1011000 \end{pmatrix} \quad \begin{pmatrix} 001110110 \\ 100101110 \\ 010010111 \\ 111101000 \\ 100110101 \\ 001111001 \\ 111010010 \\ 011001011 \\ 110001101 \end{pmatrix}$$

Fig. 2. Examples of $[14, 7, 4]$ and $[18, 9, 6]$ codes over $\mathbb{F}_2$

5   Software-Friendly Examples and the Cipher PRIDE

In this section, we describe our new lightweight software-cipher PRIDE, a 64-bit block cipher that uses a 128-bit key. We refer to Appendix D of [2] for a sketch of the security analysis and to the full version for more details.

We chose to design an SPN block cipher because it seems that this structure is better understood than e.g. ARX designs. We are, unsurprisingly, making use of the construction given in Theorem 1. We here decided on a linear layer with high dependency and a linear & differential branch number of 4. One key-observation is that the construction of Theorem 1 fits naturally with a bit-sliced implementation of the cipher, in particular with the Sbox layer. As a bit-sliced implementation of the Sbox layer is advantageous on 8-bit micro-controllers, in any case this is a nice match.

The target platform of PRIDE is Atmel's AVR micro-controller [4], as it is dominating the market along with PIC [44] (see [45]). Furthermore, many implementations in literature are also implemented in AVR, we therefore opt for this platform to provide a better comparison to other ciphers (including SIMON and SPECK [10]). However, the reconfigurable nature of our search architecture (cf. Section 5.1) to find the basic layers of the cipher allows us to extend the search to various platforms in the future.

5.1   The Search for the Linear Layer

A natural choice in terms of Theorem 1 is to choose $k = 4$ and $b_1 = b_2 = b_3 = b_4 = 1$. Thus, the task reduces to find four 16×16 matrices forming one 64×64 matrix (to permute the whole state) of the following form:

$$\begin{pmatrix} L_0 & 0 & 0 & 0 \\ 0 & L_1 & 0 & 0 \\ 0 & 0 & L_2 & 0 \\ 0 & 0 & 0 & L_3 \end{pmatrix}$$

Each of these four 16×16 matrices should provide branch number 4 and together achieve high dependency with the least possible number of instructions. Instead of searching for an efficient implementation for a given matrix, we decided to search for the most efficient solution fulfilling our criteria.

To find such matrices (L_i) that could be implemented very efficiently given the AVR instruction set, we performed an extensive and hardware-aided tree search. Our search engine was optimized to look for AVR assembly code segments utilizing a limited set of instructions that would result in linear behaviour at matrix level. These are namely CLC, EOR, MOV, MOVW, CLR, SWAP, ASR, ROR, ROL, LSR, and LSL instructions. As we are looking for 16×16 matrices, the state to be multiplied with each L_i is stored in two 8-bit registers, which we call X and Y. We also allowed utilization of four temporary registers, namely $T0$, $T1$, $T2$, and $T3$. We designed and optimized our search engine according to these registers. Our search engine checks the resulting matrix L_i after N instructions to see if it provides the desired characteristics. While trying to reach instruction N, we try all possible instruction-register combinations in each step. This of course comes with an impractical time complexity, especially when N is increased further. To deal with this time complexity, we came up with several optimizations. As a first step, we limited the utilization of certain instruction-register combinations. For example, we excluded CLC and CLR instructions from the combinations for the first and last instructions. Also, EOR is not considered in the first instruction. Again, for the first and last instructions, SWAP, ASR, ROR, ROL, LSR, and LSL instructions are only used with X and Y. Furthermore, we did not allow temporary registers as the destination while trying MOV and MOVW instructions in the last instruction and $X - Y$ registers as the destination while trying MOV and MOVW instructions in the first instruction.

However, such optimizations were not enough to reduce the time complexity. We therefore applied further optimizations, i.e., when the matrices of all registers do not give full rank, we stop the search as we know that we cannot find an invertible linear layer any more.

In the end, we found matrices that fulfil all of our criteria starting from 7 instructions.

We implemented our search architecture on a Xilinx ML605 (Virtex-6 FPGA) evaluation board. The reconfigurable nature of the FPGA allowed us to change easily between different parameters, i.e. the number of instructions. The details of this search engine can be found in [33].

5.2   An Extremely Efficient Linear Layer

As a result of the search explained in Section 5.1, we achieved an extremely efficient linear layer. The cheapest solution provided by our search needed 36 cycles for the complete linear layer, which is what we opted for. The optimal matrices forming the linear layer are given in the Appendix G of [2]. Of these four matrices, L_0 and L_3 are involutions with the cost of 7 instructions (in turn, clock cycles), while L_1 and L_2 require 11 and 13 instructions for true and inverse matrices, respectively. The assembly codes are given in Appendix H of [2] to show the claimed number of instructions.

Comparing to linear layers of other SPN-based ciphers clearly demonstrated the benefit of our approach. Note however, that these comparisons have to be taken with care as not all linear layers operate on the same state size and do not offer the same security level. The linear layer of the ISO-standard lightweight cipher PRESENT [15] costs 144 cycles (derived from the total cycle count given in [24]). MixColumns operation of NIST-standard AES[6] costs 117 instructions (but 149 cycles because of 3-cycle data load instruction utilizations, as MixColumns constants are implemented as look-up table – which means additional 256 bytes of memory, too) [6]. Note that ShiftRows operation was merged with the look-up table of Sbox in this implementation, so we take only MixColumns cost as the linear layer cost. The linear layer of another ISO-standard lightweight cipher CLEFIA [49] (again 128-bit cipher) costs 146 instructions and 668 cycles. Bit-sliced oriented design Serpent (AES finalist, 128-bit cipher) linear layer costs 155 instructions and 158 cycles. Other lightweight proposals, KLEIN [26] and mCrypton linear layers cost 104 instructions (100 cycles) and 116 instructions (342 cycles), respectively [23]. Finally, the linear layer cost of PRINCE is 357 instructions and 524 cycles[7], which is even worse than AES. One of the reasons for this high cost is the non-cyclic 4×4 matrices forming the linear layer. The other reason is the ShiftRows operation applied on 4-bit state words, which makes coding much more complex than that of AES on an 8-bit micro-controller.

---

[6] It is of course not fair to compare a 128-bit cipher with a 64-bit cipher. However, we provide AES numbers as a reference due to the fact that it is a widely-used standard cipher and its cost is much better compared to many lightweight ciphers.

[7] We implemented this cipher on AVR, as we could not find any AVR implementations in the literature.

5.3 Sbox Selection

For our bit-sliced design, we decided to use a very simple (in terms of software-efficiency – the formulation is given in Appendix I of [2]) 10-instruction Sbox (which makes $10 \times 2 = 20$ clock cycles in total for the whole state). It is at the same time an involution Sbox, which prevents the encryption/decryption overhead. Besides being very efficient in terms of cycle count, this Sbox is also optimal with respect to linear and differential attacks. The maximal probability of a differential is $1/4$ and the best correlation of any linear approximation is $1/2$. The PRIDE Sbox is given below.

x	0x0 0x1 0x2 0x3 0x4 0x5 0x6 0x7 0x8 0x9 0xa 0xb 0xc 0xd 0xe 0xf
$\mathcal{S}(x)$	0x0 0x4 0x8 0xf 0x1 0x5 0xe 0x9 0x2 0x7 0xa 0xc 0xb 0xd 0x6 0x3

The assembly codes are given in Appendix H of [2] to show the claimed number of instructions.

5.4 Description of PRIDE

Similar to PRINCE, the cipher makes use of the FX construction [34,13]. A pre-whitening key k_0 and post-whitening key k_2 are derived from one half of k, while the second half serves as basis k_1 for the round keys, i.e.,

$$k = k_0 || k_1 \quad \text{with} \quad k_2 = k_0.$$

Moreover, in order to allow an efficient bit-sliced implementation, the cipher starts and ends with a bit-permutation. This clearly does not influence the security of PRIDE in any way. Note that in a bit-sliced implementation, none of the permutations P nor P^{-1} used in PRIDE has to be actually implemented explicitly. The cipher has 20 rounds, of which the first 19 are identical. Subkeys are different for each round, i.e., the subkey for round i is given by $f_i(k_1)$. We define

$$f_i(k_1) = k_{1_0} || g_i^{(0)}(k_{1_1}) || k_{1_2} || g_i^{(1)}(k_{1_3}) || k_{1_4} || g_i^{(2)}(k_{1_5}) || k_{1_6} || g_i^{(3)}(k_{1_7})$$

as the subkey derivation function with four byte-local modifiers of the key as

$$g_i^{(0)}(x) = (x + 193i) \bmod 256, \quad g_i^{(1)}(x) = (x + 165i) \bmod 256,$$
$$g_i^{(2)}(x) = (x + 81i) \bmod 256, \quad g_i^{(3)}(x) = (x + 197i) \bmod 256,$$

which simply add one of four constants to every other byte of k_1. The overall structure of the cipher is depicted here:

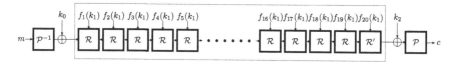

The round function $\mathcal{R}$ of the cipher shows a classical substitution-permutation network: The state is XORed with the round key, fed into 16 parallel 4-bit Sboxes and then permuted and processed by the linear layer.

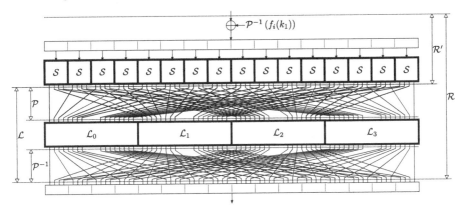

The difference between $\mathcal{R}$ and $\mathcal{R}'$ is that in the latter no more diffusion is necessary, therefore the last round ends after the substitution layer. With the software-friendly matrices we have found as described above, the linear layer is defined as follows (cf. Theorem 1 and Appendix G of [2]):

$$L := P^{-1} \circ (L_0 \times L_1 \times L_2 \times L_3) \circ P \quad \text{where} \quad P := P^{16}_{1,1,1,1}.$$

The test vectors for the cipher are provided in the Appendix J of [2].

5.5   Performance Analysis

As depicted above, one round of our proposed cipher PRIDE consists of a linear layer, a substitution layer, a key addition, and a round constant addition (key update). In a software implementation of PRIDE on a micro-controller, we also perform branching in each round of the cipher in addition to the previously listed layers. Adding up all these costs gives us the total implementation cost for one round of the cipher. The total cost can roughly be calculated by multiplying the number of rounds with the cost of each round. Note that we should subtract the cost of one linear layer from the overall cost, as PRIDE has no linear layer in the last round. The software implementation cost of the round function of PRIDE on Atmel AVR ATmega8 8-bit micro-controller [4] is presented in the following:

	Key update	Key addition	Sbox Layer	Linear Layer	Total
Time (cycles)	4	8	20	36	**68**
Size (bytes)	8	16	40	72	**136**

Comparing PRIDE to existing ciphers in literature, we can see that it outperforms many of them significantly both in terms of cycle count and code size. Note that we are not using any look-up tables in our implementation, in turn no RAMs[8]. The comparison with existing implementations is given below:

	AES-128 [24]	SERPENT-128 [24]	PRESENT-128 [24]	CLEFIA-128 [24]	SEA-96 [50]	NOEKEON-128 [23]	PRINCE-128	ITUbee-80 [32]	SIMON-64/128 [10]	SPECK-64/96 [10]	SPECK-64/128 [10]	PRIDE
t(cyc)	3159	49314	10792	28648	17745	23517	3614	2607	2000	1152	1200	**1514**
bytes	1570	7220	660	3046	386	364	1108	716	282	182	186	**266**
eq.r.	5/10	1/32	4/31	1/18	8/92	1/16	5/12	12/20	33/44	34/26	34/27	

In the table, the first row is the time (performance) in clock cycles, the second row is the code size in bytes, and the third row is the *equivalent rounds*. The third row expresses the number of rounds for the given ciphers that would result in a total running time similar to PRIDE.

Note that, as we did not come across to any reference implementations in the literature, we implemented PRINCE in AVR for comparison. We also do not list the RAM utilization for the ciphers under comparison in the table.

In the implementation of PRIDE, our target was to be fast and at the same time compact. Note that we do not exclude data & key read and data write back as well as the whitening steps in our results (these are omitted in SIMON and SPECK numbers). Although the given numbers are just for encryption, decryption overhead is also acceptable: It costs 1570 clock cycles and 282 bytes. A cautionary note is indicated for the above comparison for several reasons. AES, SERPENT, CLEFIA, and NOEKOEN are working on 128-bit blocks; so, for a cycle per byte comparison, their cycle count has to be divided by a factor of two. Moreover, the ciphers differ in the claimed security level and key-size. PRIDE does not claim any resistance against related-key attacks (and actually can be distinguished trivially in this setting) and also generic time-memory trade-offs are possible against PRIDE in contrast to most other ciphers. Besides those restrictions, the security margin in PRIDE in terms of the number of rounds is (in our belief) sufficient.

One can see that PRIDE is comparable to SPECK-64/96 and SPECK-64/128 (members of NSA's *software*-cipher family), which are based on a Feistel structure and use modular additions as the main source of non-linearity.

In addition to the above table, the recent work of Grosso et al. [28] presents LS-Designs. This is a family of block ciphers that can systematically take advantage

---

[8] Which has the additional advantage of increased resistance against cache-timing attacks.

of bit-slicing in a principled manner. In this paper, the authors make use of look-up tables. Therefore, a direct comparison with PRIDE is not fair as the use of look-up tables does not minimize the linear layer cost. However, to have an idea, we can try to estimate the cost of the 64-bit case of this family. They suggest two options: The first uses 4-bit Sbox with 16-bit Lbox, and the second uses 8-bit Sbox with 8-bit Lbox. The first option has 8 rounds, which results in 64 non-linear operations, 128 XORs, and 128 table look-ups in total. The second one has 6 rounds, which takes 72 non-linear operations, 144 XORs, and 48 table look-ups. For linear layer cost, we consider the XOR cost together with table look-ups. Unfortunately, it is not easy to estimate the overall cost of the given two options on AVR platform as the table look-ups take more than one cycle compared to the non-linear and linear operations. Another important point here to mention is that the use of look-up tables result in a huge memory utilization.

Finally, we note that, despite its target being software implementations, PRIDE is also efficient in hardware. It can be considered a hardware-friendly design, due to its cheap linear and Sbox layers.

6 Conclusion

In this work, we have presented a framework for constructing linear layers for block ciphers which allows to trade security against efficiency. For a given security level, in our case we focused on the branch number, we demonstrated techniques to find very efficient linear layers satisfying this security level. Using this framework, we presented a family of linear layers that are efficient in hardware. Furthermore, we presented a new cipher PRIDE dedicated for 8-bit micro-controllers that offers competitive performance due to our new techniques for finding linear layers.

One important question is on the optimality of a given construction for a linear layer. In particular, in the case of our construction, the natural question is if the reduction of the search space excludes optimal solutions and only sub-optimal solutions remain. For the hardware-friendly examples presented in Section 4 and Appendix C of [2], it is easy to argue that those constructions are optimal. Thus, in this case the reduction of the search space clearly did not have a negative influence on the results. In general, and for the linear layer constructed in Section 5 in particular, the situation is less clear. The main reason is that, again, the construction of linear layers is understudied and hence we do not have enough prior work to answer this question satisfactorily at the moment. Instead we view the PRIDE linear layer as a strong benchmark for efficient linear layers with the given parameters and encourage researchers to try to beat its performance.

Along these lines, we see this work as a step towards a more rigorous design process for linear layers. Our hope is that this framework will be extended in future. In particular, we would like to mention the following topic for further investigations. It seems that using an Sbox with a non-trivial branch number has the potential to significantly increase the number of active Sboxes when combined with a linear layer based on Theorem 1. Finding ways to easily prove such a result is worth investigating.

Finally, regarding PRIDE, we obviously encourage further cryptanalysis.

References

1. AES. Advanced Encryption Standard. FIPS PUB 197, Federal Information Processing Standards Publication (2001)
2. Albrecht, M.R., Driessen, B., Kavun, E.B., Leander, G., Paar, C., Yalçın, T.: Block Ciphers – Focus On The Linear Layer (feat. PRIDE): Full Version. IACR Cryptology ePrint Archive, 2014:453 (2014)
3. Anderson, R., Biham, E., Knudsen, L.: Serpent: A Proposal for the Advanced Encryption Standard (1998)
4. Atmel AVR. ATmega8 Datasheet, http://www.atmel.com/images/doc8159.pdf
5. Augot, D., Finiasz, M.: Direct Construction of Recursive MDS Diffusion Layers using Shortened BCH Codes. In: Fast Software Encryption (FSE). LNCS. Springer (to appear, 2014)
6. AVRAES: The AES block cipher on AVR controllers, http://point-at-infinity.org/avraes/
7. Barreto, P.S.L.M., Nikov, V., Nikova, S., Rijmen, V., Tischhauser, E.: Whirlwind: A New Cryptographic Hash Function. Des. Codes Cryptography 56(2-3), 141–162 (2010)
8. Barreto, P.S.L.M., Rijmen, V.: The Anubis Block Cipher. Submission to the NESSIE project (2001)
9. Barreto, P.S.L.M., Rijmen, V.: The Khazad Legacy-level Block Cipher. Submission to the NESSIE project (2001)
10. Beaulieu, R., Shors, D., Smith, J., Treatman-Clark, S., Weeks, B., Wingers, L.: The SIMON and SPECK Families of Lightweight Block Ciphers. IACR Cryptology ePrint Archive, 2013:414 (2013)
11. Bertoni, G., Daemen, J., Peeters, M., Van Assche, G.: Keccak Specifications (2009)
12. Biham, E., Shamir, A.: Differential Cryptanalysis of DES-like Cryptosystems. In: Menezes, A., Vanstone, S.A. (eds.) CRYPTO 1990. LNCS, vol. 537, pp. 2–21. Springer, Heidelberg (1991)
13. Biryukov, A.: DES-X (or DESX). In: Encyclopedia of Cryptography and Security, 2nd edn., p. 331. Springer (2011)
14. Biryukov, A., De Cannière, C., Braeken, A., Preneel, B.: A Toolbox for Cryptanalysis: Linear and Affine Equivalence Algorithms. In: Biham, E. (ed.) EUROCRYPT 2003. LNCS, vol. 2656, pp. 33–50. Springer, Heidelberg (2003)
15. Bogdanov, A., Knudsen, L.R., Leander, G., Paar, C., Poschmann, A., Robshaw, M.J.B., Seurin, Y., Vikkelsø, C.: PRESENT: An Ultra-Lightweight Block Cipher. In: Paillier, P., Verbauwhede, I. (eds.) CHES 2007. LNCS, vol. 4727, pp. 450–466. Springer, Heidelberg (2007)
16. Borghoff, J., et al.: PRINCE – A Low-Latency Block Cipher for Pervasive Computing Applications - Extended Abstract. In: Wang, X., Sako, K. (eds.) ASIACRYPT 2012. LNCS, vol. 7658, pp. 208–225. Springer, Heidelberg (2012)
17. Brinkmann, M., Leander, G.: On the Classification of APN Functions Up to Dimension Five. Des. Codes Cryptography 49(1-3), 273–288 (2008)
18. Carlet, C.: Vectorial Boolean Functions for Cryptography. In: Boolean Methods and Models. Cambridge University Press (2010)
19. Daemen, J.: Cipher and Hash Function Design, Strategies Based On Linear and Differential Cryptanalysis. PhD thesis, Katholieke Universiteit Leuven (1995)
20. Daemen, J., Knudsen, L., Rijmen, V.: The Block Cipher SQUARE. In: Biham, E. (ed.) FSE 1997. LNCS, vol. 1267, pp. 149–165. Springer, Heidelberg (1997)

21. Daemen, J., Rijmen, V.: The Wide Trail Design Strategy. In: Honary, B. (ed.) Cryptography and Coding 2001. LNCS, vol. 2260, pp. 222–238. Springer, Heidelberg (2001)

22. DES: Data Encryption Standard. FIPS PUB 46, Federal Information Processing Standards Publication (1977)

23. Eisenbarth, T., et al.: Compact Implementation and Performance Evaluation of Block Ciphers in ATtiny Devices. In: Mitrokotsa, A., Vaudenay, S. (eds.) AFRICACRYPT 2012. LNCS, vol. 7374, pp. 172–187. Springer, Heidelberg (2012)

24. Engels, S., Kavun, E.B., Mihajloska, H., Paar, C., Yalçın, T.: A Non-Linear/Linear Instruction Set Extension for Lightweight Block Ciphers. In: ARITH'21: 21st IEEE Symposium on Computer Arithmetics. IEEE Computer Society (2013)

25. Gauravaram, P., Knudsen, L., Matusiewicz, K., Mendel, F., Rechberger, C., Schläer, M., Thomsen, S.: Grøstl. SHA-3 Final-round Candidate (2009)

26. Gong, Z., Nikova, S., Law, Y.W.: KLEIN: A New Family of Lightweight Block Ciphers. In: Juels, A., Paar, C. (eds.) RFIDSec 2011. LNCS, vol. 7055, pp. 1–18. Springer, Heidelberg (2012)

27. Grassl, M.: Bounds On the Minimum Distance of Linear Codes and Quantum Codes (2007), http://www.codetables.de

28. Grosso, V., Leurent, G., Standaert, F.-X., Varıcı, K.: LS-Designs: Bitslice Encryption for Efficient Masked Software Implementations. In: Fast Software Encryption (FSE). LNCS. Springer (to appear, 2014)

29. Guo, J., Peyrin, T., Poschmann, A.: The **PHOTON** Family of Lightweight Hash Functions. In: Rogaway, P. (ed.) CRYPTO 2011. LNCS, vol. 6841, pp. 222–239. Springer, Heidelberg (2011)

30. Guo, J., Peyrin, T., Poschmann, A., Robshaw, M.J.B.: The LED Block Cipher. In: Preneel, B., Takagi, T. (eds.) CHES 2011. LNCS, vol. 6917, pp. 326–341. Springer, Heidelberg (2011)

31. Intel. Advanced Encryption Standard Instructions, Intel AES-NI (2008)

32. Karakoç, F., Demirci, H., Harmancı, A.E.: ITUbee: A Software Oriented Lightweight Block Cipher. In: Avoine, G., Kara, O. (eds.) LightSec 2013. LNCS, vol. 8162, pp. 16–27. Springer, Heidelberg (2013)

33. Kavun, E.B., Leander, G., Yalçın, T.: A Reconfigurable Architecture for Searching Optimal Software Code to Implement Block Cipher Permutation Matrices. In: International Conference on ReConFigurable Computing and FPGAs (ReConFig). IEEE Computer Society (2013)

34. Kilian, J., Rogaway, P.: How to Protect DES Against Exhaustive Key Search (An Analysis of DESX). J. Cryptology 14(1), 17–35 (2001)

35. Knežević, M., Nikov, V., Rombouts, P.: Low-Latency Encryption – Is "Lightweight = Light + Wait"? In: Prouff, E., Schaumont, P. (eds.) CHES 2012. LNCS, vol. 7428, pp. 426–446. Springer, Heidelberg (2012)

36. Leander, G., Poschmann, A.: On the Classification of 4 Bit S-Boxes. In: Carlet, C., Sunar, B. (eds.) WAIFI 2007. LNCS, vol. 4547, pp. 159–176. Springer, Heidelberg (2007)

37. Lee, R.B., Fışkıran, M., Wang, M., Hilewitz, Y., Chen, Y.-Y.: PAX: A Cryptographic Processor with Parallel Table Lookup and Wordsize Scalability. Princeton University Department of Electrical Engineering Technical Report CE-L2007-010 (2007)

38. Lee, R.B., Shi, Z., Yang, X.: Efficient Permutation Instructions for Fast Software Cryptography. IEEE Micro 21(6), 56–69 (2001)

39. Lim, C.H., Korkishko, T.: mCrypton – A Lightweight Block Cipher for Security of Low-Cost RFID Tags and Sensors. In: Song, J.-S., Kwon, T., Yung, M. (eds.) WISA 2005. LNCS, vol. 3786, pp. 243–258. Springer, Heidelberg (2006)
40. Lin, S., Costello, D.J. (eds.): Error Control Coding, 2nd edn. Prentice Hall (2004)
41. Matsui, M.: Linear Cryptanalysis Method for DES Cipher. In: Helleseth, T. (ed.) EUROCRYPT 1993. LNCS, vol. 765, pp. 386–397. Springer, Heidelberg (1994)
42. McGregor, J.P., Lee, R.B.: Architectural Enhancements for Fast Subword Permutations with Repetitions in Cryptographic Applications. In: 19th International Conference on Computer Design (ICCD 2001), pp. 453–461 (2001)
43. Nyberg, K.: Differentially Uniform Mappings for Cryptography. In: Helleseth, T. (ed.) EUROCRYPT 1993. LNCS, vol. 765, pp. 55–64. Springer, Heidelberg (1994)
44. PIC. 12-Bit Core Instruction Set
45. PIC vs. AVR, http://www.ladyada.net/library/picvsavr.html
46. Saarinen, M.-J.O.: Cryptographic Analysis of All 4×4-Bit S-Boxes. In: Miri, A., Vaudenay, S. (eds.) SAC 2011. LNCS, vol. 7118, pp. 118–133. Springer, Heidelberg (2012)
47. Sajadieh, M., Dakhilalian, M., Mala, H., Sepehrdad, P.: Recursive Diffusion Layers for Block Ciphers and Hash Functions. In: Canteaut, A. (ed.) FSE 2012. LNCS, vol. 7549, pp. 385–401. Springer, Heidelberg (2012)
48. Shi, Z.J., Yang, X., Lee, R.B.: Alternative Application-Specific Processor Architectures for Fast Arbitrary Bit Permutations. IJES 3(4), 219–228 (2008)
49. Shirai, T., Shibutani, K., Akishita, T., Moriai, S., Iwata, T.: The 128-Bit Blockcipher CLEFIA (Extended Abstract). In: Biryukov, A. (ed.) FSE 2007. LNCS, vol. 4593, pp. 181–195. Springer, Heidelberg (2007)
50. Standaert, F.-X., Piret, G., Gershenfeld, N., Quisquater, J.-J.: SEA: A Scalable Encryption Algorithm for Small Embedded Applications. In: Domingo-Ferrer, J., Posegga, J., Schreckling, D. (eds.) CARDIS 2006. LNCS, vol. 3928, pp. 222–236. Springer, Heidelberg (2006)
51. Suzaki, T., Minematsu, K., Morioka, S., Kobayashi, E.: TWINE: A Lightweight Block Cipher for Multiple Platforms. In: Knudsen, L.R., Wu, H. (eds.) SAC 2012. LNCS, vol. 7707, pp. 339–354. Springer, Heidelberg (2013)
52. Ullrich, M., De Cannière, C., Indesteege, S., Küçük, Ö., Mouha, N., Preneel, B.: Finding Optimal Bitsliced Implementations of 4×4-Bit S-boxes. In: Symmetric Key Encryption Workshop (2011)
53. Wu, S., Wang, M., Wu, W.: Recursive Diffusion Layers for (Lightweight) Block Ciphers and Hash Functions. In: Knudsen, L.R., Wu, H. (eds.) SAC 2012. LNCS, vol. 7707, pp. 355–371. Springer, Heidelberg (2013)
54. Wu, W., Zhang, L.: LBlock: A Lightweight Block Cipher. In: Lopez, J., Tsudik, G. (eds.) ACNS 2011. LNCS, vol. 6715, pp. 327–344. Springer, Heidelberg (2011)

Related-Key Security for Pseudorandom Functions Beyond the Linear Barrier

Michel Abdalla[1], Fabrice Benhamouda[1],
Alain Passelègue[1], and Kenneth G. Paterson[2]

[1] Département d'Informatique, École normale supérieure, Paris, France
{michel.abdalla,fabrice.ben.hamouda,alain.passelegue}@ens.fr
http://www.di.ens.fr/users/{mabdalla,fbenhamo,passeleg}
[2] Information Security Group, Royal Holloway, University of London, Surrey, UK
kenny.paterson@rhul.ac.uk
http://www.isg.rhul.ac.uk/~kp/

Abstract. Related-key attacks (RKAs) concern the security of cryptographic primitives in the situation where the key can be manipulated by the adversary. In the RKA setting, the adversary's power is expressed through the class of related-key deriving (RKD) functions which the adversary is restricted to using when modifying keys. Bellare and Kohno (Eurocrypt 2003) first formalised RKAs and pin-pointed the foundational problem of constructing RKA-secure pseudorandom functions (RKA-PRFs). To date there are few constructions for RKA-PRFs under standard assumptions, and it is a major open problem to construct RKA-PRFs for larger classes of RKD functions. We make significant progress on this problem. We first show how to repair the Bellare-Cash framework for constructing RKA-PRFs and extend it to handle the more challenging case of classes of RKD functions that contain claws. We apply this extension to show that a variant of the Naor-Reingold function already considered by Bellare and Cash is an RKA-PRF for a class of affine RKD functions under the DDH assumption, albeit with an exponential-time security reduction. We then develop a second extension of the Bellare-Cash framework, and use it to show that the same Naor-Reingold variant is actually an RKA-PRF for a class of degree d polynomial RKD functions under the stronger decisional d-Diffie-Hellman inversion assumption. As a significant technical contribution, our proof of this result avoids the exponential-time security reduction that was inherent in the work of Bellare and Cash and in our first result.

Keywords: Related-Key Security, Pseudorandom Functions.

1 Introduction

Background and Context. A common approach to prove the security of a cryptographic scheme, known as provable security, is to relate its security to one of its underlying primitives or to an accepted hard computational problem. While this approach is now standard and widely accepted, there is still a significant gap

J.A. Garay and R. Gennaro (Eds.): CRYPTO 2014, Part I, LNCS 8616, pp. 77–94, 2014.

between the existing models used in security proofs and the actual environment in which these cryptosystems are deployed. For example, most of the existing security models assume that the adversary has no information about the user's secret key. However, it is well known that this is not always true in practice: the adversary may be able to learn partial information about the secrets using different types of side-channel attacks, such as the study of energy consumption, fault injection, or timing analysis. In the particular case of fault injection, for instance, an adversary can learn not only partial information about the secret key, but he may also be able to force a cryptosystem to work with different but related secret keys. Then, if he can observe the outcome of this cryptosystem, he may be able to break it. This is what is known in the literature as a related-key attack (RKA).

Most primitives are designed without taking related-key attacks into consideration so their security proofs do not provide any guarantee against such attacks. Hence, a cryptographic scheme that is perfectly safe in theory may be completely vulnerable in practice. Indeed, many such attacks were found during the last decade, especially against practical blockciphers [10–14, 18]. Inspired by this cryptanalytic work, some years ago, theoreticians started to develop appropriate security models and search for cryptographic primitives which can be proven RKA secure.

Formal Foundations of RKA Security. Though RKAs were first introduced by Biham and Knudsen [9, 19] in the early 1990s, it was only in 2003 that Bellare and Kohno [6] began the formalisation of the theoretical foundations for RKA security. We recall their security definition for RKA security of PRFs here. Let $F\colon \mathcal{K} \times \mathcal{D} \to \mathcal{R}$ be a family of functions for a security parameter k, and let $\Phi = \{\phi\colon \mathcal{K} \to \mathcal{K}\}$ be a set of functions on the key space $\mathcal{K}$, called a related-key deriving (RKD) function set. We say that F is a Φ-RKA-PRF if for any polynomial-time adversary, its advantage in the following game is negligible. The game starts by picking a random challenge bit b, a random target key $K \in \mathcal{K}$ and a random function $G\colon \mathcal{K} \times \mathcal{D} \to \mathcal{R}$. The adversary can repeatedly query an oracle that, given a pair $(\phi, x) \in \Phi \times \mathcal{D}$, returns either $F(\phi(K), x)$, if $b = 1$, or $G(\phi(K), x)$, if $b = 0$. Finally, the adversary outputs a bit b', and its advantage is defined by $2\Pr[\,b = b'\,] - 1$. Note that if the class Φ of RKD functions contains only the identity function, then this notion matches standard PRF security.

Bellare and Cash [3] designed the first RKA-PRFs secure under standard assumptions, by adapting the Naor-Reingold PRF [21]. Their RKA-PRFs are secure for RKA function classes consisting of certain multiplicative and additive classes. To explain their results, let us begin by recalling the definition of the Naor-Reingold PRF. Let $\mathbb{G} = \langle g \rangle$ be a group of prime order p. Let NR$\colon (\mathbb{Z}_p^*)^{n+1} \times \{0,1\}^n \to \mathbb{G}$ denote the Naor-Reingold PRF that given a key $\mathbf{a} = (\mathbf{a}[0], \ldots, \mathbf{a}[n]) \in (\mathbb{Z}_p^*)^{n+1}$ and input $x = x[1] \ldots x[n] \in \{0,1\}^n$ returns

$$\mathsf{NR}(\mathbf{a}, x) = g^{\mathbf{a}[0]\prod_{i=1}^{n} \mathbf{a}[i]^{x[i]}} .$$

The keyspace of the Naor-Reingold PRF is $\mathcal{K} = (\mathbb{Z}_p^*)^{n+1}$, which has a group structure under the operation of component-wise multiplication modulo p, denoted $*$. Now let Φ_* denote the class of component-wise multiplicative functions on $(\mathbb{Z}_p^*)^{n+1}$, that is $\Phi_* = \{\phi\colon \mathbf{a} \in (\mathbb{Z}_p^*)^{n+1} \mapsto \mathbf{b} * \mathbf{a} \mid \mathbf{b} \in (\mathbb{Z}_p^*)^{n+1}\}$. It is easy to see that NR is not itself a Φ_*-RKA-PRF, since it suffers from simple algebraic attacks, but using a collision-resistant hash function $h\colon \{0,1\}^n \times \mathbb{G}^{n+1} \to \{0,1\}^{n-2}$, Bellare and Cash were able to show that a simple modification of the Naor-Reingold PRF does yield a Φ_*-RKA-PRF under the DDH assumption. Specifically, they defined $F\colon (\mathbb{Z}_p^*)^{n+1} \times \{0,1\}^n \to \mathbb{G}$ by:

$$F(\mathbf{a}, x) = \mathsf{NR}(\mathbf{a}, 11 \| h(x, (g^{\mathbf{a}[0]}, g^{\mathbf{a}[0]\mathbf{a}[1]}, \dots, g^{\mathbf{a}[0]\mathbf{a}[n]})))$$

and showed that this F is indeed a Φ_*-RKA-PRF under the DDH assumption. A second construction in [3] uses similar techniques to build an RKA-PRF under the DLIN assumption.

In the original version of their paper, Bellare and Cash also used a variant of the Naor-Reingold PRF, $\mathsf{NR}^*\colon (\mathbb{Z}_p)^n \times \{0,1\}^n \backslash \{0^n\} \to \mathbb{G}$, defined by:

$$\mathsf{NR}^*(\mathbf{a}, x) = g^{\prod_{i=1}^{n} \mathbf{a}[i]^{x[i]}},$$

to obtain a third RKA-PRF, this one for additive RKD functions. In more detail, the keyspace $\mathcal{K} = (\mathbb{Z}_p)^n$ of NR^*, has a natural group structure under the operation of component-wise addition modulo p. We define Φ_+ to be the class of functions, $\Phi_+ = \{\phi\colon \mathbf{a} \in (\mathbb{Z}_p)^n \mapsto \mathbf{a} + \mathbf{b} \mid \mathbf{b} \in (\mathbb{Z}_p)^n\}$. Then, Bellare and Cash claimed that the function $F\colon (\mathbb{Z}_p)^n \times (\{0,1\}^n \setminus 0^n) \to \mathbb{G}$ with

$$F(\mathbf{a}, x) = \mathsf{NR}^*(\mathbf{a}, 11 \| h(x, (g^{\mathbf{a}[1]}, g^{\mathbf{a}[2]}, \dots, g^{\mathbf{a}[n]})))$$

is a Φ_+-RKA-PRF under the DDH assumption, when the function $h\colon \{0,1\}^n \times \mathbb{G}^n \to \{0,1\}^{n-2}$ is a collision-resistant hash function. The running time of their security reduction in this case was exponential in the input size.

These foundational results of [3] were obtained by applying a single, elegant, general framework to the Naor-Reingold PRFs. The framework hinges on two main tools, key-malleability and key-fingerprints for PRFs and associated RKD function classes Φ. The former property means that there is an efficient deterministic algorithm, called a *key-transformer*, that enables one to transform an oracle for computing $F(K, x)$ into one for computing $F(\phi(K), x)$ for any $\phi \in \Phi$ and any input x (the technical requirements are in fact somewhat more involved than these). The latter provides a means to ensure that, in the Bellare-Cash construction for an RKA-PRF from a (normal) PRF F, all adversarial queries to the putative Φ-RKA-PRF get appropriately separated before being processed by F. In combination, these two features enable a reduction to be made to the PRF security of the underlying function F.

Unfortunately, it was recently discovered that the original framework of [3] has a bug, in that a technical requirement on the key-transformer, called hash function compatibility, was too weak to enable the original security proof of the Bellare-Cash construction to go through. When hash function compatibility is appropriately strengthened to enable a proof, it still holds for the key-transformers used in the analysis of their two main constructions, the multiplicative DDH and DLIN-based RKA-PRF constructions. However, the new compatibility definition no longer holds for the key-transformer used in their additive, DDH-based RKA-PRF construction. With respect to their framework and, specifically, their additive, DDH-based RKA-PRF construction, Bellare and Cash note in the latest version of their paper [4]: *We see no easy way to fill the gap within our current framework and accordingly are retracting our claims about this construction and omitting it from the current version.*

Main Question. A natural question that arises from the work of Bellare-Cash is whether it is possible to go further, to obtain RKA-PRFs for larger classes of RKD function than Φ_* and Φ_+. This is important in understanding whether there are yet to be discovered fundamental barriers in achieving RKA security for PRFs, as well as bringing the current state of the art for RKA security closer to practical application. This question becomes even more relevant in the light of the results of Bellare, Cash and Miller [5], who showed that RKA-security can be transferred from PRFs to several other primitives, including identity-based encryption (IBE), signatures, as well as symmetric (SE) and public-key encryption (PKE) secure against chosen-ciphertext attacks. Their results illustrate the central role that RKA-PRFs play in related-key security more generally: any advance in constructing RKA-PRFs for broader classes would immediately transfer to these other primitives via the results of [5]. A subsidiary question is whether it is possible to repair the Bellare-Cash framework without requiring stronger hash compatibility conditions on the key-transformer. This, if achievable, would reinstate their Φ_+-RKA-PRF.

A partial answer to the first question was provided by Goyal, O'Neill and Rao [17], who proposed RKA-secure weak-PRF and symmetric encryption schemes for polynomial functions using the Decisional Truncated q-ADHE problem. RKA-secure weak-PRFs, however, are significantly weaker than standard RKA-PRFs since their security only holds with respect to random inputs. Wee [22] provided RKA-secure PKE for linear functions, while Bellare, Paterson, and Thomson [7] proposed a framework for obtaining RKA-secure IBE for affine and polynomial RKD function sets, from which RKA security for signatures, PKE (and more) for the same RKD function sets follows using the results of [5] and extensions thereof. However, in respect of these works, it should be noted that achieving RKA security for randomised primitives appears to be substantially easier than for PRFs which are deterministic objects. An extended discussion on this point can be found in [3, Section 1].

In parallel work to ours, Lewi et al. [20] showed that the key homomorphic PRFs from Boneh et al. [15] (and slight extensions of them) are RKA-secure. Specifically, they show RKA-security for a strict subset of Φ_+ for the PRF of [15] that is based on the Learning with Error (LWE) problem, and against a claw-free class of affine functions for the PRF of [15] that is based on multilinear maps. They also showed that, if the adversary's queries are restricted to unique inputs, these two PRFs are RKA-secure for larger classes, namely a class of affine RKD functions (with a low-norm for the "linear" part) for the LWE-based

PRF and a class of polynomial RKD functions for the PRF based on multilinear maps. These classes are not really comparable to our classes Φ_{aff} and Φ_d of affine and polynomial functions defined below, because the secret-key structures are slightly different. However, we remark that Lewi et al. [20] do not deal with claw-free classes and do not show ways to leverage unique-input RKA security to full RKA security. We handle both of these issues in our paper, and it may be possible to extend our solutions to their setting. It should also be remarked that the construction of Barnahee and Peikert [2] may also yield another RKA-secure PRF based on LWE.

Our Contributions. In this paper, we make substantial progress on the main question above, obtaining RKA-PRFs for substantially larger classes of RKD functions than were previously known. Along the way, we recover the original Bellare-Cash framework, showing that their original technical conditions on the key-transformer are in fact *already* sufficient to enable a (different) proof of RKA security to go through. Let us first introduce our main results on specific RKA-PRFs, and then explain the technical means by which they are obtained.

For p prime and $n, d \geq 1$, let Φ_d denote the class of functions from $\mathbb{Z}_p^n$ to $\mathbb{Z}_p^n$ each of whose component functions is a non-constant polynomial in one variable of degree at most d. That is, we have:

$$\Phi_d = \left\{ \phi \colon \mathbb{Z}_p^n \to \mathbb{Z}_p^n \,\middle|\, \begin{array}{l} \phi = (\phi_1, \ldots, \phi_n); \phi_i \colon A_i \mapsto \sum_{j=0}^{d} \alpha_{i,j} \cdot A_i^j, \\ \forall i = 1, \ldots, n, \ (\alpha_{i,1}, \ldots, \alpha_{i,d}) \neq 0^d \end{array} \right\}.$$

For the special case $d = 1$, we denote Φ_1 by Φ_{aff} (aff for affine functions). Note that $\Phi_+ \subset \Phi_{\mathsf{aff}}$.

We will construct RKA-PRFs for the RKD-function classes Φ_{aff} and Φ_d for each d. To this end, let $\mathbb{G} = \langle g \rangle$ be a group of prime order p, let $\overline{\mathcal{D}} = \{0,1\}^n \times \mathbb{G}^n$ and let $h \colon \overline{\mathcal{D}} \to \{0,1\}^{n-2}$ be a hash function. Let $\mathbf{w}[i] = 0^{i-1} \| 1 \| 0^{n-i}$, for $i = 1, \ldots, n$. Define $F \colon \mathbb{Z}_p^n \times (\{0,1\}^n \setminus 0^n) \to \mathbb{G}$ by:

$$F(\mathbf{a}, x) = \mathsf{NR}^*(\mathbf{a}, 11 \| h(x, \mathsf{NR}^*(\mathbf{a}, \mathbf{w})))$$

for all $\mathbf{a} \in \mathbb{Z}_p^n$ and $x \in \{0,1\}^n$. This is the same F as in the withdrawn construction of [3]. Theorems 7 and 13 show that this function is an RKA-PRF for both the RKD-function classes Φ_{aff} and Φ_d (for each d), under reasonable hardness assumptions.

For our first result on the Φ_{aff}-RKA-PRF security of F, we recover and extend the withdrawn result of Bellare and Cash [3], under the same hardness assumption that they required, namely the standard DDH assumption. Here our proof, like that in [3], requires an exponential-time reduction. We then develop a further extension of the Bellare-Cash framework enabling us to circumvent their use of key-transformers having a key malleability property. We use this framework to modularise our proof that F is also a Φ_d-RKA-PRF. As part of this proof, we require the decisional d-Diffie-Hellman Inversion (d-DDHI) assumption, introduced in [17]. Informally, the d-DDHI problem in a group $\mathbb{G}$ of prime order p consists of deciding, given inputs $(g, g^a, \ldots, g^{a^d})$ and z, where g is a generator of $\mathbb{G}$, whether z is equal to $g^{\frac{1}{a}}$ or to a random group element. Notably, in our analysis of the Φ_d-RKA-PRF security of F, we are able to avoid an exponential-time reduction. This puts the RKA-PRF F on the same footing as the surviving constructions in [3].

Let us now expand on the technical aspects of our contributions.

Proof Barriers and Techniques. We first show how the Bellare-Cash framework can be modified to deal with RKD functions that are *not* claw-free, meaning that there exist pairs of different RKD functions ϕ_1 and ϕ_2 and a key $K \in \mathcal{K}$, such that $\phi_1(K) = \phi_2(K)$. Up to now, only claw-free classes have been considered for RKA-PRFs. But classes Φ underlying practical attacks such as fault injections have no reason to be claw-free, so dealing with non-claw-free classes of RKD functions is important in advancing RKA security towards practice. Moreover, both our RKD function classes of interest, Φ_{aff} and Φ_d, do contain claws. The lack of claw-freeness poses a problem in security proofs because, if an adversary is able to find two RKD functions which lead to the same derived key, he can detect this via his queries, and then the equation $\phi_1(K) = \phi_2(K)$ may leak information on K sufficient to enable the adversary to break RKA-PRF security in a particular construction.

We overcome the lack of claw-freeness in our adaptation of the Bellare-Cash framework by introducing two new concepts, Φ-Key-Collision Security for PRFs and Φ-Statistical-Key-Collision Security. The former is a property similar to the identity-collision-resistance property defined in [5] in the context of pseudorandom generators and refers to the non-existence of an adversary who can find a colliding key (i.e. $\phi_1(K) = \phi_2(K)$ for $\phi_1, \phi_2 \in \Phi$), when given oracle access to the PRF under related keys $\phi(K)$. The latter concept is essentially the same, but now oracle access to the PRF is replaced by oracle access to a random function. These properties are just the right ingredients necessary to generalise the Bellare-Cash framework to the non-claw-free case.

At the same time as dealing with claws, we are able to repair the gap in the proof for the original Bellare-Cash framework, showing that their original hash function compatibility condition required of the key-transformer is already strong enough to enable an alternative proof of RKA security. Our new proof introduces a slightly different sequence of game hops in order to avoid the apparent impasse in the original proof. Our main theorem establishing the RKA-PRF security of functions arising from this framework is Theorem 1. It repairs and extends the corresponding main theorem in [3]. Our theorem is then combined with an analysis of the specific function NR^* to obtain Theorem 7 concerning the Φ_{aff}-RKA-PRF security of F.

To show that F is also an RKA-PRF for Φ_d, we still have a second major difficulty to overcome. While Φ_d-Key-Collision Security and Φ_d-Statistical-Key-Collision Security can still be proven for F, we no longer have the key-transformer component that is critical to the Bellare-Cash framework. Instead, in Section 5, we introduce a further extension of their framework, replacing the key-transformer with a stronger pseudorandomness condition on the base PRF M used in the construction, which we call (S, Φ)-unique-input-prf-rka security. The new requirement essentially states that M should already act as a Φ-RKA-PRF on a restricted domain S, provided the queries $(\phi_1, x_1), \ldots, (\phi_k, x_k)$ made by the Φ-RKA-PRF adversary to its oracle with $x_i \in S$ are all for *distinct* x_i. Under this condition, we are able to prove Theorem 8 establishing the security of RKA-PRFs arising from our further extension of the Bellare-Cash framework. This theorem then enables us to prove in a modular fashion that F is also an RKA-PRF for Φ_d.

The final technical challenge is in proving that NR*, playing the role of M, satisfies the relevant (S, Φ)-unique-input-prf-rka security property so as to allow the application of Theorem 8. This is done in a crucial lemma, Lemma 12, whose proof involves a delicate series of hybrids in which we gradually replace the oracle responses to queries (ϕ_i, x_i) for x_i in a suitable set S with random values. We exploit the algebraic nature of the function NR* to ensure that the hybrids are close under a particular pair of hardness assumptions the (N, d)-Polynomial DDH and (N, d)-EDDH assumptions). We also make use of an efficient, approximate (but close to perfect) procedure to detect linear dependencies arising in the simulation from the adversary's oracle queries. This procedure is key to making the entire proof efficient (rather than exponential-time). Finally, we provide a series of reductions relating our pair of hardness assumptions to the d-DDHI assumption. Examining the details of the proof shows that we can recover our result concerning Φ_{aff}-RKA-PRF security of F under DDH (rather than 1-DDHI), but now without an exponential-time reduction.

2   Definitions

Notations and Conventions. Let $\mathsf{Fun}(\mathcal{K} \times \mathcal{D}, \mathcal{R})$ be the set of all families of functions $F \colon \mathcal{K} \times \mathcal{D} \to \mathcal{R}$. A family of functions $F \colon \mathcal{K} \times \mathcal{D} \to \mathcal{R}$ takes a key $K \in \mathcal{K}$ and an input $x \in \mathcal{D}$ and returns an output $F(K, x) \in \mathcal{R}$. If $\mathbf{x}$ is a vector then $|\mathbf{x}|$ denotes its length, and $\mathbf{x} = (\mathbf{x}[1], \ldots, \mathbf{x}[|\mathbf{x}|])$. A binary string x is identified with a vector over $\{0, 1\}$ so that $|x|$ denotes its length, $x[i]$ its i-th bit and, for $i, j \in \{1, \ldots, n\}$, $i \leq j$, $x[i, \ldots, j]$ the binary string $x[i]\| \ldots \|x[j]$. For a binary string $x \in \{0, 1\}^n$ and an integer d, we denote by $d \cdot x$ the string $y = y[1]\| \ldots \|y[n] \in \{0, d\}^n$ defined by $y[i] = d \cdot x[i]$ for $i = 1, \ldots, n$. For two strings $x, y \in \{0, \ldots, d\}^n$, we denote by $y \preceq x$ the fact that $y[i] \leq x[i]$, $\forall i = 1, \ldots, n$ and we denote by $S(x)$ the set $\{i \mid x[i] \neq 0\}$. If ϕ is a vector of functions from S_1 to S_2 with $|\phi| = n$ and $\mathbf{a} \in S_1^n$ then we denote by $\phi(\mathbf{a})$ the vector $(\phi[1](\mathbf{a}[1]), \ldots, \phi[n](\mathbf{a}[n])) \in S_2^n$. If $F \colon \mathcal{K} \times \mathcal{D} \to \mathcal{R}$ is a family of functions and $\mathbf{x}$ is a vector over $\mathcal{D}$ then $F(K, \mathbf{x})$ denotes the vector $(F(K, \mathbf{x}[1]), \ldots, F(K, \mathbf{x}[|\mathbf{x}|]))$. If S is a set, then $|S|$ denotes its size. We denote by $s \xleftarrow{\$} S$ the operation of picking at random s in S. If A is a randomized algorithm, we denote by $y \xleftarrow{\$} \mathsf{A}(x_1, x_2, \ldots)$ the operation of running A on inputs $(x_1, x_2, \ldots)$ with fresh coins and letting y denote the output.

Games. Some of our definitions and proofs use code-based game-playing [8]. Recall that a game has an **Initialize** procedure, procedures to respond to adversary's oracle queries, and a **Finalize** procedure. A game G is executed with an adversary A as follows. First, **Initialize** executes and its outputs are the inputs to A. Then A executes, its oracle queries being answered by the corresponding procedures of G. When A terminates, its outputs become the input to the **Finalize** procedure. The output of the latter, denoted G^A is called the output of the game, and we let "$G^A \Rightarrow 1$", abbreviated W in the proofs, denote the event that this game output takes the value 1. Boolean flags are assumed initialized to false. Games G_i, G_j are identical until flag if their code differs only in statements that follow the setting of flag to true. The running time of an adversary by convention is the worst case time for the execution of the adversary with any of the games defining its security, so that the time of the called game procedures is included.

PRFs. PRFs were introduced by [16]. A PRF is a family of functions $F \colon \mathcal{K} \times \mathcal{D} \to \mathcal{R}$ which is efficiently computable and so that it is hard to distinguish a function chosen randomly from the PRF family from a random function, which is formally defined as the fact that the advantage of any efficient adversary in attacking the standard prf security of F is negligible. The advantage of an adversary A in attacking the standard prf security of a family of functions $F \colon \mathcal{K} \times \mathcal{D} \to \mathcal{R}$ is defined via

$$\mathbf{Adv}_F^{\mathsf{prf}}(A) = \Pr\left[\,\mathrm{PRFReal}_F^A \Rightarrow 1\,\right] - \Pr\left[\,\mathrm{PRFRand}_F^A \Rightarrow 1\,\right].$$

Game $\mathrm{PRFReal}_F$ begins by picking $K \xleftarrow{\$} \mathcal{K}$ and responds to query $\mathrm{FN}(x)$ via $F(K, x)$. Game $\mathrm{PRFRand}_F$ begins by picking $f \xleftarrow{\$} \mathsf{Fun}(\mathcal{D}, \mathcal{R})$ and responds to oracle query $\mathrm{FN}(x)$ via $f(x)$.

RKA-PRFs. We recall the definitions from [6]. Let $F \colon \mathcal{K} \times \mathcal{D} \to \mathcal{R}$ be a family of functions and $\Phi \subseteq \mathsf{Fun}(\mathcal{K}, \mathcal{K})$. The members of Φ are called RKD (Related-Key Deriving) functions. An adversary is said to be Φ-restricted if its oracle queries (ϕ, x) satisfy $\phi \in \Phi$. The advantage of a Φ-restricted adversary A in attacking the prf-rka security of F is defined via

$$\mathbf{Adv}_{\Phi, F}^{\mathsf{prf\text{-}rka}}(A) = \Pr\left[\,\mathrm{RKPRFReal}_F^A \Rightarrow 1\,\right] - \Pr\left[\,\mathrm{RKPRFRand}_F^A \Rightarrow 1\,\right].$$

Game $\mathrm{RKPRFReal}_F$ begins by picking $K \xleftarrow{\$} \mathcal{K}$ and then responds to oracle query $\mathrm{RKFn}(\phi, x)$ via $F(\phi(K), x)$. Game $\mathrm{RKPRFRand}_F$ begins by picking $G \xleftarrow{\$} \mathsf{Fun}(\mathcal{K} \times \mathcal{D}, \mathcal{R})$ and responds to oracle query $\mathrm{RKFn}(\phi, x)$ via $G(\phi(K), x)$. We say that F is a Φ-RKA-secure PRF if for any Φ-restricted, efficient adversary, its advantage in attacking the prf-rka security is negligible.

Strong Key Fingerprint. A strong key fingerprint is a tool used in proofs to detect whether a key arises more than once in a simulation, even if we do not have any information about the key itself. We recall the definition from [3]. Suppose $F \colon \mathcal{K} \times \mathcal{D} \to \mathcal{R}$ is a family of functions. Let $\mathbf{w}$ be a vector over $\mathcal{D}$ and let $n = |\mathbf{w}|$. We say that $\mathbf{w}$ is a *strong key fingerprint* for F if

$$(F(K, \mathbf{w}[1]), \dots, F(K, \mathbf{w}[n])) \neq (F(K', \mathbf{w}[1]), \dots, F(K', \mathbf{w}[n]))$$

for all distinct $K, K' \in \mathcal{K}$.

Key-Malleability. As defined in [3], let $F\colon \mathcal{K} \times \mathcal{D} \to \mathcal{R}$ be a family of functions and Φ be a class of RKD functions. Suppose T is a deterministic algorithm that, given an oracle $f\colon \mathcal{D} \to \mathcal{R}$ and inputs $(\phi, x) \in \Phi \times \mathcal{D}$, returns a point $\mathsf{T}^f(\phi, x) \in \mathcal{R}$. T is said to be a *key-transformer* for (F, Φ) if it satisfies the *correctness* and *uniformity* conditions. *Correctness* asks that $\mathsf{T}^{F(K, \cdot)}(\phi, x) = F(\phi(K), x)$ for every $(\phi, K, x) \in \Phi \times \mathcal{K} \times \mathcal{D}$. Let us say that a Φ-restricted adversary is unique-input if, in its oracle queries $(\phi_1, x_1), \ldots, (\phi_q, x_q)$, the points $x_1, \ldots, x_q$ are always distinct. *Uniformity* requires that for any (even inefficient) Φ-restricted, unique-input adversary U,

$$\Pr\left[\, \mathrm{KTReal}_\mathsf{T}^U \Rightarrow 1 \,\right] = \Pr\left[\, \mathrm{KTRand}_\mathsf{T}^U \Rightarrow 1 \,\right],$$

where game $\mathrm{KTReal}_\mathsf{T}$ is initialized by picking $f \xleftarrow{\$} \mathsf{Fun}(\mathcal{D}, \mathcal{R})$ and responds to query $\mathrm{KTFn}(\phi, x)$ via $\mathsf{T}^f(\phi, x)$, while $\mathrm{KTRand}_\mathsf{T}$ has no initialization and responds to oracle query $\mathrm{KTFn}(\phi, x)$ by returning a value $y \xleftarrow{\$} \mathcal{R}$ chosen uniformly at random in $\mathcal{R}$. If such a key-transformer exists, we say that F is a Φ-key-malleable PRF.

Compatible Hash Function. Let $F\colon \mathcal{K} \times \mathcal{D} \to \mathcal{R}$ be a family of functions and Φ be a class of RKD functions, such that there is a key-transformer T for (F, Φ). Let $\mathbf{w} \in \mathcal{D}^m$ and let $\overline{\mathcal{D}} = \mathcal{D} \times \mathcal{R}^m$. We denote by $\mathsf{Qrs}(\mathsf{T}, F, \Phi, \mathbf{w})$ the set of all $w \in \mathcal{D}$ such that there exists $(f, \phi, i) \in \mathsf{Fun}(\mathcal{D}, \mathcal{R}) \times \Phi \times \{1, \ldots, m\}$ such that the computation of $\mathsf{T}^f(\phi, \mathbf{w}[i])$ makes oracle query w. Then, we say that a hash function $H\colon \overline{\mathcal{D}} \to S$ is *compatible* with $(\mathsf{T}, F, \Phi, \mathbf{w})$, if $S = \mathcal{D} \backslash \mathsf{Qrs}(\mathsf{T}, F, \Phi, \mathbf{w})$. Note that this definition is the same as that given in the original Bellare-Cash framework [3] rather than the stronger one used in the authors' repaired version [4].

CR hash functions. The advantage of C in attacking the collision-resistance security of $H\colon \mathcal{D} \to \mathcal{R}$ is

$$\mathbf{Adv}_H^{\mathsf{cr}}(C) = \Pr\left[\, x \neq x' \text{ and } H(x) = H(x') \,\right]$$

where the probability is over $(x, x') \xleftarrow{\$} C$.

Hardness Assumptions. Our proofs make use of the d-Strong Discrete Logarithm (d-SDL) and Decisional d-Diffie-Hellman Inversion (d-DDHI) problems given in [17] and described in Fig. 1. We define the advantage of an adversary D against the d-SDL problem in $\mathbb{G}$ as

$$\mathbf{Adv}_\mathbb{G}^{d\text{-}\mathsf{sdl}}(D) = \Pr\left[\, d\text{-}\mathrm{SDL}_\mathbb{G}^D \Rightarrow \mathsf{true} \,\right]$$

where the probability is over the choices of $a \in \mathbb{Z}_p$, $g \in \mathbb{G}$, and the random coins used by the adversary. The advantage of an adversary D against the d-DDHI problem in $\mathbb{G}$ is defined to be

$$\mathbf{Adv}_\mathbb{G}^{d\text{-}\mathsf{ddhi}}(D) = \Pr\left[\, d\text{-}\mathrm{DDHI\text{-}Real}_\mathbb{G}^D \Rightarrow 1 \,\right] - \Pr\left[\, d\text{-}\mathrm{DDHI\text{-}Rand}_\mathbb{G}^D \Rightarrow 1 \,\right]$$

proc Initialize // d-SDL	**proc Finalize**(a') // d-SDL
$g \xleftarrow{\$} \mathbb{G}$; $a \xleftarrow{\$} \mathbb{Z}_p^*$	Return ($g^a = g^{a'}$)
Return ($g, g^a, \ldots, g^{a^d}$)	

proc Initialize // d-DDHI-Real	**proc Initialize** // d-DDHI-Rand
$g \xleftarrow{\$} \mathbb{G}$; $a \xleftarrow{\$} \mathbb{Z}_p^*$	$g \xleftarrow{\$} \mathbb{G}$; $a \xleftarrow{\$} \mathbb{Z}_p^*$; $z \xleftarrow{\$} \mathbb{Z}_p^*$
Return ($g, g^a, \ldots, g^{a^d}, g^{1/a}$)	Return ($g, g^a, \ldots, g^{a^d}, g^z$)
proc Finalize(b) // d-DDHI-Real	**proc Finalize**(b) // d-DDHI-Rand
Return b	Return b

Fig. 1. Games defining the d-SDL and d-DDHI problems in $\mathbb{G}$

where the probabilities are over the choices of $a, z \in \mathbb{Z}_p$, $g \in \mathbb{G}$, and the random coins used by the adversary. We have two assumptions corresponding to the hardness of these problems, the d-SDL assumption and the d-DDHI assumption. Setting $d = 1$ in the d-SDL problem, we recover the usual definition of the DL problem in $\mathbb{G}$.

3 Repairing and Extending the Bellare-Cash Framework

Here, we give a method to deal with classes of RKD functions that are not claw-free, such as affine classes, by repairing and extending the general framework of Bellare and Cash from [3]. Our approach still relies on key-malleability, meaning that it is not generally applicable since almost all the known PRFs are not key-malleable for interesting classes of functions. However, as we shall see, it *does* provide an easy way to obtain a Φ_{aff}-RKA-secure PRF, using the variant NR* of the Naor-Reingold PRF. In Section 5, we will present a further extension of the Bellare-Cash approach that enables us to deal with PRFs that are not key-malleable.

To deal with non-claw-freeness, we first introduce two new notions. The first one is called Φ-*Key-Collision Security* and captures the likelihood that an adversary finds two RKD functions which lead to the same derived key in a given PRF construction. The second one, called Φ-*Statistical-Key-Collision Security*, is similar, but replaces the oracle access to the PRF with an oracle access to a random function.

Φ-Key-Collision (Φ-kc) Security. Let Φ be a class of RKD functions. We define the advantage of an adversary A against the Φ-key-collision security of a PRF $M \colon \mathcal{K} \times \mathcal{D} \to \mathcal{R}$, denoted by $\mathbf{Adv}_{\Phi, M}^{\mathsf{kc}}(A)$, to be the probability of success in the game on the left side of Fig. 2, where the functions ϕ appearing in A's queries are restricted to lie in Φ.

Φ-Statistical-Key-Collision (Φ-skc) Security. Let Φ be a class of RKD functions. We define the advantage of an adversary A against the Φ-statistical-key-collision security for $\mathsf{Fun}(\mathcal{K} \times \mathcal{D}, \mathcal{R})$, denoted by $\mathbf{Adv}_{\Phi}^{\mathsf{skc}}(A)$, to be the probability of success in the game on the right side of Fig. 2. Here the functions ϕ appearing in A's queries are again restricted to lie in Φ.

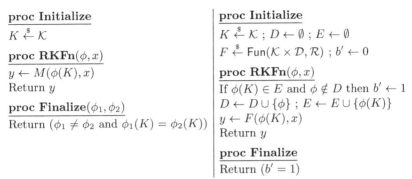

Fig. 2. Game defining the Φ-key-collision security of a PRF M on the left and Φ-statistical-key-collision security for $\mathsf{Fun}(\mathcal{K} \times \mathcal{D}, \mathcal{R})$ on the right

Using these notions, we can now prove the following theorem, which both repairs and extends the main result of [3].

Theorem 1. *Let* M: $\mathcal{K} \times \mathcal{D} \to \mathcal{R}$ *be a family of functions and* Φ *be a class of RKD functions that contains the identity function* id. *Let* T *be a key-transformer for* (M, Φ) *making* Q_{T} *oracle queries, and let* $\mathbf{w} \in \mathcal{D}^m$ *be a strong key fingerprint for* M. *Let* $\overline{\mathcal{D}} = \mathcal{D} \times \mathcal{R}^m$ *and let* H: $\overline{\mathcal{D}} \to S$ *be a hash function that is compatible with* $(\mathsf{T}, M, \Phi, \mathbf{w})$. *Define* F: $\mathcal{K} \times \mathcal{D} \to \mathcal{R}$ *by*

$$F(K, x) = M(K, H(x, M(K, \mathbf{w})))$$

for all $K \in \mathcal{K}$ *and* $x \in \mathcal{D}$. *Let* A *be a* Φ-restricted adversary against the prf-rka security of F that makes $Q_A \leq |S|$ oracle queries. Then we can construct an adversary B against the standard prf security of M, an adversary C against the cr security of H, an adversary D against the Φ-kc security of M and an adversary E against Φ-skc security for $\mathsf{Fun}(\mathcal{K} \times \mathcal{D}, \mathcal{R})$ such that*

$$\mathbf{Adv}_{\Phi,F}^{\mathsf{prf\text{-}rka}}(A) \leq \mathbf{Adv}_{M}^{\mathsf{prf}}(B) + \mathbf{Adv}_{H}^{\mathsf{cr}}(C) + \mathbf{Adv}_{\Phi,M}^{\mathsf{kc}}(D) + \mathbf{Adv}_{\Phi}^{\mathsf{skc}}(E). \quad (1)$$

Adversaries C, D *and* E *have the same running time as* A. *Adversary* B *has the same running time as* A *plus the time required for* $Q_A \cdot (m + 1)$ *executions of the key-transformer* T.

Note that if the class Φ is claw-free, then the advantage of any adversary in breaking Φ-kc security of M or Φ-skc security for $\mathsf{Fun}(\mathcal{K} \times \mathcal{D}, \mathcal{R})$ is zero. In this case Theorem 1 matches exactly the main theorem of [3], under the original and weaker definition of hash function compatibility from [3]. This justifies our claim of repairing the Bellare-Cash framework.

Overview of the Proof. The proof of the above theorem is detailed in the full version [1]. Here we provide a brief overview. Since the RKD functions that we consider in our case may have claws, we start by dealing with possible collisions on the related-keys in the RKPRFReal case, using the key-collision notion. Then, we deal with possible collisions on hash values in order to ensure that the hash values h used to compute the output y are pairwise distinct so the attacker is unique-input. Then, using the properties of the key-transformer and the compatibility condition, we show that it is hard to distinguish the output from a uniformly random output based on the standard prf security of M. Finally, we use the statistical-key-collision security notion to deal with possible key collisions in the RKPRFRand case so that the last game matches the description of the RKPRFRand game.

Remark 2. It is worth noting that we deviate from the original proof of [3] in games G_5–G_7, filling the gap in their original proof, but under the same technical conditions on compatibility. Unlike in their proof, we are able to show that the output of F is already indistinguishable from a uniformly random output as soon as one replaces the underlying PRF M with a random function f due to the uniformity condition of the transformer. In order to build a unique-input adversary against the uniformity condition, the main trick is to precompute the values of $f(w)$ for all $w \in \mathsf{Qrs}(\mathsf{T}, M, \Phi, \mathbf{w})$ and use these values to compute $\mathsf{T}^f(\phi, \mathbf{w}[i])$, for $i = 1, \ldots, |\mathbf{w}|$ and $\phi \in \Phi$, whenever needed. This avoids the need to query the oracle in the uniformity game twice on the same input when computing the fingerprint.

4   Related-Key Security for Affine RKD Functions

In this section, we apply the above framework to the variant NR^* of the Naor-Reingold PRF. Recall that $\mathsf{NR}^*: (\mathbb{Z}_p)^n \times \{0,1\}^n \backslash \{0^n\} \to \mathbb{G}$ was defined in [3] by:

$$\mathsf{NR}^*(\mathbf{a}, x) = g^{\prod_{i=1}^n \mathbf{a}[i]^{x[i]}}$$

for all $\mathbf{a} \in \mathbb{Z}_p^n$ and $x \in \{0,1\}^n \backslash \{0^n\}$. We recall the definition of Φ_{aff} $(= \Phi_1)$ from the introduction. Using the above theorem, we prove that NR^* can be used to build a Φ_{aff}-RKA-secure PRF under the DDH assumption, thereby recovering and strengthening the withdrawn result from [3]. We first recall the following lemma from [3].

Lemma 3. *Let* $\mathbb{G} = \langle g \rangle$ *be a group of prime order* p *and let* NR^* *be defined via* $\mathsf{NR}^*(\mathbf{a}, x) = g^{\prod_{i=1}^n \mathbf{a}[i]^{x[i]}}$, *where* $\mathbf{a} \in \mathbb{Z}_p^n$ *and* $x \in \{0,1\}^n \setminus \{0^n\}$. *Let* A *be an adversary against the standard prf security of* NR^* *that makes* Q_A *oracle queries. Then we can construct an adversary* B *against the DDH problem such that*

$$\mathbf{Adv}_{\mathsf{NR}^*}^{\mathsf{prf}}(A) \le n \cdot \mathbf{Adv}_{\mathbb{G}}^{\mathsf{ddh}}(B) .$$

The running time of B *is equal to the running time of* A, *plus the time required to compute* $O(Q_A)$ *exponentiations in* $\mathbb{G}$.

In what follows, we prove the properties needed to apply Theorem 1 to NR^*. The proofs of the above lemmas are detailed in the full version [1].

Strong Key Fingerprint. Let $\mathbf{w}[i] = 0^{i-1}\|1\|0^{n-i}$, for $i = 1, \ldots, n$. Then $\mathbf{w}$ is a strong key fingerprint for NR^*. Indeed, we have $(\mathsf{NR}^*(\mathbf{a}, \mathbf{w}[1]), \ldots, \mathsf{NR}^*(\mathbf{a}, \mathbf{w}[n]))$

$= (g^{\mathbf{a}[1]}, \ldots, g^{\mathbf{a}[n]})$, so if $\mathbf{a} \neq \mathbf{a}'$ are two distinct keys in $\mathcal{K} = \mathbb{Z}_p^n$, then there exists $i \in \{1, \ldots, n\}$ such that $\mathbf{a}[i] \neq \mathbf{a}'[i]$, so $g^{\mathbf{a}[i]} \neq g^{\mathbf{a}'[i]}$.

Compatible Hash Function. We have $\mathsf{Qrs}(\mathsf{T}_{\mathsf{aff}}, \mathsf{NR}^*, \Phi_{\mathsf{aff}}, \mathbf{w}) = \{\mathbf{w}[1], \ldots, \mathbf{w}[n]\}$, so let $\overline{\mathcal{D}} = \{0,1\}^n \times \mathbb{G}^n$ and let $h \colon \overline{\mathcal{D}} \to \{0,1\}^{n-2}$ be a collision resistant hash function. Then the hash function defined by $H(x, \mathbf{z}) = 11 \| h(x, \mathbf{z})$ is a collision resistant hash function that is compatible with $(\mathsf{T}_{\mathsf{aff}}, \mathsf{NR}^*, \Phi_{\mathsf{aff}}, \mathbf{w})$ since every element of $\mathsf{Qrs}(\mathsf{T}_{\mathsf{aff}}, \mathsf{NR}^*, \Phi_{\mathsf{aff}}, \mathbf{w})$ has at most one 1 bit and every output of H has at least two 1 bits. Note that in particular the output of H is never 0^n, so it is always in the domain of NR^*.

Lemma 4. *Let $\mathbb{G} = \langle g \rangle$ be a group of prime order p and let NR^* be defined via $\mathsf{NR}^*(\mathbf{a}, x) = g^{\prod_{i=1}^{n} \mathbf{a}[i]^{x[i]}}$, where $\mathbf{a} \in \mathbb{Z}_p^n$ and $x \in \{0,1\}^n \setminus \{0^n\}$. Let D be an adversary against the Φ_{aff}-key-collision security of NR^* that makes Q_D oracle queries. Then we can construct an adversary C against the DL problem in $\mathbb{G}$ with the same running time as that of D such that*

$$\mathbf{Adv}_{\Phi_{\mathsf{aff}}, \mathsf{NR}^*}^{\mathsf{kc}}(D) \leq n \cdot \mathbf{Adv}_{\mathbb{G}}^{\mathsf{dl}}(C).$$

Since the hardness of DDH implies the hardness of DL, the above lemma does not introduce any additional hardness assumptions beyond DDH.

Lemma 5. *Let $\mathbb{G} = \langle g \rangle$ be a group of prime order p and let NR^* be defined via $\mathsf{NR}^*(\mathbf{a}, x) = g^{\prod_{i=1}^{n} \mathbf{a}[i]^{x[i]}}$, where $\mathbf{a} \in \mathbb{Z}_p^n$ and $x \in \{0,1\}^n \setminus \{0^n\}$. Let $\mathsf{T}_{\mathsf{aff}}$ be defined via*

$$\mathsf{T}_{\mathsf{aff}}^f(\phi, x) = g^{\prod_{i \in S(x)} \mathbf{c}[i]} \cdot \prod_{y \preceq x, y \neq 0^n} f(y)^{\prod_{j \in S(y)} \mathbf{b}[j] \prod_{k \in S(x) \setminus S(y)} \mathbf{c}[k]}$$

where $\phi = (\phi_1, \ldots, \phi_n) \in \Phi_{\mathsf{aff}}$, with $\phi_i \colon a \to \mathbf{b}[i]a + \mathbf{c}[i]$, $\mathbf{b}[i] \neq 0$, for $i = 1, \ldots, n$. Then $\mathsf{T}_{\mathsf{aff}}$ is a key-transformer for $(\mathsf{NR}^*, \Phi_{\mathsf{aff}})$. Moreover, the worst-case running time of this key-transformer is the time required to compute $O(2^n)$ exponentiations in $\mathbb{G}$.

Lemma 6. *Let $\mathbb{G} = \langle g \rangle$ be a group of prime order p. Let A be an adversary against the Φ_{aff}-statistical-key-collision security for $\mathsf{Fun}(\mathbb{Z}_p^n \times \{0,1\}^n, \mathbb{G})$ making Q_A queries. Then we have*

$$\mathbf{Adv}_{\Phi_{\mathsf{aff}}}^{\mathsf{skc}}(A) \leq \frac{Q_A^2}{2p}.$$

We now have everything we need to apply Theorem 1 to NR^*. Combining Theorem 1, Lemmas 3–6 and the above properties, we obtain the following theorem.

Theorem 7. *Let $\mathbb{G} = \langle g \rangle$ be a group of prime order p and let NR^* be defined via $\mathsf{NR}^*(\mathbf{a}, x) = g^{\prod_{i=1}^{n} \mathbf{a}[i]^{x[i]}}$, where $\mathbf{a} \in \mathbb{Z}_p^n$ and $x \in \{0,1\}^n \setminus \{0^n\}$. Let $\overline{\mathcal{D}} = \{0,1\}^n \times \mathbb{G}^n$ and let $h \colon \overline{\mathcal{D}} \to \{0,1\}^{n-2}$ be a hash function. Let $\mathbf{w}[i] = 0^{i-1} \| 1 \| 0^{n-i}$, for $i = 1, \ldots, n$. Define $F \colon \mathbb{Z}_p^n \times \{0,1\}^n \to \mathbb{G}$ by*

$$F(\mathbf{a}, x) = \mathsf{NR}^*(\mathbf{a}, 11 \| h(x, \mathsf{NR}^*(\mathbf{a}, \mathbf{w})))$$

for all $\mathbf{a} \in \mathbb{Z}_p^n$ *and* $x \in \{0,1\}^n$. *Let* A *be a* Φ_{aff}-*restricted adversary against the prf-rka security of* F *that makes* Q_A *oracle queries. Then we can construct an adversary* B *against the DDH problem in* $\mathbb{G}$, *an adversary* C *against the cr security of* h, *and an adversary* D *against the DL problem in* $\mathbb{G}$, *such that*

$$\mathbf{Adv}_{\Phi_{\mathsf{aff}},F}^{\mathsf{prf-rka}}(A) \leq n \cdot \mathbf{Adv}_{\mathbb{G}}^{\mathsf{ddh}}(B) + \mathbf{Adv}_h^{\mathsf{cr}}(C) + n \cdot \mathbf{Adv}_{\mathbb{G}}^{\mathsf{dl}}(D) + \frac{Q_A^2}{2p} .$$

The running time of B *is that of* A *plus the time required to compute* $O(Q_A \cdot (n+1) \cdot 2^n)$ *exponentiations in* $\mathbb{G}$. *The running times of* C *and* D *are the same as that of* A.

5   Further Generalisation of the Bellare-Cash Framework

We introduce a new type of PRF, called an (S, Φ)-Unique-Input-RKA-PRF. We then use this notion as a tool in a further extension of the Bellare-Cash framework that can be applied to *non*-key-malleable PRFs and *non*-claw-free classes of RKD functions. This new framework provides in particular a route to proving that the variant of the Naor-Reingold PRF introduced in Section 3 is actually Φ_d-RKA-secure.

(S, Φ)-**Unique-Input-RKA-PRF.** Let $M: \mathcal{K} \times \mathcal{D} \to \mathcal{R}$ be a family of functions. Let S be a subset of $\mathcal{D}$ and Φ be a class of RKD functions. We consider the class of adversaries A in Fig. 3 such that all queries (ϕ, x) with $x \in S$ made by A to its oracle are for distinct values of x. That is, for any sequence of A's queries $(\phi_1, x_1), \ldots, (\phi_k, x_k)$ with $x_i \in S$ for all $i = 1, \ldots, k$, we require all the x_i to be distinct (no such restriction is made for queries (ϕ_i, x_i) with $x_i \notin S$). We denote the advantage of such an adversary A by $\mathbf{Adv}_{\Phi,S,M}^{\mathsf{ui-prf-rka}}(B)$. We then say that M is an (S, Φ)-unique-input-RKA-secure PRF if the advantage of any such Φ-restricted, efficient adversary A in attacking (S, Φ)-unique-input-prf-rka security is negligible.

proc **Initialize**	proc **RKFn**(ϕ, x)
$K \xleftarrow{\$} \mathcal{K} \,;\, b \xleftarrow{\$} \{0,1\}$	If $x \in S$ then
	$\quad$ If $b = 0$ then $y \leftarrow M(\phi(K), x)$
proc **Finalize**(b')	$\quad$ Else $y \xleftarrow{\$} \mathcal{R}$
Return $b' = b$	Else $y \leftarrow M(\phi(K), x)$
	Return y

Fig. 3. Game defining the (S, Φ)-unique-input-prf-rka security of a PRF M

The following theorem is an analogue of Theorem 1 in which the roles of key malleability and hash function compatibility are replaced by our new notion, (S, Φ)-unique-input-prf-rka security.

Theorem 8. *Let* $M: \mathcal{K} \times \mathcal{D} \to \mathcal{R}$ *be a family of functions and* Φ *be a class of RKD functions. Let* $\mathbf{w} \in \mathcal{D}^m$ *be a strong key fingerprint for* M. *Let* $\overline{\mathcal{D}} = \mathcal{D} \times \mathcal{R}^m$

and let $H\colon \overline{\mathcal{D}} \to S$ be a hash function, where $S \subseteq \mathcal{D}\backslash\{\mathbf{w}[1], \ldots, \mathbf{w}[m]\}$. Define $F\colon \mathcal{K} \times \mathcal{D} \to \mathcal{R}$ by

$$F(K, x) = M(K, H(x, M(K, \mathbf{w})))$$

for all $K \in \mathcal{K}$ and $x \in \mathcal{D}$. Let A be a Φ-restricted adversary against the prf-rka security of F that makes $Q_A \leq |S|$ oracle queries. Then we can construct an adversary B against the (S, Φ)-unique-input-prf-rka security of M, an adversary C against the cr security of H, an adversary D against the Φ-kc security of M and an adversary E against Φ-skc security for $\mathsf{Fun}(\mathcal{K} \times \mathcal{D}, \mathcal{R})$ such that

$$\mathbf{Adv}_{\Phi,F}^{\mathsf{prf\text{-}rka}}(A) \leq \mathbf{Adv}_{\Phi,S,M}^{\mathsf{ui\text{-}prf\text{-}rka}}(B) + \mathbf{Adv}_{H}^{\mathsf{cr}}(C) + \mathbf{Adv}_{\Phi,M}^{\mathsf{kc}}(D) + \mathbf{Adv}_{\Phi}^{\mathsf{skc}}(E) \; . \quad (2)$$

Adversaries C, D and E have the same running time as A. Adversary B makes $(m + 1) \cdot Q_A$ oracle queries and has the same running time as A.

Overview of the Proof. The proof of the above theorem is detailed in the full version [1]. Here we provide a brief overview. Since the RKD functions that we consider in our case may have claws, we start by dealing with possible collisions on the related-keys in the RKPRFReal case, using the key-collision notion. Then, we deal with possible collisions on hash values in order to ensure that the hash values h used to compute the output y are distinct. Then, in contrast to the proof of Theorem 1, we use the new (S, Φ)-Unique-Input-RKA-PRF notion and the compatibility condition to show that it is hard to distinguish the output of F from a uniformly random output. Finally, we use the statistical-key-collision security notion to deal with possible key collisions in the RKPRFRand case so that the last game matches the description of the RKPRFRand Game.

Remark 9. In the full version [1], we explore the relationship between key-malleable PRFs and unique-input-RKA-secure PRFs. Specifically, we show that the (S, Φ)-unique-input-prf-rka security of a Φ-key-malleable PRF M is implied by its regular prf security if the key-transformer T associated with M satisfies a new condition that we call S-uniformity. This condition demands that the usual uniformity condition for T should hold on the subset S of $\mathcal{D}$ rather than on all of $\mathcal{D}$. Whether S-uniformity is implied by (regular) uniformity is an open question.

6   Related-Key Security for Polynomial RKD Functions

We apply Theorem 8 to the variant NR^* of the Naor-Reingold PRF for the class of RKD functions $\Phi_d = \{\phi\colon \mathcal{K} \to \mathcal{K}|\phi = (\phi_1, \ldots, \phi_n); \phi_i \colon A \mapsto \sum_{j=0}^{d} \alpha_{i,j} \cdot A^j, (\alpha_{i,1}, \ldots, \alpha_{i,d}) \neq 0^d; \forall i = 1, \ldots, n\}$. Specifically, we prove that NR^* can be used to build a Φ_d-RKA-secure PRF, under the d-DDHI assumption. Remarkably, our proof provides an efficient reduction, avoiding an exponential running time like that seen in Theorem 7. The key step in establishing our result is Lemma 12. Its proof involves at its core the construction of a bespoke key-transformer to handle Φ_d and a delicate analysis of it using sequences of hybrid games.

In what follows, we prove the various properties needed to apply Theorem 8 to NR*. The proofs of Lemmas 10–12 can be found in the full version [1].

Strong Key Fingerprint. Let $\mathbf{w}[i] = 0^{i-1}\|1\|0^{n-i}$, for $i = 1, \ldots, n$. Then, as before, $\mathbf{w}$ is a strong key fingerprint for NR*.

Hash Function. Let $\overline{\mathcal{D}} = \{0,1\}^n \times \mathbb{G}^n$ and let $h\colon \overline{\mathcal{D}} \to \{0,1\}^{n-2}$ be a collision resistant hash function. Then, as previously, the hash function defined by $H(x, \mathbf{z}) = 11\|h(x, \mathbf{z})$ is a collision resistant hash function with range S satisfying the relation $S \subseteq \{0,1\}^n\backslash(\{\mathbf{w}[1], \ldots, \mathbf{w}[n]\} \cup \{0^n\})$.

Lemma 10. *Let* $\mathbb{G} = \langle g \rangle$ *be a group of prime order* p *and let* NR* *be defined via* $\mathsf{NR}^*(\mathbf{a}, x) = g^{\prod_{i=1}^n \mathbf{a}[i]^{x[i]}}$, *where* $\mathbf{a} \in \mathbb{Z}_p^n$ *and* $x \in \{0,1\}^n \setminus \{0^n\}$. *Let* D *be an adversary against the* Φ_d-*key-collision security of* NR* *that makes* Q_D *oracle queries. Then we can construct an adversary* C *against the* d-*SDL problem in* $\mathbb{G}$ *such that*

$$\mathbf{Adv}_{\Phi_d, \mathsf{NR}^*}^{\mathsf{kc}}(D) \leq n \cdot \mathbf{Adv}_{\mathbb{G}}^{d\text{-sdl}}(C) \ .$$

The running time of C *is that of* D *plus the time required to factorize a polynomial of degree at most* d *in* $\mathbb{F}_p$ *(sub-quadratic in* d *and logarithmic in* p*) plus* $O(Q_D \cdot d)$ *exponentiations in* $\mathbb{G}$.

Lemma 11. *Let* $\mathbb{G}$ *be a group of prime order* p. *Let* $\mathsf{Fun}(\mathbb{Z}_p^n \times \{0,1\}^n, \mathbb{G})$ *be the set of functions from which the random function in the* Φ_d-*statistical-key-collision security game is taken. Let* A *be an adversary against the* Φ_d-*statistical-key-collision security that makes* Q_A *queries. Then we have*

$$\mathbf{Adv}_{\Phi_d}^{\mathsf{skc}}(A) \leq \frac{d \cdot Q_A^2}{2p} \ .$$

Lemma 12. *Let* $\mathbb{G} = \langle g \rangle$ *be a group of prime order* p *and let* NR* *be defined via* $\mathsf{NR}^*(\mathbf{a}, x) = g^{\prod_{i=1}^n \mathbf{a}[i]^{x[i]}}$, *where* $\mathbf{a} \in \mathbb{Z}_p^n$ *and* $x \in \{0,1\}^n \setminus \{0^n\}$. *Let* S *denote the set* $\{0,1\}^n\backslash(\{0^n\} \cup \{\mathbf{w}[1], \ldots, \mathbf{w}[n]\})$. *Let* A *be an adversary against the* (S, Φ_d)-*unique-input-prf-rka security of* NR* *that makes* Q_A *oracle queries. Then, assuming* $nd \leq \sqrt{p}$, *we can design an adversary* B *against the* d-*DDHI problem in* $\mathbb{G}$ *such that*

$$\mathbf{Adv}_{\Phi_d, S, \mathsf{NR}^*}^{\mathsf{ui\text{-}prf\text{-}rka}}(B) \leq \left(n \cdot d \cdot \left(\frac{p}{p-1} \right)^2 + n \cdot (d-1) \right) \cdot \mathbf{Adv}_{\mathbb{G}}^{d\text{-ddhi}}(A) + \frac{2n \cdot Q_A}{p} \ .$$

The running time of B *is that of* A *plus the time required to compute* $O(d \cdot (n + Q_A))$ *exponentiations in* $\mathbb{G}$ *and* $O(Q_A^3 \cdot (nd + Q_A))$ *operations in* $\mathbb{Z}_p$.

Finally, by combining the results in Lemmas 10–12 with Theorem 8, we can prove the following theorem.

Theorem 13. *Let* $\mathbb{G} = \langle g \rangle$ *be a group of prime order* p *and let* NR* *be defined via* $\mathsf{NR}^*(\mathbf{a}, x) = g^{\prod_{i=1}^n \mathbf{a}[i]^{x[i]}}$, *where* $\mathbf{a} \in \mathbb{Z}_p^n$ *and* $x \in \{0,1\}^n \setminus \{0^n\}$. *Let* $\mathcal{D} = \{0,1\}^n \times$

$\mathbb{G}^n$ and let $h \colon \mathcal{D} \to \{0,1\}^{n-2}$ be a hash function. Let $\mathbf{w}[i] = 0^{i-1} \| 1 \| 0^{n-i}$, for $i = 1, \ldots, n$. Define $F \colon \mathbb{Z}_p^n \times \{0,1\}^n \to \mathbb{G}$ by

$$F(\mathbf{a}, x) = \mathsf{NR}^*(\mathbf{a}, 11 \| h(x, \mathsf{NR}^*(\mathbf{a}, \mathbf{w})))$$

for all $\mathbf{a} \in \mathbb{Z}_p^n$ and $x \in \{0,1\}^n$. Let A be a Φ_d-restricted adversary against the prf-rka security of F that makes $Q_A \leq |\{0,1\}^{n-2}|$ oracle queries. Then, assuming $nd \leq \sqrt{p}$, we can construct an adversary B against the d-DDHI problem in $\mathbb{G}$, an adversary C against the cr security of h, and an adversary D against the d-SDL problem in $\mathbb{G}$ such that

$$\mathbf{Adv}_{\Phi_d, F}^{\mathsf{prf\text{-}rka}}(A) \leq \left(n \cdot d \cdot (1 - 1/p)^2 + n \cdot (d-1)\right) \cdot \mathbf{Adv}_{\mathbb{G}}^{d\text{-}\mathsf{ddhi}}(B)$$

$$+ \mathbf{Adv}_h^{\mathsf{cr}}(C) + n \cdot \mathbf{Adv}_{\mathbb{G}}^{d\text{-}\mathsf{sdl}}(D) + \left(d \cdot Q_A^2 + 4n \cdot Q_A\right) / (2p) . \quad (3)$$

The running time of B is that of A plus $O(d \cdot (n + Q_A))$ exponentiations in $\mathbb{G}$ and $O(Q_A^3 \cdot (nd + Q_A))$ operations in $\mathbb{Z}_p$. C has the same running time as A. The running time of D is that of A plus the time required to factorize a polynomial of degree at most d in $\mathbb{F}_p$, which is sub-quadratic in d, logarithmic in p.

Acknowledgements. We thank Susan Thomson for bringing the issues in the original Bellare-Cash framework to our attention, and for useful comments on the paper. Michel Abdalla, Fabrice Benhamouda, and Alain Passelègue were supported by the French ANR-10-SEGI-015 PRINCE Project, the *Direction Générale de l'Armement* (DGA), the CFM Foundation, the European Commission through the FP7-ICT-2011-EU-Brazil Program under Contract 288349 SecFuNet, and the European Research Council under the European Union's Seventh Framework Programme (FP7/2007-2013 Grant Agreement 339563 – CryptoCloud). Kenneth G. Paterson was supported by an EPSRC Leadership Fellowship, EP/H005455/1.

References

1. Abdalla, M., Benhamouda, F., Passelègue, A., Paterson, K.G.: Related-key security for pseudorandom functions beyond the linear barrier, full version of this paper available at Cryptology ePrint Archive, http://eprint.iacr.org/
2. Banerjee, A., Peikert, C.: New and improved key-homomorphic pseudorandom functions. In: Garay, J.A., Gennaro, R. (eds.) CRYPTO 2014, Part I. LNCS, vol. 8616, pp. 353–370. Springer, Heidelberg (2014)
3. Bellare, M., Cash, D.: Pseudorandom functions and permutations provably secure against related-key attacks. In: Rabin, T. (ed.) CRYPTO 2010. LNCS, vol. 6223, pp. 666–684. Springer, Heidelberg (2010)
4. Bellare, M., Cash, D.: Pseudorandom functions and permutations provably secure against related-key attacks. Cryptology ePrint Archive, Report 2010/397 (2010), http://eprint.iacr.org/ (last updated October 27, 2013)
5. Bellare, M., Cash, D., Miller, R.: Cryptography secure against related-key attacks and tampering. In: Lee, D.H., Wang, X. (eds.) ASIACRYPT 2011. LNCS, vol. 7073, pp. 486–503. Springer, Heidelberg (2011)

6. Bellare, M., Kohno, T.: A theoretical treatment of related-key attacks: RKA-PRPs, RKA-PRFs, and applications. In: Biham, E. (ed.) EUROCRYPT 2003. LNCS, vol. 2656, pp. 491–506. Springer, Heidelberg (2003)

7. Bellare, M., Paterson, K.G., Thomson, S.: RKA security beyond the linear barrier: IBE, encryption and signatures. In: Wang, X., Sako, K. (eds.) ASIACRYPT 2012. LNCS, vol. 7658, pp. 331–348. Springer, Heidelberg (2012)

8. Bellare, M., Rogaway, P.: The security of triple encryption and a framework for code-based game-playing proofs. In: Vaudenay, S. (ed.) EUROCRYPT 2006. LNCS, vol. 4004, pp. 409–426. Springer, Heidelberg (2006)

9. Biham, E.: New types of cryptanalytic attacks using related keys. In: Helleseth, T. (ed.) EUROCRYPT 1993. LNCS, vol. 765, pp. 398–409. Springer, Heidelberg (1994)

10. Biham, E., Dunkelman, O., Keller, N.: Related-key boomerang and rectangle attacks. In: Cramer, R. (ed.) EUROCRYPT 2005. LNCS, vol. 3494, pp. 507–525. Springer, Heidelberg (2005)

11. Biham, E., Dunkelman, O., Keller, N.: A unified approach to related-key attacks. In: Nyberg, K. (ed.) FSE 2008. LNCS, vol. 5086, pp. 73–96. Springer, Heidelberg (2008)

12. Biryukov, A., Dunkelman, O., Keller, N., Khovratovich, D., Shamir, A.: Key recovery attacks of practical complexity on AES-256 variants with up to 10 rounds. In: Gilbert, H. (ed.) EUROCRYPT 2010. LNCS, vol. 6110, pp. 299–319. Springer, Heidelberg (2010)

13. Biryukov, A., Khovratovich, D.: Related-key cryptanalysis of the full AES-192 and AES-256. In: Matsui, M. (ed.) ASIACRYPT 2009. LNCS, vol. 5912, pp. 1–18. Springer, Heidelberg (2009)

14. Biryukov, A., Khovratovich, D., Nikolić, I.: Distinguisher and related-key attack on the full AES-256. In: Halevi, S. (ed.) CRYPTO 2009. LNCS, vol. 5677, pp. 231–249. Springer, Heidelberg (2009)

15. Boneh, D., Lewi, K., Montgomery, H.W., Raghunathan, A.: Key homomorphic PRFs and their applications. In: Canetti, R., Garay, J.A. (eds.) CRYPTO 2013, Part I. LNCS, vol. 8042, pp. 410–428. Springer, Heidelberg (2013)

16. Goldreich, O., Goldwasser, S., Micali, S.: How to construct random functions. In: 25th FOCS, pp. 464–479. IEEE Computer Society (October 1984)

17. Goyal, V., O'Neill, A., Rao, V.: Correlated-input secure hash functions. In: Ishai, Y. (ed.) TCC 2011. LNCS, vol. 6597, pp. 182–200. Springer, Heidelberg (2011)

18. Kim, J., Hong, S., Preneel, B.: Related-key rectangle attacks on reduced AES-192 and AES-256. In: Biryukov, A. (ed.) FSE 2007. LNCS, vol. 4593, pp. 225–241. Springer, Heidelberg (2007)

19. Knudsen, L.R.: Cryptanalysis of LOKI91. In: Zheng, Y., Seberry, J. (eds.) AUSCRYPT 1992. LNCS, vol. 718, pp. 196–208. Springer, Heidelberg (1993)

20. Lewi, K., Montgomery, H., Raghunathan, A.: Improved constructions of PRFs secure against related-key attacks. In: Boureanu, I., Owesarski, P., Vaudenay, S. (eds.) ACNS 2014. LNCS, vol. 8479, pp. 44–61. Springer, Heidelberg (2014)

21. Naor, M., Reingold, O.: Number-theoretic constructions of efficient pseudo-random functions. In: 38th FOCS, pp. 458–467. IEEE Computer Society (October 1997)

22. Wee, H.: Public key encryption against related key attacks. In: Fischlin, M., Buchmann, J., Manulis, M. (eds.) PKC 2012. LNCS, vol. 7293, pp. 262–279. Springer, Heidelberg (2012)

Automated Analysis of Cryptographic Assumptions in Generic Group Models

Gilles Barthe[1], Edvard Fagerholm[1,2], Dario Fiore[1], John Mitchell[3],
Andre Scedrov[2], and Benedikt Schmidt[1]

[1] IMDEA Software Institute, Madrid, Spain
{gilles.barthe,dario.fiore,benedikt.schmidt}@imdea.org
[2] University of Pennsylvania, USA
{edvardf,scedrov}@math.upenn.edu
[3] Stanford University, USA
mitchell@cs.stanford.edu

Abstract. We initiate the study of principled, automated, methods for analyzing hardness assumptions in generic group models, following the approach of symbolic cryptography. We start by defining a broad class of generic and symbolic group models for different settings—symmetric or asymmetric (leveled) k-linear groups—and by proving "computational soundness" theorems for the symbolic models. Based on this result, we formulate a very general master theorem that formally relates the hardness of a (possibly interactive) assumption in these models to solving problems in polynomial algebra. Then, we systematically analyze these problems. We identify different classes of assumptions and obtain decidability and undecidability results. Then, we develop and implement automated procedures for verifying the conditions of master theorems, and thus the validity of hardness assumptions in generic group models. The concrete outcome of this work is an automated tool which takes as input the statement of an assumption, and outputs either a proof of its generic hardness or shows an algebraic attack against the assumption.

1 Introduction

Sophisticated abstractions have often been instrumental in recent breakthroughs in the design of cryptographic schemes. Bilinear maps are perhaps the most striking instance of such an abstraction; over the last fifteen years, they have been used for building advanced and previously unknown cryptographic schemes. Now it is believed that multilinear maps will lead to similar breakthroughs. Compared to the "classical" algebraic settings based on the purported hardness of the Factoring/RSA or Discrete-log/Diffie-Hellman problems, bilinear and multilinear maps indeed provide richer and more versatile algebraic structures that are particularly suitable for new constructions. At the same time, one unsettling consequence of using such sophisticated abstractions is a significant growth in the number of hardness assumptions used in security proofs. Moreover, these assumptions are not as well studied as their classical and standard counterparts.

J.A. Garay and R. Gennaro (Eds.): CRYPTO 2014, Part I, LNCS 8616, pp. 95–112, 2014.

While it is widely acknowledged that this situation is far from ideal, relying on non-standard assumptions is sometimes the *only* known way to construct some new (or some efficient) cryptographic scheme, and hence it cannot be completely disregarded. A common view to resolving this dilemma is to develop principled, rigorous approaches for analyzing and comparing non-standard hardness assumptions.

This question has been previously considered in the literature, in which we identify at least two approaches. One approach is to devise assumptions that are general enough to be reused and allow for simple security proofs, and at the same time are shown to hold under more classical assumptions (e.g., [15,32]). A second approach is to develop idealized models, such as the Generic Group [31,33,28] and the Generic Bilinear Group [10] models, and to provide (in the form of so-called master theorems) necessary and sufficient conditions for the security of an assumption in these models. Proving the hardness of an assumption in these models is essentially a way to rule out the possibility of algebraic attacks against the underlying algorithmic problem, and it can be considered the minimal level of guarantee we need to gain confidence in an assumption. Two prominent examples along this direction are the "Uber assumption" (aka "Master theorem") of Boneh, Boyen and Goh [10,14] and the Matrix Decisional Diffie-Hellman assumption family recently proposed by Escala et al. [17].

However, although these results are quite general, they can be quite difficult to apply. Indeed, in order to argue the hardness of an assumption using the Uber assumption in [10,14] (resp. the Matrix-DDH assumption in [17]) one has to show the independence (resp. irreducibility) of certain polynomials contained in the statement of the assumption. A similar problem arises in the context of interactive assumptions such as [27,2], in which the hardness crucially relies on the restrictions posed on the queries performed by the adversary. In summary, applying these general results to verify the validity of a given assumption is far from being a trivial task, and may be error-prone, as witnessed by unfortunate failures [35,23].

In this paper, we initiate the study of principled, automated methods for analyzing hardness assumptions in generic group models. Our main contribution is essentially threefold. First, we reformulate master theorems in the style of the celebrated "computational soundness" theorem of Abadi and Rogaway [1], and formally show that the problem of analyzing assumptions in the generic group reduces to solving problems in polynomial algebra. Second, we systematically analyze these problems: while we show that the most general problem is undecidable, we distill a set of properties (capturing most interesting cases) for which the problem is decidable. Finally, by applying tools from linear algebra, we develop and implement automated procedures for verifying the conditions of master theorems, and thus the validity of hardness assumptions in generic group models. The concrete outcome of this work is an automated tool[1] which takes as input an assumption and outputs either a proof of its generic hardness (along with concrete bounds) or shows an algebraic attack against the assumption.

---

[1] The tool is available at http://www.easycrypt.info/GGA

1.1   An Overview of Our Contribution

The key contribution of our work is the development of automated decision procedures for testing the validity of hardness assumptions in generic group models. Towards this goal, we first settle a rigorous framework for carrying out this analysis. Basically, this framework consists of formalizing a class of generic group models and then stating a general master theorem. Finally, our decision procedures will be aimed at verifying the side conditions of our master theorem.

GENERIC GROUP MODELS. We formalize a broad class of generic group models capturing many interesting cases used in cryptography: symmetric and asymmetric k-linear groups, with both leveled and non-leveled maps, and with the possibility of modeling efficiently computable isomorphisms between the groups. For any experiment stated in these generic models, we generalize the commonly-used step of applying the Schwartz-Zippel Lemma, and obtain a generic transformation (cf. Theorem 1) for switching from the generic group model experiment, in which variables are uniformly sampled in the underlying field, to a completely deterministic experiment that works in a corresponding symbolic group model.

A GENERAL MASTER THEOREM. We give a general version of the Master theorem in [10] which can be stated in any of the generic group models mentioned above. As in [10], we formulate an assumption as a list L of polynomials in $\mathbb{F}_p[X_1, \ldots, X_n]$ where $X_1, \ldots, X_n$ is a set of random variables. In particular, a decisional (aka left-or-right) assumption is defined by two lists of polynomials L and L' (one for the "left" and one for the "right" distribution), and the assumption is said to hold if the adversary cannot distinguish whether it receives polynomials from L or L'. Very informally, our Master theorem states that viewing L and L' as the generating sets of two vector spaces[2], then the linear dependencies within L and within L' are the same. Previous master theorems [10,17] considered only decisional assumptions with the real-or-random formulation in which the adversary is given a list of polynomials L and either a "challenge" polynomial f or a fresh random variable Z.

Beyond obtaining a theorem that works in (leveled) k-linear groups, our general formulation allows us to capture virtually all decisional assumptions, based on k-linear groups (for any $k \geq 1$), that are used in cryptography. To mention some examples, assumptions captured by our theorem include the Matrix-DDH assumption [17], the k-BDH assumption [5], and recently proposed assumptions such as (n, k)-MMDHE [22].

AUTOMATED METHODS. Once we have settled the above framework, our goal is to develop a collection of automated methods to verify the side condition of the Master theorem for any given assumption stated in the framework. While the statement of the above side condition already suggests how to use linear algebra to make these checks, a crucial challenge is that in many important cases (e.g., ℓ-BDHI, k-Lin, etc.) the size of the lists L and L' is a variable parameter. That

---

[2] We are oversimplifying. More precisely, one has to consider lists C and C' containing all polynomials computable by doing multiplications over L and L' respectively, and then look at linear dependencies in C and C'.

Assumption Type	Algorithm	Examples
Non-parametric	D, C	DBDH [12], 2-lin, 3-lin, Freeman assm. 3&4 [18]
Parametric (real-or-random, monomials inputs) Fixed #vars, Par. linear degree and Par. arity Fixed #vars, Par. linear degree, Fixed arity Parametric #vars, Par. arity, Fixed degree	U, I D, C I	(ℓ, k)-MMDHE [22] ℓ-DHI [9], ℓ-DHE [13] (k)-BDH [5], k-Lin in k-linear groups
Interactive bounded	I,C	LRSW [27], CDDH 1&2 [2], M-LRSW [7], IBSAS-CDH [8]
Interactive unbounded	I	LRSW [27], Strong-LRSW [3], s-LRSW [20]

Fig. 1. Summary of our automated analysis methods. U=undecidable problem, D=decision procedure, I = incomplete procedure, C=find counterexample for invalid assumptions.

is, to check that the side condition holds, one would have to do computations on a vector space of variable dimension: a challenging problem for automation.

We study this problem for three main categories of hardness assumptions: (1) non-parametric, (2) parametric, and (3) interactive. *Non-parametric* assumptions are non-interactive assumptions in which the number of inputs is fixed, no input is quantified over a variable and the number of levels is fixed (examples include DDH, DBDH [12], as well as assumptions in k-linear groups for fixed k, e.g., 3-Lin in 3-linear groups). Conversely, an assumption is *parametric* if one or more of the above restrictions do not hold. Finally, *interactive* assumptions are those ones where the adversary is granted access to additional oracles (in addition to the oracles for the algebraic operations). By carefully analyzing each of these categories, we obtain the following results summarized in Fig. 1.

For non-parametric assumptions, we show how to reduce the check on the side condition to computing the kernels of certain matrices (of fixed dimension) that are derived from the lists of polynomials in the assumption's definition. Using computer algebra tools (SAGE [34]), we implement a decision procedure that shows a concrete hardness bound in the corresponding generic group model in the positive case, and an algebraic attack if the assumption does not hold.

Our methods for non-parametric assumptions offer a complete decision procedure to verify arbitrary instances of parametric assumptions where all the parameters have been fixed. This might be sufficient to test quickly a new assumption (and find attacks if any), but it is often desirable to obtain stronger guarantees that hold for *all* parameters. We show that, contrary to the non-parametric case, the side condition becomes undecidable in general. However, we identify classes of assumptions for which we develop automated methods. Interestingly, these classes still contain most cryptographic assumptions. Considering the class of real-or-random assumptions, we develop two different methods. The first method focuses on the case in which the number of random variables is fixed, and the input elements are monomials. Our method shows how to reduce the check of the side condition to an integer programming problem. Interestingly, we can show the following: if the degree of the monomials is *not* a linear polynomial, or the arity of the map is variable, then the problem is *undecidable*; otherwise (if the monomials have linear degree and the arity of the map is fixed) the problem is decidable. We implemented the translation procedure to integer programming problems and use SMT solvers to check satisfiability. For the decidable fragment of assumptions mentioned above, we obtain a complete decision procedure that also shows an attack if the assumption is invalid. For the undecidable fragment, our procedure successfully analyzes all significant examples from the literature.

Our second method focuses on the case where the number of random variables is parametric. As in the previous case, our method provides a way to reduce the side condition to a system of equations. However, the same idea as before does not work since a parametric number of variables would lead to an infinite number of equations. Therefore, we focus on a restricted, but significant, class of assumptions (one restriction is that inputs are expressed as monomials). Our method is incomplete but successfully analyzes all relevant examples in this class.

Finally, we study interactive assumptions such as LRSW [27]. To analyze interactive assumptions, we first formulate an interactive version of our master theorem. Interestingly, once applying our general "computational soundness" theorem and switching to the symbolic model, our interactive master theorem essentially becomes a variant of the non-interactive master theorem for parametric computational assumptions. This allow us to apply similar techniques as for parametric assumptions. More specifically, we use SMT solvers and Gröbner bases computations as an incomplete method to show the validity of such assumptions and find attacks. For instance, our tool automatically proves the validity of LRSW [27] and exhibits attacks for m-LRSW [7] and CDDH [2].

EXTENSIONS AND ADDITIONAL MATERIAL. We extend our results to composite-order groups. Precisely, we formulate the generic group model and our master theorem in a general way that captures also composite-order groups, and we show how to extend our decision procedures for non-parametric assumptions to this setting. Another extension of our results is handling assumptions in which the adversary receives rational values in the exponent. These extensions, full detailed proofs and some running examples appear only in the full version [4].

LIMITATIONS. While our master theorem is very general, our automated methods require to specify the assumptions in a concrete language, essentially to describe the distribution of the polynomials defining the assumption. Such language cannot support the expression of very abstract properties, and thus rules out a few examples. For instance, the definition of the Decision Multilinear No-Exact-Cover Assumption [19] is parametrized by an instance (with no solution) of the Exact-Cover NP-complete problem. Although fixing a specific Exact-Cover instance yields lists of polynomials which can be analyzed using our methods, a definition *for any instance* is too general. For a similar reason, our tool cannot handle the Matrix-DDH assumption in its full generality, unless one fixes a specific distribution for the matrix (e.g., k-Lin).

Discussion. Although well-studied standard assumptions should always be preferred when designing cryptographic schemes, the use of non-standard ones is not likely to stop. In this sense, we believe the study and development of rigorous methods for analyzing cryptographic assumptions is relevant, and that automated analysis tools can support cryptographers in multiple directions. Mainly, they provide a rigorous, fast way to test the validity of candidate assumptions in generic models by delegating this task to a machine. This is especially relevant in the recent setting of leveled multilinear maps, that have a rich algebraic structure and for which even simple assumptions may become difficult to analyze. We believe that the importance of such tools is motivated by the fact that proofs validating the hardness of an assumption in the generic group model fall exactly in the so-called "mundane part"[3] of cryptographic proofs mentioned by Halevi [21], and constitute a perfect candidate of a proof to be delegated to a machine.

Our work shows the feasibility and relevance of developing automated methods to analyze assumptions in generic group models. It can also be seen as the first step towards analyzing cryptographic protocols directly in the generic model; we expect that such analyses would allow to discover subtle flaws in protocols and supplant existing methods based on symbolic cryptography.

1.2   Related Work

The problem of analyzing and comparing hardness assumptions has been earlier considered in the literature, e.g., [30]. In particular, we identify two main approaches in previous work. The first approach aims to define generalized assumptions that reduce to standard ones. Examples of works in this direction include: the Square Diffie-Hellman assumption, shown to be equivalent to CDH by Maurer and Wolf [29]; the (P, Q)-Decisional Diffie-Hellman assumption of Bresson et al. [15] which is shown to reduce to DDH; and the decisional subspace problems of Okamoto-Takashima [32] that are reduced to DLin.

The other approach aims at directly analyzing assumptions by means of idealized models, such as the generic group model. This model was introduced by Nechaev [31] and further refined and generalized by Shoup [33], and Maurer [28]. Our work follows closely Maurer's model, in which the main difference compared to previous proposals is to model the adversary's access to group elements via handles instead of random bitstrings as in [31,33]. These two models have been proven equivalent in [25]. Worth mentioning in this context is the semi-generic group model of Jager and Rupp [24]. This is a weaker version of the bilinear generic group model, and its basic idea is to model the base groups of pairings as generic groups, whereas the target group is given in the standard model.

Two works that address the problem of devising general assumptions in the generic group are the Master theorem of Boneh, Boyen and Goh [10] (generalized by Boyen [14]), and the Matrix DDH assumption of Escala et al. [17]. Roughly speaking, the former provides a framework for arguing about the validity of several pairing-based assumptions in the generic group model, and it captures a significant fraction of assumptions in the literature. The latter is an assumption that subsumes classical problems like DDH or DLin and also introduces

---

[3] In [21], Halevi informally divides proofs in two categories (quoting): "*Most (or all) cryptographic proofs have a creative part (e.g., describing the simulator or the reduction) and a mundane part (e.g., checking that the reduction actually goes through). It often happens that the mundane parts are much harder to write and verify, and it is with these parts that we can hope to have automated help.*"

assumptions, such as k-Casc, that are proven hard in the generic k-linear group model. Also worth mentioning is the work of Freeman [18] which extends the BBG Master theorem to challenges in the source group and uses the computer algebra system Magma to verify the side conditions required to prove two of the assumptions. Our work is also close to the line of work on automation of cryptographic proofs in both the computational and symbolic models, see [6] for an overview.

1.3 Preliminaries

In our work, we denote by λ the security parameter. We use $\mathbb{G}_i$ to denote additive cyclic groups of prime order and P_i to denote a generator of $\mathbb{G}_i$. For any element $Q = xP_i$, we denote with $x = dlog(Q)$ its discrete logarithm. We use $\boldsymbol{a}$ or $\boldsymbol{v}$ to denote vectors, $\boldsymbol{a}\|\boldsymbol{b}$ for the concatenation of two vectors, and $\boldsymbol{a} \cdot \boldsymbol{b}$ to denote their inner product. We denote the power set of S with $\mathcal{P}(S)$, the i-th element of a list with $L[i]$, the range $\{n, \ldots, n+l\}$ with $[n, n+l]$, and $[1, n]$ with $[n]$.

A *symmetric k-linear group* is a pair of groups $\mathbb{G}_1$ and $\mathbb{G}_2$ together with an admissible k-linear map $e : \mathbb{G}_1^k \to \mathbb{G}_2$. An *asymmetric k-linear group* is a sequence of groups $\mathbb{G}_1, \ldots, \mathbb{G}_k, \mathbb{G}_{k+1}$ together with an admissible k-linear map $e : \mathbb{G}_1 \times \cdots \times \mathbb{G}_k \to \mathbb{G}_{k+1}$. For a k-linear map $e : \mathbb{G}_1 \times \cdots \times \mathbb{G}_k \to \mathbb{G}_{k+1}$, we call $\mathbb{G}_{k+1}$ the *target group* and other groups $\mathbb{G}_i$ *source groups*. We can further assume existence of isomorphisms $\mathbb{G}_i \to \mathbb{G}_j$ between source groups.

A *symmetric leveled k-linear group* is a sequence of groups $\mathbb{G}_1, \ldots, \mathbb{G}_k$ together with bilinear maps $e : \mathbb{G}_i \times \mathbb{G}_j \to \mathbb{G}_{i+j}$ for $i, j \in [1, k]$ and $i + j \leq k$. We say that $\mathbb{G}_n$ is the group at level n and call $\mathbb{G}_k$ the target group. An *asymmetric leveled k-linear group* is a collection of groups $\{\mathbb{G}_S\}$ for $S \in \mathcal{P}([k])$ together with bilinear maps $e_{S,T} : \mathbb{G}_S \times \mathbb{G}_T \to \mathbb{G}_{S \cup T}$ for all $S \cap T = \emptyset$.

2 Generic Group Models and Symbolic Group Models

In this section, we define a class of generic group models that captures the previously described group settings. Afterwards, we define a symbolic group model where instead of computing with (randomly sampled) group elements, the challenger computes with (fixed) polynomials. We prove that this model is equivalent to the generic group model up to some usually small error.

Generic Group Models. A generic group model for a concrete group setting captures all operations that an adversary with black-box access can perform.

Definition 1. *A group setting is a tuple $\mathcal{GS} = (p, \mathcal{G}, \Phi, \mathcal{E})$ where $\mathcal{G} = \{\mathbb{G}_i\}_{i \in \mathcal{I}}$ is a set of cyclic groups of prime order p indexed by a totally ordered set $\mathcal{I}$, Φ is a set of isomorphisms $\phi : \mathbb{G}_i \to \mathbb{G}_j$, and $\mathcal{E}$ is a set of maps, where for each $e \in \mathcal{E}$, there is a k s.t. $e : \mathbb{G}_{i_1} \times \ldots \times \mathbb{G}_{i_k} \to \mathbb{G}_{i_{k+1}}$ is an admissible k-linear map.*

The generic model for a group setting $(p, \mathcal{G}, \Phi, \mathcal{E})$ and a distribution $\mathcal{D}$ on indexed sets $\{L_i\}_{i \in \mathcal{I}}$ of lists of elements of $\mathbb{G}_i$ is defined as follows. The challenger maintains lists $\boldsymbol{L} = \{L_i\}_{i \in \mathcal{I}}$ where each list L_i contains elements from $\mathbb{G}_i$. The *lists are initialized by sampling from $\mathcal{D}$ and the adversary can apply the group operations, isomorphisms, and k-linear maps to list elements by providing the indices of elements as handles. For an operation $o : \mathbb{G}_{i_1} \times \ldots \times \mathbb{G}_{i_k} \to \mathbb{G}_{i_{k+1}}$, the corresponding oracle takes handles $h_1, \ldots, h_k$, computes $a = o(a_1, \ldots, a_k)$ for $a_j = L_{i_j}[h_j]$, appends a to $L_{i_{k+1}}$ and returns a's handle $h = |L_{i_{k+1}}|$. Note that handles are not unique, but the challenger provides an equality oracle to check if two handles refer to the same group element. A formal definition of the game appears in the full version.*

Remark 1. As mentioned in Section 1.2, our generic group model closely follows Maurer's model [28]. We provide the adversary with access to the internal state variables of the challenger via handles, and we assume that the equality queries are "free", in the sense that they do not count when measuring the computational complexity of the adversary.

Example 1. To model a asymmetric leveled k-linear map, we use the index set $\mathcal{I} = \mathcal{P}([k])$, $\Phi = \emptyset$, and $\mathcal{E} = \{e_{T,R} : \mathbb{G}_T \times \mathbb{G}_R \to \mathbb{G}_{T \cup R} \mid T, R \in \mathcal{I} \wedge T \cap R = \emptyset\}$.

Definition 2. *For a list of lists $\boldsymbol{L} = L_1, \ldots, L_k$ of polynomials over $\mathbb{F}_p[X_1, .., X_n]$, we define the distribution $\mathcal{D}_{\boldsymbol{L}}$ by the following procedure. Uniformly sample a point $\boldsymbol{x} \in \mathbb{F}_p^n$ and return the list of lists $\boldsymbol{L}' = L_1', \ldots, L_k'$ where $L_i' = [f_1(\boldsymbol{x})P_i, \ldots, f_{|L_i|}(\boldsymbol{x})P_i]$ for $f_j = L_i[j]$. A distribution $\mathcal{D}$ is* polynomially induced *if $\mathcal{D} = \mathcal{D}_{\boldsymbol{L}}$ for some $\boldsymbol{L}$.*

Most hardness assumptions in generic group models belong to the following classes of decisional, computational, or generalized extraction problems stated with respect to a group setting $\mathcal{GS}$:

- Decisional problem for $\mathcal{D}_{\boldsymbol{L}}$ and $\mathcal{D}_{\boldsymbol{L}'}$:
  Return $b \in \{0, 1\}$ to distinguish the corresponding generic group models.
- Computational problem for $\mathcal{D}_{\boldsymbol{L}}$, polynomial f, and group index i:
  Return handle to $f(\boldsymbol{x})P_i$, where $\boldsymbol{x}$ is the random point sampled by $\mathcal{D}_{\boldsymbol{L}}$.
- Generalized extraction problem for $\mathcal{D}_{\boldsymbol{L}}, n, m, i_1, \ldots, i_m, H$:
  Return $\boldsymbol{a} \in \mathbb{F}_p^n$ and handles $h_1, \ldots, h_m$ such that the random point $\boldsymbol{x}$ sampled by $\mathcal{D}_{\boldsymbol{L}}$ satisfies $H(\boldsymbol{x}, \boldsymbol{a}, dlog(L_{i_1}[h_1]), \ldots, dlog(L_{i_m}[h_m])) = 0$.

The above classification generalizes the one proposed by Maurer [28]. Precisely, in addition to decisional and computational assumptions, Maurer considered "straight" extraction problems (such as discrete logarithm) in which the adversary has to extract the random value x of a handle. Our class of *generalized extraction problems* captures extraction problems like discrete logarithm, but also captures problems like the Strong Diffie-Hellman Problem [9].[4] Moreover, note that our class of generalized extraction problems contains the class of computational problems.

From Generic to Symbolic Group Models. The *symbolic group model* for a group setting $(p, \mathcal{G}, \Phi, \mathcal{E})$ and a distribution $\mathcal{D}_{\boldsymbol{L}}$ provides the same adversary

---

[4] Set $n = 1$, $m = 0$, $H(X, a_1) = X - a_1$ for DLOG and $n = m = 1, H(X, a_1, Y) = (X - a_1)Y - 1$ for SDH.

interface as the corresponding generic group model. The difference is that, internally, the challenger now stores lists of polynomials in $\mathbb{F}_p[X_1, \ldots, X_n]$ where $X_1, \ldots, X_n$ are the variables occurring in L. The oracles perform addition, negation, and equality checks in the polynomial ring. To define the polynomial operations corresponding to applications of isomorphisms and n-linear maps, observe that for all isomorphisms ϕ there is an $a \in \mathbb{F}_p^{\times}$ such that $\phi(g_i) = g_j^a$. We therefore define the oracle $\mathsf{isom}_\phi(h)$ such that it computes $a \cdot L_i[h]$. Similarly, we define the oracle $\mathsf{map}_e(h_1, \ldots, h_k)$ such that it computes $a \cdot (L_{i_1}[h_1] \cdots L_{i_k}[h_k])$. We also define a symbolic version $S(E)$ of a generic winning condition E. For decisional problems and computational problems, the symbolic event is equal to the generic event, i.e., $S(E) = E$. For generalized extraction problems, the event E is translated to checking whether $H(X_1, \ldots, X_n, a, L_{i_1}[h_1], \ldots, L_{i_m}[h_m]) = 0$ holds in the polynomial ring. We denote the symbolic group model for a group setting $\mathcal{GS}$ and a distribution $\mathcal{D}_L$ with $Sym_{\mathcal{GS}}^{\mathcal{D}_L}$ and the corresponding generic group model with $Gen_{\mathcal{GS}}^{\mathcal{D}_L}$.

Theorem 1. *Let $(p, \mathcal{G}, \Phi, \mathcal{E})$ denote a group setting, $\mathcal{D}_L$ a distribution, $\mathcal{A}$ an adversary performing at most q queries, and E the winning event of a decisional, computational, or generalized extraction assumption. If d is an upper bound on the degrees of the polynomials occurring in the internal state of $Sym_{\mathcal{GS}}^{\mathcal{D}_L}(\mathcal{A})$ and $S(E)$, s is the sum of the sizes of the lists in L, and the event $S(E)$ contains at most e equality tests, then*

$$|Pr[\, Gen_{\mathcal{GS}}^{\mathcal{D}_L}(\mathcal{A}) : E \,] - Pr[\, Sym_{\mathcal{GS}}^{\mathcal{D}_L}(\mathcal{A}) : S(E) \,]| \leq (s+q)^2 * d/2p + ed/p$$

where the probability is taken over the coins of $Gen_{\mathcal{GS}}^{\mathcal{D}_L}$ and $\mathcal{A}$.

By applying this theorem, we can therefore analyze the hardness of assumptions in the simpler symbolic model. We note that existing master theorems usually include a similar step in their proofs. Here we explicitly prove the equivalence of the *Gen* and *Sym* experiments. This stronger result is required for our decidability results.

3   Master Theorem for Non-interactive Assumptions

In this section we state our master theorem for decisional, non-interactive problems. In Section 5, we give a master theorem for interactive assumptions which cover generalized extraction problems (and computational ones per Section 2).

To state our theorem, we first define the completion $\mathcal{C}(L)$ of a list L with respect to the group setting $(p, \mathcal{G}, \Phi, \mathcal{E})$. This notion will be instrumental to define the side condition of our master theorem. Intuitively speaking, given a list L, its completion $\mathcal{C}(L)$ is the list of all polynomials that can be computed by the adversary by applying isomorphisms and maps to polynomials in L.

We compute the completion $\mathcal{C}(L)$ of L in two steps. In the first step, we compute the *recipe lists* $\{R_i\}_{i \in \mathcal{I}}$ using the algorithm given in Figure 2. The elements of the recipe lists are monomials over the variables $W_{i,j}$ for $(i, j) \in \mathcal{I} \times [|L_i|]$.

$$\underline{\text{foreach } i \in \mathcal{I} : S'_i = \emptyset \; ; \; S_i = \{W_{i,1}, \dots, W_{i,|L_i|}\}}$$
$$\underline{\text{while } S \neq S' :}$$
$$\quad S' := S$$
$$\quad \underline{\text{foreach } e : \mathbb{G}_{j_1} \times \dots \times \mathbb{G}_{j_n} \to \mathbb{G}_{j_{n+1}} \in \mathcal{E} :}$$
$$\qquad S_{j_{n+1}} := S_{j_{n+1}} \cup \{f_1 \cdots f_n \mid f_i \in S_{j_i}, \; i \in [n]\}$$
$$\quad \underline{\text{foreach } \phi : \mathbb{G}_i \to \mathbb{G}_j \in \Phi : S_j := S_j \cup S_i}$$
$$\underline{\text{foreach } i \in \mathcal{I} : R_i := setToList(S_i)}$$

Fig. 2. Computation of lists of recipes R_i for input lists L_i.

The monomials characterize which products of elements in L the adversary can compute by applying isomorpisms and maps. The result of the first step is independent of the elements in the lists L and only depends on the lengths of the lists. In the second step, we compute the actual polynomials from the recipes as

$$\mathcal{C}(L)_i = [m_1(L), \dots, m_{|R_i|}(L)] \text{ for } [m_1, \dots, m_{|R_i|}] = R_i$$

where every m_i is a monomial over the variables $W_{i,j}$ and $m_i(L)$ denotes the result of evaluating the monomial m_i for the values $L_i[j_i]$.

To ensure that the computation of the recipes terminates, we restrict ourselves to group settings without cycles. We also assume that the group setting contains a target group. Formally, for a group setting $(p, \mathcal{G}, \Phi, \mathcal{E})$, we define the weighted directed graph $G = (V, E)$ with $V = \mathcal{G}$ and E defined as follows. For each isomorphism $\mathbb{G}_i \to \mathbb{G}_j \in \Phi$, there is an edge from $\mathbb{G}_i$ to $\mathbb{G}_j$ of weight 0. Similarly, given any $\mathbb{G}_{i_1} \times \dots \times \mathbb{G}_{i_n} \to \mathbb{G}_{i_{n+1}} \in \mathcal{E}$, there are edges from $\mathbb{G}_{i_j}$ to $\mathbb{G}_{i_{n+1}}$ of weight 1 for $j \in [n]$. We assume that the graph G contains no loops of *positive* weight. Furthermore, we assume there is a unique $\mathbb{G}_t \in V$ called the *target group*, such that from any $\mathbb{G}_i \in V$ there is a path to $\mathbb{G}_t$ and $\mathbb{G}_t$ does not have any outgoing edges.

Theorem 2. *Let $\mathcal{GS} = (p, \{\mathbb{G}_i\}_{i \in \mathcal{I}}, \Phi, \mathcal{E})$ denote a group setting, and $\mathcal{D}_L, \mathcal{D}_{L'}$ be polynomially-induced distributions such that $|L_i| = |L'_i|$ for all $i \in \mathcal{I}$. Let t denote the index of the target group, $s = \sum_{i \in \mathcal{I}} |L_i|$, $r = |\mathcal{C}(L)_t|$, and let d denote an upper bound for the total degrees of the polynomials in the completions of the lists. If*

$$\{a \in \mathbb{F}_p^r \mid a \cdot \mathcal{C}(L)_t = 0\} = \{a \in \mathbb{F}_p^r \mid a \cdot \mathcal{C}(L')_t = 0\},$$

then

$$|Pr[\, Gen_{\mathcal{D}_L}^{\mathcal{GS}}(\mathcal{A}) = 1 \,] - Pr[\, Gen_{\mathcal{D}_{L'}}^{\mathcal{GS}}(\mathcal{A}) = 1 \,]| \leq (s+q)^2 * d/p$$

for all adversaries $\mathcal{A}$ that perform at most q operations.

Note that deciding the side condition is sufficient for deciding the hardness of the corresponding decisional problem for a fixed group setting and fixed distributions. Either the side condition is satisfied or there exists an $a \in \mathbb{F}_p^r$ that is included in one of the sets, but not in the other one. In the first case, the distinguishing advantage is upper-bounded by the ϵ given above. In the second case, we can construct an adversary that distinguishes the two symbolic models with probability 1, which implies that it distinguishes the corresponding generic models with probability $1 - \epsilon$. Note that for real-or-random assumptions where the adversary is given $\hat{\boldsymbol{L}}$ and must distinguish f from a fresh variable Z in the target group $\mathbb{G}_t$, our side condition simplifies to $\sum_{j=1}^{r} a_j \mathcal{C}(\hat{\boldsymbol{L}})_t[j] \neq f$ for all $\boldsymbol{a} \in \mathbb{F}_p^r$. This is similar to the independence condition in the BBG master theorem [11].

4   Automated Analysis of Non-interactive Assumptions

In this section, we present methods to automatically verify or falsify the hardness of decisional assumptions. As mentioned earlier, our master theorem is stated with respect to a fixed group setting and fixed distributions. To consider multiple group settings or distributions at once, we define a decisional assumption $\mathbb{A}$ as a possibly infinite set of triples $(\mathcal{GS}, \mathcal{D}_{\boldsymbol{L}}, \mathcal{D}_{\boldsymbol{L}'})$. $\mathbb{A}$ is *generically hard* if the distinguishing probability is upper-bounded by ϵ in Theorem 2 for all triples in $\mathbb{A}$.

We distinguish between *non-parametric* assumptions and *parametric* assumptions. An assumption is non-parametric if only the concrete groups, isomorphisms, and n-linear maps vary, but the structure of the group setting and the lists $\boldsymbol{L}$ and $\boldsymbol{L}'$ defining the distributions remain fixed. This captures assumptions such as "3-lin is hard in all groups with a symmetric 3-linear map". Conversely, an assumption is parametric if one or more of these restrictions do not hold.

4.1   Non-parametric Assumptions

We perform the following computations over $\mathbb{Z}$ to decide the hardness of a decisional assumption defined by lists $\boldsymbol{L}$ and $\boldsymbol{L}'$ for all group settings $\mathcal{GS}$ with a given index set and types of isomorphisms and n-linear maps.

1. Initialize the set T of distinguishing tests and the set E of exceptional primes to the empty set $\emptyset$.
2. Compute the completions $\mathcal{C}(\boldsymbol{L})$ and $\mathcal{C}(\boldsymbol{L}')$ and set $\overline{L}_t := \mathcal{C}(\boldsymbol{L})_t$, $\overline{L}'_t := \mathcal{C}(\boldsymbol{L}')_t$
3. Compute a generating set K of the $\mathbb{Z}$-module $\{\boldsymbol{a} \in \mathbb{Z}^{|\overline{L}_t|} \mid \boldsymbol{a} \cdot \overline{L}_t = 0\}$ as follows:
    (a) Represent all polynomials $g \in \overline{L}_t$ as vectors $\boldsymbol{v}_1, \ldots, \boldsymbol{v}_n$ and denote by M the matrix, where row i is $\boldsymbol{v}_i$ with respect to the basis $monomials(\overline{L}_t)$.
    (b) Compute the Hermite Normal Form N of M and read off a generating set K of the left kernel from N and the transformation matrix. Set $E := E \cup F$ where F is the set of factors of pivots of N.
    Perform the same steps for $\overline{L}'_t$ to obtain M' and K'.
4. Check for every $\boldsymbol{k} \in K$ if $\boldsymbol{k}M' = 0$. If $\boldsymbol{k}M' = c \neq 0$, then set $T := T \cup \boldsymbol{k}$ and $E := E \cup F$ where F denotes the set of common factors of c. Perform the same steps for K' and M.
5. Compute distinguishing probability ϵ from degrees in $\overline{L}_t$ and $\overline{L}'_t$.

6. If T is empty, return that distinguishing probability is upper-bounded by ϵ except (possibly) for primes in E. If T is nonempty, return that using the tests in T, an adversary can distinguish with probability $1 - \epsilon$ except (possibly) for primes in E.

Note that performing division-free computations over $\mathbb{Z}$ allows us to track the set of exceptional primes, which we return. We have implemented this algorithm in a tool that takes a group setting and two sequences of group elements as input and decides if the corresponding decisional assumption is hard returning ϵ, E, and the distinguishing tests T (if nonempty).

4.2   Parametric Assumptions

For parametric decisional assumptions, we restrict ourselves to the real-or-random case. The approach can also be adapted to handle computational assumptions. We distinguish parametricity in two dimensions. First, an assumption may be parameterized by *range limits* $l_1, \ldots, l_m$ (ranging over $\mathbb{N}$) that determine the size of the adversary input. We use *range expressions* $\forall r \in [\alpha, \beta]. \, h_r$, where α and β are polynomials over range limits, to express such assumptions. The polynomials h_r can use the *range index* r in the exponent or as the index of an indexed variable X_r. We will denote range expressions with capital letters R. Second, the group setting of an assumption may be parameterized by an *arity* k that captures the maximum number of multiplications that can be performed.

Parametricity in the input size allows us to analyze assumptions such as "l-DHE is hard for all l". Parametricity in the arity allows us to analyze assumptions such "2-BDH is hard for all k-linear groups". Combining both types of parametricity allows us to analyze assumptions such as "k-lin is hard in k-linear groups" or "(l, k)-MMDHE is hard for all l and $k \geq 3$". In the following, we will present two methods that deal with both parametricity in the input size and parametricity in the arity. The first method assumes a fixed number of random variables. The second method allows for indexed random variables, but assumes that the degree of adversary input and challenge is fixed.

Fixed Number of Variables. We assume a real-or-random decisional assumption in a (leveled) k-linear group where the challenge polynomial g is in the target group, and the adversary input is expressed using range expressions $R_1, \ldots, R_n$ on the levels $\lambda_1, \ldots, \lambda_n$. Here λ_i is either of the form c or of the form $k - c$ for a constant $c \in \mathbb{N}$. Furthermore, we assume that the assumption uses random variables $\boldsymbol{X}$ and range limits $\boldsymbol{l}$. To simplify the presentation, we will use the notation $\boldsymbol{X^f} = X_1^{f_1} \cdots X_m^{f_m}$. Then the ranges are of the form

$$R_i = \forall r_{i,1} \in [\alpha_{i,1}, \beta_{i,1}], \ldots, r_{i,t_i} \in [\alpha_{i,t_i}, \beta_{i,t_i}]. \, \boldsymbol{X^{f_i}}$$

where every $\alpha_{i,j}$ and $\beta_{i,j}$ is a polynomial over $\boldsymbol{l}$ and every $f \in \boldsymbol{f}_i$ is a polynomial over k, $\boldsymbol{l}$, and $r_{i,1}, \ldots, r_{i,t_i}$. The challenge polynomial is of the form $g = \sum_{i=1}^{w} c_i \boldsymbol{X^{u_i}}$. Using the independence condition derived from Theorem 2, it follows that real distribution and the random distribution are indistinguishable iff there is a monomial $\boldsymbol{X^{u_i}}$ that is not an element of the completion of the R_i.

To check this condition, we proceed in two steps. In the first step, we compute a single range expression $\overline{R}$ that denotes the completion of the R_i in the target group. In the second step, we check for each X^{u_i} whether $X^{u_i} \in \overline{R}$, by encoding the required equalities of the exponent-polynomials into a set of diophantine (in)equalities. We then show that satisfiability checking for such constraints is undecidable in general. Nevertheless, we identify two decidable fragments and demonstrate that SMT solvers can handle most instances derived from practical cryptographic assumptions, even those that are not in the decidable fragments.

If $R_1, \ldots, R_n$ denote the sets $S_1, \ldots, S_n$, then the completion $\overline{R}$ of $R_1, \ldots, R_n$ in the target group must denote the set

$$\bigcup_{\delta \in \mathbb{N}^n \text{ s.t. } \sum_{i=1}^{n} \delta_i \cdot \lambda_i = k} S_1^{\delta_1} \cdots S_n^{\delta_n}$$

where $SS' = \{ss' \mid s \in S \wedge s' \in S'\}$ and $S^\delta = \{\prod_{i=1}^{\delta} s_i \mid s_1 \in S \wedge \ldots \wedge s_\delta \in S\}$. We therefore define multiplication of range expressions with distinct range indices as

$$(\forall r_1 \in [\alpha_1, \beta_1], \ldots, r_t \in [\alpha_t, \beta_t]. X^f)(\forall r_1' \in [\alpha_1', \beta_1'], \ldots, r_s' \in [\alpha_{t'}', \beta_{t'}']. X^{f'})$$
$$= \forall r_1 \in [\alpha_1, \beta_1], \ldots, r_t \in [\alpha_t, \beta_t], r_1' \in [\alpha_1', \beta_1'], \ldots, r_s' \in [\alpha_{t'}', \beta_{t'}']. X^{f+f'}.$$

To define the δ-fold product of a range expression, we restrict ourselves to exponent-polynomials that can be expressed as $\hat{f} + \tilde{f}$ such that $\hat{f} = \sum_{j=1}^{t} r_j \, \phi_j(l, k)$ for polynomials ϕ_j in $\mathbb{Z}[l, k]$ and such that $\tilde{f}$ is a polynomial in $\mathbb{Z}[l, k]$. The δ-fold product is then defined as

$$(\forall r_1 \in [\alpha_1, \beta_1], \ldots, r_m \in [\alpha_t, \beta_t]. X^{\hat{f}+\tilde{f}})^\delta$$
$$= \forall r_1 \in [\delta\alpha_1, \delta\beta_1], \ldots, r_m \in [\delta\alpha_t, \delta\beta_t]. X^{\hat{f}+\delta\tilde{f}}.$$

Given range expressions $R_1, \ldots, R_n$, we can now compute $\overline{R}$ by introducing fresh variables $\delta_1, \ldots, \delta_n$, computing the range expressions $R_i^{\delta_i}$, and then computing the product of these range expressions.

The remaining task is now to check if

$$X^u \in (\forall r_1 \in [\alpha_1, \beta_1], \ldots, r_t \in [\alpha_t, \beta_t]. X^f) = \overline{R}$$

where $u \in \mathbb{Z}[l, k]^m$, $\alpha_i, \beta_i \in \mathbb{Z}[\delta, l]$, $f \in \mathbb{Z}[l, k, r_1, \ldots, r_t]^m$, and $\sum_{i=1}^{n} \delta_i \cdot \lambda_i = k$. To achieve this, we compute the following set of integer constraints that is satisfiable iff $X^u \in \overline{R}$:

$$\begin{cases} 0 \le \delta_i & \text{for } i \in [1, n] \\ \alpha_i \le r_i \le \beta_i & \text{for } i \in [1, t] \\ u_i = f_i, & \text{for } i \in [1, m] \\ \sum_{i=1}^{n} \delta_i \lambda_i = k \end{cases}$$

If we allow for both types of parametricity, it is possible to reduce Hilbert's 10th problem to the generic hardness of cryptographic assumptions expressed as previously described. This yields the following theorem.

Theorem 3. *Deciding hardness of parametric assumptions with a fixed number of variables in the generic group model is undecidable, even if all exponent-polynomials are linear in range limits, range indices, and the arity.*

However, for a restricted class of assumptions, the problem is decidable.

Theorem 4. *For all parametric assumptions with a fixed number of variables such that all exponent-polynomials $f_{i,j}$ and range bounds $\alpha_{i,j}$ and $\beta_{i,j}$ in the input are linear, and either (1) the arity k is fixed or (2) the assumption does not contain range limits l_i and the input exponent-polynomials do not use k, deciding hardness in the generic group model is decidable.*

Proof (Sketch). In both cases, we transform the constraint system into a system of linear constraints. Note that the first type of constraint is already linear. In the first case, the arity k is fixed and we can eliminate the variables δ_i by performing a case distinction since there are only finitely many possible values. Then, the constraints of the first and fourth type are constant and the constraints of the second and third type are linear. If there are no range limits, then the range bounds are constants and we can eliminate the range indices by expanding all range expressions into finite sets of monomials. Then the constraints of the second type are constant and we can linearize the constraints of the last type since λ_i is either a constant c or of the form $k - c$. For constraints of the third type, every u_i is a linear polynomial in $\mathbb{Z}[k]$ and every f_i is a linear polynomial in $\mathbb{Z}[\delta, k]$.

We have implemented this method in our tool and use Z3 [16] to check the constraints. Our experiments confirm that Z3 can prove most assumptions taken from the literature, even those outside the decidable fragment.

Indexed Random Variables. For the case of indexed random variables, we have developed an (incomplete) constraint solving procedure that deals with assumptions parametric in the arity k and a range limit l. Let M denote monomials built from indexed variables and M' denote monomials built from non-indexed variables. Our procedure supports all assumptions where the challenge is of the form $\sum_{i \in [0,l]} MM'$ and the input consist of ranges $\forall i \in [0,l].\, MM'$ and non-indexed monomials M'.

5   Interactive Assumptions

In this section, we present our methods for the analysis of interactive assumptions such as LRSW [27]. To simplify the presentation, we focus on assumptions where exactly *one* additional oracle $\mathcal{O}$ is provided to the adversary and the problem is a generalized extraction problem. In the remainder, we fix a group setting $\mathcal{GS} = (p, \{\mathbb{G}\}_{i \in \mathcal{I}}, \Phi, \mathcal{E})$ and a distribution $\mathcal{D}_L$. We use X to denote the variables occurring in L and x to denote the point sampled by $\mathcal{D}_L$.

Generalizing *Gen* and *Sym*. Our first step is generalizing the generic group and symbolic group models to the interactive setting. Let q', n, m, l denote positive integers, let $i \in \mathcal{I}^l$, and let F denote an l-dimensional vector of polynomials in $\mathbb{F}_p[\boldsymbol{X}, Y_1, \ldots, Y_m, A_1, \ldots, A_n]$. We say $\mathcal{O}$ is defined by $(q', n, m, l, \boldsymbol{i}, \boldsymbol{F})$ if $\mathcal{O}$ answers at most q' queries and answers queries for parameter $\boldsymbol{a} \in \mathbb{F}_p^n$ by sampling a point $\boldsymbol{y} \in \mathbb{F}_p^m$ and returning handles to the group elements $F_j(\boldsymbol{x}, \boldsymbol{y}, \boldsymbol{a}) P_{i_j} \in \mathbb{G}_{i_j}$ for $j \in [l]$ where P_{i_j} is the generator of $\mathbb{G}_{i_j}$. Similarly, the symbolic version of $\mathcal{O}$ answers queries for $\boldsymbol{a} \in \mathbb{F}_p^n$ by choosing m fresh variables $\boldsymbol{Y}$, adding the polynomials $F_j(\boldsymbol{X}, \boldsymbol{Y}, \boldsymbol{a})$ to the lists L_{i_j} for $j \in [l]$, and returning their handles. To formalize winning conditions of interactive assumptions, we extend the previously given definition of generalized extraction problem with inequalities. Concretely, the winning condition is formalized by polynomials $H_1, \ldots, H_{d_1}, G_1, \ldots, G_{d_2}$ that capture the required equalities and inequalities for the field elements $\boldsymbol{b}$ and the handles $\boldsymbol{h}$ returned by the adversary. These polynomials are elements of $\mathbb{F}_p[\boldsymbol{X}, (\boldsymbol{Y_i})_{i \in [q']}, (\boldsymbol{A_i})_{i \in [q']}, \boldsymbol{B}, \boldsymbol{Z}]$. Intuitively, $\boldsymbol{X}$ and $\boldsymbol{Y_i}$ model random variables sampled initially and by $\mathcal{O}$, $\boldsymbol{A_i}$ and $\boldsymbol{B}$ model parameters chosen by the adversary, and $\boldsymbol{Z}$ models group elements referenced by the handles $\boldsymbol{h}$. An adversary, that queries the oracle with $\boldsymbol{a}_1, \ldots, \boldsymbol{a}_{q'}$ and returns $\boldsymbol{b}$ and $\boldsymbol{h}$, wins if the following conditions are satisfied for $\boldsymbol{y}_j$ sampled in the j-th oracle call:

$$H_j(\boldsymbol{x}, \boldsymbol{y}_1, \ldots, \boldsymbol{y}_{q'}, \boldsymbol{a}_1, \ldots, \boldsymbol{a}_{q'}, \boldsymbol{b}, dlog(L_{i_1}[h_1]), \ldots, dlog(L_{i_m}[h_m])) = 0 \ , j \in [d_1]$$

$$G_j(\boldsymbol{x}, \boldsymbol{y}_1, \ldots, \boldsymbol{y}_{q'}, \boldsymbol{a}_1, \ldots, \boldsymbol{a}_{q'}, \boldsymbol{b}, dlog(L_{i_1}[h_1]), \ldots, dlog(L_{i_m}[h_m])) \neq 0 \ , j \in [d_2]$$

Since Theorem 1 captures generalized extraction problems (with inequalities) in such an interactive setting, we can analyze such assumptions in the symbolic group model. As mentioned earlier, the symbolic version of the winning event can be obtained by plugging in the polynomials $L_{i_j}[h_j]$ for the variables Z_j instead of using the discrete logarithm.

Interactive Master Theorem. To define the interactive master theorem, we introduce the notion of parametric completion. The *parametric completion* of $\boldsymbol{L}$ with respect to a group setting $\mathcal{GS}$ and an oracle $\mathcal{O}$ defined by $(q', n, m, l, \boldsymbol{i}, \boldsymbol{F})$ is a family L_i of lists of polynomials in $\mathbb{F}_p[\boldsymbol{X}, \boldsymbol{Y}, \boldsymbol{A}]$. Here, the variables $Y_{u,v}$ range over $u \in [m]$ and $v \in [q']$ and the variables $A_{u,v}$ range over $u \in [n]$ and $v \in [q']$. They model the random values sampled by $\mathcal{O}$ and the parameters given to $\mathcal{O}$. The parametric completion first extends the lists L_{i_j} with

$$\{F_j(\boldsymbol{X}, Y_{1,v}, \ldots, Y_{m,v}, A_{1,v}, \ldots, A_{n,v}) \mid v \in [q']\}$$

for $j \in [l]$. Then, it performs the previously defined completion with respect to the isomorphisms and n-linear maps in $\mathcal{GS}$. We denote the result with $\mathcal{C}^{\mathcal{O}}(\boldsymbol{L})$.

To state our interactive master theorem, we exploit that in the symbolic model, we can translate a generalized extraction problem to an equivalent generalized extraction problem where the adversary returns only elements in $\mathbb{F}_p$ and no handles. Let $\mathcal{C}^{\mathcal{O}}(\boldsymbol{L}) = \overline{L_{i_1}}, \ldots, \overline{L_{i_l}}$ denote the lists in the completion. Then, we can translate $H(\boldsymbol{X}, (\boldsymbol{Y_i})_{i \in [q']}, (\boldsymbol{A_i})_{i \in [q']}, \boldsymbol{B}, Z_1, \ldots, Z_l)$ to

$$H'(\boldsymbol{X}, \overrightarrow{\boldsymbol{Y}}, \overrightarrow{\boldsymbol{A}}, \boldsymbol{B}, \boldsymbol{C}_1, \ldots, \boldsymbol{C}_l) = H(\boldsymbol{X}, \overrightarrow{\boldsymbol{Y}}, \overrightarrow{\boldsymbol{A}}, \boldsymbol{V}, \boldsymbol{C}_1 \cdot \overline{L_{i_1}}, \ldots, \boldsymbol{C}_l \cdot \overline{L_{i_l}}).$$

The two problems are equivalent since the adversary can return a handle to a polynomial f in L_{i_j} if and only if f is in the span of $\overline{L_{i_j}}$.

Theorem 5. *Let $\mathcal{GS}$ denote a group setting and let $\mathcal{D_L}$ denote a polynomially-induced distribution. Consider the $(\hat{n}, \hat{m}, \boldsymbol{j}, \boldsymbol{H}, \boldsymbol{G})$-extraction problem in the generic and symbolic group models for $\mathcal{GS}$, $\mathcal{D_L}$, and the oracle defined by $(q', n, m, l, \boldsymbol{i}, \boldsymbol{F})$. Let $\boldsymbol{H'}$ and $\boldsymbol{G'}$ denote the translations of $\boldsymbol{H}$ and $\boldsymbol{G}$ with respect to this model that do not use handles. Then the problem is symbolically hard if there exist no vectors $\boldsymbol{a}$, $\boldsymbol{b}$, and $\boldsymbol{c}$ in $\mathbb{F}_p$ such that*

$$\left(\bigwedge_{j=1}^{|\boldsymbol{H'}|} H'_j(\boldsymbol{X}, \boldsymbol{Y}, \boldsymbol{a}, \boldsymbol{b}, \boldsymbol{c}) = 0 \right) \wedge \left(\bigwedge_{j=1}^{|\boldsymbol{G'}|} G'_j(\boldsymbol{X}, \boldsymbol{Y}, \boldsymbol{a}, \boldsymbol{b}, \boldsymbol{c}) \neq 0 \right).$$

In this case, the winning probability for the generic version is upper-bounded by $(s + q + q'\, l)^2 * d/2p + ed/p$ where p is the group order, s is the sum of the sizes of the lists in $\boldsymbol{L}$, q the number of queries to the group-oracles, q' the number of queries to $\mathcal{O}$, d an upper bound on the degrees (in $\boldsymbol{X}$ and $\boldsymbol{Y}$) stored by the corresponding symbolic model and occuring in $\boldsymbol{H'}$ and $\boldsymbol{G'}$, and $e = |\boldsymbol{H'}| + |\boldsymbol{G'}|$.

In the proof of this theorem, we use Theorem 1 to switch to the symbolic model. In the symbolic model, the winning condition is equivalent to our side condition.

Automated Analysis. We have developed two methods for the automated analysis of interactive assumptions. Our first method deals with the bounded case, i.e., where the number of oracle queries q' is fixed. Informally, we use Gröbner basis techniques and SMT solvers to prove that there is (1) no solution for all primes, (2) no solution for all primes except for some bad primes, (3) a solution over the rationals which can be converted into an attack for almost all primes, or (4) a solution over $\mathbb{C}$. Even though we only encountered cases (1-3) in practice, case (4) is the reason for the incompleteness of our algorithm since the existence of a solution over $\mathbb{C}$ does not imply the existence of solutions over $\mathbb{F}_p$. In the unbounded case, we perform most steps symbolically to obtain results that are valid for all possible values of q'. Concretely, we encode the hardness of the assumption into a formula in the theory of non-linear arithmetic over $\mathbb{C}$ with uninterpreted function symbols, which we use to encode parameters used in queries and returned by the adversary. We use Z3 to prove the unsatisfiability of these formulas exploiting the support for nonlinear arithmetic over the reals [26] by encoding complex numbers as pairs of reals. In our experiments, Z3 can prove the unsatisfiability of formulas obtained from most valid assumptions in seconds.

Acknowledgements. This work is supported in part by ONR grant N00014-12-1-0914, Madrid regional project S2009TIC-1465 PROMETIDOS, and Spanish projects TIN2009-14599 DESAFIOS 10 and TIN2012-39391-C04-01 Strongsoft. Additional support for Mitchell, Scedrov, and Fagerholm is from the AFOSR MURI "Science of Cyber Security: Modeling, Composition, and Measurement" and from NSF Grants CNS-0831199 (Mitchell) and CNS-0830949 (Scedrov and Fagerholm). The research of Fiore and Schmidt has received funds from the European Commission's Seventh Framework Programme Marie Curie Cofund Action AMAROUT II (grant no. 291803).

References

1. Abadi, M., Rogaway, P.: Reconciling two views of cryptography (the computational soundness of formal encryption). Journal of Cryptology 20(3), 395 (2007)
2. Abdalla, M., Pointcheval, D.: Interactive Diffie-Hellman assumptions with applications to password-based authentication. In: S. Patrick, A., Yung, M. (eds.) FC 2005. LNCS, vol. 3570, pp. 341–356. Springer, Heidelberg (2005)
3. Ateniese, G., Camenisch, J., de Medeiros, B.: Untraceable RFID tags via insubvertible encryption. In: Atluri, V., Meadows, C., Juels, A. (eds.) ACM CCS 2005, pp. 92–101. ACM Press (November 2005)
4. Barthe, G., Fagerholm, E., Fiore, D., Mitchell, J., Scedrov, A., Schmidt, B.: Automated analysis of cryptographic assumptions in generic group models. Cryptology ePrint Archive 2014 (2014)
5. Benson, K., Shacham, H., Waters, B.: The k-BDH assumption family: Bilinear map cryptography from progressively weaker assumptions. In: Dawson, E. (ed.) CT-RSA 2013. LNCS, vol. 7779, pp. 310–325. Springer, Heidelberg (2013)
6. Blanchet, B.: Security protocol verification: Symbolic and computational models. In: Degano, P., Guttman, J.D. (eds.) POST 2012. LNCS, vol. 7215, pp. 3–29. Springer, Heidelberg (2012)
7. Boldyreva, A., Gentry, C., O'Neill, A., Yum, D.H.: Ordered multisignatures and identity-based sequential aggregate signatures, with applications to secure routing. In: Ning, P., di Vimercati, S.D.C., Syverson, P.F. (eds.) ACM CCS 2007, pp. 276–285. ACM Press (October 2007)
8. Boldyreva, A., Gentry, C., O'Neill, A., Yum, D.H.: Ordered multisignatures and identity-based sequential aggregate signatures, with applications to secure routing. Cryptology ePrint Archive, Report 2007/438 (2007) (revised February 21, 2010)
9. Boneh, D., Boyen, X.: Short signatures without random oracles. In: Cachin, C., Camenisch, J.L. (eds.) EUROCRYPT 2004. LNCS, vol. 3027, pp. 56–73. Springer, Heidelberg (2004)
10. Boneh, D., Boyen, X., Goh, E.-J.: Hierarchical identity based encryption with constant size ciphertext. In: Cramer, R. (ed.) EUROCRYPT 2005. LNCS, vol. 3494, pp. 440–456. Springer, Heidelberg (2005)
11. Boneh, D., Boyen, X., Goh, E.-J.: Hierarchical identity based encryption with constant size ciphertext. Cryptology ePrint Archive, Report 2005/015 (2005)
12. Boneh, D., Franklin, M.: Identity-based encryption from the weil pairing. In: Kilian, J. (ed.) CRYPTO 2001. LNCS, vol. 2139, pp. 213–229. Springer, Heidelberg (2001)
13. Boneh, D., Gentry, C., Waters, B.: Collusion resistant broadcast encryption with short ciphertexts and private keys. In: Shoup, V. (ed.) CRYPTO 2005. LNCS, vol. 3621, pp. 258–275. Springer, Heidelberg (2005)
14. Boyen, X.: The uber-assumption family. In: Galbraith, S.D., Paterson, K.G. (eds.) Pairing 2008. LNCS, vol. 5209, pp. 39–56. Springer, Heidelberg (2008)
15. Bresson, E., Lakhnech, Y., Mazaré, L., Warinschi, B.: A generalization of DDH with applications to protocol analysis and computational soundness. In: Menezes, A. (ed.) CRYPTO 2007. LNCS, vol. 4622, pp. 482–499. Springer, Heidelberg (2007)
16. de Moura, L., Bjørner, N.: Z3: An efficient SMT solver. In: Ramakrishnan, C.R., Rehof, J. (eds.) TACAS 2008. LNCS, vol. 4963, pp. 337–340. Springer, Heidelberg (2008)
17. Escala, A., Herold, G., Kiltz, E., Ràfols, C., Villar, J.: An algebraic framework for Diffie-Hellman assumptions. In: Canetti, R., Garay, J.A. (eds.) CRYPTO 2013, Part II. LNCS, vol. 8043, pp. 129–147. Springer, Heidelberg (2013)

18. Freeman, D.M.: Converting pairing-based cryptosystems from composite-order groups to prime-order groups. In: Gilbert, H. (ed.) EUROCRYPT 2010. LNCS, vol. 6110, pp. 44–61. Springer, Heidelberg (2010)
19. Garg, S., Gentry, C., Sahai, A., Waters, B.: Witness encryption and its applications. In: Boneh, D., Roughgarden, T., Feigenbaum, J. (eds.) 45th ACM STOC, pp. 467–476. ACM Press (ACM Press)
20. Gjøsteen, K., Thuen, Ø.: Password-based signatures. In: Petkova-Nikova, S., Pashalidis, A., Pernul, G. (eds.) EuroPKI 2011. LNCS, vol. 7163, pp. 17–33. Springer, Heidelberg (2012)
21. Halevi, S.: A plausible approach to computer-aided cryptographic proofs. Cryptology ePrint Archive, Report 2005/181 (2005)
22. Hohenberger, S., Sahai, A., Waters, B.: Full domain hash from (Leveled) multilinear maps and identity-based aggregate signatures. In: Canetti, R., Garay, J.A. (eds.) CRYPTO 2013, Part I. LNCS, vol. 8042, pp. 494–512. Springer, Heidelberg (2013)
23. Hwang, J.Y., Lee, D.H., Yung, M.: Universal forgery of the identity-based sequential aggregate signature scheme. In: Li, W., Susilo, W., Tupakula, U.K., Safavi-Naini, R., Varadharajan, V. (eds.) ASIACCS 2009, Mar. 2009, pp. 157–160. ACM Press (March 2009)
24. Jager, T., Rupp, A.: The semi-generic group model and applications to pairing-based cryptography. In: Abe, M. (ed.) ASIACRYPT 2010. LNCS, vol. 6477, pp. 539–556. Springer, Heidelberg (2010)
25. Jager, T., Schwenk, J.: On the equivalence of generic group models. In: Baek, J., Bao, F., Chen, K., Lai, X. (eds.) ProvSec 2008. LNCS, vol. 5324, pp. 200–209. Springer, Heidelberg (2008)
26. Jovanović, D., de Moura, L.: Solving non-linear arithmetic. In: Gramlich, B., Miller, D., Sattler, U. (eds.) IJCAR 2012. LNCS, vol. 7364, pp. 339–354. Springer, Heidelberg (2012)
27. Lysyanskaya, A., Rivest, R.L., Sahai, A., Wolf, S.: Pseudonym systems (Extended abstract). In: Heys, H.M., Adams, C.M. (eds.) SAC 1999. LNCS, vol. 1758, pp. 184–199. Springer, Heidelberg (2000)
28. Maurer, U.M.: Abstract models of computation in cryptography. In: Smart, N.P. (ed.) Cryptography and Coding 2005. LNCS, vol. 3796, pp. 1–12. Springer, Heidelberg (2005)
29. Maurer, U.M., Wolf, S.: Diffie-Hellman oracles. In: Koblitz, N. (ed.) CRYPTO 1996. LNCS, vol. 1109, pp. 268–282. Springer, Heidelberg (1996)
30. Naor, M.: On cryptographic assumptions and challenges. In: Boneh, D. (ed.) CRYPTO 2003. LNCS, vol. 2729, pp. 96–109. Springer, Heidelberg (2003)
31. Nechaev, V.I.: Complexity of a determinate algorithm for the discrete logarithm. Mathematical Notes 55(2), 165–172 (1994)
32. Okamoto, T., Takashima, K.: Fully secure functional encryption with general relations from the decisional linear assumption. In: Rabin, T. (ed.) CRYPTO 2010. LNCS, vol. 6223, pp. 191–208. Springer, Heidelberg (2010)
33. Shoup, V.: Lower bounds for discrete logarithms and related problems. In: Fumy, W. (ed.) EUROCRYPT 1997. LNCS, vol. 1233, pp. 256–266. Springer, Heidelberg (1997)
34. Stein, W., et al.: Sage Mathematics Software (Version 5.12). The Sage Development Team (2013), http://www.sagemath.org
35. Szydlo, M.: A note on chosen-basis decisional diffie-hellman assumptions. In: Di Crescenzo, G., Rubin, A. (eds.) FC 2006. LNCS, vol. 4107, pp. 166–170. Springer, Heidelberg (2006)

The Exact PRF-Security of NMAC and HMAC

Peter Gaži, Krzysztof Pietrzak, and Michal Rybár

IST Austria

Abstract. NMAC is a mode of operation which turns a fixed input-length keyed hash function f into a variable input-length function. A practical single-key variant of NMAC called HMAC is a very popular and widely deployed message authentication code (MAC). Security proofs and attacks for NMAC can typically be lifted to HMAC.

NMAC was introduced by Bellare, Canetti and Krawczyk [Crypto'96], who proved it to be a secure pseudorandom function (PRF), and thus also a MAC, assuming that (1) f is a PRF and (2) the function we get when cascading f is weakly collision-resistant. Unfortunately, HMAC is typically instantiated with cryptographic hash functions like MD5 or SHA-1 for which (2) has been found to be wrong. To restore the provable guarantees for NMAC, Bellare [Crypto'06] showed its security based solely on the assumption that f is a PRF, albeit via a non-uniform reduction.

- Our first contribution is a simpler and uniform proof for this fact: If f is an ε-secure PRF (against q queries) and a δ-*non-adaptively* secure PRF (against q queries), then NMAC^f is an $(\varepsilon + \ell q\delta)$-secure PRF against q queries of length at most ℓ blocks each.
- We then show that this $\varepsilon + \ell q\delta$ bound is basically tight. For the most interesting case where $\ell q\delta \geq \varepsilon$ we prove this by constructing an f for which an attack with advantage $\ell q\delta$ exists. This also violates the bound $O(\ell\varepsilon)$ on the PRF-security of NMAC recently claimed by Koblitz and Menezes.
- Finally, we analyze the PRF-security of a modification of NMAC called NI [An and Bellare, Crypto'99] that differs mainly by using a compression function with an additional keying input. This avoids the constant rekeying on multi-block messages in NMAC and allows for a security proof starting by the standard switch from a PRF to a random function, followed by an information-theoretic analysis. We carry out such an analysis, obtaining a tight $\ell q^2/2^c$ bound for this step, improving over the trivial bound of $\ell^2 q^2/2^c$. The proof borrows combinatorial techniques originally developed for proving the security of CBC-MAC [Bellare et al., Crypto'05].

Keywords: Message authentication codes, pseudorandom functions, NMAC, HMAC, NI.

1 Introduction

NMAC is a mode of operation which transforms a keyed fixed input-length function f : $\{0,1\}^c \times \{0,1\}^b \to \{0,1\}^c$ (with $b \geq c$) into a keyed variable input-length

J.A. Garay and R. Gennaro (Eds.): CRYPTO 2014, Part I, LNCS 8616, pp. 113–130, 2014.

function $\mathsf{NMAC}^\mathsf{f} : \{0,1\}^{2c} \times \{0,1\}^{b*} \to \{0,1\}^c$ (where $\{0,1\}^{b*}$ denotes all bit strings whose length is a multiple of b) as

$$\mathsf{NMAC}^\mathsf{f}((K_1, K_2), M) := \mathsf{f}(K_2, \mathsf{Casc}^\mathsf{f}(K_1, M)\|0^{b-c})$$

where $\mathsf{Casc}^\mathsf{f} : \{0,1\}^c \times \{0,1\}^{b*} \to \{0,1\}^c$ is the cascade (also known as Merkle-Damgård) construction

$$\mathsf{Casc}^\mathsf{f}(K_1, m_1\|\ldots\|m_\ell) := \mathsf{f}(\ldots\mathsf{f}(\mathsf{f}(K_1, m_1), m_2)\ldots m_\ell) \,.$$

HMAC is a variant of NMAC (we postpone its exact definition to Section 2.2) tweaked for applicability in practice. As security proofs for NMAC can typically be lifted to HMAC, it is usually sufficient to analyse the security of the cleaner NMAC construction, we will discuss this point further in Section 1.2.

NMAC and HMAC were introduced by Bellare, Canetti and Krawczyk in 1996 [4] and later standardized [18]. HMAC has also become very popular and widely used, being implemented in SSL, SSH, IPsec and TLS amongst other places. Although originally designed as a MAC, it is also often employed more broadly, as a pseudorandom function (PRF). This is the case for example when used for key-derivation in TLS and IKE (the Internet Key Exchange protocol of IPsec). This proliferation into practice motivates the need for a good understanding of the exact security guarantees provided by NMAC and HMAC when used as a PRF.

PRF-SECURITY OF NMAC. Bellare *et al.* [4] prove that NMAC is a secure PRF if (1) f is a PRF and (2) Casc^f is weakly collision-resistant (WCR). This is a relaxed notion of collision resistance, where one requires that it is hard to find a pair of messages $M \neq M'$ such that $\mathsf{Casc}^\mathsf{f}(K, M) = \mathsf{Casc}^\mathsf{f}(K, M')$ under a random key K, given oracle access to $\mathsf{Casc}^\mathsf{f}(K, .)$ (but not K, as in the standard definition of collision resistance).

HMAC is typically instantiated with cryptographic hash functions like MD5 or SHA-1 playing the role of Casc^f. However, both of these have been found not to satisfy the WCR notion [26,27], which renders the security proof from [4] irrelevant for this case. Despite that, no attacks (better than standard birthday attacks) are known for NMAC or HMAC when instantiated with MD5 or SHA-1 (though attacks on reduced round versions exist [16]).

SECURITY WITHOUT COLLISION-RESISTANCE. To restore the provable security of NMAC, Bellare [3] investigates the security of NMAC dropping assumption (2), that is, assuming only that f is a secure PRF. The exact security statement from [3] is a bit technical, but it roughly states that if f is an ε-secure PRF (against an adversary running in time t and asking q queries) and a γ-secure PRF (against time $O(\ell)$ and 2 queries), then NMAC^f is an $(\varepsilon + \ell q^2 \gamma)$-secure PRF against time t and q queries of length at most ℓ (in b-bit blocks). The security reduction is non-uniform, which means one has to be careful when deducing what this bound exactly means when instantiated in practice, we will discuss this further in Section 1.2.[1]

1.1 Our Contributions

PRF-SECURITY PROOF FOR NMAC. Our first contribution is a simpler, uniform, and as we will show, basically tight proof for the PRF-security of NMAC^f assuming only that f is a PRF: If f is an ε-secure PRF against q queries, then NMAC^f is roughly $\ell q\varepsilon$-secure against q queries of length at most ℓ blocks each.

Our actual result is more fine-grained, and expresses the security in terms of both the adaptive and non-adaptive security of f. Let δ denote the PRF-security of f against q *non-adaptive* queries. Then our Theorem 1 states that NMAC^f is roughly $(\varepsilon + \ell q\delta)$-secure (against q queries, each at most ℓ blocks). As non-adaptive adversaries are a subset of adaptive ones we have $\delta \leq \varepsilon$, and if $\delta \ll \varepsilon$, then our fine-grained bound is much better than the simpler $\ell q\varepsilon$ bound. The reduction works in the best running time one could hope for, its overhead being $\tilde{O}(\ell q)$.

The main technical part of our proof closely follows a proof by Bellare *et al.* [5] who show that if f is a secure fixed input-length PRF, then Casc^f is a secure PRF if queried on prefix-free queries. We first observe that their proof also holds in the non-adaptive setting. Then we reduce the security of NMAC^f against arbitrary adaptive queries to the security of Casc^f against non-adaptive prefix-free queries.

MATCHING ATTACK FOR NMAC. In Section 3.2 we prove that the above lower bound is basically tight. From any PRF, we construct another PRF f for which NMAC^f can be broken with advantage $\Theta(\ell q\delta)$. This shows that our bound is tight for the practically most important case when $\ell q\delta$ is larger (or at least comparable) to ε.

We also consider the case where $\varepsilon \gg \ell q\delta$, that is, when the PRF has much better security against non-adaptive than adaptive distinguishers. We observe that for any ε, we can use a result due to Pietrzak [23] who shows that cascading non-adaptively secure PRFs does not give an adaptively secure PRF in general, to construct an ε-secure f where NMAC^f can be broken with advantage $\Theta(\varepsilon^2)$. This only shows the ε term is necessary if ε is constant as then $\Theta(\varepsilon) = \Theta(\varepsilon^2) = \Theta(1)$. We conjecture that $\Theta(\varepsilon^2)$ is the correct value, and the ε term in the lower bound can be improved to $\Theta(\varepsilon^2)$ using security amplification techniques along the lines of [22,25].

PRF-SECURITY PROOF FOR NI. The main difficulty in security analyses of NMAC^f and HMAC^f based on the PRF-security of the underlying compression function f is that both these constructions are constantly rekeying f during the evaluation of Casc^f, using the output from the last invocation as the key for the

---

[1] We note that in a very recent update of the ePrint version of [3], Bellare observes that the proof in [3] can also give a uniform reduction, differing from the non-uniform case only in the running time of the 2-query adversary which then becomes t. The uniform bound given in this paper is better for most reasonable parameters.

next one. This prevents the proof approach typically applied to constructions that use a PRF f under a fixed random secret key, where the analysis starts by replacing the PRF with an ideal random function (introducing an error that is upper-bounded by the PRF-security of f) and proceeds by a fully information-theoretic argument.

To circumvent this issue, as our third contribution we investigate the PRF-security of the nested iterated (NI) construction introduced in [2]. The construction NI^h is very similar to $NMAC^f$, but is based on a compression function h that (compared to f) takes an additional k-bit input which is used for keying instead of the chaining input: NI^h uses h under the same key throughout the whole cascade. Additionally, it includes the length of the message in the input to the final, outer h-call. The modified keying allows for the simple switching argument from PRF to a random function. We focus on enhancing the information-theoretic analysis that follows this switch and prove an essentially tight $\ell q^2/2^c$ bound for this step, improving significantly over the trivial bound of $\ell^2 q^2/2^c$. For completeness, we also consider the modification of NI that does not include the message length in the last h-call and show a security bound of $\ell d'(\ell)q^2/2^c$ for this case, where $d'(\ell) \approx \ell^{1/\ln\ln\ell}$ denotes the maximum number of divisors of any positive integer not greater than ℓ. Our proofs employ combinatorial techniques originally developed for proving the security of CBC-MAC [7], considerably adapted for our setting.

1.2   More Related Work

INDIFFERENTIABILITY. In practice, the HMAC construction is sometimes used in a setting where stronger guarantees than PRF-security are needed. Motivated by this, recent work [12] investigates the indifferentiability [21,10] of HMAC from a (keyed) random oracle. This result is incomparable to ours: While the stronger notion of indifferentiability covers the settings where HMAC is not used as a PRF, the bound achieved in [12] is understandably much weaker, being $\Theta(\ell^2 q^2/2^c)$.

ANOTHER LOOK AT [17]. As already mentioned, Bellare [3] proved that $NMAC^f$ is an $(\varepsilon + \ell q^2 \gamma)$-secure PRF against q queries if f is ε-secure against q queries, and γ-secure against 2 queries. In a recent paper [17], Koblitz and Menezes present a criticism of the way [3] discusses the practical implications of this result. In a nutshell, Bellare estimates that for a well-designed PRF the γ term is roughly $t/2^c$ (for a 2-query adversary running in time t), but as this γ is derived in a non-uniform way, it is in the order of $2^{-c/2}$ already for constant t.

At the time when [3] appeared, the fact that non-uniform attacks can distinguish any pseudorandom object generated using a c-bit key with advantage $2^{-c/2}$ in constant time was not widely known in the crypto community[2] and overoptimistic estimates for the exact security implied by non-uniform

---

[2] Let us stress that this only holds for pseudorandom objects which do not require additional *public* randomness, such as PRFs. This does not extend to weak PRFs, which are defined like PRFs but the adversary only sees the output on random inputs.

reductions have appeared in numerous papers.[3] This changed at the latest with the Crypto 2010 paper [11], who discuss this issue in detail and attribute such generic non-uniform attacks to the 1992 paper by Alon *et al.* [1].

The paper [17] also claimed that HMAC is an $\varepsilon\ell$-secure PRF, a bound that is falsified by an attack given in this paper. In response, [17] was updated to take account of this by employing a non-standard definition of a PRF for the underlying compression function. We believe that the updated claim can be obtained via a simpler proof from [5].

HMAC VS NMAC. The proofs in this paper consider NMAC. There is a standard reduction of HMAC-to-NMAC PRF-security given by Bellare [3], albeit under some additional requirements on the underlying compression function f. Informally, one needs to assume that f is a PRF even when keyed through the b-bit data input, as opposed to being keyed by the c-bit chaining variable. Moreover, security of the single-key version of HMAC requires the PRF to be secure under a specific class of related-key attacks. Formally, the reductions are given in Lemmas 5.1 and 5.2 in the full version of [3] for the case of double- and single-keyed HMAC, respectively. Since these reductions only relate to NMAC via its PRF-security, they apply to our result in a blackbox way, thus giving clear statements also for HMAC.

2 Preliminaries

BASIC DEFINITIONS. We reserve the letter λ do denote the empty string. With $\{0,1\}^{b*} := \bigcup_{z \geq 0} \{0,1\}^{bz}$ we denote the set of all bitstrings whose length is a multiple of b. $\mathcal{F}(b,c)$ (resp. $\mathcal{F}(b*,c)$) denotes the sets of all functions from $\{0,1\}^b$ to $\{0,1\}^c$ (resp. from $\{0,1\}^{b*}$ to $\{0,1\}^c$). We denote by $\mathsf{Pow}(\mathcal{S})$ the power set of the set $\mathcal{S}$. For an integer n, $d(n) = |\{i \in \mathbb{N} : i \mid n\}|$ is the number of its positive divisors and $d'(n) := \max_{n' \in \{1,\ldots,n\}} |\{d \in \mathbb{N} : d \mid n'\}| \approx n^{1/\ln\ln n}$ is the maximum, over all positive integers $n' \leq n$, of the number of positive divisors of n'. More precisely, we have $\forall \varepsilon > 0\ \exists n_0\ \forall n > n_0 : d(n) < n^{(1+\varepsilon)/\ln\ln n}$ [13]. All logarithms considered in the paper are base 2 unless indicated otherwise.

RANDOM VARIABLES AND EXPERIMENTS. Random variables and concrete values they can take are usually denoted by upper-case letters $X, Y, \ldots$ and lower-case letters $x, y, \ldots$, respectively. If $\mathcal{M}$ is a distribution (respectively, a set), then we denote by $X \leftarrow \mathcal{M}$ sampling the random variable X according to $\mathcal{M}$ (respectively, choosing it uniformly at random from $\mathcal{M}$). For events A and B and random variables U and V with ranges $\mathcal{U}$ and $\mathcal{V}$, respectively, we denote

---

[3] This should not be confused with the (less trivial, but in the crypto community long well-known) fact that non-uniform generic attacks beating simple brute-force key search exist for "large" running times, as shown in a classical result by Hellman [14]. Hellman's result for example implies that there almost certainly exist key-recovery attacks against AES with a k bit key (k being 128, 192 or 256) which succeed with probability at least $1/2$ and run in time $\approx 2^{2k/3}$, and in particular much less than 2^k required for brute-force key search.

by $\mathsf{P}_{UA|VB}$ the corresponding conditional probability distribution, seen as a (partial) function $\mathcal{U} \times \mathcal{V} \to [0,1]$. The value $\mathsf{P}_{UA|VB}(u,v) = \mathsf{P}[U = u \wedge A|V = v \wedge B]$ is well-defined for all $u \in \mathcal{U}$ and $v \in \mathcal{V}$ such that $\mathsf{P}_{VB}(v) > 0$ and undefined otherwise. Two probability distributions P_U and $\mathsf{P}_{U'}$ on the same set $\mathcal{U}$ are equal, denoted $\mathsf{P}_U = \mathsf{P}_{U'}$, if $\mathsf{P}_U(u) = \mathsf{P}_{U'}(u)$ for all $u \in \mathcal{U}$. Conditional probability distributions are equal if the equality holds for all arguments for which both of them are defined. To emphasize the random experiment $\mathcal{E}$ in consideration, we sometimes write it in the superscript, e.g. $\mathsf{P}^{\mathcal{E}}_{U|V}(u,v)$. If the distribution of a random variable U is clear from the context, we also sometimes write P^U to refer to the random experiment where U is chosen according to its distribution.

2.1   Random Systems

To present our results we make use of Maurer's random systems framework [20], which we now introduce in a self-contained exposition sufficient to follow the rest of the paper. This choice is a matter of authors' taste, we believe that the results could also be obtained using the game-playing framework [8].

We start by observing that the input-output behavior of any kind of reactive discrete system with inputs in $\mathcal{X}$ and outputs in $\mathcal{Y}$ can be described by an infinite family of functions specifying, for each $i \geq 1$, the probability distribution of the system's i-th output $Y_i \in \mathcal{Y}$, given the values of the first i inputs $X^i \in \mathcal{X}^i$ and the previous $i-1$ outputs $Y^{i-1} \in \mathcal{Y}^{i-1}$. Using this viewpoint, we say that an $(\mathcal{X},\mathcal{Y})$-*(random) system* $\mathbf{F}$ is an infinite sequence of functions $\mathsf{p}^{\mathbf{F}}_{Y_i|X^iY^{i-1}} \colon \mathcal{Y} \times \mathcal{X}^i \times \mathcal{Y}^{i-1} \to [0,1]$ such that $\sum_{y_i} \mathsf{p}^{\mathbf{F}}_{Y_i|X^iY^{i-1}}(y_i,x^i,y^{i-1}) = 1$ for all $i \geq 1$, $x^i \in \mathcal{X}^i$ and $y^{i-1} \in \mathcal{Y}^{i-1}$. Note that $\mathsf{p}^{\mathbf{F}}_{Y_i|X^iY^{i-1}}$ by itself does not represent a (conditional) probability distribution in any particular random experiment with well-defined random variables Y_i, X^i, Y^{i-1} until the system is connected to a distinguisher (see below), in which case these random variables will exist and take the role of the transcript. We shall typically define discrete systems by a high level description, as long as the resulting conditional probability distributions could be derived easily from this description. Two systems $\mathbf{F}$ and $\mathbf{G}$ are called *equivalent* (denoted $\mathbf{F} \equiv \mathbf{G}$) if their input-output behaviors are the same, i.e., $\mathsf{p}^{\mathbf{F}}_{Y_i|X^iY^{i-1}} = \mathsf{p}^{\mathbf{G}}_{Y_i|X^iY^{i-1}}$ for all $i \geq 1$.

A system $\mathbf{F}$ might often be used as a component (subsystem) in a construction $\mathbf{C}^{(\cdot)}$, resulting in the composed system $\mathbf{C}^{\mathbf{F}}$. $\mathbf{F} \triangleright \mathbf{G}$ denotes the serial composition of systems: every input to $\mathbf{F} \triangleright \mathbf{G}$ is fed to $\mathbf{F}$, its output is fed to $\mathbf{G}$ and the output of $\mathbf{G}$ is used as the output of $\mathbf{F} \triangleright \mathbf{G}$. In case $\mathbf{G}$ takes as inputs longer bitstrings than $\mathbf{F}$ outputs (as will be the case in the definition of NMAC), the construction $\mathbf{F} \triangleright \mathbf{G}$ pads the outputs of $\mathbf{F}$ with trailing zeroes before passing them to $\mathbf{G}$.

EXAMPLES. We denote by $\mathbf{R}$ a system that provides access to a function chosen uniformly at random from the set of all functions with domain $\{0,1\}^{b*}$ and range $\{0,1\}^c$. (This unusual domain slightly deviates from the standard definition of $\mathbf{R}$ in the random-systems literature, but will be advantageous for our exposition.) Similarly, for a finite domain $\{0,1\}^b$ we denote by $\mathbf{r}$ a system realizing a function chosen uniformly from $\mathcal{F}(b, c)$. Finally, we also consider a system $\mathbf{f}$ realizing a function chosen uniformly from $\mathcal{F}(c+b, c)$. We refer to $\mathbf{R}$, $\mathbf{r}$ and $\mathbf{f}$ as a uniformly random function (URF), a fixed input-length URF, and an ideal compression function, respectively. In each case the parameters b and c will be clear from the context.

DISTINGUISHERS AND ADVERSARIES. A *distinguisher* $\mathbf{D}$ for an $(\mathcal{X}, \mathcal{Y})$-random system asking q queries is a $(\mathcal{Y}, \mathcal{X})$-random system which is "one query ahead:" its input-output behavior is defined by the conditional probability distributions of its queries $\mathsf{p}^{\mathbf{D}}_{X_i \mid X^{i-1} Y^{i-1}}$ for all $1 \leq i \leq q$. (Its first query is determined by $\mathsf{p}^{\mathbf{D}}_{X_1}$.) After the distinguisher asks all q queries, it outputs a bit W_q depending on the transcript (X^q, Y^q). Given a random system $\mathbf{F}$ and a distinguisher $\mathbf{D}$, we denote by $\mathbf{DF}$ the random experiment where $\mathbf{D}$ interacts with $\mathbf{F}$, with the distributions of the transcript (X^q, Y^q) and of the bit W_q being uniquely defined by their conditional probability distributions. For two $(\mathcal{X}, \mathcal{Y})$-random systems $\mathbf{F}$ and $\mathbf{G}$, the *distinguishing advantage* of $\mathbf{D}$ in distinguishing systems $\mathbf{F}$ and $\mathbf{G}$ by q queries is the quantity $\Delta^{\mathbf{D}}(\mathbf{F}, \mathbf{G}) = |\mathsf{P}^{\mathbf{DF}}_{W_q}(1) - \mathsf{P}^{\mathbf{DG}}_{W_q}(1)|$ and the maximal distinguishing advantage over all distinguishers asking q queries is denoted by $\Delta_q(\mathbf{F}, \mathbf{G}) = \max_{\mathbf{D}} \Delta^{\mathbf{D}}(\mathbf{F}, \mathbf{G})$ (with $\mathbf{D}$ ranging over all such distinguishers).

As opposed to the information-theoretic notion of a distinguisher, we often need to consider an attacker with restricted computational resources. Although such an attacker also participates in a distinguishing experiment, to emphasize this restriction we call it an *adversary* and denote using a sans-serif symbol (e.g. A). Note that a computationally restricted adversary implicitly defines a random system by its input-output behavior and hence any notation defined for information-theoretic distinguishers is also well-defined for such an adversary. We often restrict the computational power of an adversary by its running time, for this we assume some reasonable fixed model of computation.

MONOTONE CONDITIONS. For a random system $\mathbf{F}$, we often consider an internal *monotone condition* defined on it. Such a condition is initially satisfied (true), but once it gets violated, it cannot become true again (hence the name monotone). We use such conditions to capture whether the behavior of the system meets some additional requirement (e.g. distinct outputs, consistent outputs) or this was already violated during the interaction that occurred so far. A monotone condition is formalized by a sequence of events $\mathcal{A} = A_0, A_1, \ldots$ such that A_0 always holds, and A_i holds if the condition holds after answering the i-th query. The probability that a distinguisher $\mathbf{D}$ issuing q queries to $\mathbf{F}$ makes a monotone condition $\mathcal{A}$ fail in the random experiment $\mathbf{DF}$ is denoted by $\nu^{\mathbf{D}}(\mathbf{F}, \overline{A_q}) = \mathsf{P}^{\mathbf{DF}}(\overline{A_q})$ and maximum over all such distinguishers is denoted by $\nu(\mathbf{F}, \overline{A_q}) = \max_{\mathbf{D}} \nu^{\mathbf{D}}(\mathbf{F}, \overline{A_q})$. We also define $\mu(\mathbf{F}, \overline{A_q}) = \max_{x^q} \mathsf{P}^{\mathbf{F}}_{\overline{A_q} \mid X^q}(x^q)$ to be the maximal probability of violating the condition $\mathcal{A}$ by a sequence of q non-adaptive queries.

For a random system $\mathbf{F}$ with a monotone condition $\mathcal{A} = A_0, A_1, \ldots$ and a random system $\mathbf{G}$, we say that $\mathbf{F}$ *conditioned on $\mathcal{A}$ is equivalent to* $\mathbf{G}$, denoted $\mathbf{F}|\mathcal{A} \equiv \mathbf{G}$, if $\mathsf{p}^{\mathbf{F}}_{Y_i \mid X^i Y^{i-1} A_i} = \mathsf{p}^{\mathbf{G}}_{Y_i \mid X^i Y^{i-1}}$ for $i \geq 1$, for all arguments for which

$\mathsf{p}^{\mathbf{F}}_{Y_i|X^iY^{i-1}A_i}$ is defined. Intuitively, this captures the fact that as long as the condition $\mathcal{A}$ holds in $\mathbf{F}$, it behaves the same as $\mathbf{G}$. The following useful claims were given in [20], see also [15] for the proof of claim (ii) and [19] for further discussion.

Lemma 1. *Let $\mathbf{F}$ and $\mathbf{G}$ be random systems, let $\mathcal{A}$ be a monotone condition defined on $\mathbf{F}$, let $\mathbf{D}$ be a distinguisher asking q queries. Then:*

(i) [20, Lemma 7] If $\mathbf{F}|\mathcal{A} \equiv \mathbf{G}$ then $\Delta^{\mathbf{D}}(\mathbf{F}, \mathbf{G}) \leq \nu^{\mathbf{D}}(\mathbf{F}, \overline{A_q})$.
(ii) [20, Theorem 2] If $\mathsf{p}^{\mathbf{F}}_{A_i|X^iY^{i-1}A_{i-1}} = \mathsf{p}^{\mathbf{F}}_{A_i|X^iA_{i-1}}$ for all $i \geq 1$, then $\nu(\mathbf{F}, \overline{A_q}) = \mu(\mathbf{F}, \overline{A_q})$.

2.2   Message Authentication Codes and PRFs

The standard security requirement for a MAC is *unforgeability under chosen-message attack*. However, it is well-known that any PRF attains this property [6], hence in this paper we focus on PRF-security of the analyzed constructions.

If the first component of the input to a function f is to be seen as a key, we sometimes call f a *keyed* function to emphasize this. For a keyed function $f: \mathcal{K} \times \mathcal{D} \to \mathcal{R}$ under a key $k \in \mathcal{K}$ we often write $f_k(\cdot)$ instead of $f(k, \cdot)$. A variable input-length keyed function $\mathsf{G} : \{0,1\}^c \times \{0,1\}^{b*} \to \{0,1\}^c$ is an:

- $(\varepsilon, t, q, \ell)$-*secure PRF*, if for any adversary A running in time t and making at most q queries, each of length at most ℓ (in b-bit blocks), a URF $\mathbf{R}: \{0,1\}^{b*} \to \{0,1\}^c$ and a uniformly random key $K \leftarrow \{0,1\}^c$, we have $\Delta^{\mathsf{A}}(\mathsf{G}_K, \mathbf{R}) \leq \varepsilon$.
- $(\varepsilon, t, q, \ell)$-*NA-secure PRF*, if the above is true for all adversaries A that choose their queries non-adaptively (i.e., A has to choose its q queries before seeing any of the outputs).
- $(\varepsilon, t, q, \ell)$-*PF-secure PRF*, if the above is true for all adversaries A that choose their queries to be prefix-free (i.e., no query is a prefix of another query).
- $(\varepsilon, t, q, \ell)$-*NA-PF-secure PRF*, if the above is true for all adversaries A that choose queries *both* non-adaptively and prefix-free.

For fixed input-length functions, we define analogous notions by omitting the parameter ℓ and distinguishing from $\mathbf{r}$ instead of $\mathbf{R}$. Moreover, we refer to an adversary A as an $(\varepsilon, t, q, \ell)$-PRF adversary against G if it runs in time t, asks at most q queries each consisting of at most ℓ blocks, and achieves the advantage $\Delta^{\mathsf{A}}(\mathsf{G}_K, \mathbf{R}) = \varepsilon$. We refer analogously to adversaries for the other PRF-notions defined above.

For a keyed function $f : \{0,1\}^c \times \{0,1\}^b \to \{0,1\}^c$ we denote with $\mathsf{Casc}^f : \{0,1\}^c \times \{0,1\}^{b*} \to \{0,1\}^c$ the cascade construction (also known as Merkle-Damgård) built from f as $\mathsf{Casc}^f(K, m_1\|\ldots\|m_\ell) := y_\ell$ where $y_0 := K$ and for $i \geq 1$ we have $y_i := f(y_{i-1}, m_i)$, in particular $\mathsf{Casc}^f(K, \lambda) := K$.

The construction $\mathsf{NMAC}^f : (\{0,1\}^c)^2 \times \{0,1\}^{b*} \to \{0,1\}^c$ is derived from Casc^f by adding an additional, independently keyed application of f at the end. It assumes that the domain sizes of f satisfy $b \geq c$ and the output of the cascade is padded with zeroes before the last f-call. Formally,

$$\mathsf{NMAC}^{\mathsf{f}}((K_1, K_2), M) := \mathsf{f}(K_2, \mathsf{Casc}^{\mathsf{f}}(K_1, M) \| 0^{b-c})$$

or $\mathsf{NMAC}^{\mathsf{f}}_{K_1, K_2} := \mathsf{Casc}^{\mathsf{f}}_{K_1} \triangleright \mathsf{f}_{K_2}$. Note that practical MD-based hash functions take as input arbitrary-length bitstrings and then pad them to a multiple of the block length, often including the message length in the so-called MD-strengthening. This padding then also appears in NMAC (and HMAC) but since it does not affect any of our arguments, we take the customary shortcut and our definition above actually corresponds to the generalized construction denoted as GNMAC in [3] where this step is also justified in detail.

HMAC^{f} is a practice-oriented version of NMAC^{f}, where the two keys (K_1, K_2) are derived from a single key $K \in \{0,1\}^b$ by xor-ing it with two fixed b-bit strings ipad and opad. In addition, the keys are not given through the key-input of the compression function f, but are prepended to the message instead. This allows for the usage of existing implementations of hash functions that contain a hard-coded initialization vector IV. Formally:

$$\mathsf{HMAC}^{\mathsf{f}}(K, m) := \mathsf{Casc}^{\mathsf{f}}(\mathsf{IV}, K_2 \| \mathsf{Casc}^{\mathsf{f}}(\mathsf{IV}, K_1 \| m) \| \mathsf{fpad})$$

$$\text{where } (K_1, K_2) := (K \oplus \mathsf{ipad}, K \oplus \mathsf{opad})$$

and fpad is a fixed $(b - c)$-bit padding not affecting the security analysis. (Technically, [18] allows for arbitrary length of the key K: a key shorter than b bits is padded with zeroes before applying the xor transformations, a longer key is first hashed.) As discussed in Section 1.2, we can focus on the PRF-security of NMAC as it translates to analogous results for HMAC under the assumptions stated in [3].

Finally, we also introduce the nested iterated (NI) construction defined in [2]. For this, we consider a keyed compression function $\mathsf{h} \colon \{0,1\}^k \times \{0,1\}^c \times \{0,1\}^b \to \{0,1\}^c$. When such h is used in a cascading construction, its c-bit and b-bit inputs are used for the chaining value and the next block, respectively. In contrast to the function f considered above, h has an additional k-bit input that is used for keying. Formally, for such h we define the *nested iterated* construction $\mathsf{NI}^{\mathsf{h}} \colon (\{0,1\}^k)^2 \times \{0,1\}^{b*} \to \{0,1\}^c$ as

$$\mathsf{NI}^{\mathsf{h}}_{K_1, K_2}(m) := \mathsf{h}_{K_2}(\mathsf{Casc}^{\mathsf{h}_{K_1}}_0(m), |m|)$$

where $\mathbf{0}$ denotes the all zero bitstring 0^c and $|m|$ is the length of m encoded as a b-bit string. Alternatively, for a function $\mathsf{f} \colon \{0,1\}^c \times \{0,1\}^b \to \{0,1\}^c$ and a key K we will denote by $\mathsf{LenCasc}^{\mathsf{f}}_K$ a system that given a message m outputs the pair $(\mathsf{Casc}^{\mathsf{f}}_K(m), |m|)$. This allows us to describe NI equivalently as $\mathsf{NI}^{\mathsf{h}}_{K_1, K_2} := \mathsf{LenCasc}^{\mathsf{h}_{K_1}}_0 \triangleright \mathsf{h}_{K_2}$. For a detailed discussion of the relationship of NI to NMAC, see [2].

3   PRF-Security of NMAC

In this section we analyze the PRF security of NMAC^{f} in terms of the PRF-security of the underlying function f.

3.1   Security Lower Bound

Before moving to the NMAC^f construction, we start by stating a lower bound on the security of the cascade Casc^f when queried on prefix-free inputs. A similar statement has already been proven in [5], and we follow their proof, modifying it where necessary to obtain security against *non-adaptive* adversaries, assuming only *non-adaptive security* of the underlying compression function f. The proof of Proposition 1 is postponed to the full version due to space constraints.

Proposition 1 (Casc^f as a NA-PF-PRF). *Let* $f\colon \{0,1\}^c \times \{0,1\}^b \to \{0,1\}^c$ *be a compression function. There exists an explicit reduction* T *(described in the proof) such that for any* (ε',t',q,ℓ)*-NA-PF-PRF adversary* A *against* Casc^f, T^A *is an* (ε_{na}, t, q)*-NA-PRF adversary against* f *such that*

$$\varepsilon' \leq \ell q \varepsilon_{na} \qquad \text{and} \qquad t = t' + \tilde{O}(\ell q) .$$

This allows us to present our main result in this section, which relates the adaptive PRF-security of the construction NMAC^f to both the adaptive and non-adaptive PRF-security of f.

Theorem 1 (NMAC^f as a PRF). *Let* $f\colon \{0,1\}^c \times \{0,1\}^b \to \{0,1\}^c$ *be a compression function. There exist explicit reductions* T_1 *and* T_2 *(described in the proof) such that for any* (ε',t',q,ℓ)*-PRF adversary* A *against* NMAC^f,

1. T_1^A *is an* (ε,t,q)*-PRF adversary against* f,
2. T_2^A *is an* (ε_{na},t,q)*-NA-PRF adversary against* f,

and their parameters satisfy

$$\varepsilon' \leq \varepsilon + (\ell+1)q\varepsilon_{na} + \frac{q^2}{2^c} \qquad \text{and} \qquad t = t' + \tilde{O}(\ell q) .$$

Proof. Let A be a PRF-adversary running in time t' and asking q queries, each of length at most ℓ blocks. Let $\mathbf{r}\colon \{0,1\}^b \to \{0,1\}^c$, $\mathbf{R}\colon \{0,1\}^{b*} \to \{0,1\}^c$ and $K = (K_1, K_2) \leftarrow \{0,1\}^c \times \{0,1\}^c$ denote a fixed input-length URF, a URF and a key pair chosen independently at random, respectively.

We turn A into an adversary T_1^A against the PRF-security of f_K as follows: Given access to g (which is either f_K or $\mathbf{r}$), sample some key K_1 at random, and then invoke A, answering its queries with $\mathsf{Casc}^f_{K_1} \triangleright g$. Finally, output the decision bit of A. Clearly we have $\Delta^A(\mathsf{Casc}^f_{K_1} \triangleright f_{K_2}, \mathsf{Casc}^f_{K_1} \triangleright \mathbf{r}) = \Delta^{\mathsf{T}_1^A}(f_K, \mathbf{r})$ and if we denote $\Delta^{\mathsf{T}_1^A}(f_K, \mathbf{r})$ by ε then using triangle inequality we get

$$\Delta^A(\mathsf{NMAC}^f_K, \mathbf{R}) = \Delta^A(\mathsf{Casc}^f_{K_1} \triangleright f_{K_2}, \mathbf{R}) \leq \varepsilon + \Delta^A(\mathsf{Casc}^f_{K_1} \triangleright \mathbf{r}, \mathbf{R}) .$$

In the experiment where A interacts with $\mathsf{Casc}^f_{K_1} \triangleright \mathbf{r}$, let C_i denote the event that during the first i queries to $\mathsf{Casc}^f_{K_1} \triangleright \mathbf{r}$, for any two distinct queries M and M' the values $\mathsf{Casc}^f_{K_1}(M)$ and $\mathsf{Casc}^f_{K_1}(M')$ (inputs to the final $\mathbf{r}$-call) are also distinct. As long as the monotone condition $\mathcal{C} = C_0, C_1, \dots$ remains satisfied, the responses of $\mathsf{Casc}_{K_1}^{\mathsf{f}} \rhd \mathbf{r}$ to distinct queries are equivalent to outputs of $\mathbf{r}$ on distinct inputs, and thus independent, uniformly random values, in particular $(\mathsf{Casc}_{K_1}^{\mathsf{f}} \rhd \mathbf{r})|\mathcal{C} \equiv \mathbf{R}$. We can therefore apply Lemma 1(i) to conclude that distinguishing $\mathsf{Casc}^{\mathsf{f}} \rhd \mathbf{r}$ from a URF $\mathbf{R}$ is at least as hard as making the condition $\mathcal{C}$ fail, i.e.,

$$\Delta^{\mathsf{A}}(\mathsf{Casc}_{K_1}^{\mathsf{f}} \rhd \mathbf{r}, \mathbf{R}) \leq \nu^{\mathsf{A}}(\mathsf{Casc}_{K_1}^{\mathsf{f}} \rhd \mathbf{r}, \overline{C_q}) .$$

Below we explain how to use the adversary A to construct[4] a *non-adaptive* adversary $\mathsf{A_{na}}$ such that

$$\nu^{\mathsf{A}}(\mathsf{Casc}_{K_1}^{\mathsf{f}} \rhd \mathbf{r}, \overline{C_q}) = \nu^{\mathsf{A_{na}}}(\mathsf{Casc}_{K_1}^{\mathsf{f}} \rhd \mathbf{r}, \overline{C_q}) . \qquad (1)$$

$\mathsf{A_{na}}$ simply runs A and responds to all its fresh queries by fresh random values, while answering repeated queries consistently. In the end, $\mathsf{A_{na}}$ (non-adaptively) asks all the queries that A asked during this simulated interaction. The equation (1) follows from the fact that the simulation for A is perfect as long as its queries do not violate $\mathcal{C}$. Since $\mathcal{C}$ is defined on $\mathsf{Casc}_{K_1}^{\mathsf{f}}$ and $\mathsf{A_{na}}$ is non-adaptive, we additionally have

$$\nu^{\mathsf{A_{na}}}(\mathsf{Casc}_{K_1}^{\mathsf{f}} \rhd \mathbf{r}, \overline{C_q}) = \nu^{\mathsf{A_{na}}}(\mathsf{Casc}_{K_1}^{\mathsf{f}}, \overline{C_q}) .$$

Next, for $\mathsf{A_{na}}$ we can construct another non-adaptive adversary $\mathsf{A_{pf}}$ that violates the condition $\mathcal{C}$ (i.e., creates a collision in the outputs of $\mathsf{Casc}_{K_1}^{\mathsf{f}}$) with the same probability as $\mathsf{A_{na}}$, but all its queries are *prefix-free*. This can be done, for example, by simply appending an additional block to all queries asked by $\mathsf{A_{na}}$, such that this block does not appear in the original queries. Hence we have

$$\nu^{\mathsf{A_{na}}}(\mathsf{Casc}_{K_1}^{\mathsf{f}}, \overline{C_q}) = \nu^{\mathsf{A_{pf}}}(\mathsf{Casc}_{K_1}^{\mathsf{f}}, \overline{C_q})$$

for a non-adaptive adversary $\mathsf{A_{pf}}$ asking prefix-free queries of length at most $\ell+1$.

Finally, consider the non-adaptive adversary A^* that simply asks the same prefix-free queries as $\mathsf{A_{pf}}$ and then outputs 1 if and only if the responses to these queries contain a collision. Then A^* interacting with $\mathsf{Casc}_{K_1}^{\mathsf{f}}$ outputs 1 with probability $\nu^{\mathsf{A_{pf}}}(\mathsf{Casc}_{K_1}^{\mathsf{f}}, \overline{C_q})$, while in an interaction with $\mathbf{R}$ it outputs 1 with probability at most $q^2/2^c$ via the well-known birthday bound. Hence, by the definition of $\Delta^{\mathsf{A}^*}(\mathsf{Casc}_{K_1}^{\mathsf{f}}, \mathbf{R})$, we have

$$\nu^{\mathsf{A_{pf}}}(\mathsf{Casc}_{K_1}^{\mathsf{f}}, \overline{C_q}) \leq \Delta^{\mathsf{A}^*}(\mathsf{Casc}_{K_1}^{\mathsf{f}}, \mathbf{R}) + \frac{q^2}{2^c} .$$

Since A^* is non-adaptive and prefix-free, we can now employ the reduction T guaranteed by Proposition 1 to obtain an NA-PRF adversary $\mathsf{T}^{\mathsf{A}^*}$ against f such that

$$\Delta^{\mathsf{A}^*}(\mathsf{Casc}_{K_1}^{\mathsf{f}}, \mathbf{R}) \leq (\ell+1)q \cdot \Delta^{\mathsf{T}^{\mathsf{A}^*}}(\mathsf{f}, \mathbf{r}) .$$

Putting $\mathsf{T}_2^{\mathsf{A}} := \mathsf{T}^{\mathsf{A}^*}$ hence concludes the proof of Theorem 1. $\qquad \square$

---

[4] One could use a lemma from the random system framework [20] in the spirit of Lemma 1(ii) to switch to non-adaptivity. We prefer to spell out the actual construction to emphasize the uniformity of our reduction.

Corollary 1. *If* $f\colon \{0,1\}^c \times \{0,1\}^b \to \{0,1\}^c$ *is an* (ε, t, q)-*secure PRF and an* (ε_{na}, t, q)-*NA-secure PRF, then* NMAC^f *is an* $(\varepsilon', t', q, \ell)$-*secure PRF with*

$$\varepsilon' = \varepsilon + (\ell+1)q\varepsilon_{na} + \frac{q^2}{2^c} \qquad \text{and} \qquad t = t' + \tilde{O}(\ell q) \,.$$

3.2   Matching Attacks

We now argue that the bound obtained in Theorem 1 is essentially tight. First, we show that the term $\ell q \varepsilon_{na}$ is unavoidable (up to a constant factor) by constructing a particular compression function f, which is an (ε_{na}, t, q)-NA-secure PRF, yet there is a simple attack against the PRF-security of NMAC^f achieving advantage roughly $\ell q \varepsilon_{na}$.

Proposition 2. *Let* b, c, ℓ *be positive integers such that* $b \geq c$, *let* $\varepsilon_{na} \in (0,1)$, *and moreover, assume that pseudo-random functions exist. Then there exists a function* $f\colon \{0,1\}^c \times \{0,1\}^b \to \{0,1\}^c$ *and an adversary* A *against* NMAC^f *such that for any* q *that satisfies* $\varepsilon_{na} = \omega(q^2 2^{-b}, 2^{-c})$, *we have:*

- f *is* (ε_{na}, t, q)-*NA-secure PRF;*
- *the adversary* A, *when asking* q *queries of length* ℓ *blocks each, runs in time* $\tilde{O}(\ell q)$ *and achieves distinguishing advantage*

$$\Delta^A(\mathsf{NMAC}^f_K, \mathbf{R}) = \Theta(\ell q \varepsilon_{na}) \,.$$

In particular, NMAC^f *is not an* $(o(\ell q \varepsilon_{na}), \tilde{O}(\ell q), q, \ell)$-*secure PRF.*

Proof (sketch). Here we only describe the high-level idea for constructing f and A and defer the discussion of the technical obstacles in implementing this idea to the full version.

Roughly speaking, we construct an (ε_{na}, t, q)-NA-secure PRF f that behaves pseudo-randomly for all keys except for a small, $\varepsilon_{na}/2$-fraction of them. We denote the set of these keys by $\mathcal{K}$ and refer to them as the *weak keys*. Under any weak key k, the function $f(k, \cdot)$ outputs some constant value $w \in \mathcal{K}$ irrespective of its input.

To attack the NA-PRF security of $\mathsf{NMAC}^f_{K=(K_1,K_2)}$, consider a pair of messages M_1, M_2 chosen by sampling $M \leftarrow \{0,1\}^{b(\ell-1)}$ at random and then setting $M_1 = M\|x_1$ and $M_2 = M\|x_2$ for some distinct blocks $x_1, x_2 \in \{0,1\}^b$. If some of the $\ell-1$ intermediate values in the evaluation of the inner function $\mathsf{Casc}^f(K_1, M)$ is in $\mathcal{K}$, then all following intermediate values are w, and in particular we have $\mathsf{Casc}^f(K_1, M_i) = w$ for both $i \in \{1,2\}$, and hence also $\mathsf{NMAC}^f(K, M_1) = \mathsf{NMAC}^f(K, M_2) = f_{K_2}(w)$. This implies that it is much more likely to get a collision for a pair of messages as described above for NMAC^f_K than for $\mathbf{R}$. Our adversary A simply choses $q/2$ message pairs at random as above, and it outputs 1 if it observes a collision for at least one of those pairs. As there are $q/2$ message pairs, each of length ℓ, we have a total of $\ell q/2$ possibilities to "hit" a weak key, each having probability ε_{na}. By the union bound this gives us a total probability of $\Theta(\ell q \varepsilon_{na})$ for observing a collision when querying NMAC^f_K. On the other hand the probability of observing a colliding pair in $\mathbf{R}$ is only $O(q/2^c)$.   □

We now consider the tightness of the bound in Theorem 1 when $\varepsilon \gg \ell q \varepsilon_{na}$ is the dominating term. This is the case when the best adaptive attack against f is by more than a factor ℓq better than any non-adaptive attack.

In [23] a pair $\mathsf{g}_1, \mathsf{g}_2$ of PRFs is constructed such that g_1 and g_2 are ε_{na}-secure *non-adaptive* PRFs for some negligible ε_{na}, and the serial composition $\mathsf{g}_1 \triangleright \mathsf{g}_2$ with independent keys can be broken by an *adaptive* attack (in a constant number of queries) with advantage almost 1.[5] From such $\mathsf{g}_1, \mathsf{g}_2$ we can get a single PRF f which is an ε_{na}-secure NA-PRF for a negligible ε_{na}, an ε-secure PRF for any ε of our choice, and where $f \triangleright f$ is not $\Theta(\varepsilon^2)$-secure, by setting $f := \mathsf{g}_1$ and $f := \mathsf{g}_2$ with probability $\varepsilon/2$, respectively, and some strong standard PRF with probability $1 - \varepsilon$ (over the choice of the key). We now observe that NMAC^f_K computed on single-block messages is simply a cascade of two f's with independent keys. Thus, when using the above ε-secure PRF f, we can break NMAC^f_K with advantage $\Theta(\varepsilon^2)$. This shows that the ε term in Theorem 1 is necessary if ε is constant as then $\Theta(\varepsilon) = \Theta(\varepsilon^2) = \Theta(1)$. We conjecture that $\Theta(\varepsilon^2)$ is the correct value, and the ε term in the lower bound can be improved to $\Theta(\varepsilon^2)$ using security amplification techniques along the lines of [22,25].

4 PRF-Security of the NI Construction

In this section we analyze the PRF-security of the NI^h construction under the assumption that the keyed compression function h is a PRF (when keyed via its k-bit input).

Theorem 2. *If* $\mathsf{h}: \{0,1\}^k \times \{0,1\}^c \times \{0,1\}^b \to \{0,1\}^c$ *is an* (ε_1, t, q)-*secure PRF and an* $(\varepsilon_2, t, \ell q)$-*secure PRF, then* NI^h *is an* $(\varepsilon', t', q, \ell)$-*secure PRF with*

$$\varepsilon' = \varepsilon_1 + \varepsilon_2 + \frac{q^2}{2^c} \cdot \left(\ell + \frac{64\ell^4}{2^c} \right) \qquad \text{and} \qquad t = t' + \tilde{O}(\ell q) \ .$$

Proof. We prove Theorem 2 in four consecutive steps. First, we use the PRF-security of h to replace it by an ideal compression function, making the rest of our analysis information-theoretic. Second, we observe that the resulting system behaves identically to $\mathbf{R}$ as long as no non-trivial collision occurs in the outputs of the initial cascade. Third, we reduce estimating the probability of such a collision to a counting problem of upper-bounding the number of graphs satisfying certain properties (modeling the computation of the cascade). Finally, we give a bound on the number of these graphs, hence concluding the argument.

---

[5] The NA-PRF security of this construction relies on the DDH assumption, [9] construct such a PRF under the weaker assumption that "uniform transcript key-agreement" exists, and this assumption is necessary [24].

From a PRF to a Random Function. Let A be a PRF-adversary against NI_K^h running in time t and asking q queries, each of length at most ℓ blocks. To simplify the notation let $\mathbf{0} := 0^c$. By a standard argument as in the proof of Theorem 1, we have

$$\Delta^A(NI_K^h, \mathbf{R}) = \Delta^A\left(\mathsf{LenCasc}_0^{h_{K_1}} \triangleright h_{K_2}, \mathbf{R}\right) \leq \varepsilon_1 + \varepsilon_2 + \Delta^A\left(\mathsf{LenCasc}_0^{f_1} \triangleright f_2, \mathbf{R}\right) \tag{2}$$

where $K = (K_1, K_2) \leftarrow (\{0,1\}^k)^2$ is a uniformly random key and f_1 and f_2 are two independent ideal compression functions. Interestingly, the system $\mathsf{LenCasc}_0^{f_1} \triangleright f_2$ is very similar to NMAC with an ideal compression function and $(\mathbf{0}, |m|)$ being used instead of the key pair.

Bound via Collision Probability. Let $\mathsf{CColl}(\ell)$ denote the probability that a random choice of the compression function f_1 results in a collision in $\mathsf{Casc}_0^{f_1}$, maximized over the choice of the two distinct, equal-length inputs m_1, m_2 consisting of at most ℓ blocks each. (Note that we require length equality $|m_1| = |m_2|$ to obtain a collision also for $\mathsf{LenCasc}_0^{f_1}$.) Formally, for uniformly random $f_1 \leftarrow \mathcal{F}(c+b, c)$ we define

$$\mathsf{CColl}(\ell) := \max_{\substack{m_1 \neq m_2 \\ |m_1| = |m_2| \leq \ell b}} \mathsf{P}^{f_1}\left[\mathsf{Casc}_0^{f_1}(m_1) = \mathsf{Casc}_0^{f_1}(m_2)\right] . \tag{3}$$

In the experiment where A interacts with $\mathsf{LenCasc}_0^{f_1} \triangleright f_2$, let E_i denote the event that during the first i queries to $\mathsf{LenCasc}_0^{f_1} \triangleright f_2$, for any two distinct queries M and M' the values $\mathsf{LenCasc}_0^{f_1}(M)$ and $\mathsf{LenCasc}_0^{f_1}(M')$ (inputs to the final f_2-call) were also distinct. As long as the monotone condition $\mathcal{E} = E_0, E_1, \ldots$ remains satisfied, the responses of $\mathsf{LenCasc}_0^{f_1} \triangleright f_2$ to distinct queries are clearly independent, uniformly random values thanks to f_2. Hence, we have $(\mathsf{LenCasc}_0^{f_1} \triangleright f_2)|\mathcal{E} \equiv \mathbf{R}$ and $\mathsf{p}_{E_i|X^iY^{i-1}E_{i-1}}^{\mathsf{LenCasc}_0^{f_1} \triangleright f_2} = \mathsf{p}_{E_i|X^iE_{i-1}}^{\mathsf{LenCasc}_0^{f_1} \triangleright f_2}$ and can therefore consecutively apply Lemma 1(i), Lemma 1(ii), and finally the union bound to get

$$\Delta^A(\mathsf{LenCasc}_0^{f_1} \triangleright f_2, \mathbf{R}) \leq \nu(\mathsf{LenCasc}_0^{f_1} \triangleright f_2, \overline{E_q}) \leq \mu(\mathsf{LenCasc}_0^{f_1} \triangleright f_2, \overline{E_q}) \leq q^2 \cdot \mathsf{CColl}(\ell) . \tag{4}$$

Graph-Based Representation of Casc. The probability $\mathsf{CColl}(\ell)$ could trivially be upper-bounded by $O(\ell^2/2^c)$ using a union-bound argument, achieving a non-trivial and significantly better bound on $\mathsf{CColl}(\ell)$ is the central part of our proof. To this end, we use an approach inspired by [7] and represent the computation of $\mathsf{Casc}_0^{f_1}$ on various inputs by directed graphs.

Let m_1 and m_2 be two distinct, equal-length messages that can be parsed into b-bit blocks as $m_i = m_i^1 \| \cdots \| m_i^{\ell'}$ for some $\ell' \leq \ell$, and let $\Lambda := 2\ell'$. For convenience, we use the notation $m^{(i)}$ as a reference to the block m_1^i if $i \leq \ell'$, otherwise it denotes the block $m_2^{i-\ell'}$. For any fixed compression function $f \in \mathcal{F}(c+b, c)$ and a pair of such messages $\mathcal{M} = (m_1, m_2)$, we define the *structure graph* $G_f^{\mathcal{M}}$ to be the triple $G_f^{\mathcal{M}} = (\mathcal{V}, \mathcal{E}, \mathcal{L})$, such that:

– $(\mathcal{V}, \mathcal{E})$ is a directed graph. To describe it, let

$$
s_i := \begin{cases}
\mathbf{0} & \text{for } i = 0 \\
f(s_{i-1}, m_1^i) & \text{for } 1 \leq i \leq \ell' \\
f(\mathbf{0}, m_2^1) & \text{for } i = \ell' + 1 \\
f(s_{i-1}, m_2^{i-\ell'}) & \text{for } \ell' + 2 \leq i \leq \Lambda
\end{cases}
\tag{5}
$$

and consider the mappings $[\cdot]_G$ and $[\cdot]_G'$ defined on $\{0, \ldots, \Lambda\}$ such that $[i]_G := \min\{j : s_i = s_j\}$ (so $[i]_G = i$ if and only if s_i is "fresh") and $[i]_G' := [i]_G$ for $i \neq \ell'$, while $[\ell']_G' := 0$. Now we let

$$
\mathcal{V} := \{[i]_G : 0 \leq i \leq \Lambda\} \quad \text{and} \quad \mathcal{E} := \{([i-1]_G', [i]_G) : 1 \leq i \leq \Lambda\} .
$$

– $\mathcal{L} : \mathcal{V}^2 \to \mathsf{Pow}(\{0,1\}^b)$ is a labeling function that labels every edge $(u, v) \in \mathcal{E}$ with the set $\{m^{(i)} : [i-1]_G' = u \wedge [i]_G = v\}$ and every pair of vertices that do not form an edge with the empty set $\emptyset$ (to simplify our notation later).

Intuitively, if all the values s_i are distinct, $G_f^{\mathcal{M}}$ simply consists of two directed paths starting in the root vertex 0, representing the evaluation of $\mathsf{Casc}_0^{\mathsf{f}_1}$ on the messages m_1 and m_2 (the edges are labeled by the corresponding blocks). If some collisions among the values s_i occur, one can obtain the graph $G_f^{\mathcal{M}}$ by collapsing every pair of vertices i, j where $s_i = s_j$ into one vertex labeled $\min\{i, j\}$, as well as merging the edge labels in the natural way.

Let $\mathcal{G}(\mathcal{M}) := \{G_f^{\mathcal{M}} : f \in \mathcal{F}(c+b, c)\}$ denote the set of all structure graphs associated with the message pair $\mathcal{M}$. Note that the uniformly distributed random variable $F \leftarrow \mathcal{F}(c+b, c)$ also induces a distribution on $\mathcal{G}(\mathcal{M})$, therefore we denote by $G_F^{\mathcal{M}}$ the resulting random variable (taking on structure graphs as values). Similarly, F also induces a distribution on the values s_i defined above and we denote the resulting random variables S_i.

For a fixed structure graph $G = G_f^{\mathcal{M}}$ we denote by $G_i = (\mathcal{V}_i, \mathcal{E}_i, \mathcal{L}_i)$ the graph that is obtained after processing only the first i out of Λ blocks of $\mathcal{M}$. More formally, $G_i := G_f^{\mathcal{M}'}$ where $\mathcal{M}' := (m_1^1 \| \cdots \| m_1^i, \lambda)$ if $i \leq \ell'$ and $\mathcal{M}' := (m_1, m_2^1 \| \cdots \| m_2^{i-\ell'})$ otherwise. Building on this notion, we call $\mathsf{fColl}(G)$ the *set of f-collisions* that occurred in G:

$$
\mathsf{fColl}(G) := \left\{ (i, [i]_G) : [i]_G < i \wedge m^{(i)} \notin \mathcal{L}_{i-1}([i-1]_G', [i]_G) \right\} .
\tag{6}
$$

Informally, imagine we reveal the structure graph G step by step, i.e., by a sequence of transitions from G_{i-1} to G_i, for $i = 1, \ldots, \Lambda$. The pair $(i, [i]_G)$ belongs to $\mathsf{fColl}(G)$ (and we say that the i-th step caused an f-collision), if during this step, instead of adding a new vertex, we arrive at a vertex already visited, while not following an existing edge already labeled with $m^{(i)}$ (i.e., not repeating a step we have made before).

PROPERTIES OF STRUCTURE GRAPHS. We first upper-bound the probability of $G_F^{\mathcal{M}}$ taking the form of any particular fixed structure graph $g \in \mathcal{G}(\mathcal{M})$. The following result is inspired by Lemma 8 from [7]. Due to space constraints, we postpone the proofs of all technical lemmas below to the full version of this paper.

Lemma 2. *Let* $F \leftarrow \mathcal{F}(c + b, c)$ *be chosen uniformly at random. For a fixed graph* $g \in \mathcal{G}(\mathcal{M})$ *we have*

$$\mathsf{P}^F \left[G_F^{\mathcal{M}} = g \right] \leq 2^{-c \cdot |\mathsf{fColl}(g)|} .$$

Using Lemma 2, it is easy to see that the event that at least two f-collisions occur in G is highly unlikely.

Lemma 3. *Let* $F \leftarrow \mathcal{F}(c + b, c)$ *be chosen uniformly at random. Then*

$$\mathsf{P}^F \left[|\mathsf{fColl} \left(G_F^{\mathcal{M}} \right)| \geq 2 \right] \leq \frac{4\Lambda^4}{2^{2c}} .$$

FROM COLLISION PROBABILITY TO COUNTING GRAPHS. We can now proceed to upper-bounding the value $\mathsf{CColl}(\ell)$. Let $\mathcal{M} := (m_1, m_2)$ be the two distinct, equal-length messages of length at most ℓ blocks that maximize the probability $\mathsf{CColl}(\ell) := \max_{m_1 \neq m_2} \mathsf{P}^F \left[\mathsf{Casc}_0^F(m_1) = \mathsf{Casc}_0^F(m_2) \right]$. For $j \in \{1, 2\}$ let V_j^i be the random variable denoting the i-th vertex (counting from 0) in the path corresponding to m_j in $G_F^{\mathcal{M}}$ (randomness taken over the uniform choice of F). Formally, $V_1^i := [i]_G$ and $V_2^i := [\ell' + i]_G$. Using this notation, we have $\mathsf{CColl}(\ell) = \mathsf{P}[V_1^{\ell'} = V_2^{\ell'}]$. Since $m_1 \neq m_2$, $V_1^{\ell'} = V_2^{\ell'}$ cannot occur without any f-collision, hence we can split $\mathsf{CColl}(\ell)$ into

$$\mathsf{P} \left[V_1^{\ell'} = V_2^{\ell'} \wedge |\mathsf{fColl}(G_F^{\mathcal{M}})| = 1 \right] + \mathsf{P} \left[V_1^{\ell'} = V_2^{\ell'} \wedge |\mathsf{fColl}(G_F^{\mathcal{M}})| \geq 2 \right] . \quad (7)$$

The latter probability can be readily upper-bounded by $4\Lambda^4/2^{2c}$ using Lemma 3. As for the former, let us denote by $\mathcal{H}(\mathcal{M})$ the set of structure graphs for $\mathcal{M}$ that contain exactly one f-collision and where the vertices $V_1^{\ell'}$ and $V_2^{\ell'}$ coincide. The first term in (7) can then be upper-bounded by $|\mathcal{H}(\mathcal{M})|/2^c$ using Lemma 2, hence it remains to bound the size of the set $\mathcal{H}(\mathcal{M})$.

COUNTING THE STRUCTURE GRAPHS. We give such a bound in the following lemma, proven in the full version of this paper.

Lemma 4. *For two distinct, equal-length messages* $\mathcal{M} = \{m_1, m_2\}$ *each of length at most* ℓ *blocks, we have* $|\mathcal{H}(\mathcal{M})| \leq \ell$.

Finally, combining the equations (2), (4), (7), and the bounds obtained in Lemma 3 and Lemma 4, we get

$$\Delta^A(\mathsf{NI}_K^h, \mathbf{R}) \leq \varepsilon_1 + \varepsilon_2 + q^2 \cdot \left(\frac{\ell}{2^c} + \frac{4\Lambda^4}{2^{2c}} \right) \leq \varepsilon_1 + \varepsilon_2 + \frac{q^2}{2^c} \cdot \left(\ell + \frac{64\ell^4}{2^c} \right)$$

and conclude the proof of Theorem 2. □

In the full version we also show that Lemma 4 is tight, and discuss the implications for the tightness of Theorem 2. Moreover, we show a generalization of Lemma 4 that does not require the messages in $\mathcal{M}$ to have the same length, in which case we prove $|\mathcal{H}(\mathcal{M})| \leq \ell d'(\ell)$. This translates directly into a PRF-security statement for a variant of NI that does not include the message length in its last h-call, giving a bound that is equivalent to Theorem 2 except for the term $\ell q^2/2^c$ that is replaced by $\ell d'(\ell)q^2/2^c$.

Acknowledgements. We thank the anonymous reviewers for useful comments and suggestions. This work was partly funded by the European Research Council under an ERC Starting Grant (259668-PSPC).

References

1. Alon, N., Goldreich, O., Håstad, J., Peralta, R.: Simple construction of almost k-wise independent random variables. Random Struct. Algorithms 3(3), 289–304 (1992)
2. An, J.H., Bellare, M.: Constructing VIL-mACs from FIL-mACs: Message authentication under weakened assumptions. In: Wiener, M. (ed.) CRYPTO 1999. LNCS, vol. 1666, pp. 252–269. Springer, Heidelberg (1999)
3. Bellare, M.: New proofs for NMAC and HMAC: Security without collision-resistance. In: Dwork, C. (ed.) CRYPTO 2006. LNCS, vol. 4117, pp. 602–619. Springer, Heidelberg (2006)
4. Bellare, M., Canetti, R., Krawczyk, H.: Keying hash functions for message authentication. In: Koblitz, N. (ed.) CRYPTO 1996. LNCS, vol. 1109, pp. 1–15. Springer, Heidelberg (1996)
5. Bellare, M., Canetti, R., Krawczyk, H.: Pseudorandom functions revisited: The cascade construction and its concrete security. In: 37th Annual Symposium on Foundations of Computer Science, pp. 514–523. IEEE Computer Society Press (1996)
6. Bellare, M., Kilian, J., Rogaway, P.: The security of the cipher block chaining message authentication code. Journal of Computer and System Sciences 61(3), 362–399 (2000)
7. Bellare, M., Pietrzak, K., Rogaway, P.: Improved security analyses for CBC MACs. In: Shoup, V. (ed.) CRYPTO 2005. LNCS, vol. 3621, pp. 527–545. Springer, Heidelberg (2005)
8. Bellare, M., Rogaway, P.: The security of triple encryption and a framework for code-based game-playing proofs. In: Vaudenay, S. (ed.) EUROCRYPT 2006. LNCS, vol. 4004, pp. 409–426. Springer, Heidelberg (2006)
9. Cho, C., Lee, C.-K., Ostrovsky, R.: Equivalence of uniform key agreement and composition insecurity. In: Rabin, T. (ed.) CRYPTO 2010. LNCS, vol. 6223, pp. 447–464. Springer, Heidelberg (2010)
10. Coron, J.-S., Dodis, Y., Malinaud, C., Puniya, P.: Merkle-damgård revisited: How to construct a hash function. In: Shoup, V. (ed.) CRYPTO 2005. LNCS, vol. 3621, pp. 430–448. Springer, Heidelberg (2005)
11. De, A., Trevisan, L., Tulsiani, M.: Time space tradeoffs for attacks against one-way functions and PRGs. In: Rabin, T. (ed.) CRYPTO 2010. LNCS, vol. 6223, pp. 649–665. Springer, Heidelberg (2010)
12. Dodis, Y., Ristenpart, T., Steinberger, J., Tessaro, S.: To hash or not to hash again (In)Differentiability results for h^2 and HMAC. In: Safavi-Naini, R., Canetti, R. (eds.) CRYPTO 2012. LNCS, vol. 7417, pp. 348–366. Springer, Heidelberg (2012)
13. Hardy, G.H., Wright, E.M.: An Introduction to the Theory of Numbers, 6th edn. Oxford University Press, USA (2008)
14. Hellman, M.E.: A cryptanalytic time-memory trade-off. IEEE Transactions on Information Theory 26(4), 401–406 (1980)
15. Jetchev, D., Özen, O., Stam, M.: Understanding adaptivity: Random systems revisited. In: Wang, X., Sako, K. (eds.) ASIACRYPT 2012. LNCS, vol. 7658, pp. 313–330. Springer, Heidelberg (2012)

16. Kim, J., Biryukov, A., Preneel, B., Hong, S.: On the security of HMAC and NMAC based on HAVAL, MD4, MD5, SHA-0 and SHA-1 (Extended abstract). In: De Prisco, R., Yung, M. (eds.) SCN 2006. LNCS, vol. 4116, pp. 242–256. Springer, Heidelberg (2006)

17. Koblitz, N., Menezes, A.: Another look at HMAC. Cryptology ePrint Archive, Report 2012/074 (2012)

18. Krawczyk, H., Bellare, M., Canetti, R.: HMAC: Keyed-Hashing for Message Authentication. IETF Internet Request for Comments 2104 (February 1997)

19. Maurer, U.: Conditional equivalence of random systems and indistinguishability proofs. In: 2013 IEEE International Symposium on Information Theory Proceedings (ISIT), pp. 3150–3154 (July 2013)

20. Maurer, U.: Indistinguishability of random systems. In: Knudsen, L.R. (ed.) EUROCRYPT 2002. LNCS, vol. 2332, pp. 110–132. Springer, Heidelberg (2002)

21. Maurer, U., Renner, R., Holenstein, C.: Indifferentiability, impossibility results on reductions, and applications to the random oracle methodology. In: Naor, M. (ed.) TCC 2004. LNCS, vol. 2951, pp. 21–39. Springer, Heidelberg (2004)

22. Maurer, U., Tessaro, S.: Computational indistinguishability amplification: Tight product theorems for system composition. In: Halevi, S. (ed.) CRYPTO 2009. LNCS, vol. 5677, pp. 355–373. Springer, Heidelberg (2009)

23. Pietrzak, K.: Composition does not imply adaptive security. In: Shoup, V. (ed.) CRYPTO 2005. LNCS, vol. 3621, pp. 55–65. Springer, Heidelberg (2005)

24. Pietrzak, K.: Composition implies adaptive security in minicrypt. In: Vaudenay, S. (ed.) EUROCRYPT 2006. LNCS, vol. 4004, pp. 328–338. Springer, Heidelberg (2006)

25. Tessaro, S.: Security amplification for the cascade of arbitrarily weak PRPs: Tight bounds via the interactive hardcore lemma. In: Ishai, Y. (ed.) TCC 2011. LNCS, vol. 6597, pp. 37–54. Springer, Heidelberg (2011)

26. Wang, X., Yin, Y.L., Yu, H.: Finding collisions in the full SHA-1. In: Shoup, V. (ed.) CRYPTO 2005. LNCS, vol. 3621, pp. 17–36. Springer, Heidelberg (2005)

27. Wang, X., Yu, H.: How to break MD5 and other hash functions. In: Cramer, R. (ed.) EUROCRYPT 2005. LNCS, vol. 3494, pp. 19–35. Springer, Heidelberg (2005)

Updates on Generic Attacks
against HMAC and NMAC

Jian Guo[1], Thomas Peyrin[1], Yu Sasaki[2], and Lei Wang[1]

[1] Nanyang Technological University, Singapore
{guojian,thomas.peyrin,wang.lei}@ntu.edu.sg
[2] NTT Secure Platform Laboratories, Japan
sasaki.yu@lab.ntt.co.jp

Abstract. In this paper, we present new generic attacks against HMAC and other similar MACs when instantiated with an n-bit output hash function maintaining a l-bit internal state. Firstly, we describe two types of selective forgery attacks (a forgery for which the adversary commits on the forged message beforehand). The first type is a tight attack which requires $O(2^{l/2})$ computations, while the second one requires $O(2^{2l/3})$ computations, but offers much more freedom degrees in the choice of the committed message. Secondly, we propose an improved universal forgery attack which significantly reduces the complexity of the best known attack from $O(2^{5l/6})$ to $O(2^{3l/4})$. Finally, we describe the very first time-memory tradeoff for key recovery attack on HMAC. With $O(2^l)$ precomputation, the internal key K_{out} is firstly recovered with $O(2^{2l/3})$ computations by exploiting the Hellman's time-memory tradeoff, and then the other internal key K_{in} is recovered with $O(2^{3l/4})$ computations by a novel approach. This tends to indicate an inefficiency in using long keys for HMAC.

Keywords: HMAC, NMAC, selective forgery, universal forgery, key recovery.

1 Introduction

A message authentication code (MAC) ensures the integrity of messages trans-ferred between two parties sharing a secret key K in advance. When the sender would like to send a message $\mathcal{M}$, he first generates the tag T computed by $T = \text{MAC}(K, \mathcal{M})$, and then sends the pair $(\mathcal{M}, T)$ to the other party. The re-ceiver computes the tag value with the received message using his own key K and checks if this value matches the received tag value T. If they do match, he knows that the message $\mathcal{M}$ received was indeed sent by the other party.

A classical method to build a MAC algorithm is to use a hash function, and the well-known example is HMAC [1] designed by Bellare *et al.*, which has been standardized by ANSI, IETF, ISO and NIST, and is widely implemented in various security protocols such as SSL/TLS and IPSec.

There are several security requirements one expects a secure MAC to verify. Informally, it should resist key recovery attacks, any type of forgery attacks, as

J.A. Garay and R. Gennaro (Eds.): CRYPTO 2014, Part I, LNCS 8616, pp. 131–148, 2014.

well as any distinguishing attacks. We note that key recovery and forgeries are arguably the most important as they have the greatest impact in practice. We provide below definitions of the attacks which are related to this paper. It is assumed that the adversary can interact with an oracle that outputs the valid tag $T = \texttt{MAC}(K, \mathcal{M})$ when queried with a message $\mathcal{M}$.

Key recovery: the adversary recovers the secret-key K used in the MAC algorithm.

Selective forgery: the adversary first commits on a message $\mathcal{M}$ before interacting with the oracle, and then builds a valid pair $(\mathcal{M}, T)$, without having queried $\mathcal{M}$.

Universal forgery: the adversary first receives a message $\mathcal{M}$ sent as challenge, and then builds a valid pair $(\mathcal{M}, T)$, without having queried $\mathcal{M}$.

The security of a MAC construction is discussed in terms of lower-bound and upper-bound on the complexity of the attack for each notion. Regarding the lower-bound, many hash based MACs including HMAC are proven to be indistinguishable from a PRF (pseudo-random function) up to $O(2^{l/2})$ queries, where l is the internal state size of the underlying hash function with n-bit hash digests.

On the other hand, the upper-bound on the complexity is shown by demonstrating a generic attack for each notion. Concerning the notions existential forgery and the distinguishing-R, Preneel *et al.* proposed a tight generic attack, *i.e.*, the attack complexity matches the proven lower-bound [17]. Their method is based on an internal collision generated with a birthday complexity of $O(2^{l/2})$. Following a similar collision-detection based approach, Naito *et al.* proposed a distinguishing-H attack attack with a complexity of $O(2^l/l)$ [14]. Although this approach is powerful, its direct application to other notions, particularly the above three defined notions, seems difficult.

Recently, cryptographers have proposed new attack approaches by studying the cycle property of functional graphs and by studying entropy loss of sequential iterations, and applied them to find new generic attacks on hash based MACs [15,4,12,16], as well as dedicated attacks on instances of specific designs [8,7]. Interestingly, these approaches have even been used to analyze the notions selective forgery and universal forgery. In [16], Peyrin and Wang showed a universal forgery with a complexity of $O(2^{5l/6})$, which is based on cycle property of functional graph. At the same time with our paper, in [3], Dinur and Leurent found another universal forgery with a complexity of $O(2^{6l/7})$ but with shorter queries and thus with wider applications (Peyrin and Wang's attack inherently needs to use queries of $O(2^{l/2})$ blocks long), which is based on collision entropy loss of iterations. We note that selective forgery is a weaker security notion than universal forgery (an attacker can use a universal forgery producing oracle to generate selective forgeries), and hence these universal forgery attacks can be directly used to obtain a selective forgery with the same complexity.

As we can see, the current best known universal forgery and selective forgery attacks on hash-based MAC [16] are not tight and it remains an open problem if attacks and/or proofs can be improved. Moreover, as of today, key recovery remains the only security notion for which no generic attack was proposed on hash-based MAC.

Our Contributions. In this article, we present improved selective forgery attacks against HMAC, NMAC, and other similar MACs, as well as an improved universal forgery attack and the very first time-memory tradeoff for key recovery attack.

More precisely, we first describe two types of selective forgery attacks. The first type offers rather limited choice to the adversary regarding the committed message and this message must consist of at least $O(2^{l/2})$ blocks, but the overall complexity is only $O(2^{l/2})$ computations, which matches the proven lower-bound for hash-based MAC algorithms. On the other hand, the second type permits a much broader choice of committed message (the scenario is actually quite close to a universal forgery attack), and its complexity depends on the block length of the committed message. Particularly, the committed message must consist of at least $O(2^{l/3})$ blocks in order to obtain the optimal complexity of $O(l \cdot 2^{2l/3})$ computations. Giving an example with the specifications of widely used hash functions, such as SHA-1 or SHA-256 [20], the adversary can freely choose the committed message, except about 12.5% of it. The former type is a direct application of the distinguishing-H attacks from [12], while the latter is obtained by devising an expandable message technique in the keyed scenario, which was originally proposed by Kelsey and Schneier for keyless hash functions [10]. The obvious main difficulty in the keyed scenario is that the adversary cannot access the internal state values anymore.

Secondly, we improve the complexity of the best known universal forgery attack for hash-based MAC algorithms, which is reduced to $O(\max(2^{l-s}, 2^{3l/4}, 2^s))$ for a challenge message composed of 2^s blocks. Roughly, the complexity has been significantly reduced from $O(2^{5l/6})$ to $O(2^{3l/4})$. Previous universal forgery attacks [16] are based on the analysis of nodes' *height* in the MAC functional graph, the height of a node x being the number of nodes linking x to the cycle of its own component in the functional graph. The basic principle of this attack is that the adversary will first collect offline many values and their exact height in the MAC functional graph, and then use this information to perform the forgery. Unfortunately, the authors failed to estimate the height distributions for the nodes collected offline, which essentially prevents the attack complexity to go below $2^{5l/6}$ computations. In order to overcome this, we performed experiments in order to investigate these height distributions, and we finally observed a very interesting property. This observation remains a conjecture as of today, but confidence in its validity is backed up by our experiments. Based on this conjecture, we managed to improve the universal forgery attack.

Lastly, we propose the first time-memory tradeoff for key recovery attack against HMAC and NMAC. Before discussing our attacks, the key size of HMAC and NMAC needs to be specified. For NMAC instantiated with a l-bit internal state hash function, the key size is l bits for the first key and l bits for the second key, for a total key size of $2l$ bits. HMAC is defined to accept a secret key of an arbitrary size. If the key size is longer than the block size, the key is first hashed by using the underlying hash function, and then the corresponding digest is used as the key. As later explained in Section 2, keys longer than n bits are quite common in industry implementations. Our key recovery attack will target these keys that are larger than n bits. We show that by performing a clever precomputation phase, the second key in NMAC or the equivalent key K_{out} in HMAC can be recovered with a time complexity of $O(2^{2l/3})$ computations and a memory to store $O(2^{2l/3})$ states by applying the Hellman's time-memory tradeoff. After that, the first key in NMAC or the equivalent key K_{in} in HMAC can be recovered with a time complexity of $O(2^{3l/4})$ computations and a memory to store $O(2^{3l/4})$ states.

Paper Outline. We recall the HMAC and NMAC specifications in Section 2 and properties of functional graphs in Section 3. Then, we explain the two types of selective forgery attacks in Section 4 and the improved universal forgery attack in Section 5. Finally, we describe the generic key recovery attack in Section 6 and we conclude the paper in Section 7.

2   Description of NMAC and HMAC

A Hash Function. H maps arbitrarily long messages to an n-bit digest. It is usually built by iterating a compression function f, which maps inputs of $l + b$ bits to outputs of l bits. In details, H first pads an input message $\mathcal{M}$ to be a multiple of b bits, then splits it into blocks of b bits each, i.e. $pad(\mathcal{M}) = M_1\|M_2\|\cdots\|M_s$, where $\|$ denotes the concatenation operation. It then calls the compression function f iteratively to process these blocks. Finally, H may use a finalization function g that maps l bits to n bits to produce the hash digest. Namely, set $X_0 \leftarrow IV$, compute $X_i \leftarrow f(X_{i-1}, M_i)$ for $i = 1, 2, \ldots, s$, and produce $g(X_s)$ as the final digest, with some finalization function g. Each internal state word X_i is l-bit long, and IV (initial value) is a public constant.

NMAC **Algorithm [1]** keys a hash function H by replacing the public IV with a secret key K, which is denoted as H_K. It then uses two l-bit secret keys K_{in} and K_{out} referred to as the inner and the outer keys respectively, and makes two calls to the hash function H. NMAC is simply defined to process an input message $\mathcal{M}$

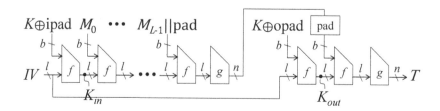

Fig. 1. HMAC with a narrow-pipe hash function

as $\text{NMAC}(K_{out}, K_{in}, \mathcal{M}) = H_{K_{out}}(H_{K_{in}}(\mathcal{M}))$. Keyed functions $H_{K_{in}}$ and $H_{K_{out}}$ are referred to as the inner and the outer (hash) functions respectively.

HMAC Algorithm [1] is a single-key variant of NMAC, depicted in Figure 1. It derives K_{in} and K_{out} from the single secret key K as $K_{in} = f(IV, K \oplus \texttt{ipad})$ and $K_{out} = f(IV, K \oplus \texttt{opad})$, where ipad and opad are two distinct public constants. HMAC is then simply defined to process an input message $\mathcal{M}$ as $\text{HMAC}(K, \mathcal{M}) = H(K \oplus \texttt{opad} \| H(K \oplus \texttt{ipad} \| \mathcal{M}))$. HMAC accepts any key size. If the key K is shorter than b bits, then it is padded with 0 bits to reach the size b of an entire compression function message block. Otherwise, if the key K is longer than b bits, then it is hashed and padded with 0 bits: $K \leftarrow H(K) \| 0^{b-n}$.

Regarding the use of keys which are longer than the tag size (n bits), there are both positive and negative decisions by standardization bodies. Indeed, RFC [11] only specifies that n bits is the minimum recommended key size. Though it does not specify the maximum key size, it explains that keys longer than n bits are acceptable, but the extra length would not significantly increase the function strength. However, it recommends longer key sizes when the randomness of the key is considered weak. FIPS [21] specifies that the effective security strength of the HMAC key is the minimum of the security strength of the key and the value of $2l$, where l is the internal state size. Hence, it seems natural to use $2l$-bit keys if that is possible, so as to maximize the security of the construction. Finally, we observe that in fact industry often implements HMAC with much longer key sizes than n bits. This is the case for example in MonoCrypt, which is a cryptographic library currently operated in commerce developed by SBI Net Systems [19]. MonoCrypt supports 80-bit, 128-bit, 512-bit, 576-bits, and 640-bit keys for HMAC-SHA-1 and 160-bit, 192-bit, 512-bit, 576-bit, and 640-bit keys for HMAC-SHA-256.

For simplicity, hereafter we will describe the attacks based on HMAC. However, we emphasize that our methods apply similarly to hash function based MACs such as NMAC [1] and Sandwich-MAC [22].

3 Functional Graph

In this article and in previous works on HMAC cryptanalysis [15,12,16], the analysis of properties of functional graphs for random functions is very important. We recall a few results in this section.

The functional graph $\mathcal{G}_f$ of a function $f : \{0,1\}^l \rightarrow \{0,1\}^l$ is simply the directed graph in which the vertices (or nodes) are all the values in $\{0,1\}^l$ and where the directed edges are the iterations of f (i.e. a directed edge from a vertex a to a vertex b exists iff $f(a) = b$). The functional graph of a function is composed of one or several components, each having its own internal cycle.

The following Theorems 1, 2 and 3 state several statistical properties of the functional graph of a random function.

Theorem 1 ([5, Th. 2]). *The expectations of the number of components, number of cyclic nodes (a node belonging to the cycle of its component), number of* *terminal nodes (a node without a preimage), and number of image nodes (a node with a preimage) in a random mapping of size N have the asymptotic forms, as $N \to \infty$:*

(i) #*Components:* $\frac{1}{2} \log N$ (iii) #*Terminal nodes:* $e^{-1} N$

(ii) #*Cyclic nodes:* $\sqrt{\pi N / 2}$ (iv) #*Image nodes:* $(1 - e^{-1}) N$

Starting from any node x, the iteration structure of f is described by a simple path that connects to a cycle. The length of the path (measured by the number of edges) is called the tail length of x (or the height of x) and is denoted by $\lambda(x)$. The length of the cycle is called the cycle length of x and is denoted $\mu(x)$. Finally, the rho-length of x is denoted $\rho(x)$ and represents the length of the non repeating trajectory of x: $\rho(x) = \lambda(x) + \mu(x)$.

Theorem 2 ([5, Th. 3]). *Seen from a random node in a random mapping of size N, the expectations of the tail length, cycle length, rho length, tree size, component size, and predecessors size have the following asymptotic forms:*

(i) *Tail length* (λ): $\sqrt{\pi N / 8}$ (iv) *Tree size:* $N/3$

(ii) *Cycle length* (μ): $\sqrt{\pi N / 8}$ (v) *Component size:* $2N/3$

(iii) *Rho length* ($\rho = \lambda + \mu$): $\sqrt{\pi N / 2}$ (vi) *Predecessors size:* $\sqrt{\pi N / 8}$

Moreover, the asymptotic expectations of the giant component and its giant tree have been provided in [6].

Theorem 3 ([6, VII.14]). *In a random mapping of size N, the largest tree and the largest component have expectations asymptotic, respectively, of $0.48 * N$ and $0.7582 * N$.*

In this article, we will study the functional graph of a compression function f, when a constant value M is used as message block input (*i.e.* the function f is iterated with fixed messages block all equal to M). We will denote f_M such a function $f_M : \{0,1\}^l \to \{0,1\}^l$, and $\mathcal{G}_{f_M}$ its corresponding functional graph.

4   Selective Forgery Attacks

In this section, we show two types of generic selective forgery attacks against HMAC. The attacker first commits on some message $\mathcal{M}$, and then can interact with the MAC oracle to output the valid tag T corresponding to $\mathcal{M}$ without querying $\mathcal{M}$. Note that the offline phase refers to the computations done before committing on $\mathcal{M}$, while the online phase refers to the computations done and the queries sent after committing on $\mathcal{M}$. Moreover, we denote $M^{(i)}$ the i successive concatenation of M.

4.1   Attack with a Very Constrained Target Message

The adversary will have to choose a long message, composed of $O(2^{l/2})$ blocks. Our method is a direct application of the distinguishing-H technique by [12].

Committing Phase (Offline)
1. As done in [12], draw the functional graph $\mathcal{G}_{f_M}$ of the underlying compression function f where a fixed message block M is used as the input message block, and compute the size γ of the cycle of the largest component.
2. Select as target message for the selective forgery the message $\mathcal{M} = M^{(2^{l/2})} \|$ $M' \| M^{(2^{l/2+\gamma})}$, where M' can be any message block such that $M' \neq M$.

Challenging Phase (Online)
3. Query $M^{(2^{l/2+\gamma})} \| M' \| M^{(2^{l/2})}$ to obtain the tag T. Output the pair $(\mathcal{M}, T)$.

The complexity and success probability evaluation is exactly the same as in [12] and we refer to the original article for more details. Informally, since a least $2^{l/2}$ identical message blocks M are used as prefix and suffix, there is a good chance that we enter in the main cycle of the functional graph $\mathcal{G}_{f_M}$ before and after processing message block M'. If this is true for $\mathcal{M}$ then it will be true for the queried message $M^{(2^{l/2+\gamma})} \| M' \| M^{(2^{l/2})}$ as well, and both will be fully synchronized inside the cycle, which means that they will end up to the same tag value. The overall success probability is equal to 0.14, but if needed it can be improved by iterating the fixed message block M a little bit more. The overall attack complexity is $O(2^{l/2})$ computations, which matches the proven lower-bound of the HMAC construction. This attack is therefore tight and it closes the discussions on the security gap of HMAC with regards to selective forgery notion.

Concerning the choice of the target message $\mathcal{M}$ by the adversary, we note that he can freely choose the values of M and M', but he can also append any prefix and suffix while preserving the validity of the attack. However, the target message $\mathcal{M}$ eventually contains quite a long iteration of an identical message block M, which constrains a lot the adversary's freedom to choose it.

4.2   Attack with More Freedom Degrees on the Target Message

The padding scheme used in the underlying hash function is heavily related to this analysis. Here, we suppose the MD-strengthening padding, which is widely used in practice e.g. by SHA-1 or SHA-256 [20]. Its essence is appending the message length information to the end of the message. See [20] for details.

In Section 4.2, we suppose that the underlying hash function is narrow-pipe, i.e. $l \leftarrow n$.

Our selective forgery attack uses a strategy similar to the previous generic second-preimage attack for hash functions [2,10], which can generate a second-preimage with a complexity of 2^{n-c} for a target message of size 2^c blocks. We briefly recall previous second-preimage attacks on hash functions. The initial idea is to try 2^{n-c} random messages in order to find one that collides with one

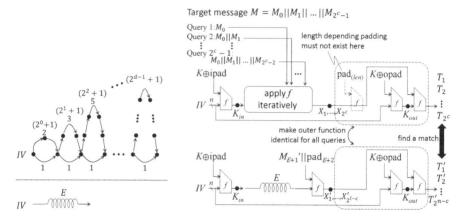

Fig. 2. (Top) Expandable message ranging from d to $d+2^d-1$ blocks. (Bottom) Simplified representation

Fig. 3. Strategy for computing a selective forgery with more freedom degrees on the target message

value of the 2^c internal chaining variables of the target message. However, the pre-specified message length in the padding string prevents this naive attack. Kelsey and Schneier showed that this issue can be solved with a multi-collision consisting of messages with different block lengths [10]. The generated multi-collision structure is called an *expandable message*. Informally, it generates a collision between a 1-block message and a $1+2^0=2$-block message, followed by a collision between an 1-block message and a $1+2^1=3$-block message. Similarly, a collision between an 1-block message and a $(1+2^i)$-block message is generated for $i=0,1,\ldots,d-1$. Then, any block length from d to $d+2^d-1$ can be reached by choosing the appropriate combination of the message blocks. An example is shown in Figure 2.

Adapting the generic second-preimage attacks for hash functions to compute selective forgery in the setting of MAC's presents several main difficulties:

1. Due to the two equivalent secret keys in HMAC, the adversary cannot access the internal state values after each message block is processed.
2. If the length of an input message changes, the MD-strengthening padding for the inner function will be different and thus the result of the outer function will change as well. This makes it difficult to cut an input message and analyze only up to the exact middle (i-th) message block.

Attack Overview. Before going into details, we explain our strategy, which is illustrated in Figure 3. To begin, we solve the first issue. In our attack, instead of storing the internal state value after each message block (which is unknown because of the secret key used in the MAC), the adversary queries the first i blocks of the 2^c-block target message for $i = 1, 2, \ldots, 2^c - 1$, and stores the corresponding tags. Namely, the first query is M_0, the second query is $M_0 \| M_1$, the third query is $M_0 \| M_1 \| M_2$, and so on. Let X_i be the unknown internal state value after processing the i-th message block and T_i its corresponding tag value. This is illustrated in top of Figure 3. The adversary later searches for a connection from the target message to the 2^{n-c}-block second message. Let X'_j be the output of the inner function (the internal state) and T'_j its corresponding tag value for the second message, where $j = 1, 2, \ldots, 2^{n-c}$. This is illustrated in bottom of Figure 3. If we can make the function from X_i to T_i and the function from X'_j to T'_j identical, a collision on the tag (i.e. $T_i = T'_j$) suggests a collision of the internal state (i.e. $X_i = X'_j$) with a good probability. Therefore, a connection from the second message to the target message can be found just by looking at the tag outputs, without even knowing the internal state values.

We then explain our strategy to solve the second issue. As indicated in Figure 3, the function from X_i to T_i may process the padding block, which depends on the input message length. This padding block issue can be avoided by selecting the target message so that the padding string is embedded inside each message block. Namely, message block M_i is chosen to be composed of $M_i = m_i \| p_i$ where m_i can be any value, and where p_i is the padding string corresponding to a message of $i + 1$ blocks plus $|m_i|$ bits.

With these few tricks, we can adapt the second-preimage attacks to the MAC setting and the selective forgery attack can eventually be carried out successfully.

Attack Procedure. Our attack is divided into five steps: 1) selecting the target message, 2) obtaining T_i, 3) building an expandable message, 4) obtaining T'_j, and 5) forging the tag. Only the first step is done offline, the rest being online.

1. **Selecting the Target Message.** The attacker must commits on the target message $\mathcal{M}$ of length $c + 2^c + 1$ blocks[1], where c is a parameter for the attack that we will determine later. More precisely, he will choose $\mathcal{M} = M_{-c} \| \cdots \| M_{-1} \| M_0 \| \cdots \| M_{2^c-1} \| M_{2^c}$, where the first c blocks $M_{-c} \| \cdots \| M_{-1}$ and the last block M_{2^c} can take any value of his choice. For the middle 2^c blocks from M_0 to M_{2^c-1}, He will set each block as $M_i \leftarrow m_i \| p_i$, where m_i can be set to any value of his choice and where p_i is the padding string corresponding to $1 + c + i$ blocks plus $|m_i|$ bits (here the "$1 + c$" corresponds to the first $1 + c$ blocks $(K \oplus \mathtt{ipad}) \| M_{-c} \| \cdots \| M_{-1}$ that will be handled by the internal hash function).

2. **Obtaining T_i.** After committing on $\mathcal{M}$, the online part can start:
   1. Query $M_{-c} \| \cdots \| M_{-1} \| m_0$ and store the tag T_1 in a list L.
   2. Query $M_{-c} \| \cdots \| M_{-1} \| M_0 \| m_1$ and store the tag T_2 in L.
   i. Query $M_{-c} \| \cdots \| M_{-1} \| M_0 \| \cdots \| M_{i-2} \| m_{i-1}$ and store the tag T_i in L for $i = 3, 4, \ldots, 2^c$.

---

[1] "c" comes from the minimum length of the expandable message and "$+1$" comes from the last message block. The detailed reasoning for the "$+1$" is explained later.

For the query at step i, due to the padding process the last message block becomes $m_{i-1}\|p_{i-1}$ which is in fact M_{i-1}.

3. **Building an Expandable Message.** The adversary builds an expandable message ranging from c blocks to $c + 2^c - 1$ blocks in order to later have the possibility to freely adjust the length. E_{prev} will denote the shortest colliding message discovered so far and is naturally initialized to a `null` string. Then, the following procedure is iterated for $i = 0, 1, \ldots, c - 1$:

3.1 Choose $2^{n/2}$ distinct 1-block messages $E_i[u]$. Query $E_{prev}\|E_i[u]$ and store the tags in a list L_u.

3.2 Choose $2^{n/2}$ distinct $(2^i + 1)$-block messages $E_i'[v]$. Query $E_{prev}\|E_i'[v]$ and store the tags in a list L_v.

3.3 Find a match between L_u and L_v. Let $\hat{u}$ and $\hat{v}$ be the matched indices.

3.4 To eliminate the false positives, find a collision by appending $2^{n/2}$ distinct single block messages after $E_{prev}\|E_i[\hat{u}]\|p_{\hat{u}}$, where $p_{\hat{u}}$ is the corresponding padding bits. Let E_x and E_x' be two messages that lead to a collision.

3.5 Query $E_{prev}\|E_i'[\hat{v}]\|p_{\hat{v}}\|E_x$ and $E_{prev}\|E_i'[\hat{v}]\|p_{\hat{v}}\|E_x'$. If their tags collide, $E_i[\hat{u}]$ and $E_i[\hat{v}]$ are internal collisions. Store $E_i[\hat{u}]\|p_{\hat{u}}$ and $E_i'[\hat{v}]\|p_{\hat{v}}$ as the i-th colliding pair of the expandable message. Otherwise, they are false positive, and we continue the search.

3.6 Update $E_{prev} \leftarrow E_{prev}\|E_i[\hat{u}]\|p_{\hat{u}}$.

The number of queries for Step 3.1 and Step 3.2 for the i-th block is $(i+1) \cdot 2^{n/2}$ and $(i + 1 + 2^i) \cdot 2^{n/2}$ respectively, which is unbalanced. For optimization, we generate more shorter messages $E_{prev}\|E_i[u]$ than longer messages $E_{prev}\|E_i'[v]$. Let α and β be $i + 1$ and $2^i + i + 1$, respectively. We get balance between the two query costs by generating $2^{n/2+(\log \beta - \log \alpha)/2}$ choices of $E_{prev}\|E_i[u]$ and $2^{n/2-(\log \beta - \log \alpha)/2}$ choices of $E_{prev}\|E_i'[v]$. The entire cost is the sum of two costs over the c iterations, $\sum_{i=0}^{c-1} 2^{(n/2+\log \beta+\log \alpha)/2+1}$, which amounts to $O(c \cdot 2^{n/2+c/2})$ blocks of queries. The memory cost is for storing tag values for $E_{prev}\|E_i'[v]$ in which the number of generation is smaller than the tag values for $E_{prev}\|E_i[u]$. Hence, the memory cost is $2^{n/2-(\log(2^i+i+1)-\log(i+1))/2}$ When $i \leftarrow c$, the memory cost is $O(2^{n/2-c/2})$. The cost for eliminating false positives at Step 3.4 is $(i + 1) \cdot 2^{n/2}$, which is smaller than Steps 3.1 and 3.2.

4. **Obtaining T_j'.** The length of the expandable message is at minimum c blocks, and we let M_E' denote this shortest c-block instance of the expandable message. Then, $(K_{in} \oplus \text{ipad})\|M_E'$ fits in $1 + c$ blocks. The adversary generates 2^j distinct 1-block message $M_j' = m_j'\|p'$ for $j = 1, 2, \ldots, 2^{n-c}$, where p' is the padding string for messages of $1 + c$ blocks plus $|m_j'|$ bits long. Query $M_E'\|m_j'$ for $j = 1, 2, \ldots, 2^{n-c}$, and store the received tag T_j' in a list L'.

5. **Forging the Tag.** Because 2^c T_i values are stored in L and 2^{n-c} T_j' values are stored in L', we expect to find a match between T_i and T_j'. With a good probability, the corresponding X_i and X_j' are also colliding.

Then, the length of the expandable message is adjusted to be equal to $c + i - 1$ blocks so that the length of the expandable message followed by the block M'_j can be the same as the length of $M_{-c}\| \cdots \|M_{-1}\|M_0\| \cdots \|M_{i-1}$. We denote $\overline{M'}$ the message chunk build by concatenating the length-adjusted expandable message and M'_j.

These two messages have the same length and result in the same internal state value. Thus, the adversary can append $M_i\|M_{i+1}\| \cdots \|M_{2^c}$ to the end of $\overline{M'}$, and query to the oracle this newly formed message. The received tag value $\overline{T}$ is also a valid tag for the selected target message $\mathcal{M}$.

Note that we need to ensure that the match is done before the last message block of $\mathcal{M}$, so that we have at least 1 block appended to $\overline{M'}$. That is the reason why we add the last block M_{2^c}.

Complexity Evaluation. We proceed the complexity evaluation of our attack. First, one can see that Step 1 is negligible, while Step 2 requires to query $c + 1, c + 2, \ldots, c + 2^c$ blocks, which amounts $c \cdot 2^c + 2^c \cdot (2^c + 1)/2 \approx 2^{2c}$ blocks in total. It also requires a memory sufficient to store 2^c tags. In Step 3, the cost is $O(c \cdot 2^{n/2+c/2})$ queries and $O(2^{n/2-c/2})$ tags as explained previously, and the memory cost is equivalent to $O(2^{n/2-c/2})$. tags. Step 4 requires to query $(c+1) \cdot 2^{n-c}$ blocks and a memory to store 2^{n-c} tags, while Step 5 is negligible (one can combine Step 4 and Step 5 so that values generated at Step 4 are tested immediately, which would render this part memoryless).

In total, the number of queries is about $2^{2c} + c \cdot 2^{n/2+c/2} + (c + 1) \cdot 2^{n-c}$ blocks, which is minimized to $O(n \cdot 2^{2n/3})$ blocks when $c = n/3$. The memory requirement is $O(2^c + 2^{n/2-c/2})$, which would become $O(2^{n/3})$ when $c = n/3$.

5   Improved Universal Forgery Attacks

In this section, we show an improved generic universal forgery attack against HMAC. We recall that for universal forgeries, the attacker is first challenged with a message $\mathcal{M}$, and after interacting with the MAC oracle he must output the valid tag T corresponding to $\mathcal{M}$ (without querying $\mathcal{M}$ to the oracle).

5.1   Revisiting Previous Universal Forgery Attacks on HMAC and NMAC

Recently, Peyrin and Wang published a universal forgery attack on iterative hash function-based MACs [16]. Their attack use a special property: the height of a node in a functional graph. We recall that in a functional graph each node x has a unique path connecting it with a cycle node, and the length of this path is called the *height* of x and is denoted as $\lambda(x)$. A brief description of their attack is provided below.

Let $\mathcal{M} = M_1\|M_2\| \cdots \|M_{2^s}$ be the given challenge message (after padding) for the universal forgery, and $X = \{X_1, X_2, \ldots, X_{2^s}\}$ be the successive internal state values during the processing of $\mathcal{M}$ in inner hash function, where X_i denotes the internal state after $M_1\| \cdots \|M_i$ has been processed.

The attacker computes the height in a functional graph of 2^{s_1} ($s_1 < s$) unknown internal state values during the processing of the challenge message.

Meanwhile he also collects 2^{l-s_1} offline values with their heights in the same functional graph. Note there is a good probability that one unknown internal state value collides with one offline collected value, which of course have the same height value. Then the attacker deduces the exact value of one of the unknown internal state values by identifying such a collision pair. In details, the attacker first matches the height values between the unknown internal state values and the offline collected values, which exponentially reduce the candidate pairs, and then examines each remaining pair individually. Finally, Once an internal state value is recovered, a classical second-preimage-like attack trivially allows to compute a universal forgery for the challenge.

However, Peyrin and Wang left an open problem, that is the height distribution in the set of the offline collected values. It is essential in order to obtain *tighter* upper bound of the attack complexity, and thus deserves further investigation. Due to the limited space, we refer to [16] for detailed argument. Here we mainly recall the procedure of collecting offline values and computing their heights, and then illustrate why it is hard to analyze their height distribution. Let $\mathcal{G}_{f_M}$ be the functional graph used in the attack, where V is a random message block value chosen by the attacker. The procedure is described below.

1. Initialize a table Y to be empty.
2. Select a random node y_1 such that $y_1 \notin Y$.
3. Iteratively compute $y_i = f_V(y_{i-1})$ until either of two cases occur:
   - y_i collides with a previously stored nodes in Y; or
   - y_i collides with a previous node y_j ($1 \leq j \leq i-1$) in the currently computed chain, namely a new cycle is generated.
4. Compute the height values for all nodes $y_1, \ldots, y_i$ in the chain, and store them in Y.
5. Repeat Steps $2-4$ until the number of nodes in Y becomes 2^{l-s_1}.

As we see, it is quite a difficult task to analyze the height distribution in the set Y because the nodes are not chosen uniformly (the process does not pick each node individually and randomly, but it picks a node y_1 and then picks all the nodes in the chain from y_1 to the cycle of y_1's component in the functional graph $\mathcal{G}_{f_M}$).

5.2   Our Observations

We have experimentally investigated the height distributions in the set Y, which is generated by the procedure in Section 5.1. Denote by Y_λ a subset of nodes in Y that have the height value λ, and by $|Y_\lambda|$ the number of nodes in Y_λ. In our experiment, we mainly pay attention on finding the smallest height value λ such that $|Y_\lambda|$ is *asymptotically* less than $2^{l/2-s}$, and observed an interesting phenomenon. Although we did not manage to prove formally this observation, we state this reasonable conjecture below.

Conjecture 1. If in total 2^t distinct nodes, where $l/2 \leq t \leq l$ holds, are collected following the procedure in Section 5.1, then for any integer λ satisfying $1 \leq \lambda \leq 2^{l/2}/l$, there are $\Theta(2^{t-l/2})$ nodes collected with the height value λ.

The extreme cases $t = l/2$ and $t = l$ are easy to analyze. For the case $t = l/2$, a randomly selected starting node has a height value $\Theta(2^{l/2})$ on average, and then for each height value λ such that $1 \leq \lambda \leq 2^{l/2}/l$ holds, $\Theta(1)$ nodes will be collected. For the case $t = l$, researchers have already carried out extensive studies on this topic. The set of all nodes with the same height λ is usually called the λ-th *stratum* of the functional graph, and we denote it as S_λ. Particularly, Mutafchiev [13] has proven the following theorem.

Theorem 4 ([13, Lemma 2]). *If $l \to \infty$ and $\lambda = o(2^{l/2})$, the mean value of the λ-th stratum S_λ is $\sqrt{\pi/2} * 2^{l/2}$.*

Since $2^{l/2}/l = o(2^{l/2})$ indeed holds, we get that Conjecture 1 has actually already been proven for the case $t = l$.

To further verify Conjecture 1, we performed experiments for small values of l (namely we computed the smallest value of the subset size $|Y_i|$ for $1 \leq i \leq 2^{l/2}/l$), which will be reported in full version of the paper.

5.3 Improved Universal Forgery Attacks

We present an improved universal forgery attack based on Conjecture 1. We divide $\mathcal{M}$ into two parts: $M_1\|\cdots\|M_{2^{s_1}}$ and $M_{2^{s_1}+1}\|\cdots\|M_{2^s}$ with $s_1 \leq s - 1$.

1. **(online)** Recover the height value $\lambda(X_i)$ of each X_i with $1 \leq i \leq 2^{s_1}$ in the functional graph $\mathcal{G}_{f_M}$. For the interested reader, the exact procedure of this step is referred to [16]. For each X_i the complexity of evaluating its height is $O(2^{l/2})$.
2. **(online)** Find a pair of 1-block message (m, m') with a birthday-like collision attack, such that $M_1\|\cdots\|M_{2^{s_1}}\|m$ and $M_1\|\cdots\|M_{2^{s_1}}\|m'$ is a collision on the inner hash function. The complexity is upper-bounded by $O(2^{s_1+l/2})$. Moreover, it is important to notice that (m, m') is a filter for all X_i with $1 \leq i \leq 2^{s_1}$ as the relation below holds:

$$f(f(\cdots f(X_i, M_{i+1})\cdots, M_{2^{s_1}}), m) = f(f(\cdots f(X_i, M_{i+1})\cdots, M_{2^{s_1}}), m').$$

3. **(offline)** Use the same collection procedure with previous attacks [16] to select 2^{l-s_1} nodes and obtain their respective height in the functional graph $\mathcal{G}_{f_M}$. However, in contrary to the previous attack, we only store the nodes with height λ satisfying $0 \leq \lambda \leq 2^{l/2}/l$. Moreover, for each such height λ, we store exactly $2^{l/2-s_1}$ nodes in Y in the end. According to Conjecture 1, we just need to repeat the collection procedure by at most a constant number of times. Thus, the complexity of this step is upper-bounded by $O(2^{l-s_1})$. It is important to recall that we now know the height distribution for the selected nodes in the set Y. More precisely, for each height λ such that $0 \leq \lambda \leq 2^{l/2}/l$ hold, there are $2^{l/2-s_1}$ nodes in Y that have height λ.
4. **(offline)** Recover the value of some X_i by matching the elements between X and Y. In details, for each X_i, if $\lambda(X_i) \leq 2^{l/2}/l$ holds, then:

4.1 Obtain the elements in Y that have the height value $\lambda(X_i)$. Let them be a subset of Y denoted as $Y_{\lambda(X_i)}$. We know that $|Y_{\lambda(X_i)}| = 2^{l/2 - s_1}$ holds.

4.2 For each node y in Y with height $\lambda(X_i)$, check if the following holds

$$f(f(\cdots f(y, M_{i+1})\cdots, M_{2^{s_1}}), m) = f(f(\cdots f(y, M_{i+1})\cdots, M_{2^{s_1}}), m').$$

and if it does, then output the value of y as the value of X_i.

The complexity of this step for a single X_i is computed as $(2^{s_1} - i) \cdot |Y_{\lambda(X_i)}| = (2^{s_1} - i) \cdot 2^{l/2 - s_1} = 2^{l/2} - i \cdot 2^{l/2 - s_1}$ and so the total complexity of this step is given by

$$\sum_{i=1}^{2^{s_1}} (2^{l/2} - i \cdot 2^{l/2 - s_1}) = O(2^{s_1 + l/2})$$

5. **(offline)** Based on the knowledge of some intermediate hash value X_i, construct a second-preimage $\mathcal{M}'$ of the challenge message $\mathcal{M}$ with respect to the inner hash function. Note that once X_i is known, the following intermediate hash values X_j with $i \leq j \leq 2^s$ are also known. Then, previous generic second-preimage attacks [10] can be applied to find $\mathcal{M}'$ and the complexity is known to be upper-bounded by $O(2^{l-s})$.

6. **(online)** Query $\mathcal{M}'$ to MAC and receive the tag T. Output T as the valid tag for the challenge message $\mathcal{M}$. The complexity of this step is obviously upper-bounded by the block length of $\mathcal{M}'$, that is $O(2^s)$.

Note that there are in total $2^{l-s_1}/l$ nodes in Y, and 2^{s_1} intermediate hash values in X. So a collision between an element in X and an element in Y occurs with a probability around $1/l$. Thus, we need to repeat the attack procedure $\Theta(l)$ times in order to increase the success probability to a constant value.

Now, we can eventually summarize the complexity of the entire universal forgery attack. We recall that $s_1 \leq s - 1$.

Step 1:	$O(2^{s_1 + l/2})$	**Step 2:**	$O(2^{s_1 + l/2})$	**Step 3:**	$O(2^{l-s_1})$
Step 4:	$O(2^{s_1 + l/2})$	**Step 5:**	$O(2^{l-s})$	**Step 6:**	$O(2^s)$

- For the case $0 < s < l/4$, the complexity is dominated by Step 3. Set $s_1 = s - 1$ and then the total complexity is upper-bounded by $O(l \cdot 2^{l-s})$.
- For the case $l/4 \leq s \leq 3l/4$, set $s_1 = l/4$ to make the complexities at Steps 1 and 3 equal, which optimizes the overall complexity. The complexity is upper-bounded by $O(l \cdot 2^{3l/4})$.
- For the case $s > 3l/4$, set $s_1 = l/4$, and the complexity is dominated by Step 6, which is upper-bounded by $O(l \cdot 2^s)$.

Overall, our attacks have significantly decreased the complexity of universal forgery attack on iterated hash-based MACs from $O(2^{5l/6})$ (attack complexity in [16]) to $O(2^{3l/4})$ by ignoring the polynomial factors.

6   Time-Memory Tradeoff for Key Recovery Attacks

In this section, we discuss time-memory tradeoff for key recovery attacks on NMAC or for the equivalent key recovery attacks on HMAC. To start with, it has been known that the complexity of the brute-force key recovery attack can be reduced to 2^l although the key size is $2l$ bits, by following a divide-and-conquer approach. In short, the adversary firstly generates an inner collision, and then brute force recovers the inner key by detecting if the collision can be reached for each key candidate. After the inner key is recovered, the adversary moves to recover the outer key by using the trivial brute-force attack based on the knowledge of the inner key. While this attack does not use any precomputation, surprisingly it is even more efficient than the straightforward application of Hellman's time-memory tradeoff [9], which uses 2^{2l} precomputation, and for key recovery phase $2^{4l/3}$ computations and $2^{4l/3}$ memory. This motivates us to investigate if there are more efficient time-memory tradeoff for the key recovery attacks on HMAC and NMAC with the usage of precomputation.

In the following, we will present our new time-memory tradeoff. Roughly speaking, our tradeoff utilizes both the divide-and-conquer approach and the Hellman's time-memory tradeoff. With a precomputation, we firstly recover the outer key K_{out} and then recover the inner key K_{in}. It is important to note that both of the precomputation for K_{out} and K_{in} are performed before launching any key recovery attacks for K_{out} and K_{in}.

6.1   Recovering K_{out}

Hellman's tradeoff approach is not trivially applicable to recover K_{out} because the input from the inner hash function is unknown due to the inner key. To overcome this problem, we preset the output of the inner hash function to a constant X_e. Thanks to the recent internal state recovery attack on hash-based MAC [12] and the second preimage attack on hash function [10], we can always successfully constructed a message which will produce an output of the inner hash function, which is the fixed X_e.

The attack procedure is described as below.

Precomputation Phase
1. Randomly pick a chaining value X_0 and iteratively compute $X_i = f_M(X_{i-1})$ for $i = 1, \ldots, O(2^{l/2})$ while storing all the X_i's in a lookup table. Denote the final internal state value as X_e.
2. Build Hellman's precomputed lookup tables for the function f_{X_e}, *i.e.* the compression function with X_e as the message block.

Key Recovery Phase
1. Recover the unknown internal state for a message m with $O(2^{l/2})$ blocks using the technique from [12] with $2^{l/2}$ time complexity and $2^{l/2}$ memory requirement.
2. Append m with an expandable message M_E of range $[l/2, l/2 + 2^{l/2} - 1]$.

3. Find a message block M_L that links the expandable message to one of the precomputed X_i's.

4. Query the MAC oracle with message $\mathcal{M}_q = m\|M_E\|M_L\|M\|\cdots\|M$ to obtain the tag T, where M_E's length is chosen in the way that the overall length of $\mathcal{M}_q$ becomes $O(2^{l/2})$. Note that we shall choose the last block M so that $\mathcal{M}_q$ is already a valid padded message, and this message $\mathcal{M}_q$ ensures that the output of the inner layer will be X_e.

5. Use T as the input of Hellman's key recovery phase to recover K_{out}.

In this attack, the first step of the precomputation phase and the second and third steps of the key recovery phase are essentially performed to find a second-preimage of the hash function for the given message $M\|\cdots\|M$ with prefix m, with length $2^{l/3}$ and with the initial value changed to X_0. The entire process costs $2^{2l/3}$ computation and $2^{l/3}$ memory. The second step of the precomputation phase and the fifth step of the key recovery phase are exactly Hellman's tradeoff costing 2^l precomputation, and $2^{2l/3}$ online computations and memory. Overall, K_{out} can be recovered with $2^{2l/3}$ time and $2^{2l/3}$ memory (both dominated by the fifth step) with 2^l precomputation.

6.2 Recovering K_{in}

Our time-memory tradeoff for recovering K_{in} is based on the height of nodes in the functional graph. In short, during the precomputation phase, we collect a set of nodes in a functional graph $\mathcal{G}_{f_M}$ with a certain pattern of heights. Then during the key recovery phase, we first recover the height of K_{in} in $\mathcal{G}_{f_M}$ following the procedure in [16], then derive a set of nodes, which have the same height with K_{in}, from the collected nodes of the prcomputation phase, and checks if K_{in} is inside these nodes or not. Moreover, we need to utilize more than one functional graph in order to amplify the success probability to a constant value.

The attack procedure is described as below.

Precomputation Phase

1. Randomly pick an internal state value X_0 and iteratively compute $X_j = f_{M_i}(X_{j-1})$ until some X_j collides with a previous one. This allows to deduce the height of X_0 in the functional graph $\mathcal{G}_{f_{M_i}}$. Store in table T_i the pair $(X_j, \lambda(X_j))$ with $\lambda(X_j)$ being a multiple of $2^{l/4}$ and $\lambda(X_j) < 2^{l/2}/l$ (omit if the pair is already in T_i). Repeat the process for $2^{l/4}$ random X_0 and sort the table T_i according to the heights, and save the final T_i together with M_i.

2. Repeat the process for random M_i so as to obtain $l \times 2^{l/4}$ structures of (T_i, M_i)'s.

Key Recovery Phase

1. Obtain the height of K_{in} using the technique from [16] using the functional graph $\mathcal{G}_{f_{M_i}}$. Let λ be the smallest multiple of $2^{l/4}$ greater than $\lambda(K_{in})$. Retrieve all X_j's whose height in $\mathcal{G}_{f_{M_i}}$ is equal to λ. Test if $f^{\lambda - \lambda(K_{in})}(X_j)$ is the correct guess of K_{in} for all X_j in the collection of T_i. Repeat for all M_i until K_{in} is recovered.

Following *Conjecture* 1, in the range that interests us (*i.e.* $[1, 2^{l/2}/l]$), there will be $\Theta(2^{l/4})$ nodes with the same height collected in each table. Since the overall number of nodes at each height of interest is $O(2^{l/2})$, the chance for a collision to happen at each height is $o(2^{-l/4} = 2^{l/4}/2^{l/2})$, and we covered $1/l$ portion of all possible nodes, so the chance to find a match in one table is $o(l^{-1} \cdot 2^{-l/4})$. Since there are $l \cdot 2^{l/4}$ independent tables, our key recovery phase will be successful with a non-negligible probability. The time and memory complexity for this attack is eventually $2^{3l/4}$ with 2^l precomputation.

7   Conclusion

In this paper, we presented selective forgery attacks, improved universal forgery attacks, and time-memory tradeoff for key recovery attacks against the most popular MAC constructions built upon iterative hash functions, such as HMAC and NMAC. Our cryptanalysis methods are based on the extension of various techniques including expandable messages, second-preimage attack, functional graph-based forgery attacks, etc. Our work provides the community with a better understanding of the security margin of iterative hash-based MACs.

Acknowledgments. The authors would like to thank the anonymous referees for their helpful comments, especially for suggesting the conversion from the previous distinguishing-H attack into the selective forgery attack. Jian Guo, Thomas Peyrin and Lei Wang were supported by the Singapore National Research Foundation Fellowship 2012 (NRF-NRFF2012-06).

References

1. Bellare, M., Canetti, R., Krawczyk, H.: Keying Hash Functions for Message Authentication. In: Koblitz, N. (ed.) CRYPTO 1996. LNCS, vol. 1109, pp. 1–15. Springer, Heidelberg (1996)
2. Dean, R.D.: Formal Aspects of Mobile Code Security. Ph.D Dissertation, Princeton University (January 1999)
3. Dinur, I., Leurent, G.: Improved Generic Attacks Against Hash-Based MACs and HAIFA. In: Garay, J., Gennaro, R. (eds.) CRYPTO 2014, Part I. LNCS, vol. 8616, pp. 149–168. Springer, Heidelberg (2014)
4. Dodis, Y., Ristenpart, T., Steinberger, J., Tessaro, S.: To Hash or Not to Hash Again (In)Differentiability Results for H^2 and HMAC. In: Safavi-Naini, R., Canetti, R. (eds.) CRYPTO 2012. LNCS, vol. 7417, pp. 348–366. Springer, Heidelberg (2012)
5. Flajolet, P., Odlyzko, A.M.: Random Mapping Statistics. In: Quisquater, J.-J., Vandewalle, J. (eds.) EUROCRYPT 1989. LNCS, vol. 434, pp. 329–354. Springer, Heidelberg (1990)
6. Flajolet, P., Sedgewick, R.: Analytic Combinatorics. Cambridge University Press (2009)
7. Guo, J., Sasaki, Y., Wang, L., Wang, M., Wen, L.: Equivalent Key Recovery Attacks against HMAC and NMAC with Whirlpool Reduced to 7 Rounds. In: Cid, C., Rechberger, C. (eds.) Fast Software Encryption. LNCS. Springer (to appear, 2014)

148 J. Guo et al.

8. Guo, J., Sasaki, Y., Wang, L., Wu, S.: Cryptanalysis of HMAC/NMAC-Whirlpool. In: [18], pp. 21–40

9. Hellman, M.E.: A Cryptanalytic Time-Memory Trade-Off. IEEE Transactions on Information Theory 26(4), 401–406 (1980)

10. Kelsey, J., Schneier, B.: Second Preimages on n-Bit Hash Functions for Much Less Than 2^n Work. In: Cramer, R. (ed.) EUROCRYPT 2005. LNCS, vol. 3494, pp. 474–490. Springer, Heidelberg (2005)

11. Krawczyk, H., Bellare, M., Canetti, R.: HMAC: Keyed-Hashing for Message Authentication. Internet Engineering Task Force, IETF (1997), http://www.rfc-editor.org/rfc/rfc2104.txt

12. Leurent, G., Peyrin, T., Wang, L.: New Generic Attacks against Hash-Based MACs. In: [18], pp. 1–20

13. Mutafchiev, L.R.: The limit distribution of the number of nodes in low strata of a random mapping. Statistics & Probability Letters 7(3), 247–251 (1988)

14. Naito, Y., Sasaki, Y., Wang, L., Yasuda, K.: Generic State-Recovery and Forgery Attacks on ChopMD-MAC and on NMAC/HMAC. In: Sakiyama, K., Terada, M. (eds.) IWSEC 2013. LNCS, vol. 8231, pp. 83–98. Springer, Heidelberg (2013)

15. Peyrin, T., Sasaki, Y., Wang, L.: Generic Related-Key Attacks for HMAC. In: Wang, X., Sako, K. (eds.) ASIACRYPT 2012. LNCS, vol. 7658, pp. 580–597. Springer, Heidelberg (2012)

16. Peyrin, T., Wang, L.: Generic Universal Forgery Attack on Iterative Hash-Based MACs. In: Nguyen, P.Q., Oswald, E. (eds.) EUROCRYPT 2014. LNCS, vol. 8441, pp. 147–164. Springer, Heidelberg (2014)

17. Preneel, B., van Oorschot, P.C.: On the Security of Two MAC Algorithms. In: Maurer, U.M. (ed.) EUROCRYPT 1996. LNCS, vol. 1070, pp. 19–32. Springer, Heidelberg (1996)

18. Sako, K., Sarkar, P. (eds.): ASIACRYPT 2013, Part II. LNCS, vol. 8270, pp. 2013–2019. Springer, Heidelberg (2013)

19. SBI Net Systems: MonoCrypt home page, http://capg.sbins.co.jp/products/monocrypt/index.html.

20. U.S. Department of Commerce, National Institute of Standards and Technology: Secure Hash Standard (SHS) (Federal Information Processing Standards Publication 180-3) (2008), http://csrc.nist.gov/publications/fips/fips180-3/fips180-3_final.pdf

21. U.S. Department of Commerce, National Institute of Standards and Technology: Recommendation for Applications Using Approved Hash Algorithms (Federal Information Processing Standards Publication 800-107) (2012), http://csrc.nist.gov/publications/nistpubs/800-107-rev1/sp800-107-rev1.pdf

22. Yasuda, K.: "Sandwich" Is Indeed Secure: How to Authenticate a Message with Just One Hashing. In: Pieprzyk, J., Ghodosi, H., Dawson, E. (eds.) ACISP 2007. LNCS, vol. 4586, pp. 355–369. Springer, Heidelberg (2007)

Improved Generic Attacks
against Hash-Based MACs and HAIFA[*]

Itai Dinur[1] and Gaëtan Leurent[2]

[1] Département d'Informatique, École Normale Supérieure, Paris, France
Itai.Dinur@ens.fr
[2] Inria, EPI SECRET, France
Gaetan.Leurent@inria.fr

Abstract. The security of HMAC (and more general hash-based MACs) against state-recovery and universal forgery attacks was very recently shown to be suboptimal, following a series of surprising results by Leurent *et al.* and Peyrin *et al.*. These results have shown that such powerful attacks require much less than 2^ℓ computations, contradicting the common belief (where ℓ denotes the internal state size). In this work, we revisit and extend these results, with a focus on properties of concrete hash functions such as a limited message length, and special iteration modes.

We begin by devising the first state-recovery attack on HMAC with a HAIFA hash function (using a block counter in every compression function call), with complexity $2^{4\ell/5}$. Then, we describe improved trade-offs between the message length and the complexity of a state-recovery attack on HMAC. Consequently, we obtain improved attacks on several HMAC constructions used in practice, in which the hash functions limit the maximal message length (e.g., SHA-1 and SHA-2). Finally, we present the first universal forgery attacks, which can be applied with short message queries to the MAC oracle. In particular, we devise the first universal forgery attacks applicable to SHA-1 and SHA-2.

Keywords: Hash functions, MAC, HMAC, Merkle-Damgård, HAIFA, state-recovery attack, universal forgery attack, GOST, Streebog, SHA family.

1 Introduction

MAC algorithms are an important symmetric cryptography primitive, used to verify the integrity and authenticity of messages. First, the sender appends to the message a tag, computed from the message and a key. The receiver can recompute the tag using the key and reject the message when the computed tag does not match the received one. The main security requirement of a MAC is the resistance to existential forgery. Namely, after querying the MAC oracle to obtain the tags of some carefully chosen messages, it should be hard to forge a valid tag for a different message.

---

[*] Some of the work presented in this paper was done during Dagstuhl Seminar 14021.

J.A. Garay and R. Gennaro (Eds.): CRYPTO 2014, Part I, LNCS 8616, pp. 149–168, 2014.

One of the most widely used MAC algorithms in practice is HMAC, a MAC construction using a hash function designed by Bellare, Canetti and Krawczyk in 1996 [4]. The algorithm has been standardized by ANSI, IETF, ISO and NIST, and is widely deployed to secure internet communications (*e.g.* SSL, SSH, IPSec). As these protocols are widely used, the security of HMAC has been extensively studied, and several security proofs [3,4] show that it gives a secure MAC and a secure PRF up to the birthday bound (assuming good properties of the underlying compression function). At the same time, there is a simple existential forgery attack on any iterative MAC with an ℓ-bit state, with complexity $2^{\ell/2}$, matching the security proof. Nevertheless, security beyond the birthday bound for stronger attacks (such as state-recovery and universal forgery) is still an important topic.

Surprisingly, the security of HMAC beyond the birthday bound has not been thoroughly studied until 2012, when Peyrin and Sasaki described an attack on HMAC in the related-key setting [19]. Later work focused on single-key security, and included a paper by Naito, Sasaki, Wang and Yasuda [17], which described state-recovery attacks with complexity $2^\ell/\ell$. At Asiacrypt 2013, Leurent, Peyrin and Wang [16] gave state-recovery attacks with complexity $2^{\ell/2}$, closing the gap with the security proof. More recently, at Eurocrypt 2014, Peyrin and Wang [20] described a universal forgery attack with complexity as low as $2^{5\ell/6}$. The complexity of the universal forgery attack was further improved to $2^{3\ell/4}$ in [9], showing that even this very strong attack is possible with less than 2^ℓ work.

These generic attacks have also been used as a first step to build specific attacks against HMAC with the concrete hash function Whirlpool [11,10].

These very recent and surprising results show that more work is needed to better understand the exact security provided by HMAC and hash-based MACs.

1.1   Our Results

In this paper, we provide several important contributions to the security analysis of HMAC and similar hash-based MAC constructions. In particular, we devise improved attacks when HMAC is used with many popular concrete hash functions, and in several cases our attacks are the first to be applicable to HMAC with the given hash function. Some results with concrete instantiations are summarized in Table 1.

As a first contribution, we focus on the HAIFA [5] mode of operation, used in many recent designs such as BLAKE [1,2], Skein [8], or Streebog [7]. The HAIFA construction uses a block counter to tweak the compression functions, such that they resemble independent random functions, in order to thwart some narrow-pipe attacks (*e.g.* the second-preimage attack of Kelsey and Schneier [14]). Indeed, the recent attacks against HMAC [16,20] use in a very strong way the assumption that the same compression function is applied to all the message blocks, and thus they cannot be applied to HAIFA. In this work, we present the first state-recovery attack on HMAC using these hash functions, whose optimal complexity is $2^{4\ell/5}$.

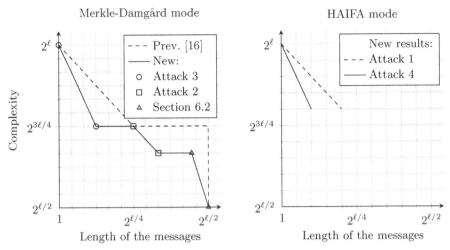

Fig. 1. Trade-offs between the message length and the complexity

In an interesting application of our state-recovery attack on HAIFA (given in the full version of this paper [6]), we show how to extend it into a key-recovery attack on the new Russian standard Streebog, recovering the 512-bit key of HMAC-Streebog with a complexity of 2^{410}. This key recovery attack is similar to the one of [16] for Merkle-Damgård, and confirms its surprising observation: adding an internal checksums in a hash function (such as Streebog) *weakens* the design when used in HMAC, even for hash functions based on the HAIFA mode.

As a second contribution of this paper, we revisit the results of [16], and give a formal proof of the conjectures used in its short message attacks. Some of our proofs are of broad interest, as they give insight into the behavior of classical collision search algorithms for random functions. These proofs explain for the first time an interesting phenomenon experimentally observed in several previous works (such as [18]), namely, that the collisions found by such algorithms are likely to belong to a restricted set of a surprisingly small size.

Then, based on our proofs, we describe several new algorithms with various improved trade-offs between the message length and the complexity as shown in Figure 1. As many concrete hash functions restrict the message size, we obtain improved attacks in many cases: for instance, we reduce the complexity of a state-recovery attack against HMAC-SHA-1 from 2^{120} to 2^{107} (see Table 1).

Finally, we focus on universal forgery attacks, and devise attacks using techniques which are different from those of [20]. While the attack of [20] (and its improvement in [9]) is much more efficient than exhaustive search, it requires, in an inherent way, querying the MAC oracle with very long messages of about $2^{\ell/2}$ blocks, and thus has limited impact in practice. On the other hand, our attacks can be efficiently applied with much shorter queries to the MAC oracle, and thus have many more applications. In particular, we devise the first universal forgery attack applicable to HMAC with SHA-1 and SHA-2 (see Table 1).

Table 1. Complexity of attacks on HMAC instantiated with some concrete hash functions. The state size is denoted as ℓ, and the maximum message length as 2^s. For the new results, we give a reference to the Attack number, or the full version of the paper.

Function	Mode	ℓ	s	State-recovery		Universal forgery	
				[16]	New	[20]	New
SHA-1	MD	160	2^{55}	2^{120}	2^{107} (2)	N/A	2^{132} (6)
SHA-256	MD	256	2^{55}	2^{201}	2^{152} (3)	N/A	2^{228} (5,6)
SHA-512	MD	512	2^{118}	2^{394}	2^{282} (3)	N/A	2^{453} (5,6)
HAVAL	MD	256	2^{54}	2^{202}	2^{154} (3)	N/A	2^{229} (5,6)
BLAKE-256	HAIFA	256	2^{55}	N/A	2^{213} (4)	N/A	N/A
BLAKE-512	HAIFA	512	2^{118}	N/A	2^{419} (4)	N/A	N/A
Skein-512	HAIFA	512	2^{90}	N/A	2^{419} (4)	N/A	N/A
						Key recovery	
						[16]	New[6]
Streebog	HAIFA+σ	512	∞	N/A	2^{419} (4)	N/A	2^{419}

1.2 Framework of the Attacks

In order to recover an internal state, computed by the MAC oracle during the processing of some message, we use a framework which is similar to the framework of [16]. Namely, we match states that are computed offline with (unknown) states that are computed online (during the processing of messages by the MAC oracle). However, as arbitrary states match with low probability (which does not lead to efficient attacks), we only match *special states*, which have a higher probability to be equal. These special states are the result of iterating random functions using chains, computed by applying the compression function on a fixed message from arbitrary initial states. In this paper, we exploit special states of two types, which were also exploited in [16]: states on which two evaluated chains collide, and states on which a single chain collides with itself to form a cycle. Additionally, some of our attacks (and in particular our attacks on HAIFA) use special states which are a result of the reduction of the image space that occurs when applying a fixed sequence of random functions.

As described above, after we compute special states both online and offline, we need to match them in order to recover an online state. However, since the online states are unknown, the matching cannot be performed directly, and we are forced to match the nodes indirectly using *filters*. A filter for a node (state) is a property that identifies it with high probability, i.e., once the filters of two nodes match, then the nodes themselves match with high probability. Since the complexity of the matching steps in a state-recovery attack depend on the complexity on building a filter for a node and testing a filter on a node, we are interested in building filters efficiently. In this paper, we use two types of filters: collision filters (which were also used in [16]) and diamond filters, which exploit the diamond structure (proposed in [13]) in order to build filters for a large set of nodes with reduced average complexity. Furthermore, in this paper we use a novel online construction of the diamond structure via the MAC oracle, whereas such a structure is typically computed offline. In particular, we show that despite the fact that the online diamond filter increases the complexity of building the filter, the complexity of the actual matching phase is significantly reduced, and gives improved attacks in many cases.

Outline. The paper is organized as follows: we begin with a description of HMAC in Section 2. We then describe and analyze the algorithms we use to compute special states in Section 3, and the filters we use in our attacks in Section 4. Next, we present a simple attack against HMAC with a HAIFA hash function in Section 5, and revisit the results of [16] in Section 6, presenting new trade-offs for attacks on Merkle-Damgård hash functions. In Section 7, we give more complex attacks for shorter messages. Finally, in Section 8, we present our universal forgery attacks with short queries, and conclude in Section 9.

2   HMAC and Hash-Based MACs

In this paper we study MAC algorithms based on a hash function, such as HMAC. HMAC is defined using a hash function H as $\mathtt{HMAC}(K, M) = H(K \oplus \mathtt{opad} \| H(K \oplus \mathtt{ipad} \| M))$. More generally, we consider a class of designs defined as:

$$x_0 = I_K \qquad\qquad x_{i+1} = h_i(x_i, m_i) \qquad\qquad t = g(K, x_p, |M|).$$

The message processing updates an internal state of size ℓ, starting from a key-dependant value I_K, and the output is produced with a key-dependant finalization function g. In particular, we note that the state update does not depend on the key. Our description covers HMAC [4], Sandwich-MAC [23] and envelope-MAC with any common hash function. The hash function can use the message length in the finalization process, which is a common practice, and the rounds function can depend on a block counter, as in the HAIFA mode. If the hash function uses the plain Merkle-Damgård mode, the round functions h_i are all identical (this is the model of previous attacks [16,20]).

In this work, we assume that the tag length n is larger than ℓ, so that collision in the tag result from collisions in the internal state with very high probability. This greatly simplifies the description of the attacks, and does not restrict the scope of our results. Indeed from a function $\mathtt{MAC}_1(K, M)$ with an output of n bits, we can build a function $\mathtt{MAC}_2(K, M)$ with a $2n$-bit output by appending message blocks $[0]$ and $[1]$ to M, as $\mathtt{MAC}_2(K, M) = \mathtt{MAC}_1(K, M \| [0]) \| \mathtt{MAC}_1(K, M \| [1])$. Our attacks applied to $\mathtt{MAC}_2$ can immediately be turned to attacks on $\mathtt{MAC}_1$.

3   Description and Analysis of Collision Search Algorithms

In this section, we describe and analyze the collision search algorithms which are used in our state-recovery attacks in order to compute special states. We then analyze these algorithms and prove the conjectures of [16]. Lemma 1 proves the first conjecture, while Lemma 3 proves the second conjecture. We also give further results in the full version of the paper [6].

3.1   Collision Search Algorithms

We use standard collision search algorithms, which evaluate chains starting from arbitrary points. Namely, a chain $\overrightarrow{x}$ starts from x_0, and is constructed iteratively by the equation $x_i = f_i(x_{i-1})$ up to $i = 2^s$ for a fixed value of $s \leq \ell/2$. We consider two different types of collisions between two chains $\overrightarrow{x}$ and $\overrightarrow{y}$: free-offset collisions ($x_i = y_j$ for any i, j, with all the f_i's being equal), and same-offset collisions ($x_i = y_i$).

Free-Offset Collision Search. When searching offline for collisions in iterations of a *fixed random function* f, we evaluate 2^t chains starting from arbitrary points, and extended to length 2^s for $s < \ell/2$.

Assuming that $2^t \cdot 2^{t+s} \leq 2^\ell$ (i.e., $2t + s \leq \ell$), then each of the chains is not expected to collide with more than one other chain in the structure. This implies that the structure contains a total of about 2^{t+s} distinct points, and (according to the birthday paradox) we expect it to contain a total of $2^c = 2^{2(t+s)-\ell}$ collisions. We can easily recover all of these collisions in $O(2^{t+s}) = O(2^{(c+\ell)/2})$ time by storing all the evaluated points and checking for collisions in memory.

We note that we can reduce the memory requirements of the algorithm by using the parallel collision search algorithm of van Oorschot and Wiener [18]. However, in this paper, we generally focus on time complexity and do not try to optimize the memory complexity of our attacks.

Same-Offset Collision Search. While free-offset collisions are the most general form of collisions, they cannot always be efficiently detected and exploited by our attacks. In particular, they cannot be efficiently detected in queries to the online oracle (as a collision between messages of different lengths would lead to different values after the finalization function). Furthermore, if the hash function uses the HAIFA iteration mode, it is also not clear how to exploit free-offset collisions offline, as the colliding chains do not merge after the collision (and thus we do not have any easily detectable non-random property).

In the cases above, we are forced to only use collisions that occur at the same-offset. When computing 2^t chains of length 2^s (for t not too large), a pair of chains collide at a fixed offset i with probability $2^{-\ell}$, and thus a pair of chains collide with probability $2^{s-\ell}$. As we have 2^{2t} pairs of chains, we expect to find about $2^{2t+s-\ell}$ fixed-offset collisions.

Locating collisions online. Online collisions are detected by sorting and comparing the tags obtained by querying the MAC oracle with chains of a fixed length 2^s. If we find two massages such that $MAC(M) = MAC(M')$, we can easily compute the message prefix that gives the (unknown) collision state, as described in [16]. Namely, if we denote by $M_{|i}$ the i-block prefix of M, then we find the smallest i such that $MAC(M_{|i}) = MAC(M'_{|i})$ using binary search. This algorithm queries the

MAC oracle with $O(s)$ messages of length $O(2^s)$, and thus the time complexity of locating a collision online is $s \cdot 2^s = \tilde{O}(2^s)$.

3.2   Analysis of the Collision Search Algorithms

In this section, we provide useful lemmas regarding the collision search algorithms described above. These lemmas are used in order to estimate the collision probability of special states that are calculated by our attacks and thus to bound their complexity. Lemmas 1 and 2 can generally be considered as common knowledge in the field, and their proofs are given in the full version of this paper [6]. Perhaps, the most interesting results in this section are lemmas 3 and 4. These lemmas show that the probability that our collision search algorithms reach the same collision twice from different arbitrary starting points, is perhaps higher than one would expect. This phenomenon was already observed in previous works such as [18], but to the best of our knowledge, this is the first time that this lemma is formally proven. As the proof of lemma 4 is very similar to that of lemma 3, it is given in the full version of this paper [6].

Lemma 1. *Let $s \le \ell/2$ be a non-negative integer. Let $f_1, f_2, \ldots, f_{2^s}$ be a sequence of random functions over the set of 2^ℓ elements, and $g_i \triangleq f_i \circ \ldots \circ f_2 \circ f_1$ (with the f_i being either all identical, or independently distributed). Then, the images of two arbitrary inputs to g_{2^s} collide with probability of about $2^{s-\ell}$, i.e. $\Pr_{x,y}[g_{2^s}(x) = g_{2^s}(y)] = \Theta(2^{s-\ell})$.*

Lemma 2. *Let $f_1, f_2, \ldots, f_{2^s}$ be a sequence of random functions, then the image of the function $g_{2^s} \triangleq f_{2^s} \circ \ldots \circ f_2 \circ f_1$ contains at most $\tilde{O}(2^{\ell-s})$ points.*

Lemma 3. *Let $\hat{x}$ and $\hat{y}$ be two random collisions (same-offset or free-offset) found by a collision search algorithm using chains of length 2^s, with a **fixed ℓ-bit function f** such that $s < \ell/2$. Then $\Pr[\hat{x} = \hat{y}] = \Theta(2^{2s-\ell})$.*

Proof. First, we note that we generally have 4 cases to analyze, according to whether $\hat{x}$ and $\hat{y}$ were found using a free-offset, or a same-offset collision search algorithm. However, the number of cases can be easily reduced to 3, as we have 2 symmetric cases, where one collision is free-offset, and the other is same-offset. In this proof, we assume that $\hat{x}$ is a same-offset collision and $\hat{y}$ is a free-offset collision (this is the configuration used in our attacks). However, the proof can easily be adapted to the 2 other different settings.

We denote the starting points of the chains which collide on $\hat{x}$ by (x_0, x'_0), and the actual corresponding colliding points of the chains by (x_i, x'_i), and thus $f(x_i) = f(x'_i) = \hat{x}$. In the following, we assume that $0.25 \cdot 2^s \le i \le 0.75 \cdot 2^s$, which occurs with probability about $1/2$ since the offset of the collision $\hat{x}$ is roughly uniformly distributed in the interval $[0, 2^s]$.[1] This can be shown using Lemma 1, as increasing the length of the chains, increases the collision probability by the same multiplicative factor.

---

[1] The assumption simplifies the proof of the lower bound on the collision probability.

Fixing (x_0, x_0'), we now calculate the probability that 2 chains of length 2^s, starting from arbitrary points (y_0, y_0'), collide on $\hat{x}$. This occurs if $y_0, y_1, \ldots, y_{2^s-i}$ collides with $x_0, x_1, \ldots, x_i$, and $y_0', y_1', \ldots, y_{2^s-i}'$ collides with $x_0', x_1', \ldots, x_i'$ (or vise-versa), which happens with probability $\Theta(2^{2(2s-\ell)})$ (assuming $0.25 \cdot 2^s \leq i \leq 0.75 \cdot 2^s$, all chains are of length $\Theta(2^s)$). This lower bounds the collision probability on $\hat{x}$ by $\Omega(2^{2(2s-\ell)})$. At the same time, the collision on $\hat{x}$ is also upper bounded by $O(2^{2(2s-\ell)})$, as all 4 chains are of length $O(2^s)$. We conclude that the collision probability on $\hat{x}$ is $\Theta(2^{2(2s-\ell)})$.

On the other hand, the probability that the chains starting from (y_0, y_0') collide on any point is $\Theta(2^{2s-\ell})$. Assuming that the collision search algorithm evaluates 2^t chains such that $2t+s \leq \ell$, then each evaluated chain is not expected to collide with more than one different chain, and the pairs of chains can essentially be analyzed independently.

We denote by A the event that the chains starting from (y_0, y_0') collide on $\hat{x}$, and by B the event that the chains starting from (y_0, y_0') collide. We are interested in calculating the conditional probability $\Pr[A|B]$, and we have $\Pr[A|B] = \Pr[A \cap B]/\Pr[B] = \Pr[A]/\Pr[B] = \Theta(2^{2(2s-\ell)-(2s-\ell)}) = \Theta(2^{2s-\ell})$, as required. □

Lemma 4. *Let $\hat{x}$ and $\hat{y}$ be two arbitrary same-offset collisions found, respectively, at offsets i and j by a collision search algorithm using chains of fixed length 2^s, with **independent ℓ-bit functions** f_i, such that $s < \ell/2$. Then $\Pr[(\hat{x}, i) = (\hat{y}, j)] = \Theta(2^{s-\ell})$. Furthermore, given that $i = j$, we have $\Pr[\hat{x} = \hat{y}] = \Theta(2^{2s-\ell})$.*

4   Filters

We describe the two types of filters that we use in our attacks in order to match (known) states computed offline with unknown states computed online.

4.1   Collision Filters

A simple filter that we use in some of our attacks was also used in the previous work of [16]. We build a collision filter $([b], [b'])$ for a state x offline by finding message blocks $([b], [b'])$ such that the states, obtained after processing these blocks from x, collide. In order to build this filter, we find a collision in the underlying hash function by evaluating its compression function for about $2^{\ell/2}$ different messages blocks from the state x. In order to test this filter online on the unknown node x' obtained after processing a message m', we simply check whether the tags of $m' \| [b]$ and $m' \| [b']$ collide. As the tags of $m' \| [b]$ and $m' \| [b']$ collide with probability $2^{-n} < 2^{-\ell}$ if the state obtained after processing m' is not x, we can conclude that the collision filter identifies the state x with high probability.

The complexity of building a collision filter offline is $O(2^{\ell/2})$. Testing the filter online requires querying the MAC oracle with $m' \| [b]$ and $m' \| [b']$, and assuming that the length of m' is $2^{s'}$, then it requires $O(2^{s'})$ time.

4.2 Diamond Filters

In order to build filters for 2^t nodes, we can build a collision filter for each one of them separately, requiring a total of $O(2^{t+\ell/2})$ time. However, this process can be optimized using the diamond structure, introduced by Kelsey and Kohno in the herding attack [13]. We now recall the details of this construction.

The diamond structure is built from a set of 2^t states x_i, constructing a set of messages m_i of length $O(t)$, such that iterating the compression function from any state x_i using message m_i leads to a fixed final state y. The structure is built in $O(t)$ iterations, where each iteration processes a layer of nodes, and outputs a smaller layer to be processed by the next iteration. This process terminates once the layer contains only one node, which is denoted by y.

Starting from the first layer with 2^t points, we evaluate the compression function from each point x_i with about $2^{(\ell-t)/2}$ random message blocks. This gives a total of about $2^{(\ell+t)/2}$ random values, and we expect them to contain about 2^t collisions. Each collision allows to match two different values x_i, x_j and to send them to a common value in the next layer, such that its size is reduced to about $1/2$. The message m_i for a state x_i is constructed by concatenating the $O(t)$ message blocks on its path leading to y. According to the detailed analysis of [15], the time complexity of building the structure is is $\Theta(2^{(\ell+t)/2})$.

Once we finish building the diamond structure, we construct a standard collision filter for the final node y, using some message blocks $([b], [b'])$. Thus, building a diamond filter offline for 2^t states requires $O(2^{(\ell+t)/2})$ time, which is faster than the $O(2^{t+\ell/2})$ time required to build a collision filter for each node separately.

In order to test the filter for a state x_i (in the first layer of the diamond structure), on the unknown node x' obtained after processing a message m' online, we simply check whether the tags of $m' \| m_i \| [b]$ and $m' \| m_i \| [b']$ collide. Assuming that the length of m' is $2^{s'}$, then the online test requires $O(t + 2^{s'})$ time.

Online Diamond Filter. A novel observation that we use in this paper, is that in some attacks it is more efficient to build the diamond structure online by calling the MAC oracle. Namely, we construct a diamond structure for the set of 2^t states x_i, where (the unknown) x_i is a result of querying the MAC oracle with a message M_i. Note the online construction is indeed possible, as the construction algorithm does not explicitly require the value of x_i, but rather builds the corresponding m_i by testing for collisions between the states (which can be detected according to collisions in the corresponding tags). However, testing for collisions online requires that all the messages M_i, for which we build the online diamond filter, are of the same length. Assuming that the messages M_i are of length 2^s, this construction requires $O(2^{s+(t+\ell)/2})$ calls to the compression function.

In order to test the filter for an unknown online state x_i, on a known state x'_i, we simply evaluate the compression function from x' on $m_i \| [b]$ and $m_i \| |[b']$, and check whether the resulting two states are equal. Thus, the offline test requires $O(t)$ time.

5   Internal State-Recovery for NMAC and HMAC with HAIFA

In this section, we describe the first internal state-recovery attack applicable to HAIFA (which can also be used as a distinguishing-H attack). Our optimized attack has a complexity of $\tilde{O}(2^{\ell-s})$ using messages of length 2^s, but this only applies with $s \leq \ell/5$; the lowest complexity we can reach is $2^{4\ell/5}$. We note that attacks against HAIFA can also be used to attack a Merkle-Damgård hash function; this gives more freedom in the queried messages by removing the need for long series of identical blocks as in [16].

In this attack, we fix a long sequence of random functions in order to reduce the entropy of the image states, based on Lemma 1. We then use an online diamond structure to match the states computed online with states that are compute offline. The detailed attack is as follows:

Attack 1: State-recovery attack against HMAC with HAIFA
Complexity $\tilde{O}(2^{\ell-s})$, with $s \leq \ell/5$ (min: $2^{4\ell/5}$)

1. (online) Fix a message C of length 2^s. Query the oracle with 2^u messages $M_i = [i]\|C$. Build an online diamond filter for the set of unknown states X, obtained after M_i.
2. (offline) Starting from 2^t arbitrary starting points, iterate the compression function with the fixed message C.
3. (offline) Test each image point x', obtained in Step 2, against each of the unknown states of X. If a match is found, then with high probability the state reached after the corresponding M_i is x'.

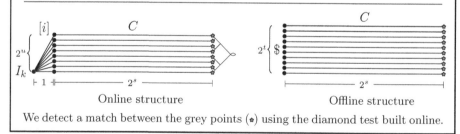

Online structure   Offline structure

We detect a match between the grey points (•) using the diamond test built online.

Complexity Analysis. In Step 3, we match the set X of size 2^u (implicitly computed during Step 1), and a set of size 2^t (computed during Step 2). We compare 2^{t+u} pairs of points, and each pair collides with probability $2^{s-\ell}$ according to Lemma 1. Therefore, the attack is successful with high probability if $t + u \geq \ell - s$. We now assume that $t = \ell - s - u$, and evaluate the complexity of each step of the attack:

Step 1: $2^{s+u/2+\ell/2}$ **Step 2:** $2^{s+t} = 2^{\ell-u}$ **Step 3:** $2^{t+u} \cdot u = 2^{\ell-s} \cdot u$

The lowest complexity is reached when all the steps of the attack have the same complexity, with $s = \ell/5$. More generally, we assume that $s \leq \ell/5$ and we set

$u = s$. This give an attack with complexity $\tilde{O}(2^{\ell-s})$ since $s + u/2 + \ell/2 = 3s/2 + \ell/2 \leq 4\ell/5 \leq \ell - s$.

6 New Tradeoffs for Merkle-Damgård

In this section, we revisit the results of [16], and give more flexible tradeoffs for various message lengths.

6.1 Trade-Off Based on Iteration Chains

In this attack, we match special states obtained using collision, based on Lemma 3. This attack extends the original tradeoff of [16] by using two improved techniques: first, while [16] used a fixed-offset offline collision search, we use a more general, free-offset offline collision search, which enables us to find collisions more efficiently. Second, while [16] used collision filters, we use a more efficient diamond filter.

Attack 2: Chain-based trade-off for HMAC with Merkle-Damgård
Complexity $O(2^{\ell-s})$, with $s \leq \ell/3$ (min: $2^{2\ell/3}$)

1. (offline) Use free-offset collision search from $2^{\ell-2s}$ starting points with chains of length 2^s, and find 2^c collisions (denoted by the set $\hat{X}$).
2. (offline) Build a diamond filter for the points in $\hat{X}$.
3. (online) Query the oracle with 2^t messages $M_i = [i] \parallel [0]^{2^s}$. Sort the tags, and locate 1 collision.
4. (online) Use a binary search to find the message prefix giving the unknown online collision state $\hat{y}$.
5. (online) Match the unknown online state $\hat{y}$ with each offline state in $\hat{X}$ using the diamond filter. If a match with $\hat{x} \in \hat{X}$ is found, then with very high probability $\hat{y} = \hat{x}$.

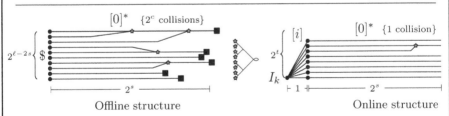

Offline structure   Online structure

We generate collisions offline using free-offset collision search, build a diamond filter for the collision points (•), and recover the state of an online collision.

Complexity Analysis. In Step 1, we use free-offset collision search with $2^{\ell-2s}$ starting points and chains of length 2^s, and thus according to Section 3.1, we find $2^{\ell-2s}$ collisions (*i.e.* $c = \ell - 2s$). Furthermore, according to Lemma 3, $\hat{y} \in \hat{X}$ with high probability, in which case the attack succeeds.

In Step 3, we use fixed-offset collision search with 2^t starting points and chains of length 2^s, and thus according to Section 3.1, we find $2^{2t+s-\ell}$ collisions. As we require one collision, we have $t = (\ell - s)/2$. We now compute the complexity of each step of the attack:

Step 1:	$2^{\ell/2+c/2} = 2^{\ell-s}$	**Step 2:**	$2^{\ell/2+c/2} = 2^{\ell-s}$
Step 3:	$2^{t+s} = 2^{(\ell+s)/2}$	**Step 4:**	$s \cdot 2^s$
Step 5:	$2^{c+s} = 2^{(\ell+s)/2}$		

With $s \leq \ell/3$, we have $(\ell + s)/2 \leq 2/3 \cdot \ell \leq \ell - s$, and the complexity of the attack is $O(2^{\ell-s})$.

6.2   Trade-Off Based on Cycles

We also generalize the cycle-based state-recovery attack of [16], which uses messages of length $2^{\ell/2}$ and has a complexity of $2^{\ell/2}$. Our attack uses (potentially) shorter messages of length 2^s for $s \leq \ell/2$, and has a complexity of $2^{2\ell-3s}$. The full attack and its analysis is given in the full version of this paper [6].

7   Shorter Message Attacks

In this section, we describe more complex attacks that can reach a tradeoff of $2^{\ell-2s}$, for relatively small values of s. These attacks are useful in cases where the message length of the underlying hash function is very restricted (*e.g.* the SHA-2 family). In order to reach a complexity of $2^{\ell-2s}$, we combine the idea of building filters in the online phase with lemmas 3 and 4.

In the case of Merkle-Damgård with identical compression functions, we reach a complexity of $2^{\ell-2s}$ for $s \leq \ell/8$, *i.e.* the optimal complexity of this attack is $2^{3/4\cdot\ell}$. With the HAIFA mode of operation, we reach a complexity of $2^{\ell-2s}$ for $s \leq \ell/10$ *i.e.* the optimal complexity of $2^{4/5\cdot\ell}$, matching the optimal complexity of the attack of Section 5.

7.1   Merkle-Damgård

Attack 3: Short message attack for HMAC with Merkle-Damgård
Complexity $\tilde{O}(2^{\ell-2s})$, with $s \leq \ell/8$ (min: $2^{3\ell/4}$)

---

1. (online) Query the oracle with 2^u messages $M_i = [i] \parallel [0]^{2^s}$, and locate 2^{c_1} collisions.
2. (online) For each collision (i, j), use a binary search to find the distance (offset) μ_{ij} from the starting point to the collision, and denote the (unknown) state reach after M_i (or M_j) by y_{ij}.
   Denote the set of all y_{ij} (containing about 2^{c_1} states) by Y. Build an online diamond filter for all the states in Y.

3. (offline) Run a free-offset collision search algorithm from 2^t starting points with chains of length 2^s, and locate 2^{c_2} collisions.
4. (offline) For each offline collision $\hat{x}$, match its iterates with all points $y_{ij} \in Y$: iterate the compression function with a zero message starting from $\hat{x}$ (up to 2^s times), and match iterate $2^s - \mu_{ij}$ (i.e., $f^{2^s - \mu_{ij}}(\hat{x})$) with y_{ij} using the diamond filter. If a match is found, then with high probability $y_{ij} = f^{2^s - \mu_{ij}}(\hat{x})$.

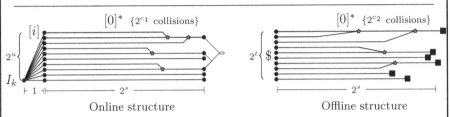

Online structure   Offline structure

We generate collisions and build a diamond filter online, and match them with collisions found offline.

Complexity Analysis. Using similar analysis to Section 6.1, we have $c_1 = 2u + s - \ell$ (as a pair of chains collide at the same offset with probability $2^{s-\ell}$, and we have 2^{2u} such pairs) and $c_2 = 2t + 2s - \ell$. The attack succeeds if the sets of collisions found online and offline intersect. According to Lemma 3, this occurs with high probability if $c_1 + c_2 \geq \ell - 2s$. In the following, we assume $c_1 + c_2 = \ell - 2s$.

Step 1: $2^{u+s} = 2^{s/2 + c_1/2 + \ell/2}$ **Step 2:** $2^{s + c_1/2 + \ell/2} = 2^{\ell - c_2/2}$

Step 3: $2^{t+s} = 2^{\ell/2 + c_2/2}$ **Step 4:** $2^{c_2 + s} + 2^{c_1 + c_2} \cdot c_1 = 2^{c_2 + s} + 2^{\ell - 2s} \cdot c_1$

The best tradeoffs are achieved by balancing steps 2 and 3, *i.e.* with $c_2 = \ell/2$. This reduces the complexity to:

Step 1: $2^{3\ell/4 - s/2}$ **Step 2:** $2^{3\ell/4}$

Step 3: $2^{3\ell/4}$ **Step 4:** $2^{\ell/2 + s} + 2^{\ell - 2s} \cdot \ell/2$

With $s \leq \ell/8$, we have $\ell/2 + s \leq 5\ell/8$ and $3\ell/4 \leq \ell - 2s$; therefore the complexity of the attack is $\tilde{O}(2^{\ell - 2s})$.

7.2   HAIFA

Since the attack is very similar to the previous attack on Merkle-Damgård, we only specify the differences between the attacks.

Attack 4: Short message attack for HMAC with HAIFA
Complexity $\tilde{O}(2^{\ell - 2s})$, with $s \leq \ell/10$ (min: $2^{4\ell/5}$)

- In Step 1 of Attack 3, we fix an arbitrary suffix C of length 2^s, and use $M_i = [i] \parallel C$.
- Correspondingly, in Step 3, we use a fixed-offset collision search by iterating the compression function with C from 2^t starting points.
- In Step 4, we match each offline collision $\hat{x}$, only with online collisions that occur at the same offset as $\hat{x}$. Thus, for each $\hat{x}$, we test only the end point of its chain (at offset 2^s) with the corresponding states in Y. Note that each $\hat{x}$ is matched with $2^{c_1} \cdot 2^{-s}$ states in Y on average.

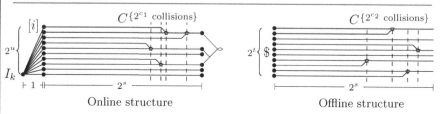

Online structure   Offline structure

We generate collisions and build an online diamond filter, and match them with offline collisions using the collision offset as a first filter.

Analysis. The attack succeeds in case there is a match between the set of collisions detected online and offline, that occurs at the same offset. According to Lemma 4, this match occurs with high probability when $c_1 + c_2 \geq \ell - s$, and thus we assume that $c_1 + c_2 = \ell - s$.

Complexity Analysis. Similar to the analysis of the previous attacks, we have $c_1 = 2u + s - \ell$ and $c_2 = 2t + s - \ell$.

Step 1: $2^{u+s} = 2^{s/2+c_1/2+\ell/2}$ **Step 2:** $2^{s+c_1/2+\ell/2} = 2^{\ell-c_2/2+s/2}$

Step 3: $2^{s+t} = 2^{s/2+c_2/2+\ell/2}$ **Step 4:** $2^{c_1+c_2-s} \cdot u = 2^{\ell-2s} \cdot u$

The best tradeoffs are achieved by balancing steps 2 and 3, *i.e.* with $c_2 = \ell/2$. This reduces the complexity to:

Step 1: $2^{3\ell/4}$ **Step 2:** $2^{3\ell/4+s/2}$ **Step 3:** $2^{3\ell/4+s/2}$ **Step 4:** $2^{\ell-2s} \cdot 3\ell/4$

With $s \leq \ell/10$, we have $3\ell/4 + s/2 \leq 4\ell/5 \leq \ell - 2s$; therefore the complexity of the attack is $\tilde{O}(2^{\ell-2s})$.

8   Universal Forgery Attacks with Short Queries

We now revisit the universal forgery attack of Peyrin and Wang [20]. In this attack, the adversary receives a challenge message of length 2^t at the beginning of the game, and interacts with the oracle in order to predict the tag of the challenge. The attack of [20] has two phases, where in the first phase, the adversary recovers the internal state of the MAC at some step during the computation on the challenge. In the second phase, the adversary uses a second-preimage attack on long messages in order to generate a different message with the same tag as the challenge.

The main draw back of the attack of Peyrin and Wang (as well as its recent improvement [9]) is that its first phase uses very long queries to the MAC oracle, regardless of the length of the challenge. In this section, we use the tools developed in this paper to devise two universal forgery attacks which use shorter queries to the MAC oracle. Our first universal forgery attack has a complexity of $2^{\ell-t}$ for $t \leq \ell/7$, using queries to the MAC oracle of length of at most 2^{2t} (which is much smaller than $2^{\ell/2}$ for any $t \leq \ell/7$). Thus, the optimal complexity of this attack is $2^{6\ell/7}$, obtained with a challenge of length at least $2^{\ell/7}$. Our second universal forgery attack has a complexity of only $2^{\ell-t/2}$. However, it is applicable for any $t \leq 2\ell/5$, using queries to the MAC oracle of length of at most 2^t. Thus, this attack has an improved optimal complexity of $2^{4\ell/5}$, which is obtained with a challenge of length at least $2^{2\ell/5}$.

In order to devise our attacks, we construct different state-recovery algorithms than the one used in [20], but reuse its second phase (i.e., the second-preimage attack) in both of the attacks. Thus, in the following, we concentrate of the state-recovery algorithms, and note that since the complexity of the second phase of the attack is $2^{\ell-t}$ for any value of t, it does not add a significant factor to the time complexity.

8.1 A Universal Forgery Attack Based on the Reduction of the Image-Set Size

Directly matching the 2^t states of the challenge message with some states evaluated offline is too expensive. Thus, we first reduce the number of nodes we match by computing and matching the images of the states under iterations of a fixed function. After matching the images, we can efficiently match and recover on the states of the challenge message.

We denote the challenge message as C, and the first κ blocks of C as $C_{|\kappa}$. The details of the first phase of the attack are as follows.

Attack 5: Universal forgery attack based on the reduction of the image-set size (first phase)
Complexity $\tilde{O}(2^{\ell-t})$, with $t \leq \ell/7$ (min: $2^{6\ell/7}$)

1. (online) Build a collision filter for the last (unknown) state z obtained during the computation of MAC(C).
2. (online) Query the oracle with 2^t messages $M_i = C_{|i} \parallel [0]^{2^{2t}-i}$. Denote the set of (unknown) final states of the chains by Y. Build a diamond filter for all states in Y.
3. (offline) Compute a structure of chains containing a total of $2^{\ell-t}$ points. Each chain is extended to a maximal length of 2^{2t+1}, or until it collides with a previous chain. Consider the set X of the 2^{2t} final states of all

---

the chains. According to Lemma 2, this set contains (no more than) about $2^{\ell-2t}$ distinct points, as all the points are in the image of $f^{2^{2t}}$.

4. (offline) Match all the points $x \in X$ with the 2^t points in Y. For each match between $x \in X$ and an online state in Y (obtained using M_i), restart the chains that merge into x, in order to locate all the points at a (backward) distance of $2^{2t} - i$ from x. Denoted this set by $\mathrm{CAND}(x)$.

5. (offline) Test the candidates: for each $x' \in \mathrm{CAND}(x)$, compute the state obtained by following the last $2^t - i$ blocks of the challenge message, and match this state with z using the collision filter. When a match is found, the state obtained after $C_{|i}$ is x' with high probability.

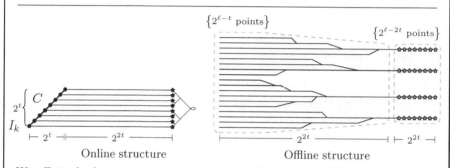

We efficiently detect a match between the challenge points ($\bullet$) and the first part of the offline structure, by first matching X ($\bullet$) and Y ($\bullet$).

Analysis. The structure of Step 3 contains $2^{\ell-t}$ points, and thus according to the birthday paradox, it covers one of the 2^t points of the challenge with high probability. In this case, the attack will successfully recover the state of the covered point with high probability, as we essentially have in the offline set X almost all $2^{\ell-2t}$ images of $f^{2^{2t}}$ (including the image of the covered point).

As X contains almost all $2^{\ell-2t}$ images of $f^{2^{2t}}$, we expect a match for every point in Y in Step 3 (a total of 2^t matches). In order to calculate the expected size of $\mathrm{CAND}(x)$ in Step 5, we first calculate the expected number of the endpoints of the chains computed in Step 3 of the attack. As the chains are of length of 2^{2t+1}, we expect that after evaluating the first $2^{\ell-4t}$ chains, a constant fraction of the chains will not collide with any other chain, and thus we have (at least) about $2^{\ell-4t}$ endpoints. Since the structure contains a total of $2^{\ell-t}$ points, each endpoint is a root of a tree of average size of (at most) $2^{\ell-t-(\ell-4t)} = 2^{3t}$. This gives about $2^{3t-2t} = 2^t$ candidates nodes $\mathrm{CAND}(x)$ at a fixed (backwards) distance (note that each $x' \in \mathrm{CAND}(x)$ is extended with a message of length about 2^t, according to the challenge).

Complexity.

Step 1:	$2^{\ell/2+t}$	**Step 2:**	$2^{2t+t/2+\ell/2} = 2^{\ell/2+5t/2}$
Step 3:	$2^{\ell-t}$	**Step 4:**	$t \cdot 2^{\ell-t}$
Step 5:	2^{3t}		

With $t \leq \ell/7$, we have $\ell/2 + 5t/2 \leq 6\ell/7 \leq \ell - t$; the complexity of the first phase of the universal forgery attack is $\tilde{O}(2^{\ell-t})$, and as the second phase has a similar complexity, this is also the complexity of the full attack.

8.2   A Universal Forgery Attack Based on Collisions

In this attack, we devise a different algorithm which recovers one of the states computed during the execution of the challenge message. The main idea here is to find collisions between chains evaluated online, and directly match them with collisions obtained offline. This is different from the previous algorithm, which matched the endpoints of the chains, rather than nodes on which the collisions occur. Once the collisions are matched, similarly to the previous algorithm, we obtain a small set of candidate nodes, which we match with the actual challenge nodes.

Attack 6: Universal forgery attack based on collisions (first phase)
Complexity $O(2^{\ell-t/2})$, with $t \leq 2\ell/5$ (min: $2^{4\ell/5}$)

1. (online) Query the oracle with 2^t messages $M_i = C_{|i} \parallel [0]^{2^{t+1}-i}$, and sort the tags.
2. (online) Execute state-recovery Attack 2 using messages of length $min(2^t, 2^{\ell/3})$, and denote by W a message of length 2^t whose last computed state is recovered.[a]
3. (online) Query the oracle with 2^v messages $W_j = W \parallel [j] \parallel 0^{2^t-1}$, sort the tags, and locate 2^c collisions with the tags computed using the messages M_i. For each collision of tags between M_i and W_j, find the first collision point using binary search (note that the state of the collision is known, as the state obtained after processing W is known). Store all the collision states $\hat{x}_{ij}$ in a sorted list, each one next to its distance d_{ij} from $C_{|i}$.
4. (offline) Compute a structure of chains containing a total of $2^{\ell-c}$ points. Each chain is extended to a maximal length of 2^{t+1}, or until it collides with a previous chain.
5. (offline) For each offline point in the structure y which collides with an online collision $\hat{x}_{ij}$ (i.e., $y = \hat{x}_{ij}$), retrieve candidate points $\text{CAND}(y)$ for the state obtained after processing $C_{|i}$. This is done by computing the d_{ij}-preimage points of y in the structure (i.e., the points which are at distance d_{ij} backwards from y). Assume that for each $y = \hat{x}_{ij}$, we have an average of 2^u candidate points, and thus we have a total of at most 2^{c+u} candidate points to test in the set $\bigcup_{ij}(\text{CAND}(y = \hat{x}_{ij}))$. Build a diamond filter for all the 2^{c+u} candidate points.
6. (online) For each $(\hat{x}_{ij}, y)$, match the state obtained after $C_{|i}$ with all the corresponding 2^u candidate points in $\text{CAND}(y)$ using the diamond filter. If a match is found, then with high probability the state obtained after processing $C_{|i}$ is equal to the tested candidate.

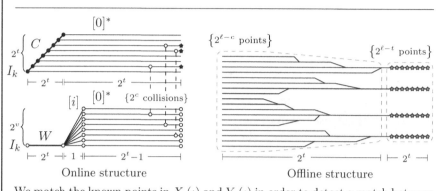

Online structure   Offline structure

We match the known points in X (•) and Y (•) in order to detect a match between the challenge points (•) and the first part of the offline structure.

[a] In case $t > \ell/3$, we first recover the last computed state of a message of size $2^{\ell/3}$, and then complement it arbitrarily to a length of 2^t.

Analysis. In Step 3 of the attack, we find 2^c collisions between pairs of chains, where the prefix of one chain in each pair is some challenge prefix $C_{|i}$. Thus, the 2^c collisions cover 2^c such challenge prefixes, and moreover, the offline structure, computed in Step 4, contains $2^{\ell-c}$ points. Thus, according to the birthday paradox, with high probability, the offline structure covers one of the states obtained after the computation of a prefix $C_{|i}$, such that the message M_i collides with some W_j on a point $\hat{x}_{ij}$ in Step 3. Since the state obtained after the computation of $C_{|i}$ is covered by the offline structure, then $\hat{x}_{ij}$ is also covered by the offline structure, and thus the state corresponding to $C_{|i}$ will be matched as a candidate and recovered in Step 6.

In order to calculate the value of c, note that the online structure, computed in Step 1, contains 2^t chains, each of length at least 2^t, and thus another arbitrary chain of length 2^t collides with one of the chains in this structure at the same offset with probability of about $2^{2t-\ell}$. Since the structure computed in Step 3 contains 2^v such chains, the expected number of detected collisions between the structures is $2^c = 2^{2t+v-\ell}$, i.e., $c = 2t + v - \ell$.

In order to calculate the value of u, we first calculate the expected number of the endpoints of the chains computed in Step 3 of the attack. As the chains are of length of 2^{t+1}, after evaluating the first $2^{\ell-2t}$ chains, a constant fraction of the chains will not collide with any other chain, and thus we have (at least) about $2^{\ell-2t}$ endpoints. Since the structure contains a total of $2^{\ell-c}$ points, each endpoint is a root of a tree of average size of (at most) $2^{\ell-c-(\ell-2t)} = 2^{2t-c}$. This gives about $2^{2t-c-t} = 2^{t-c}$ candidates nodes at a fixed depth, i.e., $u = t - c = \ell - t - v$.

We note that the last argument we use here is heuristic, as we assume that the average number of preimages at a certain distance for the specific collision points y is similar to the average for arbitrary points. However, steps 4 and 5 are not bottlenecks of the attack (as described in the complexity analysis below), and thus even if their complexity is somewhat higher, it will not effect the complexity of the full attack. Furthermore, we can perform a more complicated matching phase, in which we iteratively build filters for the offline structure at depths about $2^{t-1}, 2^{t-2}, \ldots$, and match them with the online structure. This guarantees that the expected complexity of the attack is as claimed below.

Step 1:	2^{2t}	**Step 2:**	$max(2^{\ell-t}, 2^{2\ell/3})$
Step 3:	2^{v+t}	**Step 4:**	$2^{\ell-c} = 2^{2\ell-2t-v}$
Step 5:	$(c+u) \cdot 2^{(c+u+l)/2} = t \cdot 2^{\ell/2+t/2}$	**Step 6:**	$2^{c+u+t} = 2^{2t}$

We balance steps 3 and 4 by setting $v + t = 2\ell - 2t - v$, or $v = \ell - 3t/2$. This gives a total complexity of $O(2^{\ell-t/2})$ for any $t \le 2\ell/5$.

9   Conclusions and Open Problems

In this paper, we provided improved analysis of HMAC and similar hash-based MAC constructions. More specifically, we devised the first state-recovery attacks on HMAC built using hash functions based on the HAIFA mode, and provided improved trade-offs between the message length and the complexity of state-recovery attacks for HMAC built using Merkle-Damgård hash functions. Finally, we presented the first universal forgery attacks which can be applied with short queries to the MAC oracle. Since it is widely deployed, our attacks have many applications to HMAC constructions used in practice, built using GOST, the SHA family, and other concrete hash functions.

Our results raise several interesting future work items such as devising efficient universal forgery attacks on HMAC built using hash functions based on the HAIFA mode, or proving that this mode provides resistance against such attacks. At the same time, there is also a wide gap between the complexity of the best known attacks and the security proofs for HMAC built using Merkle-Damgård hash functions. For example, the best universal forgery attacks on these MACs are still significantly slower than the birthday bound, which is the security guaranteed by the proofs.

References

1. Aumasson, J.P., Henzen, L., Meier, W., Phan, R.C.W.: SHA-3 proposal BLAKE. Submission to NIST (2008/2010), http://131002.net/blake/blake.pdf
2. Aumasson, J.-P., Neves, S., Wilcox-O'Hearn, Z., Winnerlein, C.: BLAKE2: Simpler, Smaller, Fast as MD5. In: Jacobson, M., Locasto, M., Mohassel, P., Safavi-Naini, R. (eds.) ACNS 2013. LNCS, vol. 7954, pp. 119–135. Springer, Heidelberg (2013)
3. Bellare, M.: New Proofs for NMAC and HMAC: Security Without Collision-Resistance. In: Dwork, C. (ed.) CRYPTO 2006. LNCS, vol. 4117, pp. 602–619. Springer, Heidelberg (2006)
4. Bellare, M., Canetti, R., Krawczyk, H.: Keying Hash Functions for Message Authentication. In: Koblitz, N. (ed.) CRYPTO 1996. LNCS, vol. 1109, pp. 1–15. Springer, Heidelberg (1996)

5. Biham, E., Dunkelman, O.: A Framework for Iterative Hash Functions - HAIFA. IACR Cryptology ePrint Archive, Report 2007/278 (2007)
6. Dinur, I., Leurent, G.: Improved Generic Attacks Against Hash-based MACs and HAIFA. IACR Cryptology ePrint Archive, Report 2014/441 (2014)
7. Dolmatov, V., Degtyarev, A.: GOST R 34.11-2012: Hash Function. RFC 6986 (Informational) (August 2013), http://www.ietf.org/rfc/rfc6986.txt
8. Ferguson, N., Lucks, S., Schneier, B., Whiting, D., Bellare, M., Kohno, T., Callas, J., Walker, J.: The Skein hash function family. Submission to NIST (2008/2010), http://skein-hash.info
9. Guo, J., Peyrin, T., Sasaki, Y., Wang, L.: Updates on Generic Attacks against HMAC and NMAC. In: Garay, J., Gennaro, R. (eds.) CRYPTO 2014, Part I. LNCS, vol. 8616, pp. 131–148. Springer, Heidelberg (2014)
10. Guo, J., Sasaki, Y., Wang, L., Wang, M., Wen, L.: Equivalent Key Recovery Attacks against HMAC and NMAC with Whirlpool Reduced to 7 Rounds. In: FSE (2014)
11. Guo, J., Sasaki, Y., Wang, L., Wu, S.: Cryptanalysis of HMAC/NMAC-Whirlpool. In: Sako, Sarkar (eds.) [21], pp. 21–40.
12. Joux, A.: Multicollisions in Iterated Hash Functions. Application to Cascaded Constructions. In: Franklin, M. (ed.) CRYPTO 2004. LNCS, vol. 3152, pp. 306–316. Springer, Heidelberg (2004)
13. Kelsey, J., Kohno, T.: Herding Hash Functions and the Nostradamus Attack. In: Vaudenay, S. (ed.) EUROCRYPT 2006. LNCS, vol. 4004, pp. 183–200. Springer, Heidelberg (2006)
14. Kelsey, J., Schneier, B.: Second Preimages on n-Bit Hash Functions for Much Less than 2^n Work. In: Cramer, R. (ed.) EUROCRYPT 2005. LNCS, vol. 3494, pp. 474–490. Springer, Heidelberg (2005)
15. Kortelainen, T., Kortelainen, J.: On Diamond Structures and Trojan Message Attacks. In: Sako, Sarkar (eds.) [21], pp. 524–539
16. Leurent, G., Peyrin, T., Wang, L.: New Generic Attacks against Hash-Based MACs. In: Sako, Sarkar (eds.) [21], pp. 1–20.
17. Naito, Y., Sasaki, Y., Wang, L., Yasuda, K.: Generic State-Recovery and Forgery Attacks on ChopMD-MAC and on NMAC/HMAC. In: Sakiyama, K., Terada, M. (eds.) IWSEC 2013. LNCS, vol. 8231, pp. 83–98. Springer, Heidelberg (2013)
18. van Oorschot, P.C., Wiener, M.J.: Parallel Collision Search with Cryptanalytic Applications. J. Cryptology 12(1), 1–28 (1999)
19. Peyrin, T., Sasaki, Y., Wang, L.: Generic Related-Key Attacks for HMAC. In: Wang, X., Sako, K. (eds.) ASIACRYPT 2012. LNCS, vol. 7658, pp. 580–597. Springer, Heidelberg (2012)
20. Peyrin, T., Wang, L.: Generic Universal Forgery Attack on Iterative Hash-based MACs. In: Nguyen, P.Q., Oswald, E. (eds.) EUROCRYPT 2014. LNCS, vol. 8441, pp. 147–164. Springer, Heidelberg (2014)
21. Sako, K., Sarkar, P. (eds.): ASIACRYPT 2013, Part II. LNCS, vol. 8270. Springer, Heidelberg (2013)
22. Tsudik, G.: Message authentication with one-way hash functions. SIGCOMM Comput. Commun. Rev. 22(5), 29–38 (1992), http://doi.acm.org/10.1145/141809.141812
23. Yasuda, K.: "Sandwich" Is Indeed Secure: How to Authenticate a Message with Just One Hashing. In: Pieprzyk, J., Ghodosi, H., Dawson, E. (eds.) ACISP 2007. LNCS, vol. 4586, pp. 355–369. Springer, Heidelberg (2007)

Cryptography from Compression Functions: The UCE Bridge to the ROM

Mihir Bellare, Viet Tung Hoang, and Sriram Keelveedhi

Dept. of Computer Science and Engineering, University of California San Diego, USA

Abstract. This paper suggests and explores the use of UCE security for the task of turning VIL-ROM schemes into FIL-ROM ones. The benefits we offer over indifferentiability, the current leading method for this task, are the ability to handle multi-stage games and greater efficiency. The paradigm consists of (1) Showing that a VIL UCE function can instantiate the VIL RO in the scheme, and (2) Constructing the VIL UCE function given a FIL random oracle. The main technical contributions of the paper are domain extension transforms that implement the second step. Leveraging known results for the first step we automatically obtain FIL-ROM constructions for several primitives whose security notions are underlain by multi-stage games.Our first domain extender exploits indifferentiability, showing that although the latter does not work directly for multi-stage games it can be used indirectly, through UCE, as a tool for this end. Our second domain extender targets performance. It is parallelizable and shown through implementation to provide significant performance gains over indifferentiable domain extenders.

1 Introduction

Two forms of the random oracle model (ROM) of BR [9] have emerged, namely the VIL-ROM and FIL-ROM. In the VIL-ROM, the random oracle, denoted RO, is variable input length (VIL), meaning takes inputs of arbitrary length. In the FIL-ROM, the random oracle, denoted ro, is fixed input length (FIL), meaning only takes inputs of one, particular length. The VIL-ROM is preferable for the design and analysis of ROM schemes and reflects the original view of BR [9] that random oracles would be instantiated by cryptographic hash functions that, like SHA-256, take variable length inputs. However hash functions are built in a very structured way from their underlying compression functions. This lead researchers beginning with Coron, Dodis, Malinaud and Puniya [14] to suggest that it should be the compression function, rather than the hash function, that is treated as "ideal," leading to the FIL-ROM. Indeed, SHA-256 is built from its compression function sha-256 in a way that renders SHA-256 subject to the extension attack, which can lead to attacks when SHA-256 is used to instantiate a VIL random oracle. Treating the compression function (rather than the full hash function) as the ideal object is more reflective of the design goals and intuition of practitioners and leads to better security.

J.A. Garay and R. Gennaro (Eds.): CRYPTO 2014, Part I, LNCS 8616, pp. 169–187, 2014.

The consensus then is that we should design schemes in the FIL-ROM. The question is how best to do this. One option is to directly design and analyze schemes in this model, but this is difficult and ad hoc. A better option is to provide a construction E^{ro} of a VIL function that can substitute a VIL RO, meaning we would design schemes secure in the VIL-ROM as usual and then automatically replace RO with E^{ro} to obtain security in the FIL-ROM. We refer to such an E as a domain extension construction or domain extender.

For this to work in some broad and useful way, we need a *definition* of some property, call it X, that, if satisfied by E^{ro}, allows the latter to securely replace RO in the VIL-ROM and thus provide security in the FIL-ROM, for some useful and hopefully large set of schemes that are proven secure in the VIL-ROM. The leading proposal for X is "indifferentiability from a random oracle" as defined by Maurer, Renner and Holenstein [19] and advocated by [14].

This paper suggests, and explores, an alternative X. We suggest that X be the notion of UCE (Universal Computational Extractor) security defined by BHK [6]. Our results will show both theoretical and practical benefits of X= UCE over X= indifferentiability in this role. On the theoretical side, UCE allows us to move from the VIL-ROM to the FIL-ROM for primitives whose security is defined via multi-stage games, a setting where indifferentiability fails [23,15]. On the practical side, we exhibit UCE domain extenders E that are significantly more efficient than known indifferentiability ones, in particular parallelizable to take advantage of modern multi-core machines, our efficiency claims being not just asymptotic but supported by implementations and experiments. Conceived as a way to remove random oracles, UCE now becomes a bridge to better security in the ROM.

LIMITATIONS OF INDIFFERENTIABILITY. While indifferentiability works well in some settings, it has two major limitations. The first is that indifferentiable-from-RO functions do not suffice to securely replace a VIL random oracle for primitives whose security definition is underlain by multi-stage games [23,15]. This gap is more than academic, for we are seeing the emergence of numerous primitives and security notions of practical importance whose definitions are inherently multi-stage. Examples include Deterministic PKE (D-PKE) [5], Message-Locked Encryption (MLE) [8], and proofs of storage. In each case there are natural, efficient and canonical solutions in the VIL-ROM that we would like to implement in the FIL-ROM, but indifferentiability offers no way to do this.

The second limitation of indifferentiability is performance. Typical indifferentiable domain extenders iterate the compression function sequentially. This means that instantiations are left unable to take advantage of modern multi-core processors to provide performance gains. This reduces the potential for high volume usage and deployment of cryptography based on compression functions.

OUR PERSPECTIVE. We conceptualize the goal that motivated the use of indifferentiability as aiming to design an X-secure domain extender —this being a construction E^{ro} that, given the FIL random oracle ro, computes a VIL, X-secure function— for a "good" choice of X, meaning one that allows E^{ro} to securely replace RO in the VIL-ROM for some significant set of applications.

Method	Notions	Performance	Applications
Keyed-Indiff	UCE[$\mathcal{S}^{\text{srs}}$] UCE[$\mathcal{S}^{\text{crs}}$]	About $m/(m-n)$ times the speed of M	All schemes in [6]
AU-then-Hash	UCE[$\mathcal{S}^{\text{sup}}$]	Parallelizable ~ 0.4 cycles per byte	MLE, key derivation, storage auditing

Fig. 1. Our UCE domain extension constructions and their properties. The second column gives the UCE notion that is achieved. M is the indifferentiable domain extender used in the first construction. The numbers n and m are the key length of the hash function and the input length of the ideal compression function, respectively. Typically, $n = 128$ and $m = 512$.

While X=indifferentiability has been very successful in some domains, it also, as discussed above, has important limitations. We ask if there are alternative definitions X that can overcome these limitations and complement indifferentiability in its role.

The core limitation of indifferentiability is the inability to handle multi-stage games. We suggest that a natural route around this is that X-*security itself be multi-stage*. The particular candidate X we suggest is the UCE notion of [6], which is indeed multi-stage. Our suggested UCE-based paradigm to move schemes from the VIL-ROM to the FIL-ROM has two steps: (1) Show that instantiating the VIL random oracle in the scheme with a VIL UCE function preserves security, and (2) Implement the VIL UCE function as E^{ro} to obtain a FIL-ROM scheme. Prior work has already given us the first step for many constructions: UCE-secure hash functions are shown in [6] to be able to securely instantiate VIL random oracles for diverse multi-stage applications including the important practical ones noted above and all examples of multi-stage schemes listed in [23]. The missing element is UCE domain extenders E for the second step. If we had those, we could immediately harvest the existing results to get FIL-ROM constructions for many multi-stage primitives. The concrete quest that emerges, then, is for UCE domain extenders.

OUR RESULTS. Our core contribution is two new domain extenders for UCE that together allow us to reach the above goals of security and speed. These are constructions E that take a FIL random oracle ro and return a VIL, keyed function E^{ro} that meets UCE security notions of BHK in the FIL-ROM. (UCE hash functions are keyed, whence the introduction of a key in this setting. Also, UCE is not a monolithic or single security notion, but rather a framework in which one parameterizes notions of security by classes of "sources." Applications rely on different choices of the starting class. The framework is recalled in Section 3. Here we will avoid the details beyond noting for which classes each of our constructions is secure and what this entails for applications.) See Fig. 1 for a summary of the two domain extenders and their properties.

Our first construction is generic, turning any indifferentiable domain extender into a UCE domain extender. Given an indifferentiable domain extender M, we show that the hash family $\mathsf{H}_{hk} = \mathsf{M}^{\text{ro}(hk\,\|\,\cdot)}$ is UCE-secure. The forms of UCE for which this works are enough to prove security for all schemes listed in [6], for example the EwH D-PKE scheme of [5], or the storage-auditing scheme used in [23] as a counterexample for the failure of the indifferentiability framework in multi-stage settings.

This construction illustrates what we believe is an interesting relation between UCE and indifferentiability. Indifferentiability cannot *directly* yield the applications we have obtained for multi-stage primitives. However, it can be used, in a blackbox way, to create a domain extender that meets a *particular* multi-stage notion of security, namely UCE. Then, exploiting known UCE results, we can obtain FIL-ROM security for *many* multi-stage primitives. Thus our construction shows how to use UCE to leverage indifferentiability to solve a problem that indifferentiability could not solve directly.

While our first construction delivers, we believe, important advances on the theoretical front, its performance is that of the underlying indifferentiable construction. Our second construction targets speed. It follows the Carter-Wegman paradigm [13], first using an almost-universal hash to condense the input, and then running $\mathrm{ro}(K \,\|\, \cdot)$ on the result, where K is the hash key. This gives us highly efficiently, fully parallelizable hash constructions that are not achievable if the target is indifferentiability. In more detail, we show that if F is almost-universal, then the hash family $\mathsf{H}_{hk}(x) = \mathrm{ro}(K \,\|\, \mathsf{F}_{fk}(x))$, with $hk = (fk, K)$, is UCE-secure. The most important application here is the message-locked encryption (MLE) scheme CE of [8]. Due to the space constraint, we leave the proofs of our theorems to the full version [7].

GENERAL DOMAIN EXTENSION. Above we presented the domain extension problem for notion X as being to design E such that E^{ro} is a VIL X-secure function in the FIL-ROM. More generally, the problem is to design E such that if $\overline{\mathsf{H}}$ is a FIL X-secure function then $\mathsf{E}^{\overline{\mathsf{H}}}$ is a VIL X-secure function. Here $\overline{\mathsf{H}}$ can be a FIL-ROM function, and thus the prior formulation is the special case $\overline{\mathsf{H}}_{hk}(\cdot) = \mathrm{ro}(hk\|\cdot)$. Our first construction discussed above generalizes to solve this problem, letting $\mathsf{H}_{hk} = \mathsf{M}^{\overline{\mathsf{H}}(hk\,\|\,\cdot)}$ where M, as before, is an indifferentiable domain extender. Setting $\overline{\mathsf{H}}_{hk}(\cdot) = \mathrm{ro}(hk\|\cdot)$ recovers the result stated above. The generalization however yields something new, namely a *standard model* domain extender for UCE. This follows by letting $\overline{\mathsf{H}}$ be a standard model FIL UCE function. This is interesting because it shows that indifferentiability, which so far has been a ROM notion and tool, can be leveraged to get results purely in the standard model.

INSTANTIATION AND EXPERIMENTAL RESULTS. We give a very fast instantiation of F based on reduced-round AES and polynomial-based evaluation. Our construction makes use of the fact that four-round AES, with the four subkeys chosen uniformly and independently, is an almost-xor-universal hash [18]. We stress that our universal hashing construction is unconditional, making no assumption on AES. This leads to a highly efficient, parallelizable UCE-secure hash FastHash. Our experiments show that even in the sequential setting, FastHash is about 5.3 times faster than SHA-256. When parallelism is employed, FastHash achieves a much better speedup, about 24 times faster than SHA-256. Finally, we demonstrate the utility of FastHash by giving an extremely fast MLE scheme.

RELATED WORK. Mittelbach [21] defines restrictions on a multi-stage game so that the indifferentiability composition theorem still holds for a subclass of indifferentiable domain extenders called iterative domain extenders, and is thereby able to show that the latter suffice for applications like D-PKE and MLE. He also shows that if M is an iterative domain extender then M^{ro} is UCE-secure. In comparison, our first construction is more general in the following ways: It is able to use any indifferentiable domain extender, and as a result our applications are able to use a broader class of domain extenders; it turns any FIL UCE function into a VIL one; it works both in the standard model and the ROM. On the other hand, Mittelbach's construction is about $m/(m-n)$ times faster than ours, where m is the input length of the compression function, and n is the key length.

Dodis, Ristenpart, and Shrimpton [16] define preimage-awareness (PrA) as a strengthening of collision resistance and show that the plain Merkle-Damgård is a PrA extender. PrA can also be used in multi-stage games: Ristenpart, Shacham, and Shrimpton [23] show how to compose a PrA-secure hash with a FIL RO to achieve D-PKE.

Some versions of UCE are shown by [12] to be unachievable in the standard mode if indistinguishability obfuscation for all circuits exists, but most of the applications in [6] only need weaker versions of UCE where our domain extenders work but the attacks in [12] do not. All versions of UCE in [6] are shown by the latter to be achievable in the VIL-ROM, so our domain extenders achieve all the applications in the FIL-ROM.

2   Preliminaries

Concrete security bounds are important for applications. However, notions in the current domain, involving simulators and multiple conditions and adversaries, are complex. The result is that when theorems are stated purely concretely, it is hard to understand the (much more simple) conceptual import. We will try to achieve the "best of both worlds." We formulate definitions asymptotically. The first cut theorem statements are asymptotic so that one can quickly see the core implication and result. This is followed by a concrete statement with bounds.

NOTATION. By $\lambda \in \mathbb{N}$ we denote the security parameter. If $n \in \mathbb{N}$ then 1^n denotes its unary representation. We denote the size of a finite set X by $|X|$, the number of coordinates of a vector $\mathbf{x}$ by $|\mathbf{x}|$, and the length of a string $x \in \{0,1\}^*$ by $|x|$. We let ε denote the empty string. If x is a string then $x[i]$ is its i-th bit and $x[1, \ell] = x[1] \dots x[\ell]$. By $x \| y$ we denote the concatenation of strings x, y. If X is a finite set, we let $x \leftarrow_\$ X$ denote picking an element of X uniformly at random and assigning it to x. Algorithms may be randomized unless otherwise indicated. Running time is worst case. "PT" stands for "polynomial-time," whether for randomized algorithms or deterministic ones. If A is an algorithm, we let $y \leftarrow A(x_1, \ldots; r)$ denote running A with random coins r on inputs $x_1, \ldots$ and assigning the output to y. We let $y \leftarrow_{\$} A(x_1, \ldots)$ be the resulting of picking r at random and letting $y \leftarrow A(x_1, \ldots; r)$. We let $[A(x_1, \ldots)]$ denote the set of all possible outputs of A when invoked with inputs $x_1, \ldots$. We say that $f : \mathbb{N} \to \mathbb{R}$ is negligible if for every positive polynomial p, there exists $n_p \in \mathbb{N}$ such that $f(n) < 1/p(n)$ for all $n > n_p$.

GAMES. We use the code based game playing framework of [10]. (See Fig. 3 for an example.) By $\mathrm{G}^{A_1, A_2, \cdots}(\lambda) \Rightarrow y$ we denote the event that the execution of game G with adversaries $A_1, A_2, \ldots$ and security parameter λ results in output y. We abbreviate $\mathrm{G}^{A_1, A_2, \cdots}(\lambda) \Rightarrow$ true by $\mathrm{G}^{A_1, A_2, \cdots}(\lambda)$, the occurrence of this event meaning that $A_1, A_2, \ldots$ win the game.

For concrete security assessments, let the *number of queries* of A to an oracle Proc be the function $\mathbf{Q}_A^{\mathsf{Proc}}$ that on input λ returns the maximum number of queries that A makes to Proc when executed with security parameter λ, the maximum over all coins and all possible replies to queries to all oracles of A. Time assessments are simplified by the convention that running time is that of the game rather than merely the adversary, and we let $\mathbf{T}(\mathrm{G}^{A_1, A_2, \cdots})$ denote the function of λ that returns the maximum execution time of game G with adversaries $A_1, A_2, \ldots$ and security parameter λ, the maximum over all coins, and the time being all inclusive, meaning the time taken by game procedures to compute replies is included.

RANDOM ORACLES. A random oracle $\mathsf{RO} : U \to \{0,1\}^n$ is a procedure that maintains a table H, initially empty, and is defined by

$\underline{\mathsf{RO}(x)}$

If $H[x] \neq \bot$ then $H[x] \leftarrow_{\$} \{0,1\}^n$; Return $H[x]$

We say that RO is variable-input length (VIL) if $U = \{0,1\}^*$ and fixed-input length (FIL) if there is $m \in \mathbb{N}$ such that $U = \{0,1\}^m$. Formally, any random oracle referred to in a game should appear explicitly in the game as a procedure defined as above, but for the same of brevity of game descriptions, we omit writing it explicitly, instead only indicating the domain and range of each random oracle. By convention, RO indicates a VIL random oracle, and ro a FIL random oracle.

3   UCE Framework

The Universal Computational Extractor (UCE) framework of BHK [6] is intended to define security notions for families of hash functions in the standard model, but BHK also lift this to the ROM to show its achievability there. We use the latter with the random oracle being FIL. We note that the standard-model definition is the special case where parties and algorithms make no queries to the random oracle.

BHK first give a single-key version of the definition and then extend it to a multi-key one. We will work directly with the multi-key version, calling it UCE rather than mUCE as in [6].

FUNCTION FAMILIES. Our syntax for function families follows [6], in particular allowing variable output lengths. A family of functions H specifies the following. On input the unary representation 1^λ of the security parameter $\lambda \in \mathbb{N}$, key generation algorithm H.Kg returns a key $hk \in \{0,1\}^{\text{H.kl}(\lambda)}$, where H.kl: $\mathbb{N} \to \mathbb{N}$ is the keylength function associated to H. The deterministic, PT evaluation algorithm H.Ev takes 1^λ, a key $hk \in [\text{H.Kg}(1^\lambda)]$, an input $x \in \{0,1\}^*$ with $|x| \in \text{H.IL}(\lambda)$, and a unary encoding 1^ℓ of an output length $\ell \in \text{H.OL}(\lambda)$ to return $\text{H.Ev}(1^\lambda, hk, x, 1^\ell) \in \{0,1\}^\ell$. Here H.IL is the input-length function associated to H, so that $\text{H.IL}(\lambda) \subseteq \mathbb{N}$ is the set of allowed input lengths, and similarly H.OL is the output-length function associated to H, so that $\text{H.OL}(\lambda) \subseteq \mathbb{N}$ is the set of allowed output lengths. The latter allows us to cover functions of variable output length. If H has fixed input length then let H.il denote the function such that $\text{H.IL}(\lambda) = \{\text{H.il}(\lambda)\}$ for every $\lambda \in \mathbb{N}$. If H has fixed output length, define H.ol likewise. In the ROM, we allow H.Ev access to a FIL random oracle denoted ro. We write H.Ev^{ro} to indicate explicitly that H.Ev needs access to a FIL random oracle ro.

FRAMEWORK. Let H be a family of functions. Let S be an adversary called the *source* and D an adversary called the *distinguisher*. We associate to them and H the game $\text{UCE}_H^{S,D}(\lambda)$ in the left panel of Fig. 2. Initially, the source specifies a unary-encoded integer $n \geq 1$ to indicate the number of hash keys that it wants to use. The game then chooses a secret vector $\mathbf{hk}$ of n uniformly random hash keys and grants the source access to an oracle Hash. We require that any query $(x, 1^\ell, i)$ made to this oracle satisfy $|x| \in \text{H.IL}(\lambda)$, $\ell \in \text{H.OL}(\lambda)$ and $i \in \{1, \ldots, n\}$. When the challenge bit b is 1 (the "real" case) the oracle responds via H.Ev under $\mathbf{hk}[i]$. When $b = 0$ (the "random" case) it responds via the ith random-oracle procedure. The source then leaks a string L to its accomplice distinguisher. The latter *does* get the keys $\mathbf{hk}$ as input and must now return its guess $b' \in \{0,1\}$ for b. The game returns true iff $b' = b$, and the uce-advantage of (S, D) is defined for $\lambda \in \mathbb{N}$ via

$$\text{Adv}_{H,S,D}^{\text{uce}}(\lambda) = 2\Pr[\text{UCE}_H^{S,D}(\lambda)] - 1 .$$

If S is a class (set) of sources, we say that H is UCE[$\mathcal{S}$]-secure if $\text{Adv}_{H,S,D}^{\text{uce}}(\cdot)$ is negligible for all sources $S \in \mathcal{S}$ and all PT distinguishers D. Trivial attacks from [6] show that UCE[$\mathcal{S}$]-security is not achievable if $\mathcal{S}$ is the class of all PT sources. To obtain meaningful notions of security, BHK [6] impose restrictions on the source. There are many ways to do this; below we'll focus on what they call unpredictable and reset-secure sources. To discuss the concrete security of constructions it will be useful to say that S is a N-key source if we always have $n \leq N(\lambda)$ when $(1^n, t) \leftarrow_\$ S(1^\lambda, \varepsilon)$.

UNPREDICTABLE SOURCES. A source is unpredictable if it is hard to guess the source's Hash queries even given the leakage, in the *random case* of UCE game. Formally, let S be a source and P an adversary called a predictor. Consider game $\text{Pred}_S^P(\lambda)$ in the middle panel of Fig. 2 associated to S, P. Given 1^n and

Game $\text{UCE}_H^{S,D}(\lambda)$	$\text{Hash}(x, 1^\ell, i)$
$(1^n, t) \leftarrow_\$ S(1^\lambda, \varepsilon)$	If $T[x, \ell, i] = \bot$ then
For $i = 1, \ldots, n$ do $\mathbf{hk}[i] \leftarrow_\$ \text{H.Kg}(1^\lambda)$	If $b = 0$ then $T[x, \ell, i] \leftarrow_\$ \{0, 1\}^\ell$
$b \leftarrow_\$ \{0, 1\}$; $L \leftarrow_\$ S^{\text{Hash,ro}}(1^\lambda, t)$	Else $T[x, \ell, i] \leftarrow \text{H.Ev}^{\text{ro}}(1^\lambda, \mathbf{hk}[i], x, 1^\ell)$
$b' \leftarrow_\$ D^{\text{ro}}(1^\lambda, \mathbf{hk}, L)$; Return $(b' = b)$	Return $T[x, \ell, i]$

Game $\text{Pred}_S^P(\lambda)$	Game $\text{Reset}_S^R(\lambda)$
$(1^n, t) \leftarrow_\$ S(1^\lambda, \varepsilon)$	$U \leftarrow \emptyset$; $(1^n, t) \leftarrow_\$ S(1^\lambda, \varepsilon)$
$Q \leftarrow \emptyset$; $L \leftarrow_\$ S^{\text{Hash,ro}}(1^n, t)$	$L \leftarrow_\$ S^{\text{Hash,ro}}(1^n, t)$; $b \leftarrow_\$ \{0, 1\}$
$Q' \leftarrow_\$ P^{\text{ro}}(1^\lambda, 1^n, L)$	If $b = 0$ then // reset the array T
Return $(Q' \cap Q \neq \emptyset)$	For $(x, \ell, i) \in U$ do $T[x, \ell, i] \leftarrow_\$ \{0, 1\}^\ell$
	$b' \leftarrow_\$ R^{\text{Hash,ro}}(1^\lambda, L)$; Return $(b = b')$
$\text{Hash}(x, 1^\ell, i)$	
$Q \leftarrow Q \cup \{x\}$	$\text{Hash}(x, 1^\ell, i)$
If $T[x, \ell, i] = \bot$ then $T[x, \ell, i] \leftarrow_\$ \{0, 1\}^\ell$	If $T[x, \ell, i] = \bot$ then $T[x, \ell, i] \leftarrow_\$ \{0, 1\}^\ell$
Return $T[x, \ell, i]$	$U \leftarrow U \cup \{(x, \ell, i)\}$; Return $T[x, \ell, i]$

Fig. 2. Games UCE **(top),** Pred **(bottom left), and** Reset **(bottom right) to define UCE security.** Here $\text{ro} : \{0, 1\}^{\text{ro.il}(\lambda)} \to \{0, 1\}^{\text{ro.ol}(\lambda)}$ is a random oracle.

the leakage, the predictor outputs a set Q'. The predictor wins if Q' contains a Hash-query of the source. For $\lambda \in \mathbb{N}$ we let

$$\text{Adv}_{S,P}^{\text{pred}}(\lambda) = \Pr[\text{Pred}_S^P(\lambda)] .$$

We require that the size of Q', as well as the number of queries that P makes to ro, be bounded by a polynomial (allowed to depend on P) in λ. We say that S is computationally (respectively, statistically) unpredictable if $\text{Adv}_{S,P}^{\text{pred}}(\cdot)$ is negligible for all PT (respectively, all, even computationally unbounded) predictors P. We let $\mathcal{S}^{\text{cup}}$ be the class of computationally unpredictable PT sources, and $\mathcal{S}^{\text{sup}}$ the class of statistically unpredictable PT sources. The corresponding security notions for H are $\text{UCE}[\mathcal{S}^{\text{cup}}]$ and $\text{UCE}[\mathcal{S}^{\text{sup}}]$.

RESET-SECURE SOURCES. We recall the second restriction on sources from [6], called reset security. Let S be a source and R an adversary called a reset adversary. The source again is executed with its Hash being a random oracle. The reset adversary is either given access to the same random oracle or to an *independent* one. The requirement is that it should not be able to tell which. Formally, consider game $\text{Reset}_S^R(\lambda)$ at the right panel of Fig. 2 associated to S, R. For $\lambda \in \mathbb{N}$ we let

$$\text{Adv}_{S,R}^{\text{reset}}(\lambda) = 2 \Pr[\text{Reset}_S^R(\lambda)] - 1 .$$

We require that the number of queries that P makes to Hash and ro be bounded by a polynomial (allowed to depend on R) in λ. We say S is computationally (respectively, statistically) reset-secure if $\text{Adv}_{S,R}^{\text{reset}}(\cdot)$ is negligible for all PT (respectively, all, even computationally unbounded) reset adversaries R. We let $\mathcal{S}^{\text{crs}}$ be the class of all PT computationally reset-secure sources, and $\mathcal{S}^{\text{srs}}$ the class of

Game $\text{Indiff}^A_{M,\overline{M}}(\lambda)$	$\text{Func}(x)$	$\text{Prim}(x)$
$b \leftarrow\!\!\$ \{0,1\}$; $st \leftarrow \varepsilon$	If $b = 1$ then return $M^{\text{ro}}(1^\lambda, x)$	If $b = 1$ then return $\text{ro}(x)$
$b' \leftarrow\!\!\$ A^{\text{Prim},\text{Func}}(1^\lambda)$	Else return $\text{RO}(x)$	$(y, st) \leftarrow\!\!\$ \overline{M}^{\text{RO}}(1^\lambda, st, x)$
Return $(b = b')$		Return y

Fig. 3. Game Indiff **defining indifferentiability.** Here $\text{RO} : \{0,1\}^* \to \{0,1\}^{M.\text{fol}(\lambda)}$ and $\text{ro} : \{0,1\}^{M.\text{pil}(\lambda)} \to \{0,1\}^{M.\text{pol}(\lambda)}$ are random oracles.

all PT statistically reset-secure sources. The corresponding security notions for H are $\text{UCE}[\mathcal{S}^{\text{crs}}]$ and $\text{UCE}[\mathcal{S}^{\text{srs}}]$.

RELATIONS AND ACHIEVABILITY. Reset security is a relaxation of unpredictability. In particular BHK [6] show that $\text{UCE}[\mathcal{S}^{\text{crs}}]$-security of H implies $\text{UCE}[\mathcal{S}^{\text{cup}}]$-security of H and $\text{UCE}[\mathcal{S}^{\text{srs}}]$-security of H implies $\text{UCE}[\mathcal{S}^{\text{sup}}]$-security of H. The converses are not necessarily true. BFM [12] show that if indistinguishability obfuscation for all circuits is possible then $\text{UCE}[\mathcal{S}^{\text{crs}}]$-security is not achievable in the standard model. In the ROM however BHK [6] show that both $\text{UCE}[\mathcal{S}^{\text{crs}}]$-security and $\text{UCE}[\mathcal{S}^{\text{srs}}]$-security are achievable.

4   UCE from Indifferentiability

We first review necessary definitions of the indifferentiability framework [19].

INDIFFERENTIABILITY. We consider an algorithm M that, given a FIL random oracle ro, attempts to have input-output behavior approximating that of a VIL random oracle. Indifferentiability provides one definition of what it means for M to succeed at this task. Consider game $\text{Indiff}^A_{M,\overline{M}}(\lambda)$ of Fig. 3 associated to M, an algorithm $\overline{M}$ called a simulator, and an adversary A. In the first world ($b = 1$), oracle Prim implements the FIL random oracle ro while oracle Func implements the construction, namely M^{ro}, that aims to approximate a VIL random oracle. In the second world ($b = 0$), oracle Func implements a true VIL random oracle RO while replies to Prim queries are determined by the simulator that itself has access to RO. The simulator is stateful, its state st being maintained by the game. The input x to M has arbitrary length, the oracle provided to M maps $M.\text{pil}(\lambda)$-bit inputs to $M.\text{pol}(\lambda)$-bit outputs, and M returns outputs of length $M.\text{fol}(\lambda)$, where $M.\text{pil}, M.\text{pol}, M.\text{fol} : \mathbb{N} \to \mathbb{N}$ are functions associated to M called the input-length of M's primitive, output-length of M's primitive, and output-length of M's functionality, respectively. For $\lambda \in \mathbb{N}$ we let

$$\text{Adv}^{\text{indiff}}_{M,\overline{M},A}(\lambda) = 2 \Pr[\text{Indiff}^A_{M,\overline{M}}(\lambda)] - 1 .$$

We require that the number of queries that A makes to its oracles be bounded by a polynomial (allowed to depend on A) in λ. Then we say that M is a *pseudorandom oracle* (PRO) if there is a PT simulator $\overline{M}$ such that $\text{Adv}^{\text{indiff}}_{M,\overline{M},A}(\cdot)$ is negligible for every (even computationally unbounded) adversary A.

For concrete security assessments we let $Q_{\overline{M},q}$ be the function that on input λ returns the maximum, over all $x_1, \ldots, x_q \in \{0,1\}^{M.\text{pil}(\lambda)}$, of the total number of oracle queries that $\overline{M}$ makes when run sequentially on inputs $x_1, \ldots, x_q$, starting from state ε. Also let $T_{\overline{M},q}$ be the function that on input λ returns the maximum, over all $x_1, \ldots, x_q \in \{0,1\}^{M.\mathsf{pil}(\lambda)}$, of the total running time of $\overline{M}$ when run sequentially on inputs $x_1, \ldots, x_q$, starting from state ε, the time for an oracle query being taken as linear in the length of the query and reply.

THE Keyed-Indiff EXTENDER. Let $\overline{H}$ be a FIL function family that is $\mathsf{UCE}[\mathcal{S}^{\mathsf{xxx}}]$-secure for some xxx. We want to build a VIL family of functions H that is also $\mathsf{UCE}[\mathcal{S}^{\mathsf{xxx}}]$-secure. Our construction uses as a tool any PRO M with $M.\mathsf{pil} = \overline{H}.\mathsf{il}$ and $M.\mathsf{pol} = \overline{H}.\mathsf{ol}$. We associate to M and $\overline{H}$ the family of functions $H = $ Keyed-Indiff$[M, \overline{H}]$ defined as follows. We let $H.\mathsf{IL} = \mathbb{N}$, meaning H is VIL. The output length of H is $H.\mathsf{ol} = M.\mathsf{fol}$. We let $H.\mathsf{Kg} = \overline{H}.\mathsf{Kg}$, meaning keys for H are the same as for $\overline{H}$. Finally for any $\lambda \in \mathbb{N}$, any $hk \in [H.\mathsf{Kg}(1^\lambda)]$ and any $x \in \{0,1\}^*$ we let

$$H.\mathsf{Ev}^{\mathsf{ro}}(1^\lambda, hk, x, 1^{H.\mathsf{ol}(\lambda)}) = M^{\overline{H}.\mathsf{Ev}^{\mathsf{ro}}(1^\lambda, hk, \cdot, 1^{\overline{H}.\mathsf{ol}(\lambda)})}(1^\lambda, x) \,. \tag{1}$$

This needs some explanation. Begin by ignoring ro, so that we are looking at a standard-model construction. Recall that M takes an oracle mapping $\{0,1\}^{M.\mathsf{pil}(\lambda)}$ to $\{0,1\}^{M.\mathsf{pol}(\lambda)}$. In the indifferentiability setting, this is a random oracle. Our construction however does something different. It implements M's oracle via the given $\mathsf{UCE}[\mathcal{S}^{\mathsf{xxx}}]$-secure family $\overline{H}$. The key hk is held fixed. Our claim will be that H is itself $\mathsf{UCE}[\mathcal{S}^{\mathsf{xxx}}]$-secure for xxx $\in \{\mathsf{crs}, \mathsf{srs}\}$. Something we consider interesting is that this result is entirely standard model, yet uses ROM theory, in the form of a PRO, for the construction and proof. Finally the ro in the construction simply reflects that the result lifts to the ROM. In case $\overline{H}$ was a ROM family of functions, H will be as well. This extension, together with known applications of $\mathsf{UCE}[\mathcal{S}^{\mathsf{xxx}}]$-security, allow us to implement in the FIL-ROM many constructions given in the VIL-ROM.

RESULT. We view Keyed-Indiff$[M, \cdot]$ as a domain extension transform taking a FIL family $\overline{H}$ and returning a VIL family $H = $ Keyed-Indiff$[M, \overline{H}]$. The following says that this transform preserves $\mathsf{UCE}[\mathcal{S}^{\mathsf{xxx}}]$-security for xxx $\in \{\mathsf{crs}, \mathsf{srs}\}$.

Theorem 1. *Let $\overline{H}$ be a hash function family. Let M be a PRO such that $M.\mathsf{pil} = \overline{H}.\mathsf{il}$ and $M.\mathsf{pol} = \overline{H}.\mathsf{ol}$. Let $H = $ Keyed-Indiff$[M, \overline{H}]$. Let xxx $\in \{\mathsf{crs}, \mathsf{srs}\}$.*

Asymptotic result: If $\overline{H}$ is $\mathsf{UCE}[\mathcal{S}^{\mathsf{xxx}}]$-secure then so is H.

Concrete result: Let $\overline{M}$ be a simulator for M. Let S be an N-key source, D a distinguisher and $\overline{R}$ a reset adversary. Then we construct an N-key source $\overline{S}$, indifferentiability adversaries A, B and a reset adversary R such that

$$\mathsf{Adv}^{\mathsf{uce}}_{H,S,D}(\lambda) \leq \mathsf{Adv}^{\mathsf{uce}}_{\overline{H},\overline{S},D}(\lambda) + N(\lambda) \cdot \mathsf{Adv}^{\mathsf{indiff}}_{M,\overline{M},A}(\lambda) \tag{2}$$

$$\mathsf{Adv}^{\mathsf{reset}}_{\overline{S},\overline{R}}(\lambda) \leq \mathsf{Adv}^{\mathsf{reset}}_{S,R}(\lambda) + 3N(\lambda) \cdot \mathsf{Adv}^{\mathsf{indiff}}_{M,\overline{M},B}(\lambda) \tag{3}$$

for all $\lambda \in \mathbb{N}$. *Furthermore:*

$$\mathbf{Q}_A^{\mathsf{Prim}} = 0; \; \mathbf{Q}_A^{\mathsf{Func}} = \mathbf{Q}_B^{\mathsf{Func}} = \mathbf{Q}_S^{\mathsf{Hash}}; \; \mathbf{Q}_B^{\mathsf{Prim}} = \mathbf{Q}_{\overline{R}}^{\mathsf{Hash}}$$

$$\mathbf{Q}_R^{\mathsf{ro}} = \mathbf{Q}_{\overline{R}}^{\mathsf{ro}}; \; \mathbf{Q}_R^{\mathsf{Hash}} = Q_{\overline{\mathsf{M}},q} \; where \; q = \mathbf{Q}_{\overline{R}}^{\mathsf{Hash}}; \; \mathbf{Q}_{\overline{S}}^{\mathsf{ro}} = \mathbf{Q}_S^{\mathsf{ro}}$$

$\mathbf{Q}_{\overline{S}}^{\mathsf{Hash}}$ *is the number of oracle queries of* M *in the execution of* $\mathsf{UCE}_{\mathsf{H}}^{S,D}$

$$\mathbf{T}(\mathsf{Indiff}_{\mathsf{M},\overline{\mathsf{M}}}^A) = \mathbf{T}(\mathsf{UCE}_{\mathsf{H}}^{S,D}); \; \mathbf{T}(\mathsf{UCE}_{\mathsf{H}}^{\overline{S},D}) = \mathbf{T}(\mathsf{UCE}_{\mathsf{H}}^{S,D})$$

$$\mathbf{T}(\mathsf{Reset}_S^R) = \mathbf{T}(\mathsf{Reset}_{\overline{S}}^{\overline{R}}) + T_{\overline{\mathsf{M}},q} \; where \; q = \mathbf{Q}_{\overline{R}}^{\mathsf{Hash}}$$

$$\mathbf{T}(\mathsf{Indiff}_{\mathsf{M},\overline{\mathsf{M}}}^B) = \mathbf{T}(\mathsf{Reset}_S^R) + \mathbf{T}(\mathsf{Reset}_{\overline{S}}^{\overline{R}}) \quad \square$$

We emphasize that Keyed-Indiff works in both the standard and the random oracle models. In particular if FIL family $\overline{\mathsf{H}}$ is $\mathsf{UCE}[\mathcal{S}^{\mathrm{xxx}}]$-secure in the standard model, then so is Keyed-Indiff$[\mathsf{M}, \overline{\mathsf{H}}]$, for $\mathrm{xxx} \in \{\mathrm{crs}, \mathrm{srs}\}$. This resolves an open problem from [6] to construct UCE domain extenders in the standard model.

INSTANTIATION. To obtain a concrete result that can be used in applications, we now instantiate $\overline{\mathsf{H}}$ above in a simple way, namely (1) $\overline{\mathsf{H}}.\mathsf{Kg}(1^\lambda)$ returns $hk \leftarrow_{\$} \{0,1\}^\lambda$, and (2) $\overline{\mathsf{H}}.\mathsf{Ev}^{\mathrm{ro}}(1^\lambda, hk, x, 1^{\overline{\mathsf{H}}.\mathsf{ol}(\lambda)})$ returns $\mathrm{ro}(hk \,\|\, x)$. This is shown by BHK [6] to be UCE secure in the FIL-ROM for all forms of UCE they define. From Theorem 1 we obtain the following.

Theorem 2. *Let* $\overline{\mathsf{H}}$ *be constructed as above. Let* M *be a PRO such that* $\mathsf{M}.\mathsf{pil} = \overline{\mathsf{H}}.\mathsf{il}$ *and* $\mathsf{M}.\mathsf{pol} = \overline{\mathsf{H}}.\mathsf{ol}$. *Let* $\mathsf{H} = \mathsf{Keyed\text{-}Indiff}[\mathsf{M}, \overline{\mathsf{H}}]$. *Let* $\mathrm{xxx} \in \{\mathrm{crs}, \mathrm{srs}\}$.

Asymptotic result: H *is* $\mathsf{UCE}[\mathcal{S}^{\mathrm{crs}}]$-secure.

Concrete result: Let $\overline{\mathsf{M}}$ *be a simulator for* M. *Let* S *be an* N-key source and D *a distinguisher. We can construct a reset adversary* R *and an indifferentiability adversary* A *such that*

$$\mathsf{Adv}_{\mathsf{H},S,D}^{\mathsf{uce}}(\lambda) \leq \mathsf{Adv}_{S,R}^{\mathsf{reset}}(\lambda) + 4N(\lambda) \cdot \mathsf{Adv}_{\mathsf{M},\overline{\mathsf{M}},A}^{\mathsf{indiff}}(\lambda) + \frac{2N(\lambda) \cdot q(\lambda) + N^2(\lambda)}{2^\lambda}$$

for every $\lambda \in \mathbb{N}$. *Furthermore,*

$$\mathbf{Q}_A^{\mathsf{Prim}} = \mathbf{Q}_S^{\mathsf{Hash}}; \; \mathbf{Q}_A^{\mathsf{Func}} = \mathbf{Q}_R^{\mathsf{ro}} = \mathbf{Q}_D^{\mathsf{ro}}; \; and \; \mathbf{Q}_R^{\mathsf{Hash}} = Q_{\overline{\mathsf{M}},q}, \; where \; q = \mathbf{Q}_D^{\mathsf{ro}}$$

$$\mathbf{T}(\mathsf{Indiff}_{\mathsf{M},\overline{\mathsf{M}}}^A) = \mathbf{T}(\mathsf{Reset}_S^R) = \mathbf{T}(\mathsf{UCE}_{\mathsf{H}}^{S,D}) + T_{\overline{\mathsf{M}},q}, \; where \; q = \mathbf{Q}_D^{\mathsf{ro}} \quad \square$$

Theorem 2 is the one that can be used for the applications, namely to obtain FIL-ROM constructions for (possibly multi-stage) primitives that have been constructed using a VIL UCE function, such as those in BHK [6]. We simply instantiate the VIL UCE function with H given by Theorem 2. The broader paradigm to move from the VIL-ROM to the FIL-ROM is thus the following. Take a primitive with a VIL-ROM proof, and show that the random oracle can be UCE-instantiated. Then apply Theorem 2.

5 UCE from Universal Hashing

In this section, we show how almost universal hash functions can be used to build a domain extender for UCE.

$H.Kg(1^\lambda)$	$H.Ev^{ro}(1^\lambda, hk, x, 1^\ell)$
$fk \leftarrow\!\!{\scriptstyle\$}\ F.Kg(1^\lambda)$; $\overline{hk} \leftarrow\!\!{\scriptstyle\$}\ \overline{H}.Kg(\lambda)$	$(\overline{hk}, fk) \leftarrow hk$; $u \leftarrow F.Ev(1^\lambda, fk, x, 1^{F.ol(\lambda)})$
$hk \leftarrow (\overline{hk}, fk)$; Return hk	$y \leftarrow \overline{H}.Ev^{ro}(1^\lambda, \overline{hk}, u, 1^\ell)$; Return y

Fig. 4. The $H = \text{AU-then-Hash}[F, \overline{H}]$ construction, built from a AU hash F and a FIL UCE-secure hash $\overline{H}$.

AU HASH FAMILIES. For any function family F let

$$\textbf{Coll1}_F(\lambda, m) = \max_{|y|=F.ol(\lambda), |x|\leq m} \left\{ \Pr_{fk \leftarrow\!\!{\scriptstyle\$}\ F.Kg(1^\lambda)}[y = F.Ev(1^\lambda, fk, x, 1^{F.ol(\lambda)})] \right\},$$

and define $\textbf{Coll2}_F(\lambda, m_0, m_1)$ as

$$\max\left\{ \Pr_{fk \leftarrow\!\!{\scriptstyle\$}\ F.Kg(1^\lambda)}[F.Ev(1^\lambda, fk, x_0, 1^{F.ol(\lambda)}) = F.Ev(1^\lambda, fk, x_1, 1^{F.ol(\lambda)})] \right\};$$

the maximum is taken over distinct strings x_0, x_1 such that each $|x_i| \leq m_i$. Let

$$\textbf{Coll}_F(\lambda, m_0, m_1) = \max\{\textbf{Coll2}_F(\lambda, m_0, m_1), \textbf{Coll1}_F(\lambda, \min\{m_0, m_1\})\} \ .$$

A hash family F is *almost universal* (AU) if $f(\lambda) = \textbf{Coll}_F(\lambda, m_0, m_1)$ is negligible for all polynomials m_0, m_1. This generalizes the Carter-Wegman notion of universal hashing [13].

A similar definition is given in [11], which is very useful when one needs to work with arbitrarily large input and short hash keys. In Section 6, we'll show how to concretely instantiate a very fast AU hash for $\lambda = 128$, from reduced-round AES and a classic polynomial-based universal hash. Define

$$\text{Adv}_F^{\text{coll}}(\lambda, p, \sigma) = \max_{\ell \leq p, \ell' \leq p, m_1 + \cdots + m_\ell \leq \sigma, m'_1 + \cdots + m'_{\ell'} \leq \sigma} \left\{ \sum_{i=1}^{\ell} \sum_{j=1}^{\ell'} \textbf{Coll}_F(\lambda, m_i, m'_j) \right\} \ .$$

If F is AU then $\text{Adv}_F^{\text{coll}}(\lambda, p, \sigma)$ is negligible for all polynomials p and σ: since $\textbf{Coll}(\lambda, \cdot, \cdot)$ is increasing in both arguments, it follows that $\text{Adv}_F^{\text{coll}}(\lambda, p, \sigma) \leq p^2 \textbf{Coll}_F(\lambda, \sigma, \sigma)$.

UCE EXTENDER FROM AN AU HASH. We now describe a UCE extender from AU hash. Intuitively, one first uses the AU hash to condense the input, and then applies the resulting string to the (keyed) compression function. Formally, let $\overline{H}$ be a hash function family of fixed input length, and F be a universal hash function family with $F.ol = \overline{H}.il$ and $F.IL = \mathbb{N}$. Consider the hash function family $H = \text{AU-then-Hash}[F, \overline{H}]$ as given in Fig. 4, with $H.OL = \overline{H}.OL$ and $H.IL = \mathbb{N}$. The construction essentially follows the widely used Carter-Wegman paradigm [24] Below, we show that $\text{AU-then-Hash}[F, \cdot]$ is also a domain extender for $UCE[\mathcal{S}^{\text{sup}}]$ security.

Theorem 3. *Let $\overline{H}$ be a function family of fixed input length, and F be an AU hash function family with $F.ol = \overline{H}.il$ and $F.IL = \mathbb{N}$. Let $H = \text{AU-then-Hash}[F, \overline{H}]$.*

Asymptotic result: If $\overline{H}$ is UCE$[\mathcal{S}^{sup}]$-secure then so is H.

Concrete result: Let S be a N-key source, D a distinguisher, and $\overline{P}$ a predictor. We can construct a source $\overline{S}$, a distinguisher $\overline{D}$, and a predictor P such that

$$\mathsf{Adv}^{uce}_{H,S,D}(\lambda) \leq \mathsf{Adv}^{uce}_{H,\overline{S},\overline{D}}(\lambda) + \mathsf{Adv}^{coll}_{F}(\lambda,p,\sigma) \qquad (4)$$

$$\mathsf{Adv}^{pred}_{\overline{S},\overline{P}}(\lambda) \leq \sqrt{2q\mathsf{Adv}^{coll}_{F}(\lambda,p,\sigma)} + \sqrt{q\mathsf{Adv}^{pred}_{S,P}(\lambda)} \qquad (5)$$

where $p = \mathbf{Q}^{Hash}_{S}$, q is the maximum of the size of $\overline{P}$'s output in the execution of $\mathrm{Pred}^{\overline{P}}_{\overline{S}}$, and σ is the maximum of the total length of Hash queries that S makes in $\mathrm{UCE}^{S,D}_{H}$. Furthermore,

$$\mathbf{Q}^{ro}_{\overline{S}} = \mathbf{Q}^{ro}_{S}; \ \mathbf{Q}^{Hash}_{\overline{S}} = \mathbf{Q}^{Hash}_{S}; \ \mathbf{Q}^{ro}_{\overline{D}} = \mathbf{Q}^{ro}_{D}$$

$$\mathbf{T}(\mathrm{UCE}^{\overline{S},\overline{D}}_{\overline{H}}) = \mathbf{T}(\mathrm{UCE}^{S,D}_{H}), \ and \ P \ outputs \ a \ set \ of \ size \ at \ most \ \mathbf{Q}^{Hash}_{S} \quad \square$$

We emphasize that AU-then-Hash works in both the standard and the random-oracle models. In particular If FIL family $\overline{H}$ is UCE$[\mathcal{S}^{sup}]$-secure in the standard model then so is AU-then-Hash$[F,\overline{H}]$.

The intended applications for the AU-then-Hash$[F,\cdot]$ transform, as listed in Fig. 1, use only a single hash key, that is, they only need UCE$[\mathcal{S}^{sup} \cap \mathcal{S}^{one}]$ security, where $\mathcal{S}^{one}$ is the class of 1-key sources. AU-then-Hash$[F,\cdot]$ is also a domain extender for UCE$[\mathcal{S}^{sup} \cap \mathcal{S}^{one}]$ security because the value of N is preserved.

INSTANTIATION. So far we have assumed the existence of a fixed-input-length UCE-secure hash $\overline{H}$. In the full version, we'll construct hash family H_{rom}, of variable output length, in the ROM, by using a pseudorandom permutation (PRP) E, which will be instantiated by AES. We conclude the following.

Theorem 4. *Let F be an AU hash function family with $F.ol = H_{rom}.il$ and $F.IL = \mathbb{N}$. Let $H = AU\text{-}then\text{-}Hash[F, H_{rom}]$.*

Asymptotic result: H is UCE$[\mathcal{S}^{sup}]$-secure.

Concrete result: Let S be an N-key source and D a distinguisher. We can construct a predictor P and a PRP adversary A such that

$$\mathsf{Adv}^{uce}_{H,S,D}(\lambda) \leq 2\sqrt{q(\lambda)\mathsf{Adv}^{coll}_{F}(\lambda,p(\lambda),\sigma(\lambda))} + \sqrt{q(\lambda)\mathsf{Adv}^{pred}_{S,P}(\lambda)} +$$

$$2p(\lambda) \cdot \mathsf{Adv}^{prp}_{E,A}(\lambda) + \frac{2s^2(\lambda) + N^2(\lambda) + q^2(\lambda)}{2^\lambda}$$

for every $\lambda \in \mathbb{N}$, where $p = \mathbf{Q}^{Hash}_{S}$; $q = \mathbf{Q}^{ro}_{S} + \mathbf{Q}^{ro}_{D}$; σ and s are the maximum of the total length of the first components and the total number of λ-bit blocks in the second components, respectively, of Hash queries in the execution of $\mathrm{UCE}^{S,D}_{H}$. Furthermore

$\mathbf{Q}^{LR}_{A}$ *is maximum of the number of λ-bit blocks in the second component of a Hash query in $\mathrm{UCE}^{S,D}_{H}$*

$\mathbf{T}(\mathrm{PRP}^{A}_{E}) = \mathbf{T}(\mathrm{UCE}^{S,D}_{H})$, *and P outputs a set of size at most $\mathbf{Q}^{Hash}_{S}$* $\quad \square$

6 Fast, Parallelizable AU Hash from Reduced-Round AES

We now show how to construct a fast parallelizable AU hash, which we call F_{aes4}. In this section, let $n = 128$, $C = 2^{15}$, and let r be a small integer, say $r = 5$. All function families in this section are concrete; the security parameter λ is hidden in the formulas, but implicitly, it is $\lambda = 128$. For any integer m, let $\|m\|_n$ denote $\lfloor m/n \rfloor + 1$. We'll first describe two building blocks: F_{poly}, a polynomial-based AU hash that operates on $\{0,1\}^*$, and F_{tree}, a highly efficient AU hash based on reduced-round AES that operates on $\{x \in (\{0,1\}^n)^+ : |x| \le 2^r n\}$. We then show how to combine them to produce a highly efficient AU hash F_{aes4} whose domain is $\{0,1\}^*$.

THE F_{poly} CONSTRUCTION. We now describe a variant of a classic polynomial-based universal hash [13], which we call F_{poly}. Let $F_{poly}.ol = n$. As described in the pseudocode below, the key fk is picked as a random element of $GF(2^n)$. To hash, we parse the input string $x \in \{0,1\}^*$ to a unique sequence $(w_0, \ldots, w_m)$, where each $w_i \in GF(2^n)$ and w_m is not the zero element. This is performed by (i) parse $v_0 \| \cdots \| v_m \leftarrow x \| 10^s 1$, where $s \in \mathbb{N}$ is the smallest number such that $s + |x| \equiv -2 \pmod{n}$ and each $|w_i| = n$, and (ii) let each w_i be the encoding of v_i in $GF(2^n)$. Then, the hash is computed as $\sum_{i=0}^m w_i \cdot fk^i$.

$F_{poly}.Kg()$	$F_{poly}.Ev(fk, x, 1^n)$
$fk \leftarrow\!\!\$ \; GF(2^n)$	$(w_0, \ldots, w_m) \leftarrow x \; ; \; y \leftarrow w_0$
Return fk	For $i = 1$ to m do $y \leftarrow y + w_i \cdot fk^i$
	Return y

Proposition 5. *(a) For any $m \in \mathbb{N}$, we have $\mathbf{Coll1}_{F_{poly}}(m) \le \|m\|_n/2^n$, and (b) for any $m_0, m_1 \in \mathbb{N}$, we have $\mathbf{Coll2}_{F_{poly}}(m_0, m_1) \le \max\{\|m_0\|_n, \|m_1\|_n\}/2^n$.*

THE F_{tree} CONSTRUCTION. Let $E : \{0,1\}^{4n} \times \{0,1\}^n \to \{0,1\}^n$ denote a function based on 4-round AES which works as follows. Parse the key K as the concatenation of n-bit substrings S_0, S_1, S_2, S_3, and let $S_4 = 0^n$. The input is initially xored with S_0, and each S_i is used as the subkey of the i-th AES round, for $i \in \{1, 2, 3, 4\}$. One can build from E a hash of domain $\{n, 2n, 3n, \ldots, 2^r n\}$ as follows. Let Halve denote the following operation. On input $(K, x) \in \{0,1\}^{4n} \times (\{0,1\}^n)^*$, we partition x into n-bit blocks $x_1 \cdots x_m$. For every two consecutive blocks x_{2i-1} and x_{2i}, we compute $y_i \leftarrow E_K(x_{2i-1}) \oplus x_{2i}$. If m is odd then let $y_{\lceil m/2 \rceil} \leftarrow x_m$. Finally output $y_1 \| \cdots \| y_{\lceil m/2 \rceil}$. Consider the following tree-hash construction F_{tree}, with $F_{tree}.IL = \{n, 2n, 3n, \ldots, 2^r n\}$ and $F_{tree}.ol = n$.

$F_{tree}.Kg()$	$F_{tree}.Ev(fk, x, 1^n)$
For $i = 1$ to r do $K_i \leftarrow\!\!\$ \; \{0,1\}^{4n}$	$z_0 \leftarrow x \; ; \; (K_1, \ldots, K_r) \leftarrow fk$
$hk \leftarrow (K_1, \ldots, K_r) \; ;$ Return fk	For $i = 1$ to r do $z_i \leftarrow \mathsf{Halve}(K_i, z_{i-1})$
	Return z_r

Minematsu and Tsunoo [20] show that

$$\mathbf{Coll2}_{\mathsf{F_{tree}}}(m_0, m_1) \leq \frac{Cr}{2^n} \quad (6)$$

for any $m_0, m_1 \leq 2^r$. We stress that the result in [20] makes no assumption on AES. This is based on the fact that four-round AES, with the subkeys chosen uniformly and independently, is an almost-xor-universal hash [18].

COMBINING $\mathsf{F_{tree}}$ AND $\mathsf{F_{poly}}$. One can "cascade" $\mathsf{F_{tree}}$ and $\mathsf{F_{poly}}$ to produce a hash $\mathsf{F_{fast}}$ of domain $\{0, 1\}^*$ as follows.

$\mathsf{F_{fast}.Kg}()$	$\mathsf{F_{fast}.Ev}(fk, x, 1^n)$	$\mathsf{Shrink}(fk_1, x)$
$fk_1 \leftarrow\!\!{\$}\ \mathsf{F_{tree}.Kg}()$	$(fk_1, fk_2) \leftarrow fk$	$w_1 w_2 \cdots w_k \leftarrow x\,;\ u_k \leftarrow w_k$
$fk_2 \leftarrow\!\!{\$}\ \mathsf{F_{poly}.Kg}()$	$y \leftarrow \mathsf{Shrink}(fk_1, x)$	For $i = 1$ to $k - 1$ do
Return (fk_1, fk_2)	$z \leftarrow \mathsf{F_{poly}.Ev}(fk_2, y, 1^n)$	$\quad u_i \leftarrow \mathsf{F_{tree}.Ev}(fk_1, w_i, 1^n)$
	Return z	$y \leftarrow u_1 \| \cdots \| u_k\,;$ Return y

In the procedure Shrink above, we parse a string x as the concatenation of substrings $w_1, \ldots, w_k$, where the length of each w_i, with $i \leq k-2$, is exactly $2^r n$, and $|w_{k-1}| > 0$ is a multiple of n but does not exceed $2^r n$, and $0 \leq |w_k| < n - 1$. Note that on a large input x, the hash F will make at most $(1 - 2^{-r})\lceil x/n \rceil$ calls on E, and then run $\mathsf{F_{poly}}$ on a string of length about $|x|/2^r$.

Proposition 6. *For any $m_0, m_1 \in \mathbb{N}$, we have*

$$\mathbf{Coll}_{\mathsf{F_{fast}}}(m_0, m_1) \leq \frac{Cr + \max\{\|m_0\|_n, \|m_1\|_n\}}{2^n}$$

USING WITH AU-then-Hash. The hash $\mathsf{F_{fast}}$ can't be used directly with the AU-then-Hash transform in Section 5, because the term $(q\mathsf{Adv}_{\mathsf{F_{fast}}}^{\mathsf{coll}}(p, \sigma))^{1/2}$ in Theorem 3 is about $(\sqrt{qp\sigma} + Crp\sqrt{q})/2^{n/2}$, which is inferior. The reason for this is that the output length of this hash is only n bits, which is too short. We therefore need to "double" the output length. Formally, given a hash family $\overline{\mathsf{F}}$, the family $\mathsf{F} = \mathsf{Double}[\overline{\mathsf{F}}]$, with $\mathsf{F.IL} = \overline{\mathsf{F}}.\mathsf{IL}$ and $\mathsf{F.ol} = 2\overline{\mathsf{F}}.\mathsf{ol}$, is constructed as follows.

$\mathsf{F.Kg}()$	$\mathsf{F.Ev}(fk, x, 1^{\mathsf{F.ol}})$
$fk_1, fk_2 \leftarrow\!\!{\$}\ \overline{\mathsf{F}}.\mathsf{Kg}()$	$(fk_1, fk_2) \leftarrow fk$
$fk \leftarrow (fk_1, fk_2)\,;$ Return fk	For $i = 1$ to 2 do $y_i \leftarrow \overline{\mathsf{F}}.\mathsf{Ev}(fk_i, x, 1^{\overline{\mathsf{F}}.\mathsf{ol}})$
	Return $y_1 \| y_2$

Let $\mathsf{F_{aes4}}$ denote $\mathsf{Double}[\mathsf{F_{fast}}]$. In Proposition 7 below, the term $(q\mathsf{Adv}_{\mathsf{F_{fast}}}^{\mathsf{coll}}(p, \sigma))^{1/2}$ in Theorem 3 is bounded by $(Crp\sqrt{2q} + 2(\|\sigma\|_n + p)\sqrt{pq})/2^n$, which is good.

Proposition 7. *For any p and σ, we have* $\mathsf{Adv}_{\mathsf{F_{aes4}}}^{\mathsf{coll}}(p, \sigma) \leq \frac{2C^2 r^2 p^2 + 4p(\|\sigma\|_n + p)^2}{2^{2n}}$.

KEY LENGTH. The key material of $\mathsf{FastHash} = \mathsf{AU\text{-}then\text{-}Hash}[\mathsf{F_{aes4}}, \mathsf{H_{rom}}]$ is relatively large: 672B for $r = 5$. It's slightly bigger than that of some widely used schemes such as RSA [22] (256B). This is acceptable because the key is used as a public parameter.

Hash function	Setting	Speed (cycles per byte)		
		1MB	16MB	128MB
SHA-256 [1]		11.5	12.0	12.0
FastHash	sequential	2.1	2.2	2.2
	parallel - 12 threads	0.4	0.4	0.5

Fig. 5. Running time of the hash constructions. The first column lists the hash names, the second column lists the setting, namely sequential or parallel, along with the number of threads, and the last three columns list the running time on messages of sizes 1MB, 16MB, and 128MB respectively.

7   Implementation

In this section, we'll describe how to instantiate the AU hash F_{aes4} in Section 6, and the FIL UCE-secure hash H_{rom} in Section 5. We then compare the speed of FastHash, the resulting instantiation of AU-then-Hash[F_{aes4}, H_{rom}], with a standard hash function, SHA-256. We first describe our choices for components and parameters to instantiate the construction, and then provide an overview of the implementation, before outlining the testing environment and test specifications. We also compare the convergent encryption (CE) MLE scheme [1] from FastHash and SHA-256. Our results indicate a speedup of 5.3x for our hash function over SHA-256 and 6.3x for CE in the sequential setting, and 24x and 20x speedups, respectively, once parallelism is enabled.

INSTANTIATIONS. To instantiate F_{aes4}, we use the standard irreducible polynomial $p(x) = x^{127} + x^7 + x^2 + x + 1$ for multiplication over GF(2^{128}). For H_{rom}, the FIL RO is instantiated by the compression function of SHA-256, and the PRP by AES128.

IMPLEMENTATION. We implemented FastHash in C with inline assembly. We used Intel's library for multiplication over $\mathbb{GF}(2^{128})$ [3], Intel's optimized SHA256 implementation [1], and Intel's AES-NI library [2] for the code involving AES operations. We used the pthreads library for implementing threads for parallelization.

SETUP. We performed experiments on an Intel Core i7-970 processor clocking at 3201 MHz with a 12288 KB L1 cache. The machine provides hardware support for SSE4 vector instructions, AES operations (AES-NI), and multiplication in $\mathbb{GF}(2^{128})$. Tests were compiled with gcc version 4.6 optimization level -O3, with support for SSE4 via -msse4 flag, AES-NI instructions through the -maes flag, $\mathbb{GF}(2^{128})$ multiplications via the -mpcmulqdq flag, and parallelization via the -pthread flag. We ran the tests in isolation, after turning off processor frequency scaling. We used the rdtsc instruction to count cycles.

---

[1] In CE [8], one first hashes the message x to derive a key K, and then runs AES-CTR on key K to encrypt x. To use FastHash on CE, one needs to use the CE variant of [6], in which AES-CTR on message m is replaced by FastHash($hk, K, 1^{|x|}$)$\oplus x$. Note that this doesn't give us any speed advantage over the standard version of CE, as the masking via FastHash is essentially AES-CTR. The only thing we gain is the abstraction of AES as part of the hash, so that one can apply UCE[$\mathcal{S}^{sup}$].

MLE Scheme	Setting	Speed (cycles per byte)		
		1MB	16MB	128MB
CE implementation in [8]		22.1	22.3	22.6
CE[FastHash]	sequential	3.5	3.6	3.7
	parallel - 12 threads	1.2	1.1	1.1

Fig. 6. Running time of CE instantiations. The first column lists the instantiations, the second column lists the setting, namely sequential or parallel, along with the number of threads, and the last three columns list the running time (key generation + encryption) on messages of size 1MB, 16MB, and 128MB respectively.

EXPERIMENTS. We measured the performance of instantiations of the hash functions (i.e. FastHash and SHA-256) as well as CE schemes based on these hash functions on messages of lengths 1MB, 16MB and 128MB. In each case, we measured the median running times of the different hash functions over 100 iterations, repeated this process 100 times and obtained the mean of the medians.

In the case of parallelizable constructions, viz. FastHash and CE[FastHash], we ran tests with multiple levels of parallelism, starting from single-threaded, serial constructions, and increasing the number of threads until we reached a point of thrashing where the performance starts to deteriorate because of other bottlenecks in the system. We report both the single-thread sequential running time, and the optimal parallel running time along with the optimal number of threads. In the latter case, the reported time does not include the time to create and destroy the threads.

In Fig. 5, we report the median running times of the hash function instantiations, in cycles per byte. We compare these times with the best times reported for SHA-256 on similar processors [1]. Our construction achieves substantially better running times. On messages of 1MB, SHA runs at 11.5 cycles per byte, but our instantiation runs more than 5.3 times faster, at a cost of 2.1 cycles per byte. With parallelism, we achieve much better speeds, below one cycle per byte.

In Fig. 6, we demonstrate the benefits of having faster hash functions by comparing the speeds of CE implemented with FastHash with the implementation of CE by SHA-256 and AES-CTR in [8]. Our experiments show that CE[FastHash], even in the sequential setting, is about 6.3x faster than the speeds reported in [8]. When parallelism enabled, we achieve about 20x speedup.

Acknowledgments. Work done while Keelveedhi was a PhD student at UCSD. The authors were supported in part by NSF grants CNS-1116800 and CNS-1228890.

References

1. Fast SHA-256 Implementations on Intel Architecture Processors, goo.gl/Hh81eB.
2. Intel AESNI Library, goo.gl/l2czm1.

3. Intel Carry-Less Multiplication Instruction and its Usage for Computing the GCM Mode, http://goo.gl/qJLrF1
4. Barak, B., Dodis, Y., Krawczyk, H., Pereira, O., Pietrzak, K., Standaert, F.-X., Yu, Y.: Leftover hash lemma, revisited. In: Rogaway, P. (ed.) CRYPTO 2011. LNCS, vol. 6841, pp. 1–20. Springer, Heidelberg (2011)
5. Bellare, M., Boldyreva, A., O'Neill, A.: Deterministic and efficiently searchable encryption. In: Menezes, A. (ed.) CRYPTO 2007. LNCS, vol. 4622, pp. 535–552. Springer, Heidelberg (2007)
6. Bellare, M., Hoang, V.T., Keelveedhi, S.: Instantiating random oracles via UCEs. Cryptology ePrint Archive, Report 2013/424 (2013); Preliminary version appeared in Canetti, R., Garay, J.A. (eds.) CRYPTO 2013, Part II. LNCS, vol. 8043, pp. 398–415. Springer, Heidelberg (2013)
7. Bellare, M., Hoang, V.T., Keelveedhi, S.: Cryptography from compression functions: The UCE bridge to the ROM. Cryptology ePrint Archive (2014)
8. Bellare, M., Keelveedhi, S., Ristenpart, T.: Message-locked encryption and secure deduplication. In: Johansson, T., Nguyen, P.Q. (eds.) EUROCRYPT 2013. LNCS, vol. 7881, pp. 296–312. Springer, Heidelberg (2013)
9. Bellare, M., Rogaway, P.: Random oracles are practical: A paradigm for designing efficient protocols. In: ACM CCS 1993. ACM (1993)
10. Bellare, M., Rogaway, P.: The security of triple encryption and a framework for code-based game-playing proofs. In: Vaudenay, S. (ed.) EUROCRYPT 2006. LNCS, vol. 4004, pp. 409–426. Springer, Heidelberg (2006)
11. Black, J., Rogaway, P.: CBC MACs for arbitrary-length messages: The three-key constructions. Journal of Cryptology 18(2), 111–131 (2005)
12. Brzuska, C., Farshim, P., Mittelbach, A.: Indistinguishability obfuscation and UCEs: The case of computationally unpredictable sources. Cryptology ePrint Archive, Report 2014/099. To appear in Garay, J.A., Gennaro, R. (eds.) CRYPTO 2014. LNCS, vol. 8616, pp. 188–205. Springer, Heidelberg (2014)
13. Carter, L., Wegman, M.: Universal classes of hash functions. Journal of Computer and System Sciences 18(2), 143–154 (1979)
14. Coron, J.-S., Dodis, Y., Malinaud, C., Puniya, P.: Merkle-Damgård revisited: How to construct a hash function. In: Shoup, V. (ed.) CRYPTO 2005. LNCS, vol. 3621, pp. 430–448. Springer, Heidelberg (2005)
15. Demay, G., Gaži, P., Hirt, M., Maurer, U.: Resource-restricted indifferentiability. In: Johansson, T., Nguyen, P.Q. (eds.) EUROCRYPT 2013. LNCS, vol. 7881, pp. 664–683. Springer, Heidelberg (2013)
16. Dodis, Y., Ristenpart, T., Shrimpton, T.: Salvaging Merkle-Damgård for practical applications. In: Joux, A. (ed.) EUROCRYPT 2009. LNCS, vol. 5479, pp. 371–388. Springer, Heidelberg (2009)
17. Håstad, J., Impagliazzo, R., Levin, L.A., Luby, M.: A pseudorandom generator from any one-way function. SIAM Journal on Computing 28(4), 1364–1396 (1999)
18. Keliher, L., Sui, J.: Exact maximum expected differential and linear probability for two-round advanced encryption standard. IET Information Security 1(2), 53–57 (2007)
19. Maurer, U., Renner, R., Holenstein, C.: Indifferentiability, impossibility results on reductions, and applications to the random oracle methodology. In: Naor, M. (ed.) TCC 2004. LNCS, vol. 2951, pp. 21–39. Springer, Heidelberg (2004)
20. Minematsu, K., Tsunoo, Y.: Provably secure MACs from differentially-uniform permutations and AES-based implementations. In: Robshaw, M. (ed.) FSE 2006. LNCS, vol. 4047, pp. 226–241. Springer, Heidelberg (2006)

21. Mittelbach, A.: Salvaging indifferentiability in a multi-stage setting. In: Nguyen, P.Q., Oswald, E. (eds.) EUROCRYPT 2014. LNCS, vol. 8441, pp. 603–621. Springer, Heidelberg (2014)
22. PKCS #1: RSA cryptography standard. RSA Data Security, Inc, Version 2.0. (September 1998)
23. Ristenpart, T., Shacham, H., Shrimpton, T.: Careful with composition: Limitations of the indifferentiability framework. In: Paterson, K.G. (ed.) EUROCRYPT 2011. LNCS, vol. 6632, pp. 487–506. Springer, Heidelberg (2011)
24. Wegman, M.N., Carter, L.: New hash functions and their use in authentication and set equality. Journal of Computer and System Sciences 22, 265–279 (1981)

Indistinguishability Obfuscation and UCEs: The Case of Computationally Unpredictable Sources

Christina Brzuska[1], Pooya Farshim[2], and Arno Mittelbach[3]

[1] Tel Aviv University, Israel
[2] Royal Holloway, University of London, UK
[3] Darmstadt University of Technology, Germany
brzuska@post.tau.ac.il, pooya.farshim@rhul.ac.uk,
arno.mittelbach@cased.de

Abstract. Random oracles are powerful cryptographic objects. They facilitate the security proofs of an impressive number of practical cryptosystems ranging from KDM-secure and deterministic encryption to point-function obfuscation and many more. However, due to an uninstantiability result of Canetti, Goldreich, and Halevi (STOC 1998) random oracles have become somewhat controversial. Recently, Bellare, Hoang, and Keelveedhi (BHK; CRYPTO 2013 and ePrint 2013/424, August 2013) introduced a new abstraction called Universal Computational Extractors (UCEs), and showed that they suffice to securely replace random oracles in a number of prominent applications, including all those mentioned above, without suffering from the aforementioned uninstantiability result. This, however, leaves open the question of constructing UCEs in the standard model.

We show that the existence of indistinguishability obfuscation (iO) implies (non-black-box) attacks on all the definitions that BHK proposed within their UCE framework in the original version of their paper, in the sense that no concrete hash function can satisfy them. We also show that this limitation can be overcome, to some extent, by restraining the class of admissible adversaries via a *statistical* notion of unpredictability. Following our attack, BHK (ePrint 2013/424, September 2013), independently adopted this approach in their work.

In the updated version of their paper, BHK (ePrint 2013/424, September 2013) also introduce two other novel source classes, called *bounded parallel sources* and *split sources*, which aim at recovering the computational applications of UCEs that fall outside the statistical fix. These notions keep to a computational notion of unpredictability, but impose structural restrictions on the adversary so that our original iO attack no longer applies. We extend our attack to show that indistinguishability obfuscation is sufficient to also break the UCE security of any hash function against bounded parallel sources. Towards this goal, we use the *randomized encodings* paradigm of Applebaum, Ishai, and Kushilevitz (STOC 2004) to parallelize the obfuscated circuit used in our attack, so that it can be computed by a bounded parallel source whose second stage consists of constant-depth circuits. BHK, in the latest version of their paper (ePrint 2013/424, May 2014), have subsequently replace

J.A. Garay and R. Gennaro (Eds.): CRYPTO 2014, Part I, LNCS 8616, pp. 188–205, 2014.

bounded parallel sources with new source classes. We conclude by discussing the composability and feasibility of hash functions secure against split sources.

Keywords: Randomized encodings, obfuscation, UCE, random oracle.

1   Introduction

Since their formal introduction in the seminal paper of Bellare and Rogaway [13], random oracles have found extensive use across a wide spectrum of cryptographic protocols. Their versatility has lead researchers to seek for a unified formalization of their useful properties, hoping that such a definition could be eventually realized. Canetti, Goldreich, and Halevi [20] proposed such a definition, but somewhat disappointingly, also proved a negative result which ruled out instantiations of random oracles in *arbitrary* (perhaps artificial) cryptographic protocols by *any* keyed hash functions. This negative result was subsequently extended in a number of works [35,25,22,32,7,21].

UCE security. Bellare, Hoang, and Keelvedhi (BHK) [8,9,10,12][1] revisited the above question and formulated an attractive new security notion called *Universal Computational Extractor* (UCE). They were able to apply their framework to an interesting and diverse set of security goals, which included among other things, security under key-dependent attacks, security under related-key attacks, simultaneous hardcore bits, point-function obfuscation, garbling schemes, proofs of storage, and deterministic encryption. Recently, Matsuda and Hanaoka [33] used UCEs to also build CCA-secure public-key encryption schemes.

The UCE framework comes in two versions: a single-key version (UCE) and a multi-key version (mUCE). For a keyed hash function H, single-key UCE security is defined via a two-stage security game consisting of algorithms S and D, called the *source* and the *distinguisher*, respectively. In the first stage, the source is given access to an oracle HASH that, depending on a challenge bit b, implements either a random oracle or the concrete hash function with a randomly chosen key hk. The source terminates with some leakage L, which is then communicated together with hk to the distinguisher D. The distinguisher's goal is to guess the bit b, i.e., guess whether the source interacted with the random oracle or the hash function. The UCE advantage of the pair (S, D) is defined as the probability of returning the correct answer scaled away from one-half. (The stronger multi-key version is defined analogously by introducing HASH oracles for multiple keys and providing the keys together with leakage to the distinguisher.) We summarize this interaction schematically in Figure 1, and give the pseudocode in Figure 2. We refer the reader to the original work for an excellent philosophical perspective on this framework.

---

[1] Citation [8] refers to the CRYPTO 2013 proceedings version, [9] refers to its full version on Cryptology ePrint Archive from August 2013 prior to communicating our basic iO attack (presented in this paper), and [10] refers to the version from September/October 2013, and [12] refers to the latest version from May 2014.

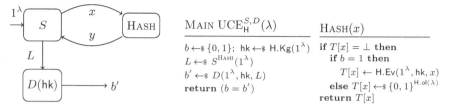

Fig. 1. Schematic of the UCE game

Fig. 2. Pseudocode for the UCE game. Here $\mathsf{H.ol}(\lambda)$ is a function which specifies the length of hash values.

$$\underline{\text{MAIN UCE}_{\mathsf{H}}^{S,D}(\lambda)}$$
$b \leftarrow_\$ \{0,1\}; \; \mathsf{hk} \leftarrow_\$ \mathsf{H.Kg}(1^\lambda)$
$L \leftarrow_\$ S^{\text{HASH}}(1^\lambda)$
$b' \leftarrow_\$ D(1^\lambda, \mathsf{hk}, L)$
return $(b = b')$

$$\underline{\text{HASH}(x)}$$
if $T[x] = \bot$ **then**
  if $b = 1$ **then**
    $T[x] \leftarrow \mathsf{H.Ev}(1^\lambda, \mathsf{hk}, x)$
  else $T[x] \leftarrow_\$ \{0,1\}^{\mathsf{H.ol}(\lambda)}$
  return $T[x]$

Without any restrictions UCE security cannot be achieved: the source can simply leak one of its oracle queries together with the corresponding answer to the distinguisher, which then can locally compute the hash value on the queried point (the distinguisher knows the hash key) and compare it to the leaked hash value. Thus, the source needs to be somehow restricted, and this restriction forms the actual UCE definition: for a source class S we denote the UCE assumption with sources restricted to S by UCE[S]. Prior to our work, BHK proposed two source classes via *unpredictability* and *reset security* conditions, which in turn gave rise to two notions called UCE1 and UCE2, respectively.

The UCE1 notion [8,9] is defined using an unpredictability game which requires that when the source is run with a *random oracle*, its leakage does not *computationally* reveal any of its queries. This is formalized by requiring that the probability that an efficient predictor P can guess a query of S when given L is negligible. Such a source is then called unpredictable, and leads to the following definition of UCE1 security: a hash function is UCE1 secure if the advantage of all efficient, unpredictable sources S, and all efficient distinguishers D in the UCE game is negligible. The stronger notion of UCE2 security is defined analogously by requiring that the source satisfies the weaker requirement of reset security.

Following our obfuscation-based attack (that we describe next and that we communicated to BHK in August 2013 [11]), the UCE1 and UCE2 notions were revised in [10] and additional restrictions on sources were imposed. We will be discussing these shortly, after presenting our first attack.

An obfuscation-based attack on UCE1. Our first attack, described in Section 3, targets the original UCE notions UCE1 and UCE2, and is based on a recent breakthrough in the construction of obfuscation schemes. Garg et al. [24] give a candidate construction for the so-called notion of indistinguishability obfuscation [6] based on intractability assumptions related to multi-linear maps. Our attack shows that any UCE1 construction would need to falsify one of these assumptions. Put differently, if indistinguishability obfuscation exists, then UCE1 security (and hence also the stronger UCE2 security) cannot be achieved.

Roughly speaking, a secure indistinguishability obfuscation (iO) scheme assures that the obfuscations of any two circuits that implement the same function are computationally indistinguishable. Our attack uses this primitive as follows. The source picks a random point x, and queries it to HASH to get y. It then prepares an iO of the Boolean circuit $(H(\cdot, x) = y)$, and leaks it to the distinguisher as L. The distinguisher now plugs the hash key hk into this obfuscated circuit and returns whatever the circuit outputs. It is easy to see that the distinguisher recovers the challenge bit correctly with an overwhelming probability. What is less clear, however, is whether or not the source is unpredictable. Recall that the unpredictability game operates with respect to a random oracle. Let us now assume, for simplicity, that $|hk| < |y|/2$ (we will not need to rely on this assumption in our full attack). For any x, there are at most $2^{|hk|}$ possible values for $H(hk, x)$, and a random y would be one of them with probability at most $2^{|hk|}/2^{|y|} < 2^{-|y|/2}$, which is negligible. Consequently, the obfuscated circuit implements the constant *zero* function with overwhelming probability. This allows us to apply the security of the obfuscator to conclude the attack: the obfuscated circuit does not leak any more information about x than the zero function would, and since x was chosen randomly, it remains hidden from the view of any efficient predicator.

Salvaging UCE. Assuming the existence of indistinguishability obfuscation, we ask to what extent UCE can be salvaged. That is, do there exist other UCE assumptions that allow recovering (some of) the originally presented applications? We partially salvage UCEs by modifying the unpredictability condition and letting the predictor run in *unbounded* time. This statistical notion of unpredictability restricts the class of admissible sources such that the source implementing the iO attack falls outside it: an unbounded predictor can reverse-engineer the *computationally* secure obfuscator. This modification is validated by the work of Goldwasser and Rothblum [26] who show that a statistical analogue of iO is impossible unless the polynomial hierarchy collapses to its second level. As we discuss in Section 3.2, a large number of interesting applications (such as KDM and RKA security) survives under this definition.

After communicating our attack, BHK independently suggested the statistical patch [10]. In the revised version of their paper [10], they recast their proofs of security to rely only on statistical unpredictability for all applications where this is possible. We refer to [10] for details on the applications that can be salvaged by statistical UCE1. As mentioned earlier, not all applications can be salvaged by statistical unpredictability. Hence, BHK also present two additional UCE notions based on computational unpredictability, which together with the statistical patch allowed them to fully recover their original set of applications in light of the aforementioned iO attack. We discuss these next.

Computational UCE. Some applications discussed in [8,9], specifically hardcore functions, deterministic public-key encryption (D-PKE), message-locked encryption (MLE), and OAEP rely on computational unpredictability in an intrinsic way; that is, the reduction only works if the predictor is bound to run in polynomial time. For instance, the source presented in [8,9] for D-PKEs produces leakage which contains encryptions of messages that have been sent to the HASH oracle. An unbounded predictor can easily decipher the ciphertexts and predict HASH queries of the source.

Following the above attack, in the updated version of their paper, BHK [10] propose two novel UCE notions by imposing additional restrictions on the way the source operates, while keeping the original *computational* unpredictability game. The goal here is that these restrictions are sufficiently strong to circumvent our attack, but weak enough so that successful security reductions can be established.

To recover D-PKEs, MLEs, and OAEP, BHK propose a new UCE assumption based on computational unpredictability restricted to so-called *bounded parallel sources*. Such a source splits into two stages S_0 and S_1. In the first phase, algorithm S_0 prepares a vector of strings. In the second phase, independent instances of S_1 for each entry in the previously prepared vector are run in parallel. Each instance gets access to the HASH oracle and their combined outputs make up the final leakage. To circumvent our attack two restrictions on S_1 are imposed: its runtime and number of HASH queries (per instance). The idea here is that computing the obfuscation of a hash function is "too costly," and hence the attack cannot be mounted.

In Section 4 we show that this refined notion still falls prey to a similar, but somewhat more complex attack. The idea is to split the iO attack into two stages consisting of a high-complexity first stage and a parallelizable second stage. To this end, we use the powerful *randomized encodings* paradigm of Applebaum, Ishai, and Kushilevitz [2] to bring down the complexity of the second stage of the attack. The randomized encoding $\hat{f}(x; r)$ of $f(x)$ is simply an encoding of $f(x)$ such that a decoder dec can retrieve the original value $f(x)$ from it, i.e., $\text{dec}(\hat{f}(x; r)) = f(x)$. In addition, a randomized encoding specifies an efficient simulator Sim such that for all x the distributions $\hat{f}(x; r)$ over uniformly chosen r and $\text{Sim}(f(x))$ are computationally indistinguishable. These properties combined allow us to show that we can adapt our original attack such that the source does not leak the obfuscated circuit but rather a randomized encoding of it. This alone, however, is still not enough for an attack with the restrictions of bounded parallel sources. Finally, we utilize a special form of *decomposable* randomized encodings [31] to realize an attack. Such encodings have the property that each output bit of $\hat{f}(x; r)$ depends on at most a *single* bit of x (but possibly on the entire string r). The randomized encoding of Applebaum, Ishai, and Kushilevitz [3] is decomposable and supports all functions in $\mathcal{P}/poly$. We show how to use such an encoding scheme to split the computation of the encoding into two phases: a complex first preprocessing phase which does not depend on the actual input and a very simple second stage which can be parallelized and where each parallel instance essentially only has to drop one of two bits. We show that this second stage (which will correspond to S_1) can be implemented by constant-depth circuits consisting only of very few gates. This application of decomposable randomized encodings could be of interest also in other scenarios where efficiently computing an encoding is important and preprocessing is possible. In the latest version of their paper [12] BHK has removed bounded parallel sources and replaced them by new source classes to recover the original applications.

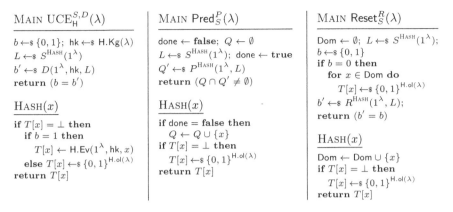

Fig. 3. The UCE security game together with the unpredictability and reset-security games

While bounded parallel sources suffice to also recover simultaneous hardcore functions, BHK propose a second, simpler UCE assumption based on *split sources*. A split source consists of two parts S_0 and S_1, in which each part independently contributes to the leakage sent to the distinguisher. The idea is that none of these sub-sources gets direct access to the HASH oracle. Rather, algorithm S_0 defines the queries (without access to any hash values) and algorithm S_1 gets to see the hash values but not the queries. As for our attack, note that the associated source needs to know both the query x and its hash value $y \leftarrow \text{HASH}(x)$ in order to compute the circuit $(\mathsf{H}(\cdot, x) = y)$. We discuss split sources in a larger context and present necessary conditions for a hash function to achieve split-source UCE security in the full version of this work [19]. For example, we show that in order to prove the security of a hash function H, one needs to show that the function that maps x to the obfuscation of the circuit $\mathsf{H}(\cdot, x)$ must not be one way. We also discuss intricacies regarding composition of such functions with one-way permutations, and show that such a composition does not harm standard notions such as collision resistance, pseudorandomness (and indeed statistical UCE1 security) but provably fails for split-source security. We present this discussion in the full version.

To conclude, although UCEs strengthen our confidence in the security of many practical schemes in the random-oracle model, our attacks highlight the need for a thorough assessment of definitional choices that can be made within the UCE framework. This assessment, in addition to instantiability questions, should also include studying concrete instantiations of UCEs such as the SHA family [34] in HMAC mode, as suggested by BHK [8].

2   Preliminaries

Notation. We denote by $\lambda \in \mathbb{N}$ the security parameter, which is implicitly given to all algorithms (if not explicitly stated so) in the unary representation 1^λ.

By $\{0,1\}^\ell$ we denote the set of all bit-strings of length ℓ, and by $\{0,1\}^*$ the set of all bit-strings of finite length. For two strings $x_1, x_2 \in \{0,1\}^*$ their concatenation is written as $x_1\|x_2$. The length of x is denoted by $|x|$ and $x[i]$ is the i-th bit of x. For a finite set X, we denote the action of sampling x uniformly at random from X by $x \leftarrow_\$ X$, and denote the cardinality of X by $|X|$. Algorithms are assumed to be randomized, unless otherwise stated. We call an algorithm efficient or PPT if it runs in time polynomial in the security parameter. By $y \leftarrow \mathcal{A}(x; r)$ we denote that y was output by algorithm $\mathcal{A}$ on input x and randomness r. If $\mathcal{A}$ is randomized and no randomness is specified, then we assume that $\mathcal{A}$ is run with freshly sampled uniform random coins, and write this is as $y \leftarrow_\$ \mathcal{A}(x)$. We often refer to algorithms, or tuples of algorithms, as adversaries. We say a function $\mathrm{negl}(\lambda)$ is negligible if $|\mathrm{negl}(\lambda)| \in \lambda^{-|\omega(1)|}$. In this paper we deploy the game-playing framework of Bellare and Rogaway [14] with the augmented game procedures described in [36].

Syntax of hash functions. In line with [8], we consider the following formalization of hash functions. A function family H is a five tuple of PPT algorithms $(\mathsf{H.Kg}, \mathsf{H.Ev}, \mathsf{H.kl}, \mathsf{H.il}, \mathsf{H.ol})$ as follows. The algorithms $\mathsf{H.kl}$, $\mathsf{H.il}$, and $\mathsf{H.ol}$ are deterministic and on input 1^λ define the key length, input length, and output lengths, respectively. (We have adopted the simplified notion from [8] here.) The key generation algorithm $\mathsf{H.Kg}$ gets the security parameter 1^λ as input and outputs a key $\mathsf{hk} \in \{0,1\}^{\mathsf{H.kl}(\lambda)}$. The deterministic evaluation algorithm $\mathsf{H.Ev}$ takes as input the security parameter 1^λ, a key hk, a message $x \in \{0,1\}^{\mathsf{H.il}(\lambda)}$ and generates a hash value $\mathsf{H.Ev}(1^\lambda, \mathsf{hk}, x) \in \{0,1\}^{\mathsf{H.ol}(\lambda)}$.

UCE *game.* Let $\mathsf{H} = (\mathsf{H.Kg}, \mathsf{H.Ev}, \mathsf{H.kl}, \mathsf{H.il}, \mathsf{H.ol})$ be a hash function and (S, D) be a pair of PPT algorithms. We define the UCE advantage of (S, D) against H through

$$\mathsf{Adv}^{\mathsf{uce}}_{\mathsf{H},S,D}(\lambda) := 2 \cdot \Pr\left[\mathrm{UCE}^{S,D}_{\mathsf{H}}(\lambda)\right] - 1 \,,$$

where game $\mathrm{UCE}^{S,D}_{\mathsf{H}}(\lambda)$ is shown in Figure 3 on the left.

Unpredictability. A source S is called *computationally unpredictable* if the advantage of any PPT predictor P defined by

$$\mathsf{Adv}^{\mathsf{pred}}_{S,P}(\lambda) := \Pr\left[\mathsf{Pred}^P_S(\lambda)\right]$$

is negligible, where game $\mathsf{Pred}^P_S(\lambda)$ is shown in Figure 3 in the middle. We denote the class of all computationally unpredictable sources by $\mathcal{S}^{\mathrm{cup}}$.

UCE *security.* We say a hash function H is UCE1 secure if for all computationally unpredictable PPT sources S and all PPT distinguishers D the advantage $\mathsf{Adv}^{\mathsf{uce}}_{\mathsf{H},S,D}(\lambda)$ is negligible. In the later version of their paper [10], BHK refer to UCE1 as $\mathrm{UCE}[\mathcal{S}^{\mathrm{cup}}]$. BHK introduce a stronger version called UCE2 which is based on the reset-security game $\mathsf{Reset}^R_S(1^\lambda)$ shown in Figure 3 on the right. We refer the reader to [9] for the details, but note here that UCE2 security implies UCE1 security and, thus, any attack on UCE1 also applies to UCE2.

We discuss the revised UCE assumptions introduced in [10], namely those for *bounded parallel sources* and *split sources*, in Section 4 and in the full version [19], respectively.

Indistinguishability obfuscation. Roughly speaking, an indistinguishability obfuscation (iO) scheme ensures that the obfuscations of any two functionally equivalent circuits are computationally indistinguishable. Indistinguishability obfuscation was originally proposed by Barak et al. [6] as a potential weakening of virtual-black-box obfuscation. We recall the definition from [24]. A PPT algorithm iO is called an *indistinguishability obfuscator* for a circuit class $\{\mathcal{C}_\lambda\}_{\lambda \in \mathbb{N}}$ if the following conditions are satisfied:

- CORRECTNESS. For all security parameters $\lambda \in \mathbb{N}$, for all $C \in \mathcal{C}_\lambda$, and for all inputs x we have that

$$\Pr\left[C'(x) = C(x) : C' \leftarrow_\$ \mathsf{iO}(1^\lambda, C) \right] = 1 .$$

- SECURITY. For any PPT distinguisher D, for all pairs of circuits $C_0, C_1 \in \mathcal{C}_\lambda$ such that $C_0(x) = C_1(x)$ on all inputs x the following distinguishing advantage is negligible:

$$\mathsf{Adv}^{\mathsf{io}}_{\mathsf{iO}, D, C_0, C_1}(\lambda) := \Pr\left[D(\mathsf{iO}(1^\lambda, C_1)) = 1 \right] - \Pr\left[D(\mathsf{iO}(1^\lambda, C_0)) = 1 \right] .$$

With their recent candidate construction for indistinguishability obfuscation, Garg et al. [24] have revived interest in the study of obfuscation schemes (see, for example, [37,16,28,17,18,23,5,15] and the references therein). Garg et al. prove that under an intractability assumption related to multi-linear maps their construction yields an indistinguishability obfuscator for all circuits in $\mathcal{NC}^1$. Additionally, assuming a perfectly correct fully homomorphic encryption scheme and a perfectly sound non-interactive witness-indistinguishable proof system, they also show how their obfuscation scheme can be bootstrapped to support any polynomial-size circuit. In a recent work, Barak et al. [5] have further simplified the construction and showed that it is secure against all generic multi-linear attacks.

3   UCE1 and UCE2 Security

In this section we formalize our iO attack on the UCE1 (and hence the stronger UCE2) security of *any* concrete hash function. We also propose a fix to these notions which avoids the attack while still being applicable to a number of cryptosystems.

3.1   The iO Attack

Our attack uses an indistinguishability obfuscation scheme in a black-box way, but is non-black-box as it relies on the code of the hash function for obfuscation. (Therefore, the attack does not contradict the positive feasibility of BHK in the random-oracle model.) We stress that the complexity of running our attack, although high, is polynomial, and will benefit from future advances in the construction of iO schemes.

Theorem 1 (UCE1 infeasibility). *If indistinguishability obfuscation exists, then UCE1 security cannot be achieved in the standard model.*

We now present a sketch of the proof and defer the full proof to the full version [19].

Proof (sketch). Let H be a UCE1-secure hash function family. Let us assume for now that $H.ol(\lambda) \geq 2 \cdot H.kl(\lambda)$, that is, the output length of the hash function is at least twice the size of a hash key. (We will be dropping this condition shortly.) Define a source S which generates a random value $x \leftarrow_\$ \{0,1\}^{H.il(\lambda)}$ and computes $y \leftarrow \text{HASH}(x)$. It then constructs the Boolean circuit

$$C_{\lambda,H,x,y}(\cdot) := (H.Ev(1^\lambda, \cdot, x) = y),$$

that returns 1 on input hk if, and only if, $H.Ev(1^\lambda, hk, x)$ equals y. The source S passes on an encoding of circuit $C_{\lambda,H,x,y}(\cdot)$ as leakage L to the distinguisher. We will later use obfuscation to ensure that x is not leaked by the encoding of $C_{\lambda,H,x,y}(\cdot)$ (this is needed for unpredictability). The distinguisher D recovers circuit $C_{\lambda,H,x,y}(\cdot)$ from the leakage L, and computes $b' \leftarrow C_{\lambda,H,x,y}(hk)$ using the given hash key hk, and returns b'. The UCE1 adversary (S, D) has advantage $1 - 2^{-H.ol(\lambda)}$: when the source is run with oracle access to $H.Ev(1^\lambda, hk, \cdot)$, the circuit always returns 1. When S interacts with a random oracle, y coincides with $H.Ev(1^\lambda, hk, x)$ with probability $2^{-H.ol(\lambda)}$.

Now let iO be an indistinguishability obfuscator. Instead of leaking circuit $C_{\lambda,H,x,y}(\cdot)$, we let S compute an obfuscation of the circuit and output $L \leftarrow_\$ iO$ $(C_{\lambda,H,x,y}(\cdot))$. By the correctness property of the obfuscator, distinguisher D, as before, has an overwhelming advantage in guessing the challenge bit correctly. It remains to show that the adapted source S is unpredictable.

In the unpredictability game $\text{Pred}_S^P(\lambda)$ oracle HASH is always implemented by a random oracle. Thus, with high probability the circuit $C_{\lambda,H,x,y}(\cdot)$ is the constant zero circuit: for any $x \in \{0,1\}^n$, there are at most $2^{|H.kl((\lambda))|}$ possible values for $H.Ev(hk, x)$, and a random y would be one of the image values with probability at most $2^{|H.kl(\lambda)|}2^{-\ell}$ which by assumption is less than $2^{-H.ol(\lambda)/2}$. Now, to see that the source S is unpredictable note that the zero function and $C_{\lambda,H,x,y}(\cdot)$ are functionally equivalent. This means that an indistinguishability obfuscation of $C_{\lambda,H,x,y}$ will not leak any more information about x than the zero function would. Since x was chosen randomly, it remains hidden from the view of any PPT predicator P.

It remains to argue how we can drop the requirement on the size of hash keys. For this note that we can simply choose a t such that $t \geq 2 \cdot \lceil H.kl(\lambda)/H.ol(\lambda) \rceil$ and let the source leak an obfuscation of the circuit $(H.Ev(1^\lambda, \cdot, x_1) = y_1 \wedge \cdots \wedge H.Ev(1^\lambda, \cdot, x_t) = y_t)$. $\square$

In the above proof, we relied on the source being able to make multiple queries to its hash oracle. Bellare, Hoang, and Keelveedhi [11] point out that the theorem can be extended to a single-query source by applying a pseudorandom generator to the output of the hash function. This result is noteworthy as several applications only require the source to make a single query.

3.2   Statistical Unpredictability

The iO attack immediately gives rise to the following question: can the UCE1 and/or UCE2 notions be somehow patched so that they avoid the attack while maintaining (part of) their wide applicability? Fortunately, we show that this is indeed the case. We start by observing that the security guarantee of the indistinguishability obfuscator is only computational. Consequently, the attack can be directly ruled out by demanding the source to be *statistically* unpredictable, i.e., by letting a potential predictor run in unbounded time (but still impose polynomial query complexity). More formally, we say a source S is *statistically unpredictable* if the advantage of any (possibly unbounded) predictor P with polynomial query complexity in the $\mathsf{Pred}_S^P(\lambda)$ game shown in Figure 3 (middle) is negligible. Statistical UCE2 security can be defined analogously, where we let the reset distinguisher run in unbounded time and only place a polynomial bound on the number of its queries.

The above definition, in turn, leads to the following two questions: (1) Is a statistically secure variant of indistinguishability obfuscation possible? (2) Are there any application scenarios which only rely on this weaker property? Goldwasser and Rothblum [26] provide a negative answer to the first question by showing that the existence of a statistically secure iO scheme implies the collapse of the polynomial hierarchy to its second level. This impossibility result reinforces our confidence in the soundness of the above definition. For the second question, recall that the unpredictability game is always defined with respect to a random oracle, and hence statistical unpredictability may be (non-trivially) achievable. Indeed, consider a source which samples a random point x, queries it to its oracle, and leaks the result to the distinguisher. It is easy to see that this source is statistically unpredictable as a random oracle is one-way against unbounded adversaries. Indeed, many of the cryptosystems considered by BHK admit security proofs with sources that essentially take this simple form [8,9]. We present a brief discussion of these in the full version of this work.

After we communicated our attack [11], BHK in the revised version of their paper [10] also independently suggested the statistical notion of unpredictability. They denote by $\mathcal{S}^{\mathrm{sup}}$ the class of all statistically unpredictable sources and recast their proofs of the above to use UCE[$\mathcal{S}^{\mathrm{sup}}$]. We refer to [10] for details on the applications that can be salvaged with statistical UCE1 aka UCE[$\mathcal{S}^{\mathrm{sup}}$] (resp. statistical UCE2 aka UCE[$\mathcal{S}^{\mathrm{srs}}$]).

We end this section by noting that for the hardcore predicate, BR93 encryption, D-PKE, MLE and OAEP application scenarios discussed in [8,9], the leakage contains auxiliary information related to a query x that only computationally hides x (e.g., it might contain a one-way image $f(x)$, or an encryption of x). Consequently, an unbounded predictor might well be able to guess the point x, and in these cases our statistical patch is no longer useful. Despite this, we observe that UCE-secure hash functions with regard to statistical unpredictability are hardcore for highly non-injective one-way functions. (The proof is essentially equivalent to that in [9] and relies on the fact that any (even an unbounded) predictor cannot recover the *exact* query if the preimage space is super-polynomially large.)

4   Bounded Parallel Sources

In version [10] of their paper, BHK introduce novel UCE-type security notions to recover applications where statistical unpredictability is of no help. The main idea behind these new UCE assumptions is that, in order to keep the unpredictability condition computational, the source needs to operate in a restricted way so that the iO attack cannot be mounted any longer.

A new restricted source class that BHK introduce to recover the deterministic public-key encryption (D-PKE), message-locked encryption (MLE), and OAEP applications is that of *bounded parallel sources*. In parallel sources the source splits into two parts S_0 and S_1 as follows. The first part of the source S_0 does not get oracle access to HASH, and simply outputs some preliminary leakage L_0 and a vector $\mathbf{L}'$ of arbitrary bit strings. For each entry in $\mathbf{L}'$ an independent instance of the second part of the source S_1 is run. This can be done in parallel as the several invocations do not share any coins or state. Instance i of S_1 is given $\mathbf{L}'[i]$ as input which then produces leakage $\mathbf{L}[i]$. As opposed to S_0, the second part S_1 of parallel sources has oracle access to HASH. The final leakage of the source $S := \mathsf{Prl}[S_0, S_1]$ is set to be $L := (L_0, \mathbf{L})$. The details of a parallel source $S = \mathsf{Prl}[S_0, S_1]$ are given in Figure 4 on the left.

Without any further restrictions, parallel sources are as powerful as regular sources: simply ignore S_0 and let a single S_1 generate the entire leakage. Thus, in order to circumvent the iO attack, further restrictions are necessary. To this end, BHK restrict the resources of S_0 and S_1 via polynomials τ, σ, and q as follows: (1) the running time (circuit size) of each invocation of S_1 is at most $\tau(\cdot)$; (2) each invocation of S_1 makes at most $q(\cdot)$ oracle queries; and (3) the length of initial leakage L_0 output by S_0 is at most $\sigma(\cdot)$. BHK then consider the class $\mathcal{S}_{\tau,\sigma,q}^{\mathrm{prl}}$ consisting of all parallel sources satisfying these bounds, and define UCE for computationally unpredictable, bounded parallel sources by considering UCE[$\mathcal{S}^{\mathrm{cup}} \cap \mathcal{S}_{\tau,\sigma,q}^{\mathrm{prl}}$].

For their results on D-PKE and MLE schemes, the parameters τ, σ, and q need to be fine-tuned according to the underlying encryption scheme. More precisely, BHK set q to 1 (each instance of S_1 makes a single hash query), σ to the size of a key-pair (0 in the case of MLEs), and τ to the runtime of the encryption operation plus the input and key sizes of the encryption scheme. It is easily seen that our basic attack does not fall into this class as long as the computation of the obfuscated circuit takes longer than what is granted by τ.

Prl SOURCE $S^{\text{HASH}}(1^\lambda)$	Splt SOURCE $S^{\text{HASH}}(1^\lambda)$
$(L_0, \mathbf{L}') \leftarrow_{\$} S_0(1^\lambda)$	$(L_0, \mathbf{x}) \leftarrow_{\$} S_0(1^\lambda)$
for $i = 1, \ldots, \|\mathbf{L}'\|$ do	for $i = 1, \ldots, \|\mathbf{x}\|$ do
$\quad \mathbf{L}[i] \leftarrow_{\$} S_1^{\text{HASH}}(1^\lambda, \mathbf{L}'[i])$	$\quad \mathbf{y}[i] \leftarrow_{\$} \text{HASH}(\mathbf{x}[i])$
$L \leftarrow (L_0, \mathbf{L})$	$L_1 \leftarrow_{\$} S_1(1^\lambda, \mathbf{y}); L \leftarrow (L_0, L_1)$
return L	return L

Fig. 4. The parallel source $S = \text{Prl}[S_0, S_1]$ on the left and the split source $S = \text{Splt}[S_0, S_1]$ on the right as defined in the updated version of [10]. In both cases the source consists of two parts S_0 and S_1 that jointly generate leakage L. For split sources neither part gets direct oracle access to HASH. For parallel sources additional restrictions on the runtime and the number of queries of S_1, and the length of leakage L_0 are imposed. Note that the invocations of S_1 are parallelizable and independent of one another.

In choosing the parameters for bounded parallel sources, one has to strike a delicate balance between the complexity of obfuscating a hash function and the cost of encryption (resp. the application in question). Indeed, suppose that a bounded parallel source assumption with parameters as above is used to prove an MLE scheme secure in the standard model. Now if the complexity of the encryption scheme is high (e.g., because it is implemented based on iO [37] or because it includes (artificial) redundant code), then the assumption can be broken by the iO attack, as described in the previous section. Similarly, if one could reduce the complexity of obfuscating the hash function, an attack would become feasible. However, considering the current state of research, obfuscation is a very costly operation and thus, intuitively, computing the obfuscation of a hash function should be harder than encrypting a message.

Interestingly, as we show in this section, it is the *parallel* complexity of obfuscating a hash function (after a possibly complex preprocessing phase) that matters for the attack, and we can show that the latter can lie in a complexity class which is dramatically below that of computing the obfuscation of the hash function. More precisely, we show how to combine our iO attack with the *randomized encodings* of Applebaum, Ishai, and Kushilevitz [2] to split the attack into two stages such that the second stage is highly parallelizable. Before describing our attack, let us briefly recall the notion of randomized encodings.

4.1 Randomized Encodings

Randomized encodings allow one to substantially reduce the complexity of computing a function f by instead computing an *encoding* of it. This technique was first introduced by Ishai and Kushilevitz [29,30] in the context of multi-party computation and has since found many applications [2,3,31,27,4,1]. The formalization of randomized encodings that we use here is due to Applebaum, Ishai, and Kushilevitz (AIK) [2] and is adapted to the setting of perfect correctness and computational privacy. Informally, we say that $\hat{f}(x; r)$ is a randomized encoding of some function $f(x)$ if (1) given $\hat{f}(x;r)$ one can efficiently recover function value $f(x)$, and (2) given $f(x)$, one can efficiently sample from the distribution $\hat{f}(x;r)$ induced by uniformly choosing r.

More precisely, a randomized encoding scheme RE consists of three efficient algorithms (enc, dec, Sim) as follows: (1) a probabilistic encoding algorithm enc which on input a security parameter 1^λ, a circuit computing $f_\lambda : \{0,1\}^{n(\lambda)} \to \{0,1\}^{\ell(\lambda)}$ (of size polynomial in λ) and an $x \in \{0,1\}^{n(\lambda)}$ outputs an encoding $z \in \{0,1\}^{s(\lambda)}$; (2) a deterministic decoder algorithm dec which on input the security parameter 1^λ and an encoding $z \in \{0,1\}^{s(\lambda)}$ outputs an image point $y \in \{0,1\}^{\ell(\lambda)}$; and (3) a probabilistic simulation algorithm Sim which on input 1^λ and an image point $y \in \{0,1\}^{\ell(\lambda)}$ outputs an encoding $z \in \{0,1\}^{s(\lambda)}$. To keep our notation consistent with the previous literature on randomized encoding, for a given circuit f_λ, we will refer to the the mapping $\mathsf{enc}(1^\lambda, f_\lambda, \cdot; \cdot)$ by $\hat{f}_\lambda : \{0,1\}^{n(\lambda)} \times \{0,1\}^{m(\lambda)} \to \{0,1\}^{s(\lambda)}$, where $\{0,1\}^{m(\lambda)}$ is the randomness space of enc. We say scheme RE is a perfectly correct, computationally private randomized encoding for a circuit class $\{\mathcal{F}_\lambda\}_{\lambda \in \mathbb{N}}$ if it satisfies the following two conditions.

- CORRECTNESS. For any $f_\lambda \in \mathcal{F}_\lambda$ and any input $x \in \{0,1\}^{n(\lambda)}$ we have that

$$\Pr\left[\mathsf{dec}(1^\lambda, \hat{f}_\lambda(x; r_{\mathsf{enc}})) = f_\lambda(x) : r_{\mathsf{enc}} \leftarrow\!\!\!{}_\$ \{0,1\}^{m(\lambda)} \right] = 1 .$$

- PRIVACY. For any PPT distinguisher D, any $f_\lambda \in \mathcal{F}_\lambda$, and any input $x \in \{0,1\}^{n(\lambda)}$ the distinguishing advantage $\mathsf{Adv}^{\mathsf{re}}_{\mathsf{RE},D,x}$ is negligible, where advantage $\mathsf{Adv}^{\mathsf{re}}_{\mathsf{RE},D,x}(\lambda)$ is defined as:

$$\Pr[D(1^\lambda, \hat{f}_\lambda(x; r_{\mathsf{enc}})) = 1 : r_{\mathsf{enc}} \leftarrow\!\!\!{}_\$ \{0,1\}^{m(\lambda)}] - \Pr[D(1^\lambda, \mathsf{Sim}(1^\lambda, f_\lambda(x))) = 1] .$$

Functions n, ℓ, s, and m are polynomials, however, we will be dropping the explicit dependency on λ in order to simplify notation, and set $n := n(\lambda)$, $\ell := \ell(\lambda)$, $s := s(\lambda)$, and $m := m(\lambda)$.

AIK used randomized encodings to construct cryptography in $\mathcal{NC}^0$. For us, the complexity of the encoding is not important. Rather, we will make use of encodings with small locality, where each bit in the randomized encoding $\hat{f}(x;r)$ only depends on at most a single bit of x (but possibly many bits of r). We will return to the topic of locality in Section 4.3.

4.2   Composing iO with Randomized Encodings

To ease readability, we present our attack in two stages. First, we show that our iO attack can be composed with any randomized encoding scheme in a way which neither affects the adversary's advantage nor the unpredictability of its implicit source. Then, in the next subsection, we use a special type of RE scheme known as *decomposable randomized encodings* [31] to split and parallelize the adversary's source in order to meet the (minimal) bounds of $q(\lambda) = 1$, $\sigma(\lambda) = 0$, and $\tau(\lambda) \in \mathcal{O}(\lambda)$. Consequently, our attack will rule out bounded parallel sources for these parameters. Since the bounds that our attacks achieves are very stringent, and an encryption scheme has to at least run in time $\mathcal{O}(\lambda)$ (and make a single HASH query), assuming indistinguishability obfuscation, it is unlikely that bounded parallel sources can be used to instantiate ROs in any meaningful application scenario.

Let H be a $\mathsf{UCE}[\mathcal{S}^{\mathrm{cup}} \cap \mathcal{S}^{\mathrm{prl}}_{\tau,\sigma,q}]$-secure hash function, iO be an indistinguishability obfuscator, and let us assume once again that $\mathsf{H.ol}(\lambda) \geq 2 \cdot \mathsf{H.kl}(\lambda)$. (As in the proof of Theorem 1, this assumption will be without loss of generality.)

The attacker. Define $C_{\lambda,\mathsf{H},x,y}(\cdot) := (\mathsf{H.Ev}(1^\lambda, \cdot, x) = y)$, and compute a randomized encoding of the circuit

$$ f : (x, y, r_{\mathsf{io}}) \mapsto \mathsf{iO}(C_{\lambda,\mathsf{H},x,y}(\cdot); r_{\mathsf{io}}), $$

where r_{io} is the randomness used by the obfuscator. As in the proof of Theorem 1, we consider the source S which chooses random values x, r_{io}, and r_{enc}, queries x to its oracle to obtain $y \leftarrow \mathrm{HASH}(x)$, and leaks the randomized encoding

$$ L := \hat{f}(x, y, r_{\mathsf{io}}; r_{\mathsf{enc}}) \ . $$

The distinguisher D gets as input a hash key hk and an encoding $\hat{f}(x, y, r_{\mathsf{io}}; r_{\mathsf{enc}})$. It uses the decoder dec of the randomized encoding scheme to recover

$$ f(x, y, r_{\mathsf{io}}) \leftarrow \mathsf{dec}(\hat{f}(x, y, r_{\mathsf{io}}; r_{\mathsf{enc}})) \ . $$

It then interprets the result as a circuit, runs it on on hk, and returns whatever the circuit outputs.

By correctness of the randomized encoding, the advantage of the adversary is identical to the one in our original iO-attack. Moreover, the source is computationally unpredictable which follows when combining the analysis of the previous section with the privacy of the randomized encoding. We give the formal analysis of advantage and success probability in the full version [19].

4.3   Splitting and Parallelizing S Using Decomposable REs

The attack described in the previous subsection works for any randomized encoding scheme. In particular, now, we will use a *decomposable* randomized encoding scheme to instantiate the attack; this allows us to recast the above source as a bounded parallel source. Let us begin with the definition decomposable randomized encodings.

Decomposable encodings. In a decomposable randomized encoding (DRE) scheme, every output bit of the encoding $\hat{f}(x; r)$ depends on at most a single bit of x (but possibly on arbitrarily many bits of r). More precisely, a decomposable randomized encoding scheme DRE consists of a four tuple of algorithms (idx, enc, dec, Sim) as follows. Algorithm idx on input a circuit f and an index $i \in [s]$ outputs an index $j \in [n] \cup \{0\}$. The decomposable encoding algorithm enc operates based on a local encoding algorithm $\overline{\mathsf{enc}}$ as follows. On input a circuit f, a point x, and random coins r_{enc}, for each $i \in [s]$

compute $z_i \leftarrow \overline{\mathsf{enc}}(f, i, x[\mathsf{idx}(f, i)]; r_{\mathsf{enc}})$, where we define $x[0] := \perp$, and return $z \leftarrow (z_1, \ldots, z_s)$. Algorithms dec and Sim play the same roles as those in a conventional RE scheme. As before, we denote $\overline{\mathsf{enc}}(f, i, b; r_{\mathsf{enc}})$ by $\hat{f}_i(b; r_{\mathsf{enc}})$. Thus we may write

$$\hat{f}(x; r_{\mathsf{enc}}) = \hat{f}_1(x[\mathsf{idx}(1)]; r_{\mathsf{enc}}) \| \hat{f}_2(x[\mathsf{idx}(2)]; r_{\mathsf{enc}}) \| \cdots \| \hat{f}_s(x[\mathsf{idx}(s)]; r_{\mathsf{enc}}) .$$

As Ishai et al. [31] point out, several constructions of randomized encodings are decomposable. For example, AIK's construction based on garbled circuits [3] is a decomposable, perfectly correct, and computationally private randomized encoding for any function in $\mathcal{P}/poly$. Their construction relies only on the existence of secure pseudorandom generators.

Using decomposable encodings, we show that our attack can be parallelized. The idea is that each instance of S_1 is responsible for computing a single bit of $\hat{f}(x; r_{\mathsf{enc}})$. However, potentially, computing even a single bit of $\hat{f}(x; r_{\mathsf{enc}})$ can be a computationally heavy task. We thus outsource pre-computation to S_0 such that for S_1, computing a single bit of $\hat{f}(x; r_{\mathsf{enc}})$ becomes easy. For concreteness, let us think about the instance of S_1 that computes the first bit of $\hat{f}(x; r_{\mathsf{enc}})$. As the encoding is decomposable, the first bit of $\hat{f}(x; r_{\mathsf{enc}})$ only depends on a single bit x_i of x. S_0 now picks r_{enc} and computes the first bit of $\hat{f}(x; r_{\mathsf{enc}})$ simply for both cases, $x_i = 0$ and if $x_i = 1$. It obtains two values and passes these two values to S_1. Now, as source S_1 has access to HASH it can compute the actual x_i and its task is thus merely picking the right precomputed bit as output. We give the full description of the attack in the full version [19].

Theorem 2 (Bounded parallel UCE infeasibility). *If indistinguishability obfuscation (and PRGs) exist, then* $\mathrm{UCE}[\mathcal{S}^{\mathrm{cup}} \cap \mathcal{S}^{\mathrm{prl}}_{\tau, \sigma, q}]$ *security cannot be achieved in the standard model for* $q \neq 0$, *any* $\sigma \geq 0$, *and* $\tau \in \Omega(\lambda)$.

Following the above attack, BHK [12] retracted bounded parallel sources and replaced them by new source classes that are specifically designed according to each application scenario.

5 Split Sources

In principle, bounded parallel sources would also suffice to recover the application of UCEs to hardcore functions. However, for this purpose, BHK [10] introduce a second, simpler UCE notion which is based on computational unpredictability and so-called *split sources*. A split source S is composed of two algorithms S_0 and S_1, where neither gets direct access to the HASH oracle. Algorithm S_0 outputs L_0 together with a vector of points $\mathbf{x}$. For each entry of $\mathbf{x}$, the corresponding HASH value is computed, and the vector of hash values $\mathbf{y}$ is formed. Algorithm S_1 is then run on $\mathbf{y}$ produces leakage L_1. The leakage of the split source $S := \mathsf{Splt}[S_0, S_1]$ then equals $L := (L_0, L_1)$. We give the pseudocode in Figure 4 on the right.

Split sources avoid our original attack, as well as its generalized version, as neither component of the source gets direct access to the HASH oracle. In the full version of this work [19], we discuss the composition of split-source UCE-secure functions with one-way permutations and also study the implications of existence of certain forms of obfuscators on their feasibility.

For example, consider a hash function where its inputs are first run through a one-way permutation before being hashed. Intuitively, this application of a one-way permutation should not harm UCE security. Indeed, this can be easily seen to be the case for the standard notions of one-wayness, collision resistance, and pseudorandomness. We show that statistical UCE1 security also enjoys this property. However, when composing a $\text{UCE}[\mathcal{S}^{\text{cup}} \cap \mathcal{S}^{\text{splt}}]$ hash functions with a one-way permutation, the resulting function fails to be $\text{UCE}[\mathcal{S}^{\text{cup}} \cap \mathcal{S}^{\text{splt}}]$ secure.

We also show that certain levels of unobfuscatability are necessary for a hash function to achieve UCE security with respect to split sources. For instance, the function that maps x to an obfuscation of the circuit $\mathsf{H}(\cdot, x)$ must *not* be one way. Further, this must also be the case for obfuscators that are specially designed to support H. For example, as a practical instantiation of UCEs, BHK suggest to use the SHA family [34] in HMAC mode [8]. Our results imply that in order to obtain confidence in the security of this construction, its extractability properties in conjunction with, say, the candidate obfuscator of Garg et al. [24] should be studied. We note that due to their simplicity, our results potentially also apply to other UCE notions which rely on a computational unpredictability notion. We refer to [19] for a more detailed discussion of split sources.

Acknowledgments. We thank Mihir Bellare, Viet Tung Hoang, and Sriram Keelveedhi for their personal communication [11]. Christina Brzuska was supported by the Israel Science Foundation (grant 1076/11 and 1155/11), the Israel Ministry of Science and Technology grant 3-9094), and the German-Israeli Foundation for Scientific Research and Development (grant 1152/2011). Pooya Farshim was supported by grant Fi 940/4-1 of the German Research Foundation (DFG). Arno Mittelbach was supported by CASED (www.cased.de) and DFG SPP 1736.

References

1. Applebaum, B.: Bootstrapping obfuscators via fast pseudorandom functions. Cryptology ePrint Archive, Report 2013/699 (2013),
   http://eprint.iacr.org/2013/699
2. Applebaum, B., Ishai, Y., Kushilevitz, E.: Cryptography in NC^0. In: 45th FOCS, pp. 166–175. IEEE Computer Society Press (October 2004)
3. Applebaum, B., Ishai, Y., Kushilevitz, E.: Computationally private randomizing polynomials and their applications. Computational Complexity 15(2), 115–162 (2006)
4. Applebaum, B., Ishai, Y., Kushilevitz, E., Waters, B.: Encoding functions with constant online rate or how to compress garbled circuits keys. In: Canetti, R., Garay, J.A. (eds.) CRYPTO 2013, Part II. LNCS, vol. 8043, pp. 166–184. Springer, Heidelberg (2013)

5. Barak, B., Garg, S., Kalai, Y.T., Paneth, O., Sahai, A.: Protecting obfuscation against algebraic attacks. In: Nguyen, P.Q., Oswald, E. (eds.) EUROCRYPT 2014. LNCS, vol. 8441, pp. 221–238. Springer, Heidelberg (2014)
6. Barak, B., Goldreich, O., Impagliazzo, R., Rudich, S., Sahai, A., Vadhan, S.P., Yang, K.: On the (im)possibility of obfuscating programs. In: Kilian, J. (ed.) CRYPTO 2001. LNCS, vol. 2139, pp. 1–18. Springer, Heidelberg (2001)
7. Bellare, M., Boldyreva, A., Palacio, A.: An uninstantiable random-oracle-model scheme for a hybrid-encryption problem. In: Cachin, C., Camenisch, J.L. (eds.) EUROCRYPT 2004. LNCS, vol. 3027, pp. 171–188. Springer, Heidelberg (2004)
8. Bellare, M., Hoang, V.T., Keelveedhi, S.: Instantiating random oracles via uCEs. In: Canetti, R., Garay, J.A. (eds.) CRYPTO 2013, Part II. LNCS, vol. 8043, pp. 398–415. Springer, Heidelberg (2013)
9. Bellare, M., Hoang, V.T., Keelveedhi, S.: Instantiating random oracles via UCEs. Cryptology ePrint Archive, Report 2013/424 (August 1, 2013), http://eprint.iacr.org/2013/424/20130801:043135 (Latest version prior to our attack [11])
10. Bellare, M., Hoang, V.T., Keelveedhi, S.: Instantiating random oracles via UCEs. Cryptology ePrint Archive, Report 2013/424 (October 17, 2013), http://eprint.iacr.org/2013/424/20131017:000316
11. Bellare, M., Hoang, V.T., Keelveedhi, S.: Personal communication (September 2013)
12. Bellare, M., Hoang, V.T., Keelveedhi, S.: Instantiating random oracles via UCEs. Cryptology ePrint Archive, Report 2013/424 (May 20, 2014), http://eprint.iacr.org/2013/424/20140520:182716 (Latest version at the time of writing)
13. Bellare, M., Rogaway, P.: Random oracles are practical: A paradigm for designing efficient protocols. In: Ashby, V. (ed.) ACM CCS 1993, pp. 62–73. ACM Press (November 1993)
14. Bellare, M., Rogaway, P.: The security of triple encryption and a framework for code-based game-playing proofs. In: Vaudenay, S. (ed.) EUROCRYPT 2006. LNCS, vol. 4004, pp. 409–426. Springer, Heidelberg (2006)
15. Boyle, E., Chung, K.M., Pass, R.: On extractability obfuscation. In: Lindell, Y. (ed.) TCC 2014. LNCS, vol. 8349, pp. 52–73. Springer, Heidelberg (2014)
16. Brakerski, Z., Rothblum, G.N.: Obfuscating conjunctions. In: Canetti, R., Garay, J.A. (eds.) CRYPTO 2013, Part II. LNCS, vol. 8043, pp. 416–434. Springer, Heidelberg (2013)
17. Brakerski, Z., Rothblum, G.N.: Black-box obfuscation for d-CNFs. In: Naor, M. (ed.) ITCS 2014, pp. 235–250. ACM (January 2014)
18. Brakerski, Z., Rothblum, G.N.: Virtual black-box obfuscation for all circuits via generic graded encoding. In: Lindell, Y. (ed.) TCC 2014. LNCS, vol. 8349, pp. 1–25. Springer, Heidelberg (2014)
19. Brzuska, C., Farshim, P., Mittelbach, A.: Indistinguishability obfuscation and UCEs: The case of computationally unpredictable sources. Cryptology ePrint Archive, Report 2014/099 (2014), http://eprint.iacr.org/2014/099
20. Canetti, R., Goldreich, O., Halevi, S.: The random oracle methodology, revisited (preliminary version). In: 30th ACM STOC, pp. 209–218. ACM Press (May 1998)
21. Canetti, R., Goldreich, O., Halevi, S.: On the random-oracle methodology as applied to length-restricted signature schemes. In: Naor, M. (ed.) TCC 2004. LNCS, vol. 2951, pp. 40–57. Springer, Heidelberg (2004)

22. Dodis, Y., Oliveira, R., Pietrzak, K.: On the generic insecurity of the full domain hash. In: Shoup, V. (ed.) CRYPTO 2005. LNCS, vol. 3621, pp. 449–466. Springer, Heidelberg (2005)
23. Garg, S., Gentry, C., Halevi, S., Raykova, M.: Two-round secure MPC from indistinguishability obfuscation. In: Lindell, Y. (ed.) TCC 2014. LNCS, vol. 8349, pp. 74–94. Springer, Heidelberg (2014)
24. Garg, S., Gentry, C., Halevi, S., Raykova, M., Sahai, A., Waters, B.: Candidate indistinguishability obfuscation and functional encryption for all circuits. In: 54th FOCS, pp. 40–49. IEEE Computer Society Press (October 2013)
25. Goldwasser, S., Kalai, Y.T.: On the (in)security of the Fiat-Shamir paradigm. In: 44th FOCS, pp. 102–115. IEEE Computer Society Press (October 2003)
26. Goldwasser, S., Rothblum, G.N.: On best-possible obfuscation. In: Vadhan, S.P. (ed.) TCC 2007. LNCS, vol. 4392, pp. 194–213. Springer, Heidelberg (2007)
27. Goyal, V., Ishai, Y., Sahai, A., Venkatesan, R., Wadia, A.: Founding cryptography on tamper-proof hardware tokens. In: Micciancio, D. (ed.) TCC 2010. LNCS, vol. 5978, pp. 308–326. Springer, Heidelberg (2010)
28. Hohenberger, S., Sahai, A., Waters, B.: Replacing a random oracle: Full domain hash from indistinguishability obfuscation. In: Nguyen, P.Q., Oswald, E. (eds.) EUROCRYPT 2014. LNCS, vol. 8441, pp. 201–220. Springer, Heidelberg (2014)
29. Ishai, Y., Kushilevitz, E.: Randomizing polynomials: A new representation with applications to round-efficient secure computation. In: 41st FOCS, pp. 294–304. IEEE Computer Society Press (November 2000)
30. Ishai, Y., Kushilevitz, E.: Perfect constant-round secure computation via perfect randomizing polynomials. In: Widmayer, P., Triguero, F., Morales, R., Hennessy, M., Eidenbenz, S., Conejo, R. (eds.) ICALP 2002. LNCS, vol. 2380, pp. 244–256. Springer, Heidelberg (2002)
31. Ishai, Y., Kushilevitz, E., Ostrovsky, R., Sahai, A.: Cryptography with constant computational overhead. In: Ladner, R.E., Dwork, C. (eds.) 40th ACM STOC, pp. 433–442. ACM Press (May 2008)
32. Kiltz, E., Pietrzak, K.: On the security of padding-based encryption schemes – or – why we cannot prove OAEP secure in the standard model. In: Joux, A. (ed.) EUROCRYPT 2009. LNCS, vol. 5479, pp. 389–406. Springer, Heidelberg (2009)
33. Matsuda, T., Hanaoka, G.: Chosen ciphertext security via UCE. In: Krawczyk, H. (ed.) PKC 2014. LNCS, vol. 8383, pp. 56–76. Springer, Heidelberg (2014)
34. National Institute of Standards and Technology: FIPS 180-4, Secure Hash Standard (SHS). Tech. rep. (March 2012)
35. Nielsen, J.B.: Separating random oracle proofs from complexity theoretic proofs: The non-committing encryption case. In: Yung, M. (ed.) CRYPTO 2002. LNCS, vol. 2442, pp. 111–126. Springer, Heidelberg (2002)
36. Ristenpart, T., Shacham, H., Shrimpton, T.: Careful with composition: Limitations of the indifferentiability framework. In: Paterson, K.G. (ed.) EUROCRYPT 2011. LNCS, vol. 6632, pp. 487–506. Springer, Heidelberg (2011)
37. Sahai, A., Waters, B.: How to use indistinguishability obfuscation: Deniable encryption, and more. Cryptology ePrint Archive, Report 2013/454 (2013), http://eprint.iacr.org/2013/454

Low Overhead Broadcast Encryption
from Multilinear Maps

Dan Boneh[1], Brent Waters[2], and Mark Zhandry[1]

[1] Stanford University, CA, USA
{dabo,zhandry}@cs.stanford.edu
[2] University of Texas at Austin, TX, USA
bwaters@cs.utexas.edu

Abstract. We use multilinear maps to provide a solution to the long-standing problem of public-key broadcast encryption where all parameters in the system are small. In our constructions, ciphertext overhead, private key size, and public key size are all poly-logarithmic in the total number of users. The systems are fully collusion-resistant against any number of colluders. All our systems are based on an $O(\log N)$-way multilinear map to support a broadcast system for N users. We present three constructions based on different types of multilinear maps and providing different security guarantees. Our systems naturally give identity-based broadcast systems with short parameters.

1 Introduction

Broadcast encryption [FN94] is an important generalization of public-key encryption to the multi-user setting. In a broadcast encryption scheme, a broadcaster encrypts a message for a subset S of users who are listening on a broadcast channel. The broadcaster can encrypt to any set S of its choice, and any user in S can decrypt the broadcast using its secret key. The system is said to be *fully collusion resistant* if even a coalition of all users outside of S learns nothing about the plaintext. Broadcast systems are regularly used in TV and radio subscription services where broadcasts are encrypted for currently active subscribers. They are also used in encrypted file systems where a file is encrypted so that only users who have access to the file can decrypt it.

The efficiency of a broadcast system is measured in the ciphertext overhead: the number of bits in the ciphertext beyond what is needed for the description of the recipient set S and the symmetric encryption of the plaintext payload. The shorter the overhead, the better. We say that the system has *low overhead* if the ciphertext overhead depends at most logarithmically on the number of users N in the system.

Existing constructions with low ciphertext overhead. Several broadcast systems are fully collusion resistant with low ciphertext overhead. The first such system by Boneh, Gentry, and Waters [BGW05] is built from bilinear maps. It has

J.A. Garay and R. Gennaro (Eds.): CRYPTO 2014, Part I, LNCS 8616, pp. 206–223, 2014.

constant ciphertext overhead and short secret keys, but the public encryption key size is *linear* in the number of users N. Other systems using bilinear-maps achieve adaptive security [GW09, DPP07] and some are even identity-based [GW09, Del07, SF07], but the public encryption key is always large.

Multilinear maps give *secret-key* broadcast systems with optimal ciphertext overhead [BS03, GGH13a, FHPS13, CLT13, BW13]. However, in these systems the broadcaster's key must be kept secret, and they require an N-way multilinear map to support N users. Current constructions of N-linear maps [GGH13a, CLT13] have group elements of size $O(N^2)$ bits, resulting in large space requirements. While these broadcast systems can be made public-key by including a few group elements in the ciphertext, their dependence on N-linear maps leads to an $O(N^2)$ ciphertext overhead, which is worse than the trivial broadcast system. Until this work, it has not been known how to use multilinear maps to construct low overhead broadcast systems with a short public encryption key.

A third class of constructions employs the powerful candidates for indistinguishability obfuscation (iO) [BGI+01, GGH+13b]. Using iO it is possible to build a *public-key* broadcast system with optimal ciphertext overhead and short private keys, though public keys are large [BZ14]. The resulting systems have several other remarkable properties. However, current iO candidates add considerable complexity on top of multilinear maps. Our goal here is to construct broadcast systems using only simple assumptions on multilinear maps, namely, without relying on iO.

Our results. We describe three broadcast systems for N users that use an $O(\log N)$-way multilinear map. The systems have ciphertext overhead and decryption key of only $O(1)$ group elements which is $O(\log^2 N)$ bits using the current multilinear map candidates. The public encryption key contains $O(\log N)$ group elements which is $O(\log^3 N)$ bits. The first system uses an *asymmetric* multilinear map and follows the [BGW05] construction closely. It uses the $O(\log N)$-way multilinear map to compress the public key of that system from $O(N)$ group elements to $O(\log N)$ elements while keeping the ciphertext overhead and secret key short. We prove static security under a multivariate equivalent of the [BGW05] assumption.

The second system uses a general *symmetric* $O(\log N)$-way multilinear map to similarly compress the public key in [BGW05]. The added flexibility of a symmetric map has both positive and negative consequences. On the negative side, this flexibility allows the adversary to combine extra elements together. To maintain security we must ensure that all user indexes $u \in [N]$ are mapped to integers $\hat{u} \in [O(N \log N)]$ where all $\hat{u}$ have the same Hamming weight. This mapping does not affect ciphertext or private key size. On the positive side, this flexibility allows us to obtain slightly better parameters and base static security on a slightly simpler, though similar, complexity assumption.

The third system is built from a symmetric $O(\log N)$-way map, but we can prove *adaptive* security of the scheme in generic multilinear groups. This system has secret keys of length $O(\log^3 N)$ bits, which is longer than the previous two schemes, but has a tighter security proof in generic groups.

Because the parameters of these systems are logarithmic in N, we can let N be exponential, and in particular be as large as the range of a collision resistant hash function (e.g., $N = 2^{256}$). This, in effect, turns all our broadcast systems into efficient identity-based schemes. A user with identity id $\in \{0,1\}^*$ is given the secret key associated with index number $H(\text{id}) \in [N]$ where H is a collision resistant hash whose range is $[N]$. A broadcaster can then transmit to a set of recipients simply by hashing their public identities. For this reason, we describe all our broadcast systems as identity-based broadcast schemes.

Additional related work. Collusion resistant broadcast encryption has been widely studied. Revocation systems (e.g., [NNL01, HS02, GST04, DF02, LSW10]) can encrypt to $N - r$ users with ciphertext size of $O(r)$. Further combinatorial solutions (e.g., [NP00, DF03]) achieve similar parameters. A broadcast encryption system is said to be recipient-private if broadcast ciphertexts reveal nothing about the intended set of recipients [BBW06, LPQ12, FP12]. Our broadcast systems are not recipient private, and it is a long-standing open problem to build a low-overhead recipient-private broadcast system. Such a system was recently built using indistinguishability obfuscation [BZ14], but constructing such systems under weaker assumptions remains open.

2   Preliminaries

2.1   Broadcast Encryption

We begin by defining broadcast encryption. A (public key) identity-based broadcast encryption scheme consists of three randomized algorithms:

Setup($\mathcal{ID}$): Sets up a broadcast scheme for identity space $\mathcal{ID}$. It outputs public
    parameters params as well as a master secret key msk

KeyGen(msk, u): Takes the master secret key and a user $u \in \mathcal{ID}$ and outputs a
    secret key sk_u for user u.

Enc(params, S): The encryption algorithm takes the public parameters and a
    polynomial sized set $S \subseteq \mathcal{ID}$ of recipients, and produces a pair (Hdr, K). We
    refer to Hdr as the header, and K as the message encryption key.
    The message is encrypted using a symmetric encryption scheme with the key
    K to obtain a ciphertext c. The overall ciphertext is (Hdr, c).

Dec(params, u, sk_u, S, Hdr): The decryption algorithm takes the header Hdr and
    the secret key for user u, and if $u \in S$, outputs the message encryption key
    K. If $u \notin S$, the decryption algorithm outputs $\perp$.
    To actually decrypt the overall ciphertext (Hdr, c), user u runs Dec to obtain
    K, and then decryption c using K to obtain the message.

For correctness, we require that the decryption algorithm always succeeds when it is supposed to. That is, for every (params, msk) output by Setup($\mathcal{ID}$), every set $S \subseteq \mathcal{ID}$, every sk_u output by $\mathsf{KeyGen}(\mathsf{msk}, u)$, and (Hdr, K) outputted by $\mathsf{Enc}(\mathsf{params}, S)$ where $u \in S$, that $\mathsf{Dec}(\mathsf{params}, u, \mathsf{sk}_u, S, \mathsf{Hdr}) = K$.

For security, several notions of security are possible. We start by defining active chosen ciphertext security. For any adversary $\mathcal{A}$, let $\mathtt{EXP}(b)$ denote the following experiment on $\mathcal{A}$:

Setup: The challenger runs $(\mathsf{params}, \mathsf{msk}) \leftarrow \mathsf{Setup}(\mathcal{ID})$, and gives $\mathcal{A}$ the public key params.

Secret Key Queries: $\mathcal{A}$ may adaptively make secret key queries for user u. In response, the challenger runs $\mathsf{sk}_u \leftarrow \mathsf{KeyGen}(\mathsf{msk}, u)$ and gives sk_u to $\mathcal{A}$.

CCA Queries: $\mathcal{A}$ may make chosen ciphertext queries on tuples (u, S, Hdr). The challenger responds with $\mathsf{Dec}(\mathsf{params}, u, \mathsf{sk}_u, S, \mathsf{Hdr})$ where $\mathsf{sk}_u \leftarrow \mathsf{KeyGen}(\mathsf{msk}, u)$[1].

Challenge: $\mathcal{A}$ submits a set $S^* \subset \mathcal{ID}$, subject to the restriction that $u \notin S^*$ for any user u requested in a secret key query. The challenger lets $(\mathsf{Hdr}^*, K_0^*) \leftarrow \mathsf{Enc}(\mathsf{params}, S^*)$. If $b = 0$, the challenger gives (Hdr^*, K_0^*) to the adversary. If $b = 1$, the challenger computes a random key K_1^* and gives (Hdr^*, K_1^*) to the adversary.

More Secret Key Queries: $\mathcal{A}$ may continue making secret key queries for users $u \notin S^*$

More CCA Queries: $\mathcal{A}$ may continue making CCA queries on headers $\mathsf{Hdr} \neq \mathsf{Hdr}^*$[2].

Guess: $\mathcal{A}$ produces a guess b' for b.

Using a simple hybrid argument, we can assume the adversary makes only a single challenge query. Let W_b be the event that $\mathcal{A}$ outputs 1 in $\mathtt{EXP}(b)$. We define the adaptive CCA advantage of $\mathcal{A}$, as

$$\mathrm{BE}^{(\mathrm{adv})}{}_{\mathcal{A}} = |\Pr[W_0] - \Pr[W_1]|$$

Definition 1. *A broadcast encryption scheme is adaptively secure under a chosen ciphertext attack (adaptively CCA-secure) if, for all polynomial time adversaries $\mathcal{A}$, $\mathrm{BE}^{(adv)}{}_{\mathcal{A}}$ is negligible.*

We will also consider several weaker notions of security. For example, we get static security if we require $\mathcal{A}$ to commit to the challenge set S^* before seeing the public parameters. We also get CPA security if we do not allow chosen ciphertext queries. In this paper, we will be focusing on the following notion of static CPA security, but will also discuss the other variants:

Definition 2. *A broadcast encryption scheme is statically secure under a chosen plaintext attack (statically CPA-secure) if, for all polynomial time adversaries $\mathcal{A}$ that must commit to S^* before seeing the public parameters and cannot make CCA queries, $\mathrm{BE}^{(adv)}{}_{\mathcal{A}}$ is negligible.*

---

[1] Another variation is to have the challenger maintain a table of (u, sk_u) pairs, and only run KeyGen once for a particular user, using a single sk_u to answer multiple secret key and CCA queries. Note that the correctness of a broadcast scheme implies that this does not affect CCA queries.

[2] Another potentially stronger variation is to require $(S, \mathsf{Hdr}) \neq (S^*, \mathsf{Hdr}^*)$

2.2   Multilinear Maps

We now review multilinear maps [BS03, GGH13a, CLT13]. A multilinear map consists of two algorithms:

Setup(n): Sets up an n-linear map. It outputs n groups $\mathbb{G}_1, \ldots, \mathbb{G}_n$ of prime order p, along with generators $g_i \in \mathbb{G}_i$. We call $\mathbb{G}_1$ the source group, $\mathbb{G}_n$ the target group, and $\mathbb{G}_2, \ldots, \mathbb{G}_{n-1}$ intermediate groups.

$e_{i,j}(g, h)$: Takes in two elements $g \in \mathbb{G}_i$ and $h \in \mathbb{G}_j$ with $i + j \leq n$, and outputs an element of $\mathbb{G}_{i+j}$ such that

$$e_{i,j}(g_i^a, g_j^b) = g_{i+j}^{ab}$$

We often omit the subscripts and just write e. We can also generalize e to multiple inputs as $e(h^{(1)}, \ldots, h^{(k)}) = e(h^{(1)}, e(h^{(2)}, \ldots, h^{(k)}))$.

We sometimes call g_i^a as a level-i *encoding* of a. The scalar a itself could be referred to as a level-0 encoding of a. Then the map e combines a level i encoding and a level j encoding, and produces a level $i + j$ encoding of the product.

We will make use of *asymmetric* multilinear maps. In such maps, groups are indexed by integer vectors rather than integers. The pairing operations maps $\mathbb{G}_{\mathbf{v}_1} \times \mathbb{G}_{\mathbf{v}_2}$ into $\mathbb{G}_{\mathbf{v}_1+\mathbf{v}_2}$. More precisely, we have the following algorithms:

Setup($\mathbf{n}$) Sets up an $\mathbf{n}$-linear map, where $\mathbf{n} \in \mathbb{Z}^\ell$ is some positive integer vector. It outputs a description of groups $\mathbb{G}_{\mathbf{v}}$ of prime order p where $\mathbf{v}$ are non-negative integer vectors and $\mathbf{v} \leq \mathbf{n}$ (that is, the comparison must hold component-wise). It also outputs a description of generators $g_{\mathbf{v}} \in \mathbb{G}_{\mathbf{v}}{}^3$. Let $\mathbf{e}_i$ be the ith standard basis vector, with a 1 at position i and a 0 elsewhere. We call $\mathbb{G}_{\mathbf{e}_i}$ the ith source group, $\mathbb{G}_{\mathbf{n}}$ the target group, and the rest of the $\mathbb{G}_{\mathbf{v}}$ groups are intermediate groups.

$e_{\mathbf{v}_1, \mathbf{v}_2}(g, h)$ Takes in two elements $g \in \mathbb{G}_{\mathbf{v}_1}$ and $h \in \mathbb{G}_{\mathbf{v}_2}$ with $\mathbf{v}_1 + \mathbf{v}_2 \leq \mathbf{n}$, and outputs an element of $\mathbb{G}_{\mathbf{v}_1+\mathbf{v}_2}$ such that

$$e_{\mathbf{v}_1, \mathbf{v}_2}(g_{\mathbf{v}_1}^a, g_{\mathbf{v}_2}^b) = g_{\mathbf{v}_1+\mathbf{v}_2}^{ab}$$

We often omit the subscripts and just write e. We can also generalize e to multiple inputs as $e(h^{(1)}, \ldots, h^{(k)}) = e(h^{(1)}, e(h^{(2)}, \ldots, h^{(k)}))$.

We note that current candidates of multilinear maps [GGH13a, CLT13] depart from the ideal notions of multilinear maps described above. In particular, in these candidates, representations of group elements are not unique and contain a noise term that can cause errors during group and multilinear operations. While we present our constructions using ideal multilinear maps for simplicity, we stress that our constructions can easily be instantiated using current non-ideal candidates. We need multilinear maps with the following properties:

---

[3] There may be an exponential number of groups and generators. The setup algorithm outputs a set of parameters from which the groups $\mathbb{G}_{\mathbf{v}}$ and generators $g_{\mathbf{v}}$ can be derived. In particular, each $g_{\mathbf{v}}$ can be derived from the pairing operation and $\{g_{\mathbf{e}_i}\}$, where $\mathbf{e}_i$ is the ith standard basis vector

- A way to hide the group and multilinear operations that lead to a particular element. In current multilinear maps, this is obtained by performing a re-randomization procedure which makes the representation of an element statistically independent of the operations that lead to that element.
- A way, given any representation of an element in the target group, to "extract" a canonical representation of that element. This is handled by a "zero-test parameter" in current maps.
- The ability for the person who sets up the map to compute elements of the form g^{α^x} for exponentially-large x. In ideal multilinear maps, this would be accomplished by computing $z = \alpha^x$ in $\mathbb{Z}_p$, and then computing g^z. However, with current multilinear maps, it is not possible for normal users to compute g^z for a specific z of their choice[4]. However, the person who sets up the map knows a trapdoor that *does* allow computing g^z for any $z \in \mathbb{Z}_p$.
- The ability to generate asymmetric multilinear maps for any positive integer vector $\mathbf{n} \in \mathbb{Z}^\ell$. Section 4.3 of [GGH13a] shows how to do this.
- A way to make sure the noise growth does not cause any errors during normal execution of our protocols. Since there is no circular dependence between the parameters of the multilinear map and the number of operations our protocols require, we can set the parameters so the noise stays small enough to avoid errors.

3 Our Asymmetric Multilinear Map Construction

In this section, we give our first construction of identity-based broadcast encryption from multilinear maps. Our starting point is the scheme of Boneh, Gentry, and Waters [BGW05], henceforth referred to as the BGW scheme. Recall in their scheme, the public parameters consist of $O(N)$ source group elements (where N is the number of users), secret keys and headers are a constant number of source group elements, and the message encryption key is a group element in the target group. Our goal is to shrink the public key size to $O(\log N)$ group elements. We accomplish this by embedding the BGW scheme in a multilinear map, where the BGW parameters lie in an intermediate group. The BGW public parameters can then be derived from a small number of elements in the source group of the map — these few source group elements are the new public key.

In more detail, the significant component of the BGW public key are the elements $Z_1 = g_1^\alpha, Z_2 = g_1^{\alpha^2}, \ldots, Z_N = g_1^{\alpha^N}, Z_{N+2} = g_1^{\alpha^{N+2}}, \ldots, Z_{2N} = g_1^{\alpha^{2N}}$. The rest of the BGW public keys, secret keys, and header components are also element in $\mathbb{G}_1$, whereas the message encryption key is an element in the group $\mathbb{G}_2$.

Let $N = 2^n - 1$ for some integer n, and let $\mathbf{n} = \overbrace{(1, \ldots, 1)}^{n+1 \text{ 1s}}$ be the vector of $n + 1$ 1s. Our idea is to use an asymmetric multilinear map, where the target group is $\mathbb{G}_{2\mathbf{n}}$. We note that pairing two elements in $\mathbb{G}_\mathbf{n}$ gives an element in $\mathbb{G}_{2\mathbf{n}}$. Thus, while the entire multilinear map is asymmetric, the pairing operation acts symmetrically on the group $\mathbb{G}_\mathbf{n}$. Now we replace the groups $\mathbb{G}_1$ and $\mathbb{G}_2$ in the

---

[4] Instead, users can compute g^z for a random, but unknown, z.

BGW scheme with $\mathbb{G}_\mathbf{n}$ and $\mathbb{G}_{2\mathbf{n}}$. Thus $Z_u = g_\mathbf{n}^{\alpha^u}$. Rather than explicitly include the Z_u in the public parameters, we give a few group elements in the groups $\mathbb{G}_{\mathbf{e}_i}$ where $\mathbf{e}_i$ are the standard basis vectors. Specifically, we provide the parameters $X_i = g_{\mathbf{e}_i}^{\alpha^{(2^i)}}$ for $i = 0, \ldots, n-1$. By pairing various subsets of these X_i together, we can build all of the Z_u for $u \le 2^n - 1 = N$. In particular, if $u = \sum_{i=0}^{n-1} u_i 2^i$ is the binary representation of u, then

$$Z_u = e(\, X_0^{u_0} g_{\mathbf{e}_0}^{1-u_0} \,,\, X_1^{u_1} g_{\mathbf{e}_1}^{1-u_1} \,,\, \ldots, X_{n-1}^{u_{n-1}} g_{\mathbf{e}_{n-1}}^{1-u_{n-1}} \,,\, g_{\mathbf{e}_n} \,)$$

where $X_i^0 g_{\mathbf{e}_i}^1 = g_{\mathbf{e}_i}$ and $X_i^1 g_{\mathbf{e}_i}^0 = X_i$

To allow computation of Z_u for $u \ge 2^n + 1 = N+2$, we might decide to publish $g_{\mathbf{e}_n}^{\alpha^{(2^n)}}$. However, this would allow computation of Z_{N+1}, which will break the security of the BGW scheme. Therefore, we instead publish $X_n = g_{\mathbf{e}_n}^{\alpha^{(2^n+1)}}$. Then, for $u \in [N+2, 2N]$, let $u' = u - (2^n + 1) = \sum_{i=0}^{n-1} u_i' 2^i$. Then we can write

$$Z_u = e(\, X_0^{u_0'} g_{\mathbf{e}_0}^{1-u_0'} \,,\, X_1^{u_1'} g_{\mathbf{e}_1}^{1-u_1'} \,,\, \ldots, X_{n-1}^{u_{n-1}'} g_{\mathbf{e}_{n-1}}^{1-u_{n-1}'} \,,\, X_n \,)$$

Now we make the observation that $O(\log N)$ graded encodings remain efficient even up to exponential N. Therefore, we can actually make our scheme identity-based, where identities are bit strings of length n with the caveat that the 0^n is not a valid identity. Now we give our entire construction:

Construction 1. *Let* Setup$'$ *be the setup algorithm for a multilinear map, where groups have order* p. *Our first identity-based broadcast scheme consists of the following algorithms:*

Setup(n): *Takes as input the length* n *of identities. Let* $\mathcal{ID} = \{0,1\}^n \setminus \{0^n\}$ *be the identity space. Let* $\mathbf{n}$ *be the all-ones vector of length* $n + 1$. *Run* Setup$'$ *on* $2\mathbf{n}$, *obtaining the public parameters* params$'$ *for a multilinear map with target group* $\mathbb{G}_{2\mathbf{n}}$.

Choose a random $\alpha \in \mathbb{Z}_p$ *and let* $X_i = g_{\mathbf{e}_i}^{\alpha^{(2^i)}}$ *for* $i = 0, \ldots, n-1$ *and let* $X_n = g_{\mathbf{e}_n}^{\alpha^{(2^n+1)}}$. *Also choose a random* $\gamma \in \mathbb{Z}_p$ *and let* $Y = g_\mathbf{n}^\gamma$. *Lastly, let* $W = g_{2\mathbf{n}}^{\alpha^{(2^n)}}$. *The public key is*

$$\text{params} = (\text{params}', W, \{X_i\}_{i \in \{0, \ldots, n\}}, Y)$$

The master secret key is (α, γ).

KeyGen(params, α, γ, u): *The secret key for identity* $u \in [1, 2^n - 1]$ *is* $\mathsf{sk}_u = g_\mathbf{n}^{\gamma \alpha^u}$.

Enc(params, S): *Recall that we can compute* Z_j *for* $j \in [1, 2^n - 1]$ *from the public parameters* $\{X_i\}_{i \in \{0, \ldots, n-1\}}$. *Pick a random* $t \in \mathbb{Z}_p$ *and compute the key and header as*

$$K = W^t = g_{2\mathbf{n}}^{t\alpha^{(2^n)}} \quad \text{and}$$

$$\mathsf{Hdr} = \left(\, g_\mathbf{n}^t \,,\, \left(Y \cdot \prod_{u \in S} Z_{2^n - u}\right)^t \,\right) = \left(\, g_\mathbf{n}^t \,,\, g_\mathbf{n}^{t\left(\gamma + \sum_{u \in S} \alpha^{(2^n - u)}\right)} \,\right)$$

Dec(params, u, sk_u, S, Hdr): *If $u \notin S$, output $\bot$. Otherwise, write* Hdr *as* (C_0, C_1). *Recall that we can compute Z_j for $j \in [2^n + 1, 2^{n+1}]$. Output*

$$
K = \frac{e(Z_u \,,\, C_1)}{e\left((\text{sk}_u \cdot \prod_{j \in S, j \neq u} Z_{2^n - j + u}) \,,\, C_0\right)}
$$

If C_0 and C_1 are as above, then we can write $K = g_{2\mathbf{n}}^c$ where

$$
c = \alpha^u t \left(\gamma + \sum_{j \in S} \alpha^{(2^n - j)}\right) - \left(\gamma \alpha^u + \sum_{j \in S, j \neq u} \alpha^{(2^n - j + u)}\right) t
$$

Most of the terms cancel, leaving $c = t\alpha^{2^n}$ as desired.

Implementation details. As mentioned in Section 2, there are some minor complications with implementing our scheme using current multilinear map constructions [GGH13a, CLT13], but we stress that these complications do not affect the semantics of our scheme. First, during normal operations, computing g_1^α for a random α involves computing a "level-0" encoding of a random (unknown) α, and then pairing with g_1. In order to compute $g_1^{\alpha^2}$, we would pair g_1 with the level-0 encoding twice. However, the noise growth with repeated pairing operations would prevent us from computing $g_1^{\alpha^{(2^i)}}$ for sufficiently high powers of i. Instead, the setup algorithm must choose an explicit (known) $\alpha \in \mathbb{Z}_p$, compute the various α^{2^i}, encode these powers as level-0 encodings, and only then pair with g_1. This requires knowing the secrets used to set up the multilinear map, meaning the broadcaster must set up the map himself and cannot rely on maps generated by trusted parties. Note, however, that this exponentiation is only required during setup, and not encryption or decryption, meaning the secrets can be discarded immediately after setup, and anyone can still broadcast using the public parameters.

To make sure the header does not leak any important information, we also need to re-randomize the header components. This means re-randomization parameters need to be included for the group $\mathbb{G}_\mathbf{n}$. No other re-randomization parameters are necessary.

Before discussing security, we must discuss our new security assumption, which is closely related to the bilinear Diffe-Hellman Exponent assumption (BDHE) as used in BGW.

3.1 The Hybrid Diffie-Hellman Exponent Assumption (HDHE) Assumption

We define the (computational) n-Hybrid Diffie-Hellman Exponent problem as follows: Let params' $\leftarrow$ Setup'$(2\mathbf{n})$ where $\mathbf{n}$ is the all-ones vector of length $n + 1$. Choose $\alpha \in \mathbb{Z}_p$ at random, and let $X_i = g_{e_i}^{\alpha^{(2^i)}}$ for $i = 0, \ldots, n - 1$ and

$X_n = g_{\mathbf{e}_n}^{\alpha^{(2^n+1)}}$. Choose a random $t \in \mathbb{Z}_p$ and let $V = g_{\mathbf{n}}^t$. Given $(\{X_i\}_{i \in \{0,\dots,n-1\}}, V)$, the goal is to compute $K = g_{\mathbf{2n}}^{t\alpha^{(2^n)}}$.

We now define the decisional n-Hybrid Diffie-Hellman Expoent problem as, given the tuple $(\{X_i\}_{i \in \{0,\dots,n-1\}}, V, K)$ where K is either $g_{\mathbf{2n}}^{t\alpha^{(2^n)}}$ or a random element of $\mathbb{G}_{\mathbf{2n}}$, to distinguish the two cases.

Definition 3. *We say the decisional n-Hybrid Diffie-Hellman Exponent assumption holds for* Setup' *if, for any polynomial n and probabilistic polynomial time algorithm $\mathcal{A}$, $\mathcal{A}$ has negligible advantage in solving the n-Hybrid Diffie-Hellman Exponent problem.*

Given the X_i for $i = 0, \dots, n-1$, it is straightforward to compute $g_{\mathbf{n}}^{\alpha^j}$ for any $j \in [0, 2^n - 1]$. Moreover, including X_n, it is straightforward to extend this to $j \in [2^n + 1, 2^{n+1}]$. However, computing $K = g_{\mathbf{2n}}^{t\alpha^{(2^n)}}$ from the X_i and V appears difficult. The reason is that we only have one term that depends on t, namely V. So to compute K, we would need to pair V with some combination of the X_i. In other words, we need to be able to compute $g_{\mathbf{n}}^{\alpha^{(2^n)}}$ from the X_i. However, since $\mathbf{n}$ has a one in each component, we can never pair any of the X_i with itself. This means we can only compute products of terms of the form $e(X_0^{s_0}, X_1^{s_1}, \dots, X_n^{s_n})$ for $s_i \in \{0, 1\}$, where we take $X_i^0 = g_{\mathbf{e}_i}$. Notice that we can never include an X_n, since then we would already exceed the desired degree of 2^n. Put another way, we can only compute products of terms of the form

$$g_{\mathbf{n}}^{\prod_{i \in S} \alpha^{(2^i)}}$$

where $S \subseteq [0, n-1]$. However, $\prod_{i \in S} \alpha^{2^i} = \alpha^{\sum_{i \in S} 2^i}$, and $\sum_{i \in S} 2^i < 2^n$ for all subsets $S \subseteq [0, n-1]$. This is the basis for our assumption that the n-HDHE assumption is hard. In the full version [BWZ14], we discuss the difficulty of our assumption in the generic multilinear map model.

3.2   Security of Our Construction

With our assumption defined, we can now state and prove the security of our scheme:

Theorem 2. *Let* Setup' *be the setup algorithm for a multilinear map, and suppose that the decisional n-Hybrid Diffie-Hellman Exponent assumption holds for* Setup'. *Then the scheme in Construction 1 is a statically secure identity-based broadcast encryption scheme.*

Proof. Our proof closely follows the security proof for BGW [BGW05]. Suppose we have an adversary $\mathcal{A}$ that breaks the security of the scheme. We use $\mathcal{A}$ to build an adversary $\mathcal{B}$ that breaks the decisional n-HDHE problem for Setup'. $\mathcal{B}$ works as follows:

- $\mathcal{B}$ obtains a challenge tuple $(\mathsf{params}', \{X_i\}_{i\in[0,n]}, V, K)$ where:
  - $\mathsf{params}' \leftarrow \mathsf{Setup}'(2\mathbf{n})$ where $\mathbf{n}$ is the all-ones vector of length $n+1$.
  - $X_i = g_{\mathbf{e}_i}^{\alpha^{2^i}}$ for $i = 0, \ldots, n-1$ for a random $\alpha \in \mathbb{Z}_p$.
  - $X_n = g_{\mathbf{e}_n}^{\alpha^{2^n+1}}$
  - $V = g_{\mathbf{n}}^t$ for a random $t \in \mathbb{Z}_p$
  - $K = g_{2\mathbf{n}}^{t\alpha^{2^n}}$ or K is a random group element in $\mathbb{G}_{2\mathbf{n}}$.
- $\mathcal{B}$ simulates $\mathcal{A}$ until $\mathcal{A}$ submits a subset $S \subseteq [1, 2^n - 1]$ of users that $\mathcal{A}$ will challenge.
- $\mathcal{B}$ chooses a random $r \in \mathbb{Z}_p$. It sets

$$Y = \frac{g_{\mathbf{n}}^r}{\prod_{u \in S} Z_{2^n - u}}$$

where the Z_j are calculated from the X_i as before. This amounts to setting

$$\gamma = r - \sum_{u \in S} \alpha^{2^n - u}$$

Since r is uniform in $\mathbb{Z}_p$ and independent of α, so is γ. $\mathcal{B}$ also computes

$$W' = e(g_{\mathbf{e}_0}, g_{\mathbf{e}_1}, \ldots, g_{\mathbf{e}_{n-2}}, X_{n-1}, g_{\mathbf{e}_n})$$

and

$$W = e(W', W')$$

Observe that $W = g_{2\mathbf{n}}^{\alpha^{2^n}}$.
- $\mathcal{B}$ gives $\mathcal{A}$ the public parameters $(W, \{X_i\}_{i\in[0,n]}, Y)$
- Now $\mathcal{A}$ is allowed to ask for private keys for users $u \notin S$. $\mathcal{B}$ computes

$$\mathsf{sk}_u = \frac{Z_u^r}{\prod_{j \in S} Z_{2^n - j + u}}$$

Observe that

$$\mathsf{sk}_u = g_{\mathbf{n}}^{r\alpha^u - \sum_{j \in S} \alpha^{(2^n - j + u)}} = g_{\mathbf{n}}^{\gamma\alpha^u + \sum_{j \in S} \alpha^{(2^n - j + u)} - \sum_{j \in S} \alpha^{(2^n - j + u)}} = g_{\mathbf{n}}^{\gamma\alpha^u}$$

as desired.
- When $\mathcal{A}$ asks for the challenge, $\mathcal{B}$ lets $\mathsf{Hdr} = (V, V^r)$ and responds with (Hdr, K). Observe that

$$V^r = g_{\mathbf{n}}^{rt} = g_{\mathbf{n}}^{t\left(\gamma + \sum_{u \in S} \alpha^{(2^n - u)}\right)}$$

which means (V, V^r) is a valid header for the set S. Also observe that if $K = g_{2\mathbf{n}}^{t\alpha^{(2^n)}}$, then K is the correct key for this header.
- When $\mathcal{A}$ returns a guess b for which K it is given, $\mathcal{B}$ returns b as its guess.

As shown above, $\mathcal{B}$ perfectly simulates the view of $\mathcal{A}$ in the broadcast encryption security game. Therefore, $\mathcal{B}$ has the same advantage as $\mathcal{A}$, which must therefore be negligible, as desired.

4   Our Symmetric Multilinear Map Construction

In this section, we give our second construction of broadcast encryption, this time from traditional symmetric multilinear maps. That is, we do not require the more complicated asymmetric structure of Construction 1, but can use a basic multilinear map. The idea, however, is very similar. We implement BGW [BGW05] in middle levels of the multilinear map, and use elements in the bottom level to generate the BGW public parameters. Similar to the graded encoding scheme, the BGW parameters will have the form $Z_u = g_n^{\alpha^u}$, which can be computed from the public parameters $X_i = g_1^{\alpha^{(2^i)}}$.

However, we run into a problem. With asymmetric maps, we could enforce that X_i could not be paired itself. This was used to ensure that Z_{2^n} was not computable given X_i for $i = 0, \ldots, n$. However, in the symmetric multilinear map setting, X_{n-1} could be paired with itself, giving Z_{2^n}. Instead, we create a hole by limiting the total number of X_i that can be paired together. If we allow only $n - 1$ of them to be paired together, the first hole occurs at Z_{2^n-1}. We therefore set $N = 2^n - 2$ so that the hole is at $N + 1$ as in BGW.

Notice that a second hole occurs at $Z_{2^n+2^{n-1}-1}$, and since $2^n + 2^{n-1} - 1 < 2(2^n - 2) = 2N$, we can not yet compute all the Z_u needed by BGW. One possible fix is to include extra X_i that can be used to fill in the unwanted holes. Instead, we opt to restrict the bit representations of all identities in the system to having the same Hamming weight. We show that this allows the computation of all the necessary Z_u.

We now describe our scheme:

Construction 3. *Let* Setup' *be the setup algorithm for a multilinear map, where groups have order p. Our second identity-based broadcast scheme consists of the following algorithms:*

Setup(n, ℓ) *Sets up a broadcast scheme for n-bit identities with Hamming weight ℓ. Run* Setup' *on $n + \ell - 1$, obtaining the public parameters* params' *for a multilinear map with target group $\mathbb{G}_{n+\ell-1}$. Let $\alpha, \gamma \in \mathbb{Z}_p$ be chosen at random. Let $W = g_{n+\ell-1}^{\alpha^{(2^n-1)}}$. Compute $X_i = g_1^{\alpha^{(2^i)}}$ for $i = 0, \ldots, n$. Lastly, let $Y = g_{n-1}^\gamma$. Output*

$$\mathsf{params} = (\mathsf{params}', W, \{X_i\}_{i \in [0,n]}, Y)$$

KeyGen$(\mathsf{params}, \alpha, \gamma, u)$ *The secret key for an identity $u \in \{0, 1\}^n$ of Hamming weight ℓ is*

$$\mathsf{sk}_u = g_{n-1}^{\gamma \alpha^u}$$

Enc(params, S) *Let $Z_j = g_{n-1}^{\alpha^j}$. We will show shortly that we can compute all of the necessary Z_j from the X_i. Pick a random $t \in \mathbb{Z}_p$ and compute the key and header as*

$$K = W^t = g_{n+\ell-1}^{t\alpha^{(2^n-1)}} \quad and$$

$$\mathsf{Hdr} = \left(g_\ell^t, \, (Y \prod_{u \in S} Z_{2^n-1-u})^t \right) = \left(g_\ell^t, \, g_{n-1}^{t\left(\gamma + \sum_{u \in S} \alpha^{(2^n-1-u)}\right)} \right)$$

$\mathsf{Dec}(\mathsf{params}, u, \mathsf{sk}_u, S, \mathsf{Hdr})$ *If $u \notin S$, output $\perp$. Otherwise, write $\mathsf{Hdr} = (C_0, C_1)$.*
Also let $Z'_u = g_\ell^{\alpha^u}$. We will shortly show that Z'_u can be computed from the
X_i*. Compute*

$$K = \frac{e(Z'_u, C_1)}{e(\mathsf{sk}_u \cdot \prod_{j \in S, j \neq u} Z_{2^n - 1 - j + u}, \ C_0)}$$

If C_0, C_1 are as above, notice that we can write $K = g_{n+\ell-1}^c$ where

$$c = \alpha^u t \left(\gamma + \sum_{j \in S} \alpha^{2^n - 1 - j} \right) - \left(\gamma \alpha^u + \sum_{j \in S, j \neq u} \alpha^{2^n - 1 - j + u} \right) t = t \alpha^{2^n - 1}$$

as desired.

We need to show how to compute Z_j and Z'_j.

Claim. Let $Z_j = g_{n-1}^{\alpha^j}$ and $Z'_j = g_\ell^{\alpha^j}$. Let $X_i = g_1^{\alpha^{(2^i)}}$ for $i = 0, \ldots, n$. Then,
using group multiplications and paring operations on the X_i, it is possible to
compute:

- Z'_j for $j \in [1, 2^n - 2]$ of weight exactly ℓ.
- $Z_{2^n - 1 - j}$ for $j \in [1, 2^n - 2]$ of weight exactly ℓ.
- $Z_{2^n - 1 - j + u}$ for $j, u \in [1, 2^n - 2], j \neq u$ of weight exactly ℓ.

Proof. Let $h(j)$ denote the Hamming weight of j. First, observe that we can
easily compute $g_{h(j)}^{\alpha^j}$ for $j \in [2^n - 1]$ by paring together X_i where the ith bit of j
is 1. This allows us to compute the Z'_j. We can also compute $g_{n-\ell}^{\alpha^{(2^n - 1 - j)}}$ for any
j of weight exactly ℓ. Thus, we can pair with $g_{\ell-1}$ to get $Z_{2^n - 1 - j}$.

Now we show how to compute $Z_{2^n - 1 - j + u}$. $2^n - 1 - j$, written as a bit string,
has Hamming weight $n - \ell$. Therefore, write $2^n - 1 - j = \sum_{i \in T} 2^i$ for some
subset $T \subseteq [0, n-1]$ of size $n - \ell$. Similarly, write $u = \sum_{i \in U} 2^i$ for some subset
$U \subseteq [0, n-1]$ of size ℓ. Notice that U and T are only disjoint if $2^n - 1 - j + u = 2^n - 1$, in which case $j = u$. Since we do not allow this case, there must be some
$\hat{i} \in [0, n-1]$ inside U and T. Then we can write

$$2^n - 1 - j + u = \left(\sum_{i \in T \setminus \{\hat{i}\}} 2^i \right) + \left(\sum_{i \in U \setminus \{\hat{i}\}} 2^i \right) + 2^{\hat{i}+1}$$

which is the sum of $n + \ell - 1$ powers of two. This means we can write

$$Z_{2^n - 1 - j + u} = e \left(\{X_i\}_{i \in T \setminus \{\hat{i}\}}, \ \{X_i\}_{i \in U \setminus \{\hat{i}\}}, \ X_{\hat{i}+1} \right)$$

which is the pairing of $n + \ell - 1$ of the X_i, as desired.

Setting n and ℓ. Suppose we want to handle λ-bit identities. We would map those identities to bit strings of length n and weight ℓ. Therefore, we need

$$\lambda \geq \log_2 \binom{n}{\ell}$$

A simple solution which minimizes n (and hence the number of elements in the public parameters) is to set $n = \lambda + \lceil (\log_2 \lambda)/2 \rceil + 1$ and $\ell = \lfloor n/2 \rfloor$. However, for existing multilinear map constructions, the multilinearity itself is expensive, so we might try to minimize the total multilinearity $n + \ell - 1$. When $\ell = \lfloor n/2 \rfloor$, the total multilinear is roughly $1.5(\lambda + (\log_2 \lambda)/2)$. However, setting $n \approx 1.042(\lambda + (\log_2 \lambda)/2)$ and $\ell \approx 0.398(\lambda + (\log \lambda)/2)$ gives us roughly 2^λ identities with total multilinearity about $1.440(\lambda + (\log \lambda)/2)$, slightly beating the trivial construction. The following table gives the settings of n and ℓ which minimize the total multilinearity for common identity lengths:

Length of identities (λ)	n	ℓ	Total Multilinearity ($n + \ell - 1$)
128	138	52	189
160	175	62	236
256	272	103	374
512	545	200	744

Implementation. As with Construction 1, we must take advantage of the secrets used to construct the multilinear map to compute the X_i. We also need to re-randomize the header components. This time, however, there are two groups that need re-randomization terms: $\mathbb{G}_\ell$ and $\mathbb{G}_{n-1}$. No other re-randomization parameters are necessary.

4.1   The Multilinear Diffie-Hellman Exponent Assumption

We define the computational (n, ℓ)-multilinear Diffie-Hellman Exponent $((n, \ell)$-MDHE) Problem as follows: Let params $\leftarrow$ Setup$'(n + \ell - 1)$. Choose random $\alpha, t \in \mathbb{Z}_p$, and let $X_i = g_1^{\alpha^{(2^i)}}$ for $i = 0, \dots, n$. Let $V = g_\ell^t$. Given $(\{X_i\}_{i \in [0,n]}, V)$, the goal is to compute $K = g_{n+\ell-1}^{t\alpha^{(2^n-1)}}$.

As before, we define the decisional version as the problem of distinguishing K from a random element in $G_{n+\ell-1}$.

Definition 4. *We say the decisional (n, ℓ)-multilinear Diffie-Hellman Exponent assumption holds for Setup$'$ if, for any polynomial n and probabilistic polynomial time algorithm $\mathcal{A}$, $\mathcal{A}$ has negligible advantage in solving the (n, ℓ)-multilinear Diffie-Hellman Exponent problem.*

This problem appears difficult for the same reasons as the n-HDHE assumption from Section 3. Computing $K = g_{n+\ell-1}^{t\alpha^{(2^n-1)}}$ requires pairing $V = g_\ell^t$ with a term $g_{n-1}^{\alpha^{(2^n-1)}}$, which must in turn be computed from the X_i. However, there is no way to pair at most $n - 1$ of the X_i to create the desired exponent $2^n - 1$. In the full version [BWZ14], we discuss the difficulty of the (n, ℓ)-MDHE problem in the generic multilinear map model.

4.2   Security of Our Construction

With our assumption defined, we can now state the security of our scheme:

Theorem 4. *Let* Setup′ *be the setup algorithm for a multilinear, and suppose that the decisional* (n, ℓ)*-multilinear Diffie-Hellman Exponent assumption holds for* Setup′. *Then the scheme in Construction 3 is a secure identity-based broadcast encryption scheme.*

Proof. Again, our proof follows BGW [BGW05]. Suppose we have an adversary $\mathcal{A}$ that breaks the security of the scheme. We use $\mathcal{A}$ to build an adversary $\mathcal{B}$ that breaks the decisional MDHE problem for Setup′. $\mathcal{B}$ works as follows:

- $\mathcal{B}$ obtains a challenge tuple $(\mathsf{params}', \{X_i\}_{i \in [0, n+1]}, V, K)$ where:
  - $\mathsf{params}' \leftarrow \mathsf{Setup}'(n + \ell - 1)$
  - $X_i = g_1^{\alpha^{(2^i)}}$ for $i = 0, \dots, n$ for a random $\alpha \in \mathbb{Z}_p$
  - $V = g_\ell^t$ for a random $t \in \mathbb{Z}_p$
  - $K = g_{n+\ell-1}^{t\alpha^{(2^n-1)}}$ or K is a random element in $\mathbb{G}_{n+\ell-1}$.
- $\mathcal{B}$ simulates $\mathcal{A}$ until $\mathcal{A}$ submits a subset $S \subseteq [1, 2^n - 2]$ of users that all have Hamming weight ℓ.
- $\mathcal{B}$ chooses a random $r \in \mathbb{Z}_p$. It sets

$$Y = g_{n-1}^r / \prod_{u \in S} Z_{2^n - 1 - s}$$

where the Z_j are calculated from the X_i as before. This amounts to setting

$$\gamma = r - \sum_{u \in S} \alpha^{2^n - 1 - u}$$

Since r is uniform in $\mathbb{Z}_p$ and independent of α, so is γ. $\mathcal{B}$ computes

$$W = e(X_0, X_1, \dots, X_{n-1}, g_{\ell-1})$$

Observe that $W = g_{n+\ell-1}^{2^n-1}$.
- $\mathcal{B}$ gives α the public parameters $(W, \{X_i\}_{i \in [0, n+1]}, Y)$
- Now $\mathcal{A}$ is allowed to ask for private keys for users $u \notin S$ of Hamming weight ℓ. $\mathcal{B}$ computes

$$\mathsf{sk}_u = Z_u^r / \prod_{j \in S, j \neq u} Z_{2^n - 1 - j + u}$$

Observe that

$$\mathsf{sk}_u = g_{n-1}^{r\alpha^u - \sum_{j \in S} \alpha^{(2^n - 1 - j + u)}} = g_{n-1}^{\left(r - \sum_{j \in S} \alpha^{(2n-1-j)}\right)\alpha^u} = g_{n-1}^{\gamma \alpha^u}$$

as desired.

– When $\mathcal{A}$ asks for the challenge, $\mathcal{B}$ lets $\mathsf{Hdr} = (V, e(V, g_{n-1-\ell})^r)$ and responds with (Hdr, K). Observe that

$$e(V, g_{n-1-\ell})^r = g_{n-1}^{rt} = g_{n-1}^{t\left(\gamma + \sum_{u \in S} \alpha^{(2^n - 1 - u)}\right)}$$

which means $(V, e(V, g_{n-1-\ell})^r)$ is a valid header for the set S. Also, observe that if $K = g_{n+\ell-1}^{t\alpha^{(2^n-1)}}$, then K is the correct key for this header.

– When $\mathcal{A}$ returns a guess b for which K it is given, $\mathcal{B}$ returns b as its guess.

As shown above, $\mathcal{B}$ perfectly simulates the view of $\mathcal{A}$ in the broadcast encryption security game. Therefore, $\mathcal{B}$ has the same advantage as $\mathcal{A}$, which must therefore be negligible, as desired.

5   Our Third Construction

In this section, we give our third and final broadcast scheme. This scheme is based on the basic broadcast scheme of Gentry and Waters [GW09], henceforth called the GW scheme. Like the BGW scheme, the GW scheme has public keys consisting of $O(N)$ elements, where N is the number of users. Our idea is to, similar to Constructions 1 and 3, run the GW scheme in the higher levels of a multilinear map, and derive the public key elements from $O(\log N)$ low-level elements.

However, unlike the BGW public parameters, which are all derived from a single scalar $\alpha \in \mathbb{Z}_p$, each of the GW public key elements are derived from a separate random scalar. Therefore, we cannot possibly hope to simulate the GW public key elements exactly. Instead, we we generate them using a Naor-Reingold-style PRF [NR97].

Also, unlike the BGW scheme, the secret keys in the GW scheme have $O(N)$ group elements. To make our scheme more efficient, and more importantly to make our scheme identity-based, we need to shrink the secret keys to $O(\log N)$ elements. To accomplish this, we observe that the secret key components are actually some of the outputs of another Naor-Reingold-style PRF, and we can allow the secret key holder to compute just those outputs by puncturing the PRF, similar to Boneh and Waters [BW13].

We now present out scheme:

Construction 5. *Let* Setup' *be the setup algorithm for a multilinear map, where groups have order* p. *Our final identity-based broadcast scheme consists of the following algorithms:*

$\mathsf{Setup}(n)$ *Takes as input the length* n *of identities. Run the setup algorithm for a multilinear map,* Setup', *to construct an* $n + 1$-*linear map with parameters* params'. *Draw a random* $\alpha \in \mathbb{Z}_p$. *For* $i = 0, \ldots, n-1$ *and* $b = 0, 1$, *draw random* $\beta_{i,b} \in \mathbb{Z}_p$. *The public key is*

$$\mathsf{pk} = (\mathsf{params}', \{X_{i,b} = g_1^{\beta_{i,b}}\}_{i \in [0,n-1], b \in \{0,1\}}, W = g_{n+1}^{\alpha})$$

For any user $\mathbf{u} \in \{0,1\}^n$, *note that we can compute*

$$Z_{\mathbf{u}} \equiv g_n^{\prod_{i=1}^{n} \beta_{i,u_i}} = e(X_{1,u_1}, X_{2,u_2}, \ldots, X_{n,u_n})$$

KeyGen(params, α, $\{\beta_{i,b}\}$, $\mathbf{u}$) *Pick a random* $r_{\mathbf{u}} \in \mathbb{Z}_p$. *Let*

$$U_0^{(\mathbf{u})} = g_1^{r_{\mathbf{u}}}$$

$$U_i^{(\mathbf{u})} = X_{i,1-u_i}^{r_{\mathbf{u}}} = g_1^{r_{\mathbf{u}}\beta_{i,1-u_i}} \quad for \ i = 1, \dots, n$$

$$U_{n+1}^{(\mathbf{u})} = g_n^{\alpha} Z_{\mathbf{u}}^{r_{\mathbf{u}}} = g_n^{\alpha + r_{\mathbf{u}} \cdot \prod_{i=1}^{n} \beta_{i,u_i}}$$

The secret key for user $\mathbf{u}$ *is* $\mathsf{sk}_{\mathbf{u}} = \{U_i^{(\mathbf{u})}\}_{i \in [0, n+1]}$.

Observe that for $\mathbf{v} \neq \mathbf{u}$, *we can compute* $Z_{\mathbf{v}}^{r_{\mathbf{u}}}$ *by finding an* i^* *where* $v_{i^*} = 1 - u_{i^*}$, *and computing*

$$e(X_{1,v_1}, \dots, X_{i^*-1,v_{i^*-1}}, U_{i^*}^{(\mathbf{u})}, X_{i^*+1,v_{i^*+1}}, \dots, X_{n,v_n}) = g_n^{r_{\mathbf{u}}\beta_{i^*,v_{i^*}} \cdot \prod_{i \neq i^*} \beta_{i,v_i}}$$

$$= g_n^{r_{\mathbf{u}} \cdot \prod_{i=1}^{n} \beta_{i,v_i}} = Z_{\mathbf{v}}^{r_{\mathbf{u}}}$$

Enc(params, S) *Choose a random* $t \in \mathbb{Z}_p$ *and compute the key and header as*

$$K = W^t = g_{n+1}^{t\alpha} \quad and \quad \mathsf{Hdr} = \left(g_1^t, \ \left(\prod_{\mathbf{u} \in S} Z_{\mathbf{u}}\right)^t\right) = \left(g_1^t, \ g_n^{t \sum_{\mathbf{u} \in S} \prod_{i=1}^{n} \beta_{i,u_i}}\right)$$

where $Z_{\mathbf{u}}$ *are computed as above.*

Dec(params, $\mathbf{u}$, $\mathsf{sk}_{\mathbf{u}}$, S, Hdr) *If* $\mathbf{u} \notin S$, *output* $\perp$. *Otherwise, write* $\mathsf{Hdr} = (C_0, C_1)$. *Compute*

$$k = \frac{e(U_{n+1}^{(\mathbf{u})} \cdot \prod_{\mathbf{v} \in S, \mathbf{v} \neq \mathbf{u}} Z_{\mathbf{v}}^{r_{\mathbf{u}}}, \ C_0)}{e(U_0^{(\mathbf{u})}, \ C_1)}$$

Observe that if (C_0, C_1) *are as above, we can write* k *as* g_{n+1}^c *where*

$$c = (\alpha + r_{\mathbf{u}} \prod \beta_{i,u_i} + \sum_{\mathbf{v} \in S, \mathbf{v} \neq \mathbf{u}} r_{\mathbf{u}} \prod \beta_{i,v_i}) \cdot t - r_{\mathbf{u}} \cdot (t \sum_{\mathbf{v} \in S} \prod \beta_{i,v_i}) = \alpha t$$

as desired.

Correctness follows from the comments above.

Differences from GW. In the Gentry and Waters scheme [GW09], the $Z_{\mathbf{u}}$ are generated independently and given explicitly in the public parameters (as elements of the source group $\mathbb{G}_1$). In our scheme, the $Z_{\mathbf{u}}$ are generated pseudorandomly by means of a Naor-Reingold PRF. Similarly, in the GW scheme, the $Z_{\mathbf{v}}^{r_{\mathbf{u}}}$ for $\mathbf{v} \neq \mathbf{u}$ are also given explicitly to user $\mathbf{u}$. In our scheme, we note that the $Z_{\mathbf{v}}^{r_{\mathbf{u}}}$ for fixed $\mathbf{u}$ actually form another Naor-Reignold PRF, which we puncture at the point $\mathbf{u}$ to allow user $\mathbf{u}$ to compute the necessary values without learning $Z_{\mathbf{u}}^{r_{\mathbf{u}}}$. Our puncturing follows the puncturing used by Boneh and Waters [BW13].

Comparison to Constructions 1 and 3. Construction 5 has a couple advantages and disadvantages over our previous schemes:

- Unlike the BGW-based schemes, there are no high-degree terms being generated. This means we do not need the secret parameters for the multilinear map to set up our scheme. Therefore, we can use a map from some trusted third party. We do, however, need to make sure re-randomization parameters are available in the groups $\mathbb{G}_1$ and $\mathbb{G}_n$ to re-randomize the header elements. If we are using a map that we did not set up, we also need to re-randomize the user secret keys.
- To handle identities of length λ, the total multilinearity of Construction 5 is $\lambda + 1$. Compare this to 2λ and $1.440(\lambda + (\log_2 \lambda)/2)$ from the previous constructions.
- On the negative side, secret keys in Construction 5 consist of $O(\log N)$ group elements, compared to the single element secret keys of the previous schemes.
- For security, we unfortunately are unable to prove security relative to a non-interactive assumption. In the original GW scheme, the security proof involved manipulating the $Z_{\mathbf{u}}$ for $\mathbf{u} \notin S$. Since each of the $Z_{\mathbf{u}}$ are independent in the GW scheme, this is achievable. For our scheme, however, the $Z_{\mathbf{u}}$ are generated from $O(\log N)$ parameters, meaning we cannot modify them independently. Instead, in the full version [BWZ14], we opt to prove security in the generic multilinear map model. We note, however, that we obtain a better generic security theorem than is possible for Constructions 1 and 3.

Acknowledgments. This work is supported by NSF, DARPA, IARPA, and others, as listed in the full version.

References

[BBW06]   Barth, A., Boneh, D., Waters, B.: Privacy in encrypted content distribution using private broadcast encryption. In: Di Crescenzo, G., Rubin, A. (eds.) FC 2006. LNCS, vol. 4107, pp. 52–64. Springer, Heidelberg (2006)

[BGI+01] Barak, B., Goldreich, O., Impagliazzo, R., Rudich, S., Sahai, A., Vadhan, S.P., Yang, K.: On the (Im)possibility of Obfuscating Programs. In: Kilian, J. (ed.) CRYPTO 2001. LNCS, vol. 2139, p. 1. Springer, Heidelberg (2001)

[BGW05]   Boneh, D., Gentry, C., Waters, B.: Collusion resistant broadcast encryption with short ciphertexts and private keys. In: Shoup, V. (ed.) CRYPTO 2005. LNCS, vol. 3621, pp. 258–275. Springer, Heidelberg (2005)

[BS03]   Boneh, D., Silverberg, A.: Applications of multilinear forms to cryptography. Contemporary Mathematics 324, 71–90 (2003)

[BW13]   Boneh, D., Waters, B.: Constrained pseudorandom functions and their applications. In: Sako, K., Sarkar, P. (eds.) ASIACRYPT 2013, Part II. LNCS, vol. 8270, pp. 280–300. Springer, Heidelberg (2013)

[BWZ14]   Boneh, D., Waters, B., Zhandry, M.: Low overhead broadcast encryption from multilinear maps. Full version available at the Cryptology ePrint Archives, Report 2014/195

[BZ14]   Boneh, D., Zhandry, M.: Multiparty key exchange, efficient traitor tracing, and more from indistinguishability obfuscation. In: Garay, J.A., Gennaro, R. (eds.) CRYPTO 2014, Part I. LNCS, vol. 8616, pp. 480–499. Springer, Heidelberg (2014)

[CLT13]   Coron, J.-S., Lepoint, T., Tibouchi, M.: Practical multilinear maps over the integers. In: Canetti, R., Garay, J.A. (eds.) CRYPTO 2013, Part I. LNCS, vol. 8042, pp. 476–493. Springer, Heidelberg (2013)

[Del07]   Delerablée, C.: Identity-Based Broadcast Encryption with Constant Size
            Ciphertexts and Private Keys 2, 200–215 (2007)
[DF02]   Dodis, Y., Fazio, N.: Public Key Broadcast Encryption for Stateless Re-
            ceivers. In: Feigenbaum, J. (ed.) DRM 2002. LNCS, vol. 2696, pp. 61–80.
            Springer, Heidelberg (2003)
[DF03]   Dodis, Y., Fazio, N.: Public key trace and revoke scheme secure against
            adaptive chosen ciphertext attack. In: Desmedt, Y.G. (ed.) PKC 2003.
            LNCS, vol. 2567, pp. 100–115. Springer, Heidelberg (2002)
[DPP07]   Delerablée, C., Paillier, P., Pointcheval, D.: Fully collusion secure dynamic
            broadcast encryption with constant-size ciphertexts or decryption keys.
            In: Takagi, T., Okamoto, T., Okamoto, E., Okamoto, T. (eds.) Pairing
            2007. LNCS, vol. 4575, pp. 39–59. Springer, Heidelberg (2007)
[FHPS13]   Freire, E.S.V., Hofheinz, D., Paterson, K.G., Striecks, C.: Programmable
            Hash Functions in the Multilinear Setting. In: Canetti, R., Garay, J.A. (eds.)
            CRYPTO 2013, Part I. LNCS, vol. 8042, pp. 513–530. Springer, Heidelberg
            (2013)
[FN94]   Fiat, A., Naor, M.: Broadcast Encryption. In: Stinson, D.R. (ed.)
            CRYPTO 1993. LNCS, vol. 773, pp. 480–491. Springer, Heidelberg (1994)
[FP12]   Fazio, N., Perera, I.M.: Outsider-Anonymous Broadcast Encryption with
            Sublinear Ciphertexts. In: Fischlin, M., Buchmann, J., Manulis, M. (eds.)
            PKC 2012. LNCS, vol. 7293, pp. 225–242. Springer, Heidelberg (2012)
[GGH13a]   Garg, S., Gentry, C., Halevi, S.: Candidate multilinear maps from ideal
            lattices. In: Johansson, T., Nguyen, P.Q. (eds.) EUROCRYPT 2013.
            LNCS, vol. 7881, pp. 1–17. Springer, Heidelberg (2013)
[GGH+13b]   Garg, S., Gentry, C., Halevi, S., Raykova, M., Sahai, A., Waters, B.: Can-
            didate indistinguishability obfuscation and functional encryption for all
            circuits. In: Proc. of FOCS 2013 (2013)
[GST04]   Goodrich, M.T., Sun, J.Z., Tamassia, R.: Efficient tree-based revocation in
            groups of low-state devices. In: Franklin, M. (ed.) CRYPTO 2004. LNCS,
            vol. 3152, pp. 511–527. Springer, Heidelberg (2004)
[GW09]   Gentry, C., Waters, B.: Adaptive security in broadcast encryption systems
            (with short ciphertexts). In: Joux, A. (ed.) EUROCRYPT 2009. LNCS,
            vol. 5479, pp. 171–188. Springer, Heidelberg (2009)
[HS02]   Halevy, D., Shamir, A.: The lsd broadcast encryption scheme. In: Yung,
            M. (ed.) CRYPTO 2002. LNCS, vol. 2442, pp. 47–60. Springer, Heidelberg
            (2002)
[LPQ12]   Libert, B., Paterson, K.G., Quaglia, E.A.: Anonymous broadcast encryp-
            tion: Adaptive security and efficient constructions in the standard model.
            In: Fischlin, M., Buchmann, J., Manulis, M. (eds.) PKC 2012. LNCS,
            vol. 7293, pp. 206–224. Springer, Heidelberg (2012)
[LSW10]   Lewko, A.B., Sahai, A., Waters, B.: Revocation systems with very small pri-
            vate keys. In: IEEE Symposium on Security and Privacy, pp. 273–285 (2010)
[NNL01]   Naor, D., Naor, M., Lotspiech, J.: Revocation and tracing schemes for
            stateless receivers. In: Kilian, J. (ed.) CRYPTO 2001. LNCS, vol. 2139,
            pp. 41–62. Springer, Heidelberg (2001)
[NP00]   Naor, M., Pinkas, B.: Efficient trace and revoke schemes. In: Frankel, Y.
            (ed.) FC 2000. LNCS, vol. 1962, pp. 1–20. Springer, Heidelberg (2001)
[NR97]   Naor, M., Reingold, O.: Number-theoretic constructions of efficient
            pseudo-random functions. In: FOCS, pp. 458–467 (1997)
[SF07]   Sakai, R., Furukawa, J.: Identity-Based Broadcast Encryption. IACR
            Cryptology ePrint Archive (2007)

Security Analysis of Multilinear Maps
over the Integers

Hyung Tae Lee[1] and Jae Hong Seo[2]

[1] Division of Mathematical Sciences,
School of Physical and Mathematical Sciences,
Nanyang Technological University, Singapore
hyungtaelee@ntu.edu.sg
[2] Myongji University, Korea
jaehongseo@mju.ac.kr

Abstract. At Crypto 2013, Coron, Lepoint, and Tibouchi (CLT) proposed a practical Graded Encoding Scheme (GES) over the integers, which has very similar cryptographic features to ideal multilinear maps. In fact, the scheme of Coron *et al.* is the second proposal of a secure GES, and has advantages over the first scheme of Garg, Gentry, and Halevi (GGH). For example, unlike the GGH construction, the subgroup decision assumption holds in the CLT construction. Immediately following the elegant innovations of the GES, numerous GES-based cryptographic applications were proposed. Although these applications rely on the security of the underlying GES, the security of the GES has not been analyzed in detail, aside from the original papers produced by Garg *et al.* and Coron *et al.*

We present an attack algorithm against the system parameters of the CLT GES. The proposed algorithm's complexity $\tilde{\mathcal{O}}(2^{\rho/2})$ is exponentially smaller than $\tilde{\mathcal{O}}(2^{\rho})$ of the previous best attack of Coron *et al.*, where ρ is a function of the security parameter. Furthermore, we identify a flaw in the generation of the zero-testing parameter of the CLT GES, which drastically reduces the running time of the proposed algorithm. The experimental results demonstrate the practicality of our attack.

1 Introduction

In 2003, Boneh and Silverberg [2] introduced the concept of cryptographic multilinear maps by generalizing cryptographic bilinear maps. They proposed interesting applications based on the concept, such as the multipartite Diffie-Hellman key exchange and an efficient broadcast encryption. Until recently, it was an important, yet hard-to-achieve open problem to construct multilinear maps satisfying cryptographic requirements. At Eurocrypt 2013, Garg, Gentry, and Halevi [18] proposed the first candidate multilinear maps, called *Graded Encoding Scheme (GES)*, having very similar cryptographic features to ideal multilinear maps. At Crypto 2013, Coron, Lepoint, and Tibouchi [10] proposed the second GES over the integers. The CLT construction has an advantage over the GGH construction; specifically, it allows one to use a desirable assumption

J.A. Garay and R. Gennaro (Eds.): CRYPTO 2014, Part I, LNCS 8616, pp. 224–240, 2014.

such as the subgroup decision assumption, which does not hold with the GGH construction. Thus, the CLT construction has broader applications. Very recently, Langlois, Stehlé, and Steinfeld [24] improved the GGH construction in terms of the bit size of the public parameters. Immediately following the elegant inventions of the GES, they received significant attention from the cryptography community, and numerous cryptography applications based on the GES inventions were built; for example, programmable hash [17], full-domain hash [22], functional encryption [19,20], witness encryption [21], and indistinguishability obfuscation [5,19,6]. Although these applications rely on the security of the underlying GES, the security of the GES itself has not been analyzed in detail, aside from the original papers produced by Garg *et al.* and Coron *et al.*

1.1 Our Contributions

n-**Masked Partial Approximate Common Divisors (n-MPACD).** We begin by introducing a new number theoretic problem, called n-*Masked Partial Approximate Common Divisors (n-MPACD)*, which is a generalization of the system parameters (such as the zero-testing parameter [10] and the rerandomization parameter [10,8]) from integer-based schemes such as multilinear maps [10] and Fully Homomorphic Encryptions (FHE) [8]. Roughly speaking, a problem instance is a product of η-bit primes $x_0 = \prod_i p_i$ and polynomially-many samples x_j such that $x_j \equiv Q \cdot r_{ij} \pmod{p_i}$ where $Q \xleftarrow{\$} \mathbb{Z}_{x_0}$, $r_{ij} \xleftarrow{\$} (-2^\rho, 2^\rho)$ and $\rho \ll \eta$. Because of the unknown Q, it is unlikely to directly apply the meet-in-the-middle attack of Chen and Nguyen [7]; therefore, it appears to be harder than the Partial Approximate Common Divisors (PACD) problem [23]. In fact, the attack algorithm of Coron, Lepoint, and Tibouchi (CLT) [10], which is the most efficient currently known algorithm for n-MPACD, has $\tilde{\mathcal{O}}(2^\rho)$ complexity, although it employs the technique used in the Chen-Nguyen attack.

Exponentially Faster Attack for n-MPACD. We present an attack algorithm for n-MPACD, which is exponentially faster than the CLT attack. The proposed algorithm follows the basic flow of the strategy of the Chen-Nguyen attack [7]. However, several tricks are required to manage the unknown Q and several moduli. Our attack is based on the following observation for subset-sums of integers in the same interval $(-2^\rho, 2^\rho)$: given $2m$ integers, there are 2^{2m}

Table 1. Algorithms for n-MPACD

Algorithm	Error Type	Computation ($\mathbb{Z}_{x_0}$ op.)	Space
(Corrected) CLT [10]	arbitrary errors†	$\mathcal{O}(\rho^2 2^\rho)$	$\mathcal{O}(\rho^2 2^\rho)$
This paper	arbitrary errors†	$\mathcal{O}(\sqrt{\rho \log \rho} \cdot \rho^2 2^{\rho/2})$	$\mathcal{O}(\sqrt{\rho \log \rho} \cdot \rho^2 2^{\rho/2})$
	uniform errors	$\mathcal{O}(\sqrt{\frac{\rho \log \rho}{n}} \cdot \rho^2 2^{\rho/2})$	$\mathcal{O}(\sqrt{\frac{\rho \log \rho}{n}} \cdot \rho^2 2^{\rho/2})$

An instance of n-MPACD consists of x_0 (product of n primes) and polynomially many samples with errors chosen from $(-2^\rho, 2^\rho)$.

†: Mild assumptions are necessary, which are specified in the paper.

different subset-sums (ignoring duplications), but such subset-sums range from $(-2m2^\rho, 2m2^\rho)$. That is, the number of subset-sums increases exponentially in m; however, those ranges increase only polynomially in m. Therefore, by slightly increasing m, we can find a collision among subset-sums. This observation is essential to our exponentially faster algorithm, as compared to the CLT attack. We summarize the comparison in Table 1.

A Flaw in the Generation of the Zero-Testing Parameter. We apply the proposed attack algorithm to the system parameters of multilinear maps over the integers; in particular, the zero-testing parameter [10]. The complexity of both our attack algorithm and the CLT attack primarily depend on ρ, the size of errors r_{ij}; therefore, it is necessary to enlarge the size of errors. In the generation of the zero-testing parameter, the matrix $\mathbf{H} = (h_{ij}) \in \mathbb{Z}^{n \times n}$ plays the role of (r_{ij}) in n-MPACD, indicating that the size of h_{ij} is very important for the security of the CLT GES. For the functionality of the multilinear maps, the matrix $\mathbf{H}$ is defined to be unimodular, and to satisfy two bounds $\|\mathbf{H}^\top\|_\infty \leq 2^\beta$ and $\|(\mathbf{H}^{-1})^\top\|_\infty \leq 2^\beta$. In [10], the authors provided a method for generating $\mathbf{H}$. However, we point out that the given method does not provide sufficient randomness in $\mathbf{H}$; that is, the average size of each entry h_{ij} in $\mathbf{H}$ is much less than expected. Eventually, this will weaken the security of multilinear maps over the integers.

Experimental Results. We provide several experimental results for our algorithm. In particular, we apply our attack algorithm to the implementation parameters on Small size for 52-bit security and Medium size for 62-bit security in [10] with a slight modification; the implementation in [10] used only a single zero-testing *integer*. However, we assume that a zero-testing *vector* is given, as in the original CLT GES. Our experimental results demonstrate that our algorithm requires less than $2^{34.84}$ and $2^{37.23}$ clock cycles on average for Small size and Medium size, respectively.

We remark that a part of this paper was made public through [28] and the missing details can be found in the full version of this paper [25].

1.2 Outline

In the following section, we provide some preliminary information that should be helpful for reading this paper. In Section 3, we define our new problem, and investigate a relation between it and the system parameters of multilinear maps. Section 4 provides our attack algorithm along with a detailed analysis. We describe how to speed our basic algorithm up and provide implementation results of our algorithm on the parameters of multilinear maps over the integers in Section 5. In Section 6, we discuss additional issues related to multilinear maps and our attack algorithms.

2   Preliminaries

Notation. Throughout the paper, λ is the security parameter, and we consider only discrete values; the interval notation $[a, b]$ indicates all integers between a and b, containing a and b. Similarly, (a, b) and $(a, b]$ notations also indicate respective sets of all integers contained in the corresponding continuous intervals. For integers a and p, the reduction of a modulo p is denoted by $a \pmod{p} \in (-p/2, p/2]$. Problem instances are defined by Chinese Remaindering with respect to n co-prime integers $p_1, \ldots, p_n$, making it convenient to use the notation $\mathsf{CRT}_{p_1, \ldots, p_n}(r_1, \ldots, r_n)$ (abbreviated as $\mathsf{CRT}_{(p_i)}(r_i)$) to denote the unique integer x in $(-\frac{1}{2} \prod_{i \in [1,n]} p_i, \frac{1}{2} \prod_{i \in [1,n]} p_i]$ with $x \equiv r_i \pmod{p_i}$ for all $i \in [1, n]$.

2.1 Fast Polynomial Algorithms

We consider polynomials with integer coefficients modulo x_0. There are classic algorithms for fast polynomial arithmetic, which use the Fast Fourier Transformation (FFT) [15,3,4] and have been used in various areas of cryptography, in particular, cryptanalysis [9,7,16]. In this paper, we use two fast polynomial arithmetic algorithms, each denoted by $\mathsf{Alg}_{Poly}^{FFT}$ and Alg_{MPE}^{FFT}, as subroutines; the algorithm $\mathsf{Alg}_{Poly}^{FFT}$ takes ℓ points as inputs and outputs a monic degree-ℓ polynomial over $\mathbb{Z}_{x_0}$ having ℓ input points as roots. The algorithm Alg_{MPE}^{FFT} takes a degree-ℓ polynomial $f(x)$ over $\mathbb{Z}_{x_0}$ and ℓ points as inputs, and then it evaluates $f(x)$ at ℓ input points and outputs the results. $\mathsf{Alg}_{Poly}^{FFT}$ (Alg_{MPE}^{FFT}, resp.) has quasi-linear complexity in the number of the input points (the degree of the input polynomial, resp.). We summarize the basic information regarding these fast polynomial algorithms in Table 2. We omit details of these classical algorithms; instead, we refer to [31,27].

Table 2. Fast polynomial algorithms using FFT

	$\mathsf{Alg}_{Poly}^{FFT}$	Alg_{MPE}^{FFT}
Input	x_0 and $\{a_0, \ldots, a_{\ell-1}\}$	x_0, $f(X)$ of ℓ-deg., and $\{\mathsf{pt}_i\}_{i \in [0, \ell-1]}$
Output	$f(X) = \prod_{i=0}^{\ell-1}(X - a_i) \pmod{x_0}$	$f(\mathsf{pt}_0), \ldots, f(\mathsf{pt}_{\ell-1}) \pmod{x_0}$
Comp. cost	$\mathcal{O}(\ell \log^2 \ell)$ operations modulo x_0	$\mathcal{O}(\ell \log^2 \ell)$ operations modulo x_0
Space cost	$\mathcal{O}(\ell \log^2 \ell)$ polynomially many bits	$\mathcal{O}(\ell \log^2 \ell)$ polynomially many bits

3 Masked Partial Approximate Common Divisors

Before providing our algorithm, we first generalize the problem instances for both the re-randomization parameter and the zero-testing parameter in the CLT GES. We believe that the following generalization will help readers to understand the security of the multilinear maps; specifically, both the hardness and weakness of the problem. We introduce a new number theoretic problem, which is a variant of *(Partial) Approximate Common Divisors* [23]. First, we describe the new hardness problem, then discuss its relationship with the system parameters of CLT GES in the following subsection.

Definition 1 (n-Masked Partial Approximate Common Divisors). *Given integers $Q, q_0, p_1, \ldots, p_n$, we state that x_j is sampled from the distribution $\mathcal{D}_\rho^M(Q, q_0, p_1, \ldots, p_n)$ if*

$$x_j = Q \cdot \mathsf{CRT}_{q_0, (p_i)}(q_j, r_{1j}, \ldots, r_{nj})(\mathrm{mod}\ q_0 \prod_{i \in [1,n]} p_i),$$

where $q_j \leftarrow [0, q_0)$ and $r_{ij} \leftarrow (-2^\rho, 2^\rho)$.

We define the (ρ, η, γ, n)-Masked Partial Approximate Common Divisors (abbreviated as n-MPACD) problem as follows. Choose η-bit random primes p_i for $i \in [1, n]$ and let π be their product. Set $x_0 := q_0 \cdot \pi$, where q_0 is a randomly chosen 2^{λ^2}-rough integer from $[0, 2^\gamma/\pi)$. Choose $Q \leftarrow [0, x_0)$. Given x_0 and polynomially many samples x_j from $\mathcal{D}_\rho^M(Q, q_0, p_1, \ldots, p_n)$, find a non-trivial factor of (x_0/q_0).

Note that we do not restrict the distribution of r_{ij}'s and Q in Definition 1 explicitly to cover various variants; in addition, our attack algorithm provided in the following section succeeds regardless of the distributions of Q and r_{ij}'s. We require only mild restrictions satisfied by both the zero-parameters and the re-randomization parameters of multilinear maps, which are the primary targets of our algorithm.

Hardness of n-MPACD: This paper mainly proposes attack algorithms against n-MPACD; however, it would be interesting to precisely understand the hardness of n-MPACD as well. To this end, we prove that n-MPACD is hard if PACD [23,13,14,7] is also hard. The reduction is provided in the full version.

Asymptotic Parameters: When we consider algorithms for n-MPACD, we basically assume that parameters are set to thwart various lattice-based attacks and factoring algorithms; that is, γ (x_0's bit size) must be large enough to prevent lattice-based attacks, so that $\gamma = \omega(\eta^2 \log \lambda)$ [30,13,10] and $\eta = \omega(\lambda^2)$, to prevent an efficient factorization algorithm such as ECM from having sub-exponential complexity in the size of factors. In this paper, we focus on the size of errors $r_{ij} \in (-2^\rho, 2^\rho)$ and the complexities of all algorithms associated with ρ.

3.1 Parameters as an Instance of the MPACD Problem

We demonstrate that the system parameters in the CLT GES can be considered as instances of n-MPACD.

Zero-testing Parameter: The zero-testing parameters $(x_0, (\boldsymbol{p}_{zt})_j$ for $j \in [1, n])$ are of form

$$(\boldsymbol{p}_{zt})_j = \sum_{i=1}^n h_{ij} \cdot (z^\kappa \cdot g_i^{-1} \mod p_i) \cdot \prod_{i' \neq i} p_{i'} \ (\mathrm{mod}\ x_0)$$
$$= Q \cdot \mathsf{CRT}_{(p_i)}(h_{ij})\ (\mathrm{mod}\ x_0)$$

where $Q = \mathsf{CRT}_{(p_i)}(z^\kappa \cdot g_i^{-1} \cdot \prod_{i' \neq i} p_{i'})$. Here, h_{ij} is distributed in a small bounded set $(-2^\beta, 2^\beta)$, where $2^\beta \ll p_i$. Therefore, we can regard the zero-testing parameters as an instance of n-MPACD.

Re-randomization Parameter: The re-randomization parameters are of form

$$\Pi_j = \mathsf{CRT}_{(p_i)}\left(\frac{\varpi_{ij} \cdot g_i}{z}\right) \equiv Q \cdot \mathsf{CRT}_{(p_i)}(\varpi_{ij}) \mod x_0,$$

where $Q = \mathsf{CRT}_{(p_i)}(\frac{g_i}{z})$. Note that the ϖ_{ij}'s of the errors are not chosen from the same set, unlike those in n-MPACD; non-diagonal entries are chosen from

$(-2^\rho, 2^\rho)$, while the diagonal entries are chosen from $(n2^\rho, n2^\rho + 2^\rho)$. Although errors are chosen from two different sets, the sizes of both sets are almost equal to 2^ρ. This is sufficient for our attack algorithm provided in Section 4.

Remark 1. In fact, by excluding some parts that have entries chosen from $(n2^\rho, n2^\rho + 2^\rho)$, the re-randomization parameters generated by n primes may be considered as an instance of $(n-k)$-MPACD as well for $k < n$. That is, $\{\Pi_j\}_{j\in[1,k]}$ for $k \in [1,n]$ can be re-written by

$$\Pi_j \equiv \mathrm{CRT}_{(p_i)}(\frac{g_i}{z}) \cdot \mathrm{CRT}_{(p_i)}(\varpi_{ij}) \equiv \mathrm{CRT}_{q_0,(p_i)_{i\in[k+1,n]}}(q', \frac{\varpi_{ij} \cdot g_i}{z}) \bmod x_0,$$

where $q_0 = \prod_{i=1}^k p_i$ and $q' = \mathrm{CRT}_{p_1,\dots,p_k}(\varpi_{1j},\dots,\varpi_{kj})$. Subsequently, all errors ϖ_{ij} for $i \in [k+1,n]$ and $j \in [1,k]$ are chosen from $(-2^\rho, 2^\rho)$, so that $(x_0, \{\Pi_j\}_{j\in[1,k]})$ is an instance of $(n-k)$-MPACD.

4 Our Algorithms for the n-MPACD Problem

We present an exponentially faster algorithm for solving n-MPACD problems; our (basic) algorithm requires $\mathcal{O}((\log \rho)^{0.5}\rho^{2.5}2^{\rho/2})$ $\mathbb{Z}_{x_0}$ operations. In [10], the attack algorithm for n-MPACD is roughly sketched and details are omitted. We present the detailed description of the CLT attack based on our speculation in the full version, which achieves the complexity Coron *et al.* claimed. Our analysis of the CLT algorithm for n-MPACD requires two mild assumptions about the distribution of samples. Similarly, the proposed algorithm also requires two mild assumptions about samples satisfied by our main application, multilinear maps over the integers.

4.1 Overview

We provide an overview of our algorithm for solving n-MPACD problems. Our strategy follows the basic flow of the Chen-Nguyen attack; however, we require several additional ideas to manage the unknown masking Q and several moduli in the n-MPACD problem, in contrast to the Chen-Nguyen attack for the PACD problem.

Consider an instance of an n-MPACD problem: $x_0 = q_0 \prod_{i=1}^n p_i$ and $x_j \equiv r_{ij} \bmod p_i$ where p_i's are η-bit primes, $r_{ij} \in (-2^\rho, 2^\rho)$ for $1 \leq j \leq 2m$, and $2^\rho \ll p_i$. For randomly chosen bits b'_j's, if m is sufficiently large, then for each p_i there is a high probability that

$$p_i \mid \gcd\left(x_0, \prod_{\substack{(b_1,\dots,b_{2m})\in\{0,1\}^{2m} \\ (b_1,\dots,b_{2m})\neq(b'_1,\dots,b'_{2m})}} (\sum_{j=1}^{2m} b_jx_j - \sum_{j=1}^{2m} b'_jx_j) \ (\bmod\ x_0)\right). \tag{1}$$

For each p_i, there are 2^{2m} possible sums $\sum_{j=1}^{2m} b_jx_j$ such that $\sum_{j=1}^{2m} b_jx_j \ (\bmod\ p_i)$ is contained in the relatively small range $(-2m2^\rho, 2m2^\rho)$, which is contained in $(-p_i/2, p_i/2]$. If the number of samples m satisfies an inequality $2^{2m} \geq m2^{\rho+3}$ (e.g., $2m = \rho + \log\rho + \log\log\rho$ for sufficiently large ρ), then there are many collisions in the range. In fact, at least a half of all possible elements have a collision in the range $(-2m2^\rho, 2m2^\rho)$ according to the pigeonhole principle. Therefore, for such an m, we have $\prod_{\substack{(b_1,\dots,b_{2m})\in\{0,1\}^{2m} \\ (b_1,\dots,b_{2m})\neq(b'_1,\dots,b'_{2m})}} (\sum_{j=1}^{2m} b_j x_j - \sum_{j=1}^{2m} b'_j x_j) \equiv 0 \pmod{p_i}$ with at least $1/2$ probability, depending on the choice of b'_j's.

To solve an n-MPACD problem using the relation (1), two remaining issues must be considered, in terms of efficiency and correctness. First, $2^{2m} > 2^\rho$ modulus multiplications, which are quite large, are required for naive computation of the above product. To reduce the complexity, we follow the concept of the meet-in-the-middle approach, similar to the Chen-Nguyen attack. Second, it is likely that the result of the gcd computation in (1) is not a non-trivial factor of x_0, but just x_0. To overcome this obstacle, we additionally equip our algorithm with the concept of the *divide-and-conquer* technique.

Let us address the efficiency issue first. We define the 2^d-degree polynomial $f_{d,(b'_j)}(X)$ over $\mathbb{Z}_{x_0}$ as follows:

$$f_{d,(b'_j)}(X) = \prod_{(b_1,\dots,b_d)\in\{0,1\}^d} ((X + \sum_{j=1}^{d} b_j x_j) - \sum_{j=1}^{2m} b'_j x_j) \pmod{x_0} \qquad (2)$$

Using this new notation, we can rewrite (1) as[1]

$$p_i \mid \gcd\left(x_0, \prod_{(b_{m+1},\dots,b_{2m})\in\{0,1\}^m} f_{m,(b'_j)}(\sum_{k=m+1}^{2m} b_k x_k) \pmod{x_0}\right). \qquad (3)$$

We can compute the 2^m-degree polynomial $f_{m,(b'_j)}(X)$ via Alg_{Poly}^{FFT} and evaluate $f_{m,(b'_j)}(X)$ at 2^m points $\{\sum_{k=m+1}^{2m} b_k x_k\}_{(b_{m+1},\dots,b_{2m})\in\{0,1\}^m}$ via Alg_{MPE}^{FFT} so that we can solve the n-MPACD problem with $\mathcal{O}(2^m m^2)$ complexity. If we set $2m = \rho + \log\rho + \log\log\rho$, then we determine that the complexity is $\mathcal{O}((\log\rho)^{0.5}\rho^{2.5}2^{\rho/2})$.

For the second issue regarding the gcd computation result, we can apply the divide-and-conquer method. It is clear that the result should be x_0 or its divisor. If the output of the gcd computation is x_0, then we divide the product $\prod_{(b_{m+1},\dots,b_{2m})\in\{0,1\}^m} f_{m,(b'_j)}(\sum_{k=m+1}^{2m} b_k x_k) \pmod{x_0}$ into four factors and compute all factors. If there is a non-trivial factor among four factors, then the algorithm succeeds. Otherwise, we select a factor that is a multiple of x_0, and repeat the same process until a non-trivial factor is found. We can demonstrate that this process will find a non-trivial factor with overwhelming probability, and the recursive process's asymptotic complexity is still $\mathcal{O}((\log\rho)^{0.5}\rho^{2.5}2^{\rho/2})$. We provide a clear description and analysis of our algorithm in the following subsections.

If the errors r_{ij}'s are distributed (almost) uniformly, then we can reduce the complexity further by scrunching the domain of the product up; if the domain size is decreasing, we cannot expect that $(\sum_{j=1}^{2m} b'_j x_j)$ will have a collision in each modulus p_i with high probability; however, we can expect that it will have a collision in at least one modulus p_i, which is exactly what we want. In fact, we can

---

[1] Strictly speaking, (3) is not equal to (1) because (3) contains the case $(b_1,\dots,b_{2m}) = (b'_1,\dots,b'_{2m})$ therefore the product is trivially 0. We can easily change (3) to not contain the case $(b_1,\dots,b_{2m}) = (b'_1,\dots,b'_{2m})$. Because such a modification is technical, we omit it in this overview and relegate the details to the next subsection.

reduce the $\sqrt{n}$ factor further from the complexity. In Section 5, we discuss the method we used to increase the speed of our basic algorithm.

4.2   Basic Algorithm for n-MPACD

Given $2m$ samples x_j's when $2m \leq n$ and $m2^{\rho+2} \leq 2^{2m}$, we require two mild assumptions regarding samples.

Assumption 1. $2m2^{\rho+1} \leq p_i$ *for each* p_i.
Assumption 2. The rank of the integer matrix $(r_{ij})_{\substack{i \in [1,n] \\ j \in [1,2m]}} \in \mathbb{Z}^{n \times 2m}$ *is* $2m$, *where* $x_j \equiv r_{ij} \pmod{p_i}$.

Note that both the zero-testing parameter and the re-randomization parameter of multilinear maps over the integers satisfy both Assumption 1 & 2; Assumption 1 is trivial. In the zero-testing parameter, the matrix (h_{ij}) is invertible, so it can satisfy Assumption 2. For the re-randomization parameter, r_{ij}'s are distributed uniformly and independently; thus, the $rank(r_{ij})$ will be equal to $2m$ with overwhelming probability because r_{ij}'s are chosen from the exponentially large set in the security parameter.

Our n-MPACD Algorithm: We present our basic algorithm for n-MPACD in Algorithm 1. Our algorithm consists of two steps. First, the algorithm computes a product A that is a multiple of some prime factor of x_0. Second, if A is not a multiple of x_0, then the algorithm stops and outputs it. Otherwise, the algorithm runs the *while loop* to extract a non-trivial factor from the multiple of x_0; that is, we repeatedly split multiples of x_0 into four factors, until a non-trivial factor is found.

Because A is a product, we can compute A's four factors denoted by A_{00}, A_{01}, A_{10}, and A_{11} via the same process used for computing A such that $A = A_{00}A_{01}A_{10}A_{11}$, and then check if there is a non-trivial factor of x_0 among them. If not, repeat the same process until a non-trivial factor of x_0 is found. To optimize efficiency, we divide A into four factors *evenly*, that is, each A_i is also a product with the same size domain. Furthermore, we should set each domain of A_i to take full advantage of $\mathsf{Alg}_{Poly}^{FFT}$ and Alg_{MPE}^{FFT}. To this end, we define A_{00}, A_{01}, A_{10}, and A_{11} as follows: In the while loop, $A \in \mathbb{Z}_{x_0}$ is of the form

$$\prod_{\substack{\forall (b_{i_1}, \ldots, b_m), \forall (b_{i_2}, \ldots, b_{2m}) \\ (b_1, \ldots, b_{2m}) \neq (b'_1, \ldots, b'_{2m})}} \left(\sum_{j=i_1}^{m} b_j x_j + \sum_{j=i_2}^{2m} b_j x_j + C \right) \pmod{x_0},$$

where $b_1, \ldots, b_{i_1-1}, b_{m+1}, \ldots, b_{i_2-1}$ are fixed for some $1 \leq i_1 \leq m, m+1 \leq i_2 \leq 2m$, and so $C = \sum_{j=1}^{i_1-1} b_j x_j + \sum_{j=m+1}^{i_2-1} b_j x_j - \sum_{j=1}^{2m} b'_j x_j$ is a constant. Then,

$$A_{00} := \prod (\textstyle\sum_{j=i_1+1}^{m} b_j x_j + \sum_{j=i_2+1}^{2m} b_j x_j + C) \pmod{x_0},$$
$$A_{01} := \prod (\textstyle\sum_{j=i_1+1}^{m} b_j x_j + \sum_{j=i_2+1}^{2m} b_j x_j + C + x_{i_2}) \pmod{x_0},$$
$$A_{10} := \prod (\textstyle\sum_{j=i_1+1}^{m} b_j x_j + \sum_{j=i_2+1}^{2m} b_j x_j + C + x_{i_1}) \pmod{x_0},$$
$$A_{11} := \prod (\textstyle\sum_{j=i_1+1}^{m} b_j x_j + \sum_{j=i_2+1}^{2m} b_j x_j + C + x_{i_1} + x_{i_2}) \pmod{x_0},$$

Algorithm 1. n-MPACD algorithm: arbitrary distribution

Input: $(x_0, x_1, \ldots, x_{2m})$
Output: a non-trivial factor of x_0 or $\perp$

1: Choose $b'_j \overset{\$}{\leftarrow} \{0,1\}$ for $1 \leq j \leq 2m$.
2: Compute $A = \prod_{\substack{(b_1,\ldots,b_{2m}) \in \{0,1\}^{2m} \\ (b_1,\ldots,b_{2m}) \neq (b'_1,\ldots,b'_{2m})}} \left(\sum_{j=1}^{2m} b_j x_j - \sum_{j=1}^{2m} b'_j x_j \right) \pmod{x_0}$

  $\triangleright$ by using Alg. 2
3: **if** $A \not\equiv 0 \pmod{x_0}$ **then return** $\gcd(x_0, A)$.
4: **else** Set $k \leftarrow 1$.
5:     **while** $k \leq m$ **do**
6:         Compute $\gcd(x_0, A_i)$ for $i \in \{00, 01, 10, 11\}$.

       $\triangleright$ by using (a variant of) Alg. 2
7:         **if** $\gcd(x_0, A_i) \in (1, x_0)$ for some i **then return** $\gcd(x_0, A_i)$.
8:         **else** Choose an A_i s.t. $A_i \equiv 0 \pmod{x_0}$ and set $A \leftarrow A_i$, and $k \leftarrow k + 1$.
9:         **end if**
10:     **end while return** $\perp$.
11: **end if**

---

where C is defined as before and each product is defined over all $b_{i_1+1}, \ldots, b_m$, $b_{i_2+1}, \ldots, b_{2m} \in \{0,1\}$ such that $(b_1, \ldots, b_{2m}) \neq (b'_1, \ldots, b'_{2m})$. It is clear that $A = A_{00} A_{01} A_{10} A_{11}$ and each A_i has the same form as A with a different domain for the product.

Subroutine for Computing A and Its Factors: We describe how to compute $A = \prod_{\substack{(b_1,\ldots,b_{2m}) \in \{0,1\}^{2m} \\ (b_1,\ldots,b_{2m}) \neq (b'_1,\ldots,b'_{2m})}} \left(\sum_{j=1}^{2m} b_j x_j - \sum_{j=1}^{2m} b'_j x_j \right) \pmod{x_0}$.
Using the notation in (2), A can be rewritten as

$$\prod_{\substack{(b_{m+1},\ldots,b_{2m}) \\ \in \{0,1\}^m, \\ (b_{m+1},\ldots,b_{2m}) \\ \neq (b'_{m+1},\ldots,b'_{2m})}} f_{m,(b'_j)} \left(\sum_{k=m+1}^{2m} b_k x_k \right) \cdot \prod_{\substack{(b_1,\ldots,b_m) \\ \in \{0,1\}^m, \\ (b_1,\ldots,b_m) \\ \neq (b'_1,\ldots,b'_m)}} \left(\sum_{j=1}^{m} (b_j - b'_j) x_j \right) \quad (4)$$

The left term is for the case $(b_{m+1}, \ldots, b_{2m}) \neq (b'_{m+1}, \ldots, b'_{2m})$ and the right term is for the case $(b_{m+1}, \ldots, b_{2m}) = (b'_{m+1}, \ldots, b'_{2m})$ with $(b_1, \ldots, b_m) \neq (b'_1, \ldots, b'_m)$. Therefore, (4) covers all $(b_1, \ldots, b_{2m})$'s except $(b'_1, \ldots, b'_{2m})$, so that it is equal to A. We describe an algorithm for (4) in Algorithm 2. Factors A_{00}, A_{01}, A_{10} and A_{11} of A have approximately the same form as A, and hence we can compute it similarly to Algorithm 2.

Algorithm 2. Subroutine for solving n-MPACD

Input: $(x_0, x_1, \ldots, x_{2m})$ and $(b'_1, \ldots, b'_{2m})$.

Output: $A = \prod_{\substack{(b_1,\ldots,b_{2m}) \in \{0,1\}^{2m} \\ (b_1,\ldots,b_{2m}) \neq (b'_1,\ldots,b'_{2m})}} (\sum_{j=1}^{2m} b_j x_j - \sum_{j=1}^{2m} b'_j x_j) \pmod{x_0}$

1: Compute a polynomial $f_{m,(b'_j)}(X)$ over $\mathbb{Z}_{x_0}$ as follows.

   $\prod_{(b_1,\ldots,b_m) \in \{0,1\}^m} ((X + \sum_{j=1}^{m} b_j x_j) - \sum_{j=1}^{2m} b'_j x_j) \pmod{x_0}$.

   ▷ by $\mathsf{Alg}^{FFT}_{Poly}$ with x_0 and $\{(\sum_{j=1}^{2m} b'_j x_j - \sum_{j=1}^{m} b_j x_j)\}_{(b_1,\ldots,b_m) \in \{0,1\}^m}$ as inputs.

2: Perform multi-points evaluation of $f_{m,(b'_j)}(X)$ at $\{\sum_{k=m+1}^{2m} b_k x_k\}_{\forall b_k \in \{0,1\}}$ ▷ by Alg^{FFT}_{MPE}.

3: **return**

$$\prod_{\substack{(b_{m+1},\ldots,b_{2m}) \\ \in \{0,1\}^m, \\ (b_{m+1},\ldots,b_{2m}) \\ \neq (b'_{m+1},\ldots,b'_{2m})}} f_{m,(b'_j)}(\sum_{k=m+1}^{2m} b_k x_k) \cdot \prod_{\substack{(b_1,\ldots,b_m) \\ \in \{0,1\}^m, \\ (b_1,\ldots,b_m) \\ \neq (b'_1,\ldots,b'_m)}} (\sum_{j=1}^{m} (b_j - b'_j) x_j) \pmod{x_0}$$

4.3 Analysis

Success Probability: We demonstrate that Algorithm 1 correctly finds a non-trivial factor of x_0 with at least $1/2$ probability, where the probability goes over only the algorithm's random tape.[2]

Algorithm 1 begins by selecting $b'_j \in \{0,1\}$ for $1 \leq j \leq 2m$. Given an n-MPACD instance x_0 and x_j's, we state that $(b'_1, \ldots, b'_{2m}) \in \{0,1\}^{2m}$ is 'good for p_i' if there exists $(b_1, \ldots, b_{2m}) \in \{0,1\}^{2m}$ such that $(b_1, \ldots, b_{2m}) \neq (b'_1, \ldots, b'_{2m})$ and $\sum_{j=1}^{2m} b_j x_j = \sum_{j=1}^{2m} b'_j x_j \pmod{p_i}$. We can prove that if we select b'_j's uniformly and independently, then with high probability $(b'_1, \ldots, b'_{2m})$ is 'good for p_i' for each p_i. See Lemma 1 for details.

Lemma 1. *Given an n-MPACD instance x_0 and x_j's, we have that for each $i \in [1, n]$,*

$$\Pr_{b'_j \xleftarrow{\$} \{0,1\}} [(b'_1, \ldots, b'_{2m}) \text{ is good for } p_i] > 1/2 \quad \textit{under Assumption 1.}$$

Once the algorithm has a good $(b'_1, \ldots, b'_{2m})$ for p_1, then we can demonstrate that the algorithm eventually outputs a non-trivial factor of x_0. If the while loop arrives at the end before finding a non-trivial factor of x_0 (that is, it is repeated m times), then ultimately we should have an integer $\sum_{j=1}^{2m} b_j x_j - \sum_{j=1}^{2m} b'_j x_j \equiv 0 \pmod{x_0}$ for some $(b_1, \ldots, b_{2m}) \neq (b'_1, \ldots, b'_{2m})$; that is, we are not able to divide A any further. Therefore, it is sufficient to demonstrate that such a tuple $(b_1, \ldots, b_{2m})$ cannot exist, and Lemma 2 guarantees it.

---

[2] Because the success probability of our algorithm is constant, we can make that the probability of success is overwhelming by running the algorithm linear in the security parameter, with a fresh random tape.

Lemma 2. *Under Assumption 1 and 2, if* $(b_1, \ldots, b_{2m}) \neq (b'_1, \ldots, b'_{2m})$, *then there is an index* $i' \in [1, n]$ *such that*

$$\sum_{j=1}^{2m} b_j x_j \neq \sum_{j=1}^{2m} b'_j x_j \pmod{p_{i'}}$$

so that $\sum_{j=1}^{2m} b_j x_j \neq \sum_{j=1}^{2m} b'_j x_j \pmod{x_0}$.

Algorithm 1 uses the randomness only in the 1st and 8th steps. Because any A_i with correct conditions will suffice in the 8th step, it does not affect the success probability of the algorithm. Only a selection of $(b'_1, \ldots, b'_{2m})$ will determine the success of the algorithm, and we have a probability of greater than $1/2$ for a good $(b'_1, \ldots, b'_{2m})$ for p_1. Therefore, the proposed algorithm has at least a $1/2$ probability for success.

Complexity: The complexity of Algorithm 1 is dominated by computing A and its factors. The complexity of Algorithm 2 mainly depends on the domain size in the product; we require $\mathcal{O}(m^2 2^m)$ operations modulo x_0 (from $\mathsf{Alg}_{Poly}^{FFT}$ and Alg_{MPE}^{FFT}'s complexity). Similarly, for each of A's four factors, we must perform $\mathcal{O}((m-1)^2 2^{m-1})$ operations modulo x_0 because each factor of A uses a half-size degree polynomial and number of points in Algorithm 2 in comparison with A. Similarly, we require $\mathcal{O}((m-2)^2 2^{m-2})$ operations modulo x_0 for each of A_i's four factors, and so on. Overall, the computational complexity for A and all its factors is bounded by $\mathcal{O}(m^2 2^m) + 4\mathcal{O}((m-1)^2 2^{m-1}) + \cdots + 4\tilde{\mathcal{O}}(2^1) = \mathcal{O}(5m^2 2^m) = \mathcal{O}(m^2 2^m)$ operations modulo x_0. Therefore, the overall computational cost is $\mathcal{O}(m^2 2^m) + \mathcal{O}(m 2^m) = \mathcal{O}(m^2 2^m)$ $\mathbb{Z}_{x_0}$ operations. Similarly, we can demonstrate that the space complexity is bounded by $\mathcal{O}(m^2 2^m)$ polynomially many bits from the storage complexity of Alg_{MPE}^{FFT} and $\mathsf{Alg}_{Poly}^{FFT}$.

If we set $m = \frac{\rho + \log \rho + \log \log \rho}{2}$, then it asymptotically satisfies the requirement $2m \leq n$ and $2m 2^{\rho+1} < 2^{2m}$, where $\rho \geq 4$. Therefore, for $m = \frac{\rho + \log \rho + \log \log \rho}{2}$, the computational cost is $\mathcal{O}((\frac{\rho + \log \rho + \log \log \rho}{2})^2 \, 2^{\frac{\rho + \log \rho + \log \log \rho}{2}}) = \mathcal{O}((\log \rho)^{0.5} \rho^{2.5} 2^{\rho/2})$ $\mathbb{Z}_{x_0}$ operations and the space complexity is $\mathcal{O}((\log \rho)^{0.5} \rho^{2.5} 2^{\rho/2})$ polynomially many bits.

5   Attack on System Parameters of Multilinear Maps over the Integers

5.1   Speed Increase for Multilinear Maps Parameters

We introduce several techniques to increase the speed of Algorithm 1, where all of our techniques are applicable to the parameters of multilinear maps. If r_{ij}'s are uniformly distributed, we can increase the speed of the attack algorithm. For example, ϖ_{ij}'s in the re-randomization parameter are uniformly distributed in the corresponding domains. Furthermore, we know the distribution of h_{ij}'s in the zero-testing parameter. Although it is not a uniform distribution, we can consider it as a quasi-uniform distribution in an appropriate bound.

Using Shorter m**:** To guarantee exponentially many *good* $(b'_1, \ldots, b'_{2m})$ for each p_i, we select m with $2m2^{\rho+1} \leq 2^{2m}$. The sum of uniform variables follows the *bell-shaped* distribution, so that $\sum_{j=1}^{2m} b_j x_j$ has a shorter image size than its range. Furthermore, the bell-shaped distribution has more collisions around a center than uniform distributions. This fact allows us to select a shorter m, and our experimental results provided in Table 3 support our expectation.

Table 3. Shorter domains (Experimental results on average of 100 instances)

ρ	14	16	18	20
m	8	9	10	11
|domain|/|range|	0.25	0.22	0.20	0.18
|domain|/|image|	1.49	1.51	1.48	1.44

Shorter Domain in Products: Basically, Algorithm 1 becomes a brute-force attack once we select a good $(b'_1, \ldots, b'_{2m})$ for some p_i at the beginning. It is likely that $(b'_1, \ldots, b'_{2m})$ is good for several moduli p_i's. (That is the exact reason why we must have the while loop in Algorithm 1.) However, our goal is to select a vector

---

Algorithm 3. n-MPACD algorithm: speedup for the uniform distribution

---

Input: $(x_0, x_1, \ldots, x_{2m})$, $d = 2^\delta$ for $\delta \geq 1$
Output: a non-trivial factor of x_0 or $\perp$

1: Choose $(b'_1, \ldots, b'_{2m}) \xleftarrow{\$} \{0,1\}^{2m}$.

2: Choose $b_1, \ldots, b_\delta, b_{m+1}, \ldots, b_{m+\delta} \xleftarrow{\$} \{0,1\}$.

3: Compute $A = \prod_{\substack{(b_{\delta+1},\ldots,b_m)\in\mathbb{Z}_{x_0}^{m-\delta} \\ (b_{m+\delta+1},b_{2m})\in\mathbb{Z}_{x_0}^{m-\delta} \\ (b_1,\ldots,b_{2m})\neq(b'_1,\ldots,b'_{2m})}} \left(\sum_{j=1}^{2m} b_j x_j - \sum_{j=1}^{2m} b'_j x_j \right) \pmod{x_0}$

$\triangleright$ by using Alg. 2

4: The remaining process is the same as Step $3 - 11$ of Algorithm 1.

Table 4. Speedup with shorter interval (Experimental results on average of 100 instances)

Instantiation	λ	n	η	ρ	m	d	(Average) trials
Micro	≥ 34	64	1528	22	12	8	1.81 times

Parameters are set the average ratio between the domain and the image (modulus p_i) of $\sum_{1 \leq j \leq 2m} b_j x_j$ for 100 problem instances to be 1.44 for each p_i.

$(b'_1, \ldots, b'_{2m})$ that is good for only one (or a few) p_i and is not good for any others. We compute a product A' that is roughly $1/n$ of a random portion of the product A in Algorithm 1. Then, we can expect the probability, $\Pr_i$, that p_i divides A' is roughly equal to $1/2n$. Furthermore, r_{ij}'s are independent, and thus we can also expect that the probabilities $\Pr_i$'s are nearly independent. Therefore, the probability that A' is a multiple of at least one of p_i is significant, from the birthday paradox; e.g., $1 - 1/\sqrt{e}$. Applying this technique, we present an improved attack in Algorithm 3. The analysis above is heuristic, and thus to support our expectations and the heuristic analysis, we provide experimental results in Table 4.

Insufficient Entropy in Zero-testing Parameters: The matrix $\mathbf{H} = (h_{ij}) \in \mathbb{Z}^{n \times n}$ in the zero-testing parameters is selected to satisfy $\|\mathbf{H}^\top\|_\infty \leq 2^\beta$ and $\|(\mathbf{H}^{-1})^\top\|_\infty \leq 2^\beta$ where $\|\cdot\|_\infty$ is the operator norm of $n \times n$ matrices with respect to the ℓ^∞ norm on $\mathbb{R}^n$. In [10], Coron $et\ al.$ proposed an algorithm to generate such a matrix $\mathbf{H}$, with sufficient entropy. However, their approach does not rapidly increase the entropy of $\mathbf{H}$, though it satisfies the above two bounds. We will demonstrate this by providing some experimental results in this section.

Table 5. Bit-size of entries of a matrix $\mathbf{H}$ (Experimental results on average of 100 matrices for Toy and Small and 10 matrices for Medium)

	λ	n	ρ	β	Average Bit Size	Maximum Bit Size	$\beta - \log n$
Toy	42	136	26	26	1.33	8	18.91
				$42(=\lambda)$	4.66	16	34.91
				80	11.80	25	72.91
				$84(=2\lambda)$	13.99	29	76.91
				$126(=3\lambda)$	23.73	41	118.91
				$168(=4\lambda)$	32.67	51	160.91
Small	52	540	41	41	2.84	14	31.92
				$52(=\lambda)$	4.14	17	42.92
				80	9.70	29	70.92
				$104(=2\lambda)$	16.17	34	94.92
				$156(=3\lambda)$	29.07	47	146.92
				$208(=4\lambda)$	41.69	66	198.92
Medium	62	2085	56	56	5.59	17	44.97
				$62(=\lambda)$	5.63	17	50.97
				80	11.73	27	68.97

Table 5 lists the average bit size of entries in $\mathbf{H}$ generated by the algorithm of Coron $et\ al.$ on various parameters β and n. From the last three columns of Table 5, one can observe that average bit sizes are approximately 10 when $\beta = 80$ as in the implemntation parameters in [10]; moreover, the maximum bit sizes are lower than 30, and they are much smaller than the best $\beta - \log n$, which is obtained from the bound $\|\mathbf{H}^\top\|_\infty \leq 2^\beta$.

In [10, Section 3.1], the authors stated that "$One\ can\ take\ \beta = \lambda$"; however, our analysis and experimental results indicate that β should be much larger than λ. In Table 5, when $\beta \leq 3\lambda$, the expected average bit-size of $|h_{ij}|$ is still smaller than ρ, and for Small security, when $\beta \approx 4\lambda$, the expectation of the average bit-size of $|h_{ij}|$ is equal to ρ; thus, $\beta \geq 4\lambda$ would be safe for the security of the multilinear maps. We investigate the reason why the $\mathbf{H}$-generation in [10] could not increase enough entropy in the full version.

5.2   Implementation

We have implemented Algorithm 3 with various parameters in C++, using the Gnu MP library [1] and NTL library [29], on an Intel(R) Core(TM) i7-2600 CPU at 3.4 GHz with 16 GB RAM.

Attack on Zero-testing Parameter: We have implemented Algorithm 3 to attack on the zero-testing parameters; we set n, η, and ρ as in the implementation parameters for Small (52-bit) and Medium (62-bit) security [10, Section 6.4] and generated the zero-testing parameter normally by using the method described in [11, Appendix F]. We summarize the result in Table 6, and it displays that Algorithm 3 finds a non-trivial factor very quickly on the parameters for Small and Medium security levels.

Table 6. Attack on zero-testing parameter

Inst.	λ	n	η	β	$\mathrm{Exp}(|h_{ij}|)$	m	d	Time*	Security against Alg. 3
Small	52	540	1838	80	10	8	16	8.42 sec	$\leq 2^{34.84}$ clock cycles
Medium	62	2085	2043	80	12	9	32	47.28 sec	$\leq 2^{37.23}$ clock cycles

* The average running time for solving 50 problem instances

Attack on Re-randomization Parameter: We first define Toy parameters for 42-bit security. To this end, we benchmark the parameter of FHEs in [12], which is conservatively determined according to the complexity of the Chen-Nguyen attack [7]. In Table 7, we provide the average running time to solve 50 problems for Toy parameters, and the experimental result demonstrates that the expected security level is tight.

Table 7. Attack on re-randomization parameter

Inst.	λ	n	η	ρ	m	d	(Average) trials	Running time	Sec. ag. Alg 3†
Toy	42	136	1628	26	14	16	3.7 times	1979.55 sec	$2^{42.72}$

† This counts the number of clock cycles.

In fact, the complexity difference between Algorithm 3 and the Chen-Nguyen attack is $\mathcal{O}(\sqrt{\frac{\rho \log \rho}{n}})$ and $\sqrt{\frac{\rho \log \rho}{n}} \approx 1$ for 42-bit security. For Large and Extra security level parameters, $\sqrt{\frac{\rho \log \rho}{n}}$ is less than 1; therefore, Algorithm 3 will be slightly faster than the Chen-Nguyen attack algorithm. We extrapolate Algorithm 3 to be at least $2^{1.38}$ ($2^{2.12}$, resp.) times faster than the Chen-Nguyen attack for Large security (Extra security, resp.), with a similar storage advantage. Therefore, when one selects secure ρ size for large security level integer-based multilinear maps, we suggest that the performance of Algorithm 3 should be considered.

6 Discussions

Encoding-Validity Test: Zero-testing Vector vs. Zero-testing Integer:
In [10], Coron *et al.* implemented a one-round N-way Diffie-Hellman key exchange
protocol [2], based on their multilinear maps. They used heuristic optimizations
for implementation, in particular the zero-testing *integer*, instead of the
zero-testing *vector* as in the original construction. Note that both the CLT attack algorithm and our attack algorithm for n-MPACD require more than one
sample; therefore, both are inapplicable to their optimized version of multilinear
maps over the integers.

Garg *et al.* [18] pointed out a plausible threat when using a single zero-testing
element. In applications that require resilience of the zero test, including against
invalid encodings, several zero-testing elements can be utilized to prevent the use
of invalid encodings. In cryptographic applications such as the Diffie-Hellman key
exchange, it is important to test whether a given encoding is a group element.
Because GES is a substitute for ideal multilinear groups, it is also important to
test whether a given encoding is valid, and has an appropriate level. In the full
version, we present a (polynomial-time) key recovery attack on the multipartite
Diffie-Hellman key exchange protocol, based on the CLT GES with a single integer zero-testing parameter. The basic idea of the attack is analogous to the Lim-
Lee [26] key recovery attack of using invalid encodings on two-party Diffie-Hellman
key exchange based on group structures.

Applications Beyond Multilinear Maps: We note that Algorithm 3 is applicable to the public parameters of a FHE scheme in [8]. Due to the space limitation,
we relegate the detail to the full version.

Acknowledgement. This work was supported by IT R&D program of MSIP/
KEIT [No. 10047212]. Hyung Tae Lee was also supported in part by the Singapore
Ministry of Education under Research Grant MOE2013-T2-1-041. Part of work
was done while Hyung Tae Lee was with Seoul National University, Korea. Jae
Hong Seo is the corresponding author for this paper. The authors also would like
to thank Jung Hee Cheon and the anonymous reviewers of CRYPTO 2014 for their
helpful comments.

References

1. GMP: The GNU multiple precision arithmetic library ver. 5.1.3 (2013),
   http://gmplib.org
2. Boneh, D., Silverberg, A.: Applications of multilinear forms to cryptography. Contemporary Mathematics 324(1), 71–90 (2003)
3. Bostan, A., Schost, É.: On the complexities of multipoint evaluation and interpolation. Theoretical Computer Sciences 329(1-3), 223–235 (2004)
4. Bostan, A., Schost, É.: Polynomial evaluation and interpolation on special sets of
   points. Journal of Complexity 21(4), 420–446 (2005)
5. Brakerski, Z., Rothblum, G.N.: Obfuscating conjunctions. In: Canetti, R., Garay,
   J.A. (eds.) CRYPTO 2013, Part II. LNCS, vol. 8043, pp. 416–434. Springer, Heidelberg (2013)

6. Brakerski, Z., Rothblum, G.N.: Virtual black-box obfuscation for all circuits via generic graded encoding. In: Lindell, Y. (ed.) TCC 2014. LNCS, vol. 8349, pp. 1–25. Springer, Heidelberg (2014)

7. Chen, Y., Nguyen, P.Q.: Faster algorithms for approximate common divisors: Breaking fully-homomorphic-encryption challenges over the integers. In: Pointcheval, D., Johansson, T. (eds.) EUROCRYPT 2012. LNCS, vol. 7237, pp. 502–519. Springer, Heidelberg (2012)

8. Cheon, J.H., Coron, J.-S., Kim, J., Lee, M.S., Lepoint, T., Tibouchi, M., Yun, A.: Batch fully homomorphic encryption over the integers. In: Johansson, T., Nguyen, P.Q. (eds.) EUROCRYPT 2013. LNCS, vol. 7881, pp. 315–335. Springer, Heidelberg (2013)

9. Coron, J.-S., Joux, A., Mandal, A., Naccache, D., Tibouchi, M.: Cryptanalysis of the RSA subgroup assumption from TCC 2005. In: Catalano, D., Fazio, N., Gennaro, R., Nicolosi, A. (eds.) PKC 2011. LNCS, vol. 6571, pp. 147–155. Springer, Heidelberg (2011)

10. Coron, J.-S., Lepoint, T., Tibouchi, M.: Practical multilinear maps over the integers. In: Canetti, R., Garay, J.A. (eds.) CRYPTO 2013, Part I. LNCS, vol. 8042, pp. 476–493. Springer, Heidelberg (2013)

11. Coron, J.-S., Lepoint, T., Tibouchi, M.: Practical multilinear maps over the integers (2013), http://eprint.iacr.org/2013/183 (Full version of [10])

12. Coron, J.-S., Lepoint, T., Tibouchi, M.: Batch fully homomorphic encryption over the integers (2013), http://eprint.iacr.org/2013/036 (Full version of the second part of [8])

13. Coron, J.-S., Mandal, A., Naccache, D., Tibouchi, M.: Fully homomorphic encryption over the integers with shorter public keys. In: Rogaway, P. (ed.) CRYPTO 2011. LNCS, vol. 6841, pp. 487–504. Springer, Heidelberg (2011)

14. Coron, J.-S., Naccache, D., Tibouchi, M.: Public key compression and modulus switching for fully homomorphic encryption over the integers. In: Pointcheval, D., Johansson, T. (eds.) EUROCRYPT 2012. LNCS, vol. 7237, pp. 446–464. Springer, Heidelberg (2012)

15. Fiduccia, A.M.: Polynomial evaluation via the division algorithm: the fast Fourier transform revisited. In: STOC 1972, pp. 88–93 (1972)

16. Fouque, P.-A., Tibouchi, M., Zapalowicz, J.-C.: Recovering private keys generated with weak PRNGs. In: Stam, M. (ed.) IMACC 2013. LNCS, vol. 8308, pp. 158–172. Springer, Heidelberg (2013)

17. Freire, E.S.V., Hofheinz, D., Paterson, K.G., Striecks, C.: Programmable hash functions in the multilinear setting. In: Canetti, R., Garay, J.A. (eds.) CRYPTO 2013, Part I. LNCS, vol. 8042, pp. 513–530. Springer, Heidelberg (2013)

18. Garg, S., Gentry, C., Halevi, S.: Candidate multilinear maps from ideal lattices. In: Johansson, T., Nguyen, P.Q. (eds.) EUROCRYPT 2013. LNCS, vol. 7881, pp. 1–17. Springer, Heidelberg (2013)

19. Garg, S., Gentry, C., Halevi, S., Raykova, M., Sahai, A., Waters, B.: Candidate indistinguishability obfuscation and functional encryption for all circuits. In: FOCS 2013, pp. 40–49. IEEE Computer Society (2013)

20. Garg, S., Gentry, C., Halevi, S., Sahai, A., Waters, B.: Attribute-based encryption for circuits from multilinear maps. In: Canetti, R., Garay, J.A. (eds.) CRYPTO 2013, Part II. LNCS, vol. 8043, pp. 479–499. Springer, Heidelberg (2013)

21. Garg, S., Gentry, C., Sahai, A., Waters, B.: Witness encryption and its applications. In: STOC 2013, pp. 467–476. ACM (2013)

22. Hohenberger, S., Sahai, A., Waters, B.: Full domain hash from (leveled) multilinear maps and identity-based aggregate signatures. In: Canetti, R., Garay, J.A. (eds.) CRYPTO 2013, Part I. LNCS, vol. 8042, pp. 494–512. Springer, Heidelberg (2013)
23. Howgrave-Graham, N.: Approximate integer common divisors. In: Silverman, J.H. (ed.) CaLC 2001. LNCS, vol. 2146, pp. 51–66. Springer, Heidelberg (2001)
24. Langlois, A., Stehlé, D., Steinfeld, R.: GGHLite: More efficient multilinear maps from ideal lattices. In: Nguyen, P.Q., Oswald, E. (eds.) EUROCRYPT 2014. LNCS, vol. 8441, pp. 239–256. Springer, Heidelberg (2014)
25. Lee, H.T., Seo, J.H.: Security analysis of multilinear maps over the integers. IACR Cryptology ePrint Archive (2014), http://eprint.iacr.org
26. Lim, C.H., Lee, P.J.: A key recovery attack on discrete log-based schemes using a prime order subgroup. In: Kaliski Jr., B.S. (ed.) CRYPTO 1997. LNCS, vol. 1294, pp. 249–263. Springer, Heidelberg (1997)
27. Mateer, T.: Fast Fourier transform algorithms with applications. PhD thesis, Clemson University (2008)
28. Seo, J.H.: Faster algorithms for variants of approximate common divisors problem: Application to multilinear maps over the integers. In: Memoirs of the 7th Cryptology Paper Contest, arranged by Korea government organization (2013)
29. Shoup, V.: NTL: A library for doing number theory ver. 6.0.0 (2013), http://www.shoup.net/ntl/
30. van Dijk, M., Gentry, C., Halevi, S., Vaikuntanathan, V.: Fully homomorphic encryption over the integers. In: Gilbert, H. (ed.) EUROCRYPT 2010. LNCS, vol. 6110, pp. 24–43. Springer, Heidelberg (2010)
31. von zur Gathen, J., Gerhard, J.: Modern computer algebra, 2nd edn. Cambridge University Press (2003)

Converting Cryptographic Schemes
from Symmetric to Asymmetric Bilinear Groups

Masayuki Abe[1], Jens Groth[2,*], Miyako Ohkubo[3], and Takeya Tango[4]

[1] NTT Secure Platform Laboratories, Japan
abe.masayuki@lab.ntt.co.jp
[2] University College London, UK
j.groth@ucl.ac.uk
[3] Security Fundamentals Lab, NSRI, NICT, Japan
m.ohkubo@nict.go.jp
[4] Kyoto University, Japan
tkytango@ai.soc.i.kyoto-u.ac.jp

Abstract. We propose a method to convert schemes designed over symmetric bilinear groups into schemes over asymmetric bilinear groups. The conversion assigns variables to one or both of the two source groups in asymmetric bilinear groups so that all original computations in the symmetric bilinear groups go through over asymmetric groups without having to compute isomorphisms between the source groups. Our approach is to represent dependencies among variables using a directed graph, and split it into two graphs so that variables associated to the nodes in each graph are assigned to one of the source groups. Though searching for the best split is cumbersome by hand, our graph-based approach allows us to automate the task with a simple program. With the help of the automated search, our conversion method is applied to several existing schemes including one that has been considered hard to convert.

Keywords: Conversion, Symmetric Bilinear Groups, Asymmetric Bilinear Groups.

1 Introduction

It is often believed that once a scheme for a purpose is shown feasible over symmetric bilinear groupsa scheme for the same purpose should be constructable over asymmetric bilinear groups. One approach is to use different mechanisms and stronger assumptions available only in the asymmetric setting. The other, which we study in this paper, is to convert a scheme in the symmetric setting into one in the asymmetric setting keeping the original design intact.

We will given a bilinear group setting with a pairing $e : \mathbb{G} \times \tilde{\mathbb{G}} \to \mathbb{G}_T$ follow [13] in calling the symmetric setting where $\mathbb{G} = \tilde{\mathbb{G}}$ for Type-I and in the asymmetric

* The research leading to these results has received funding from the the Engineering and Physical Sciences Research Council grant EP/J009520/1 and the European Research Council under the European Union's Seventh Framework Programme (FP/2007-2013) / ERC Grant Agreement n. 307937.

J.A. Garay and R. Gennaro (Eds.): CRYPTO 2014, Part I, LNCS 8616, pp. 241–260, 2014.

setting distinguish between Type-II, where there is an efficiently computable isomorphism $\psi : \tilde{\mathbb{G}} \to \mathbb{G}$, and Type-III where there are no efficiently computable isomorphisms between the source groups. In this paper, we focus on converting schemes over Type-I groups to Type-III groups, i.e., converting to the fully asymmetric setting, so we will in general be referring to the Type-III setting when speaking of asymmetric bilinear groups.

We argue that the benefit of conversion is threefold. First, it allows designers to focus on implementing their ideas using a simpler description in the symmetric setting without being encumbered by the deployment of group elements over two source groups. Second, it is an effective way to save schemes in the symmetric setting from cryptanalytic advances. Recent progress in solving the discrete logarithm problems over groups with small characteristics [15,16,14,3] has necessitated the use of larger parameters for a major class of symmetric bilinear groups. A conversion method allows one to port these schemes to an asymmetric setting. Finally, it yields potentially more secure schemes than those dedicated to asymmetric groups because preserving the symmetric group assumptions they may remain secure even if cryptanalysis were to discover techniques to efficiently compute isomorphisms between the source groups.

There are two issues that makes conversion a non-trivial task. The first is the potential presence of symmetric pairings $e(X, X)$. A symmetric pairing can occur indirectly like $e(X, Z)$ for $Z = X \cdot Y$ and it is not necessarily easy to see if there are indirect symmetric pairings in intricate algorithms. It is in particular problematic when a function that maps a string into a source group element is used in the original scheme since it is only possible to either of the source groups at a time. The second is the security proof and the underlying assumptions. Not only should the scheme be executable, but the reduction algorithms used in the security proof must also be executable in the asymmetric setting. Furthermore, the assumptions need to be cast in the asymmetric setting as well. An instructive example is given in [18] that demonstrates how conversion without a security guarantee can yield a scheme that seems to work but is insecure.

OUR RESULT. We propose a conversion method that turns schemes designed over symmetric bilinear groups into schemes over asymmetric bilinear groups. As our method converts not only the algorithms in a scheme but also the security proofs by black-box reduction and the underpinning assumptions, the security is preserved based on the converted assumptions. We then argue that, if the original assumptions are justified in the generic Type-I group model, the converted counterparts are justified in the generic Type-III group model. Our conversion includes schemes in the random oracle model that hash a string onto a random group element and hash group elements to a random string. We present a formal model for the class of schemes our conversion method can handle.

Our conversion method takes as input a *dependency graph* that represents computation among source group elements in the scheme to convert. This is a directed graph whose nodes correspond to group elements in the scheme. Directed edges in the graph represent dependency in such a way that the destination node is computed from the source nodes through the group operation. By splitting the dependency graph into two graphs in such a way that no dependency is lost and two nodes that represent group elements input to a pairing appear in different graphs, we obtain two dependency graphs that represents computation in the source groups of asymmetric bilinear groups. There may be nodes that appear in both graphs. Their presence is necessary for consistent computation in each source group. These nodes correspond to the symmetric group elements that need to be duplicated in both source groups when converting to an asymmetric bilinear group. The most cumbersome part of our conversion procedure is to find a splitting of the dependency graph in a way that conforms to given constraints and efficiency measures.

We present an algorithm to search for the best valid split. It is implemented with Java and applied to dependency graphs for several known cryptographic schemes originally built over symmetric bilinear groups:

- Waters' Identity-based Encryption scheme [24]. We have chosen this scheme since it has a small number of parameters so that one can manually verify the result. An interesting observation is that conversion is not possible without duplicating some group elements in the assumption.
- Boneh and Shacham's verifier-local revocation group signature scheme [7]. This scheme involves hash-onto-point functions. In [23], Smart and Vercauteren observed that converting to Type-III is not possible, and [10] introduced the scheme as an typical example that cannot be converted due to their use of hash-to-point functions and homomorphisms. We present a conversion based on assumptions that duplicates elements in the original assumptions. It does not contradict [23] as they limit themselves to the case where no element duplicates in the assumption.
- Waters' dual-system encryption scheme [25]. The purpose of converting this scheme is to compare the converted result using our conversion technique with existing manually built schemes[11,20]. Our conversion yields a slightly less efficient scheme. An assumption that fully duplicates elements of the decision linear assumption is inevitable for this conversion.
- Tagged one-time signature scheme from [1]. By converting the scheme, we obtain the first tagged one-time signature scheme over asymmetric bilinear groups with minimal tag size. It answers the open question positively in [1].

The search algorithm runs in exponential time in the number of pairings and the number of bottom nodes that do not have outgoing edges as we will explain later. It takes a standard PC (Windows 7, Intel(R) Core(TM) i7-3720QM CPU @ 2.60GHz, 8.0GB RAM) 9 msec to convert the easiest case, Waters' IBE, and about 105 min to convert the most intricate case, Waters' Dual Encryption scheme. We summarize the result of our experimental conversions in Table 1.

It should be noted that our conversion is not tight, i.e., when our conversion algorithm fails to find a scheme in Type-III, it does not mean anything more than the fact that our conversion algorithm does not find a conversion for the scheme. In particular, it does not imply general impossibility. It is an open problem to show impossibility of conversion.

Table 1. Size of public parameters $(pp + pk)$ and ciphertext (ct) or signature (σ) in the number of source group elements including a default generator. "conv'd(opt=xxx)" denotes that the scheme is obtained by our conversion in terms of minimizing the size of assumption or public-keys. Assumption coXXX$^{+\alpha}$ denotes co-XXX assumption(s) that involve α duplicated elements. For TOS, σ includes a tag.

Scheme	Construction	Size and Delta		Assumptions
		$pp + pk$	ct or σ	
Waters' IBE	original [24]	$4 + \lambda$	2	DBDH
	conv'd(opt=pk)	$+2$	$+0$	coDBDH^{+3}
	conv'd(opt=$assm$)	$+3 + \lambda$	$+0$	coDBDH^{+2}
VLR Group Sig	original [7]	2	2	DLIN, q-SDH
	conv'd	$+1$	$+0$	coDLIN^{+5}, q-coSDH^{+q}
Dual System Enc	original [25]	12	9	DLIN, DBDH
	conv'd(opt=pk)	$+1$	$+0$	coDLIN$^{+4,+5}$, coDBDH^{+2}
	conv'd(opt=$assm$)	$+6$	$+0$	coDLIN$^{+3,+3}$, coDBDH^{+2}
	manually conv'd [19]	-3	-2	DDH1, DLIN, DBDH
		-6	-5	DDH1, DDH2v, DBDH
	new design [11]	-2	-5	SXDH
TOS	original [1]	$2k + 6$	4	SDP
	conv'd	$+5$	$+0$	coSDP^{+5}

Due to space limitation, most of proofs for theorems and lemmas and all details of experiments are dropped from this version of the paper.

RELATED WORKS. Chatterjee and Menezes [9,10] considered conversion from schemes over Type-II groups to Type-III groups and discuss the role of the isomorphism in Type-II groups. Their conversion shares the basic idea with ours – represent a group element by a pair of source group elements and drop one of them if unnecessary. They proposed a heuristic guideline for when a scheme allows or resists conversion. Chatterjee et al. [8] discussed relations among assumptions over Type-II and Type-III groups that include ones with or without duplicated elements in a problem instance. Smart and Vercauteren [23] explored variations of Boneh and Franklin's identity-based encryption scheme [5] and Boneh, Lynn and Shacham's signature scheme [6] based on a family of BDH assumptions over Type-II groups. They investigated which variations suffice for the security proofs and how efficient the corresponding schemes are. Chen et al. [11] presented modifications of Waters' dual-system encryption scheme over Type-III groups. They obtained a more efficient scheme than the original by a careful manual conversion.

A more general work by Akinyele, Green and Hohenberger [2] introduced a powerful software system called AutoGroup whose purpose is the same as ours.

It takes a specification of the target scheme written in a scheme description language and uses a satisfiability modulo theory solver [12] to find a valid deployment of elements over two source groups. AutoGroup is a powerful tool to find optimized computation in the resulting scheme conforming to one's design demands. On the other hand, and contrary to our tool, it does not say anything about the security of the resulting scheme; it is left as a subsequent task to check whether the resulting scheme is secure, to identify sufficient assumptions, and to provide convincing arguments that the assumptions are plausible. Our procedure requires manual work as well but only to examine if the original proof works over generic symmetric bilinear groups and to verify that the assumptions are plausible in the generic Type-I group model. Once this has been confirmed, the converted scheme retains its security under converted assumptions that also retain plausibility arguments in the generic group model.

2   Preliminaries

We follow standard definition and notations of symmetric and asymmetric bilinear groups. Let $\mathcal{G}$ be a group generator that takes security parameter 1^λ and outputs $(q, \mathbb{G}, \tilde{\mathbb{G}}, \mathbb{G}_T, e, G, \tilde{G})$ where $\mathbb{G}$, $\tilde{\mathbb{G}}$, and $\mathbb{G}_T$ are groups of prime order q, $e : \mathbb{G} \times \tilde{\mathbb{G}} \to \mathbb{G}_T$ is an efficiently computable non-degenerate bilinear map, and $G, \tilde{G}$ and $G_t = e(G, \tilde{G})$ are generators of $\mathbb{G}, \tilde{\mathbb{G}}$ and $\mathbb{G}_T$ respectively. $\mathbb{G}$ and $\tilde{\mathbb{G}}$ are called source groups and $\mathbb{G}_T$ is called the target group. When $\mathbb{G} \neq \tilde{\mathbb{G}}$ and there are no efficiently computable isomorphisms between $\mathbb{G}$ and $\tilde{\mathbb{G}}$ we call it the Type-III bilinear group setting. When $\mathbb{G} = \tilde{\mathbb{G}}$ we let $G = \tilde{G}$ and call it a symmetric bilinear group or Type-I group. For simplicity, we will often use the shortened notation $\mathcal{G}_{\mathsf{sym}}$ and $(q, \mathbb{G}, \mathbb{G}_T, e, G)$ for the symmetric case, and on the other hand write $\mathcal{G}_{\mathsf{asym}}$ when emphasizing a group generator outputs a Type-III bilinear group. Hashing to $\mathbb{G}$ is doable in Type-I groups over supersingular elliptic curves. In Type-III groups over elliptic curves, it is possible to hash to $\mathbb{G}$ and $\tilde{\mathbb{G}}$ independently with different costs. We focus on source group elements in the paper and other data are mostly ignored as they are handled equally in Type-I and III settings. Notes will be given otherwise.

Throughout the paper we will work with directed graphs using the notation (X, Y) for an edge from node X to Y. For two directed graphs $\Gamma = (V, E)$ and $\Gamma' = (V', E')$ the merger operation $\Gamma \oplus \Gamma'$ is defined by a graph $(V \cup V', E \cup E')$. For a graph Γ and a subgraph Γ', we define $\Gamma \ominus \Gamma'$ as a graph obtained by removing nodes in Γ' and edges that involves nodes in Γ' from Γ. If a graph Γ' is a subgraph of Γ, we write $\Gamma' \subseteq \Gamma$. For a node X in Γ, we define $\mathsf{Anc}(\Gamma, X)$ by the subgraph of Γ that consists of paths reaching X. Nodes in $\mathsf{Anc}(\Gamma, X)$ are called ancestors of X, where we use the convention that X is an ancestor to itself.

By $(x, y) \leftarrow (A(a), B(b))$, we denote a process where two interactive algorithms A and B take inputs a and b and output x and y, respectively, as a result of their interaction.

3   Overview with Example

This section illustrates our conversion procedure for Waters' IBE [24]. We mostly use the original notation and description although the reader may want to refer to the original paper for details.

The procedure starts by looking at the original description of Water's IBE scheme over symmetric bilinear groups. It builds a *dependency graph*, that describes how each group element appearing in the scheme depends on other group elements. Later the dependency graph is used to decide which group element is computed in which source group of the asymmetric bilinear groups. The type assignment yields an instantiation of Waters IBE scheme over asymmetric bilinear groups. As we convert the reduction algorithm in the security proof and the underlying assumptions as well, the underlying security argument also translates to the asymmetric setting.

Step 1. Building a dependency graph for each algorithm. Waters' IBE scheme consists of four algorithms: Setup, Key Generation, Encryption and Decryption. The security proof consists of a reduction algorithm that uses a purported adversary in a black-box manner to break an instance of the decisional bilinear Diffie-Hellman (DBDH) problem. The first step is to build a dependency graph for each algorithm as illustrated in Fig. 2.

- The setup algorithm takes a security parameter, and outputs default generator g and random source group elements g_1, g_2, u', and $u_1, \ldots, u_n$ where $g_1 = g^\alpha$ for a random $\alpha \in \mathbb{Z}_q$. The dependency graph includes nodes $\mathsf{g}, \mathsf{g1}, \mathsf{g2}, \mathsf{u'}$ that correspond to g, g_1, g_2, and u', respectively. Elements $u_1, \ldots, u_n$ are represented by a single node ui in the graph. We assume that the random source group elements are generated from the default generator by using group operations. Thus the dependency graph has edges from g to every other node. The algorithm also computes a master secret key $msk = g_2{}^\alpha$. We thus add a node labeled by msk and have an edge from $\mathsf{g2}$ to msk. This results in graph (1) in Fig. 2.

- The key generation algorithm takes the master secret key and the public key, and computes decryption key (d_1, d_2) for an identity $v \in \{0, 1\}^n$. Let $\mathcal{V}$ be the set of indices for which the bit-string v is set to 1.

$$d_1 := msk \cdot (u' \prod_{i \in \mathcal{V}} u_i)^r \quad \text{and} \quad d_2 := g^r. \tag{1}$$

  A corresponding dependency graph thus involves nodes $\mathsf{d1}, \mathsf{msk}, \mathsf{u'}, \mathsf{ui}$, $\mathsf{d2}, \mathsf{g}$, and edges from $\mathsf{msk}, \mathsf{u'}, \mathsf{ui}$ to $\mathsf{d1}$, and g to $\mathsf{d2}$ as illustrated in graph (2) in Fig. 2. Note that g_1 is in the public key given to the algorithm but not involved in computation. Such unused elements can be ignored and do not appear in the graphs.

- The encryption algorithm involves both group operations and pairings. It takes the public key and message M from the target group, and computes a ciphertext (C_1, C_2, C_3) as follows:

$$C_1 := e(g_1, g_2)^t \cdot M, \quad C_2 := g^t \quad \text{and} \quad C_3 := u' \prod_{i \in \mathcal{V}} u_i^t. \tag{2}$$

The pairing operation $e(g_1, g_2)$ is represented in the graph by connecting nodes g1, g2 to a pair of special nodes called *pairing nodes*, whose label looks like p1[0] and p1[1]. The trunk name p1 is unique throughout the system. A paring node indicates that its parent node corresponds to an input to the pairing operation identified by the name. C_2 and C_3 are source group elements computed from g and u', u_i, respectively, and represented in the graph in (3) in Fig. 2 accordingly. Since M and C_1 are in the target group, no corresponding nodes are included in the graph.

- The decryption algorithm computes

$$M := C_1 \, e(d_2, C_3) \, e(d_1, C_2)^{-1}. \tag{3}$$

The pairing $e(d_2, C_3)$ yields nodes d2, C3 connected to pairing nodes p2[0] and p2[1], respectively. Similarly, the pairing $e(d_1, C_2)$ yields nodes d1, C2 connected to p3[0] and p3[1], respectively. The resulting graph is (4) in Fig. 2.

- Next we consider the instance generator of DBDH that generates default generator g and random group elements A, B, and C. (Target group element Z is irrelevant and ignored here.) The graph contains nodes g, A, B, and C, and edges from g to every other node as illustrated in (5) in Fig. 2. Graphs for associated verification and random guessing algorithms are empty as they do not involve any group operations.

- Finally we consider a graph for the reduction algorithm. The whole algorithm and its analysis is intricate but group operations are only used in a few places. The reduction first takes group elements A and B from the given instance of DBDH problem and sets them to public key g_1 and g_2, respectively. The remaining parts of the public key u' and u_i are generated normally. It then simulates an individual key by

$$d_1 := g_1^{\frac{-J(v)}{F(v)}} \left(u' \prod_{i \in V} u_i\right)^r \quad \text{and} \quad d_2 := g_1^{\frac{-1}{F(v)}} g^r \tag{4}$$

where we refer to [24] for $J(v)$ and $F(v)$. It is repeated for each key query and there are many d_1 and d_2 computed in the same manner. In the graph, these keys are represented by a single pair of nodes d1 and d2 directed from g1, u', ui and g1, g, respectively. The reduction algorithm also creates a challenge ciphertext that includes

$$C_2 := C \quad \text{and} \quad C_3 := C^{J(v^*)}. \tag{5}$$

They are represented by nodes C2, C3, C and edges directed from C to C2 and C3. The resulting graph is (6) in Fig. 2.

Step 2. Merge. Merge the graphs from Step 1 into a single graph, Γ, as illustrated in Fig. 3.

Step 3. Split. Split Γ into two graphs Γ_0 and Γ_1 such that

[Waters' IBE Scheme in Type-III]

$Setup((q, \mathbb{G}, \tilde{\mathbb{G}}, \mathbb{G}_T, e, g, \tilde{g}))$: Select $g_2 \leftarrow \mathbb{G}$, $\alpha \leftarrow \mathbb{Z}_q$. Compute $g_1 := g^\alpha$, $\tilde{g}_1 := \tilde{g}^\alpha$. $u' \leftarrow \mathbb{G}$ and a random n-length vector $U = (u_i) \in \mathbb{G}^n$, those elements are chosen at random from $\mathbb{G}$. Publish the public parameters $prm = (g, \tilde{g}, g_1, \tilde{g}_1, g_2, u', U) \in \mathbb{G}^{4+n} \times \tilde{\mathbb{G}}^2$. The master secret is $msk = g^\alpha \in \mathbb{G}$.

$KeyGeneration(prm)$: $r \leftarrow \mathbb{Z}_q$ and a private key for identity v is

$$d_v := (d_1, \tilde{d}_2) = \left(g_2^\alpha \left(u' \prod_{i \in \mathcal{V}} u_i \right)^r, \tilde{g}^r \right) \in \mathbb{G} \times \tilde{\mathbb{G}} \qquad (6)$$

$Encryption(prm, M)$: Let a message $M \in \mathbb{G}_T$. $t \leftarrow \mathbb{Z}_q$. The ciphertext is

$$C = (C_1, \tilde{C}_2, C_3) = \left(e(g_2, \tilde{g}_1)^t M, \tilde{g}^t, \left(u' \prod_{i \in \mathcal{V}} u_i \right)^t \right) \in \mathbb{G}_T \times \tilde{\mathbb{G}} \times \mathbb{G} \qquad (7)$$

$Decryption(prm, d_v, C)$: Parse $C = (C_1, \tilde{C}_2, C_3) \in \mathbb{G}_T \times \tilde{\mathbb{G}} \times \mathbb{G}$. Calculate

$$C_1 \frac{e(C_3, \tilde{d}_2)}{e(d_1, \tilde{C}_2)} = \left(e(g_2, \tilde{g}_1)^t M \right) \frac{e((u' \prod_{i \in \mathcal{V}} u_i)^t, \tilde{g}^r)}{e(g_2^\alpha (u' \prod_{i \in \mathcal{V}} u_i)^r, \tilde{g}_t)} = M$$

[Decisional Co-BDH Problem]

Given $(g, \tilde{g}, A = g^a, \tilde{A} = \tilde{g}^a, B = g^b, C = g^c, \tilde{C} = \tilde{g}^c, Z) \in \mathbb{G}^4 \times \tilde{\mathbb{G}}^3 \times \mathbb{G}_T$, where $Z = e(g, \tilde{g})^{abc+\beta r}$, $r \leftarrow \mathbb{Z}_q$ and $\beta \leftarrow \{0, 1\}$, guess β.

Fig. 1. Converted Waters' IBE and underlying hard problem

- No nodes or edges are lost, i.e., merging Γ_0 and Γ_1 recovers Γ.
- For every pair of paring nodes, if one node is in Γ_0, the other node is exclusively in Γ_1.
- For every node X in each split graph, the ancestor subgraph of X in Γ is included in the same graph.

Given 3 pairs of pairing nodes in Γ, there exist 2^3 valid splits that satisfy the above conditions. Select a valid split (Γ_0, Γ_1) according to a criterion for ones purpose of conversion. In Fig. 4, we give a valid split that yields a minimal public key size. As shown in Table 1, another valid split is possible to minimize the assumptions. We give an algorithm that searches for the best split according to an arbitrary criteria in Section 5.4.

Step 4. Derive the converted scheme. Nodes in Γ_0 and Γ_1 correspond to elements in $\mathbb{G}$ and $\tilde{\mathbb{G}}$, respectively. Based on the assignment, one can derive the resulting Waters' IBE scheme over Type-III groups and its underlying assumption as illustrated in Fig. 1.

Remark 1. To preserve the security, it is required that the security of the cryptosystem is proven by a black-box reduction [21] and that the reduction algorithms are abstract as defined in Section 4.2.

Remark 2. In the formal model, we consider correctness as part of scheme and hence a dependency graph for correctness should be included. As nodes are given consistent names in this example, the dependency graph for correctness becomes trivial and is therefore omitted. In general, consistent names are given by *object identifiers* as explained in Section 4.2.

Remark 3. It is important to check group membership for every input. For instance, if an input X in the original scheme is converted into X and $\tilde{X}$, then the group membership testing on X in the original scheme is translated to checking the Diffie-Hellman relation $e(X, \tilde{G}) = e(G, \tilde{X})$ in the converted scheme. In the above example, the relation between g_1 and $\tilde{g}_1$ in the common parameter should be verified. Since the common parameters will be verified once for all in practice, it is not explicitly shown in Fig. 1.

4   Formal Model

4.1   Cryptographic System

We consider secure cryptographic schemes whose correctness and security are defined by game-like interactive algorithms, and the security is proven by black-box reductions to hardness of computational or decisional problems. Formally, we formulate a secure cryptosystem Π by sets of efficient algorithms $\Pi = \{\mathcal{F}, \mathcal{C}, \mathcal{I}, \mathcal{R}\}$ that represent the functionality, correctness, underlying problems and security reductions. Properties of these algorithms are defined in the following.

The functionality $\mathcal{F}$ is a set of algorithms $\mathcal{F} = (\mathcal{F}_1, \ldots, \mathcal{F}_t)$ where each $\mathcal{F}_i$ implements some function for the cryptosystem such as "key generation", "encryption", and so on. Correctness of $\mathcal{F}$ is defined by $\mathcal{C}$ that has black-box access to the functionalities in $\mathcal{F}$ and outputs 1 if everything works as intended.

Definition 4 (Correctness). *Π is correct if $\Pr[1 \leftarrow \mathcal{C}^{\mathcal{F}}(1^\lambda)] = 1$ for all λ.*

A problem $\mathcal{I}$ is a triple of algorithms $\mathcal{I} = (\mathcal{I}_{gen}, \mathcal{I}_{ver}, \mathcal{I}_{guess})$, where $\mathcal{I}_{gen}$ is an instance generator that generates a problem instance, $\mathcal{I}_{ver}$ is a verification algorithm that verifies a given answer, and $\mathcal{I}_{guess}$ is a guessing algorithm that returns an answer by random guessing.

Definition 5 (Hardness of $\mathcal{I}$). *Problem $\mathcal{I}$ is hard if the advantage function*

$$\mathrm{Adv}_{\mathcal{B}}^{\mathcal{I}}(\lambda) := \Pr\left[1 \leftarrow \mathcal{I}_{ver}(x, y) \mid (x, y) \leftarrow (\mathcal{I}_{gen}(1^\lambda), \mathcal{B}(1^\lambda))\right] - \mathcal{I}_{guess}(1^\lambda)$$

is negligible in λ for any probabilistic polynomial-time adversary $\mathcal{B}$.

In this work, we consider cryptographic schemes where security is proven by an efficient algorithm called a reduction, $\mathcal{R}$, that is successful in solving problem $\mathcal{I}$

given black-box access to an adversary that successfully attacks the scheme. We define security in the form of advantage functions $\mathrm{Adv}_{\mathcal{A}}^{\mathcal{S}}(\lambda) := \Pr[1 \leftarrow \mathcal{S}^{\mathcal{A}}(1^{\lambda})]$ for algorithm $\mathcal{S}$ (which is often called a challenger), which should be negligible for any probabilistic polynomial time adversary $\mathcal{A}$.

Definition 6 (Security of Π). *Cryptosystem Π is secure in the sense of $\mathcal{S}$ under the hardness assumption on $\mathcal{I}$ with black-box reduction $\mathcal{R}$ if for any $\mathcal{A}$ advantage $\mathrm{Adv}_{\mathcal{R}^{\mathcal{A}}}^{\mathcal{I}}(\lambda)$ is not negligible if $\mathrm{Adv}_{\mathcal{A}}^{\mathcal{S}}(\lambda)$ is not negligible.*

Though $\mathcal{C}$, $\mathcal{R}$, $\mathcal{I}$, and $\mathcal{S}$ are defined as single algorithms, they can be naturally extended to sets of algorithms. In particular, security is often proven by sequences of games and each game reduces to an individual hardness assumption.

4.2   Abstract Algorithms

Let $\tilde{\mathcal{A}}^{\mathcal{O}}$ denote an algorithm where $\tilde{A}$ is called an abstract algorithm that computes group operations through oracle $\mathcal{O}$. Oracle $\mathcal{O}$ is called a group operation oracle and given locally to host algorithm $\tilde{A}$. It is initialized with a description of bilinear groups that is common for all algorithms in a cryptosystem. It performs generic group operations over the bilinear groups like the generic group oracle [22]. We follow the model by Maurer [17] for group operations. It forces the host algorithm be explicit in checking equality. It is useful for our purpose as we need to know which elements the host algorithm tests equality.

Oracle $\mathcal{O}$ also works as an interface for sending and receiving group elements. When $\tilde{A}^{\mathcal{O}}$ and $\tilde{B}^{\mathcal{O}}$ interact, group elements are sent and received through the oracles and all other data are exchanged directly between the host algorithms. As mentioned above, a description of bilinear groups is common to the oracles.

We also consider $\mathcal{O}$ in the random oracle model [4] to capture functions that map arbitrary input to a random source group element and that map group elements attached by arbitrary string to a random string. $\mathcal{O}$ provides these functions by interacting with random oracles $H_{\mathbb{G}} : \{0,1\}^* \to \mathbb{G}$ and $H_{str} : \mathbb{G}^k \times \{0,1\}^* \to \{0,1\}^{\ell}$ for some k and ℓ.

Every group element (more precisely a pointer to it) is associated with an object identifier (oid for short). It is an arbitrarily prescribed string that identifies the role of the element in a cryptosystem like "the third element of a ciphertext" or "the first element of a secret-key." The way oids are assigned to group elements is a part of an algorithm and indeed as important as computations in the algorithm. We restrict that only a constant number of distinct oids is used in a cryptosystem so that a dependency graph can be described in a constant size as we explain later. In general, oids can be arbitrarily specified. We consider a conventional case where oids are named after variables used in describing algorithms in $\mathcal{F}$ and $\mathcal{I}$. It is indeed how we did for Waters' IBE in Section 3. When there are indexed variables that grows in the security parameter like the public-key of Waters' IBE, they are assigned the same oid if they are involved in the computation in the same manner. For instance, the same oid u_i is given to all group elements $u_1, \ldots, u_n$ in the example in Section 3. This convention for oids applies to all schemes considered in this paper.

Let $(q, \mathbb{G}, \mathbb{G}_T, e, G)$ be symmetric bilinear groups and let $G_t = e(G, G)$. We define an extended group operation oracle $\mathcal{O}$ for symmetric bilinear group of prime order q. In the following, pointers are taken sequentially from 1, and queries with unused pointers are rejected. We omit group operations and equality checking in the target group in the following description.

[Extended Group Operation Oracle $\mathcal{O}$]

- init: Initialize lists $\mathcal{L}_s$ and $\mathcal{L}_t$ with entries (pt, G) and (pt_t, G_t) respectively with fresh pointers pt and pt_t. Return (q, pt, pt_t).
- gop$(pt_1, a_1, \ldots, pt_k, a_k, oid)$: For every pt_i, search $\mathcal{L}_s$ for (pt_i, X_i). Compute $X := \prod X_i^{a_i} \in \mathbb{G}$ and store (pt, X) with fresh pt. Output pt.
- pair(pt_1, pt_2): Search $\mathcal{L}_s$ for (pt_1, X_1) and (pt_2, X_2). Compute $X := e(X_1, X_2)$ and store (pt, X) to $\mathcal{L}_t$ with fresh pt. Output pt.
- equal(pt_1, pt_2): Search $\mathcal{L}_s$ for (pt_1, X_1) and (pt_2, X_2). If $X_1 \equiv X_2 \in \mathbb{G}$ then return 1. Return 0, otherwise.
- hash2g(str, oid): (This query is accepted only when random oracle $H_{\mathbb{G}}$ is available.) If the same input has been queried before, return the same answer. Otherwise send (str, oid) to $H_{\mathbb{G}}$ and receive $X \in \mathbb{G}$. Store (pt, X) to $\mathcal{L}_s$ with fresh pt, and return pt.
- hash2str$(pt_1, \ldots, pt_k, str, oid)$: (This query is accepted only when random oracle H_{str} is available.) Search $\mathcal{L}_s$ for each $X_i \in \mathbb{G}$ that corresponds to pt_i. If $(X_1, \ldots, X_k, str, oid)$ has been asked before, return the same value. Otherwise, send it to H_{str} and return the resulting string.
- send(pt, oid): Search $\mathcal{L}_s$ for (pt, X) and send (X, oid) to the implicitly specified destination.
- receive(oid): On receiving this query from the host algorithm, wait to receive (X, oid') from implicitly specified entity. Reject if $X \notin \mathbb{G}$ or $oid \neq oid'$ and continue waiting. Otherwise, store (pt, X) to $\mathcal{L}_s$ with fresh pt, and send pt to the host algorithm.

We make some remarks about object identifier oid given as input for most queries. When calling gop and hash2g, the host algorithm assigns an object identifier to the resulting group element by specifying it with oid. The oracle does not use oid in handling gop query, but it is needed later to build a dependency graph. It is important to see that oid is included in the input to the random oracle in hash2g. It allows the host algorithm to virtually deal with several independent random oracles indexed by oid. For hash2str, it is assumed that every group element X_i is transformed to its canonical representation in $\mathbb{G}$ before being sent to random oracle H_{str}. In general, even if $X \equiv Y \in \mathbb{G}$ holds, hashing X and Y may yield different values. This is an important issue as we simulate a group element in Type-I groups with a pair of group elements in Type-III allowing different representations. With oid specified in the input, we can control the representation so that group elements having the same oid has the same representation.

Let $\Sigma_{\mathcal{G}_{\mathsf{sym}}}$ be the set of possible extended group operation oracles based on a group generated by $\mathcal{G}_{\mathsf{sym}}$ for all sufficiently large λ and all random coins/oracles.

We say $\mathcal{O}$ is based on $\mathcal{G}_{\mathsf{sym}}$ when we refer to an extended group operation oracle $\mathcal{O}$ in $\Sigma_{\mathcal{G}_{\mathsf{sym}}}$. We now define a cryptosystem consisting of abstract algorithms. Let $\tilde{\Pi}^{\mathcal{O}}_{\mathcal{G}_{\mathsf{sym}}} := (\tilde{\mathcal{F}}^{\mathcal{O}}, \tilde{\mathcal{C}}^{\mathcal{O}}, \tilde{\mathcal{R}}^{\mathcal{O}}, \tilde{\mathcal{I}}^{\mathcal{O}})$ be a cryptosystem obtained by giving oracle $\mathcal{O}$ based on $\mathcal{G}_{\mathsf{sym}}$ to (sets of) abstract algorithms $\tilde{\mathcal{F}}$, $\tilde{\mathcal{C}}$, $\tilde{\mathcal{R}}$, and $\tilde{\mathcal{I}}$. Let also $\tilde{\mathcal{S}}$ be an abstract challenger algorithm. Let Δ_q denote all $\mathcal{G}_{\mathsf{sym}}$ that outputs q with the same probability distribution.

Definition 7 (Correct and Secure Abstract Cryptosystem). *A set of abstract algorithms $\tilde{\Pi} = (\tilde{\mathcal{F}}, \tilde{\mathcal{C}}, \tilde{\mathcal{R}}, \tilde{\mathcal{I}})$ is an abstract cryptosystem with respect to Δ_q and it is correct and secure in the sense of $\tilde{\mathcal{S}}$ if, for any $\mathcal{G}_{\mathsf{sym}} \in \Delta_q$ and $\mathcal{O} \in \Sigma_{\mathcal{G}_{\mathsf{sym}}}$, $\tilde{\Pi}^{\mathcal{O}}_{\mathcal{G}_{\mathsf{sym}}} := (\tilde{\mathcal{F}}^{\mathcal{O}}, \tilde{\mathcal{C}}^{\mathcal{O}}, \tilde{\mathcal{R}}^{\mathcal{O}}, \tilde{\mathcal{I}}^{\mathcal{O}})$ is a cryptosystem that is correct and secure in the sense of $\tilde{\mathcal{S}}^{\mathcal{O}}$.*

5   Conversion Using Dependency Graph

5.1   Simulating Group Operation Oracle

Let $(q, \mathbb{G}, \tilde{\mathbb{G}}, \mathbb{G}_T, e, G, \tilde{G})$ be asymmetric bilinear groups generated by an asymmetric group generator $\mathcal{G}_{\mathsf{asym}}$. Let $\phi : \mathbb{G} \to \tilde{\mathbb{G}}$ be an (inefficient) isomorphism between the source groups. We use three representations, $(\mathbb{G}, -)$, $(-, \tilde{\mathbb{G}})$, and $(\mathbb{G}, \tilde{\mathbb{G}})$, for a source group element and say that they are of type left, right, and both, respectively. By type we denote {left, right, both}. We say that type t is *covered* by t' if $t' =$ both or $t' = t$, and denote by $t \subseteq t'$. If two types cover at least in one way, we say that they are *compatible*. We design operations so that they can be performed efficiently over two compatible elements. Let $\mathbb{G}'$ denote $(\mathbb{G} \cup \{\bot\}) \times (\tilde{\mathbb{G}} \cup \{\bot\}) \setminus (\bot, \bot)$ where $\bot$ represents absence of data. Let $\mathcal{J} : \mathbb{G}' \to \mathsf{type}$ be a function that takes an element of $\mathbb{G}'$ as input and outputs its type. Let $\mathsf{matchtype} : \mathbb{G}' \times \mathsf{type} \to \mathbb{G}'$ be a subroutine that takes $(X, \tilde{X}) \in \mathbb{G}'$ and $t \in \mathsf{type}$ as input, and remodeled $(X, \tilde{X})$ so that $\mathcal{J}(X, \tilde{X}) = t$ holds. It is done by computing $\tilde{X} := \phi(X)$ or $X := \phi^{-1}(\tilde{X})$ if necessary, and setting $\bot$ to X or $\tilde{X}$ if either is unnecessary.

Types are assigned by an algorithm $\mathcal{D}$ called a *deployment*. It is an algorithm that takes an object identifier as input and outputs a type to assign to the identifier. Based on the asymmetric bilinear groups and deployment $\mathcal{D}$, we construct an oracle $\mathcal{O}^*$ that simulates a symmetric group operation oracle.

[Simulated Group Operation Oracle $\mathcal{O}^*$]

- init: Initialize lists $\mathcal{L}_s$ and $\mathcal{L}_t$ with entries $(pt, G, \tilde{G})$ and (pt_t, G_t) respectively with fresh pointers pt and pt_t. Return (q, pt, pt_t).
- gop$(pt_1, a_1, \ldots, pt_k, a_k, oid)$: For every pt_i, search $\mathcal{L}_s$ for $(pt_i, X_i, \tilde{X}_i)$. For every i where $\mathcal{J}(X_i, \tilde{X}_i)$ does not cover $t := \mathcal{D}(oid)$, call $\mathsf{matchtype}(X_i, \tilde{X}_i, t)$. Then compute $X := \prod X_i^{a_i}$ for $t =$ left, or $\tilde{X} := \prod \tilde{X}_i^{a_i}$ for $t =$ right, or both for $t =$ both. Set $\bot$ to X or $\tilde{X}$ if either is not computed. Store $(pt, X, \tilde{X})$ with fresh pt, and return pt.

- pair(pt_1, pt_2): Search $\mathcal{L}_s$ for $(pt_1, X_1, \tilde{X}_1)$ and $(pt_2, X_2, \tilde{X}_2)$. If both $\mathcal{J}(X_1, \tilde{X}_1)$ and $\mathcal{J}(X_2, \tilde{X}_2)$ are left or right, then use ϕ or ϕ^{-1} to compute an element on the missing side. Then compute $Z := e(X_1, \tilde{X}_2)$ or $e(X_2, \tilde{X}_1)$ whichever possible. Store (pt, Z) to $\mathcal{L}_t$ with fresh pointer pt, and output pt.

- equal(pt_1, pt_2): Search $\mathcal{L}_s$ for $(pt_1, X_1, \tilde{X}_1)$ and $(pt_2, X_2, \tilde{X}_2)$. If $\mathcal{J}(X_1, \tilde{X}_1)$ and $\mathcal{J}(X_2, \tilde{X}_2)$ are incompatible, then compute either $\tilde{X}_2 := \phi(X_2)$ or $X_2 := \phi^{-1}(\tilde{X}_2)$ whichever missing. Then if $X_1 \equiv X_2 \in \mathbb{G}$ or $\tilde{X}_1 \equiv \tilde{X}_2 \in \tilde{\mathbb{G}}$, return 1. Return 0, otherwise.

- hash2g(str, oid): (This query is accepted only when random oracles $H_{\mathbb{G}} : \{0,1\}^* \to \mathbb{G}$ and $H_{\tilde{\mathbb{G}}} : \{0,1\}^* \to \tilde{\mathbb{G}}$ are available.) Compute $t \leftarrow \mathcal{D}(oid)$ and pick fresh pt. If (str, oid) has been queried before and answered with $(pt', X, \tilde{X}) \in \mathcal{L}_s$, store $(pt, X, \tilde{X})$ to $\mathcal{L}_s$ and return pt. Otherwise query (str, oid) to random oracle $H_{\mathbb{G}}$ for $t = $ right or $H_{\tilde{\mathbb{G}}}$ for $t = $ left. If $t = $ both, query (str, oid) to $H_{\mathbb{G}}$ and use ϕ to get the corresponding element in $\tilde{\mathbb{G}}$. Store the result with pt to $\mathcal{L}_s$. Then return pt.

- hash2str($pt_1, \ldots, pt_k, str, oid$): (This query is accepted only when random oracle $H_{str} : (\mathbb{G}')^k \times \{0,1\}^* \to \{0,1\}^{\text{poly}(\lambda)}$ is available.) Search $\mathcal{L}_s$ for each $(pt_i, X_i, \tilde{X}_i)$. Let $oid := (oid_i, \ldots, oid_k)$. For every i where $\mathcal{J}(X_i, \tilde{X}_i)$ does not cover $t_i := \mathcal{D}(oid_i)$, perform matchtype($X_i, \tilde{X}_i, t_i$). Then, if $(X_1, \tilde{X}_1, \ldots, X_k, \tilde{X}_k, str, oid)$ has been queried before, return the same value. Otherwise, send it to random oracle H_{str}, and receive a string, str'. Then return str'.

- send(pt, oid): Search $\mathcal{L}_s$ for $(pt, X, \tilde{X})$. Compute $t \leftarrow \mathcal{D}(oid)$. If $\mathcal{J}(X, \tilde{X}) \neq t$, call matchtype($X, \tilde{X}, t$). Output $((X, \tilde{X}), oid)$.

- receive(oid): On receiving this query from the host algorithm, wait for receiving $((X, \tilde{X}), oid')$ from outside. Ignore if $(X, \tilde{X}) \notin \mathbb{G}'$ or $oid' \neq oid$ or $\mathcal{J}(X, \tilde{X}) \neq \mathcal{D}(oid)$, and continue waiting. Otherwise, pick fresh pt, store $(pt, X, \tilde{X})$ to $\mathcal{L}_s$, and send pt to the host algorithm.

Observe that there are some cases where oracle $\mathcal{O}^*$ performs inefficient computations ϕ or ϕ^{-1}. Nevertheless, it is not hard to inspect that $\mathcal{O}^*$ perfectly simulates the extended symmetric group operation oracle.

For abstract cryptosystem $\tilde{\Pi}$, let $\tilde{\Pi}^{\mathcal{O}^*}_{\mathcal{G}_{\text{asym}}, \mathcal{D}}$ denote a cryptosystem where oracle $\mathcal{O}^*$ based on asymmetric groups generated by $\mathcal{G}_{\text{asym}}$ and deployment $\mathcal{D}$. Let Δ'_q denote the set of $\mathcal{G}_{\text{asym}}$ that outputs q with the same distribution as those in Δ_q. We claim that if $\tilde{\Pi}^{\mathcal{O}}_{\mathcal{G}_{\text{sym}}}$ is correct and secure, then so is $\tilde{\Pi}^{\mathcal{O}^*}_{\mathcal{G}_{\text{asym}}, \mathcal{D}}$ for any $\mathcal{G}_{\text{sym}} \in \Delta_q$, $\mathcal{G}_{\text{asym}} \in \Delta'_q$, and $\mathcal{D}$ that stops with an output for any input. Nevertheless, the notion of secure Π requires algorithms to run efficiently. For a while, we assume that ϕ and ϕ^{-1} can be computed efficiently and state the following.

Lemma 8. *If $\tilde{\Pi}$ is a cryptosystem with respect to Δ_q, and it is correct and secure in the sense of $\tilde{\mathcal{S}}$, then for any $\mathcal{G}_{\text{asym}} \in \Delta'_q$ and for any $\mathcal{D}$, cryptosystem $\tilde{\Pi}^{\mathcal{O}^*}_{\mathcal{G}_{\text{asym}}, \mathcal{D}}$ is correct and secure in the sense of $\tilde{\mathcal{S}}^{\mathcal{O}^*}$ if $\mathcal{O}^*$ computes ϕ and ϕ^{-1} efficiently.*

Proof. (sketch) Correctness can be assured by observing that views of abstract algorithms with $\mathcal{O}^*$ and $\mathcal{O}$ are identical. Regarding security, we show that if there exists $\mathcal{G}_{\sf sym} \in \Delta_q$ and $\mathcal{A}$ that successfully attacks $\tilde{\Pi}^{\mathcal{O}^*}_{\mathcal{G}_{\sf asym}, \mathcal{D}}$ in the sense of $\tilde{\mathcal{S}}^{\mathcal{O}^*}$, then there exists adversary $\mathcal{B}$ that is successful in attacking $\tilde{\Pi}^{\mathcal{O}}_{\mathcal{G}_{\sf sym}}$ in the sense of $\tilde{\mathcal{S}}^{\mathcal{O}}$ for $\mathcal{G}_{\sf sym}$. We construct $\mathcal{G}_{\sf sym}$ using $\mathcal{G}_{\sf asym}$ by representing a group element with a pair of source group elements of asymmetric groups. We then construct $\mathcal{B}$ by using $\mathcal{A}$. On receiving a group element, $\mathcal{B}$ invokes $\mathcal{D}$ and run matchtype to remodel the input for $\mathcal{A}$. Outgoing group elements from $\mathcal{A}$ are also remodeled by applying matchtype. Non-group elements are sent and received intact. We argue that if $\mathrm{Adv}^{\tilde{\mathcal{S}}^{\mathcal{O}^*}}_{\mathcal{A}}(\lambda)$ is not negligible then so is $\mathrm{Adv}^{\tilde{\mathcal{S}}^{\mathcal{O}}}_{\mathcal{B}}(\lambda)$ since the view of $\tilde{\mathcal{S}}$ is identical. Since $\mathcal{O}^*$ computes ϕ and ϕ^{-1} in matchtype efficiently by hypothesis, all algorithms are efficient here. □

In reality, however, ϕ and ϕ^{-1} are inefficient in Type-III groups and hence $\mathcal{O}^*$ is inefficient in general. Nevertheless, there may exist $\tilde{\Pi}$ and $\mathcal{D}$ where $\tilde{\Pi}^{\mathcal{O}^*}_{\mathcal{G}_{\sf asym}, \mathcal{D}}$ never performs the inefficient computation. For such $\tilde{\Pi}$ and $\mathcal{D}$, $\tilde{\Pi}^{\mathcal{O}^*}_{\mathcal{G}_{\sf asym}, \mathcal{D}}$ is correct and secure. Accordingly, the task of conversion is now reduced to find efficient $\mathcal{D}$ that never have $\mathcal{O}^*$ compute either ϕ or ϕ^{-1}. It is the main issue we address in the rest of this section.

We proceed to argue whether the assumption deduced from the converted problem is plausible or not. Consider group operation oracle $\mathcal{O}$ based on $\mathcal{G}_{\sf sym}$. Let $\mathcal{I}_{\sf sym} = (\tilde{\mathcal{I}}^{\mathcal{O}}_{gen}, \tilde{\mathcal{I}}^{\mathcal{O}}_{ver}, \tilde{\mathcal{I}}^{\mathcal{O}}_{guess})$ be a problem defined over symmetric bilinear groups. Similarly, let $\mathcal{O}^*$ be a group operation oracle based on $\mathcal{G}_{\sf asym}$ and $\mathcal{D}$. Let $\mathcal{I}_{\sf asym} = (\tilde{\mathcal{I}}^{\mathcal{O}^*}_{gen}, \tilde{\mathcal{I}}^{\mathcal{O}^*}_{ver}, \tilde{\mathcal{I}}^{\mathcal{O}^*}_{guess})$ be a problem defined over asymmetric bilinear groups. Let $\{\mathcal{D}\}$ be a set of $\mathcal{D}$ that makes $\mathcal{I}_{\sf asym}$ efficient. By $\{\mathcal{I}_{\sf asym}\}$ we denote the family of problems obtained by defining $\mathcal{I}_{\sf asym}$ for each $\mathcal{D} \in \{\mathcal{D}\}$. We call $\{\mathcal{I}_{\sf asym}\}$ a family of co*-problems.

Some restrictions apply to $\mathcal{I}_{\sf sym}$. We only consider $\mathcal{I}_{\sf sym}$ that verifies membership of all incoming group elements, and do not use hash functions hash2g and hash2str. Furthermore, the default generator is included in every problem instance. Then the following holds.

Theorem 9 (Generic Hardness of co*-Problem Family). *If there exists* $(q_s, \tilde{q}_s, q_t, q_p, \epsilon, \tau)$-*successful generic adversary for* $\mathcal{I}_{\sf asym} \in \{\mathcal{I}_{\sf asym}\}$, *then there exists an* $(q_s + \tilde{q}_s, q_t, q_p, \epsilon, \tau')$-*successful generic adversary against* $\mathcal{I}_{\sf sym}$ *where* τ' *is* $\tau + O(q_s + \tilde{q}_s)$.

5.2 Dependency Graphs

We begin by defining a dependency graph for an abstract algorithm. Let $\mathcal{O}$ be a group operation oracle for some group order q. Let $\tilde{\mathcal{A}}$ be an abstract algorithm and $\tilde{\mathcal{A}}^{\mathcal{O}}(in)$ be its execution on input in. (When $\tilde{\mathcal{A}}$ is interactive, in also represents inputs obtained through interaction.) We define a dependency graph for $\tilde{\mathcal{A}}$ through the interaction between $\tilde{\mathcal{A}}$ and $\mathcal{O}$ as follows. It also defines a list NoDup called a ban list that includes oids for group elements generated by hash2g.

[Dependency Graph for Abstract Algorithm $\tilde{A}$]

1. Initialize Γ to an empty directed graph. Also initialize lists L_{pair}, L_{eq}, and NoDup to be empty.
2. Pick $\mathcal{O}$ for order q, and select in from appropriate domain. Initialize L_{pt} to empty. Run $\tilde{A}$. For each query from $\tilde{A}$ to $\mathcal{O}$, do as follows.
   - gop($pt_1, a_1, \ldots, pt_k, a_k, oid$) $\rightarrow$ pt: Add a node labeled by oid to Γ and record (pt, oid) to L_{pt}. Then, for every node oid_i that corresponds to pt_i, add an edge (oid_i, oid) to Γ.
   - pair(pt_1, pt_2): Find (pt_1, oid_1) and (pt_2, oid_2) from L_{pt}. If (oid_1, oid_2) $\in$ L_{pair} in any order, do nothing. Otherwise, pick two nodes with unique labels, say $p[0]$ and $p[1]$, and add them to Γ. Add edges ($oid_1, p[0]$) and ($oid_2, p[1]$) to Γ as well. Add (oid_1, oid_2) to L_{pair}.
   - equal(pt_1, pt_2): Let oid_1 and oid_2 be nodes for pt_1 and pt_2, respectively. If $oid_1 = oid_2$, or (oid_1, oid_2) $\in L_{eq}$ in any order, do nothing. Otherwise, add a node with a unique label, say E, and edges (oid_1, E) and (oid_2, E) to Γ. Add (oid_1, oid_2) to L_{eq}.
   - hash2g(str, oid) $\rightarrow$ pt: Add a node oid to Γ and add (pt, oid) to L_{pt}. Store oid to NoDup if not yet stored.
   - hash2str($pt_1, \ldots, pt_k, str, oid$): Let $oid = (oid_i, \ldots, oid_k)$. For every oid_i' stored with pt_i in L_{pt}, if $oid_i \neq oid_i'$, then add node oid_i and edge (oid_i', oid_i) to Γ. (This means that the element identified by pt_i was originally associated to object identifier oid_i' but now regarded as oid_i by the host algorithm.)
   - send(pt, oid): For oid' stored with pt in L_{pt}, if $oid \neq oid'$, then add node oid' and edge (oid', oid).
   - receive(oid) $\rightarrow$ pt: Add node oid to Γ. Then record (pt, oid) to L_{pt}.

   In the above, adding nodes and edges are done if they do not exist in Γ. Also skip adding self-directing edges.
3. Go back to step 2 and repeat the above for all q and in.

The above algorithm defines how to construct a dependency graph for $\tilde{A}$. In fact, repeating for all q and in as instructed in the last step is infeasible in reality. We nevertheless expect that Γ is finite size for certain $\tilde{A}$. It is particularity the case when $\tilde{A}$ behaves independently of the security parameter, or nodes related to the security parameter are indexed and given the same object identifier as we see for public key u_i in the example in Section 3. In the real world, building a dependency graph for algorithm $\tilde{A}$ needs to look into $\tilde{A}$ rather than treating $\tilde{A}$ as black-box as above.

The nodes added in pair are called *pairing nodes*. A node is called a *regular node* if it is not either of the above.

Next we define a dependency graph for a cryptosystem, $\tilde{\Pi} = (\tilde{\mathcal{F}}, \tilde{\mathcal{C}}, \tilde{\mathcal{R}}, \tilde{\mathcal{I}})$. Basically it is a graph obtained by merging dependency graphs for all algorithms in $\tilde{\Pi}$. Yet we need to work on some details for formality as shown below.

[Dependency Graph for Cryptosystem $\tilde{\Pi}$]

1. Build a dependency graph and list NoDup for every algorithm in $\tilde{\Pi}$. It is assumed that pairing nodes, equality nodes, and local nodes are given globally unique names.
2. Merge all the graphs and NoDup obtained in the previous step.
3. If two nodes are connected to more than one pair of pairing nodes, remove all but one pair of the pairing nodes. Do the same for equality nodes.
4. Output the resulting graph and NoDup.

Let $\mathcal{C}_{const}$ be a class of abstract cryptosystems that has a constant-size dependency graph in the security parameter. In the rest of the paper, we focus on cryptosystems in $\mathcal{C}_{const}$.

5.3   Deployment Algorithm

Given a dependency graph Γ for $\tilde{\Pi} \in \mathcal{C}_{const}$, we construct a deployment $\mathcal{D}$. Recall that if the target algorithm performs hash2g(str, oid), then the result should be a group element in either $\mathbb{G}$ or $\tilde{\mathbb{G}}$ but not both simultaneously. Thus $\mathcal{D}$ must not return both for such oid. To deal with similar demand from practice that some nodes should stay in either group, we use a ban list, NoDup, which specifies oids that must not be assigned to both.

We consider splitting a dependency graph into two graphs so that each graph represents nodes and computations in $\mathbb{G}$ or $\tilde{\mathbb{G}}$. The split must meet the conditions defined below.

Definition 10 (Valid Split). *Let $\Gamma = (V, E)$ be a dependency graph for $\tilde{\Pi} \in \mathcal{C}_{const}$. Let $P = (p_1[0], \dots, p_{n_p}[1]) \subset V$ be pairing nodes. A pair of graphs $\Gamma_0 = (V_0, E_0)$ and $\Gamma_1 = (V_1, E_1)$ is a valid split of Γ with respect to NoDup $\subseteq V$ if:*

1. *merging Γ_0 and Γ_1 recovers Γ,*
2. *for each $i \in \{0, 1\}$ and every $X \in V_i \setminus P$ the subgraph Anc(Γ, X) is in Γ_i,*
3. *for each $i \in \{1, \dots, n_p\}$ paring nodes $p_i[0]$ and $p_i[1]$ are separately included in V_0 and V_1.*
4. *No node in $V_0 \cap V_1$ is included in NoDup.*

We then construct a deployment $\mathcal{D}$ based on a valid split as follows:

[Deployment Algorithm $\mathcal{D}$: oid → type]
    Given object identifier oid as input, return left or right or both if a node labeled as oid is included in Γ_0 or Γ_1 or both, respectively.

Lemma 11. *If there exists a valid split with respect to $\tilde{\Pi} \in \mathcal{C}_{const}$ and NoDup, then, for oracle $\mathcal{O}^*$ based on $\mathcal{G}_{asym}$ and $\mathcal{D}$ with a valid split, $\tilde{\Pi}_{\mathcal{G}_{asym}, \mathcal{D}}^{\mathcal{O}^*}$ is efficient.*

The above lemma can be proved by observing that for each case that $\tilde{\Pi}_{\mathcal{G}_{asym}, \mathcal{D}}^{\mathcal{O}^*}$ computes inefficient isomorphisms in gop, pair, equal, hash2g, and hash2str, the split $\mathcal{D}$ is based on must be invalid.

5.4 How to Find the Best Valid Split

Let $\Gamma = (V, E)$ be a dependency graph for $\tilde{\Pi}$ where V consists of regular nodes $\{X_1, \ldots, X_k\}$ and pairing nodes $P := \{p_1[0], \ldots, p_{n_p}[1]\}$. We construct an algorithm FindSplit that finds all valid splits.

[**Algorithm:** FindSplit(Γ, NoDup)]

1. Initialize L to empty.
2. Set $B \subseteq R$ so that every $B_i \in B$ has no outgoing edges. For each $X_i \in R$ do:
   - if X_i is not in $\mathsf{Anc}(\Gamma, B_i)$ for all $B_i \in B \cup P$, then:
     • For every $X_j \in B$ that is in $\mathsf{Anc}(\Gamma, X_i)$, do $B := B \setminus \{X_j\}$.
     • $B := B \cup \{X_i\}$.
3. Repeat the following for $\ell = 0, \ldots, 2^{n_p + n_b} - 1$ where $n_b := |B|$.
   (a) Set $\Gamma_0 = (V_0, E_0), \Gamma_1 = (V_1, E_1)$ be empty graphs.
   (b) For $i = 1, \ldots, n_p$, do $\Gamma_0 \leftarrow \Gamma_0 \oplus \mathsf{Anc}(\Gamma, p_i[\mathsf{bit}_i(\ell)])$ and $\Gamma_1 \leftarrow \Gamma_1 \oplus \mathsf{Anc}(\Gamma, p_i[1 - \mathsf{bit}_i(\ell)])$.
   (c) For $j = 1, \ldots, n_b$ and $i = \mathsf{bit}_{n_p + j}(\ell)$, do $\Gamma_i \leftarrow \Gamma_i \oplus \mathsf{Anc}(\Gamma, B_j)$.
   (d) Append (Γ_0, Γ_1) to L if $V_0 \cap V_1 \cap \mathsf{NoDup}$.
4. Output L.

Lemma 12. *List L includes all valid split of Γ.*

Proof. (sketch) We verify that every (Γ_0, Γ_1) in L satisfies the conditions in Definition 10. First we show that $\Gamma_0 \oplus \Gamma_1 = \Gamma$. Observe that FindSplit is deterministic and only the order of elements in R may impact the result through construction of B. Consider B obtained from R, and B' from permutation of R. Suppose that $X \in B$ and $X \notin B'$ happens. Then there exits $Y \in B'$ that has path from X to Y. We can argue that such Y cannot exist in B without contradicting to the presence of X in B. Similarly, as Y is not in B, there exists node, say Z, in B that has path from Y to Z. If Z is not identical to X, there exists path from X to Z that contradicts to the presence of X in B. Thus, we have $X = Z$. This means that X and Y are on a circle and thus $\mathsf{Anc}(\Gamma, X) = \mathsf{Anc}(\Gamma, Y)$. Thus procedures in further steps are not affected whichever B or B' are used. Then observe that, from step (b) and (c), we have $\Gamma_0 \oplus \Gamma_1 = \mathsf{Anc}(\Gamma, p_1[1]) \oplus \mathsf{Anc}(\Gamma, p_1[0]) \oplus \cdots \oplus \mathsf{Anc}(\Gamma, p_{n_p}[0]) \oplus \mathsf{Anc}(\Gamma, B_1) \oplus \cdots \oplus \mathsf{Anc}(\Gamma, B_{n_b}) = \Gamma$. Next, the second condition is met since, in step 3-(b) and (c), every node is included in Γ_i together with their ancestor subgraphs. By the property of $\mathsf{Anc}()$, the subgraph contains a subgraph of every node in it. The third condition is assured since in step 3-(b) every pair of pairing nodes are merged to Γ_0 and Γ_1 separately. Finally, the constraint by NoDup is met due to step 3-(d) which forces the exactly the same constraint as in the fourth condition. ∎

References

1. Abe, M., David, B., Kohlweiss, M., Nishimaki, R., Ohkubo, M.: Tagged one-time signatures: Tight security and optimal tag size. In: Kurosawa, K., Hanaoka, G. (eds.) PKC 2013. LNCS, vol. 7778, pp. 312–331. Springer, Heidelberg (2013)
2. Akinyele, J.A., Green, M., Hohenberger, S.: Using SMT solvers to automate design tasks for encryption and signature schemes. In: ACM CCS 2013, pp. 399–410 (2013)

3. Barbulescu, R., Gaudry, P., Joux, A., Thomé, E.: A quasi-polynomial algorithm for discrete logarithm in finite fields of small characteristic. IACR ePrint Archive, 2013/400 (2013)
4. Bellare, M., Rogaway, P.: Random oracles are practical: a paradigm for designing efficient protocols. In: ACM CCS 1993, pp. 62–73 (1993)
5. Boneh, D., Franklin, M.: Identity-based encryption from the weil pairing. In: Kilian, J. (ed.) CRYPTO 2001. LNCS, vol. 2139, pp. 213–229. Springer, Heidelberg (2001)
6. Boneh, D., Lynn, B., Shacham, H.: Short signatures from the weil pairing. In: Boyd, C. (ed.) ASIACRYPT 2001. LNCS, vol. 2248, pp. 514–532. Springer, Heidelberg (2001)
7. Boneh, D., Shacham, H.: Group signatures with verifier-local revocation. In: ACM CCS 2004, pp. 168–177 (2004)
8. Chatterjee, S., Hankerson, D., Knapp, E., Menezes, A.: Comparing two pairing-based aggregate signature schemes. DCC 2010 55(2-3), 141–167 (2010)
9. Chatterjee, S., Menezes, A.: On cryptographic protocols employing asymmetric pairings - the role of psi revisited. IACR ePrint Archive, 2009/480 (2009)
10. Chatterjee, S., Menezes, A.: On cryptographic protocols employing asymmetric pairings - the role of revisited. Discrete Applied Math. 159(13), 1311–1322 (2011)
11. Chen, J., Lim, H.W., Ling, S., Wang, H., Wee, H.: Shorter IBE and signatures via asymmetric pairings. In: Abdalla, M., Lange, T. (eds.) Pairing 2012. LNCS, vol. 7708, pp. 122–140. Springer, Heidelberg (2013)
12. De Moura, L., Bjørner, N.: Z3: An efficient SMT solver. In: Ramakrishnan, C.R., Rehof, J. (eds.) TACAS 2008. LNCS, vol. 4963, pp. 337–340. Springer, Heidelberg (2008)
13. Galbraith, S.D., Paterson, K.G., Smart, N.P.: Pairings for cryptographers. Discrete Applied Mathematics 156(16), 3113–3121 (2008)
14. Göloglu, F., Granger, R., McGuire, G., Zumbrägel, J.: On the function field sieve and the impact of higher splitting probabilities: Application to discrete logarithms in $f_{2^{1971}}$. IACR ePrint Archive, 2013/074 (2013)
15. Joux, A.: Faster index calculus for the medium prime case application to 1175-bit and 1425-bit finite fields. In: Johansson, T., Nguyen, P.Q. (eds.) EUROCRYPT 2013. LNCS, vol. 7881, pp. 177–193. Springer, Heidelberg (2013)
16. Joux, A.: A new index calculus algorithm with complexity $1(1/4+o(1))$ in very small characteristic. IACR ePrint Archive, 2013/095 (2013)
17. Maurer, U.M.: Abstract models of computation in cryptography. In: Smart, N.P. (ed.) Cryptography and Coding 2005. LNCS, vol. 3796, pp. 1–12. Springer, Heidelberg (2005)
18. Menezes, A.: Asymmetric pairings. Invited Talk in ECC 2009 (2009), http://math.ucalgary.ca/sites/ ecc.math.ucalgary.ca/files/u5/Menezes_ECC2009.pdf
19. Ramanna, S.C., Chatterjee, S., Sarkar, P.: Variants of waters' dual-system primitives using asymmetric pairings. IACR ePrint Archive, 2012/024 (2012)
20. Ramanna, S.C., Chatterjee, S., Sarkar, P.: Variants of waters' dual system primitives using asymmetric pairings. In: Fischlin, M., Buchmann, J., Manulis, M. (eds.) PKC 2012. LNCS, vol. 7293, pp. 298–315. Springer, Heidelberg (2012)
21. Reingold, O., Trevisan, L., Vadhan, S.P.: Notions of reducibility between cryptographic primitives. In: Naor, M. (ed.) TCC 2004. LNCS, vol. 2951, pp. 1–20. Springer, Heidelberg (2004)
22. Shoup, V.: Lower bounds for discrete logarithms and related problems. In: Fumy, W. (ed.) EUROCRYPT 1997. LNCS, vol. 1233, pp. 256–266. Springer, Heidelberg (1997)
23. Smart, N.P., Vercauteren, F.: On computable isomorphisms in efficient asymmetric pairing-based systems. Discrete Applied Mathematics 155(4), 538–547 (2007)

24. Waters, B.: Efficient identity-based encryption without random oracles. In: Cramer, R. (ed.) EUROCRYPT 2005. LNCS, vol. 3494, pp. 114–127. Springer, Heidelberg (2005)
25. Waters, B.: Dual system encryption: Realizing fully secure IBE and HIBE under simple assumptions. In: Halevi, S. (ed.) CRYPTO 2009. LNCS, vol. 5677, pp. 619–636. Springer, Heidelberg (2009)

Appendix

A Dependency Graphs for Waters' IBE

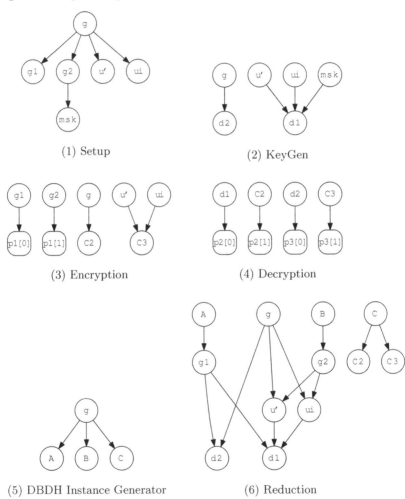

Fig. 2. Dependency graph for each algorithm in Waters' IBE scheme

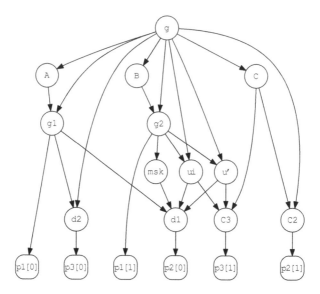

Fig. 3. Dependency graph for Waters' IBE scheme obtained by merging all graphs for individual algorithms

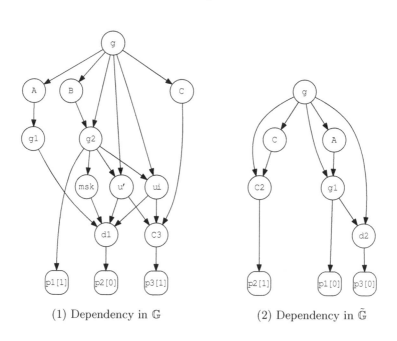

(1) Dependency in $\mathbb{G}$ (2) Dependency in $\tilde{\mathbb{G}}$

Fig. 4. A valid split for minimum public key. Nodes in graph (1) and (2) represent group elements in $\mathbb{G}$ and $\tilde{\mathbb{G}}$, respectively

Polynomial Spaces: A New Framework for Composite-to-Prime-Order Transformations

Gottfried Herold[1], Julia Hesse[2], Dennis Hofheinz[2],
Carla Ràfols[1], and Andy Rupp[2]

[1] Horst Görtz Institute for IT Security and Faculty of Mathematics,
Ruhr University Bochum, Germany
{gottfried.herold,carla.rafols}@rub.de
[2] Karlsruhe Institute of Technology, Germany
{julia.hesse,dennis.hofheinz,andy.rupp}@kit.edu

Abstract. At Eurocrypt 2010, Freeman presented a framework to convert cryptosystems based on composite-order groups into ones that use prime-order groups. Such a transformation is interesting not only from a conceptual point of view, but also since for relevant parameters, operations in prime-order groups are faster than composite-order operations by an order of magnitude. Since Freeman's work, several other works have shown improvements, but also lower bounds on the efficiency of such conversions.

In this work, we present a new framework for composite-to-prime-order conversions. Our framework is in the spirit of Freeman's work; however, we develop a different, "polynomial" view of his approach, and revisit several of his design decisions. This eventually leads to significant efficiency improvements, and enables us to circumvent previous lower bounds. Specifically, we show how to verify Groth-Sahai proofs in a prime-order environment (with a symmetric pairing) almost twice as efficiently as the state of the art.

We also show that our new conversions are optimal in a very broad sense. Besides, our conversions also apply in settings with a multilinear map, and can be instantiated from a variety of computational assumptions (including, e.g., the k-linear assumption).

Keywords: bilinear maps, composite-order groups, Groth-Sahai proofs.

1 Introduction

Motivation. Cyclic groups are a very popular platform for cryptographic constructions. Starting with Diffie and Hellman's seminal work [4], there are countless examples of cryptographic schemes that work in any finite, cyclic group G, and whose security can be reduced to a well-defined computational problem in G. In many cases, the order of the group G should be prime (or is even irrelevant). However, some constructions (e.g., [2, 10, 17, 13]) explicitly require a group G of *composite* order.

J.A. Garay and R. Gennaro (Eds.): CRYPTO 2014, Part I, LNCS 8616, pp. 261–279, 2014.

In particular in combination with a pairing (i.e., a bilinear map) e, groups of composite order exhibit several interesting properties. (For instance, $e(g_1, g_2) = 1$ for elements g_1, g_2 of coprime order. Or, somewhat more generally, the pairing operation operates on the different prime-order components of G independently.) This enables interesting technical applications (e.g., [17, 13]), but also comes at a price. Namely, to accommodate suitably hard computational problems, composite-order groups have to be chosen substantially larger than prime-order groups. Specifically, it should be hard to factor the group order. This leads to significantly slower operations in composite-order groups: [6] suggests that for realistic parameters, Tate pairings in composite-order groups are by a factor of about 50 less efficient than in prime-order groups.

Freeman's Composite-Order-to-Prime-Order Transformation. It is thus interesting to try to find substitutes for the technical features offered by composite-order groups in prime-order settings. In fact, Freeman [6] has offered a framework and tools to semi-generically convert cryptographic constructions from a composite-order to a prime-order setting. Similar transformations have also been implicit in previous works [8, 17]. The premise of Freeman's approach is that composite-order group elements "behave as" vectors over a prime field. In this interpretation, subgroups correspond to linear subspaces.

Moreover, we can think of the vector components as exponents of prime-order group elements; we can then associate, e.g., a composite-order subgroup indistinguishability problem with the problem of distinguishing vectors (chosen either from a subspace or the whole space) "in the exponent." More specifically, Freeman showed that the composite-order subgroup indistinguishability assumption can be implemented in a prime-order group with the Decisional Diffie-Hellman (or with the k-linear) assumption. A pairing operation over the composite-order group then translates into a suitable "multiplication of vectors," which can mean different things, depending on the desired properties. For instance, Freeman considers both an inner product and a Kronecker product as "vector multiplication" operations (of course with different effects).

Limitations of Freeman's Approach. Freeman's work has spawned a number of follow-up results that investigate more general or more efficient conversions of this type [13, 15, 14, 11, 12]. We note that all of these works follow Freeman's interpretation of vectors, and even his possible interpretations of a vector multiplication. Unfortunately, during these investigations, certain lower bounds for the efficiency of these transformations became apparent. For example, Seo [14] proves lower bounds both for the computational cost and the dimension of the resulting vector space of *arbitrary* transformations in Freeman's framework. More specifically, Seo reports a concrete bound on the number of required prime-order pairing operations necessary to simulate a composite-order pairing.

However, of course, these lower bounds crucially use the vector-space interpretation of Freeman's framework. Specifically, it is conceivable that a (perhaps completely different) more efficient composite-order-to-prime-order transformation exists outside of Freeman's framework. Such a more efficient transformation could also provide a way to implement, e.g., the widely used Groth-Sahai proof system [8] more efficiently.

Our Contribution: A Different View on Composite-Order-to-Prime-Order Conversions. In this work, we take a step back and question several assumptions that are implicitly made in Freeman's framework. We exhibit a different composite-order-to-prime-order conversion outside of his model, and show that it circumvents previous lower bounds. In particular, our construction leads to more efficient verification of Groth-Sahai proofs in the symmetric setting (i.e., with a symmetric pairing). Moreover, our construction can be implemented from *any* matrix assumption [5] (including the k-linear assumption) and scales better to multilinear settings than previous approaches. In the following, we give more details on our construction and its properties.

A Technical Perspective: A Polynomial Interpretation of Linear Subspaces. To explain our approach, recall that Freeman identifies a composite-order group with a vector space over a prime field. Moreover, in his work, subgroups of the composite-order group always correspond to *uniformly chosen* subspaces of a certain dimension. Of course, such "unstructured" subspaces only allow for rather generic interpretations of composite-order pairings (as generic "vector multiplications" as above).

Instead, we interpret the composite-order group as a very structured vector space. More concretely, we interpret a composite-order group element as (the coefficient vector of) a polynomial $f(X)$ over a prime field. In this view, a composite-order subgroup corresponds to the set of all polynomials with a common zero s (for a fixed and hidden s). Composite-order group operation and pairing correspond to polynomial addition and multiplication. Moreover, the hidden common zero s can be used as a trapdoor to decide subgroup membership, and thus to implement a "projection" in the sense of Freeman.

Specifically, our "vector multiplication" is very structured and natural, and there are several ways to implement it efficiently. For instance, we can apply a convolution on the coefficient vectors, or, more efficiently, we can represent f as a vector of evaluations $f(i)$ at sufficiently many fixed values i, and multiply these evaluation vectors component-wise. In particular, we circumvent the mentioned lower bound of Seo [14] by our different interpretation of composite-order group elements as vectors.

Another interesting property of our construction is that it scales better to the multilinear setting than previous approaches. For instance, while it seems possible to generalize at least Freeman's construction of a "projecting pairing" to a setting with a k-linear map (instead of a pairing), the corresponding generic vector multiplication would lead to exponentially (in k) large vectors in the target group. In our case, a k-linear map corresponds to the multiplication of k polynomials, and only requires a quadratic number of group elements in the target group.[1]

---

[1] We multiply k polynomials, and each polynomial should be of degree at least k, in order to allow for suitable subgroup indistinguishability problems that are plausible even in face of a k-linear map.

In the description above, f is always a univariate polynomial. With this interpretation, we can show that the SCasc assumption from Escala et al. [5] implies subgroup indistinguishability. However, we also provide a "multivariate" variant of our approach (with polynomials f in several variables) that can be implemented with *any* matrix assumption (such as the k-linear and even weaker assumptions). Furthermore, in the terminology of Freeman, we provide both a "projecting" and a "projecting and canceling" pairing construction (although the security of the "projecting and canceling" construction requires additional complexity assumptions).

Applications. The performance improvements of our approach are perhaps best demonstrated by the case of Groth-Sahai proofs. Compared to the most efficient previous implementations of Groth-Sahai proofs in prime-order groups with symmetric pairing [15, 5], we almost halve the number of required prime-order pairing operations (cf. Table 1). As a bonus, we also improve on the number of prime-order group elements in the target group, while retaining the small common reference string from [5]. Additionally, in the full version [9] of our paper, we show how to implement a variant of the Boneh-Goh-Nissim encryption scheme [2] in prime-order groups with a k-linear map. As already sketched, this is possible with Freeman's approach only for logarithmically small k.

Structural Results. Of course, a natural question is whether *our* results are optimal, and if so, in what sense exactly. We can settle this question in the following sense: we show that the construction sketched above is optimal in our generalized framework. We also prove a similar result for our construction from general matrix assumptions.

Open Problems. In this work, we focus on settings with a *symmetric* pairing (resp. multilinear map). It is an interesting open problem to extend our approach to asymmetric settings. Furthermore, the conversion that leads to a canceling *and* projecting map (in the terminology of Freeman) requires a nonstandard complexity assumption (that however holds generically, as we prove). It would be interesting to find constructions from more standard assumptions.

Outline. After recalling some preliminaries in Section 2, we describe our framework in Section 3. Our conversions follow in Section 4. We discuss the optimality of our conversions in Section 5, and compare them to previous conversions in Section 6. Finally, we discuss in Section 7 how our results imply more efficient Groth-Sahai proofs. We refer to the full version [9] for more detailed explanations and proofs.

2   Preliminaries

Notation. Throughout the paper we will use additive notation for all groups G. Nevertheless, we still talk about an exponentiation with exponent a considering a scalar multiplication $a\mathcal{P}$ for $\mathcal{P} \in G$ and $a \in \mathbb{Z}_{|G|}$. Let G be a cyclic group of order p generated by $\mathcal{P}$. Then by $[a] := a\mathcal{P}$ we denote the *implicit representation* of $a \in \mathbb{Z}_p$ in G. To distinguish between implicit representations in the domain

G and the target group G_T of a multilinear map we use $[\cdot]$ and $[\cdot]_T$, respectively. More generally, we also define such representations for vectors $\vec{f} \in \mathbb{Z}_p^n$ by $[\vec{f}] := ([f_i])_i \in G^n$, for matrices $\mathbf{A} = (a_{i,j})_{i,j} \in \mathbb{Z}_p^{n \times m}$ by $[\mathbf{A}] := ([a_{i,j}])_{i,j} \in G^{n \times m}$, and for sets $H \subset \mathbb{Z}_p^n$ by $[H] := \{[a] \mid a \in H\} \subset G^n$. We will often identify $\vec{f} \in \mathbb{Z}_p^n$ with the coefficients of a polynomial f in some space V with respect to a (fixed) basis $\mathfrak{q}_0, \ldots, \mathfrak{q}_{n-1}$ of V, i.e., $f = \sum_{i=0}^{n-1} f_i \mathfrak{q}_i$ (e.g., $V = \{f \mid f \in \mathbb{Z}_p[X], \deg(f) < n\}$ and $\mathfrak{q}_i = X^i$). In this case we may also write $[f] := [\vec{f}]$.

Symmetric Prime-Order k-linear Group Generators. We use the following formal definition of a k-linear prime-order group generator as the foundation for our constructions. In the scope of these constructions, we will refer to the output of such a generator as a *basic* (or, *prime-order*) k-linear map.

Definition 1 (symmetric Prime-Order k-Linear Group Generator). *A symmetric prime-order k-linear group generator is a PPT algorithm $\mathcal{G}_k$ that on input of a security parameter 1^λ outputs a tuple of the form*

$$\mathcal{MG}_k := (k, G, G_T, e, p, \mathcal{P}, \mathcal{P}_T) \leftarrow \mathcal{G}_k(1^\lambda)$$

where G, G_T are descriptions of cyclic groups of prime order p, $\log p = \Theta(\lambda)$, $\mathcal{P}$ is a generator of G, and $e: G \times \ldots \times G \to G_T$ is a map which satisfies the following properties:
- *k-linearity: For all $Q_1, \ldots, Q_k \in G$, $\alpha \in \mathbb{Z}_p$, and $i \in \{1, \ldots, k\}$ we have $e(Q_1, \ldots, \alpha Q_i, \ldots, Q_k) = \alpha e(Q_1, \ldots, Q_k)$.*
- *Non-Degeneracy: $\mathcal{P}_T = e(\mathcal{P}, \ldots, \mathcal{P})$ generates G_T.*

In our paper, one should think of $\mathcal{G}_k$ as either a generator of a bilinear group setting (for $k = 2$) defined over some group of points of an elliptic curve and the multiplicative group of a finite field or, for $k > 2$, as generator of an abstract ideal multilinear map, approximated by the recent candidate constructions [7, 3].

Matrix Assumptions. Our constructions are based on matrix assumptions as introduced in [5].

Definition 2 (Matrix Distributions and Assumptions [5]). *Let $n, \ell \in \mathbb{N}$, $n > \ell$. We call $\mathcal{D}_{n,\ell}$ a matrix distribution if it outputs (in probabilistic polynomial time, with overwhelming probability) matrices $\mathbf{A} \in \mathbb{Z}_p^{n \times \ell}$ of full rank ℓ. $\mathcal{D}_{n,\ell}$ is called polynomially induced if it is defined by picking $\vec{s} \in \mathbb{Z}_p^d$ uniformly at random and setting $a_{i,j} := \mathfrak{p}_{i,j}(\vec{s})$ for some polynomials $\mathfrak{p}_{i,j} \in \mathbb{Z}_p[X_1, \ldots, X_d]$ whose degrees do not depend on the security parameter. We define $\mathcal{D}_\ell := \mathcal{D}_{\ell+1,\ell}$. Furthermore, we say that the $\mathcal{D}_{n,\ell}$-Matrix Diffie-Hellman assumption or just $\mathcal{D}_{n,\ell}$ assumption for short holds relative to the k-linear group generator $\mathcal{G}_k$ if for all PPT adversaries D we have $\mathbf{Adv}_{\mathcal{D}_{n,\ell},\mathcal{G}_k}(\mathsf{D}) = \mathbf{Pr}[\mathsf{D}(\mathcal{MG}_k, [\mathbf{A}], [\mathbf{A}\vec{w}]) = 1] - \mathbf{Pr}[\mathsf{D}(\mathcal{MG}_k, [\mathbf{A}], [\vec{u}]) = 1] = \mathrm{negl}(\lambda)$, where the probability is taken over the output $\mathcal{MG}_k = (k, G, G_T, e, p, \mathcal{P}, \mathcal{P}_T) \leftarrow \mathcal{G}_k(1^\lambda)$, $\mathbf{A} \leftarrow \mathcal{D}_{n,\ell}$, $\vec{w} \leftarrow \mathbb{Z}_p^\ell$, $\vec{u} \leftarrow \mathbb{Z}_p^n$ and the coin tosses of the adversary D.*

We note that all of the standard examples of matrix assumptions are polynomially induced and further, in all examples we consider in this paper, the degree of $\mathfrak{p}_{i,j}$ is 1. In particular, we will refer to the following examples of matrix distributions, all for $n = \ell + 1$:

$$SC_\ell : \mathbf{A} = \begin{pmatrix} -s & 0 & \dots & 0 & 0 \\ 1 & -s & \dots & 0 & 0 \\ 0 & 1 & & 0 & 0 \\ \vdots & & \ddots & & \vdots \\ 0 & 0 & \dots & 1 & -s \\ 0 & 0 & \dots & 0 & 1 \end{pmatrix}, \quad \mathcal{L}_\ell : \mathbf{A} = \begin{pmatrix} s_1 & 0 & 0 & \dots & 0 \\ 0 & s_2 & 0 & \dots & 0 \\ \vdots & & \ddots & & \vdots \\ 0 & 0 & 0 & \dots & s_\ell \\ 1 & 1 & 1 & \dots & 1 \end{pmatrix}, \quad \mathcal{U}_\ell : \mathbf{A} \leftarrow \mathbb{Z}_p^{(\ell+1)\times\ell},$$

where $s, s_i \leftarrow \mathbb{Z}_p$. Up to sign, the SC_ℓ assumption, introduced in [5], is the ℓ-*symmetric cascade assumption* (ℓ-SCasc). The $\mathcal{L}_\ell$ assumption is actually the well-known ℓ-*linear assumption* (ℓ-Lin, [1]) in matrix language (DDH equals 1-Lin), and the $\mathcal{U}_\ell$ assumption is the ℓ-*uniform assumption*. More generally, we can also define the $\mathcal{U}_{n,\ell}$ assumption for arbitrary $n > \ell$. Note that the $\mathcal{U}_{n,\ell}$ assumption is the weakest matrix assumption (with the worst representation size) and implied by any other $\mathcal{D}_{n,\ell}$ assumption [5]. In particular ℓ-Lin implies the ℓ-uniform assumption as shown by Freeman. Moreover, ℓ-SCasc, ℓ-Lin, and the ℓ-uniform assumption hold in the generic group model [16] relative to a k-linear group generator if $k \leq \ell$ [5].

Interpolating Sets. Let $\vec{X} = (X_1, \dots, X_d)$ be a vector of variables. Let $W \subset \mathbb{Z}_p[\vec{X}]$ be a subspace of polynomials of finite dimension m. Given a set of polynomials $\{\mathfrak{r}_0, \dots, \mathfrak{r}_{m-1}\}$ which are a basis of W, we say that $\vec{x}_1, \dots, \vec{x}_m \in \mathbb{Z}_p^d$ is an *interpolating set* for W if the matrix whose (i, j)th entry is defined as $\mathfrak{r}_{j-1}(\vec{x}_i)$ has full rank. It can be easily seen that the property of being an interpolating set is independent of the basis. Further, when p is exponential (and m and the degrees of $\mathfrak{r}_i$ are polynomial) in the security parameter, any m random vectors $\vec{x}_1, \dots, \vec{x}_m$ form an interpolating set with overwhelming probability.

3 Our Framework

We now present our definitional framework for composite-to-prime-order transformations. Basically, the definitions in this section will enable us to describe how groups of prime order p with a multilinear map e can be converted into groups of order p^n for some $n \in \mathbb{N}$ with a multilinear map $\tilde{e}$. These converted groups will then "mimic" certain features of composite-order groups. Since $\tilde{e}$ is just a composition of several instances of e, we will refer to e as the *basic multilinear map*. We start with an overview of the framework of Freeman, since this is the established model for such transformations. Afterwards, we describe our framework in terms of differences to the model of Freeman.

Freeman's Model. Freeman identifies some abstract properties of bilinear composite order groups which are essential to construct some cryptographic protocols, namely subgroup indistinguishability, the projecting property and the canceling property. For Freeman, a symmetric bilinear map generator takes a bilinear group of prime order p with a pairing e and outputs some groups $\mathbb{H} \subset \mathbb{G}, \mathbb{G}_T$ of order p^n for some $n \in \mathbb{N}$ and a symmetric bilinear map $\tilde{e}: \mathbb{G} \times \mathbb{G} \to \mathbb{G}_T$, computed via the basic pairing e. Useful instances of such generators satisfy the subgroup indistinguishability assumption, which means that it should be hard to decide membership in $\mathbb{H} \subset \mathbb{G}$. Further, the pairing is projecting if the bilinear map generator also outputs some maps π, π_T defined respectively on $\mathbb{G}, \mathbb{G}_T$ which commute with the pairing and such that $\ker \pi = \mathbb{H}$. The pairing is canceling if $\tilde{e}(\mathbb{H}, \mathbb{H}') = 0$ for some decomposition $\mathbb{G} = \mathbb{H} \oplus \mathbb{H}'$.

Instantiations. Further, Freeman gives several concrete instantiations in which the subgroups $\mathbb{H}$ output by the generator are sampled uniformly. More specifically, in the language of [5], the instantiations sample subgroups according to the $\mathcal{U}_{n,\ell}$ distribution. Although his model is not specifically restricted to this case, follow-up work seems to identify "Freeman's model" with this specific matrix distribution. For instance, the results of [13] on the impossibility of achieving the projecting and canceling property simultaneously or the impossibility result of Seo [14], who proves a lower bound on the size of the image of a projecting pairing, are also in this setting.

Our Model. Essentially, we recover Freeman's original definitions for the symmetric setting, however with some additional precisions. First, we extend his model to multilinear maps and, like Seo [14], distinguish between basic multilinear map operations (e) and multilinear map operations $(\tilde{e})$, since an important efficiency measure is how many e-operations are required to compute $\tilde{e}$. The second and main block of differences is introduced with the goal of making the model compatible with several families of matrix assumptions, yielding a useful tool to prove optimality and impossibility results. For this, we extend Freeman's model to explicitly support different families of subgroup assumptions and state clearly what the dependency relations between the different outputs of the multilinear group generator are. In Section 6, we explicitly discuss the advantages of the refinement of the model.

Definition 3. *Let $k, \ell, n, r \in \mathbb{N}$ with $k > 1$ and $r \geq n > \ell$. A $(k, (r, n, \ell))$ symmetric multilinear map generator $\mathcal{G}_{k,(r,n,\ell)}$ takes as input a security parameter 1^λ and a basic k-linear map generator $\mathcal{G}_k$ and outputs in probabilistic polynomial time a tuple $(\mathcal{MG}_k, \mathbb{H}, \mathbb{G}, \mathbb{G}_T, \tilde{e})$, where*
- *$\mathcal{MG}_k := (k, G, G_T, e, p, \mathcal{P}, \mathcal{P}_T) \leftarrow \mathcal{G}_k(1^\lambda)$ is a description of a prime order symmetric k-linear group*
- *$\mathbb{G} \subset G^r$ is a subgroup of G^r with a minimal generating set of size n*
- *$\mathbb{H} \subset \mathbb{G}$ is a subgroup of $\mathbb{G}$ with a minimal generating set of size ℓ*
- *$\tilde{e}: \mathbb{G}^k \to \mathbb{G}_T$ is a non-degenerate k-linear map.*

We assume that elements in $\mathbb{H}, \mathbb{G}$ are represented as vectors in G^r. With this representation, it is natural to identify elements in these groups with vectors in $\mathbb{Z}_p^r$ in the usual way, via the canonical basis. Via this identification, any subgroup $\mathbb{H} \subset G^r$ spanned by $[\vec{b}_1], \ldots, [\vec{b}_\ell]$ corresponds to the subspace H of $\mathbb{Z}_p^r$ spanned by $\vec{b}_1, \ldots, \vec{b}_\ell$, and we write $\mathbb{H} = [H]$. Further, we may assume that $\mathbb{G}_T = G_T^m$ and elements of $\mathbb{G}_T$ are represented by m-tuples of G_T, for some fixed $m \in \mathbb{N}$, although we do not include m as a parameter of the multilinear generator.

In most constructions $n = r$, in which case we drop the index r from the definition, and we simply refer to such a generator as a $(k, (n, \ell))$ generator

$\mathcal{G}_{k,(n,\ell)}$. We always assume that membership in $\mathbb{G}$ is easy to decide.[2] In the case where $n = r$ and $\mathbb{G} = G^r$ this is obviously the case, but otherwise we assume that the description of $\mathbb{G}$ includes some auxiliary information which allows to test it (like in [15] or [12]).

Definition 4 (Properties of Multilinear Map Generators). *Let $\mathcal{G}_{k,(r,n,\ell)}$ be a $(k, (r, n, \ell))$ symmetric multilinear map generator as in Definition 3 with output $(\mathcal{MG}_k, \mathbb{H}, \mathbb{G}, \mathbb{G}_T, \tilde{e})$. We define the following properties:*

- **Subgroup indistinguishability.** *We say that $\mathcal{G}_{k,(r,n,\ell)}$ satisfies the sub-group indistinguishability property if for all PPT adversaries D,*

$$\mathbf{Adv}_{\mathcal{G}_{k,(r,n,\ell)}}(\mathsf{D}) = \mathbf{Pr}[\mathsf{D}(\mathcal{MG}_k, \mathbb{H}, \mathbb{G}, \mathbb{G}_T, \tilde{e}, x) = 1]$$
$$- \mathbf{Pr}[\mathsf{D}(\mathcal{MG}_k, \mathbb{H}, \mathbb{G}, \mathbb{G}_T, \tilde{e}, u) = 1] = \mathrm{negl}(\lambda),$$

  where the probability is taken over $(\mathcal{MG}_k, \mathbb{H}, \mathbb{G}, \mathbb{G}_T, \tilde{e}) \leftarrow \mathcal{G}_{k,(r,n,\ell)}(1^\lambda)$, $x \leftarrow \mathbb{H}$, $u \leftarrow \mathbb{G}$ and the coin tosses of the adversary D.

- **Projecting.** *We say that $(\mathcal{MG}_k, \mathbb{H}, \mathbb{G}, \mathbb{G}_T, \tilde{e})$ is projecting if there exist two non-zero homomorphisms $\pi \colon \mathbb{G} \to \mathbb{G}$, $\pi_T \colon \mathbb{G}_T \to \mathbb{G}_T$ such that $\ker \pi = \mathbb{H}$ and $\pi_T(\tilde{e}(x_1, \ldots, x_k)) = \tilde{e}(\pi(x_1), \ldots, \pi(x_k))$ for any $(x_1, \ldots, x_k) \in \mathbb{G}^k$. For the special case $r = n = \ell + 1$, $\mathbb{G} := G^n$, we can equivalently define the maps $\pi \colon G^n \to G$, $\pi_T \colon \mathbb{G}_T \to G_T$ such that $\ker \pi = \mathbb{H}$ and $\pi_T(\tilde{e}(x_1, \ldots, x_k)) = e(\pi(x_1), \ldots, \pi(x_k))$ (matching the original definition of [8]). We say that $\mathcal{G}_{k,(r,n,\ell)}$ is projecting if its output is projecting with overwhelming probability.*

- **Canceling.** *We say that $(\mathcal{MG}_k, \mathbb{H}_1, \mathbb{G}, \mathbb{G}_T, \tilde{e})$ is canceling if there exists a decomposition $\mathbb{G} = \mathbb{H}_1 \oplus \mathbb{H}_2$ such that for any $x_1 \in \mathbb{H}_{j_1}, \ldots, x_k \in \mathbb{H}_{j_k}$, $\tilde{e}(x_1, \ldots, x_k) = 0$ except for $j_1 = \ldots = j_k$. We call $\mathcal{G}_{k,(r,n,\ell)}$ canceling if its output is canceling with overwhelming probability.*

So far, the given definitions match those of Freeman (extended to the k-linear case) except that we explicitly define the basic k-linear group $\mathcal{MG}_k$ which is used in the construction. We will now introduce two aspects of our framework that are new compared to Freeman's model. First, we will define multilinear generators that sample subgroups according to a specific matrix assumptions. Then, we will define a property of the multilinear map $\tilde{e}$ that will be very useful to establish impossibility results and lower bounds.

Definition 5. *Let $k, \ell, n, r \in \mathbb{N}$ with $k > 1$, $r \geq n > \ell$ and $\mathcal{D}_{n,\ell}$ be a matrix distribution. A $(k, (r, n, \ell), \mathcal{D}_{n,\ell})$ multilinear map generator $\mathcal{G}_{k,(r,n,\ell),\mathcal{D}_{n,\ell}}$ is a $(k, (r, n, \ell))$ multilinear map generator which outputs $(\mathcal{MG}_k, \mathbb{H}, \mathbb{G}, \mathbb{G}_T, \tilde{e})$ such that the distribution of the subspaces H such that $\mathbb{H} = [H]$ equals $\mathcal{D}_{n,\ell}$ for any fixed choice of $\mathcal{MG}_k$.*

As usual, in the case where $r = n$, we just drop r and refer to a $(k, \mathcal{D}_{n,\ell})$ multilinear map generator $\mathcal{G}_{k,\mathcal{D}_{n,\ell}}$. We conclude our framework with a definition

---

[2] We note that with the recent approximate multilinear maps from [7, 3], not even group membership is efficiently recognizable. This will not affect our results, but of course hinders certain applications (such as Groth-Sahai proofs).

Table 1. Efficiency of different symmetric projecting k-linear maps. The size of the domain (n) and codomain (m) of $\tilde{e}$ is given as number of group elements of G and G_T, respectively. Costs are stated in terms of the number of applications of the basic map e, group operations (gop) including inversion in G/G_T, and ℓ-fold multi-exponentiations of the form $e_1[a_1] + \cdots + e_\ell[a_\ell]$ (ℓ-mexp) in G/G_T. Note that in this paper, for the computation of $\tilde{e}$, we use an evaluate-multiply-approach.

Construction	Ass.	Co-/Domain	Cost $\tilde{e}$	Cost π	Cost π_T
Freeman, $k = 2$ [6]	$\mathcal{U}_2$	$9/3$	$9\ e$	3 3-mexp	9 9-mexp
Seo, $k = 2$ [14]	$\mathcal{U}_2$	$6/3$	$9\ e + 3$ gop	3 3-mexp	6 6-mexp
This paper, $k = 2$	$\mathcal{SC}_2$	$5/3$	$5\ e + 22$ gop	1 2-mexp	1 5-mexp
This paper, $k = 2$	$\mathcal{U}_2$	$6/3$	$6\ e + 12$ 3-mexp[1]	1 3-mexp	1 6-mexp
Freeman, $k > 2$	$\mathcal{U}_k$	$(k+1)^k/k+1$	$(k+1)^{k+1}\ e$	$k+1$ $(k+1)$-mexp	$(k+1)^k$ $(k+1)^k$-mexp
This paper, $k > 2$	$\mathcal{U}_k$	$\binom{2k}{k}/k+1$	$\binom{2k}{k}\ e + \binom{2k}{k}k$ $(k+1)$-mexp[1]	1 $(k+1)$-mexp	1 $\binom{2k}{k}$-mexp
This paper, $k > 2$	$\mathcal{SC}_k$	$k^2+1/k+1$	$(k^2+1)\ e + (k^3+k)$ k-mexp[1]	1 k-mexp	1 k^2+1-mexp

that enables us to distinguish generators where the multilinear map $\tilde{e}$ may or may not depend on the choice of the subgroups.

Definition 6. *We say that a $(k, (r, n, \ell), \mathcal{D}_{n,\ell})$ multilinear map generator with output $(\mathcal{MG}_k, \mathbb{H}, \mathbb{G}, \mathbb{G}_T, \tilde{e})$ as in Definition 5 defines a* fixed multilinear map *if the random variable H (s.t. $\mathbb{H} = [H]$) conditioned on $\mathcal{MG}_k$ and the random variable $(\mathbb{G}, \mathbb{G}_T, \tilde{e})$ conditioned on $\mathcal{MG}_k$ are independent.*

4 Our Constructions

All of our constructions arise from the following *polynomial point of view*: The key idea is to treat $\mathbb{G} = G^n$ as an implicit representation of some space of polynomials. Polynomial multiplication will then give us a natural multilinear map. For subspaces $\mathbb{H}^{(\vec{s})}$ that correspond to polynomials sharing a common root $\vec{s}$, this multilinear map will turn out to be projecting. We will first illustrate this idea by means of a simple concrete example where subgroup decision for $\mathbb{H}^{(\vec{s})}$ is equivalent to 2-SCasc (Section 4.1). Then we show that actually any polynomially induced matrix assumption gives rise to such a polynomial space and thus allows for the construction of a k-linear projecting map (Section 4.2). Finally, by considering $\mathbb{G}$ along with the multilinear map as an implicit representation of a polynomial ring modulo some reducible polynomial, we are able to construct a multilinear map which is both projecting and canceling (see Section 4.3 for a summary). See Table 1 for an overview of the characteristics of our projecting map constructions in comparison with previous work.

4.1 A Projecting Pairing Based on the 2-SCasc Assumption

Let $(k = 2, G, G_T, e, p, \mathcal{P}, \mathcal{P}_T) \leftarrow \mathcal{G}_2(1^\lambda)$ be the output of a symmetric prime-order bilinear group generator. We set $\mathbb{G} := G^3$ and $\mathbb{G}_T := G_T^5$. For any $[\vec{f}] =$

---

[1] For the construction based on $\mathcal{SC}_k$, the involved exponents are relatively small, namely the biggest one is $(\lceil \frac{k^2+1}{2} \rceil)^k$. Also for $\mathcal{U}_k$, the involved exponents can usually be made small.

$([f_0], [f_1], [f_2]) \in \mathbb{G} = G^3$, we identify $\vec{f}$ with the polynomial $f = f_0 + f_1 X + f_2 X^2 \in \mathbb{Z}_p[X]$ of degree at most 2. Similarly, any $[\vec{f}]_T \in \mathbb{G}_T$ corresponds to a polynomial of degree at most 4. Then the canonical group operation for $\mathbb{G}$ and $\mathbb{G}_T$ corresponds to polynomial addition (in the exponent), i.e., $[\vec{f}] + [\vec{g}] = [\vec{f+g}] = [f+g]$ and $[\vec{f}]_T + [\vec{g}]_T = [f+g]_T$. Furthermore, polynomial multiplication (in the exponent) gives a map $\tilde{e}: \mathbb{G} \times \mathbb{G} \to \mathbb{G}_T$,

$$\tilde{e}([\vec{f}], [\vec{g}]) := \left(\left[\sum_{i+j=0} f_i g_j \right]_T, \ldots, \left[\sum_{i+j=4} f_i g_j \right]_T \right) = [f \cdot g]_T$$

It is easy to see that $(\mathbb{G}, \mathbb{G}_T, \tilde{e})$ is again a bilinear group setting, where the group operations and the pairing $\tilde{e}$ can be efficiently computed.

A Subgroup Decision Problem. For some fixed $s \in \mathbb{Z}_p$ let us consider the subgroup $\mathbb{H}^{(s)} \subset \mathbb{G}$ formed by all elements $[\vec{f}] \in \mathbb{G}$ such that $\vec{f}$ viewed as polynomial f has root s, i.e., $\mathbb{H}^{(s)} = \{ [f] \in \mathbb{G} \mid f(s) = 0 \}$. In other words, $\mathbb{H}^{(s)}$ consists of all $[f]$ with f of the form

$$(X - s)(f_1' X + f_0') \ , \tag{1}$$

where $f_1', f_0' \in \mathbb{Z}_p$. Thus, given $[f]$ and $[s]$, the subgroup decision problem for $\mathbb{H}^{(s)} \subset \mathbb{G}$ means to decide whether f is of this form or not. Viewing Eq. (1) as matrix-vector multiplication, we see that this is equivalent to deciding whether $\vec{f}$ belongs to the image of the 3×2 matrix

$$\mathbf{A}(s) := \begin{pmatrix} -s & 0 \\ 1 & -s \\ 0 & 1 \end{pmatrix} \tag{2}$$

Hence, our subgroup decision problem corresponds to the 2-SCasc problem (cf. Definition 2) which is hard in a generic bilinear group [5].

Projections. Given s, we can simply define projection maps $\pi: \mathbb{G} \to G$ and $\pi_T: \mathbb{G}_T \to G_T$ by polynomial evaluation at s (in the exponent), i.e., $[\vec{f}]$ is mapped to $[f(s)]$ and $[\vec{f}]_T$ to $[f(s)]_T$. Computing π, π_T requires group operations only. Obviously, it holds that $\ker(\pi) = \mathbb{H}^{(s)}$ and $e(\pi([\vec{f_1}]), \pi([\vec{f_2}])) = \pi_T(\tilde{e}([\vec{f_1}], [\vec{f_2}]))$.

Sampling from $\mathbb{H}^{(s)}$. Given $[(-s, 1, 0)], [(0, -s, 1)] \in \mathbb{G}$, a uniform element from $\mathbb{H}^{(s)}$ can be sampled by picking $(f_0', f_1') \leftarrow \mathbb{Z}_p^2$ and, as with any matrix assumption, computing the matrix-vector product

$$\left[\begin{pmatrix} -s & 0 \\ 1 & -s \\ 0 & 1 \end{pmatrix} \cdot \begin{pmatrix} f_0' \\ f_1' \end{pmatrix} \right] = \left[(-sf_0', f_0' - sf_1', f_1')^T \right] \tag{3}$$

Again, this can be done using the group operation only.

Efficiency. Computing $\tilde{e}$ in our construction corresponds to polynomial multiplication. Although this multiplication happens in the exponent (and we are "only" given implicit representations of the polynomials), we are not forced to stick to schoolbook multiplication. Instead, we propose to follow an evaluation-multiplication-interpolation approach (using small interpolation points) where the actual interpolation step is postponed to the computation of π_T.

More precisely, so far we used coefficient representation for polynomials over $\mathbb{G}$ and $\mathbb{G}_T$ with respect to the standard basis. However, other (s-independent) bases are also possible without affecting security. For efficiency, we propose to stick to this representation for $\mathbb{G}$ but to use point-value representation for polynomials over $\mathbb{G}_T$ with respect to the fixed interpolating set $M := \{-2, -1, 0, 1, 2\}$ (cf. Definition 2). This means we now identify a polynomial g in the target space with the vector $(g(-2), g(-1), g(0), g(1), g(2))$.

More concretely, to compute $\tilde{e}([f_1], [f_2]) = ([(f_1 f_2)(x)]_T)_{x \in M}$, we first evaluate f_1 and f_2 (in the exponent) with all $x \in M$, followed by a point-wise multiplication $([f_1(x) f_2(x)]_T)_{x \in M} = (e([f_1(x)], [f_2(x)]))_{x \in M}$. This way, $\tilde{e}$ can be computed more efficiently with only five pairings. Computing π is unchanged. To apply π_T, one first needs to obtain the coefficient representation by interpolation and then evaluate the polynomial at s. However, this can be done simultaneously and as the 1×5 matrix describing this operation can be precomputed (given s) it does not increase the computational cost much.

4.2   Projecting Multilinear Maps from any Matrix Assumption

In the following, we will first demonstrate that for any vector space of polynomials, the natural pairing given by polynomial multiplication is projecting for subspaces consisting of polynomials sharing a common root. We will then show that any (polynomially induced) matrix assumption can equivalently be considered as a subspace assumption in a vector space of polynomials of this type. This way, we obtain a natural projecting multilinear map for any polynomially induced matrix assumption.

A Projecting Multilinear Map on Spaces of Polynomials. Let $\mathcal{MG}_k := (k, G, G_T, e, p, \mathcal{P}, \mathcal{P}_T) \leftarrow \mathcal{G}_k(1^\lambda)$ be the output of a prime-order k-linear group generator. Let $V \subset \mathbb{Z}_p[\vec{X}]$ be a vector space of polynomials of dimension n for which we fix a basis $\mathfrak{q}_0, \ldots, \mathfrak{q}_{n-1}$. Then for any $[\vec{f}] \in \mathbb{G} := G^n$ we can identify the vector $\vec{f} = (f_0, \ldots, f_{n-1})$ with a polynomial $f = \sum f_i \mathfrak{q}_i \in V$. In the 2-SCasc example above, V corresponds to univariate polynomials of degree at most 2 and the basis is given by $1, X, X^2$. On V, we have a natural k-linear map given by polynomial multiplication: $\text{mult}_k \colon V^k \to \mathbb{Z}_p[\vec{X}], \text{mult}_k(f_1, \ldots, f_k) = f_1 \cdots f_k$. Let $W \subset \mathbb{Z}_p[\vec{X}]$ be the span of the image of mult_k and m its dimension. Then we can again fix a basis $\mathfrak{r}_0, \ldots, \mathfrak{r}_{m-1}$ of W to identify polynomials with vectors. In the 2-SCasc example above, W consists of polynomials of degree at most 4 and we chose the basis $1, X, X^2, X^3, X^4$ of W for our initial presentation. From polynomial multiplication, we then obtain a non-degenerate k-linear map

$$\tilde{e} \colon \mathbb{G}^k \to G_T^m, \tilde{e}([\vec{f_1}], \ldots, [\vec{f_k}]) = [f_1 \cdots f_k]_T .$$

Now consider a subspace $\mathbb{H}^{(\vec{s})} \in \mathbb{G}$ of the form $\mathbb{H}^{(\vec{s})} = \{[f] \in \mathbb{G} \mid f(\vec{s}) = 0\}$. It is easy to see that $\tilde{e}$ is projecting for this subspace: A projection map $\pi \colon \mathbb{G} \to G$ with $\ker(\pi) = \mathbb{H}^{(\vec{s})}$ is given by evaluation at $\vec{s}$, i.e., $\pi([\vec{f}]) = [f(\vec{s})]$. Similarly, $\pi_T \colon G_T^m \to G_T$ is defined by $\pi_T([\vec{g}]_T) = [g(\vec{s})]_T$ and by construction we have $e(\pi([\vec{f}_1]), \ldots, \pi([\vec{f}_k])) = [f_1(\vec{s}) \cdots f_k(\vec{s})]_T = [(f_1 \cdots f_k)(\vec{s})]_T = \pi_T(\tilde{e}([\vec{f}_1], \ldots, [\vec{f}_k]))$.

From a Polynomially Induced Matrix Distribution to a Space of Polynomials. Now, let $\mathcal{D}_{n-1}$ be any polynomially induced matrix distribution as defined in Definition 2 and let $\mathbf{A}(\vec{X}) \in (\mathbb{Z}_p[\vec{X}])^{n \times (n-1)}$ be the polynomial matrix describing this distribution. Then we set $\mathbb{G} := G^n$ and consider the subspace $[\mathsf{Im}\,\mathbf{A}(\vec{s})]$ for some $\vec{s}$. We now show that we can identify $\mathbb{G}$ with a vector space V of polynomials, such that the subspace $\mathsf{Im}\,\mathbf{A}(\vec{s})$ corresponds exactly to polynomials having a root at $\vec{s}$. To this end, consider the determinant of $(\mathbf{A}(\vec{X})\|\vec{F})$ as a polynomial $\mathfrak{d}$ in indeterminates $\vec{X}$ and $\vec{F}$. Since we assume that $\mathbf{A}(\vec{s})$ has generically[3] full rank a given vector $\vec{f} \in \mathbb{Z}_p^n$ belongs to the image of $\mathbf{A}(\vec{s})$ iff the determinant of the extended matrix $(\mathbf{A}(\vec{s})\|\vec{f})$ is zero, i.e., $\mathfrak{d}(\vec{s}, \vec{f}) = 0$. To obtain the desired vector space V with basis $\mathfrak{q}_0, \ldots, \mathfrak{q}_{n-1}$, we consider the Laplace expansion of this determinant to write $\mathfrak{d}$ as

$$\mathfrak{d}(\vec{X}, \vec{F}) = \sum_{i=0}^{n-1} F_i \mathfrak{q}_i(\vec{X}). \tag{4}$$

for some polynomials $\mathfrak{q}_i(\vec{X})$ depending only on $\mathbf{A}$. For $\mathcal{SC}_2$, we have $\mathfrak{q}_i = X^i$. We note that in all cases of interest the $\mathfrak{q}_i$ are linearly independent (see [9]).

Thus, we may now identify $[\vec{f}] \in \mathbb{G}$ with the implicit representation of the polynomial $f = \mathfrak{d}(\vec{X}, \vec{f}) = \sum_i f_i \mathfrak{q}_i$. As $f(\vec{s}) = \sum_i f_i \mathfrak{q}_i(\vec{s}) = 0$ iff $\vec{f} \in \mathsf{Im}\,\mathbf{A}(\vec{s})$, we have $\mathbb{H}^{(\vec{s})} = [\mathsf{Im}\,\mathbf{A}(\vec{s})] = \{[f] \in \mathbb{G} \mid f(s) = 0\}$. Hence, we may construct a projecting k-linear map from polynomial multiplication as described in the previous paragraph.

Working through the construction, one can obtain explicit coordinates as follows: let W be the span of $\{\mathfrak{q}_{i_1} \cdots \mathfrak{q}_{i_k} \mid 0 \le i_j < n\}$ and fix a basis $\mathfrak{r}_0, \ldots, \mathfrak{r}_{m-1}$ of W. This determines coefficients $\lambda_t^{(i_1, \ldots, i_k)}$ in $\mathfrak{q}_{i_1} \cdots \mathfrak{q}_{i_k} = \sum_{t=0}^{m-1} \lambda_t^{(i_1, \ldots, i_k)} \mathfrak{r}_t$.

Recall that $\tilde{e} \colon (G^n)^k \to G_T^m$ is defined as $\tilde{e}([\vec{f}_1], \ldots, [\vec{f}_k]) = [f_1 \cdots f_k]_T$, expressed as an element of G_T^m via the basis $\vec{\mathfrak{r}}$. In coordinates this reads

$$\tilde{e}([\vec{f}_1], \ldots, [\vec{f}_k]) = \Big(\sum_{j_1 \le \cdots \le j_k} \lambda_0^{(j_1, \ldots, j_k)} \cdot \sum_{\substack{(i_1, \ldots, i_k) \in \\ \tau(j_1, \ldots, j_k)}} e([f_{1,i_1}], \ldots, [f_{k,i_k}]), \ldots,$$

$$\sum_{j_1 \le \cdots \le j_k} \lambda_{m-1}^{(j_1, \ldots, j_k)} \cdot \sum_{\substack{(i_1, \ldots, i_k) \in \\ \tau(j_1, \ldots, j_k)}} e([f_{1,i_1}], \ldots, [f_{k,i_k}]) \Big) \tag{5}$$

where $[f_{1,i_1} \cdots f_{k,i_k}]_T$ simply denotes $(f_{1,i_1} \cdots f_{k,i_k})\mathcal{P}_T$ and $\tau(j_1, \ldots, j_k)$ denotes the set of permutations of $(j_1, \ldots, j_k)$. The last optimization can be done as $\mathfrak{q}_{i_1} \cdots \mathfrak{q}_{i_k} = \mathfrak{q}_{j_1} \cdots \mathfrak{q}_{j_k}$ for $(i_1, \ldots, i_k) \in \tau(j_1, \ldots, j_k)$. For the same reason,

---

[3] This means that $\mathbf{A}(\vec{s})$ will be full rank with overwhelming probability and this is indeed equivalent to $\mathfrak{d} \ne 0$. To simplify the exposition, we may assume that the sampling algorithm is changed to exclude $\vec{s}$ where $\mathbf{A}(\vec{s})$ does not have full rank.

we have $m = \binom{n+k-1}{k}$ in the worst case. In this way, the target group in our constructions is always smaller than the target group in Freeman's construction (generalized to $k \geq 2$), which is of size n^k.

The following theorem summarizes our construction and its properties:

Theorem 1. *Let $k > 1$, $n \in \mathbb{N}$, and $\mathcal{D}_{n-1}$ be a polynomially induced matrix distribution. Let $\mathcal{G}_{k,\mathcal{D}_{n-1}}$ be an algorithm that on input of a security parameter 1^λ and a symmetric prime-order k-multilinear map generator $\mathcal{G}_k$ outputs $(\mathcal{MG}_k, \mathbb{H}^{(\vec{s})}, \mathbb{G}, \mathbb{G}_T, \tilde{e})$, where*

- $\mathcal{MG}_k := (k, G, G_T, e, p, \mathcal{P}, \mathcal{P}_T) \leftarrow \mathcal{G}_k(1^\lambda)$,
- $\mathbb{G} := G^n$, $\mathbb{H}^{(\vec{s})} := [\mathsf{Im}\,\mathbf{A}(\vec{s})]$, $\mathbf{A}(\vec{s}) \leftarrow \mathcal{D}_{n-1}$,
- $\mathbb{G}_T := G_T^m$, *where m equals the dimension of*

$$W := \left\{ \sum_{0 \leq i_1, \ldots, i_k \leq n-1} \alpha_{i_1, \ldots, i_k} \mathfrak{q}_{i_1} \cdots \mathfrak{q}_{i_k} \;\middle|\; \alpha_{i_1, \ldots, i_k} \in \mathbb{Z}_p \right\}$$

(as vector space), and $\mathfrak{q}_0(\vec{X}), \ldots, \mathfrak{q}_{n-1}(\vec{X}) \in \mathbb{Z}_p[\vec{X}]$ are polynomials s.t.

$$\det(\mathbf{A}(\vec{X})\|\vec{F}) = \sum_{i=0}^{n-1} F_i \mathfrak{q}_i(\vec{X})$$

for the matrix $\mathbf{A}(\vec{X})$ describing $\mathcal{D}_{n-1}$, and
- $\tilde{e} \colon \mathbb{G}^k \to \mathbb{G}_T$ *is the map defined by Eq. (5) for a basis $\mathfrak{r}_0, \ldots, \mathfrak{r}_{m-1}$ of W.*

Then $\mathcal{G}_{k,\mathcal{D}_{n-1}}$ is a $(k, \mathcal{D}_{n-1})$ multilinear map generator. It is projecting, where the projection maps $\pi \colon \mathbb{G} \to G$ and $\pi_T \colon \mathbb{G}_T \to G_T$ defined by $\pi(\vec{f}) := \sum_{i=0}^{n-1} \mathfrak{q}_i(\vec{s})[f_i]$ and $\pi_T(\vec{g}) := \sum_{i=0}^{m-1} \mathfrak{r}_i(\vec{s})[g_i]_T$ are efficiently computable given the trapdoor $\vec{s}$. Furthermore, if the $\mathcal{D}_{n-1}$ assumption holds with respect to $\mathcal{G}_k$, then subgroup indistinguishability holds with respect to $\mathcal{G}_{k,\mathcal{D}_{n-1}}$.

Example 1. We can construct a projecting k-linear map generator satisfying subgroup indistinguishability under k-SCasc (which is hard in a k-linear generic group model). For $\mathcal{G}_{k,\mathcal{SC}_k}$, we would get $n = k + 1$ and $\mathfrak{q}_i(X) = X^i$ if k is even and $\mathfrak{q}_i(X) = -X^i$ when k is odd, where $0 \leq i \leq k$. Using the basis $\mathfrak{r}_t(X) = X^t$ for W if k is even and $\mathfrak{r}_t(X) = -X^t$ if k is odd for $0 \leq t \leq k^2$, we obtain $\lambda_t^{(i_1, \ldots, i_k)} = 1$ for $t = i_1 + \cdots + i_k$ and $\lambda_t^{(i_1, \ldots, i_k)} = 0$ else. Note that we have $m = k^2 + 1$.

Example 2. We can also construct a k-linear map generator from k-Lin. For $\mathcal{G}_{k,\mathcal{L}_k}$, we would have $n = k+1$, and polynomials $\mathfrak{q}_k(X_0, \ldots, X_{k-1}) = X_0 \cdots X_{k-1}$ and $\mathfrak{q}_i(X_0, \ldots, X_{k-1}) = -\prod_{j \neq i} X_j$ for $0 \leq i \leq k - 1$. As a basis for W we can simply take $\{\mathfrak{q}_{j_1} \cdots \mathfrak{q}_{j_k} \mid 0 \leq j_1 \leq \ldots \leq j_k \leq k\}$ yielding $m = \binom{n+k-1}{k}$.

Example 3. Like Freeman, we could also construct a k-linear map generator from the $\mathcal{U}_k$ assumption. Although the polynomials $\mathfrak{q}_i(X_{1,1}, \ldots, X_{k,k+1})$, $0 \leq i \leq k$, associated to $\mathcal{G}_{k,\mathcal{U}_k}$ have a much more complex description than in the k-Lin case, the image size of the resulting map is the same, namely $m = \binom{n+k-1}{k}$, because a basis of the image is also $\{\mathfrak{q}_{j_1} \cdots \mathfrak{q}_{j_k} \mid 0 \leq j_1 \leq \ldots \leq j_k \leq k\}$.

Efficiency. As in our setting any change of basis is efficiently computable, the security of our construction only depends on the vector space V (which in turn determines W), but not on the bases chosen. So we are free to choose bases that improve efficiency. We propose to follow the same approach as in Section 4.1: Select points $\vec{x}_0, \ldots, \vec{x}_{m-1}$ that form an interpolating set for W and represent $f \in W$ via the vector $f(\vec{x}_0), \ldots, f(\vec{x}_{m-1})$. This corresponds to choosing the basis of W consisting of polynomials $\mathfrak{r}_0, \ldots, \mathfrak{r}_{m-1} \in W$ such that $\mathfrak{r}_i(\vec{x}_j) = 1$ for $i = j$ and 0 otherwise. For the domain V, the choice is less significant and we might simply choose the $\mathfrak{q}_i$'s that the determinant polynomial gives us. Then we can compute $\tilde{e}([\vec{f}_1], \ldots, [\vec{f}_k])$ by an evaluate-multiply approach using only m applications of e. Note that the evaluation step can also be done pretty efficiently if the $\mathfrak{q}_i$'s have small coefficients (which usually is the case). For details see [9].

4.3   Canceling and Projecting k-Linear Maps From Polynomial Spaces

By considering polynomial multiplication modulo a polynomial h, which has a root at the secret s, we are able to construct a $(k, (n = \ell + 1, \ell))$ symmetric multilinear map generator with a *non-fixed* pairing that is both canceling and projecting. Our first construction relies on a $k' := k + 1$-linear prime-order map e. The one additional multiplication in the exponent is used to perform the reduction modulo h. Based on this construction, we propose another $(k, (r = 2\ell, n = \ell+1, \ell))$ symmetric multilinear map generator that requires only a $k' = k$-linear prime-order map. The security of our constructions is based on variants of the ℓ-SCasc assumption. We need to extend ℓ-SCasc by additional given group elements to allow for reduction in the exponent, e.g., in the simplest case hints of the form $[X^i \bmod h]$ are given. In the full version of this paper we give details, efficiency considerations, and show that our constructions are secure for $\ell \geq k'$ in generic k'-linear groups. We note that, to the best of our knowledge, this is the first construction of a projecting&canceling map that naturally generalizes to $k' > 2$.

5   Optimality and Impossibility Results

5.1   Optimality of Polynomial Multiplication

In this section we show that for any polynomially induced matrix assumption $\mathcal{D}_{\ell+1,\ell}$, the projecting multilinear map resulting from the polynomial viewpoint is optimal in terms of image size.

Theorem 2. *Let $k > 0$, and let $\mathcal{D}_{\ell+1,\ell}$ be a polynomially induced matrix assumption and let $\mathfrak{q}_0, \ldots, \mathfrak{q}_\ell$ be the polynomials associated to $\mathcal{D}_{\ell+1,\ell}$ as defined in Eq. (4) in Section 4.2 and let $W \subset \mathbb{Z}_p[\vec{X}]$ be the space of polynomials spanned by $\{\mathfrak{q}_{i_1} \ldots \mathfrak{q}_{i_k} \mid 0 \leq i_j \leq \ell\}$. Let $(\mathcal{MG}_k, \mathbb{H}, G^{\ell+1}, G_T^m, \tilde{e})$ be the output of any other fixed $(k, \mathcal{D}_{\ell+1,\ell})$ projecting multilinear map generator. Then, $\overline{m} := \dim W \leq m$.*

$$\mathbb{G}^k \xrightarrow{\tilde{e}} G_T^m \qquad \mathbb{G}^k \times \ldots \times \mathbb{G}^k \xrightarrow{(\tilde{e}, \ldots, \tilde{e})} G_T^m \times \ldots \times G_T^m$$

$$\downarrow (\pi^{(\vec{s})})^k \quad \downarrow \pi_T^{(\vec{s})} \qquad \downarrow \left((\pi^{(\vec{s}_1)})^k, \ldots, (\pi^{(\vec{s}_{\overline{m}})})^k \right) \qquad \downarrow \left(\pi_T^{(\vec{s}_1)}, \ldots, \pi_T^{(\vec{s}_{\overline{m}})} \right)$$

$$G^k \xrightarrow{e} G_T \qquad G^k \times \ldots \times G^k \xrightarrow{(e, \ldots, e)} G_T \times \ldots \times G_T$$

Fig. 1. Left: Projecting property. Right: The diagram repeated $\overline{m}$ times for an interpolating set $\vec{s}_1, \ldots \vec{s}_{\overline{m}}$ for W

Proof Intuition. An intuition of the proof is given by Figure 1. The first part of the proof shows that w.l.o.g. we can assume that $\pi_T^{(\vec{s})} \circ \tilde{e}$ is polynomial multiplication for all $\vec{s}$, that is, for any $[\vec{f}_1], \ldots, [\vec{f}_k] \in G^{\ell+1}$, $\pi_T(\tilde{e}([\vec{f}_1], \ldots, [\vec{f}_k])) = [(f_1 \cdots f_k)(\vec{s})]_T$. This follows from the commutative diagram on the left, i.e., the projecting property, together with the fact that, because $\mathbb{H}$ has codimension 1, the map $\pi^{(\vec{s})}$ must (up to scalar multiples) correspond to polynomial evaluation at $\vec{s}$. The intuition for the second part of the proof is given by the diagram on the right-hand side of Figure 1. Here we show that if $\vec{s}_1, \ldots \vec{s}_{\overline{m}}$ is an interpolating set for W, then the span of $\left\{ (\pi_T^{(\vec{s}_1)}(\vec{x}), \ldots, \pi_T^{(\vec{s}_{\overline{m}})}(\vec{x})) | \vec{x} \in \tilde{e}(\mathbb{G}^k) \right\} \subset G_T^{\overline{m}}$ is of dimension $\overline{m}$. This dimension can be at most the dimension of the span of $\tilde{e}(\mathbb{G}^k)$, showing $\overline{m} \leq m$. A full proof is given in [9].

5.2 Optimality of our Projecting Multilinear Map from the SCasc-Assumption

As a result of our general viewpoint, we can actually show that the projecting multilinear map based on the SCasc-assumption is optimal among *all* polynomially induced matrix assumptions $\mathcal{D}_{n,\ell}$ that are not redundant. Non-redundancy rules out the case where some components of $\vec{z}$ are no help (even information-theoretically) in distinguishing $\vec{z} \in \mathbb{G}$ from $\vec{z} \in \mathbb{H}^{(s)}$. See [9] for a formal definition.

Theorem 3. *Let $n = \ell + 1$ and $\mathcal{D}_{n,\ell}$ be a polynomially induced matrix distribution which is not redundant. Let $(\mathcal{MG}_k, \mathbb{H}, G^n, G_T^m, \tilde{e})$ be the output of some projecting $(k, \mathcal{D}_{n,\ell})$ multilinear map generator with a fixed multilinear map. Then, $m \geq \ell k + 1$.*

Note that the projecting pairing based on the polynomial viewpoint of the ℓ-SCasc-assumption reaches this bound and is hence optimal.

Proof. We may identify G^n with some subspace $V \subset \mathbb{Z}_p[\vec{X}]$ of dimension n (see [9] for details). By Theorem 2 above, we may assume w.l.o.g. that $\tilde{e}$ is polynomial multiplication, as this only makes m smaller. Hence we can also identify G_T^m with some subspace $W \subset \mathbb{Z}_p[\vec{X}]$ of dimension m. Let $>$ be any monomial ordering on $\mathbb{Z}_p[\vec{X}]$. Let $\mathfrak{q}_0, \ldots, \mathfrak{q}_\ell$ be a basis of V in echelon form with respect to $>$.

This implies that the leading monomials satisfy $\mathrm{LM}(\mathsf{q}_0) > \ldots > \mathrm{LM}(\mathsf{q}_\ell)$. Now consider the elements

$$\mathsf{q}_0^k = \mathfrak{r}_0 = \mathsf{q}_0 \cdots \mathsf{q}_0 \mathsf{q}_0$$

$$\mathfrak{r}_1 = \mathsf{q}_0 \cdots \mathsf{q}_1 \mathsf{q}_0 \qquad \mathfrak{r}_{\ell+1} = \mathsf{q}_0 \cdots \mathsf{q}_0 \mathsf{q}_1 \mathsf{q}_\ell \qquad \mathfrak{r}_{(k-1)\ell+1} = \mathsf{q}_0 \mathsf{q}_\ell \cdots \mathsf{q}_\ell$$

$$\vdots \qquad\qquad \vdots \qquad\qquad\qquad \vdots$$

$$\mathfrak{r}_\ell = \mathsf{q}_0 \cdots \mathsf{q}_0 \mathsf{q}_\ell \qquad \mathfrak{r}_{2\ell} = \mathsf{q}_0 \cdots \mathsf{q}_0 \mathsf{q}_\ell \mathsf{q}_\ell \qquad \mathfrak{r}_{\ell k} = \mathsf{q}_\ell \mathsf{q}_\ell \cdots \mathsf{q}_\ell$$

(the definition of $\mathfrak{r}_{i+1}$ differs from that of $\mathfrak{r}_i$ in one single index being greater by one). It holds that all $\mathfrak{r}_i \in W$ by construction and $\mathrm{LM}(\mathfrak{r}_0) > \mathrm{LM}(\mathfrak{r}_1) > \ldots > \mathrm{LM}(\mathfrak{r}_{\ell k})$ by the properties of a monomial order. Hence, the $\mathfrak{r}_i$ are linearly independent, showing $m = \dim W \geq \ell k + 1$.

6 Review of Previous Results in Our Framework

Let us consider some previous results using the language introduced in Section 3.

Projecting Pairings. Implicitly, in [8], Groth and Sahai were using the fact that the bilinear symmetric tensor product is a projecting map. Subsequently, Seo [14] constructed an improved symmetric projecting pairing which he claimed to be optimal in terms of image size and operations.

Theorem 4. ([14]) Let $\mathcal{G}_{2,\mathcal{U}_\ell}$ be any (symmetric) projecting $(2,\mathcal{U}_\ell)$ bilinear map generator with output $(\mathcal{MG}_2, \mathbb{H}, \mathbb{G}, G_T^m, \tilde{e})$. Then (a) we have $m \geq (\ell+1)(\ell+2)/2$, and (b) the map $\tilde{e}$ cannot be evaluated with less than $(\ell+1)^2$ prime-order pairing operations.

Using the polynomial point of view, we prove in [9] that polynomial multiplication is optimal for *any* $\mathcal{D}_\ell$ assumption, and thus cover Theorem 4 (a) as a special case when $\mathcal{D}_\ell = \mathcal{U}_\ell$. On the other hand, the polynomial viewpoint immediately suggests a method to evaluate Seo's pairing with m (less than $(\ell+1)^2$) prime-order pairing operations, refuting Theorem 4 (b).[4] Further, our results also answer in the affirmative an open question raised by Seo about the existence of more efficient pairings outside of the model. Our construction of a k-linear map based on k-SCasc beats this lower bound and is much more efficient asymptotically in k.

Cancelling and Projecting Pairings. In his original paper [6], Freeman gives several constructions of bilinear pairings which are either projecting or canceling — but not both. Subsequently, Meiklejohn *et al.* [13] give evidence that it might be hard to obtain both features simultaneously:

---

[4] In [9] we discuss in more detail Seo's construction and the reason why Theorem 4 (b) is false.

Theorem 5. *([13]) Any symmetric $(2, \mathcal{U}_\ell)$ bilinear generator with a fixed pairing cannot be simultaneously projecting and canceling, except with negligible probability (over the output of the generator).*[5]

In [9] we show that this result can be extended to any $(2, \mathcal{L}_\ell)$ and any $(2, \mathcal{SC}_2)$ bilinear generator. It remains an open question if the impossibility results extend to $(2, \mathcal{SC}_\ell)$, for $\ell > 2$.

With these impossibility results, it is not surprising that all canceling and projecting constructions are for *non-fixed* pairings in the sense of Definition 6. Indeed, in [15] Cheon and Seo construct a pairing which is both canceling and projecting but not fixed since, implicitly, the group $\mathbb{G}$ depends on the hidden subgroup $\mathbb{H}$. In our language, the pairing of Seo and Cheon is a $(2, (r = \ell^2, n = \ell + 1, \ell))$ pairing, i.e., $\mathbb{G} \subset G^{\ell^2}$ of dimension $n = \ell + 1$. Recently, Lewko and Meiklejohn [12] simplified this construction, obtaining a $(2, (r = 2\ell, n = \ell+1, \ell))$ bilinear map generator. In [9] we also construct a $(2, (r = 2\ell, n = \ell+1, \ell))$ pairing achieving both properties (and which generalizes to any $(k, (r = 2\ell, n = \ell+1, \ell))$ with $\ell \geq k$) , but using completely different techniques. A direct comparison of [15], [12] with our pairing is not straightforward, since in fact they use dual vector spaces techniques and their pairing is not really symmetric.

7   A Direct Application: More Efficient Groth-Sahai Proofs

Using our projecting pairing from Section 4.1, we can improve the performance of Groth-Sahai proofs by almost halving the number of required prime-order pairing operations for verification (cf. Table 1). Additionally, in [9], we show how to implement a k-linear variant of the Boneh-Goh-Nissim encryption scheme [2] using the projecting multilinear map generator $\mathcal{G}_{k, \mathcal{SC}_k}$.

Groth-Sahai proofs [8] are the most natural application of projecting bilinear maps. They admit various instantiations in the prime-order setting. It follows easily from the original formulation of Groth and Sahai that their proofs can be instantiated based on any $\mathcal{D}_{n,\ell}$ assumption and any fixed projecting map. Details are given in [5] but only for the projecting pairing corresponding to the symmetric bilinear tensor product. The generalization to any projecting pairing is straightforward.

The important parameters for efficiency of NIZK proofs are the size of the common reference string, the proof size and the verification cost. The proof size (for a given equation) depends only on the size of the matrix assumption, i.e., on n, ℓ, so it is omitted in our comparison. The size of the common reference string depends essentially on the size of the commitment key, which is $n + \text{Re}_\mathbb{G}(\mathcal{D}_{n,\ell})$, where $\text{Re}_\mathbb{G}(\mathcal{D}_{n,\ell})$ is the representation size of the matrix assumption $\mathcal{D}_{n,\ell}$, which is 1 for ℓ-SCasc, ℓ for ℓ-Lin and $(\ell+1)\ell$ for $\mathcal{U}_\ell$. Therefore, the ℓ-SCasc instantiation

---

[5] Their claim is that it is impossible to achieve both properties under what they call a "natural use" of the ℓ-Lin assumption, although, they are actually using the uniform assumption.

is the most advantageous from the point of view of the size of the common reference string (regardless of the pairing used), as pointed out in [5].

On the other hand, the choice of the pairing affects only the cost of verification[6]. Except for some restricted type of linear equations, verification involves several evaluations of $\tilde{e}$. In our most efficient construction, for each pairing evaluation $\tilde{e}$, we save, according to Table 1, at least 4 prime-order pairing evaluations. For instance, this leads to a saving of 12 pairing evaluations for proving that a committed value is a bit $b \in \{0, 1\}$.

Acknowledgements. We would like to thank the anonymous reviewers for very helpful and constructive comments. This work has been supported in part by DFG grant GZ HO 4534/4-1. Carla Ràfols was supported by a Sofja Kovalevskaja Award of the Alexander von Humboldt Foundation and the German Federal Ministry for Education and Research.

References

[1] Boneh, D., Boyen, X., Shacham, H.: Short group signatures. In: Franklin, M. (ed.) CRYPTO 2004. LNCS, vol. 3152, pp. 41–55. Springer, Heidelberg (2004)

[2] Boneh, D., Goh, E.-J., Nissim, K.: Evaluating 2-DNF formulas on ciphertexts. In: Kilian, J. (ed.) TCC 2005. LNCS, vol. 3378, pp. 325–341. Springer, Heidelberg (2005)

[3] Coron, J.S., Lepoint, T., Tibouchi, M.: Practical multilinear maps over the integers. In: Canetti, R., Garay, J.A. (eds.) CRYPTO 2013, Part I. LNCS, vol. 8042, pp. 476–493. Springer, Heidelberg (2013)

[4] Diffie, W., Hellman, M.E.: New directions in cryptography. IEEE Transactions on Information Theory 22(6), 644–654 (1976)

[5] Escala, A., Herold, G., Kiltz, E., Ràfols, C., Villar, J.: An algebraic framework for diffie-hellman assumptions. In: Canetti, R., Garay, J.A. (eds.) CRYPTO 2013, Part II. LNCS, vol. 8043, pp. 129–147. Springer, Heidelberg (2013)

[6] Freeman, D.M.: Converting pairing-based cryptosystems from composite-order groups to prime-order groups. In: Gilbert, H. (ed.) EUROCRYPT 2010. LNCS, vol. 6110, pp. 44–61. Springer, Heidelberg (2010)

[7] Garg, S., Gentry, C., Halevi, S.: Candidate multilinear maps from ideal lattices. In: Johansson, T., Nguyen, P.Q. (eds.) EUROCRYPT 2013. LNCS, vol. 7881, pp. 1–17. Springer, Heidelberg (2013)

[8] Groth, J., Sahai, A.: Efficient noninteractive proof systems for bilinear groups. SIAM J. Comput. 41(5), 1193–1232 (2012)

[9] Herold, G., Hesse, J., Hofheinz, D., Ràfols, C., Rupp, A.: Polynomial spaces: A new framework for composite-to-prime-order transformations. Cryptology ePrint Archive (2014), http://eprint.iacr.org/

[10] Katz, J., Sahai, A., Waters, B.: Predicate encryption supporting disjunctions, polynomial equations, and inner products. In: Smart, N.P. (ed.) EUROCRYPT 2008. LNCS, vol. 4965, pp. 146–162. Springer, Heidelberg (2008)

---

[6] This is not exactly true, in fact, with the improved pairing for SCasc the prover needs to compute an additional 4 group operations, see the discussion in [9].

[11] Lewko, A.B.: Tools for simulating features of composite order bilinear groups in the prime order setting. In: Pointcheval, D., Johansson, T. (eds.) EUROCRYPT 2012. LNCS, vol. 7237, pp. 318–335. Springer, Heidelberg (2012)

[12] Lewko, A.B., Meiklejohn, S.: A profitable sub-prime loan: Obtaining the advantages of composite-order in prime-order bilinear groups. IACR Cryptology ePrint Archive 2013, 300 (2013)

[13] Meiklejohn, S., Shacham, H., Freeman, D.M.: Limitations on transformations from composite-order to prime-order groups: The case of round-optimal blind signatures. In: Abe, M. (ed.) ASIACRYPT 2010. LNCS, vol. 6477, pp. 519–538. Springer, Heidelberg (2010)

[14] Seo, J.H.: On the (im)possibility of projecting property in prime-order setting. In: Wang, X., Sako, K. (eds.) ASIACRYPT 2012. LNCS, vol. 7658, pp. 61–79. Springer, Heidelberg (2012)

[15] Seo, J.H., Cheon, J.H.: Beyond the limitation of prime-order bilinear groups, and round optimal blind signatures. In: Cramer, R. (ed.) TCC 2012. LNCS, vol. 7194, pp. 133–150. Springer, Heidelberg (2012)

[16] Shoup, V.: Lower bounds for discrete logarithms and related problems. In: Fumy, W. (ed.) EUROCRYPT 1997. LNCS, vol. 1233, pp. 256–266. Springer, Heidelberg (1997)

[17] Waters, B.: Dual system encryption: Realizing fully secure IBE and HIBE under simple assumptions. In: Halevi, S. (ed.) CRYPTO 2009. LNCS, vol. 5677, pp. 619–636. Springer, Heidelberg (2009)

Revisiting the Gentry-Szydlo Algorithm

H. W. Lenstra[1] and A. Silverberg[2,*]

[1] Mathematisch Instituut,
Universiteit Leiden,
The Netherlands
hwl@math.leidenuniv.nl

[2] Department of Mathematics,
University of California, Irvine
Irvine, CA, USA
asilverb@uci.edu

Abstract. We put the Gentry-Szydlo algorithm into a mathematical framework, and show that it is part of a general theory of "lattices with symmetry". For large ranks, there is no good algorithm that decides whether a given lattice has an orthonormal basis. But when the lattice is given with enough symmetry, we can construct a provably deterministic polynomial time algorithm to accomplish this, based on the work of Gentry and Szydlo. The techniques involve algorithmic algebraic number theory, analytic number theory, commutative algebra, and lattice basis reduction. This sheds new light on the Gentry-Szydlo algorithm, and the ideas should be applicable to a range of questions in cryptography.

Keywords: lattices, Gentry-Szydlo algorithm, ideal lattices, lattice-based cryptography.

1 Introduction

In §7 of [6], Gentry and Szydlo introduced some powerful new ideas that combined in a clever way lattice basis reduction and number theory. They used these ideas to cryptanalyze NTRU Signatures. The recent interest in Fully Homomorphic Encryption (FHE) and in the candidate multilinear maps of Garg-Gentry-Halevi [2] bring the Gentry-Szydlo results once again to the fore. Gentry's first FHE scheme [3] used ideal lattices, as have a number of subsequent schemes. Fully Homomorphic Encryption is performed more efficiently with ideal lattices

---

* This material is based on research sponsored by DARPA under agreement numbers FA8750-11-1-0248 and FA8750-13-2-0054. The U.S. Government is authorized to reproduce and distribute reprints for Governmental purposes notwithstanding any copyright notation thereon. The views and conclusions contained herein are those of the authors and should not be interpreted as necessarily representing the official policies or endorsements, either expressed or implied, of DARPA or the U.S. Government.

J.A. Garay and R. Gennaro (Eds.): CRYPTO 2014, Part I, LNCS 8616, pp. 280–296, 2014.

than with general lattices. However, ideal lattices are special, with much structure ("symmetries") that has the potential to be exploited. In his thesis [4], Gentry mentions that the Gentry-Szydlo attack on NTRU signatures can be used to attack principal ideal lattices in the ring $\mathbb{Z}[X]/(X^n - 1)$, if the lattice has an orthonormal basis.

As Gentry pointed out [5], the Gentry-Szydlo algorithm "seems to be a rather crazy, unusual combination of LLL with more 'algebraic' techniques. It seems like it should have more applications—e.g., perhaps to breaking or weakening ideal lattices." Generalizing or improving the Gentry-Szydlo algorithm would potentially affect the security of all cryptography that is built from ideal lattices, or whose security is based on hard problems for ideal lattices. Candidate multilinear maps were recently cryptanalyzed using the Gentry-Szydlo algorithm. As remarked by Garg, Gentry, and Halevi in [2], their "new algebraic/lattice attacks are extensions of an algorithm by Gentry and Szydlo, which combines lattice reduction and Fermat's Little Theorem in a clever way to solve a relative norm equation in a cyclotomic field."

The Gentry-Szydlo algorithm has been viewed by some as magic [11]. In this paper we revisit the algorithm and put it in a mathematical framework, in order to make it easier to understand, generalize, and improve on. That should help make it more widely applicable in cryptographic applications. We embed the algorithm in a wider theory that we refer to as "lattices with symmetry".

The algorithm of Gentry and Szydlo can be viewed as a way to find an orthonormal basis (if one exists) for an ideal lattice. Determining whether a lattice has an orthonormal basis is a difficult algorithmic problem that is easier when the lattice has many symmetries. In this paper we solve this problem when the lattice comes with a sufficiently large abelian group of automorphisms, and we show how the Gentry-Szydlo algorithm is a special case of this result.

Our algorithm runs in deterministic polynomial time, whereas [6] relies on a probabilistic algorithm. Also, our setting is more general (our theory applies to arbitrary finite abelian groups, where [6] considers only cyclic groups of odd prime order), thereby covering other cases of potential cryptographic interest.

Briefly, our main result is as follows (see §2 for background information). If G is a finite abelian group and $u \in G$ has order 2, define a G-lattice to be a lattice L with a group homomorphism $G \to \text{Aut}(L)$ that takes u to -1. The "standard" G-lattice is the modified group ring $\mathbb{Z}\langle G \rangle = \mathbb{Z}[G]/(u + 1)$. A G-isomorphism is an isomorphism of lattices that respects the G-actions.

Theorem 1.1 *There is a deterministic polynomial time algorithm that, given a finite abelian group G, an element $u \in G$ of order 2, and a G-lattice L, decides whether L and $\mathbb{Z}\langle G \rangle$ are G-isomorphic, and if they are, exhibits a G-isomorphism.*

The ingredients include the technique invented by Gentry and Szydlo in [6], lattice basis reduction, commutative algebra (finite rings and tensor algebras), analytic number theory, and algorithmic algebraic number theory. The graded tensor algebra Λ introduced in §3.4 is in a sense the hero of our story. It replaces Gentry's and Szydlo's polynomial chains. In §7 of [6], taking powers of an ideal in the ring $R = \mathbb{Z}[X]/(X^n - 1)$ required complicated bookkeeping, via polynomial chains and lattice basis reduction to avoid coefficient blow-up. We do away with this, by using the module structure of the ideal, rather than its ideal structure. More precisely, an ideal in a commutative ring R is the same as an R-module M along with an embedding $M \hookrightarrow R$ of R-modules. While Gentry and Szydlo use the embedding, we observe that one can avoid coefficient blow-up by using the module structure of M but not the actual embedding. We replace ideal multiplication with tensor products of lattices.

In §2 we introduce the concept of a G-lattice, and in §2.3 we show that Theorem 1.1 implies the result of Gentry and Szydlo. In §3–§4 we introduce invertible G-lattices, of which the ideal lattices considered by Gentry and Szydlo are examples, and give the concepts and results that we use to state our new algorithm and prove its correctness. We explicitly present the algorithm in §5.

2 G-Lattices and the Modified Group Ring

In this section we explain some notation and concepts that we use in our main result.

2.1 Lattices and G-Lattices

We first give some background on lattices (see also [10]), and introduce G-lattices.

Definition 2.1 *A* **lattice** *or* **integral lattice** *is a finitely generated abelian group L with a map $\langle \cdot, \cdot \rangle : L \times L \to \mathbb{Z}$ that is*

- *bilinear: $\langle x, y + z \rangle = \langle x, y \rangle + \langle x, z \rangle$ and $\langle x + y, z \rangle = \langle x, z \rangle + \langle y, z \rangle$ for all $x, y, z \in L$,*
- *symmetric: $\langle x, y \rangle = \langle y, x \rangle$ for all $x, y \in L$, and*
- *positive definite: $\langle x, x \rangle > 0$ if $0 \neq x \in L$.*

As a group, L is isomorphic to $\mathbb{Z}^n$ for some n, which is called the **rank** of L. In algorithms, a lattice is specified by a Gram matrix $(\langle b_i, b_j \rangle)_{i,j=1}^n$ associated to a $\mathbb{Z}$-basis $\{b_1, \ldots, b_n\}$.

Definition 2.2 *The* **standard lattice** *of rank n is $L = \mathbb{Z}^n$ with $\langle x, y \rangle = \sum_{i=1}^n x_i y_i$. Its Gram matrix is the $n \times n$ identity matrix I_n.*

Definition 2.3 *A lattice L is* **unimodular** *if the map $L \to \mathrm{Hom}(L, \mathbb{Z})$ that takes each $x \in L$ to the map $y \mapsto \langle x, y \rangle$ is bijective. Equivalently, L is unimodular if its Gram matrix has determinant 1.*

Definition 2.4 *An* **isomorphism** *$L \xrightarrow{\sim} M$ of lattices is a group isomorphism $\varphi : L \xrightarrow{\sim} M$ that respects the lattice structures, i.e., $\langle \varphi(x), \varphi(y) \rangle = \langle x, y \rangle$ for all $x, y \in L$. If such a map φ exists, then L and M are* **isomorphic** *lattices. An* **automorphism** *of a lattice L is an isomorphism from L onto itself. The set of automorphisms of L is a finite group $\mathrm{Aut}(L)$ whose center contains -1 (represented by $-I_n$).*

In algorithms, isomorphisms are specified by their matrices on the given bases of L and M.

Examples 2.5 (i) *"Random" lattices have* $\mathrm{Aut}(L) = \{\pm 1\}$.
(ii) *Letting S_n denote the symmetric group on n letters and $\rtimes$ denote semidirect product, then $\mathrm{Aut}(\mathbb{Z}^n) \cong \{\pm 1\}^n \rtimes S_n$. (The standard basis vectors can be permuted, and negatives taken.)*
(iii) *If L is the equilateral triangular lattice in the plane, then $\mathrm{Aut}(L)$ is the symmetry group of the regular hexagon, which is a dihedral group of order 12.*

From now on, suppose that G is a finite abelian group, and $u \in G$ is a fixed element of order 2.

Definition 2.6 *A G-lattice is a lattice L together with a group homomorphism $f : G \to \mathrm{Aut}(L)$ such that $f(u) = -1$. For each $\sigma \in G$ and $x \in L$, define $\sigma x \in L$ by $\sigma x = f(\sigma)(x)$.*

The abelian group G is specified by a multiplication table. The G-lattice L is specified as a lattice along with, for each $\sigma \in G$, the matrix describing the action of σ on L.

Definition 2.7 *If L and M are G-lattices, then a G-isomorphism is an isomorphism $\varphi : L \xrightarrow{\sim} M$ of lattices that respects the G-actions, i.e., $\varphi(\sigma x) = \sigma \varphi(x)$ for all $x \in L$ and $\sigma \in G$. If such an isomorphism exists, we say that L and M are G-isomorphic, or isomorphic as G-lattices.*

2.2   The Modified Group Ring $\mathbb{Z}\langle G \rangle$

We define a modified group ring $A\langle G \rangle$ whenever A is a commutative ring. We will usually take $A = \mathbb{Z}$, but will also take $A = \mathbb{Z}/m\mathbb{Z}$. We consider $A\langle G \rangle$ rather than the standard group ring $A[G]$, since G-lattices become $\mathbb{Z}\langle G \rangle$-modules. Also, it allows us to include the cyclotomic rings $\mathbb{Z}[X]/(X^{2^k} + 1)$ in our theory.

The group ring $A[G]$ is the set of formal sums $\sum_{\sigma \in G} a_\sigma \sigma$ with $a_\sigma \in A$, with addition defined by

$$\sum_{\sigma \in G} a_\sigma \sigma + \sum_{\sigma \in G} b_\sigma \sigma = \sum_{\sigma \in G} (a_\sigma + b_\sigma)\sigma$$

and multiplication defined by

$$\left(\sum_{\sigma \in G} a_\sigma \sigma\right)\left(\sum_{\tau \in G} b_\tau \tau\right) = \sum_{\rho \in G}\left(\sum_{\sigma\tau = \rho} a_\sigma b_\tau\right)\rho.$$

For example, if G is a cyclic group of order m and g is a generator, then as rings $\mathbb{Z}[X]/(X^m - 1) \cong \mathbb{Z}[G]$ via the map $\sum_{i=0}^{m-1} a_i X^i \mapsto \sum_{i=0}^{m-1} a_i g^i$.

Definition 2.8 *If A is a commutative ring, then writing 1 for the identity element of the group G, we define the* **modified group ring**

$$A\langle G \rangle = A[G]/(u+1).$$

Every G-lattice is a $\mathbb{Z}\langle G \rangle$-module, where one uses the G-action on L to define ax whenever $x \in L$ and $a \in \mathbb{Z}\langle G \rangle$.

Definition 2.9 *Define the* **scaled trace function** $t : A\langle G \rangle \to A$ *by*

$$t(\sum_{\sigma \in G} a_\sigma \sigma) = a_1 - a_u.$$

Then t is the (additive) group homomorphism satisfying $t(1) = 1$, $t(u) = -1$, and $t(\sigma) = 0$ if $\sigma \in G$ and $\sigma \neq 1, u$.

Definition 2.10 *For $a = \sum_{\sigma \in G} a_\sigma \sigma \in A\langle G \rangle$, define $\bar{a} = \sum_{\upsilon \in G} a_\sigma \sigma^{-1}$.*

The map $a \mapsto \bar{a}$ is a ring automorphism of $A\langle G \rangle$. Since $\bar{\bar{a}} = a$, it is an involution. (An involution is a map that is its own inverse.) In practice, this map plays the role of complex conjugation.

Remark 2.11 *If L is a G-lattice and $x, y \in L$, then $\langle \sigma x, \sigma y \rangle = \langle x, y \rangle$ for all $\sigma \in G$. It follows that $\langle ax, y \rangle = \langle x, \bar{a}y \rangle$ for all $a \in \mathbb{Z}\langle G \rangle$.*

Definition 2.12 *For $x, y \in \mathbb{Z}\langle G \rangle$ define $\langle x, y \rangle_{\mathbb{Z}\langle G \rangle} = t(x\bar{y})$.*

Let $n = |G|/2 \in \mathbb{Z}$.

Definition 2.13 *Let S be a set of coset representatives of $G/\langle u \rangle$ (i.e., $\#S = n$ and $G = S \sqcup uS$), and for simplicity take S so that $1 \in S$.*

The following result is straightforward.

Proposition 2.14 (i) *The additive group of the ring $\mathbb{Z}\langle G \rangle$ is a G-lattice of rank n, with lattice structure defined by $\langle x, y \rangle_{\mathbb{Z}\langle G \rangle}$ and G-action defined by $\sigma x = \sigma x$ where the right-hand side is ring multiplication in $\mathbb{Z}\langle G \rangle$.*
(ii) *As lattices, $\mathbb{Z}\langle G \rangle \cong \mathbb{Z}^n$.*
(iii) *$\mathbb{Z}\langle G \rangle = \{\sum_{\sigma \in S} a_\sigma \sigma : a_\sigma \in \mathbb{Z}\} = \bigoplus_{\sigma \in S} \mathbb{Z}\sigma$ and $t(\sum_{\sigma \in S} a_\sigma \sigma) = a_1$.*

Definition 2.15 *We call $\mathbb{Z}\langle G \rangle$ the* **standard G-lattice**.

Example 2.16 *Suppose $G = H \times \langle u \rangle$ with $H \cong \mathbb{Z}/n\mathbb{Z}$. Then $\mathbb{Z}\langle G \rangle \cong \mathbb{Z}[H] \cong \mathbb{Z}[X]/(X^n - 1)$ as rings and as lattices. When n is odd (so G is cyclic), then (by sending X to $-X$) we have $\mathbb{Z}\langle G \rangle \cong \mathbb{Z}[X]/(X^n - 1) \cong \mathbb{Z}[X]/(X^n + 1)$.*

Remark 2.17 *The ring $\mathbb{Z}\langle G \rangle$ is an integral domain (i.e., no zero divisors) if and only if G is cyclic and n is a power of 2. If G is cyclic of order 2^r, then $\mathbb{Z}\langle G \rangle \cong \mathbb{Z}[\zeta_{2^r}]$.*

2.3   Ideal Lattices

Example 2.18 *Suppose I is an ideal in the ring $\mathbb{Z}\langle G \rangle$ and $w \in \mathbb{Z}\langle G \rangle$. Suppose that $I\overline{I} = \mathbb{Z}\langle G \rangle \cdot w$ and $\psi(w) \in \mathbb{R}_{>0}$ for all ring homomorphisms $\psi : \mathbb{Z}\langle G \rangle \to \mathbb{C}$. It follows that the ideal I has finite index in $\mathbb{Z}\langle G \rangle$, that $\overline{w} = w$, and that w is not a zero divisor. Define the G-lattice $L_{(I,w)}$ to be I with G-action given by multiplication in $\mathbb{Z}\langle G \rangle$, and with lattice structure defined by*

$$\langle x, y \rangle_{I,w} = t\left(\frac{x\overline{y}}{w}\right)$$

with t as in Definition 2.9. (Note that $\frac{x\overline{y}}{w} \in \mathbb{Z}\langle G \rangle$ since w generates the ideal $I\overline{I}$.) In particular, $L_{(\mathbb{Z}\langle G \rangle, 1)} = \mathbb{Z}\langle G \rangle$.

The lattice $L_{(I,w)}$ is G-isomorphic to $\mathbb{Z}\langle G \rangle$ if and only if there exists $v \in \mathbb{Z}\langle G \rangle$ such that $I = (v)$ and $w = v\overline{v}$. Further, knowing *such a G-isomorphism is equivalent to* knowing v. *More precisely, v is the image of 1 under a G-isomorphism $\mathbb{Z}\langle G \rangle \xrightarrow{\sim} L_{(I,w)}$, and $w = v\overline{v}$ if and only if $\langle av, bv \rangle_{I,w} = t(a\overline{b}) = \langle a, b \rangle_{\mathbb{Z}\langle G \rangle}$ for all $a, b \in \mathbb{Z}\langle G \rangle$. Thus, finding v from I and $v\overline{v}$ in polynomial time is equivalent to finding a G-isomorphism $\mathbb{Z}\langle G \rangle \xrightarrow{\sim} L_{(I,w)}$ in polynomial time.*

The point of dividing by w in the definition of $\langle x, y \rangle_{I,w}$ is to make the lattice L unimodular. It follows that when we take tensor powers of L over $\mathbb{Z}\langle G \rangle$, as we will do in §3 below, there will be no coefficient blow-up.

We next show how to recover the Gentry-Szydlo result from Theorem 1.1. The Gentry-Szydlo algorithm finds a generator v of an ideal I of finite index in the ring $R = \mathbb{Z}[X]/(X^n - 1)$, given $v\overline{v}$, a $\mathbb{Z}$-basis for I, and a "promise" that v exists. Here, n is an odd prime, and for $v = v(X) = \sum_{i=0}^{n-1} a_i X^i \in R$, its "reversal" is $\overline{v} = v(X^{-1}) = a_0 + \sum_{i=1}^{n-1} a_{n-i} X^i \in R$. We take G to be a cyclic group of order $2n$. Then $R \cong \mathbb{Z}\langle G \rangle$ as in Example 2.16, and we identify R with $\mathbb{Z}\langle G \rangle$. Let $w = v\overline{v} \in \mathbb{Z}\langle G \rangle$ and let $L = L_{(I,w)}$ as above. Then L is the "implicit orthogonal lattice" in §7.2 of [6]. Once you know a $\mathbb{Z}$-basis for I and w, you know L. Theorem 1.1 produces a G-isomorphism $\mathbb{Z}\langle G \rangle \xrightarrow{\sim} L$ in polynomial time, and thus gives a generator v in polynomial time.

3   Invertible G-Lattices, Short Vectors, and the Tensor Algebra Λ

In this section we give some concepts that we will use to prove Theorem 1.1.

3.1   Invertible G-Lattices

Definition 3.1 *If L is a G-lattice, then the G-lattice $\overline{L}$ is a lattice equipped with a lattice isomorphism $L \xrightarrow{\sim} \overline{L}$, $x \mapsto \overline{x}$ and a group homomorphism $G \to \mathrm{Aut}(\overline{L})$ defined by $\sigma\overline{x} = \overline{\sigma^{-1}x} = \overline{\overline{\sigma}x}$ for all $\sigma \in G$ and $x \in L$, i.e., $\overline{\sigma}\overline{x} = \overline{\sigma}\,\overline{x}$.*

Definition 3.2 *If L is a G-lattice, define the lifted inner product*

$$\cdot : L \times \overline{L} \to \mathbb{Z}\langle G \rangle \quad by \quad x \cdot \overline{y} = \sum_{\sigma \in S} \langle x, \sigma y \rangle \sigma \in \mathbb{Z}\langle G \rangle.$$

Then

$$\langle x, y \rangle = t(x \cdot \overline{y}) \tag{1}$$

and $x \cdot \overline{y} = \overline{y \cdot \overline{x}}$. This lifted inner product is $\mathbb{Z}\langle G \rangle$-bilinear, i.e., $(ax) \cdot \overline{y} = x \cdot (a\overline{y}) = a(x \cdot \overline{y})$ for all $a \in \mathbb{Z}\langle G \rangle$ and all $x, y \in L$.

Example 3.3 *If $L = \mathbb{Z}\langle G \rangle$, then $\overline{L} = \mathbb{Z}\langle G \rangle$ with $^-$ having the same meaning as in Definition 2.10 for $A = \mathbb{Z}$, and with $\cdot$ being multiplication in $\mathbb{Z}\langle G \rangle$.*

Definition 3.4 *A G-lattice L is **invertible** if the following three conditions all hold:*

(i) $\text{rank}(L) = n = |G|/2$;
(ii) *L is unimodular (see Definition 2.3);*
(iii) *for each $m \in \mathbb{Z}_{>0}$ there exists $e_m \in L$ such that $\{\sigma e_m + mL : \sigma \in G\}$ generates the abelian group L/mL.*

Example 3.5 *If a G-lattice L is G-isomorphic to the standard G-lattice then L is invertible. For (iii), observe that the group $\mathbb{Z}\langle G \rangle$ is generated by $\{\sigma 1 : \sigma \in G\}$, so the group L is generated by $\{\sigma e : \sigma \in G\}$ where e is the image of 1 under the isomorphism. Now let $e_m = e$ for all m.*

Remark 3.6 *In the full version of the paper we will show that a G-lattice L is invertible if and only if there is a $\mathbb{Z}\langle G \rangle$-module M such that $L \otimes_{\mathbb{Z}\langle G \rangle} M$ and $\mathbb{Z}\langle G \rangle$ are isomorphic as $\mathbb{Z}\langle G \rangle$-modules and L is unimodular. (See Chapter XVI of [8] for tensor products.) We will also show that this is equivalent to the map $\varphi : L \otimes_{\mathbb{Z}\langle G \rangle} \overline{L} \to \mathbb{Z}\langle G \rangle$ defined by $\varphi(x \otimes \overline{y}) = x \cdot \overline{y}$ being an isomorphism of $\mathbb{Z}\langle G \rangle$-modules. Further, L is invertible if and only if L is G-isomorphic to $L_{(I,w)}$ for some I and w as in Example 2.18.*

Definition 3.4(iii) states that L/mL is a free $(\mathbb{Z}/m\mathbb{Z})\langle G \rangle$-module of rank one for all $m > 0$. Given an ideal, it is a hard problem to decide if it is principal. But checking (iii) of Definition 3.4 is easy algorithmically; see Proposition 4.4(ii) below.

3.2   Short Vectors

Definition 3.7 *We will say that a vector e in an integral lattice L is **short** if $\langle e, e \rangle = 1$.*

Example 3.8 *The short vectors in the standard lattice of rank n are the $2n$ signed standard basis vectors $\{(0, \ldots, 0, \pm 1, 0, \ldots, 0)\}$. Thus, the set of short vectors in $\mathbb{Z}\langle G \rangle$ is G.*

Proposition 3.9 *Suppose L is an invertible G-lattice. Then:*

(i) *if e is short, then $\{\sigma \in G : \sigma e = e\} = \{1\}$;*

(ii) *if e is short, then $\langle e, \sigma e \rangle$ is 1 if $\sigma = 1$, is -1 if $\sigma = u$, and is 0 for all other $\sigma \in G$;*

(iii) *$e \in L$ is short if and only if $e \cdot \bar{e} = 1$, with inner product $\cdot$ defined in Definition 3.2.*

Proof. Suppose $e \in L$ is short. Let $H = \{\sigma \in G : \sigma e = e\}$. For all $\sigma \in G$, by the Cauchy-Schwarz inequality we have $|\langle e, \sigma e \rangle| \leq (\langle e, e \rangle \langle \sigma e, \sigma e \rangle)^{1/2} = \langle e, e \rangle = 1$, and $|\langle e, \sigma e \rangle| = 1$ if and only if e and σe lie on the same line through 0. Thus $\langle e, \sigma e \rangle \in \{1, 0, -1\}$. Then $\langle e, \sigma e \rangle = 1$ if and only if $\sigma \in H$. Also, $\langle e, \sigma e \rangle = -1$ if and only if $\sigma e = -e$ if and only if $\sigma \in Hu$. Otherwise, $\langle e, \sigma e \rangle = 0$. Thus for (i,ii), it suffices to prove $H = \{1\}$.

Let T be a set of coset representatives for G mod $H\langle u \rangle$ and let $S = T \cdot H$, a set of coset representatives for G mod $\langle u \rangle$. If $a = \sum_{\sigma \in S} a_\sigma \sigma \in (\mathbb{Z}/m\mathbb{Z})\langle G \rangle$ is fixed by H, then $a_{\tau\sigma} = a_\sigma$ for all $\sigma \in S$ and $\tau \in H$, so $a \in (\sum_{\tau \in H} \tau)(\mathbb{Z}/m\mathbb{Z})\langle G \rangle$.

Let $m = |H|$. By Definition 3.4(iii), there is a $\mathbb{Z}[H]$-module isomorphism $L/mL \cong (\mathbb{Z}/m\mathbb{Z})\langle G \rangle$. The latter is a free module over $(\mathbb{Z}/m\mathbb{Z})[H]$ with basis T. Since $e + mL \in (L/mL)^H$ we have $e = m\varepsilon_1 + (\sum_{\tau \in H} \tau)\varepsilon_2$ with $\varepsilon_1, \varepsilon_2 \in L$. Since $\langle e, \tau\varepsilon_2 \rangle = \langle \tau e, \tau\varepsilon_2 \rangle = \langle e, \varepsilon_2 \rangle$ for all $\tau \in H$, we have

$$1 = \langle e, e \rangle = m\langle e, \varepsilon_1 \rangle + \sum_{\tau \in H} \langle e, \tau\varepsilon_2 \rangle = m\langle e, \varepsilon_1 + \varepsilon_2 \rangle \equiv 0 \mod m.$$

Thus, $m = 1$ as desired. Part (iii) follows directly from (ii) and Definition 3.2.

This enables us to prove the following result.

Proposition 3.10 *Suppose L is a G-lattice. Then:*

(i) *if L is invertible, then the map*

$$\{G\text{-isomorphisms } \mathbb{Z}\langle G \rangle \to L\} \to \{\text{short vectors of } L\}$$

that sends f to $f(1)$ is bijective;

(ii) *if $e \in L$ is short and L is invertible, then $\{\sigma e : \sigma \in G\}$ generates the abelian group L;*

(iii) *L is G-isomorphic to $\mathbb{Z}\langle G \rangle$ if and only if L is invertible and has a short vector;*

(iv) *if $e \in L$ is short and L is invertible, then the map $G \to \{\text{short vectors of } L\}$ defined by $\sigma \mapsto \sigma e$ is bijective.*

Proof. For (i), that $f(1)$ is short is clear. Injectivity of the map $f \mapsto f(1)$ follows from $\mathbb{Z}\langle G \rangle$-linearity of G-isomorphisms. For surjectivity, suppose $e \in L$ is short. Proposition 3.9(ii) says that $\{\sigma e\}_{\sigma \in S}$ is an orthonormal basis for L. Parts (ii) and (i) now follow, where the G-isomorphism f is defined by $x \mapsto xe$ for all $x \in \mathbb{Z}\langle G \rangle$. Part (iii) follows from (i) and Example 3.5. For (iv), injectivity follows from Proposition 3.9(i). For surjectivity, suppose $e' \in L$ is short. Take G-isomorphisms f and f' with $f(1) = e$ and $f'(1) = e'$ as in (i), and let $\sigma = f^{-1} \circ f'(1)$. Then σ is a short vector in $\mathbb{Z}\langle G \rangle$ such that $\sigma e = e'$. By Example 3.8 we have $\sigma \in G$.

3.3   The Witt-Picard Group

If L and M are invertible G-lattices, then the $\mathbb{Z}\langle G\rangle$-module $L \otimes_{\mathbb{Z}\langle G\rangle} M$ is a G-lattice with lifted inner product $(x \otimes v) \cdot (\overline{y} \otimes \overline{w}) = (x \cdot \overline{y})(v \cdot \overline{w})$, for all $x, y \in L$ and $v, w \in M$, and with lattice structure $\langle a, b\rangle = t(a \cdot \overline{b})$ for all $a, b \in L \otimes_{\mathbb{Z}\langle G\rangle} M$. In the notation of Example 2.18 we have

$$L_{(I_1, w_1)} \otimes_{\mathbb{Z}\langle G\rangle} L_{(I_2, w_2)} = L_{(I_1 I_2, w_1 w_2)},$$

where $I_1 I_2$ is the product of ideals.

Definition 3.11 *If L is an invertible G-lattice, let $[L]$ denote its G-isomorphism class, i.e., the class of all G-lattices that are G-isomorphic to L. We define the* **Witt-Picard group** *of $\mathbb{Z}\langle G\rangle$ to be the set of all G-isomorphism classes of invertible G-lattices, with group operation defined by $[L] \cdot [M] = [L \otimes_{\mathbb{Z}\langle G\rangle} M]$, with identity element $[\mathbb{Z}\langle G\rangle]$, and with $[L]^{-1} = [\overline{L}]$.*

The Witt-Picard group is a finite abelian group. When computing in the Witt-Picard group, one can apply a lattice basis reduction algorithm whenever the numbers get too large. More precisely, algorithmically we represent an invertible G-lattice M by letting $M = \mathbb{Z}^n$ as an abelian group, specifying a group homomorphism $G \to \mathrm{GL}(n, \mathbb{Z})$ giving the action of G on M, and giving data describing the map $\cdot : M \times M \to \mathbb{Z}\langle G\rangle$; the lattice structure is then given by $\langle a, b\rangle = t(a \cdot \overline{b})$ for all $a, b \in M$. If M_1 and M_2 are invertible G-lattices, $m_1, m_2 \in \mathbb{Z}_{>0}$, and $d_i \in M_i / m_i M_i$ for $i = 1, 2$, one can compute $(M_1 \otimes_{\mathbb{Z}\langle G\rangle} M_2, d_1 \otimes d_2)$ in polynomial time. Also, there is a deterministic polynomial time algorithm that, given M and given $d \in M/mM$, produces a pair (M', d') and a G-isomorphism $(M, d) \to (M', d')$ such that the standard basis of $M' = \mathbb{Z}^n$ is LLL-reduced (and thus each entry of the Gram matrix is at most 2^{n-1} in absolute value, by Lemma 3.12 below). This in fact proves the finiteness of the Witt-Picard group.

If $L = L_{(I, w)}$ for some I and w as in Example 2.18, and $j \in \mathbb{Z}_{>0}$, then $[L]^j$ is the G-isomorphism class of $L_{(I^j, w^j)}$. One can compute $[L]^j$ in deterministic polynomial time using an addition chain for j, and LLL-reducing intermediate powers to prevent coefficient blow-up. This takes the place of the polynomial chains in §7.4 of [6].

Lemma 3.12 *If $\{b_1, \ldots, b_n\}$ is an LLL-reduced basis for an integral unimodular lattice L and $\{b_1^*, \ldots, b_n^*\}$ is its Gram-Schmidt orthogonalization, then*

$$2^{1-i} \leq |b_i^*|^2 \leq 2^{n-i}$$

and $|b_i|^2 \leq 2^{n-1}$ for all $i \in \{1, \ldots, n\}$.

Proof. Being LLL-reduced means that $b_i = b_i^* + \sum_{j=1}^{i-1} \mu_{ij} b_j^*$ with $|\mu_{ij}| \leq \frac{1}{2}$ for all $j < i \leq n$, and $|b_i^*|^2 \leq 2|b_{i+1}^*|^2$ for all $i < n$. Thus for $1 \leq j \leq i \leq n$ we have $|b_i^*|^2 \leq 2^{j-i}|b_j^*|^2$, so for all i we have

$$2^{1-i}|b_1^*|^2 \leq |b_i^*|^2 \leq 2^{n-i}|b_n^*|^2.$$

Since L is integral we have $|b_1^*|^2 = |b_1|^2 = \langle b_1, b_1 \rangle \geq 1$, so $|b_i^*|^2 \geq 2^{1-i}$. Letting $L_i = \sum_{j=1}^i \mathbb{Z}b_j$, then $|b_i^*| = \det(L_i)/\det(L_{i-1})$. Since L is integral and unimodular, $|b_n^*| = \det(L_n)/\det(L_{n-1}) = 1/\det(L_{n-1}) \leq 1$, so $|b_i^*|^2 \leq 2^{n-i}$. Since $\{b_i^*\}$ is orthogonal we have

$$|b_i|^2 = |b_i^*|^2 + \sum_{j=1}^{i-1} \mu_{ij}^2 |b_j^*|^2 \leq 2^{n-i} + \frac{1}{4}\sum_{j=1}^{i-1} 2^{n-j}$$

$$= 2^{n-i} + (2^{n-2} - 2^{n-i-1}) = 2^{n-2} + 2^{n-i-1} \leq 2^{n-1}.$$

3.4   The Extended Tensor Algebra Λ

We are now ready to introduce the extended tensor algebra Λ in which our computations take place. Suppose L is an invertible G-lattice. Letting $L^{\otimes 0} = \mathbb{Z}\langle G \rangle$ and letting $L^{\otimes m} = L \otimes_{\mathbb{Z}\langle G \rangle} \cdots \otimes_{\mathbb{Z}\langle G \rangle} L$ (m times) and $L^{\otimes(-m)} = \overline{L}^{\otimes m} = \overline{L} \otimes_{\mathbb{Z}\langle G \rangle} \cdots \otimes_{\mathbb{Z}\langle G \rangle} \overline{L}$ for all $m \in \mathbb{Z}_{>0}$, define the extended tensor algebra

$$\Lambda = \bigoplus_{i \in \mathbb{Z}} L^{\otimes i} = \ldots \oplus \overline{L}^{\otimes 3} \oplus \overline{L}^{\otimes 2} \oplus \overline{L} \oplus \mathbb{Z}\langle G \rangle \oplus L \oplus L^{\otimes 2} \oplus L^{\otimes 3} \oplus \ldots$$

("extended" because we extend the usual notion to include negative exponents $L^{\otimes(-m)}$). Each $L^{\otimes i}$ is an invertible G-lattice, and represents $[L]^i$. For simplicity, we denote $L^{\otimes i}$ by L^i. The ring structure on Λ is defined as the ring structure on the tensor algebra, supplemented with the lifted inner product $\cdot$. The following result is straightforward.

Proposition 3.13 (i) Λ is a commutative ring containing $\mathbb{Z}\langle G \rangle$ as a subring;
  (ii) the action of G on L becomes multiplication in Λ, and likewise for the action of G on $\overline{L}$;
  (iii) Λ has an involution $x \mapsto \overline{x}$ extending both the involution of $\mathbb{Z}\langle G \rangle$ and the map $L \xrightarrow{\sim} \overline{L}$;
  (iv) the lifted inner product $\cdot : L \times \overline{L} \to \mathbb{Z}\langle G \rangle$ becomes multiplication in Λ;
  (v) if $e \in L$ is short, then $\overline{e} = e^{-1}$ in Λ and $\Lambda = \mathbb{Z}\langle G \rangle[e, e^{-1}]$.

All computations in Λ and in $\Lambda/m\Lambda$ will be done with homogeneous elements only, where the set of homogeneous elements of Λ is $\bigcup_{i \in \mathbb{Z}} L^i$.

4   The Main Ingredients

We give the main results that we will use to prove Theorem 1.1. Fix as before a finite abelian group G of order $2n$ and $u \in G$ of order 2. Let k denote the exponent of G. (The exponent of a group H is the least positive integer k such that $\sigma^k = 1$ for all $\sigma \in H$. The exponent of H divides $|H|$ and has the same prime factors as $|H|$.) For all $m \in \mathbb{Z}_{>1}$, denote by $k(m)$ the exponent of the unit group $(\mathbb{Z}\langle G \rangle/(m))^*$.

Remark 4.1 *By Proposition 3.10, the G-isomorphisms $\mathbb{Z}\langle G \rangle \xrightarrow{\sim} L$ are in one-to-one correspondence with the short vectors, and if a short $e \in L$ exists, then the short vectors of L are exactly the $2n$ vectors $\{\sigma e : \sigma \in G\}$. If k is the exponent of G, then $(\sigma e)^k = \sigma^k e^k = e^k$ in Λ. Hence for invertible L, all short vectors in L have the same k-th power $e^k \in \Lambda$. At least philosophically, it is easier to find things that are uniquely determined. We look for e^k first, and then recover e from it.*

Proposition 4.2 *There is a deterministic polynomial time algorithm that, given a finite commutative ring R and an R-module M, decides whether M is a free R-module of rank one, and if it is, finds a generator.*

Proof. We sketch a proof. A complete proof will be given in the full version of the paper.

The inputs are given as follows. The ring R is given as an abelian group (say, as a sum of cyclic groups) along with all the products of pairs of generators. The finite R-module M is given as an abelian group (say, as a sum of cyclic groups), and for all generators of the abelian group R and all generators of the abelian group M, we are given the module products in M.

If $\#M \neq \#R$, output "no" and stop.

Suppose that A and B are finite commutative rings, that $R \twoheadrightarrow A \times B$ is a surjective ring homomorphism with nilpotent kernel, and that $y_B \in M$ is such that the map $B \to M_B = B \otimes_R M$, $b \mapsto b \otimes y_B$ is an isomorphism. Let I denote the kernel of the natural map $R \to B$ and let N denote the image of IM under the natural map $M \to M_A$.

Initially, take $A = R$, $B = 0$, and $y_B = 0$. As long as $A \neq 0$, do the following. If $N = 0$, output "no" and stop. Otherwise, pick $x_A \in IM$ whose image $x \in N$ is nonzero. Compute $\boldsymbol{a} = \mathrm{Ann}_A x$, where Ann_A denotes the annihilator in A. Let $\boldsymbol{b} = \mathrm{Ann}_A \boldsymbol{a}$.

If $\boldsymbol{a} = \boldsymbol{a}^2$, then $A \xrightarrow{\sim} A/\boldsymbol{a} \times A/\boldsymbol{b}$ and $M_A \xrightarrow{\sim} M_{A/\boldsymbol{a}} \times M_{A/\boldsymbol{b}}$. The image of x is of the form $(x', 0)$. If x' does not generate $M_{A/\boldsymbol{a}}$, stop with "no". Otherwise, compute $\beta \in R$ that maps to $(0, 1)$ under the map $R \twoheadrightarrow A \times B$, and replace y_B, B, A by $\beta y_B + x_A$, $(A/\boldsymbol{a}) \times B$, $A/\boldsymbol{b}$, respectively. If $\boldsymbol{a} \neq \boldsymbol{a}^2$, then $\boldsymbol{a} \cap \boldsymbol{b}$ is a nonzero nilpotent ideal, and we replace A by $A/(\boldsymbol{a} \cap \boldsymbol{b})$ and leave y_B unchanged.

When $A = 0$, then I is nilpotent; say $I^r = 0$. Then $By = M_B = M/IM$ for $y = (y_B \mod IM)$. Thus,

$$M = Ry_B + IM = Ry_B + I(Ry_B + IM) = Ry_B + I^2 M = \ldots = Ry_B + I^r M = Ry_B,$$

so output "yes". $\qquad \square$

Lemma 4.3 *Suppose that L is a G-lattice, $m \in \mathbb{Z}_{>0}$, and $e \in L$. Then*

$$\{\sigma e + mL : \sigma \in G\}$$

generates L/mL as an abelian group if and only if $L/(\mathbb{Z}\langle G \rangle \cdot e)$ is finite of order coprime to m.

Proof. The set $\{\sigma e + mL : \sigma \in G\}$ generates L/mL as an abelian group if and only multiplication by m is onto as a map from $L/(\mathbb{Z}\langle G\rangle \cdot e)$ to itself. Since $L/(\mathbb{Z}\langle G\rangle \cdot e)$ is a finitely generated abelian group, this holds if and only if $L/(\mathbb{Z}\langle G\rangle \cdot e)$ is finite of order coprime to m.

Proposition 4.4 (i) *There is a deterministic polynomial time algorithm that, given G, a G-lattice L, and $m \in \mathbb{Z}_{>0}$, decides whether there exists $e_m \in L$ such that $\{\sigma e_m + mL : \sigma \in G\}$ generates L/mL as an abelian group, and if so, finds one.*

(ii) *There is a deterministic polynomial time algorithm that, given G, u, and a G-lattice L, decides whether L is invertible.*

Proof. For (i), apply Proposition 4.2 with $R = \mathbb{Z}\langle G\rangle/(m)$ and $M = L/mL$.

For (ii), it is easy to check whether $\operatorname{rank}(L) = n$ and whether L is unimodular (check whether the Gram matrix has determinant 1). We need to check Definition 3.4(iii) for all m's in polynomial time. We show that it suffices to check two particular values of m. First take $m = 2$, and use (i) to determine if e_2 exists. If not, output "no". If there is one, use (i) to compute $e_2 \in L$. By Lemma 4.3, the group $L/(\mathbb{Z}\langle G\rangle \cdot e_2)$ is finite of odd order. Let q denote its order. Now apply (i) with $m = q$. If no e_q exists, output "no". If e_q exists, then for *all* $m \in \mathbb{Z}_{>0}$ there exists $e_m \in L$ that generates L/mL as a $\mathbb{Z}\langle G\rangle/(m)$-module, as follows. We can reduce to m being a prime power p^t, since if $\gcd(m, m') = 1$ then $L/mm'L$ is free of rank one over $\mathbb{Z}\langle G\rangle/(mm')$ if and only if L/mL is free of rank one over $\mathbb{Z}\langle G\rangle/(m)$ and $L/m'L$ is free of rank one over $\mathbb{Z}\langle G\rangle/(m')$. Lemma 4.3 now allows us to reduce to the case $m = p$. If $p \nmid q$, we can take $e_p = e_2$. If $p \mid q$, we can take $e_p = e_q$.

Proposition 4.5 *There is a deterministic polynomial time algorithm that, given a finite abelian group G of order $2n$ and $u \in G$ of order 2, determines prime powers ℓ and m such that $\ell, m \geq 2^{n/2} + 1$ and $\gcd(k(\ell), k(m)) = k$.*

Proof. One can prove that if p is prime and $p \equiv 1 \bmod k$, then

$$k(p^j) = (p-1)p^{j-1},$$

using induction on j and the facts that $(\mathbb{Z}\langle G\rangle/(p^j))^* \supset (\mathbb{Z}/p^j\mathbb{Z})^*$ and the latter group has exponent $(p-1)p^{j-1}$.

We next give an algorithm that, given $n, k \in \mathbb{Z}_{>0}$ with k even, computes $r, s \in \mathbb{Z}_{>0}$ and primes p and q such that $p \equiv q \equiv 1 \bmod k$,

$$\gcd((p-1)p^{r-1}, (q-1)q^{s-1}) = k,$$

$p^r \geq 2^{n/2} + 1$, and $q^s \geq 2^{n/2} + 1$. (We can then take $\ell = p^r$ and $m = q^s$.) Try $p = k+1, 2k+1, 3k+1, \ldots$ until the smallest prime $p \equiv 1 \bmod k$ is found. Find the least r such that $p^r \geq 2^{n/2} + 1$. Try $q = p+k, p+2k, \ldots$ until the least prime $q \equiv 1 \bmod k$ such that $\gcd((p-1)p, q-1) = k$ is found. Find the smallest s such that $q^s \geq 2^{n/2} + 1$.

This algorithm terminates, with correct output, in time $(n + k)^{O(1)}$. The key ingredient for proving this is Heath-Brown's version of Linnik's theorem [7], which implies that the prime p found by the algorithm satisfies $p \leq ck^{5.5}$ with an effective constant c. If $p - 1 = k_1 k_2$ with every prime divisor of k_1 also dividing k and with $\gcd(k_2, k) = 1$, then to have $\gcd((p - 1)p, q - 1) = k$ it suffices to have $q \equiv 2 \bmod p$ and $q \equiv 1 + k \bmod k_1$ and $q \equiv 2 \bmod k_2$. This gives a congruence $q \equiv a \bmod p(p - 1)$ for some a. Heath-Brown's version of Linnik's theorem implies that $q \leq c(p^2)^{5.5} \leq c^{12}k^{60.5}$.

Our prime powers ℓ and m play the roles that in the Gentry-Szydlo paper [6] were played by auxiliary prime numbers $P, P' > 2^{(n+1)/2}$ such that

$$\gcd(P - 1, P' - 1) = 2n.$$

Our $k(\ell)$ and $k(m)$ replace their $P - 1$ and $P' - 1$, respectively. While the Gentry-Szydlo primes P and P' are found with at best a probabilistic algorithm, we can find ℓ and m in deterministic polynomial time. (Further, the ring elements they work with were required to not be zero divisors modulo P, P' and other small auxiliary primes; we require no analogous condition on ℓ and m, since by Definition 3.4(iii), when L is invertible then for *all* m, the $(\mathbb{Z}/m\mathbb{Z})\langle G \rangle$-module L/mL is free of rank one.)

Proposition 4.6 (i) *Suppose L is an integral lattice, $3 \leq m \in \mathbb{Z}$, and $C \in L/mL$. Then C contains at most one element x with $\langle x, x \rangle = 1$.*
(ii) *There is a deterministic polynomial time algorithm that, given a rank n integral lattice L, $m \in \mathbb{Z}$ such that $m \geq 2^{n/2} + 1$, and $C \in L/mL$, finds all $x \in C$ with $\langle x, x \rangle = 1$ (and the number of them is 0 or 1).*

Proof. For (i), suppose $x, y \in C$, $\langle x, x \rangle = \langle y, y \rangle = 1$, and $x \neq y$. Since $x - y \in mL$ and L is an integral lattice, we have

$$m \leq \langle x - y, x - y \rangle^{1/2} \leq \langle x, x \rangle^{1/2} + \langle y, y \rangle^{1/2} = 1 + 1 = 2$$

by the triangle inequality. This contradicts $m \geq 3$, giving (i).

For (ii), using LLL to solve the closest vector problem, one can find (in polynomial time) $y \in C$ such that $\langle y, y \rangle < (2^n - 1)\langle x, x \rangle$ for all $x \in C$. Suppose $x \in C$ with $\langle x, x \rangle = 1$. Since $x, y \in C$, there exists $w \in L$ such that $x - y = mw$. Then

$$m\langle w, w \rangle^{1/2} = \langle x - y, x - y \rangle^{1/2} \leq \langle x, x \rangle^{1/2} + \langle y, y \rangle^{1/2} < (1 + 2^{n/2})\langle x, x \rangle^{1/2} \leq m.$$

Therefore $1 > \langle w, w \rangle^{1/2} \in \mathbb{Z}$, so $w = 0$, and thus $y = x$. Compute $\langle y, y \rangle$. If $\langle y, y \rangle = 1$, output y. If $\langle y, y \rangle \neq 1$, there is no $x \in C$ with $\langle x, x \rangle = 1$.

The n of [6] is an odd prime, so $k = 2n$ and $\mathbb{Z}\langle G \rangle$ embeds in $\mathbb{Q}(\zeta_n) \times \mathbb{Q}$. Since the latter is a product of only two number fields, the number of zeros of $X^{2n} - v^{2n}$ is at most $(2n)^2$, and the Gentry-Szydlo method for finding v from v^{2n} is sufficiently efficient. If one wants to generalize [6] to the case where n is not prime, then the smallest t such that $\mathbb{Z}\langle G\rangle$ embeds in $F_1 \times \ldots \times F_t$ with number fields F_i can be large. Given ν, the number of zeros of $X^k - \nu$ could be as large as k^t. Finding e such that $\nu = e^k$ then requires a more efficient algorithm, which we attain with Proposition 4.9 below.

An **order** is a commutative ring A whose additive group is isomorphic to $\mathbb{Z}^n$ for some $n \in \mathbb{Z}_{\geq 0}$. We specify an order by saying how to multiply any two vectors in a given basis. Let $\mu(A)$ denote the group of roots of unity in A.

Proposition 4.7 *There is a deterministic polynomial time algorithm that, given an order A, determines a set of generators for $\mu(A)$.*

Proof. The proof is a bit intricate, involving commutative algebra and algorithmic algebraic number theory. We give a sketch. See [1] for commutative algebra background.

One starts by computing the nilradical N of the $\mathbb{Q}$-algebra $A_{\mathbb{Q}} = A \otimes_{\mathbb{Z}} \mathbb{Q}$ as well as the unique subalgebra $E \subset A_{\mathbb{Q}}$ that maps isomorphically to $A_{\mathbb{Q}}/N$. One has $\mu(A) \subset E$, so replacing A by $A \cap E$ one reduces to the case in which the nilradical of A is 0, which we now assume. Next one determines the set $\mathrm{Spec}(E)$ of prime ideals $\mathfrak{m}$ of E. For each $\mathfrak{m}$ we compute $E/\mathfrak{m}$, which is an algebraic number field, and we also compute its subring $A/(\mathfrak{m} \cap A)$. One has $E \cong \prod_{\mathfrak{m} \in \mathrm{Spec}(E)} E/\mathfrak{m}$, and we identify A with a subring of finite additive index in the product ring $B = \prod_{\mathfrak{m} \in \mathrm{Spec}(E)} A/(\mathfrak{m} \cap A)$.

For each prime number p dividing $|\mu(A)|$ one has $p \leq 1 + \dim_{\mathbb{Q}} E$, so it will suffice to find, for each such p, a set of generators for the p-primary component $\mu(A)_p$ of $\mu(A)$. Fix now a prime number $p \leq 1 + \dim_{\mathbb{Q}} E$.

Since each $A/(\mathfrak{m} \cap A)$ is contained in a number field, $\mu(A/(\mathfrak{m} \cap A))_p$ is cyclic and easy to determine. This leads to a set of generators for $\mu(B)_p$.

Compute $C = \{x \in B : p^i x \in A \text{ for some } i \in \mathbb{Z}_{\geq 0}\}$; this is a subring of B containing A. The group C/A is finite of p-power order, and the group B/C is finite of order not divisible by p. We make $\mathrm{Spec}(E)$ into the set of vertices of a graph by connecting $\mathfrak{m}, \mathfrak{n} \in \mathrm{Spec}(E)$ with an edge if and only if

$$(\mathfrak{m} \cap C) + (\mathfrak{n} \cap C) \neq C.$$

For each connected component V of this graph, determine the image C_V of C in the product ring $\prod_{\mathfrak{m} \in V} A/(\mathfrak{m} \cap A)$. Then one can show that one has $C \cong \prod_V C_V$, with V ranging over the connected components, so that $\mu(C)_p \cong \prod_V \mu(C_V)_p$. In addition, one can show that for each V and each $\mathfrak{m} \in V$ the natural map $\mu(C_V)_p \to \mu(A/(\mathfrak{m} \cap A))_p$ is *injective*, so that $\mu(C_V)_p$ is cyclic; the proof also leads to an efficient algorithm for computing $\mu(C_V)_p$. Thus, at this point one knows a set of generators for $\mu(C)_p$.

To pass from $\mu(C)_p$ to $\mu(A)_p$, one starts by computing the intersection $\mathfrak{r}$ of all maximal ideals of C that contain p, as well as $\mathfrak{s} = \mathfrak{r} \cap A$. One has $\mu(C)_p \subset 1 + \mathfrak{r}$ and $\mu(A)_p = \mu(C)_p \cap (1 + \mathfrak{s})$. To compute the latter intersection, one determines $t \in \mathbb{Z}_{>0}$ with $p^t C \subset A$ as well as a presentation for the finite abelian p-group $1 + (\mathfrak{r}/p^t C)$, which is a subgroup of the unit group $(C/p^t C)^*$; to do this, one uses that $\mathbf{r}/p^t C$ is a nilpotent ideal of $C/p^t C$. The group $\mu(A)_p$ is now obtained as the kernel of the natural map $\mu(C)_p \to (1 + (\mathbf{r}/p^t C))/(1 + (\mathbf{s}/p^t C))$.

Proposition 4.8 *Suppose L is an invertible G-lattice, $r \in \mathbb{Z}_{>0}$, and ν is a short vector in the G-lattice L^r. Let $A = \Lambda/(\nu - 1)$. Identifying $\bigoplus_{i=0}^{r-1} L^i \subset \Lambda$ with its image in A, we can view $A = \bigoplus_{i=0}^{r-1} L^i$ as a $\mathbb{Z}/r\mathbb{Z}$-graded ring. Then:*

(i) $G \subseteq \mu(A) \subseteq \bigcup_{i=0}^{r-1} L^i$,
(ii) $\{e \in L : e \cdot \bar{e} = 1\} = \mu(A) \cap L$,
(iii) *$|\mu(A)|$ is divisible by $2n$ and divides $2nr$, and*
(iv) *there exists $e \in L$ for which $e \cdot \bar{e} = 1$ if and only if $|\mu(A)| = 2nr$.*

Proof. Since the ideal $(\bar{\nu} - 1) = (\nu^{-1} - 1) = (1 - \nu) = (\nu - 1)$, the map $a \mapsto \bar{a}$ induces an involution on A. Since the lattice's inner product is symmetric and positive definite, for all ring homomorphisms $\psi : A \to \mathbb{C}$ we have $\psi(\bar{a}) = \overline{\psi(a)}$ for all $a \in A$, and $\bigcap_\psi \ker \psi = 0$. Let $E = \{e \in A : e\bar{e} = 1\}$, a subgroup of A^*.

Suppose $e \in \mu(A)$. Then for all ring homomorphisms $\psi : A \to \mathbb{C}$ we have $1 = \psi(e)\overline{\psi(e)} = \psi(e)\psi(\bar{e}) = \psi(e\bar{e})$, so $e\bar{e} = 1$. Thus, $\mu(A) \subseteq E$.

Conversely, suppose $e \in E$. Write $e = \sum_{i=0}^{r-1} \varepsilon_i$ with $\varepsilon_i \in L^i$, so $\bar{e} = \sum_{i=0}^{r-1} \bar{\varepsilon}_i$ with $\bar{\varepsilon}_i \in L^{-i} = L^{r-i}$ in A. We have $1 = e\bar{e} = \sum_{i=0}^{r-1} \varepsilon_i \bar{\varepsilon}_i$ (the degree 0 piece of $e\bar{e}$). Applying the map t of Definition 2.9 and using (1) we have $1 = \sum_{i=0}^{r-1} \langle \varepsilon_i, \varepsilon_i \rangle$. It follows that there exists j such that $\langle \varepsilon_j, \varepsilon_j \rangle = 1$, and $\varepsilon_i = 0$ if $i \neq j$. Thus, $E \subseteq \bigcup_{i=0}^{r-1} \{e \in L^i : \langle e, e \rangle = 1\}$, giving (i). By Proposition 3.9(iii) and Example 3.8 we have $E \cap \mathbb{Z}\langle G \rangle = G$, so $\mu(\mathbb{Z}\langle G \rangle) = G$.

The degree map from E to $\mathbb{Z}/r\mathbb{Z}$ that takes $e \in E$ to j such that $e \in L^j$ is a group homomorphism with kernel $E \cap \mathbb{Z}\langle G \rangle = G$. Therefore, $|E|$ divides $|G| \cdot |\mathbb{Z}/r\mathbb{Z}| = 2nr$. Thus, $E \subseteq \mu(A) \subseteq E$, so $E = \mu(A)$ and we have (ii,iii). The degree map is surjective if and only if $|\mu(A)| = 2nr$, and if and only if 1 is in the image, i.e., if and only if $\mu(A) \cap L \neq \emptyset$. Part (iv) now follows from (ii). $\quad\square$

Proposition 4.9 *There is a deterministic polynomial time algorithm that, given G of exponent k, an invertible G-lattice L, and $\nu \in L^k$, determines whether there exists $e \in L$ such that $\nu = e^k$ and $e \cdot \bar{e} = 1$, and if so, finds one.*

Proof. Check whether $\nu\bar{\nu} = 1$. If so, let $A = \Lambda/(\nu - 1)$ and apply Proposition 4.7 to compute generators for $\mu(A)$. Using Proposition 4.8 with $r = k$, apply the degree map $\mu(A) \to \mathbb{Z}/k\mathbb{Z}$ to the generators, check whether the images generate $\mathbb{Z}/k\mathbb{Z}$, and if they do, compute an element $e \in \mu(A)$ whose image is 1. Then $e \in \mu(A) \cap L = \{e \in L : e \cdot \bar{e} = 1\}$. Check whether $\nu = e^k$. If any step fails, no such e exists (by Remark 4.1). The algorithm runs in polynomial time since $2nk \leq (2n)^2$. $\quad\square$

5   The Algorithm

We present the main algorithm, followed by a fuller explanation. As before, k is the exponent of the group G and $k(j)$ is the exponent of $(\mathbb{Z}\langle G \rangle/(j))^*$ if $j \in \mathbb{Z}_{>1}$.

Algorithm 5.1 *Input a finite abelian group G, an element $u \in G$ of order 2, and a G-lattice L. Output a G-isomorphism $\mathbb{Z}\langle G \rangle \xrightarrow{\sim} L$, or a proof that none exists.*

(i) *Apply Proposition 4.4(ii) to check whether L is invertible. If it is not, terminate with "no".*

(ii) *Find ℓ and m as in Proposition 4.5.*

(iii) *Compute $e_{\ell m}$ as in Proposition 4.4(i).*

(iv) *Using an addition chain for $k(m)$ and the algorithms mentioned in §3.3, compute the pair $(L^{k(m)}, e_{\ell m}^{k(m)} + mL^{k(m)})$. Use Proposition 4.6(ii) to decide whether the coset $e_{\ell m}^{k(m)} + mL^{k(m)}$ contains a short vector $\nu_m \in L^{k(m)}$, and if so, compute it. Terminate with "no" if none exists.*

(v) *Compute $s \in ((\mathbb{Z}/\ell\mathbb{Z})\langle G \rangle)^*$ such that*

$$\nu_m = s(e_{\ell m}^{k(m)} + \ell L^{k(m)})$$

in $L^{k(m)}/\ell L^{k(m)}$.

(vi) *Use the extended Euclidean algorithm to find $b \in \mathbb{Z}$ such that*

$$bk(m) \equiv k \mod k(\ell).$$

(vii) *Using an addition chain for k and the algorithms mentioned in §3.3, compute the pair $(L^k, e_{\ell m}^k + \ell L^k)$ and compute $s^b(e_{\ell m}^k + \ell L^k)$. Use Proposition 4.6(ii) to decide whether the latter coset contains a short vector $\nu \in L^k$, and if so, compute it. Terminate with "no" if none exists.*

(viii) *Apply Proposition 4.9 to find $e \in L$ such that $\nu = e^k$ and $e \cdot \bar{e} = 1$ (or to prove there is no G-isomorphism).*

We explain the algorithm in more detail. By Proposition 3.10(iii), the G-lattice L is G-isomorphic to $\mathbb{Z}\langle G \rangle$ if and only if L is invertible and has a short vector. Run the algorithm in Proposition 4.4(ii) to check whether L is invertible. If it is not, terminate with "no". If it is, we look for an $e \in L$ such that $e\bar{e} = 1$. Lattice basis reduction algorithms such as LLL can find fairly short vectors, but they are not nearly short enough for our purpose. We supplement LLL with computations modulo m. Any short e satisfies $\mathbb{Z}\langle G \rangle e = L$, which implies that for all $m \in \mathbb{Z}_{>0}$, the coset $e + mL$ generates L/mL as a $\mathbb{Z}\langle G \rangle/(m)$-module. Proposition 4.4(i) gives another generator e_m. Thus, $e_m = ye$ for some $y \in (\mathbb{Z}\langle G \rangle/(m))^*$. We have $e_m^{k(m)} \mod m = e^{k(m)} \mod m$ in $\Lambda/m\Lambda$.

Apply Proposition 4.5 to find prime powers $m, \ell \geq 2^{n/2} + 1$ such that

$$\gcd(k(\ell), k(m)) = k.$$

Compute $e_{\ell m}$ (which works as both e_m and e_ℓ) as in Proposition 4.4(i). Proposition 4.6(ii) applied to the coset $e_{\ell m} + mL^{k(m)} \in L^{k(m)}/mL^{k(m)}$ finds a short vector ν_m (if it exists). If $e \in L$ is short, then $\nu_m = e^{k(m)}$ by Proposition 4.6(i).

Since $e_{\ell m}^{k(m)}$ (by definition) and ν_m (by Proposition 3.10(ii)) each generate the $(\mathbb{Z}/\ell\mathbb{Z})\langle G \rangle$-module $L^{k(m)}/\ell L^{k(m)}$, we can find $s \in ((\mathbb{Z}/\ell\mathbb{Z})\langle G \rangle)^*$ such that

$\nu_m = s(e_{\ell m}^{k(m)} + \ell L^{k(m)})$ in $L^{k(m)}/\ell L^{k(m)}$. Since $k = \gcd(k(\ell), k(m))$, we can use the extended Euclidean algorithm to find $a, b \in \mathbb{Z}$ such that $ak(\ell) + bk(m) = k$. Compute $s^b \in ((\mathbb{Z}/\ell\mathbb{Z})\langle G \rangle)^*$ and $s^b e_{\ell m}^k \in L^k/\ell L^k$ and use Proposition 4.6(ii) to compute a short $\nu \in L^k$ in this coset or prove that none exists. If $e \in L$ is short, then $e^{k(m)} = \nu_m \equiv se_{\ell m}^{k(m)}$ mod $\ell\Lambda$, so $e^k \equiv \nu_m^b(e_{\ell m}^{k(\ell)})^a \equiv s^b e_{\ell m}^k$ mod $\ell\Lambda$, so $s^b(e_{\ell m}^k + \ell L^k)$ contains the short vector e^k of L^k, and by Proposition 4.6(i) we have $\nu = e^k$. Proposition 4.9 then finds a short vector $e \in L$, or proves none exists. The map $x \mapsto xe$ gives the desired G-isomorphism from $\mathbb{Z}\langle G \rangle$ to L. This completes the proof of Theorem 1.1.

Remark 5.2 *There is a version of the algorithm in which checking invertibility in step (i) is skipped. In this case, the algorithm may misbehave at other points, indicating that L is not invertible and thus not G-isomorphic to $\mathbb{Z}\langle G \rangle$. At the end one would check whether $\langle e, e \rangle = 1$ and $\langle e, \sigma e \rangle = 0$ for all $\sigma \neq 1, u$. If so, then $\{\sigma e\}_{\sigma \in S}$ is an orthonormal basis for L, and $x \mapsto xe$ gives the desired isomorphism; if not, no such isomorphism exists.*

Acknowledgments. We thank the participants of the August 2013 Workshop on Lattices with Symmetry, in particular Craig Gentry, René Schoof, and Mike Szydlo, and we thank the reviewers for helpful comments.

References

1. Atiyah, M.F., Macdonald, I.G.: Introduction to commutative algebra. Addison-Wesley Publishing Co., Reading (1969)
2. Garg, S., Gentry, C., Halevi, S.: Candidate Multilinear Maps from Ideal Lattices. In: Johansson, T., Nguyen, P.Q. (eds.) EUROCRYPT 2013. LNCS, vol. 7881, pp. 1–17. Springer, Heidelberg (2013)
3. Gentry, C.: Fully homomorphic encryption using ideal lattices. In: Mitzenmacher, M. (ed.) Proceedings of the 41st ACM Symposium on Theory of Computing—STOC 2009, pp. 169–178. ACM, New York (2009)
4. Gentry, C.: A fully homomorphic encryption scheme, Stanford University PhD thesis (2009), http://crypto.stanford.edu/craig/craig-thesis.pdf
5. Gentry, C.: email (May 9, 2012)
6. Gentry, C., Szydlo, M.: Cryptanalysis of the Revised NTRU Signature Scheme. In: Knudsen, L.R. (ed.) EUROCRYPT 2002. LNCS, vol. 2332, pp. 299–320. Springer, Heidelberg (2002), Full version at http://www.szydlo.com/ntru-revised-full02.pdf
7. Heath-Brown, D.R.: Zero-free regions for Dirichlet L-functions, and the least prime in an arithmetic progression. Proc. London Math. Soc. 64(3), 265–338 (1992)
8. Lang, S.: Algebra, Graduate Texts in Mathematics, 3rd edn., vol. 211. Springer-Verlag, New York (2002)
9. Lenstra, A.K., Lenstra Jr., H.W., Lovász, L.: Factoring polynomials with rational coefficients. Math. Ann. 261, 515–534 (1982)
10. Lenstra Jr., H.W.: Lattices, in Algorithmic number theory: lattices, number fields, curves and cryptography. In: Buhler, J.P., Stevenhagen, P. (eds.) Math. Sci. Res. Inst. Publ., vol. 44, pp. 127–181. Cambridge University Press, Cambridge (2008)
11. Smart, N.: personal communication

Faster Bootstrapping with Polynomial Error

Jacob Alperin-Sheriff and Chris Peikert[*]

School of Computer Science, Georgia Institute of Technology, Atlanta, GA, USA

Abstract. *Bootstrapping* is a technique, originally due to Gentry (STOC 2009), for "refreshing" ciphertexts of a somewhat homomorphic encryption scheme so that they can support further homomorphic operations. To date, bootstrapping remains the only known way of obtaining fully homomorphic encryption for arbitrary unbounded computations.

Over the past few years, several works have dramatically improved the efficiency of bootstrapping and the hardness assumptions needed to implement it. Recently, Brakerski and Vaikuntanathan (ITCS 2014) reached the major milestone of a bootstrapping algorithm based on Learning With Errors for *polynomial* approximation factors. Their method uses the Gentry-Sahai-Waters (GSW) cryptosystem (CRYPTO 2013) in conjunction with Barrington's "circuit sequentialization" theorem (STOC 1986). This approach, however, results in *very large* polynomial runtimes and approximation factors. (The approximation factors can be improved, but at even greater costs in runtime and space.)

In this work we give a new bootstrapping algorithm whose runtime and associated approximation factor are both *small* polynomials. Unlike most previous methods, ours implements an elementary and efficient *arithmetic* procedure, thereby avoiding the inefficiencies inherent to the use of boolean circuits and Barrington's Theorem. For 2^λ security under conventional lattice assumptions, our method requires only a *quasi-linear* $\tilde{O}(\lambda)$ number of homomorphic operations on GSW ciphertexts, which is optimal (up to polylogarithmic factors) for schemes that encrypt just one bit per ciphertext. As a contribution of independent interest, we also give a technically simpler variant of the GSW system and a tighter error analysis for its homomorphic operations.

1 Introduction

Gentry's *bootstrapping* paradigm [11, 10] allows for converting a "somewhat homomorphic" encryption scheme (which supports only a bounded number of homomorphic operations) into a fully homomorphic encryption one (which has no

---

[*] This material is based upon work supported by the National Science Foundation under CAREER Award CCF-1054495, by the Alfred P. Sloan Foundation, and by the Defense Advanced Research Projects Agency (DARPA) and the Air Force Research Laboratory (AFRL) under Contract No. FA8750-11-C-0098. The views expressed are those of the authors and do not necessarily reflect the official policy or position of the National Science Foundation, the Sloan Foundation, DARPA or the U.S. Government.

J.A. Garay and R. Gennaro (Eds.): CRYPTO 2014, Part I, LNCS 8616, pp. 297–314, 2014.

such bound). The bounded nature of all known somewhat-homomorphic schemes is an artifact of "error" terms in their ciphertexts, which are necessary for security. The error grows as a result of performing homomorphic operations, and if it grows too large, the ciphertext will no longer decrypt correctly.

Bootstrapping "refreshes" a ciphertext—i.e., reduces its error—so that it can support more homomorphic operations. This is accomplished by *homomorphically evaluating* the decryption function on the ciphertext. The result is a ciphertext that still encrypts the original encrypted message, and moreover, as long as the error incurred in the homomorphic evaluation is smaller than the error in the original ciphertext, the ciphertext is "refreshed." To date, the bootstrapping paradigm is the only known way of obtaining an *unbounded* FHE scheme, i.e., one that can homomorphically evaluate any efficient function using keys and ciphertexts of a fixed size. (By contrast, *leveled* FHE schemes can evaluate functions of any *a priori* bounded depth, and can be constructed without resorting to bootstrapping [4].)

Bootstrapping has received intensive study, with progress often going hand-in-hand with innovations in the design of homomorphic encryption schemes, e.g., [12, 6, 4, 14, 13, 1, 15, 7]. Of particular interest is a recent major milestone due to Brakerski and Vaikuntanathan (BV) [7], who gave a bootstrapping method that incurs only *polynomial* error in the security parameter λ. This allows security to be based on the learning with errors (LWE) problem [20] with inverse-polynomial error rates, and hence on worst-case lattice problems with polynomial approximation factors (via the reductions of [20, 19, 5]). The BV method is centered around two main components:

1. the recent homomorphic cryptosystem of Gentry, Sahai, and Waters (GSW) [15], specifically, the "*quasi-additive*" nature of its error growth under homomorphic multiplication; and
2. the "circuit sequentialization" property of Barrington's Theorem [3], which converts any depth-d circuit (of NAND gates) into a length-4^d "branching program," which is essentially a fixed sequence of conditional multiplications.

Since decryption in homomorphic cryptosystems can be implemented in circuit depth $O(\log \lambda)$, Barrington's Theorem yields an equivalent branching program of length $4^d = \text{poly}(\lambda)$. Moreover, the quasi-additive error growth of GSW multiplication means that homomorphic evaluation of the branching program incurs only $\text{poly}(\lambda)$ error, as demonstrated in [7].

The polynomial error growth of the BV bootstrapping algorithm is a terrific feature, but the method also has two significant drawbacks: it comes at a high price in *efficiency*, and the error growth is a *large* polynomial. Both issues arise from the fact that in this context, Barrington's Theorem yields a branching program of large polynomial length. Existing analyses (e.g., [6, Lemma 4.5]) of decryption circuits (for cryptosystems with 2^λ security) yield depths of $c \log \lambda$ for some unspecified but moderately large constant $c \geq 3$, which translates to a branching program of length at least $\lambda^{2c} \geq \lambda^6$. (Even if the depth were to be improved, there is a fundamental barrier of $c \geq 1$, which yields length $\Omega(\lambda^2)$.) The branching program length is of course a lower bound on the number of homomorphic operations required to bootstrap, and it also largely determines the associated error growth and final lattice approximation factors.

Separately, Brakerski and Vaikuntanathan also show how to obtain better lattice approximation factors through a kind of "dimension leveraging" technique, but this comes at an even higher price in efficiency: if the original error growth was λ^c for some constant c, then the technique involves running the bootstrapping procedure with GSW ciphertexts of dimension $n \approx \lambda^{c/\epsilon}$, where the choice of $\epsilon \in (0,1)$ yields a final approximation factor of $\tilde{O}(n^{3/2+\epsilon})$. The high cost of dimension leveraging underscores the importance of obtaining smaller error growth and approximation factors via other means.

1.1 Our Results

Our main result is a new bootstrapping method having substantially smaller runtime *and* (polynomial) error growth than the recent one from [7]. The improvements come as a result of treating decryption as an *arithmetic* function, in contrast to most earlier works which treated decryption as a boolean circuit. This avoids the circuitous and inefficient path of constructing a shallow circuit and then transforming it via Barrington's Theorem into a branching program of (large) polynomial length. Instead, we show how to *directly* evaluate the decryption function in an elementary and efficient arithmetic form, using just basic facts about cyclic groups. See the next subsection for a detailed overview.

Our method requires only a *quasi-linear* $\tilde{O}(\lambda)$ number of homomorphic operations on GSW ciphertexts, to bootstrap essentially any LWE-based encryption scheme with 2^λ security under conventional assumptions. This performance is *quasi-optimal* (i.e., ignoring polylogarithmic factors) for a system with bitwise encryption (like GSW), because the decryption function must depend on at least λ secret key bits. When instantiated with a GSW scheme based on ring-LWE [17], in which the cost of each homomorphic operation is only $\tilde{O}(\lambda)$ bit operations, the total runtime of our algorithm is a respectable $\tilde{O}(\lambda^2)$ bit operations.[1]

Regarding error growth, the security of our basic scheme can be based on LWE with error rates as large as $1/\tilde{O}(\lambda \cdot n)$, where $n = \tilde{\Omega}(\lambda)$ is the LWE dimension used in the GSW scheme. Taking $n = \tilde{O}(\lambda)$ to be asymptotically minimal, this translates to lattice approximation factors of $\tilde{O}(n^3)$, which is quite close to the $\tilde{O}(n^{3/2})$ factors that plain public-key encryption can be based upon (and quite a bit smaller than for many other applications of LWE!). We emphasize that these small factors are obtained *directly* from our construction with *small* LWE dimensions. To further improve the assumptions at a (high) cost in efficiency, we can let $n = \lambda^{1/\epsilon}$ to directly yield $\tilde{O}(n^{2+\epsilon})$ factors for any $\epsilon \in (0,1)$, or we can use the successive dimension/modulus-reduction technique from [7] to obtain $\tilde{O}(n^{3/2+\epsilon})$ factors.

Simpler GSW variant. As a contribution of independent interest, we also give a variant of the GSW cryptosystem that we believe is technically simpler, along

---

[1] Homomorphic operations in standard-LWE-based GSW are quite a bit more expensive, due to matrix multiplications of dimensions exceeding λ.

with a tighter analysis of error terms under its homomorphic operations (see Section 3). The entire scheme, security proof, and error analysis fit into just a few lines of standard linear algebra notation, and our variant enjoys additional useful properties like full "re-randomization" of error terms as a natural side effect. The error analysis is also very clean and tight, due to the use of *subgaussian* random variables instead of coarser measures like the ℓ_2 or ℓ_∞ norms. One nice consequence of this approach is that the error in a homomorphic product of d ciphertexts grows with $\sqrt{d}$, rather than linearly as in prior analyses. This is important for establishing the small error growth of our bootstrapping method.

1.2   Technical Overview

Here we give an overview of the main ideas behind our new bootstrapping method. We start by recalling in more detail the main ideas behind the work of Brakerski and Vaikuntanathan [7], which uses the Gentry-Sahai-Waters (GSW) [15] homomorphic encryption scheme to obtain FHE from LWE with inverse-polynomial error rates, and hence from lattice problems with polynomial approximation factors.

The starting point is a simple observation about the GSW encryption scheme: for encryptions $\mathbf{C}_1, \mathbf{C}_2$ of messages $\mu_1, \mu_2 \in \mathbb{Z}$, the error in the homomorphic product $\mathbf{C}_1 \boxdot \mathbf{C}_2$ of $\mu_1 \cdot \mu_2$ is *"quasi-additive"* and *asymmetric* in the ciphertexts' respective errors e_1, e_2, namely, it is $e_1 \cdot \text{poly}(n) + \mu_1 \cdot e_2$, where n is the dimension of the ciphertexts. (The error in the homomorphic sum $\mathbf{C}_1 \boxplus \mathbf{C}_2$ is simply the sum $e_1 + e_2$ of the individual errors.) This property has a number of interesting consequences. For example, Brakerski and Vaikuntanathan use it to show that the homomorphic product of many freshly encrypted 0-1 messages, if evaluated *sequentially* in a right-associative manner, has error that grows at most linearly in the number of ciphertexts. More generally, the homomorphic product of many encrypted *permutation* matrices—i.e., 0-1 matrices in which each row and column has exactly one nonzero entry—has similarly small error growth.

The next main idea from [7] is to use Barrington's Theorem to express the boolean decryption circuit of depth $d = O(\log \lambda)$ as a branching program of length $4^d = \text{poly}(\lambda)$ over the symmetric group S_5, or equivalently, the multiplicative group of 5-by-5 permutations matrices. Their bootstrapping algorithm homomorphically (and sequentially) multiplies appropriate encrypted permutation matrices to evaluate this branching program on a given input ciphertext, thereby homomorphically decrypting it. Since evaluation is just a homomorphic product of $\text{poly}(\lambda)$ permutation matrices, the error in the final output ciphertext is only polynomial, and the LWE parameters can be set to yield security assuming the hardness of lattice problems for polynomial approximation factors.

Our Approach. Our bootstrapping method retains the use of symmetric groups and permutation matrices, but it works without the "magic" of Barrington's Theorem, by treating decryption more directly and efficiently as an *arithmetic* function, not a boolean circuit. In more detail, the decryption function for essentially every LWE-based cryptosystem can without loss of generality (via standard bit-decomposition techniques) be written as a "rounded inner product" between the secret key $\mathbf{s} \in \mathbb{Z}_q^d$ and a *binary* ciphertext $\mathbf{c} \in \{0,1\}^d$, as

$$\mathsf{Dec}(\mathbf{s},\mathbf{c}) = \lfloor \langle \mathbf{s},\mathbf{c} \rangle \rceil_2 \in \{0,1\}.$$

Here the modular rounding function $\lfloor \cdot \rceil_2 \colon \mathbb{Z}_q \to \{0,1\}$ indicates whether its argument is "far from" or "close to" 0 (modulo q), and the dimension d and modulus q can both be made as small as quasi-linear $\tilde{O}(\lambda)$ in the security parameter via dimension/modulus reduction [6], while still providing provable 2^λ security under conventional lattice assumptions. Note that the inner product itself is just a subset-sum of the $\mathbb{Z}_q$-entries of $\mathbf{s}$ indicated by $\mathbf{c}$, and uses only the additive group structure of $\mathbb{Z}_q$.

Embedding $\mathbb{Z}_q$ into S_q. As a warm up, we first observe that the additive group $\mathbb{Z}_q$ embeds (i.e., has an injective homomorphism) into the symmetric group S_q, the multiplicative group of q-by-q permutation matrices. (This is just a special case of Cayley's Theorem, which says that any finite group G embeds into $S_{|G|}$.) The embedding is very simple: $x \in \mathbb{Z}_q$ maps to the permutation that cyclically rotates by x positions. Moreover, any such permutation can be represented by an indicator vector in $\{0,1\}^q$ with its 1 in the position specified by x, and its permutation matrix is obtained from the cyclic rotations of this vector. In this representation, a sum $x + y$ can be computed in $O(q^2)$ bit operations by expanding x's indicator vector into its associated permutation matrix, and then multiplying by y's indicator vector. This representation also makes the *rounding function* $\lfloor \cdot \rceil_2 \colon \mathbb{Z}_q \to \{0,1\}$ trivial to evaluate: one just sums the entries of the indicator vector corresponding to those values in $\mathbb{Z}_q$ that round to 1.

These ideas already yield a new and simple bootstrapping algorithm that appears to have better runtime and error growth than can be obtained using Barrington's Theorem. The bootstrapping key is an encryption of each coordinate of the secret key $\mathbf{s} \in \mathbb{Z}_q^d$, represented as a dimension-q indicator vector, for a total of $d \cdot q = \tilde{O}(\lambda^2)$ GSW ciphertexts. To bootstrap a ciphertext $\mathbf{c} \in \{0,1\}^d$, the inner product $\langle \mathbf{s},\mathbf{c} \rangle \in \mathbb{Z}_q$ is computed homomorphically as a subset-sum using the addition method described above, in $O(d \cdot q^2) = \tilde{O}(\lambda^3)$ homomorphic operations. The rounding function is then applied homomorphically, using just $O(q) = \tilde{O}(\lambda)$ additions.

Embedding $\mathbb{Z}_q$ into smaller symmetric groups. While the above method yields some improvements over prior work, it is still far from optimal. Our second main idea is an efficient way of embedding $\mathbb{Z}_q$ into a *much smaller* symmetric group S_r for some $r = \tilde{O}(1)$, such that the rounding function can still be efficiently evaluated (homomorphically). We do so by letting the modulus $q = \prod_i r_i$ be the product of many small prime powers r_i of distinct primes. (We can use such a q by modulus switching, as long as it remains sufficiently large to preserve correctness of decryption.) Using known bounds on the distribution of primes, it suffices to let the r_i be maximal prime powers bounded by $O(\log \lambda)$, of which there are at most $O(\log \lambda / \log \log \lambda)$.

By the Chinese Remainder Theorem, the additive group $\mathbb{Z}_q$ is isomorphic (via the natural homomorphism) to the product group $\prod_i \mathbb{Z}_{r_i}$, which then embeds into $\prod_i S_{r_i}$ as discussed above. Therefore, we can represent any $x \in \mathbb{Z}_q$ as a tuple of $O(\log \lambda)$ indicator vectors of length $r_i = O(\log \lambda)$ representing $x \pmod{r_i}$, and can perform addition by operating on the indicator vectors as described above. In this representation, the rounding function is no longer just a sum, but it can still be expressed relatively simply as

$$\lfloor x \rceil_2 = \sum_{v \in \mathbb{Z}_q \text{ s.t. } \lfloor v \rceil_2 = 1} [x = v],$$

where each equality test $[x = v]$ returns 0 for false and 1 for true.[2] In turn, each equality test $[x = v]$ is equivalent to the product of equality tests $[x = v \pmod{r_i}]$, each of which can be implemented trivially in our representation by selecting the appropriate entry of the indicator vector for $x \pmod{r_i}$. All of these operations have natural homomorphic counterparts in our representation, so we get a corresponding bootstrapping algorithm.

As a brief analysis, each coordinate of the secret key $\mathbf{s} \in \mathbb{Z}_q^d$ is encrypted as $\sum_i r_i = \tilde{O}(1)$ GSW ciphertexts, for a total of $\tilde{O}(d) = \tilde{O}(\lambda)$ ciphertexts in the bootstrapping key. Similarly, each addition or equality test over $\mathbb{Z}_q$ takes $\tilde{O}(1)$ homomorphic operations, for a total of $\tilde{O}(d + q) = \tilde{O}(\lambda)$. Both of these measures are quasi-optimal when relying on a scheme that encrypts one bit per ciphertext (like GSW). By contrast, bootstrapping using Barrington's Theorem requires at least $4^{c \log \lambda} = \lambda^{2c}$ homomorphic operations to evaluate the branching program, where $c \log \lambda$ is the depth of the decryption circuit *using NAND gates* (of fan-in 2). To our knowledge, upper bounds on the constant c have not been optimized or even calculated explicitly, but existing analyses like [6, Lemma 4.5] yield $c \gg 3$, and the necessary dependence on λ inputs bits for 2^λ security yields a fundamental barrier of $c \geq 1$.

Related Work on Branching Programs. Several works have extended and improved Barrington's Theorem for the simulation of general circuits and formulas via branching programs, e.g., [8, 9]. Of particular interest here is the thesis of Sinha [22], which gave quasi-linear-size, log-width branching programs for threshold functions (i.e., those which output 1 if at least some k of the n inputs are 1) and "mod" functions (i.e., those which output 1 if the number of 1s in the input is zero modulo some d). Similarly to our techniques, Sinha's construction uses the Chinese Remainder Theorem over many small primes in an essential way.

Because decryption in LWE-based cryptosystems involves modular addition, and can be implemented in constant depth (and polynomial size) by threshold

---

[2] Note that we are not using any special property of the rounding function here; any boolean function $f \colon \mathbb{Z}_q \to \{0, 1\}$ can be expressed similarly by summing over $f^{-1}(1)$.

gates, it might be possible to bootstrap in a quasi-linear number of homomorphic operations by using Sinha's results in place of Barrington's Theorem. However, we have not seen a way to make this work concretely.

Organization. The rest of the paper is organized as follows. In Section 2 we recall some mathematical preliminaries on subgaussian random variables and symmetric groups. In Section 3 we present our simplified GSW variant and analysis. In Section 4 we extend this to a homomorphic encryption scheme for symmetric groups. In Section 5.2 we describe and analyze our new bootstrapping algorithm.

Acknowledgments. We thank the anonymous CRYPTO reviewers for their helpful comments, and for pointers to the additional works on branching programs.

2   Preliminaries

For a nonnegative integer n, we let $[n] = \{1, \ldots, n\}$. For an integer modulus q, we let $\mathbb{Z}_q = \mathbb{Z}/q\mathbb{Z}$ denote the quotient ring of integers modulo q, and $(\mathbb{Z}_q, +)$ its additive group.

2.1   Subgaussian Random Variables

In our constructions it is very convenient to analyze the behavior of "error" terms using the standard notion of *subgaussian* random variables. (For further details and full proofs, see [23].) A real random variable X (or its distribution) is subgaussian with parameter $r > 0$ if for all $t \in \mathbb{R}$, its (scaled) moment-generating function satisfies $\mathbb{E}[\exp(2\pi tX)] \leq \exp(\pi r^2 t^2)$. By a Markov argument, X has Gaussian tails, i.e., for all $t \geq 0$, we have

$$\Pr[|X| \geq t] \leq 2\exp(-\pi t^2/r^2). \tag{1}$$

(If $\mathbb{E}[X] = 0$, then Gaussian tails also imply subgaussianity.) Any B-bounded centered random variable X (i.e., $\mathbb{E}[X] = 0$ and $|X| \leq B$ always) is subgaussian with parameter $B\sqrt{2\pi}$.

Subgaussianity is homogeneous, i.e., X is subgaussian with parameter r, then cX is subgaussian with parameter $c \cdot r$ for any constant $c \geq 0$. Subgaussians also satisfy *Pythagorean additivity*: if X_1 is subgaussian with parameter r_1, and X_2 is subgaussian with parameter r_2 conditioned on *any* value of X_1 (e.g., if X_1 and X_2 are independent), then $X_1 + X_2$ is subgaussian with parameter $\sqrt{r_1^2 + r_2^2}$. By induction this extends to the sum of any finite number of variables, each of which is subgaussian conditioned on any values of the previous ones.

We extend the notion of subgaussianity to vectors: a random real vector $\mathbf{x}$ is subgaussian with parameter r if for all fixed real unit vectors $\mathbf{u}$, the marginal $\langle \mathbf{u}, \mathbf{x} \rangle \in \mathbb{R}$ is subgaussian with parameter r. In particular, it follows directly from the definition that the concatenation of variables or vectors, each of which is subgaussian with common parameter r conditioned on any values of the prior ones, is also subgaussian with parameter r. Homogeneity and Pythagorean additivity clearly extend to subgaussian vectors as well, by linearity.

2.2   Symmetric Groups and $\mathbb{Z}_q$-Embeddings

Here we recall some basic facts about symmetric groups, which can be found in most abstract algebra textbooks, e.g., [16]. Let S_r denote the symmetric group of order r, i.e., the group of permutations (bijections) $\pi\colon \{1,\ldots,r\} \to \{1,\ldots,r\}$ with function composition as the group operation. The group S_r is isomorphic to the multiplicative group of r-by-r *permutation matrices* (i.e., 0-1 matrices with exactly one nonzero element in each row and each column), via the map that associates $\pi \in S_r$ with the permutation matrix $\mathbf{P}_\pi = [\mathbf{e}_{\pi(1)}\ \mathbf{e}_{\pi(2)}\ \cdots\ \mathbf{e}_{\pi(r)}]$, where $\mathbf{e}_i \in \{0,1\}^r$ is the ith standard basis vector. For the remainder of this work we identify permutations with their associated permutation matrices.

The additive cyclic group $(\mathbb{Z}_r, +)$ embeds into the symmetric group S_r via the injective homomorphism that sends the generator $1 \in \mathbb{Z}_r$ to the "cyclic shift" permutation $\pi \in S_r$, defined as $\pi(i) = i+1$ for $1 \le i < r$ and $\pi(r) = 1$.[3] Clearly, this embedding and its inverse can be computed efficiently. Notice also that the permutation matrices in the image of this embedding can be represented more compactly by just their first column, because the remaining columns are just the successive cyclic shifts of this column. Similarly, such permutation matrices can be multiplied in only $O(r^2)$ operations, since we only need to multiply one matrix by the first column of the other.

For our efficient bootstrapping algorithm, we need to efficiently embed a group $(\mathbb{Z}_q, +)$, for some sufficiently large q of our choice, into a symmetric group of order much smaller than q (e.g., polylogarithmic in q). This can be done as follows: suppose that $q = r_1 r_2 \cdots r_t$, where the r_i are pairwise coprime. Then by the Chinese Remainder Theorem, the ring $\mathbb{Z}_q$ is isomorphic to the direct product of rings $\mathbb{Z}_{r_1} \times \mathbb{Z}_{r_2} \times \cdots \times \mathbb{Z}_{r_t}$, and hence their additive groups are isomorphic as well. Combining this with the group embeddings of $(\mathbb{Z}_{r_i}, +)$ into S_{r_i}, we have an (efficient) group embedding from $(\mathbb{Z}_q, +)$ into $S_{r_1} \times S_{r_2} \times \cdots \times S_{r_t}$.[4]

Importantly for our purposes, q can be exponentially large in terms of $\max_i r_i$ above. This can be shown using lower bounds on the second Chebyshev function

$$\psi(x) := \sum_{p^k \le x} \log p = \log\Big(\prod_{p \le x} p^{\lfloor \log_p x \rfloor}\Big),$$

where the first summation is over all prime powers $p^k \le x$, and the second is over all primes $p \le x$; note that $p^{\lfloor \log_p x \rfloor}$ is the largest power of p not exceeding x. Therefore, the product q of all maximal prime powers $r_i = p^{\lfloor \log_p x \rfloor} \le x$ is $\exp(\psi(x))$. Asymptotically, it is known that $\psi(x) = x \pm O(x/\log x)$, and we also have the nonasymptotic bound $\psi(x) \ge 3x/4$ for all $x \ge 7$ [21, Theorem 11]. In summary:

Lemma 2.1. *For all $x \ge 7$, the product of all maximal prime powers $r_i \le x$ is at least $\exp(3x/4)$.*

---

[3] This is just a special case of Cayley's theorem, which says that any group G embeds into the symmetric group $S_{|G|}$.

[4] The latter group can be seen as a subgroup of S_r for $r = \sum_i r_i$, but it will be more efficient to retain the product structure.

For any given lower bound $q_0 \geq 191 > \exp(21/4)$, we can therefore efficiently find a $q \geq q_0$ whose maximal prime-power divisors are all at most $x = \frac{4}{3} \log q_0 \geq 7$.

3   GSW Cryptosystem

Here we present a variant of the Gentry-Sahai-Waters homomorphic encryption scheme [15] (hereafter called GSW), which we believe is simpler to understand at a technical level. We also give a tighter analysis of its error growth under homomorphic operations.

We first recall some standard background (see, e.g., [18] for further details). For a modulus q, let $\ell = \lceil \log_2 q \rceil$ and define the "gadget" (column) vector $\mathbf{g} = (1, 2, 4, \ldots, 2^{\ell-1}) \in \mathbb{Z}_q^\ell$. Note that the penultimate entry $2^{\ell-2}$ of $\mathbf{g}$ is in the interval $[q/4, q/2) \bmod q$. It will be convenient to use the following randomized "decomposition" function.

Claim (Adapted from [18]). There is a randomized, efficiently computable function $\mathbf{g}^{-1} \colon \mathbb{Z}_q \to \mathbb{Z}^\ell$ such that $\mathbf{x} \leftarrow \mathbf{g}^{-1}(a)$ is subgaussian with parameter $O(1)$, and always satisfies $\langle \mathbf{g}, \mathbf{x} \rangle = a$.

For vectors and matrices over $\mathbb{Z}_q$, define the randomized function $\mathbf{G}^{-1} \colon \mathbb{Z}_q^{n \times m} \to \mathbb{Z}^{n\ell \times m}$ by applying $\mathbf{g}^{-1}$ independently to each entry. Notice that for any $\mathbf{A} \in \mathbb{Z}_q^{n \times m}$, if $\mathbf{X} \leftarrow \mathbf{G}^{-1}(\mathbf{A})$ then $\mathbf{X}$ has subgaussian parameter $O(1)$ and

$$\mathbf{G} \cdot \mathbf{X} = \mathbf{A}, \quad \text{where} \quad \mathbf{G} = \mathbf{g}^t \otimes \mathbf{I}_n = \text{diag}(\mathbf{g}^t, \ldots, \mathbf{g}^t) \in \mathbb{Z}_q^{n \times n\ell} \qquad (2)$$

is the block matrix with n copies of $\mathbf{g}^t$ as diagonal blocks, and zeros elsewhere.

3.1   Cryptosystem and Homomorphic Operations

The GSW scheme is parameterized by a dimension n, a modulus q with $\ell = \lceil \log_2 q \rceil$, and some error distribution χ over $\mathbb{Z}$ which we assume to be subgaussian. Formally, the message space is the ring of integers $\mathbb{Z}$, though for bootstrapping we only work with ciphertexts encrypting messages in $\{0, 1\} \subset \mathbb{Z}$. The ciphertext space is $\mathcal{C} = \mathbb{Z}_q^{n \times n\ell}$. For simplicity we present just a symmetric-key scheme, which is sufficient for our purposes (it can be converted to a public-key or even attribute-based scheme, as described in [15]).

Our GSW variant differs from the original scheme described in [15] in two main ways:

1. In [15], a ciphertext is a square binary matrix $\mathbf{C} \in \{0,1\}^{n\ell}$, a secret key is a "structured" mod-q vector $\mathbf{s} \in \mathbb{Z}_q^{n\ell}$ (having large entries), and $\mathbf{s}$ is an "approximate mod-q eigenvector" of $\mathbf{C}$, in the sense that $\mathbf{s}^t \mathbf{C} \approx \mu \mathbf{s}^t \pmod{q}$, where $\mu \in \mathbb{Z}$ is the message.

   In our variant, a ciphertext is a rectangular mod-q matrix $\mathbf{C} \in \mathbb{Z}_q^{n \times n\ell}$, a secret key is some (unstructured, short) integer vector $\mathbf{s} \in \mathbb{Z}^n$, and $\mathbf{s}^t \mathbf{C} \approx \mu \cdot \mathbf{s}^t \mathbf{G} \pmod{q}$, i.e., $\mathbf{s}$ and $\mathbf{G}^t \mathbf{s}$ are corresponding left- and right- "approximate singular vectors" of $\mathbf{C}$.

The difference between these two variants turns out to be purely syntactic, in that we can efficiently and "losslessly" switch between them (without needing the secret key). However, we believe that our variant leads to simpler notation and easier-to-understand operations and analysis.

2. The second difference is more substantial: our homomorphic multiplication procedure uses the *randomized* $\mathbf{G}^{-1}(\cdot)$ operation from Claim 3. This yields a few important advantages, such as a very tight and simple error analysis using subgaussianity (see Lemma 3.3), and the ability to *completely re-randomize* the error in a ciphertext (see Corollary 3.4).

We now describe the scheme formally.

GSW.Gen(): choose $\bar{\mathbf{s}} \leftarrow \chi^{n-1}$ and output secret key $\mathbf{s} = (\bar{\mathbf{s}}, 1) \in \mathbb{Z}^n$.

GSW.Enc($(\bar{\mathbf{s}}, 1), \mu \in \mathbb{Z}$): choose $\bar{\mathbf{C}} \leftarrow \mathbb{Z}_q^{(n-1) \times n\ell}$ and $\mathbf{e} \leftarrow \chi^m$, let $\mathbf{b}^t = \mathbf{e}^t - \bar{\mathbf{s}}^t \bar{\mathbf{C}} \pmod{q}$, and output the ciphertext

$$\mathbf{C} = \begin{pmatrix} \bar{\mathbf{C}} \\ \mathbf{b}^t \end{pmatrix} + \mu \mathbf{G} \in \mathcal{C},$$

where $\mathbf{G}$ is as defined in Equation (2). Notice that $\mathbf{s}^t \mathbf{C} = \mathbf{e}^t + \mu \cdot \mathbf{s}^t \mathbf{G} \pmod{q}$.

GSW.Dec($\mathbf{s}, \mathbf{C} \in \mathcal{C}$): let $\mathbf{c}$ be the penultimate column of $\mathbf{C}$, and output $\mu = \lfloor \langle \mathbf{s}, \mathbf{c} \rangle \rceil_2$, where $\lfloor \cdot \rceil_2 \colon \mathbb{Z}_q \to \{0, 1\}$ indicates whether its argument is closer modulo q to 0 or to $2^{\ell-2}$ (the penultimate entry of $\mathbf{g}$).[5]

Homomorphic addition is defined as $\mathbf{C}_1 \boxplus \mathbf{C}_2 = \mathbf{C}_1 + \mathbf{C}_2$.

Homomorphic multiplication is defined as $\mathbf{C}_1 \boxdot \mathbf{C}_2 \leftarrow \mathbf{C}_1 \cdot \mathbf{G}^{-1}(\mathbf{C}_2)$, and is *right associative*. Notice that this is a randomized procedure, because $\mathbf{G}^{-1}$ is randomized.

The IND-CPA security of the scheme follows immediately from the assumed hardness of $\mathsf{LWE}_{n-1,q,\chi}$, where the entries of the secret are drawn from the error distribution χ (which is no easier than for a uniformly random secret; see [2, Lemma 2]). This is because a fresh ciphertext is just $\mu \mathbf{G}$ plus a matrix of $n\ell$ independent LWE samples under secret $\bar{\mathbf{s}}$, which are pseudorandom by assumption and hence hide $\mu \mathbf{G}$.

3.2 Analysis

Here we analyze the scheme's correctness and homomorphic operations.

Definition 3.1. *We say that a ciphertext $\mathbf{C}$ is designed to encrypt message $\mu \in \mathbb{Z}$ (under a secret key $\mathbf{s}$) if it is a fresh encryption of μ, or if $\mathbf{C} = \mathbf{C}_1 \boxplus \mathbf{C}_2$ where $\mathbf{C}_1, \mathbf{C}_2$ are respectively designed to encrypt $\mu_1, \mu_2 \in \mathbb{Z}$ and $\mu = \mu_1 + \mu_2$, or similarly for homomorphic multiplication.*

---

[5] Note that we can decrypt messages in $\mathbb{Z} \cap [-\frac{q}{2}, \frac{q}{2})$, or any other canonical set of representatives of $\mathbb{Z}_q$, by "decoding" $\mathbf{s}^t \mathbf{C}$ to the nearest multiple of $\mathbf{s}^t \mathbf{G}$. The above decryption algorithm will be sufficient for our purposes.

Definition 3.2. *We say that a ciphertext* $\mathbf{C}$ *that is designed to encrypt* $\mu \in \mathbb{Z}$ *(under* $\mathbf{s}$*) has* error vector $\mathbf{e}^t \in \mathbb{Z}^{n\ell}$ *if* $\mathbf{s}^t\mathbf{C} - \mu \cdot \mathbf{s}^t\mathbf{G} = \mathbf{e}^t \pmod{q}$.

For convenience later on, we also say the matrix $\mu\mathbf{G}$ is designed to encrypt μ, and has error $\mathbf{0}$. (This is essentially implied by the above definitions, since $\mu\mathbf{G}$ is indeed a fresh encryption of μ, assuming that zero is in the support of χ.) The next claim on the correctness of decryption follows immediately from the fact that $\mathbf{s} = (\bar{\mathbf{s}}, 1)$ and the penultimate column of $\mathbf{G}$ is $(0, \ldots, 0, 2^{\ell-2})$, where $2^{\ell-2} \in [q/4, q/2) \bmod q$.

Claim. If $\mathbf{C}$ is designed to encrypt some $\mu \in \{0, 1\} \subset \mathbb{Z}$, and has error vector $\mathbf{e}^t$ whose penultimate coordinate has magnitude less than $q/8$, then GSW.Dec$(\mathbf{s}, \mathbf{C})$ correctly outputs μ.

We now analyze the behavior of the error terms under homomorphic operations.

Lemma 3.3. *Suppose* $\mathbf{C}_1, \mathbf{C}_2$ *are respectively designed to encrypt* $\mu_1, \mu_2 \in \mathbb{Z}$ *and have error vectors* $\mathbf{e}_1^t, \mathbf{e}_2^t$. *Then* $\mathbf{C}_1 \boxplus \mathbf{C}_2$ *has error vector* $\mathbf{e}_1^t + \mathbf{e}_2^t$, *and* $\mathbf{C}_1 \boxdot \mathbf{C}_2$ *has error vector* $\mathbf{e}_1^t\mathbf{X} + \mu_1\mathbf{e}_2^t$, *where* $\mathbf{X} \leftarrow \mathbf{G}^{-1}(\mathbf{C}_2)$ *is the matrix used in the evaluation of* $\boxdot$. *In particular, for any values of* $\mathbf{C}_i, \mathbf{e}_i, \mu_i$, *the latter error vector is of the form* $\mathbf{e}^t + \mu_1\mathbf{e}_2^t$, *where the entries of* $\mathbf{e}$ *are independent and subgaussian with parameter* $O(\|\mathbf{e}_1\|)$.

Importantly, the error in $\mathbf{C}_1 \boxdot \mathbf{C}_2$ is *quasi-additive* and *asymmetric* with respect to the errors in $\mathbf{C}_1, \mathbf{C}_2$: while the first error vector $\mathbf{e}_1^t$ is multiplied by a short (subgaussian) matrix $\mathbf{X}$, the second error vector $\mathbf{e}_2^t$ is only multiplied by the (scalar) *message* μ_1, which we will ensure remains in $\{0, 1\}$.

Proof. The first claim is immediate, by linearity. For the second claim, because $\mathbf{G} \cdot \mathbf{X} = \mathbf{C}_2$ we have

$$
\begin{aligned}
\mathbf{s}^t(\mathbf{C}_1 \boxdot \mathbf{C}_2) &= \mathbf{s}^t\mathbf{C}_1 \cdot \mathbf{X} \\
&= (\mathbf{e}_1^t + \mu_1 \cdot \mathbf{s}^t\mathbf{G})\mathbf{X} \\
&= \mathbf{e}_1^t\mathbf{X} + \mu_1(\mathbf{e}_2^t + \mu_2 \cdot \mathbf{s}^t\mathbf{G}) \\
&= (\mathbf{e}_1^t\mathbf{X} + \mu_1\mathbf{e}_2^t) + \mu_1\mu_2 \cdot \mathbf{s}^t\mathbf{G}.
\end{aligned}
$$

As observed in [7], the asymmetric noise growth allows for performing a long chain of homomorphic multiplications while only incurring a polynomial-factor error growth, because $\boxdot$ is defined to be right associative. For convenience of analysis, in such a chain we always include the fixed ciphertext $\mathbf{G}$, which is designed to encrypt $\mu = 1$ and has zero error, as the rightmost ciphertext in the chain. This ensures that the error vector of the output ciphertext is subgaussian and essentially *independent* of the errors in the input ciphertexts (apart from their lengths), which leads to a simpler and tighter analysis. (In [7] a weaker independence guarantee was achieved by a separate "partial re-randomization" procedure, which requires additional public key material.)

Corollary 3.4. *Suppose that* $\mathbf{C}_i$ *for* $i \in [k]$ *are respectively designed to encrypt* $\mu_i \in \{0, \pm 1\}$ *and have error vectors* $\mathbf{e}_i^t$. *Then for* any *fixed values of these variables,*

$$\mathbf{C} \leftarrow \boxed{\cdot}_{i \in [k]} \mathbf{C}_i \boxdot \mathbf{G} = \mathbf{C}_1 \boxdot (\mathbf{C}_2 \boxdot (\cdots (\mathbf{C}_k \boxdot \mathbf{G}) \cdots))$$

has an error vector whose entries are mutually independent and subgaussian with parameter $O(\|\mathbf{e}\|)$, *where* $\mathbf{e}^t = (\mathbf{e}_1^t, \dots, \mathbf{e}_k^t) \in \mathbb{Z}^{kn\ell}$ *is the concatenation of the individual error vectors.*

Proof. By Lemma 3.3, the error vector in $\mathbf{C}$ is $\sum_i \mathbf{e}_i^t \mathbf{X}_i$, where each $\mathbf{e}_i^t \mathbf{X}_i$ is a fresh independent vector that has mutually independent coordinates and is subgaussian with parameter $O(\|\mathbf{e}_i\|)$. The claim then follows by Pythagorean additivity.

4 Homomomorphic Encryption for Symmetric Groups

Brakerski and Vaikuntanathan [7] showed how to use the GSW encryption scheme to homomorphically compose permutations of five elements (i.e., to homomorphically compute the group operation in the symmetric group S_5) with small additive noise growth; the use of S_5 comes from its essential role in Barrington's theorem [3]. In [7], the homomorphic composition of permutations is intertwined with the evaluation of a branching program given by Barrington's theorem. Here we give, as a "first-class object," a homomorphic cryptosystem for any symmetric group S_r. The ability to use several different small values of r, along with a homomorphic equality test that we design, will be central to our bootstrapping algorithm.

4.1 Encryption Scheme

We now describe our (symmetric-key) homomorphic encryption scheme for symmetric groups, called HEPerm. Let $\mathcal{C}$ denote the ciphertext space for an appropriate instantiation of the GSW scheme, which we treat as a "black box." A secret key sk for HEPerm is simply a secret key for the GSW scheme.

- HEPerm.Enc($sk, \pi \in S_r$): let $\mathbf{P} = (p_{i,j}) \in \{0,1\}^{r \times r}$ be the permutation matrix associated with π. Output an entry-wise encryption of $\mathbf{P}$, i.e., the ciphertext

$$\mathbf{C} = (c_{i,j}) \in \mathcal{C}^{r \times r}, \text{ where } c_{i,j} \leftarrow \mathsf{Enc}(sk, p_{i,j}).$$

(Decryption follows in the obvious manner.) As with the GSW system, we say that a ciphertext $\mathbf{C} \in \mathcal{C}^{r \times r}$ is *designed* to encrypt a permutation $\pi \in S_r$ (or its permutation matrix $\mathbf{P}_\pi$) if its $\mathcal{C}$-entries are designed to encrypt the corresponding entries of $\mathbf{P}_\pi$. For convenience, we let $\mathbf{J} \in \mathcal{C}^{r \times r}$ denote the ciphertext that encrypts the identity permutation with zero noise, which is built in the expected way from the fixed zero-error GSW ciphertexts that encrypt 0 and 1.

We now show how to homomorphically compute two operations: the standard composition operation for permutations, and an equality test.

Homomorphic composition $\mathbf{C}^\pi \boxdot \mathbf{C}^\sigma$: on ciphertexts $\mathbf{C}^\pi = (c_{i,j}^\pi), \mathbf{C}^\sigma = (c_{i,j}^\sigma)$ $\in \mathcal{C}^{r \times r}$ encrypting permutations $\pi, \sigma \in S_r$ respectively, we compute one encrypting the permutation $\pi \circ \sigma$ by homomorphically evaluating the naïve matrix-multiplication algorithm. That is, output $\mathbf{C} = (c_{i,j}) \in \mathcal{C}^{r \times r}$ where

$$c_{i,j} \leftarrow \boxplus_{\ell \in [r]} (c_{i,\ell}^\pi \boxdot c_{\ell,j}^\sigma) \in \mathcal{C}. \tag{3}$$

Just like $\boxdot$, we define $\boxdot$ to be right associative.

Homomorphic equality test $\mathsf{Eq?}(\mathbf{C}^\pi = (c_{i,j}^\pi), \sigma \in S_r)$: given a ciphertext encrypting some permutation $\pi \in S_r$ and a permutation $\sigma \in S_r$ (in the clear), output a ciphertext $c \in \mathcal{C}$ encrypting 1 if $\pi = \sigma$ and 0 otherwise, as

$$c \leftarrow \boxdot_{i \in [r]} c_{\sigma(i),i}^\pi \boxdot g,$$

where $g \in \mathcal{C}$ denotes the fixed zero-error encryption of 1. (Recall that $\boxdot$ is right associative.)

Observe that for the above two operations, the GSW ciphertext(s) in the output are designed to encrypt the appropriate $\{0,1\}$-message. For Compose this is simply by correctness of the matrix-multiplication algorithm. For $\mathsf{Eq?}$ this is because the output ciphertext is designed to encrypt 1 if and only if every $c_{\sigma(i),i}$ is designed to encrypt 1, which is the case if and only if $\mathbf{C}^\pi$ is in fact designed to encrypt σ. All that remains is to analyze the behavior of the error terms, which we do next.

4.2   Analysis

Recalling that the GSW scheme is parameterized by n and q, denote its space of error vectors by $\mathcal{E} = \mathbb{Z}^m$ where $m = n\lceil \log_2 q \rceil$. The Euclidean norm on $\mathcal{E}^r = \mathbb{Z}^{mr}$ is defined in the expected way. In what follows it is often convenient to consider vectors and matrices over $\mathcal{E}$, i.e., each entry is itself a (row) vector in $\mathcal{E} = \mathbb{Z}^m$, and we switch between $\mathcal{E}^{h \times w}$ and $\mathbb{Z}^{h \times wm}$ as is convenient.

The following lemma describes the behavior of errors under the homomorphic composition operation $\boxdot$. Note that working with vectors and matrices over $\mathcal{E}$ lets us write a statement that is syntactically very similar to the one from Lemma 3.3, with a very similar proof.

Lemma 4.1. *Let $\mathbf{C}^\pi, \mathbf{C}^\sigma \in \mathcal{C}^{r \times r}$ respectively be designed to encrypt permutation matrices $\mathbf{P}^\pi, \mathbf{P}^\sigma \in \{0,1\}^{r \times r}$ with error matrices $\mathbf{E}^\pi, \mathbf{E}^\sigma \in \mathcal{E}^{r \times r}$. Then for any fixed values of these variables, $\mathbf{C}^\pi \boxdot \mathbf{C}^\sigma$ has error matrix $\mathbf{E} + \mathbf{P}^\pi \cdot \mathbf{E}^\sigma \in \mathcal{E}^{r \times r}$, where the $\mathbb{Z}$-entries of $\mathbf{E}$ are mutually independent, and those in its ith row are subgaussian with parameter $O(\|\mathbf{e}_i^\pi\|)$, where $\mathbf{e}_i^\pi$ is the ith row of $\mathbf{E}^\pi$.*

Proof. Let $\mathbf{C} \leftarrow \mathbf{C}^\pi \boxdot \mathbf{C}^\sigma$. It suffices to show that for all i, j, its (i,j)th entry $c_{i,j} \in \mathcal{C}$ has error

$$e_{i,j} + e_{\pi^{-1}(i),j}^\sigma \in \mathcal{E} = \mathbb{Z}^m,$$

where all the $\mathbb{Z}$-entries of all the $e_{i,j} \in \mathbb{Z}^m$ are mutually independent and subgaussian with parameter $O(\|\mathbf{e}_i^\pi\|)$, and $e_{\ell,j}^\sigma$ is the (ℓ, j)th entry of $\mathbf{E}^\sigma$. This follows directly from Equation (3) and Lemma 3.3: the error in each ciphertext $c_{i,\ell}^\pi \boxdot c_{\ell,j}^\sigma$ is $p_{i,\ell}^\pi \cdot e_{\ell,j}^\sigma$ plus a fresh vector whose entries are independent and subgaussian with parameter $O(\|e_{i,\ell}^\pi\|)$. Since $p_{i,\ell}^\pi = 1$ for $\ell = \pi^{-1}(i)$ and 0 otherwise, the claim follows by Pythagorean additivity of independent subgaussians.

Similarly to a multiplication chain of GSW ciphertexts, we can perform a (right-associative) chain of compositions while incurring only small error growth. For convenience of analysis, we always include the fixed zero-error ciphertext $\mathbf{J} \in \mathcal{C}^{r \times r}$ (which encrypts the identity permutation) as the rightmost ciphertext in the chain. The following corollary follows directly from Lemma 4.1 in the same way that Corollary 3.4 follows from Lemma 3.3.

Corollary 4.2. *Suppose that $\mathbf{C}_i \in \mathcal{C}^{r \times r}$ for $i \in [k]$ are respectively designed to encrypt permutation matrices $\mathbf{P}_i \in \{0,1\}^{r \times r}$ and have error matrices $\mathbf{E}_i \in \mathcal{E}^{r \times r}$. Then for any fixed values of these variables,*

$$\mathbf{C} \leftarrow \boxed{\circ}_{i \in [k]} \mathbf{C}_i \boxdot \mathbf{J} = \mathbf{C}_1 \boxdot (\mathbf{C}_2 \boxdot (\cdots (\mathbf{C}_k \boxdot \mathbf{J}) \cdots))$$

has an error matrix whose $\mathbb{Z}$-entries are mutually independent, and those in its ith row are subgaussian with parameter $O(\|\mathbf{e}_i\|)$, where $\mathbf{e}_i^t \in \mathcal{E}^{kr}$ is the ith row of the concatenated error matrices $[\mathbf{E}_1 \mid \cdots \mid \mathbf{E}_k]$.

Finally, since the Eq? procedure simply performs a chain of (right-associative) multiplications of GSW ciphertexts, Corollary 3.4 applies.

4.3   Optimizations for $\mathbb{Z}_r$ Embeddings

For bootstrapping, we use the above scheme only to encrypt elements in the cyclic subgroup $C_r \subseteq S_r$ that embeds the additive group $(\mathbb{Z}_r, +)$. As described in the preliminaries, an element $\pi \in C_r$ can be represented more compactly as an indicator (column) vector $\mathbf{p} \in \{0,1\}^r$ (rather than a matrix in $\{0,1\}^{r \times r}$), and its associated permutation matrix $\mathbf{P}_\pi$ is made up of the r cyclic rotations of $\mathbf{p}$. In addition, the composition of two permutations represented in this way as $\mathbf{p}, \mathbf{q}$ is given by the matrix-vector product $\mathbf{P}_\pi \cdot \mathbf{q}$, which may be computed in $O(r^2)$ operations, rather than $O(r^3)$ as in the general case. All of this translates directly to *encrypted* permutations in the expected way, i.e., ciphertexts are entry-wise encryptions in $\mathcal{C}^r$ of indicator vectors, etc.

Similarly, the equality test Eq? can be performed more efficiently when we restrict to the subgroup C_r: given r ciphertexts encrypting the entries of an indicator vector in $\{0,1\}^r$ and an $s \in \mathbb{Z}_r$, just output the ciphertext in the position corresponding to s.

Since our bootstrapping scheme uses $\mathbb{Z}_r$ embeddings only for $r = O(\log \lambda)$, these optimizations lead to polylogarithmic factor improvements in runtime and error, but no more.

5   Bootstrapping

We now describe our bootstrapping procedure.

5.1   Specification and Usage

We start by specifying the abstract preconditions and output guarantees of our
bootstrapping algorithm, and describe how to use it (with some additional pre-
and post-processing) to bootstrap known LWE-based encryption schemes.

The scheme to be bootstrapped must have *binary* ciphertexts in $\{0,1\}^d$ and
secret keys in $\mathbb{Z}_q^d$ for some dimension d and modulus q that should be made as
small as possible ($q, d = \tilde{O}(\lambda)$ are possible), and a decryption function of the
form $\mathsf{Dec_s}(\mathbf{c}) = f(\langle \mathbf{s}, \mathbf{c} \rangle) \in \{0,1\}$ for some arbitrary function $f \colon \mathbb{Z}_q \to \{0,1\}$.
We rely on an appropriate instantiation of the GSW cryptosystem, as described
in further detail in Section 5.2 below.

$\mathsf{BootGen}(\mathbf{s} \in \mathbb{Z}_q^d, sk)$ takes as input a secret key vector $\mathbf{s} \in \mathbb{Z}_q^d$ from the scheme
to be bootstrapped, and a secret key sk for GSW. It outputs a bootstrapping
key bk, which appropriately encrypts $\mathbf{s}$ under sk.

$\mathsf{Bootstrap}(bk, \mathbf{c} \in \{0,1\}^d)$ takes as input the bootstrapping key bk and a ciphertext vector $\mathbf{c} \in \{0,1\}^d$ (which decrypts under the secret key $\mathbf{s}$). It outputs a
GSW ciphertext which decrypts (under sk) to the same bit as $\mathbf{c}$ does (under
$\mathbf{s}$), but with less error.

Pre- and post-processing. We can bootstrap all known LWE-based bit-encryption
schemes using the above algorithms as follows. In all LWE-based encryption
schemes, decryption can be expressed as a "rounded inner product" $\lfloor \langle \mathbf{s}, \mathbf{c} \rangle \rceil_2$ for
some appropriate rounding function $\lfloor \cdot \rceil_2 \colon \mathbb{Z}_q \to \{0,1\}$, as required. Note that a
GSW ciphertext can trivially be put in this form by just taking its penultimate
column (see GSW.Dec in Section 3.1). As for the other conditions we need (binary
ciphertexts and small d, q), LWE encryption schemes are not always presented
in a way that fulfills them, but fortunately there are standard transformations
that do so, as we now describe. (See [6, 5] for further details.)

First, since we do not need to perform any further homomorphic operations
on the ciphertext, we can use dimension- and modulus-reduction [6] to get a ci-
phertext $\bar{\mathbf{c}}$ (over $\mathbb{Z}_q$) of dimension $\tilde{O}(\lambda)$ and modulus $q = \tilde{O}(\lambda)$, while preserving
correct decryption. These steps can be implemented with 2^λ security under con-
ventional lattice assumptions.[6] Then, we can obtain a binary ciphertext $\mathbf{c}$ using
"bit decomposition:" let $\mathbf{G}$ be as defined in Section 3, and for the ciphertext $\bar{\mathbf{c}}$
over $\mathbb{Z}_q$ under secret key $\bar{\mathbf{s}}$, let $\mathbf{c}$ be a $\{0,1\}$-vector such that $\mathbf{G}\mathbf{c} = \bar{\mathbf{c}}$, and let
$\mathbf{s} = \mathbf{G}^t \bar{\mathbf{s}}$ so that $\langle \mathbf{s}, \mathbf{c} \rangle = \langle \bar{\mathbf{s}}, \bar{\mathbf{c}} \rangle \in \mathbb{Z}_q$. (The secret key $\mathbf{s}$ is therefore the one we
need to provide to BootGen.)

---

[6] To make the modulus quasi-linear, we need to use randomized (subgaussian) rounding in the modulus-reduction step.

After bootstrapping, the output is a GSW ciphertext $\mathbf{C}$ encrypted under sk (which is just an integer vector). If desired, we can convert this ciphertext back to one for the original LWE cryptosystem, simply by taking the penultimate column of $\mathbf{C}$. We can also key-switch from sk back to the original secret key $\mathbf{s}$. (As usual in bootstrapping, going "full circle" in this way requires an appropriate circular security assumption.)

5.2 Procedures

Our algorithms rely on instantiations of GSW and HEPerm with parameters n, Q, χ. Importantly, the ciphertext modulus Q is *not* the modulus q of the scheme we are bootstrapping, but rather some $Q \gg q$ that is sufficiently larger than the error in Bootstrap's output ciphertext. Let $\mathcal{C}$ denote the GSW ciphertext space.

Our procedures need q to be of the form $q = \prod_{i \in [t]} r_i$ where the r_i are small and powers of distinct primes (and hence pairwise coprime). Specifically, using Lemma 2.1 we can choose $q = \tilde{O}(\lambda)$ to be large enough by letting it be the product of all maximal prime-powers r_i that are bounded by $O(\log \lambda)$, of which there are $t = O(\log \lambda / \log \log \lambda)$. Let ϕ be the group embedding of $(\mathbb{Z}_q, +) \cong (\mathbb{Z}_{r_1} \times \cdots \times \mathbb{Z}_{r_t}, +)$ into $S = S_{r_1} \times \cdots \times S_{r_t}$ described in Section 2.2, and let ϕ_i denote the ith component of this embedding, i.e., the one from $\mathbb{Z}_q$ into S_{r_i}.

BootGen$(\mathbf{s} \in \mathbb{Z}_q^d, sk)$: given secret key $\mathbf{s} \in \mathbb{Z}_q^d$ for the scheme to be bootstrapped and a secret key sk for HEPerm, embed each coordinate $s_j \in \mathbb{Z}_q$ of $\mathbf{s}$ as $\phi(s_j) \in S$ and encrypt the components under HEPerm. That is, generate and output the bootstrapping key

$$bk = \{\mathbf{C}_{i,j} \leftarrow \text{HEPerm.Enc}(sk, \phi_i(s_j)) : i \in [t], j \in [d]\}.$$

Recalling that we are working with embeddings of $\mathbb{Z}_{r_i}$, each $\mathbf{C}_{i,j} \in \mathcal{C}^{r_i}$ can be represented as a tuple of r_i GSW ciphertexts encrypting an indicator vector (see Section 4.3). Because $t, r_i = O(\log \lambda)$ and $d = \tilde{O}(\lambda)$, the bootstrapping key consists of $\tilde{O}(\lambda)$ GSW ciphertexts.

Bootstrap$(bk, \mathbf{c} \in \{0,1\}^d)$: given a binary ciphertext $\mathbf{c} \in \{0,1\}^d$, do the following:

Inner Product: Homomorphically compute an encryption of

$$v = \langle \mathbf{s}, \mathbf{c} \rangle = \sum_{j \,:\, c_j = 1} s_j \in \mathbb{Z}_q$$

using the encryptions of the $s_j \in \mathbb{Z}_q$ as embedded into the permutation group S, via a chain of compositions. Formally, for each $i \in [t]$ compute (recalling that $\boxdot$ is right associative, and $\mathbf{J}$ is the fixed HEPerm encryption of the identity permutation)

$$\mathbf{C}_i \leftarrow \underset{j \text{ s.t. } c_j = 1}{\boxed{\circ}} \mathbf{C}_{i,j} \boxdot \mathbf{J}. \tag{4}$$

Again, because we are working with embeddings of $\mathbb{Z}_{r_i}$, each $\mathbf{C}_i \in \mathcal{C}^{r_i}$.

Round: Homomorphically map $v \in \mathbb{Z}_q$ to $f(v) \in \mathbb{Z}_2 = \{0, 1\}$: for each $x \in \mathbb{Z}_q$ such that $f(x) = 1$, homomorphically test whether $v \overset{?}{=} x$ by homomorphically multiplying the GSW ciphertexts resulting from all the equality tests $v \overset{?}{=} x \pmod{r_i}$. Then homomorphically sum the results of all the $v \overset{?}{=} x$ tests.

Formally, compute and output the GSW ciphertext (recalling that $\boxdot$ is right associative, and $\mathbf{G}$ is the fixed GSW encryption of 1)

$$\mathbf{C} \leftarrow \biguplus_{x \in \mathbb{Z}_q \text{ s.t. } f(x)=1} \left(\boxdot_{i \in [t]} \mathsf{Eq?}(\mathbf{C}_i, \phi_i(x)) \boxdot \mathbf{G} \right). \tag{5}$$

Note that since we are working with embeddings of $\mathbb{Z}_{r_i}$, each $\mathsf{Eq?}(\mathbf{C}_i, \phi_i(x))$ is just some GSW ciphertext component of $\mathbf{C}_i \in \mathcal{C}^{r_i}$ (see Section 4.3).

Because $t, r_i = O(\log \lambda)$ and $d = \tilde{O}(\lambda)$ and by Equations (4) and (5), Bootstrap performs $\tilde{O}(\lambda)$ homomorphic multiplications and additions on GSW ciphertexts.

5.3 Analysis

The following is our main theorem. The proof is deferred to the full version due to space limitations.

Theorem 5.1. *The above bootstrapping scheme can be instantiated to be correct (with overwhelming probability) and secure assuming that the decisional Shortest Vector Problem (GapSVP) and Shortest Independent Vectors Problem (SIVP) are (quantumly) hard to approximate in the worst case to within $\tilde{O}(n^2\lambda)$ factors on n-dimensional lattices.*

Because all known (quantum) algorithms for $\mathrm{poly}(n)$-factor approximations to GapSVP and SIVP on n-dimensional lattices take $2^{\Omega(n)}$ time, for 2^λ hardness we can take $n = \Theta(\lambda)$, yielding a final approximation factor of $\tilde{O}(n^3)$. This comes quite close to the $O(n^{3/2+\epsilon})$ factors obtained in [7], but *without* any expensive "dimension leveraging:" we use GSW ciphertexts of dimension only $n = O(\lambda)$, rather than some large polynomial in λ. Alternatively, at the cost of a larger dimension $n = \lambda^{1/\epsilon}$, but without using the successive dimension-reduction procedure from [7], we can obtain factors as small as $\tilde{O}(n^{2+\epsilon})$ for any constant $\epsilon > 0$.

References

[1] Alperin-Sheriff, J., Peikert, C.: Practical bootstrapping in quasilinear time. In: Canetti, R., Garay, J.A. (eds.) CRYPTO 2013, Part I. LNCS, vol. 8042, pp. 1–20. Springer, Heidelberg (2013)

[2] Applebaum, B., Cash, D., Peikert, C., Sahai, A.: Fast cryptographic primitives and circular-secure encryption based on hard learning problems. In: Halevi, S. (ed.) CRYPTO 2009. LNCS, vol. 5677, pp. 595–618. Springer, Heidelberg (2009)

[3] Barrington, D.A.M.: Bounded-width polynomial-size branching programs recognize exactly those languages in NC^1. In: STOC, pp. 1–5 (1986)

[4] Brakerski, Z., Gentry, C., Vaikuntanathan, V. (Leveled) fully homomorphic encryption without bootstrapping. In: ITCS, pp. 309–325 (2012)

[5] Brakerski, Z., Langlois, A., Peikert, C., Regev, O., Stehlé, D.: Classical hardness of learning with errors. In: STOC, pp. 575–584 (2013)

[6] Brakerski, Z., Vaikuntanathan, V.: Efficient fully homomorphic encryption from (standard) LWE. In: FOCS, pp. 97–106 (2011)

[7] Brakerski, Z., Vaikuntanathan, V.: Lattice-based FHE as secure as PKE. In: ITCS, p. 1 (2014)

[8] Cai, J.-Y., Lipton, R.J.: Subquadratic simulations of circuits by branching programs. In: 2013 IEEE 54th Annual Symposium on Foundations of Computer Science, pp. 568–573 (1989)

[9] Cleve, R.: Towards optimal simulations of formulas by bounded-width programs. Computational Complexity 1(1), 91–105 (1991)

[10] Gentry, C.: A fully homomorphic encryption scheme. PhD thesis, Stanford University (2009), http://crypto.stanford.edu/craig

[11] Gentry, C.: Fully homomorphic encryption using ideal lattices. In: STOC, pp. 169–178 (2009)

[12] Gentry, C., Halevi, S.: Fully homomorphic encryption without squashing using depth-3 arithmetic circuits. In: FOCS, pp. 107–109 (2011)

[13] Gentry, C., Halevi, S., Smart, N.P.: Better bootstrapping in fully homomorphic encryption. In: Fischlin, M., Buchmann, J., Manulis, M. (eds.) PKC 2012. LNCS, vol. 7293, pp. 1–16. Springer, Heidelberg (2012)

[14] Gentry, C., Halevi, S., Smart, N.P.: Fully homomorphic encryption with polylog overhead. In: Pointcheval, D., Johansson, T. (eds.) EUROCRYPT 2012. LNCS, vol. 7237, pp. 465–482. Springer, Heidelberg (2012)

[15] Gentry, C., Sahai, A., Waters, B.: Homomorphic encryption from learning with errors: Conceptually-simpler, asymptotically-faster, attribute-based. In: Canetti, R., Garay, J.A. (eds.) CRYPTO 2013, Part I. LNCS, vol. 8042, pp. 75–92. Springer, Heidelberg (2013)

[16] Jacobson, N.: Basic Algebra I. Dover Publications (2012)

[17] Lyubashevsky, V., Peikert, C., Regev, O.: On ideal lattices and learning with errors over rings. Journal of the ACM 60(6), 43:1–43:35 (2013); Preliminary version in Gilbert, H. (ed.) EUROCRYPT 2010. LNCS, vol. 6110, pp. 1–23. Springer, Heidelberg (2010)

[18] Micciancio, D., Peikert, C.: Trapdoors for lattices: Simpler, tighter, faster, smaller. In: Pointcheval, D., Johansson, T. (eds.) EUROCRYPT 2012. LNCS, vol. 7237, pp. 700–718. Springer, Heidelberg (2012)

[19] Peikert, C.: Public-key cryptosystems from the worst-case shortest vector problem. In: STOC 2009, pp. 333–342 (2009)

[20] Regev, O.: On lattices, learning with errors, random linear codes, and cryptography. J. ACM 56(6), 1–40 (2009); Preliminary version in STOC 2005

[21] Schoenfeld, L.: Sharper bounds for the Chebyshev functions $\theta(x)$ and $\psi(x)$. ii. Mathematics of Computation 30(134), 337–360 (1976)

[22] Sinha, R.K.: Some topics in parallel computation and branching programs. PhD thesis, University of Washington (1995)

[23] Vershynin, R.: Compressed Sensing, Theory and Applications, ch. 5, pp. 210–268. Cambridge University Press (2012), http://www-personal.umich.edu/~romanv/papers/non-asymptotic-rmt-plain.pdf

Hardness of k-LWE and Applications in Traitor Tracing

San Ling[1], Duong Hieu Phan[2], Damien Stehlé[3], and Ron Steinfeld[4]

[1] Division of Mathematical Sciences,
School of Physical and Mathematical Sciences,
Nanyang Technological University, Singapore
[2] Laboratoire LAGA (CNRS, U. Paris 8, U. Paris 13), U. Paris 8, France
[3] Laboratoire LIP (U. Lyon, CNRS, ENSL, INRIA, UCBL), ENS de Lyon, France
[4] Faculty of Information Technology,
Monash University, Clayton, Australia

Abstract. We introduce the k-LWE *problem*, a Learning With Errors variant of the k-SIS problem. The Boneh-Freeman reduction from SIS to k-SIS suffers from an exponential loss in k. We improve and extend it to an LWE to k-LWE reduction with a polynomial loss in k, by relying on a new technique involving trapdoors for random integer kernel lattices. Based on this hardness result, we present the first algebraic construction of a traitor tracing scheme whose security relies on the worst-case hardness of standard lattice problems. The proposed LWE traitor tracing is almost as efficient as the LWE encryption. Further, it achieves public traceability, i.e., allows the authority to delegate the tracing capability to "untrusted" parties. To this aim, we introduce the notion of *projective sampling family* in which each sampling function is keyed and, with a projection of the key on a well chosen space, one can simulate the sampling function in a computationally indistinguishable way. The construction of a projective sampling family from k-LWE allows us to achieve public traceability, by publishing the projected keys of the users. We believe that the new lattice tools and the projective sampling family are quite general that they may have applications in other areas.

Keywords: Lattice-based cryptography, Traitor tracing, LWE.

1 Introduction

Since the pioneering work of Ajtai [3], there have been a number of proposals of cryptographic schemes with security provably relying on the worst-case hardness of standard lattice problems, such as the decision Gap Shortest Vector Problem with polynomial gap (see the surveys [30,40]). These schemes enjoy unmatched security guarantees: Security relies on *worst-case* hardness assumptions for problems expected to be *exponentially hard* to solve (with respect to the lattice dimension n), even with quantum computers. At the same time, they often enjoy great asymptotic efficiency, as the basic operations are matrix-vector multiplications in dimension $\tilde{O}(n)$ over a ring of cardinality $\leq \mathcal{P}oly(n)$. A breakthrough result in that field was the introduction of the Learning With Errors problem (LWE) by Regev [38,39], who showed it to be at least as hard as worst-case lattice problems and exploited it to devise an elementary encryption scheme.

J.A. Garay and R. Gennaro (Eds.): CRYPTO 2014, Part I, LNCS 8616, pp. 315–334, 2014.

Gentry et al. showed in [19] that Regev's scheme may be adapted so that a master can generate a large number of secret keys for the same public key. As a result, the latter encryption scheme, called dual-Regev, can be naturally extended into a multi-receiver encryption scheme. In the present work, we build traitor tracing schemes from this dual-Regev LWE-based encryption scheme.

TRAITOR TRACING. A traitor tracing scheme is a multi-receiver encryption scheme where malicious receiver coalitions aiming at building pirate decryption devices are deterred by the existence of a tracing algorithm: Using the pirate decryption device, the tracing algorithm can recover at least one member of the malicious coalition. Such schemes are particularly well suited for fighting copyright infringement in the context of commercial content distribution (e.g., Pay-TV, subscription news websites, etc). Since their introduction by Chor et al. [15], much work has been devoted to devising efficient and secure traitor tracing schemes. The most desirable schemes are fully collusion resistant: they can deal with arbitrarily large malicious coalitions. But, unsurprisingly, the most efficient schemes are in the bounded collusion model where the number of malicious users is limited. The first non-trivial fully collusion resistant scheme was proposed by Boneh et al. [11]. However, its ciphertext size is still large ($\Omega(\sqrt{N})$, where N is the total number of users) and it relies on pairing groups of composite order. Very recently, Boneh and Zhandry [12] proposed a fully collusion resistant scheme with poly-log size parameters. It relies on indistinguishability obfuscation [18], whose security foundation remains to be studied, and whose practicality remains to be exhibited. In this paper, we focus on the bounded collusion model. The Boneh-Franklin scheme [7] is one of the earliest algebraic constructions but it can still be considered as the reference algebraic transformation from the standard ElGamal public key encryption into traitor tracing. This transformation induces a linear loss in efficiency, with respect to the maximum number of traitors. The known transformations from encryption to traitor tracing in the bounded collusion model present at least a linear loss in efficiency, either in the ciphertext size or in the private key size [7,31,23,41,6,10]. We refer to [21] for a detailed introduction to this rich topic.

OUR CONTRIBUTIONS. We describe the first algebraic construction of a public-key lattice-based traitor tracing scheme. It is semantically secure and enjoys public traceability. The security relies on the hardness of LWE, which is known to be at least as hard as standard worst-case lattice problems [39,33,13].

The scheme is the extension, described above, of the dual-Regev LWE-based encryption scheme from [19] to a multi-receiver encryption scheme, where each user has a different secret key. In the case of traitor tracing, several keys may be leaked to a traitor coalition. To show that we can trace the traitors, we extend the LWE problem and introduce the k-LWE problem, in which k hint vectors (the leaked keys) are given out.

Intuitively, k-LWE asks to distinguish between a random vector t close to a given lattice Λ and a random vector t close to the orthogonal subspace of the span of k given short vectors belonging to the dual Λ^* of that lattice. Even if we are given $(b_i^*)_{i \leq k}$ small in Λ^*, computing the inner products $\langle b_i^*, t \rangle$ will not help in solving this problem, since they are small and distributed identically in both cases. The k-LWE problem can be interpreted as a dual of the k-SIS problem introduced by Boneh and Freeman [8], which intuitively requests to find a short vector in Λ^* that is linearly independent with the k given short vectors of Λ^*. Their reduction from SIS to k-SIS can be adapted to the LWE setup, but the hardness loss incurred by the reduction is gigantic. We propose a significantly sharper reduction from LWE_α to k-LWE_α. This improved reduction requires a new lattice technique: the equivalent for kernel lattices of Ajtai's simultaneous sampling of a random q-ary lattice with a short basis [4] (see also Lemma 2). We adapt the Micciancio-Peikert framework from [28] to sampling a Gaussian $X \in \mathbb{Z}^{m \times n}$ along with a short basis for the lattice $\ker(X) = \{\boldsymbol{b} \in \mathbb{Z}^m : \boldsymbol{b}^t X = \boldsymbol{0}\}$. Kernel lattices also play an important role in the re-randomization analysis of the recent lattice-based multilinear map scheme of Garg et al. [17], and we believe that our new trapdoor generation tool for such lattices is likely find additional applications in future. We also remark that our technique can be adapted to the SIS to k-SIS reduction. We thus solve the open question left by Boneh and Freeman of improving their reduction [8]: from an exponential loss in k to a polynomial loss in k. Consequently, their linearly homomorphic signatures and ordinary signature schemes enjoy much better efficiency/security trade-offs.

Our construction of a traitor tracing scheme from k-LWE can be seen as an additive and noisy variant of the (black-box) Boneh-Franklin traitor tracing scheme [7]. While the Boneh-Franklin scheme is transformed from the ElGamal encryption with a linear loss (in the maximum number of traitors) in efficiency, our scheme is almost as efficient as standard LWE-based encryption, as long as the maximum number of traitors is bounded below $n/(c \log n)$, where n is the LWE dimension determined by the security parameter, and c is a constant. The full functionality of black-box tracing in both the Boneh-Franklin scheme and ours are of high complexity as they both rely on the black-box confirmation: given a superset of the traitors, it is guaranteed to find at least one traitor and no innocent suspect is incriminated. Boneh and Franklin left the improvement of the black-box tracing as an interesting open problem. We show that in lattice setting, the black-box tracing can be accelerated by running the tracing procedure in parallel on untrusted machines. This is a direct consequence of the property of public traceability, i.e., the possibility of running tracing procedure on public information, that our scheme enjoys. We note that almost all traitor tracing systems require that the tracing key must be kept secret. Some schemes [14,37,9,12] achieve public traceability and some others achieve a stronger notion than public traceability, namely the non-repudiation, but the setup in these schemes require some interactive protocol between the center and each user such as a secure 2-party computation protocol in [35], a commitment protocol in [36], an oblivious polynomial evaluation in [42,24,22].

To obtain public traceability and inspired from the notion of projective hash family [16], we introduce a new notion of *projective sampling family* in which each sampling function is keyed and, with a projection of the key on a well chosen space, one can simulate the sampling function in a computationally indistinguishable way. The construction of a set of projective sampling families from k-LWE allows us to publicly sample the tracing signals.

Independently, our new lattice tools may have applications in other areas. The k-LWE problem has a similar flavour to the Extended-LWE problem from [32]. It would be interesting to exhibit reductions between these problems. On a closely-related topic, it seems our sampling of a random Gaussian integer matrix X together with a short basis of $\ker(X)$ is compatible with the hardness proof of Extended-LWE from [13]. In particular, it should be possible to use it as an alternative to [13, Def 4.5] in the proof of [13, Le 4.7], to show that Extended-LWE remains hard with many hints independently sampled from discrete Gaussians.

REMARK. Due to lack of space, some background and the missing proofs of Sections 3 and 5 have been removed from this proceedings version. The full version is available on the webpages of the authors.

2   Preliminaries

If x is a real number, then $\lfloor x \rceil$ is the closest integer to x (with any deterministic rule in case x is half an odd integer). All vectors will be denoted in bold. By default, our vectors are column vectors. We let $\langle \cdot, \cdot \rangle$ denote the canonical inner product. For q prime, we let $\mathbb{Z}_q$ denote the field of integers modulo q. For two matrices A, B of compatible dimensions, we let $(A|B)$ and $(A\|B)$ respectively denote the horizontal and vertical concatenations of A and B. For $A \in \mathbb{Z}_q^{m \times n}$, we define $\mathrm{Im}(A) = \{As : s \in \mathbb{Z}_q^n\} \subseteq \mathbb{Z}_q^m$. For $X \subseteq \mathbb{Z}_q^m$, we let $\mathrm{Span}(X)$ denote the set of all linear combinations of elements of X. We let $X^\perp$ denote the linear subspace $\{b \in \mathbb{Z}_q^m : \forall c \in X, \langle b, c \rangle = 0\}$. For a matrix $S \in \mathbb{R}^{m \times n}$, we let $\|S\|$ denote the norm of its longest column. If S is full column-rank, we let $\sigma_1(S) \geq \ldots \geq \sigma_n(S)$ denote its singular values. We let $\mathbb{T}$ denote the additive group $\mathbb{R}/\mathbb{Z}$.

If D_1 and D_2 are distributions over a countable set X, their statistical distance $\frac{1}{2}\sum_{x \in X} |D_1(x) - D_2(x)|$ will be denoted by $\Delta(D_1, D_2)$. The statistical distance is defined similarly if X is measurable. If X is of finite weight, we let $U(X)$ denote the uniform distribution over X. For any invertible $S \in \mathbb{R}^{m \times m}$ and $c \in \mathbb{R}^m$, we define the function $\rho_{S,c}(b) = \exp(-\pi\|S^{-1}(b - c)\|^2)$. For $S = sI_m$, we write $\rho_{s,c}$, and we omit the subscripts S and c when $S = I_m$ and $c = 0$. We let ν_α denote the one-dimensional Gaussian distribution with standard deviation α.

2.1   Euclidean Lattices and Discrete Gaussian Distributions

A lattice is a set of the form $\{\sum_{i \leq n} x_i b_i : x_i \in \mathbb{Z}\}$ where the b_i's are linearly independent vectors in $\mathbb{R}^m$. In this situation, the b_i's are said to form a basis of the n-dimensional lattice. The n-th minimum $\lambda_n(L)$ of an n-dimensional lattice L is defined as the smallest r such that the n-dimensional closed hyperball of radius r centered in 0 contains n linearly independent vectors of L. The smoothing parameter of L is defined as $\eta_\varepsilon(L) = \min\{r > 0 : \rho_{1/r}(\widehat{L} \setminus 0) \leq \varepsilon\}$ for any $\varepsilon \in (0, 1)$, where $\widehat{L} = \{c \in \mathrm{Span}(L) : c^t \cdot L \subseteq \mathbb{Z}\}$ is the dual lattice of L. It was proved in [29, Le. 3.3] that $\eta_\varepsilon(L) \leq \sqrt{\ln(2n(1 + 1/\varepsilon))/\pi} \cdot \lambda_n(L)$ for all $\varepsilon \in (0, 1)$ and n-dimensional lattices L.

For a lattice $L \subseteq \mathbb{R}^m$, a vector $c \in \mathbb{R}^m$ and an invertible $S \in \mathbb{R}^{m \times m}$, we define the Gaussian distribution of parameters L, c and S by $D_{L,S,c}(b) \sim \rho_{S,c}(b) = \exp(-\pi\|S^{-1}(b - c)\|^2)$ for all $b \in L$. When $S = \sigma \cdot I_m$, we simply write $D_{L,\sigma,c}$.

Note that $D_{L,S,c} = S^t \cdot D_{S^{-t}L,1,S^{-t}c}$. Sometimes, for convenience, we use the notation $D_{L+c,S}$ as a shorthand for $c + D_{L,S,-c}$. Gentry et al. [19] gave an algorithm, referred to as GPV algorithm, to sample from $D_{L,S,c}$ when given as input a basis $(b_i)_i$ of L such that $\sqrt{\ln(2n+4)/\pi} \cdot \max_i \|S^{-t}b_i\| \leq 1$.

We extensively use q-ary lattices. The q-ary lattice associated to $A \in \mathbb{Z}_q^{m \times n}$ is defined as $\Lambda^{\perp}(A) = \{x \in \mathbb{Z}^m : x^t \cdot A = 0 \bmod q\}$. It has dimension m, and a basis can be computed in polynomial-time from A. For $u \in \mathbb{Z}_q^m$, we define $\Lambda_u^{\perp}(A)$ as the coset $\{x \in \mathbb{Z}^m : x^t \cdot A = u^t \bmod q\}$ of $\Lambda^{\perp}(A)$.

2.2   Random Lattices

We consider the following random lattices, called q-ary Ajtai lattices. They are obtained by sampling $A \hookleftarrow U(\mathbb{Z}_q^{m \times n})$ and considering $\Lambda^{\perp}(A)$. The following lemma provides a probabilistic bound on the smoothing parameter of $\Lambda^{\perp}(A)$.

Lemma 1 (Adapted from [19, Le. 5.3]). *Let q be prime and m, n integers with $m \geq 2n$ and $\varepsilon > 0$, then $\eta_\varepsilon(\Lambda^{\perp}(A)) \leq 4q^{\frac{n}{m}} \sqrt{\log(2m(1+1/\varepsilon))/\pi}$, for all except a fraction $2^{-\Omega(n)}$ of $A \in \mathbb{Z}_q^{m \times n}$.*

It is possible to efficiently sample a close to uniform A along with a short basis of $\Lambda^{\perp}(A)$ (see [4,5,34,28]).

Lemma 2 (Adapted from [5, Th. 3.1]). *There exists a ppt algorithm that given n, m, $q \geq 2$ as inputs samples two matrices $A \in \mathbb{Z}_q^{m \times n}$ and $T \in \mathbb{Z}^{m \times m}$ such that: the distribution of A is within statistical distance $2^{-\Omega(n)}$ from $U(\mathbb{Z}_q^{m \times n})$; the rows of T form a basis of $\Lambda^{\perp}(A)$; each row of T has norm $\leq 3mq^{n/m}$.*

For $A \in \mathbb{Z}_q^{m \times n}$, $S \in \mathbb{R}^{m \times m}$ invertible, $c \in \mathbb{R}^m$ and $u \in \mathbb{Z}_q^n$, we define the distribution $D_{\Lambda_u^{\perp}(A),S,c}$ as $\bar{c} + D_{\Lambda^{\perp}(A),S,-\bar{c}+c}$, where $\bar{c}$ is any vector of $\mathbb{Z}^m$ such that $\bar{c}^t \cdot A = u^t \bmod q$. A sample x from $D_{\Lambda_u^{\perp}(A),S}$ can be obtained using the GPV algorithm along with the short basis of $\Lambda^{\perp}(A)$ provided by Lemma 2. Boneh and Freeman [8] showed how to efficiently obtain the residual distribution of (A, x) without relying on Lemma 2.

Theorem 1 (Adapted from [8, Th. 4.3]). *Let $n, m, q \geq 2$, $k \geq 0$ and $S \in \mathbb{R}^{m \times m}$ be such that $m \geq 2n$, q is prime with $q > \sigma_1(S) \cdot \sqrt{2\log(4m)}$, and $\sigma_m(S) = q^{\frac{n}{m}} \cdot \max(\Omega(\sqrt{n \log m}), 2\sigma_1(S)^{\frac{k}{m}})$. Let $u_1, \ldots, u_k \in \mathbb{Z}_q^n$ and $c_1, \ldots, c_k \in \mathbb{R}^m$ be arbitrary. Then the residual distributions of the tuple $(A, x_1, \ldots, x_k)$ obtained with the following two experiments are within statistical distance $2^{-\Omega(n)}$.*

$\mathrm{Exp}_0:$ $A \hookleftarrow U(\mathbb{Z}_q^{m \times n}); \quad \forall i \leq k : x_i \hookleftarrow D_{\Lambda_{u_i}^{\perp}(A),S,c_i} \cdot$

$\mathrm{Exp}_1: \forall i \leq k : x_i \hookleftarrow D_{\mathbb{Z}^m,S,c_i}; A \hookleftarrow U\left(\mathbb{Z}_q^{m \times n} | \forall i \leq k : x_i^t \cdot A = u_i^t \bmod q\right).$

This statement generalizes [8, Th. 4.3] in three ways. First, the latter corresponds to the special case corresponding to taking all the u_i's and c_i's equal to 0. This generalization does not add any extra complication in the proof of [8, Th. 4.3], but is important for our constructions. Second, the condition on m is less restrictive (the corresponding assumption in [8, Th. 4.3] is that $m \geq \max(2n \log q, 2k)$). To allow for such small values of m, we refine the bound on the smoothing parameter of the $\Lambda^{\perp}(A)$ lattice (namely, we use Lemma 1). Third, we allow for a non-spherical Gaussian distribution, which seems needed in our generalized Micciancio-Peikert trapdoor gadget used in the reduction from LWE to k-LWE in Section 3.2.

We also use the following result on the probability of the Gaussian vectors x_i from Theorem 1 being linearly independent over $\mathbb{Z}_q$.

Lemma 3 (Adapted from [8, Le. 4.5]). *With the notations and assumptions of Theorem 1, the k vectors $x_1, \ldots, x_k$ sampled in $\texttt{Exp}_0$ and $\texttt{Exp}_1$ are linearly independent over $\mathbb{Z}_q$, except with probability $2^{-\Omega(n)}$.*

2.3 Rényi Divergence

We use Rényi Divergence (RD) in our analysis, relying on techniques developed in [27,25,26]. For any two probability distributions P and Q such that the support of P is a subset of the support of Q over a countable domain X, we define the RD (of order 2) by $R(P\|Q) = \sum_{x \in X} \frac{P(x)^2}{Q(x)}$, with the convention that the fraction is zero when both numerator and denominator are zero. We recall that the RD between two offset discrete Gaussians is bounded as follows.

Lemma 4 ([25, Le. 4.2]). *For any n-dimensional lattice $L \subseteq \mathbb{R}^n$ and invertible matrix S, set $P = D_{L,S,w}$ and $Q = D_{L,S,z}$ for some fixed $w, z \in \mathbb{R}^n$. If $w, z \in L$, let $\varepsilon = 0$. Otherwise, fix $\varepsilon \in (0,1)$ and assume that $\sigma_n(S) \geq \eta_\varepsilon(L)$. Then $R(P\|Q) \leq \left(\frac{1+\varepsilon}{1-\varepsilon}\right)^2 \cdot \exp\left(2\pi\|w - z\|^2/\sigma_n(S)^2\right)$.*

We use this bound and the fact that the RD between the parameter distributions of two distinguishing problems can be used to relate their hardness, if they satisfy a certain public samplability property.

Lemma 5 ([26]). *Let Φ, Φ' denote two distributions, and $D_0(r)$ and $D_1(r)$ denote two distributions determined by some parameter r. Let P, P' be two decision problems defined as follows:*

- *P: Assess whether input x is sampled from distribution X_0 or X_1, where*
  $$X_0 = \{x : r \hookleftarrow \Phi, x \hookleftarrow D_0(r)\}, \ X_1 = \{x : r \hookleftarrow \Phi, x \hookleftarrow D_1(r)\}.$$
- *P': Assess whether input x is sampled from distribution X_0' or X_1', where*
  $$X_0' = \{x : r \hookleftarrow \Phi', x \hookleftarrow D_0(r)\}, \ X_1' = \{x : r \hookleftarrow \Phi', x \hookleftarrow D_1(r)\}.$$

Assume that $D_0(\cdot)$ and $D_1(\cdot)$ have the following public samplability property: there exists a sampling algorithm S with run-time T_S such that for all r, b, given any sample x from $D_b(r)$ we have:

- *S$(0, x)$ outputs a sample distributed as $D_0(r)$ over the randomness of S.*
- *S$(1, x)$ outputs a sample distributed as $D_1(r)$ over the randomness of S.*

If there exists a T-time distinguisher A for problem P with advantage ε, then, for every $\lambda > 0$, there exists an $O(\lambda \varepsilon^{-2} \cdot (T_S + T))$-time distinguisher A' for problem P' with advantage $\varepsilon' \geq \frac{\varepsilon^3}{8R(\Phi\|\Phi')} - O(2^{-\lambda})$.

2.4 Learning with Errors

Let $s \in \mathbb{Z}_q^n$ and $\alpha > 0$. We define the distribution $A_{s,\alpha}$ as follows: Take $a \hookleftarrow U(\mathbb{Z}_q^n)$ and $e \hookleftarrow \nu_\alpha$, and return $(a, \frac{1}{q}\langle a, s \rangle + e) \in \mathbb{Z}_q^n \times \mathbb{T}$. The *Learning With Errors problem* LWE$_\alpha$, introduced by Regev in [38,39], consists in assessing whether an oracle produces samples from $U(\mathbb{Z}_q^n \times \mathbb{T})$ or $A_{s,\alpha}$ for some constant $s \hookleftarrow U(\mathbb{Z}_q^n)$. Regev [39] showed that for $q \leq \mathcal{P}oly(n)$ prime and $\alpha \in (\frac{\sqrt{n}}{2q}, 1)$, LWE is (quantumly) not easier than standard worst-case lattice problems in dimension n with approximation factors $\mathcal{P}oly(n)/\alpha$. This hardness proof was partly dequantized in [33,13], and the requirements that q should be prime and $\mathcal{P}oly(n)$ were waived.

In this work, we consider a variant LWE where the number of oracle samples that the distinguisher requests is a priori bounded. If m denotes that bound, then we will refer to this restriction as LWE$_{\alpha,m}$. In this situation, the hardness assumption can be restated in terms of linear algebra over $\mathbb{Z}_q$: Given $A \hookleftarrow U(\mathbb{Z}_q^{m \times n})$, the goal is to distinguish between the distributions (over $\mathbb{T}^m$)

$$\frac{1}{q} U\left(\text{Im}(A)\right) + \nu_\alpha^m \quad \text{and} \quad \frac{1}{q} U\left(\mathbb{Z}_q^m\right) + \nu_\alpha^m.$$

Under the assumption that $\alpha q \geq \Omega(\sqrt{n})$, the right hand side distribution is indeed within statistical distance $2^{-\Omega(n)}$ to $U(\mathbb{T}^m)$ (see, e.g., [29, Le. 4.1]). The hardness assumption states that by adding to them a small Gaussian noise, the linear spaces $\text{Im}(A)$ and $\mathbb{Z}_q^m$ become computationally indistinguishable. This rephrasing in terms of linear algebra is helpful in the security proof of the traitor tracing scheme. Note that by a standard hybrid argument, distinguishing between the two distributions given one sample from either, and distinguishing between them given Q samples (from the same distribution), are computationally equivalent problems, up to a loss of a factor Q in the distinguishing advantage.

Finally, we will also use a variant of LWE where the noise distribution ν_α is replaced by $D_{q^{-1}\mathbb{Z},\alpha}$, and where $U(\mathbb{T})$ is replaced by $U(\mathbb{T}_q)$ with $\mathbb{T}_q$ being $q^{-1}\mathbb{Z}$ with addition mod 1. This variant, denoted by LWE$'$, was proved in [34] to be no easier than standard LWE (up to a constant factor increase in α).

3 New Lattice Tools

The security of our constructions relies on the hardness of a new variant of LWE, which may be seen as the dual of the k-SIS problem from [8].

Definition 1. *Let $k \leq m$, $S \in \mathbb{R}^{m \times m}$ invertible and $C = (c_1 \| \cdots \| c_k) \in \mathbb{R}^{k \times m}$. The (k, S, C)-LWE$_{\alpha,m}$ problem (or (k, S)-LWE if $C = 0$) is as follows: Given $A \hookleftarrow U(\mathbb{Z}_q^{m \times n})$, $u \hookleftarrow U(\mathbb{Z}_q^n)$ and $x_i \hookleftarrow D_{\Lambda_{-u}^\perp(A),S,c_i}$ for $i \leq k$, the goal is to distinguish between the distributions (over $\mathbb{T}^{m+1}$)*

$$\frac{1}{q} \cdot U\left(\text{Im}\left(\frac{u^t}{A}\right)\right) + \nu_\alpha^{m+1} \quad \text{and} \quad \frac{1}{q} \cdot U\left(\text{Span}_{i \leq k}\left(\frac{1}{x_i}\right)^\perp\right) + \nu_\alpha^{m+1}.$$

The classical LWE problem consists in distinguishing the left distribution from uniform, without the hint vectors $x_i^+ = (1\|x_i)$. These hint vectors correspond to the secret keys obtained by the malicious coalition in the traitor tracing scheme. Once these hint vectors are revealed, it becomes easy to distinguish the left distribution from the uniform distribution: take one of the vectors x_i^+, get a challenge sample y and compute $\langle x_i^+, y \rangle \in \mathbb{T}$; if y is a sample from the left distribution, then the centered residue is expected to be of size $\approx \alpha \cdot (\sqrt{m}\sigma_1(S) + \|c_i\|)$, which is $\ll 1$ for standard parameter settings; on the other hand, if y is sampled from the uniform distribution, then $\langle x^+, y \rangle$ should be uniform. The definition of (k, S)-LWE handles this issue by replacing $U(\mathbb{Z}_q^{m+1})$ by $U(\mathrm{Span}_{i \le k}(x_i^+)^\perp)$.

Sampling x_i^+ from $D_{\Lambda^\perp((u^t\|A)),S,c_i}$ may seem more natural than imposing that the first coordinate of each x_i^+ is 1. Looking ahead, this constraint will prove convenient to ensure correctness of our cryptographic primitives. Theorem 3 below and its proof can be readily adapted to this hint distribution. They may also be adapted to improve the SIS to k-SIS reduction from [8]. Setting $C = 0$ is also more natural, but for technical reasons, our reduction from LWE to (k, S, C)-LWE works with unit vectors c_i. However, we show that for small $\|c_i\|$, there exist polynomial time reductions between (k, S, C)-LWE and (k, S)-LWE.

In the proof of the hardness of (k, S)-LWE problem, we rely on a gadget integral matrix G that has the following properties: its first rows have Gaussian distributions, it is unimodular and its inverse is small. Before going to this proof, we shall build such a gadget matrix by extending Ajtai's simultaneous sampling of a random q-ary lattice with a short basis [4] (see also Lemma 2) to kernel lattices. More precisely, we adapt the Micciancio-Peikert framework [28] to sampling a Gaussian $X \in \mathbb{Z}^{m \times n}$ along with a short basis for the lattice $\ker(X) = \{b \in \mathbb{Z}^m : b^t X = 0\}$.

3.1   Sampling a Gaussian X with a Small Basis of $\ker(X)$

The Micciancio-Peikert construction [28] relies on a *leftover hash lemma* stating that with overwhelming probability over $A \hookleftarrow U(\mathbb{Z}_q^{m \times n})$ and for a sufficiently large σ, the distribution of $A^t \cdot D_{\mathbb{Z}^m,\sigma} \bmod q$ is statistically close to $U(\mathbb{Z}_q^n)$. We use a similar result over the integers, starting from a Gaussian $X \in \mathbb{Z}^{m \times n}$ instead of a uniform $A \in \mathbb{Z}_q^{m \times n}$. The proof of the following lemma relies on [1], which improves over a similar result from [2]. The result would be neater with $\sigma_2 = \sigma_1$, but, unfortunately, we do not know how to achieve it. The impact of this drawback on our results and constructions is mostly cosmetic.

Lemma 6. *Let $m \ge n \ge 100$ and $\sigma_1, \sigma_2 > 0$ satisfying $\sigma_1 \ge \Omega(\sqrt{mn \log m})$, $m \ge \Omega(n \log(\sigma_1 n))$ and $\sigma_2 \ge \Omega(n^{5/2}\sqrt{m}\sigma_1^2 \log^{3/2}(m\sigma_1))$. Let $X \hookleftarrow D_{\mathbb{Z},\sigma_1}^{m \times n}$. There exists a ppt algorithm that takes $n, m, \sigma_1, \sigma_2, X$ and $c \in \mathbb{Z}^n$ as inputs and returns $x \in \mathbb{Z}^n, r \in \mathbb{Z}^m$ such that $x = c + X^t r$ with $\|r\| \le O(\sigma_2/\sigma_1)$, with probability $1 - 2^{-\Omega(n)}$, and*

$$\Delta\big((X, x), D_{\mathbb{Z},\sigma_1}^{m \times n} \times D_{\mathbb{Z}^n,\sigma_2,c}\big) \le 2^{-\Omega(n)}.$$

We now adapt the trapdoor construction from [28] to kernel lattices.

Theorem 2. *Let* $n, m_1, \sigma_1, \sigma_2$ *be as above, and* $m_2 \geq m_1$ *bounded as* $n^{O(1)}$. *There exists a ppt algorithm that given* n, m_1, m_2 *(in unary),* σ_1 *and* σ_2, *returns* $X_1 \in \mathbb{Z}^{m_1 \times n}, X_2 \in \mathbb{Z}^{m_2 \times n}$, *and* $U \in \mathbb{Z}^{m \times m}$ *with* $m = m_1 + m_2$, *such that:*

- *the distribution of* (X_1, X_2) *is within statistical distance* $2^{-\Omega(n)}$ *of* $D_{\mathbb{Z}, \sigma_1}^{m_1 \times n} \times (D_{\mathbb{Z}^{m_2}, \sigma_2, \delta_1} \times \cdots \times D_{\mathbb{Z}^{m_2}, \sigma_2, \delta_n})$, *where* δ_i *denotes the ith canonical unit vector in* $\mathbb{Z}^{m_2}$ *whose ith coordinate is* 1 *and whose remaining coordinates are* 0.
- *we have* $|\det U| = 1$ *and* $U \cdot X = (I_n \| 0)$ *with* $X = (X_1 \| X_2)$,
- *every row of* U *has norm* $\leq O(\sqrt{nm_1}\sigma_2)$ *with probability* $\geq 1 - 2^{-\Omega(n)}$.

The second statement implies that the last $m - n$ rows of U form a basis of the random lattice $\ker(X)$.

Proof. We first sample X_1 from $D_{\mathbb{Z}, \sigma_1}^{m_1 \times n}$ using the GPV algorithm. We run m_2 times the algorithm from Lemma 6, on the input $n, m_1, \sigma_1, \sigma_2, X_1$ and c running through the columns of $C = [I_n | 0_{n \times (m_2 - n)}]$. This gives $X_2 \in \mathbb{Z}^{m_2 \times n}$ and $R \in \mathbb{Z}^{m_1 \times m_2}$ such that $X_2^t = [I_n | 0_{n \times (m_2 - n)}] + X_1^t \cdot R$. One can then see that $U \cdot X = [I_n \| 0]$, where

$$U = \left[\begin{array}{c|c} 0 & I_{m_2} \\ \hline I_{m_1} & -(X_1 | 0) \end{array}\right] \cdot \left[\begin{array}{c|c} I_{m_1} & 0 \\ \hline -R^t & I_{m_2} \end{array}\right] = \left[\begin{array}{c|c} -R^t & I_{m_2} \\ \hline I_{m_1} + (X_1 | 0)R^t & -(X_1 | 0) \end{array}\right], X = \left[\begin{array}{c} X_1 \\ \hline X_2 \end{array}\right].$$

The result then follows from Gaussian tail bounds (to bound the norms of the rows of X_1) and elementary computations.   □

Our gadget matrix G is U^{-t}. In the following corollary, we summarize the properties we will use.

Corollary 1. *Let* $n, m_1, m_2, m, \sigma_1, \sigma_2$ *be as in Theorem 2. There exists a ppt algorithm that given* n, m_1, m_2 *(in unary), and* σ_1, σ_2 *as inputs, returns* $G \in \mathbb{Z}^{m \times m}$ *such that:*

- *the top* $n \times m$ *submatrix of* G *is within statistical distance* $2^{-\Omega(n)}$ *of* $D_{\mathbb{Z}, \sigma_1}^{n \times m_1} \times (D_{\mathbb{Z}^{m_2}, \sigma_2, \delta_1} \times \cdots \times D_{\mathbb{Z}^{m_2}, \sigma_2, \delta_n})^t$,
- *we have* $|\det G| = 1$ *and* $\|G^{-1}\| \leq O(\sqrt{nm_2}\sigma_2)$, *with probability* $1 - 2^{-\Omega(n)}$.

3.2 Hardness of k-LWE

The following result shows that this LWE variant, with S a specific diagonal matrix, is no easier than LWE.

Theorem 3. *There exists* $c > 0$ *such that the following holds for* $k = n/(c \log n)$. *Let* m, q, σ, σ' *be such that* $\sigma \geq \Omega(n), \sigma' \geq \Omega(n^3 \sigma^2 / \log n), q \geq \Omega(\sigma' \sqrt{\log m})$ *is prime, and* $m \geq \Omega(n \log q)$ *(e.g.,* $\sigma = \Theta(n), \sigma' = \Theta(n^5 / \log n), q = \Theta(n^5)$ *and* $m = \Theta(n \log n))$. *Then there exists a probabilistic polynomial-time reduction from* $\text{LWE}_{m+1, \alpha}$ *in dimension* n *to* $(k, S)\text{-LWE}_{m+2n, \alpha'}$ *in dimension* $4n$, *with* $\alpha' = \Omega(mn^{3/2}\sigma\sigma'\alpha)$ *and* $S = \left[\begin{array}{c|c} \sigma \cdot I_{m+n} & 0 \\ \hline 0 & \sigma' \cdot I_n \end{array}\right]$. *More concretely, using a* $(k, S)\text{-LWE}_{m+2n, \alpha'}$ *algorithm with run-time* T *and advantage* ε, *the reduction gives an* $\text{LWE}_{m+1, \alpha}$ *algorithm with advantage* $\varepsilon' \geq \frac{\varepsilon^3}{8R(\Phi \| \Phi')} - O(2^{-\lambda})$ *and advantage* $\varepsilon' = \Omega((\varepsilon - 2^{-\Omega(n/\log n)})^3) - O(2^{-n})$.

The reduction takes an LWE instance and extends it to a related k-LWE instance for which the additional hint vectors $(x_i)_{i \le k}$ are known. The major difficulty in this extension is to restrain the noise increase, as a function of k.

The existing approach for this reduction (that we improve below) is the technique used in the SIS to k-SIS reduction from [8]. In the latter approach, the hint vectors are chosen independently from a small discrete Gaussian distribution, and then the LWE matrix A is extended to a larger matrix A' under the constraint that the hint vectors are in the q-ary lattice $\Lambda^{\perp}(A') = \{b : b^t A' = 0 \bmod q\}$. Unfortunately, with this approach, the transformation from an LWE sample with respect to A, to a k-LWE sample with respect to A', involves a multiplication by the cofactor matrix $\det(G) \cdot G^{-1}$ over $\mathbb{Z}$ of a $k \times k$ full-rank submatrix G of the hint vectors matrix. Although the entries of G are small, the entries of its cofactor matrix are almost as large as $\det G$, which is exponential in k. This leads to an "exponential noise blowup," restraining the applicability range to $k \le \tilde{O}(1)$ if one wants to rely on the hardness of LWE with noise rate $1/\alpha \le \mathcal{P}oly(n)$ (otherwise, LWE is not exponentially hard to solve). To restrain the noise increase for large k, we use the gadget of Corollary 1. Ignoring several technicalities, the core idea underlying our reduction is that the latter gadget allows us to sample a small matrix $\overline{X}_2$ with $\overline{X}_2^{-1}$ also small, which we can then use to transform the given LWE matrix $A^+ = (u^t \| A) \in \mathbb{Z}_q^{(m+1) \times n}$ into a taller k-LWE matrix $A'^+ = T \cdot A^+$, using a transformation matrix T of the form

$$T = \left[\begin{array}{c} I_{m+1} \\ \hline -\overline{X}_2^{-1} X_1 \end{array} \right],$$

for some small independently sampled matrix $X_1 = [1 | \overline{X}_1]$. We can accordingly transform the given LWE sample vector $b = A^+ s + e$ for matrix A^+ into an LWE sample $b' = Tb = A'^+ s + Te$ for matrix A'^+ by multiplying the given sample by T. Since $[X_1 | \overline{X}_2] \cdot T = 0$, it follows that $[X_1 | X_2] \cdot A'^+ = 0$, so we can use k small rows of $[X_1 | \overline{X}_2]$ as the k-LWE hints x_i^+ for the new matrix A'^+, while, at same time, the smallness of T keeps the transformed noise $e' = Te$ small.

Proof. For a technical reason related to the non-zero centers δ_i in the distribution of the hint vectors produced by our gadget from Corollary 1, we decompose our reduction from $\mathrm{LWE}_{m+1,\alpha}$ to (k, S)-LWE into two subreductions. The first subreduction (outlined above) reduces $\mathrm{LWE}_{m+1,\alpha}$ in dimension n to (k, S, C)-$\mathrm{LWE}_{m+2n,\alpha'}$ in dimension $4n$, where the ith row of C is the unit vector $c_i = (0^{m+n}|\delta_i) \in \mathbb{R}^{m+2n}$ for $i = 1, \ldots, k$. The second subreduction reduces (k, S, C)-$\mathrm{LWE}_{m+2n,\alpha'}$ in dimension $4n$ to (k, S)-$\mathrm{LWE}_{m+2n,\alpha'}$ in dimension $4n$. We first describe and analyze the first subreduction, and then explain the second subreduction.

Description of the First Subreduction. Let (A^+, b) with $A^+ = (u^t \| A)$ denote the given $\mathrm{LWE}_{\alpha,m+1}$ input instance, where $A^+ \hookleftarrow U(\mathbb{Z}_q^{(m+1) \times n})$, and $b \in \mathbb{T}^{m+1}$ comes from either the "LWE distribution" $\frac{1}{q} U(\mathrm{Im}(A^+)) + \nu_\alpha^{m+1}$ or the "Uniform distribution" $\frac{1}{q} U(\mathbb{Z}_q^{m+1}) + \nu_\alpha^{m+1}$. The reduction maps (A^+, b) to (A', u', X, b') with $A' \in \mathbb{Z}_q^{(m+2n) \times 4n}$ and $u' \in \mathbb{Z}_q^{4n}$ independent and uniform, $X \in \mathbb{Z}^{k \times (m+2n)}$ with its ith row $\boldsymbol{x}_i$ independently sampled from $D_{A_{-u'}^\perp(A'),S}$ for $i \leq k$, and $\boldsymbol{b}' \in \mathbb{T}^{m+1+2n}$ coming from either the "k-LWE distribution" $\frac{1}{q}U\left(\text{Im}(A'^+)\right) + \nu_\alpha^{m+1+2n}$ if $\boldsymbol{b}$ is from the "LWE distribution," or the "k-Uniform distribution" $\frac{1}{q}U\left(\text{Span}_{i\leq k}(\boldsymbol{x}_i^+)^\perp\right)$ if $\boldsymbol{b}$ is from the "Uniform distribution." Here $A'^+ = (\boldsymbol{u}'^t\|A')$, and $\boldsymbol{x}_i^+$ denotes the vector $(1\|\boldsymbol{x}_i)$ for $i \leq k$. The reduction is as follows.

1. Sample gadget $\overline{X}_2 \in \mathbb{Z}^{2n\times 2n}$ using Corollary 1 (with parameters $n, m_1, m_2, \sigma_1,$ σ_2 set to k, n, n, σ, σ' respectively), and sample $\overline{X}_1 \hookleftarrow D_{\mathbb{Z},\sigma}^{2n\times m}$. Define $T = \begin{bmatrix} I_{m+1} \\ -\overline{X}_2^{-1}\cdot(1|\overline{X}_1) \end{bmatrix} \in \mathbb{Z}^{(m+1+2n)\times(m+1)}$, where $\mathbf{1}$ is the all-1 vector. Let $X \in \mathbb{Z}^{k\times(m+2n)}$ denote the matrix made of the top k rows of $(\overline{X}_1|\overline{X}_2)$.
2. Sample $C^+ \in \mathbb{Z}_q^{(m+1+2n)\times 3n}$ with independent columns uniform orthogonally to $\text{Im}((1|X))$ modulo q. Let $\boldsymbol{u}_C^t \in \mathbb{Z}_q^{3n}$ be the top row of C^+, and $C \in \mathbb{Z}_q^{(m+2n)\times 3n}$ denote its remaining $m + 2n$ rows.
3. Compute $\Sigma = \alpha' \cdot I_{m+1+2n} - T\cdot T^t$ and $\sqrt{\Sigma}$ such that $\sqrt{\Sigma}\cdot\sqrt{\Sigma}^t = \Sigma$; if Σ is not positive definite, abort.
4. Compute $A'^+ = (T\cdot A^+|C^+)$ and $\boldsymbol{b}' = T\boldsymbol{b} + \frac{1}{q}C^+\cdot\boldsymbol{s}' + \sqrt{\Sigma}\boldsymbol{e}'$, with $\boldsymbol{s}' \hookleftarrow U(\mathbb{Z}_q^{3n})$ and $\boldsymbol{e}' \hookleftarrow \nu_1^{m+1+2n}$. Let $(\boldsymbol{u}')^t = (\boldsymbol{u}\|\boldsymbol{u}_C)^t \in \mathbb{Z}_q^{4n}$ be the top row of A'^+.
5. Return $(A', \boldsymbol{u}', X, \boldsymbol{b}')$.

Step 1 aims at building a transformation matrix T that sends A^+ to the left n columns of A'^+. Two properties are required from this transformation. First, it must be a linear map with small coefficients, so that when we map the LWE right hand side to the k-LWE right hand side, the noise component does not blow up. Second, it must contain some vectors $(1\|\boldsymbol{x}_i)$ in its (left) kernel, with $\boldsymbol{x}_i$ normally distributed. These vectors are to be used as k-LWE hints. For this, we use the gadget of the previous subsection. This ensures that the $\boldsymbol{x}_i$'s are (almost) distributed as independent Gaussian samples from $D_{\mathbb{Z}^n,\sigma} \times D_{\mathbb{Z}^n,\sigma'}$, and that the matrix T is integral with small coefficients. We define $B \in \mathbb{Z}_q^{2n\times n}$ by $[A^+\|B] = TA^+$, so that we have:

$$[1|\overline{X}_1|\overline{X}_2]\cdot\begin{bmatrix} A^+ \\ B \end{bmatrix} = [1|\overline{X}_1|\overline{X}_2]\cdot\begin{bmatrix} I_{m+1} \\ -\overline{X}_2^{-1}\cdot(1|\overline{X}_1) \end{bmatrix}\cdot A^+ = \boldsymbol{0} \bmod q.$$

This means each row of $(\overline{X}_1|\overline{X}_2)$ belongs to $A_{-u}^\perp(A'')$, where $A'' = [A^t|B^t]^t$.

At this stage, it is tempting to define the k-LWE matrix as A'' and give away the k-LWE hint vectors $\boldsymbol{x}_i \in A_{-u}^\perp(A'')$ making up the matrix X. However, this approach does not quite work: we have extended A by $2n$ rows, but we give only k hint vectors (we cannot output them all, as the bottom rows of $\overline{X}_2$ may not be normally distributed). This creates a difficulty for mapping "Uniform" to "k-Uniform" in the reduction. Step 2 circumvents the above difficulty by sampling extra column vectors $C^+ \in \mathbb{Z}_q^{(m+1+2n)\times 3n}$ that are uniform in the subspace orthogonal to the hint vectors $\boldsymbol{x}_i^+$ modulo q. When the parameters are properly set, the columns of $[T|C^+]$ span the full subspace orthogonal to the $\boldsymbol{x}_i$'s mod q, with overwhelming probability. We finally set $A'^+ = \left[\frac{A^+}{B}\middle|C^+\right]$.

It remains to see how to map "LWE" to "k-LWE." The main problem, when multiplying $\boldsymbol{b}$ by T, is that the LWE noise gets skewed. If its covariance matrix was of the form

$\alpha^2 \cdot I_{m+1}$, then it becomes $\alpha^2 T \cdot T^t$. To compensate for that, in Step 3, we add to $T \cdot b$ an independent Gaussian noise with well-chosen covariance $\Sigma = \alpha'^2 \cdot I_{m+1+2n} - \alpha^2 T \cdot T^t$. We set α' large enough to ensure that this symmetric matrix is positive definite. This noise unskewing technique was adapted to discrete Gaussians and used in cryptography in [34].

Analysis of the First Subreduction. All steps of the reduction can be implemented in polynomial time. Its correctness follows from the following three lemmas. The proofs can be found in the full version.

Lemma 7. *The tuple* (A', u', X) *is within statistical distance* $2^{-\Omega(n/\log n)}$ *of the distribution in which* $A' \in \mathbb{Z}_q^{(m+2n)\times 4n}$ *and* $u' \in \mathbb{Z}_q^{4n}$ *are independent and uniform, and the rows of* $X \in \mathbb{Z}^{k \times (m+2n)}$ *are from* $D_{\Lambda_{-u'}^{\perp}(A'),S,c_i}$*, where* $c_i = (0^{m+n}|\delta_i) \in \mathbb{R}^{m+2n}$ *and* δ_i *denotes the ith canonical unit vector in* $\mathbb{Z}^n$ *for* $i = 1, \ldots, k$.

Next, we assume that (A'^+, X) is fixed and consider the distribution of b' in the two cases of the distribution of b. First we consider the "LWE" to "k-LWE" distribution mapping.

Lemma 8. *The following holds with probability* $1 - 2^{-\Omega(n/\log n)}$ *over the choice of* $\overline{X}_1$ *and* $\overline{X}_2$*. If* $b \in \mathbb{T}^{m+1}$ *is sampled from* $\frac{1}{q}U(\mathrm{Im}A) + \nu_\alpha^{m+1}$*, then* $b' \in \mathbb{T}^{m+1+2n}$ *is within statistical distance* $2^{-\Omega(n)}$ *of* $\frac{1}{q}U(\mathrm{Im}A'^+) + \nu_{\alpha'}^{m+1+2n}$.

Finally, we consider the "Uniform" to "k-Uniform" distribution mapping.

Lemma 9. *The following holds with probability* $1 - 2^{-\Omega(n/\log n)}$ *over the choice of* $\overline{X}_1$ *and* $\overline{X}_2$*. If* b *is sampled from* $\frac{1}{q}U(\mathbb{Z}_q^{m+1}) + \nu_\alpha^{m+1}$*, then* b' *is within statistical distance* $2^{-\Omega(n)}$ *of* $\frac{1}{q}U(\mathrm{Span}_{i\leq k}(x_i^+)^{\perp}) + \nu_{\alpha'}^{m+1+2n}$.

Overall, we have described a reduction that maps the "LWE distribution" to the "k-LWE distribution," and the "Uniform distribution" to the "k-Uniform distribution," up to statistical distance $2^{-\Omega(n/\log n)}$.

Second Subreduction. It remains to reduce the (k, S, C)-LWE with non-zero centers for the hint distribution, to (k, S)-LWE with zero-centered hints. For this, we use Lemma 5 to obtain the following.

Lemma 10. *Let* $m' = m + 2n$, $n' = 4n$, *and assume that* $\sigma_{m'}(S) \geq \omega(\sqrt{n})$. *If there exists a distinguisher against* (k, S)-$\mathrm{LWE}_{m',\alpha'}$ *in dimension* n' *with run-time* T *and advantage* ε*, then there exists a distinguisher against* (k, S, C)-$\mathrm{LWE}_{m',\alpha'}$ *with run-time* $T' = O(\mathcal{P}oly(m') \cdot (\varepsilon - 2^{-\Omega(n)})^{-2} \cdot T)$ *and advantage* $\varepsilon' = \Omega((\varepsilon - O(2^{-n}))^3 / R - O(2^{-n}))$*, where* $R = \exp(O(k \cdot (2^{-n} + \|C\|^2/\sigma_{m'}(S)^2)))$.

The main idea of the proof of Lemma 10, given in the full version, is to apply Lemma 5 with P, P' being the (k, S)-LWE and (k, S, C)-LWE problems respectively, which have instances of the form $x = (r, y)$, where $r = (A, u, \{x_i\}_{i\leq k})$ and the hints x_i for $i \leq k$ sampled from either the zero-centered distribution $\hookleftarrow D_{\Lambda_{-u}^{\perp}(A),S,0}$ (distribution Φ of r, in (k, S)-LWE) or the non-zero center distribution $\hookleftarrow D_{\Lambda_{-u}^{\perp}(A),S,c_i}$

(distribution Φ' of r, in (k, S, C)-LWE), and $\boldsymbol{y} \in \mathbb{T}^{m+1}$ is a sample from either the distribution

$$D_0(r) = \frac{1}{q} \cdot U\left(\operatorname{Im}\left(\frac{\boldsymbol{u}^t}{A}\right)\right) + \nu_\alpha^{m+1}$$

or the distribution

$$D_1(r) = \frac{1}{q} \cdot U\left(\operatorname{Span}_{i \le k}\left(\frac{1}{\boldsymbol{x}_i}\right)^\perp\right) + \nu_\alpha^{m+1}.$$

Given $x = (r, \boldsymbol{y})$, is possible to efficiently sample $\boldsymbol{y}'$ from either $D_0(r)$ or $D_1(r)$, so the public-samplability property assumed by Lemma 5 is satisfied. This Lemma gives the desired reduction between (k, S)-LWE and (k, S, C)-LWE, as long as the RD $R(\Phi \| \Phi')$ between the distribution of r in the two problems is polynomially bounded. The latter reduces to obtaining a bound on the RD between a Gaussian distribution and a small offset thereof, which is given by Lemma 4.

In our application of Lemma 10, the (k, S, C)-LWE problem resulting from the first subreduction has $\|C\| = 1$, and $\sigma_{m'}(S) = \sigma$, so that $R = \exp(O(k \cdot (2^{-n} + 1/\sigma^2))) = O(1)$ using $\sigma = \Omega(n)$ and $k \le n$. This shows that the second subreduction is probabilistic polynomial time. □

Our technique can be applied to improve the Boneh-Freeman reduction from SIS to k-SIS, from an exponential loss in k to a polynomial loss in k. In fact, we map A to A'' in the same way (except that we do not use and add $\boldsymbol{u}$ on top of the matrix A) and then also use the top k rows of $(\overline{X}_1 | \overline{X}_2)$ as the k-SIS hints for the new matrix A''. Then, whenever the adversary can output a short vector $\boldsymbol{x}_1 \| \boldsymbol{x}_2$ that is orthogonal to A'', we can also output a short vector $(\boldsymbol{x}_1 - \boldsymbol{x}_2 \cdot \overline{X}_2^{-1} \overline{X}_1)$ which is orthogonal to A. As the rows of $\overline{X}_1$ are distributed as independent Gaussian samples and the adversary is only given its first k rows, it can be shown that, if $\boldsymbol{x}_1 \| \boldsymbol{x}_2$ is linearly independent from the k-SIS hints, then the vector $(\boldsymbol{x}_1 - \boldsymbol{x}_2 \cdot \overline{X}_2^{-1} \overline{X}_1)$ is null with a negligible probability. RD may also be used to reduce k-SIS with non-zero-centered hints (with small centers) to k-SIS with zero-centered hints.

4   A Lattice-Based Public-Key Traitor Tracing Scheme

In this section, we describe and analyze our basic traitor tracing scheme. First, we give the underlying multi-user public-key encryption scheme. We then explain how to implement black-box confirmation tracing.

4.1   A Multi-user Encryption Scheme

The scheme is designed for a given security parameter n, a number of users N and a maximum malicious coalition size t. It then involves several parameters q, m, α, S. These are set so that the scheme is correct (decryption works properly on honestly generated ciphertexts) and secure (semantically secure encryption and possibility to trace members of malicious coalitions). In particular, we define S

as $\mathrm{Diag}(\sigma, \ldots, \sigma, \sigma', \ldots, \sigma') \in \mathbb{R}^{m \times m}$ where $\sigma' > \sigma$ and their respective numbers of iterations are set so that (t, S)-$\mathrm{LWE}_{m+1,\alpha}$ is hard to solve.

Setup. The trusted authority generates a master key pair using the algorithm from Lemma 2. Let $(A, T) \in \mathbb{Z}_q^{m \times n} \times \mathbb{Z}^{m \times m}$ be the output. We additionally sample u uniformly in $\mathbb{Z}_q^n$. Matrix T will be part of the tracing key tk, whereas the public key is $pk = A^+$, with $A^+ = (u^t \| A)$.

Each user $\mathcal{U}_i$ for $i \leq N$ obtains a secret key sk_i from the trusted authority, as follows. The authority executes the GPV algorithm using the basis of $\Lambda^\perp(A)$ consisting of the rows of T, and the standard deviation matrix S. The authority obtains a sample x_i from $D_{\Lambda^\perp_{-u}(A),S}$. The standard deviations $\sigma' > \sigma$ may be chosen as small as $3mq^{n/m}\sqrt{(2m+4)/\pi}$. The user secret key is $x_i^+ = (1 \| x_i) \in \mathbb{Z}^{m+1}$. Using the Gaussian tail bound and the union bound, we have $\|x_i\| \leq \sqrt{m}\sigma'$ for all $i \leq N$, with probability $\geq 1 - N \cdot 2^{-\Omega(m)}$.

The tracing key tk consists of the matrix T and all pairs $(\mathcal{U}_i, sk_i)$.

Encrypt. The encryption algorithm is exactly the 1-bit encryption scheme from [19, Se. 7.1], which we recall, for readability.[1] The plaintext and ciphertext domains are $\mathcal{P} = \{0, 1\}$ and $\mathcal{C} = \mathbb{Z}_q^{m+1}$ respectively, and:

$$\mathrm{Enc} : M \mapsto \begin{bmatrix} u^t \\ A \end{bmatrix} \cdot s + e + \begin{bmatrix} M \cdot \lfloor q/2 \rfloor \\ 0 \end{bmatrix}, \quad \text{where } s \hookleftarrow U(\mathbb{Z}_q^n) \text{ and } e \hookleftarrow \lfloor \nu_{\alpha q} \rceil^{m+1}.$$

As explained in [19], this scheme is semantically secure under chosen plaintext attacks (IND-CPA), under the assumption that $\mathrm{LWE}_{m+1,\alpha}$ is hard to solve.

Decrypt. To decrypt a ciphertext $c \in \mathbb{Z}_q^{m+1}$, user $\mathcal{U}_i$ uses its secret key x_i^+ and evaluates the following function Dec from $\mathbb{Z}_q^{m+1}$ to $\{0, 1\}$: Map c to 0 if $\langle x_i^+, c \rangle \bmod q$ is closer to 0 than $\pm \lfloor q/2 \rfloor$.

If c is an honestly generated ciphertext of a plaintext $M \in \{0, 1\}$, we have $\langle x_i^+, c \rangle = \langle x_i^+, e \rangle + M \cdot \lfloor q/2 \rfloor \bmod q$, where $e \hookleftarrow \lfloor \nu_{\alpha q} \rceil^{m+1}$. It can be shown that the latter has magnitude $\leq 2\sqrt{m}\alpha q \|x_i^+\|$ with probability $1 - 2^{-\Omega(n)}$ over the randomness of e. This is $\leq 3m\alpha q\sigma'$ for all i, with probability $\geq 1 - N \cdot 2^{-\Omega(n)}$. To ensure the correctness of the scheme, it suffices to set $q \geq 4m\alpha q\sigma'$. Note that other constraints will be added to enable tracing.

Theorem 4. *Let m, n, q, N be integers such that q is prime and $N \leq 2^{o(n)}$. Let $\alpha, \sigma, \sigma' > 0$ such that $\sigma' \geq \sigma \geq \Omega(mq^{n/m}\sqrt{\log m})$ and $\alpha \leq 1/(4m\sigma')$. Then the scheme described above is IND-CPA under the assumption that $\mathrm{LWE}_{m+1,\alpha}$ is hard. Further, the decryption algorithm is correct:*

$$\forall M \in \{0, 1\}, \forall i \leq N : \mathrm{Dec}\left(\mathrm{Enc}(M, pk), sk_i\right) = M$$

holds with probability $\geq 1 - 2^{-\Omega(n)}$ over the randomness used in Setup *and* Enc.

---

[1] As usual, the encryption algorithm may be used to encapsulate session keys which are then fed into an efficient data encapsulation mechanism to encrypt the data.

4.2   Tracing Traitors

We now present a black-box confirmation algorithm Trace.[2] It is given access to an oracle $\mathcal{O}^{\mathcal{D}}$ that provides black-box access to a decryption device $\mathcal{D}$. It takes as inputs the tracing key $tk = (T, (\mathcal{U}_i, \boldsymbol{x}_i^+)_{i \leq N})$ and a set of suspect users $\{\mathcal{U}_{i_1}, \ldots, \mathcal{U}_{i_k}\}$ of cardinality $k \leq t$, where t is the a priori bound on any coalition size. Wlog, we may consider that $k = t$ and $i_j = j$ for all $j \leq k$.

Algorithm Trace gathers information about which keys have been used to build decoder $\mathcal{D}$, by feeding different carefully designed distributions to oracle $\mathcal{O}^{\mathcal{D}}$. We consider the following $t + 1$ distributions $Tr_0, \ldots, Tr_t$ over $\mathcal{C} = \mathbb{Z}_q^{m+1}$:

$$Tr_i = U\left(\mathrm{Span}(\boldsymbol{x}_1^+, \ldots, \boldsymbol{x}_i^+)^\perp\right) + \lfloor \nu_{\alpha q} \rceil^{m+1}.$$

The first distribution Tr_0 is the uniform distribution, whereas the last distribution Tr_t is meant to be computationally indistinguishable from Enc(0). We define p_∞ as the probability $\Pr[\mathcal{O}^{\mathcal{D}}(\boldsymbol{c}, M) = 1]$ that the decoder can decrypt the ciphertexts, over the randomness of $M \hookleftarrow U(\{0,1\})$ and $\boldsymbol{c} \hookleftarrow \mathrm{Enc}(M)$. We define p_i as the probability the decoder decrypts the signals in Tr_i, for $i \in [0,t]$:

$$p_i = \Pr_{\substack{\boldsymbol{c} \hookleftarrow Tr_i \\ M \hookleftarrow U(\{0,1\})}} \left[\mathcal{O}^{\mathcal{D}}\left(\boldsymbol{c} + \begin{bmatrix} M \cdot \lfloor q/2 \rfloor \\ \boldsymbol{0} \end{bmatrix}, M \right) = 1 \right].$$

A gap between p_{i-1} and p_i is meant to indicate that $\mathcal{U}_i$ is a traitor.

The confirmation and soundness properties are proved in the full version. We now concentrate on a new feature of our scheme: public traceability.

5   Projective Sampling and Public Traceability

We now modify the scheme of the Section 4 so that the tracing signals can be publicly sampled. For this purpose, we introduce the concept of projective sampling family.

5.1   Projective Sampling

Inspired from the notion of projective hash family [16], we propose the notion of projective sampling family in which each sampling function is keyed and, with a projected key, one can simulate the sampling function in a computationally indistinguishable way. Let X be a finite non-empty set. Let $F = (\mathsf{F}_k)_{k \in K}$ be a collection of sampling functions indexed by K, so that F_k is a sampling function over X, for every $k \in K$. We call $\mathsf{Sam} = (F, K, X)$ a sampling family. We now introduce the concept of projective sampling.

Definition 2 (Projective Sampling). *Let* $\mathsf{Sam} = (F, K, X)$ *be a sampling family. Let* J *be a finite, non-empty set, and let* $\pi : K \to J$ *be a (probabilistic) function. Let also*

---

[2] Note that in our context, minimal access is equivalent to standard access: since the plaintext domain is small, plaintext messages can be tested exhaustively.

$\mathsf{P} = (\mathsf{P}_j)_{j \in J}$ be a collection of sampling functions over X, and D be a distribution over K. Then $\mathsf{PSam} = (F, K, X, \mathsf{P}, J, \pi, D)$ is called a projective sampling family if, with overwhelming probability over the choice of $k, k' \hookleftarrow D$, and given the secret key k and its projected key $\pi(k)$, 1) the distributions obtained using F_k and $\mathsf{P}_{\pi(k)}$ are computationally indistinguishable, and 2) the distributions obtained using F_k and $\mathsf{P}_{\pi(k')}$ can be efficiently distinguished.

The first condition means that for $k \hookleftarrow D$, the value $\pi(k)$ "encodes" the sampling distribution of F_k, so that when $\pi(k)$ is made public, the sampled signal F_k can be publicly simulated by $\mathsf{P}_{\pi(k)}$. The security requirement is very strong because the adversary is not only given the projected key, as in projective hashing, but also the secret key k. We require that sampling signals from the secret key and from its projected key are indistinguishable for the insiders who know the secret key. This is relevant for traitor tracing, as the traitors are system insiders and they possess secret data. The second condition (that we actually do not directly use in our cryptographic application) allows to prevent the trivial solution consisting in setting $\mathsf{P}_{\pi(k)}$ as an efficient sampling function that is independent of k: the simulation signal $\mathsf{P}_{\pi(k)}$ must be specific to k.[3]

5.2   Projective Sampling from k-LWE

We construct a set of projective sampling families $(\mathsf{PSam}_i)_{0 \le i \le t}$. The parameters are almost identical to the parameters in the Setup of the multi-user scheme of Section 4. A further difference, required for simulation purposes in the security proof, is that $\sigma' > \sigma$ must be set $\widetilde{\Omega}(\sqrt{mn} + \pi q)$.

We let $A \hookleftarrow U(\mathbb{Z}_q^{m \times n})$ and $\boldsymbol{u} \hookleftarrow U(\mathbb{Z}_q^n)$ be public parameters. For each i, we define $K_i = (\mathbb{Z}_q^m)^i$ and D_i as the distribution on K_i that samples $k = (\boldsymbol{x}_j)_{j \le i}$ with $\boldsymbol{x}_j \hookleftarrow D_{\Lambda_{-\boldsymbol{u}}^\perp(A), \sigma}$ for all $j \le i$. The sampling function $\mathsf{F}_{i,k}$ is defined as $U(\mathrm{Span}_{j \le i}(\boldsymbol{x}_j^+)^\perp) + \lfloor \nu_{\alpha q} \rceil^{m+1}$. The projected key $\pi_i(k)$ is defined as follows:

- Sample $H \in \mathbb{Z}_q^{m \times (m-n)}$ uniformly, conditioned on $\mathrm{Im}(A) \subseteq \mathrm{Im}(H)$.
- For each $j \le i$, define $\boldsymbol{h}_j^t = -\boldsymbol{x}_j^t \cdot H$.
- Finally, set $J = \mathbb{Z}_q^{m \times (m-n)} \times (\mathbb{Z}_q^{m-n})^i$ and set $\pi_i(k) = (H, (\boldsymbol{h}_j)_{j \le i})$.

We now define the sampling $\mathsf{P}_{i, \pi_i(k)}$ with projected key $\pi_i(k) = (H, (\boldsymbol{h}_j)_{j \le i})$, as follows:

- Set $H_j = (\boldsymbol{h}_j^t \| H) \in \mathbb{Z}_q^{(m+1) \times (m-n)}$. We have $\boldsymbol{x}_j^{+t} \cdot H_j = \boldsymbol{0}$ and $\mathrm{Im}(A^+) \subseteq \mathrm{Im}(H_j)$.
- Set $\mathsf{P}_{i, \pi_i(k)} = U(\cap_{j \le i} \mathrm{Im}(H_j)) + \lfloor \nu_{\alpha q} \rceil^{m+1}$, with $\cap_{j \le 0} \mathrm{Im}(H_j) = \mathbb{Z}_q^{m+1}$ by convention. Note that $\cap_{j \le i} \mathrm{Im}(H_j) \subseteq \mathrm{Span}_{j \le i}(\boldsymbol{x}_j^+)^\perp$.

Theorem 5. *For each $i = 0, \ldots, t$, PSam_i is a projective sampling family. Concretely, under the (i, S)-LWE$_{\alpha, m}$ hardness assumptions, given the uniformly sampled public parameters $(A, \boldsymbol{u})$, the secret key $k = (\boldsymbol{x}_j)_{j \le i} \hookleftarrow D_i$ and its projected key $\pi_i(k) = (H, (\boldsymbol{h}_j)_{j \le i})$, the distributions $\mathsf{F}_{i,k}$ and $\mathsf{P}_{i, \pi_i(k)}$ are indistinguishable. Moreover, they are both indistinguishable from $U(\mathrm{Im}(A^+)) + \lfloor \nu_{\alpha q} \rceil^{m+1}$. Finally, with overwhelming*

---

[3] Another trivial situation occurs when $\pi(k) = k$: the projected key leaks the full information about the original key and one cannot safely publish the projected key.

probability, the distributions $F_{i,k}$ *and* $P_{i,\pi_i(k')}$ *can be efficiently distinguished, when* k' *is independently sampled from* D_i.

Proof. For the last statement, observe that with overwhelming probability, the secret key k' contains an $x'_j \in \mathbb{Z}_q^m$ that does not belong to $\mathrm{Span}_{j \leq i}(x_j)$ (by Lemma 3). In that case, taking the inner product of all x'_j's of k' with a sample from $P_{i,\pi_i(k')}$ gives small residues modulo q, whereas one of the inner products of the x'_j's with a sample from with a sample from $F_{i,k}$ will be uniform modulo q.

We now consider the first statement. From the hardness of (i, S)-LWE$_{m,\alpha}$, given k, the distributions

$$F_{i,k} = U(\mathrm{Span}_{j \leq i}(x_j^+)^\perp) + \lfloor \nu_{\alpha q} \rceil^{m+1} \quad \text{and} \quad U(\mathrm{Im}(A^+)) + \lfloor \nu_{\alpha q} \rceil^{m+1}$$

are indistinguishable. Further, given $k = (x_j)_{j \leq i}$, the projected key $\pi_i(k) = (H, (h_j)_{j \leq i})$ can be sampled from D_i. Therefore, given both k and $\pi_i(k)$, the distributions $F_{i,k}$ and $U(\mathrm{Im}(A^+)) + \lfloor \nu_{\alpha q} \rceil^{m+1}$ remain indistinguishable.

Now, we have $\mathrm{Im}(A^+) \subseteq \cap_{j \leq i}\mathrm{Im}(H_j) \subseteq (\mathrm{Span}_{j \leq i}(x_j^+))^\perp$. Hence:

$$U(\mathrm{Im}(A^+)) + U(\cap_{j \leq i}\mathrm{Im}(H_j)) = U(\cap_{j \leq i}\mathrm{Im}(H_j)),$$
$$U(\mathrm{Span}_{j \leq i}(x_j^+)^\perp) + U(\cap_{j \leq i}\mathrm{Im}(H_j)) = U(\mathrm{Span}_{j \leq i}(x_j^+)^\perp).$$

We note that given $h_1, \ldots, h_i$, one can efficiently sample from $U(\cap_{j \leq i}\mathrm{Im}(H_j))$. Therefore, under the hardness of (i, S)-LWE$_{m,\alpha}$, this shows that $F_{i,k}$, $P_{i,\pi_i(k)}$ and $U(\mathrm{Im}(A^+)) + \lfloor \nu_{\alpha q} \rceil^{m+1}$ are indistinguishable. $\square$

5.3   Public Traceability from Projective Sampling

In the scheme of Section 4, the tracing key $tk = (T, (\mathcal{U}_i, x_i)_{i \leq N})$ must be kept secret, as it would reveal the secret keys of the users. The tracing signals are samples from $U(\mathrm{Span}_{j \leq i}(x_j^+)^\perp) + \lfloor \nu_{\alpha q} \rceil^{m+1}$, which exactly matches $F_{i,k}$. By publishing the projected key $\pi_i(k)$, anyone can use the projective sampling $P_{i,\pi_i(k)}$: by Theorem 5, given $(k, \pi_i(k))$, $F_{i,k}$ and $P_{i,\pi_i(k)}$ are indistinguishable and they are both indistinguishable from the original sampling $U(\mathrm{Im}(A^+)) + \lfloor \nu_{\alpha q} \rceil^{m+1}$. We are thus almost done with public traceability.

However, a subtle point is that we have to use all the projective samplings $(P_{i,\pi_i(k)})$ for transforming the secret tracing to the public tracing: all the projected keys $(h_j)_{j \leq N}$ should be published. Because the keys k in $F_{i,k}$ are not independent, it could occur that the adversary exploits a projected key $\pi_i(k)$ for distinguishing $P_{i',\pi_{i'}(k')}$ from the original signals. To handle this, we prove that, given $(x_j)_{j \leq i}$ and all the keys $(h_j)_{j \leq N}$, the adversary cannot distinguish $P_{i,\pi_i(k)}$ from the original signals. For this purpose, we exploit a technique from [20] to simulate $(h_j)_{i < j \leq N}$ from the public information.

Theorem 6. *Set $i \leq t$. Under the (i, S)-LWE$_{\alpha,m}$ and the LWE$'_{\alpha,m}$ hardness assumptions, given the secret key $k = (x_j)_{j \leq i}$ and the projected keys $(H, (h_j)_{j \leq N})$, the following two distributions are indistinguishable*

$$P_{i,\alpha(k)} = U(\cap_{j \leq i}\mathrm{Im}(H_j)) + \lfloor \nu_{\alpha q} \rceil^{m+1} \quad \text{and} \quad U(\mathrm{Im}(A^+)) + \lfloor \nu_{\alpha q} \rceil^{m+1}.$$

Proof. Assume a ppt attacker is given $(x_j)_{j \leq i}$ (with the x_j's independently sampled from $D_{\Lambda^{\perp}_{-u}(A),\sigma}$) and all the projected keys $(h_j)_{j \leq N}$). We are to prove that, under the (i, S)-LWE$_{\alpha,m}$ and LWE$'_{\alpha,m}$ hardness assumptions, it cannot distinguish between the distributions (over $\mathbb{Z}_q^{m+1}$)

$$U(\text{Im}(A^+)) + \lfloor \nu_{\alpha q} \rceil^{m+1} \quad \text{and} \quad P_{i,\pi_i(k)} = U(\cap_{j \leq i} \text{Im}(H_j)) + \lfloor \nu_{\alpha q} \rceil^{m+1}.$$

We proceed by a sequence of games.

Game$_0$:   This is the above distinguishing game. We let ε_0 denote the adversary's distinguishing advantage. The goal is to show that ε_0 is negligible.

Game$_1$:   In this second game, we sample $x_1, \ldots, x_i$ from $D_{\Lambda^{\perp}_{-u}(A),\sigma}$ as in **Game$_0$**, but the x_j's for $j > i$ are sampled uniformly in $\mathbb{Z}_q^n$, conditioned on $x_j^t \cdot A = -u^t$. The h_j's for $j > i$ are modified accordingly, but the rest is as in **Game$_0$**. We let ε_1 denote the adversary's distinguishing advantage.

The main point is that in **Game$_1$**, no secret information is required for sampling the projected keys h_j's for $j > i$. The proof of the following lemma may be found in the full version.

Lemma 11. *Under the LWE$'_{\alpha,m}$ hardness assumption, the quantity $|\varepsilon_1 - \varepsilon_0|$ is negligible.*

We note that, in **Game$_1$**, the h_j's can be sampled publicly from the available data. Therefore, from Theorem 5, under the (i, S)-LWE$_{\alpha,m}$ hardness assumptions, the advantage ε_1 is negligible. □

Semantic security of the updated scheme. We modify the public information of the scheme of Section 4, so that we can use the set of projective sampling families described above. For this aim, we simply add the projected key $(H, (h_i)_{i \leq N})$ to the public key. The scheme becomes publicly traceable because the tracing signals can be sampled from the projected keys, as explained above. Finally, as the public key has been modified, we should prove that the knowledge of these projected keys provides no significant advantage for an adversary towards breaking the semantic security of the encryption scheme. Fortunately, the semantic security directly follows from Theorem 6, for the particular case of $i = 0$.

Acknowledgements. We thank M. Abdalla, D. Augot, R. Bhattacharrya, L. Ducas, V. Guleria, G. Hanrot, F. Laguillaumie, K. T. T. Nguyen, G. Quintin, O. Regev, H. Wang for helpful discussions. The authors were partly supported by the LaBaCry MERLION grant, the Australian Research Council Discovery Grant DP110100628, the ANR-09-VERSO-016 BEST and ANR-12-JS02-0004 ROMAnTIC Projects, the INRIA invited researcher scheme, the Singapore National Research Foundation Research Grant NRF-CRP2-2007-03, the Singapore MOE Tier 2 research grant MOE2013-T2-1-041, the LIA Formath Vietnam and the ERC Starting Grant ERC-2013-StG-335086-LATTAC.

References

1. Aggarwal, D., Regev, O.: A note on discrete gaussian combinations of lattice vectors (2013), Draft Available at, http://arxiv.org/pdf/1308.2405v1.pdf
2. Agrawal, S., Gentry, C., Halevi, S., Sahai, A.: Discrete gaussian leftover hash lemma over infinite domains. In: Sako, K., Sarkar, P. (eds.) ASIACRYPT 2013, Part I. LNCS, vol. 8269, pp. 97–116. Springer, Heidelberg (2013)
3. Ajtai, M.: Generating hard instances of lattice problems (extended abstract). In: Proc. of STOC, pp. 99–108. ACM (1996)
4. Ajtai, M.: Generating hard instances of the short basis problem. In: Wiedermann, J., Van Emde Boas, P., Nielsen, M. (eds.) ICALP 1999. LNCS, vol. 1644, pp. 1–9. Springer, Heidelberg (1999)
5. Alwen, J., Peikert, C.: Generating shorter bases for hard random lattices. Theor. Comput. Science 48(3), 535–553 (2011)
6. Billet, O., Phan, D.H.: Efficient Traitor Tracing from Collusion Secure Codes. In: Safavi-Naini, R. (ed.) ICITS 2008. LNCS, vol. 5155, pp. 171–182. Springer, Heidelberg (2008)
7. Boneh, D., Franklin, M.K.: An efficient public key traitor scheme (Extended abstract). In: Wiener, M. (ed.) CRYPTO 1999. LNCS, vol. 1666, pp. 338–353. Springer, Heidelberg (1999)
8. Boneh, D., Freeman, D.M.: Linearly homomorphic signatures over binary fields and new tools for lattice-based signatures. In: Catalano, D., Fazio, N., Gennaro, R., Nicolosi, A. (eds.) PKC 2011. LNCS, vol. 6571, pp. 1–16. Springer, Heidelberg (2011), Full version available at, http://eprint.iacr.org/2010/453
9. Boneh, D., Waters, B.: A fully collusion resistant broadcast, trace, and revoke system. In: Proc. of ACM CCS, pp. 211–220. ACM (2006)
10. Boneh, D., Naor, M.: Traitor tracing with constant size ciphertext. In: Ning, P., Syverson, P.F., Jha, S. (eds.) ACM CCS 2008, pp. 501–510. ACM Press (2008)
11. Boneh, D., Sahai, A., Waters, B.: Fully collusion resistant traitor tracing with short ciphertexts and private keys. In: Vaudenay, S. (ed.) EUROCRYPT 2006. LNCS, vol. 4004, pp. 573–592. Springer, Heidelberg (2006)
12. Boneh, D., Zhandry, M.: Multiparty key exchange, efficient traitor tracing, and more from indistinguishability obfuscation. Cryptology ePrint Archive, Report 2013/642 (2013), http://eprint.iacr.org/
13. Brakerski, Z., Langlois, A., Peikert, C., Regev, O., Stehlé, D.: Classical hardness of learning with errors. In: STOC, pp. 575–584. ACM (2013)
14. Chabanne, H., Phan, D.H., Pointcheval, D.: Public traceability in traitor tracing schemes. In: Cramer, R. (ed.) EUROCRYPT 2005. LNCS, vol. 3494, pp. 542–558. Springer, Heidelberg (2005)
15. Chor, B., Fiat, A., Naor, M.: Tracing traitors. In: Desmedt, Y.G. (ed.) CRYPTO 1994. LNCS, vol. 839, pp. 257–270. Springer, Heidelberg (1994)
16. Cramer, R., Shoup, V.: Universal hash proofs and a paradigm for adaptive chosen ciphertext secure public-key encryption. In: Knudsen, L.R. (ed.) EUROCRYPT 2002. LNCS, vol. 2332, pp. 45–64. Springer, Heidelberg (2002)
17. Garg, S., Gentry, C., Halevi, S.: Candidate multilinear maps from ideal lattices. In: Johansson, T., Nguyen, P.Q. (eds.) EUROCRYPT 2013. LNCS, vol. 7881, pp. 1–17. Springer, Heidelberg (2013)
18. Garg, S., Gentry, C., Halevi, S., Raykova, M., Sahai, A., Waters, B.: Candidate indistinguishability obfuscation and functional encryption for all circuits. In: Proc. of FOCS, pp. 40–49. IEEE Computer Society Press (2013)
19. Gentry, C., Peikert, C., Vaikuntanathan, V.: Trapdoors for hard lattices and new cryptographic constructions. In: Proc. of STOC, pp. 197–206. ACM (2008), Full version available at, http://eprint.iacr.org/2007/432.pdf

20. Gordon, S.D., Katz, J., Vaikuntanathan, V.: A group signature scheme from lattice assumptions. In: Abe, M. (ed.) ASIACRYPT 2010. LNCS, vol. 6477, pp. 395–412. Springer, Heidelberg (2010)
21. Kiayias, A., Pehlivanglu, S.: Encryption For Digital Content. Springer, Heidelberg (2010)
22. Kiayias, A., Yung, M.: Breaking and repairing asymmetric public-key traitor tracing. In: Digital Rights Management Workshop, pp. 32–50 (2002)
23. Kiayias, A., Yung, M.: Traitor tracing with constant transmission rate. In: Knudsen, L.R. (ed.) EUROCRYPT 2002. LNCS, vol. 2332, pp. 450–465. Springer, Heidelberg (2002)
24. Komaki, H., Watanabe, Y., Hanaoka, G., Imai, H.: Efficient asymmetric self-enforcement scheme with public traceability. In: Kim, K.-c. (ed.) PKC 2001. LNCS, vol. 1992, pp. 225–239. Springer, Heidelberg (2001)
25. Langlois, A., Stehlé, D., Steinfeld, R.: GGHLite: More efficient multilinear maps from ideal lattices. In: Nguyen, P.Q., Oswald, E. (eds.) EUROCRYPT 2014. LNCS, vol. 8441, pp. 239–256. Springer, Heidelberg (2014)
26. Langlois, A., Stehlé, D., Steinfeld, R.: Improved and simplified security proofs in lattice-based cryptography: using the Rényi divergence rather than the statistical distance (2014); Available on the webpages of the authors.
27. Lyubashevsky, V., Peikert, C., Regev, O.: On ideal lattices and learning with errors over rings. J. ACM 60(6), 43 (2013)
28. Micciancio, D., Peikert, C.: Trapdoors for lattices: Simpler, tighter, faster, smaller. In: Pointcheval, D., Johansson, T. (eds.) EUROCRYPT 2012. LNCS, vol. 7237, pp. 700–718. Springer, Heidelberg (2012)
29. Micciancio, D., Regev, O.: Worst-case to average-case reductions based on gaussian measures. SIAM J. Comput 37(1), 267–302 (2007)
30. Micciancio, D., Regev, O.: Lattice-based cryptography. In: Bernstein, D.J., Buchmann, J., Dahmen, E. (eds.) Post-Quantum Cryptography, pp. 147–191. Springer, Heidelberg (2009)
31. Naor, M., Pinkas, B.: Efficient trace and revoke schemes. In: Frankel, Y. (ed.) FC 2000. LNCS, vol. 1962, pp. 1–20. Springer, Heidelberg (2001)
32. O'Neill, A., Peikert, C., Waters, B.: Bi-deniable public-key encryption. In: Rogaway, P. (ed.) CRYPTO 2011. LNCS, vol. 6841, pp. 525–542. Springer, Heidelberg (2011)
33. Peikert, C.: Public-key cryptosystems from the worst-case shortest vector problem. In: Proc. of STOC, pp. 333–342. ACM (2009)
34. Peikert, C.: An efficient and parallel gaussian sampler for lattices. In: Rabin, T. (ed.) CRYPTO 2010. LNCS, vol. 6223, pp. 80–97. Springer, Heidelberg (2010)
35. Pfitzmann, B.: Trials of traced traitors. In: Anderson, R. (ed.) IH 1996. LNCS, vol. 1174, pp. 49–64. Springer, Heidelberg (1996)
36. Pfitzmann, B., Waidner, M.: Asymmetric fingerprinting for larger collusions. In: ACM CCS 1997, pp. 151–160. ACM Press (April 1997)
37. Phan, D.H., Safavi-Naini, R., Tonien, D.: Generic construction of hybrid public key traitor tracing with full-public-traceability. In: Bugliesi, M., Preneel, B., Sassone, V., Wegener, I. (eds.) ICALP 2006. LNCS, vol. 4052, pp. 264–275. Springer, Heidelberg (2006)
38. Regev, O.: On lattices, learning with errors, random linear codes, and cryptography. In: Proc. of STOC, pp. 84–93. ACM (2005)
39. Regev, O.: On lattices, learning with errors, random linear codes, and cryptography. J. ACM 56(6) (2009)
40. Regev, O.: The learning with errors problem. In: Invited survey in CCC 2010 (2010), http://www.cims.nyu.edu/~regev/
41. Sirvent, T.: Traitor tracing scheme with constant ciphertext rate against powerful pirates. In: Augot, D., Sendrier, N., Tillich, J.-P. (eds.) Workshop on Coding and Cryptography—WCC 2007, pp. 379–388 (April 2007)
42. Watanabe, Y., Hanaoka, G., Imai, H.: Efficient asymmetric public-key traitor tracing without trusted agents. In: Naccache, D. (ed.) CT-RSA 2001. LNCS, vol. 2020, pp. 392–407. Springer, Heidelberg (2001)

Improved Short Lattice Signatures
in the Standard Model

Léo Ducas and Daniele Micciancio

University of California, San Diego, CA, USA
{lducas,daniele}@eng.ucsd.edu

Abstract. We present a signature scheme provably secure in the standard model (no random oracles) based on the worst-case complexity of approximating the Shortest Vector Problem in ideal lattices within polynomial factors. The distinguishing feature of our scheme is that it achieves *short* signatures (consisting of a single lattice vector), and *relatively short* public keys (consisting of $O(\log n)$ vectors.) Previous lattice schemes in the standard model with similarly *short* signatures, due to Boyen (PKC 2010) and Micciancio and Peikert (Eurocrypt 2012), had substantially longer public keys consisting of $\Omega(n)$ vectors (even when implemented with ideal lattices).

1 Introduction

Lattice based cryptography [3,4], originally an area of primarily theoretical interest, has seen a tremendous growth during the last decade, due both to substantial efficiency improvements obtainable using lattices with algebraic structure [16,28], and to the enormous versatility afforded by the Learning with Errors (LWE) problem [33]. One of the problems that has received most attention so far, is that of lattice based signatures [24,13,21,9,35,14,22,12,6]. From a theoretical point of view, digital signatures can be constructed from any one-way function [34,19]. So, the existence of digital signature schemes based on the hardness of lattice problems directly follows from Ajtai's seminal work [3]. But generic constructions are rather inefficient. Inputs and outputs of lattice based cryptographic functions typically consist of one or more $\tilde{\Omega}(n)$-dimensional vectors, where n is the security parameter. Generic digital signature constructions require n parallel applications of a one-way function. So, even if each one-way function takes as input a single vector, the resulting digital signatures consist of n vectors, and require $\tilde{\Omega}(n^2)$ storage even when using algebraic lattices [28]. So, finding efficient constructions of signatures directly based on hard lattice problems has been an important problem since the early days of lattice cryptography, with the main goal of finding "short" signatures, i.e., lattice signatures consisting of a single lattice vector.

The first direct constructions of lattice signatures were given in [24] and [13]. Both schemes achieved "short" signatures, consisting of a single lattice vector, but each work had its own pros and cons. On the one hand [24] gave a scheme

J.A. Garay and R. Gennaro (Eds.): CRYPTO 2014, Part I, LNCS 8616, pp. 335–352, 2014.

provably secure in the standard model of computation, and with very simple signing/verification procedures, but only provided a direct construction of one-time signatures: digital signature schemes that can be used to sign a single message. Such schemes can be turned into general purpose signature schemes with only a logarithmic loss in efficiency using standard tree constructions. However, these transformations can be quite expensive in practice, because they lead to signatures consisting of $O(\log n)$ vectors. Given that signature size is often the most critical efficiency parameter affecting the practicality of a scheme, such signatures can no longer be considered "short". On the other hand, [13] gave a scheme that allowed to produce short signatures for arbitrarily many messages, but only offered heuristic security in the random oracle model. Moreover, the scheme of [13] was not entirely practical, involving a rather complex signing algorithm based on sampling lattice vectors with gaussian distribution, a problem that only recently has found more satisfactory solutions [29].

Two lines of research have evolved from [13], trying to address either the security or efficiency limitations of that work:

- A first line of research [21,22,14,12,6,15] kept investigating lattice signature in the random oracle model, with the goal of achieving the highest possible levels of performance, and schemes that are efficient enough to be used in practice.
- A second line of work, [11,9,29] kept pursuing the important goal of obaining security in the standard model of computation (no random oracles) while at the same time improving the efficiency and potential practicality of previous schemes. Our work is part of this second line of research, which we describe in more detail.

The current state of the art, when it comes to short lattice signatures in the standard model, is given by the scheme of Boyen [9], with additional security and efficiency improvements described in [29]. This scheme achieved the important goal of "short" lattice signatures (consisting of a single lattice vector), without resorting to the random oracle model. The main drawback of this scheme was the huge public key involved. Lattice public keys, even in the random oracle model [13,21,22,14,12,6], consist of one or more $n \times m$ matrices, each of which typically requires $\tilde{\Omega}(n^2)$ storage. For the sake of comparison, we consider natural adaptations of [11,9,29] to the algebraic/ring setting, where $n \times m$ matrices can be implicitly described by a single m-dimensional vector. Going back to the signature scheme of [9,29], public keys consist of $\Omega(n)$ matrices, and therefore require at least quadratic $\tilde{\Omega}(n^2)$ total storage even when using "compact" algebraic lattices. We remark that digital signature schemes can be efficiently constructed out of identity based encryption (IBE) by using ciphertexts as signatures, and lattice based IBE with short ciphertexts are also known [11,2,1]. However, lattice IBE schemes are built on top of the signature techniques from [11,9], and bear the same limitations when it comes to public key size: lattice IBE [11,2,1] use public keys consisting of $\Omega(n)$ matrices, and result in $\tilde{\Omega}(n^2)$ or even $\tilde{\Omega}(n^3)$ pubic key size depending on the type of lattices employed.

Reducing the size of, not only the signatures, but also the public key, was the main open problem left by [11,9,29,2,1]. We remark that the last few years have seen major efficiency progress on lattice signatures in the random oracle model [13,21,22,14,12,6], leaving a substantial gap between random oracle and standard model signatures. Still, designing efficient signature schemes without random oracles is an important and well established problem, both for the theory and practice of cryptography. A recent work in this direction is the paper of Bohl *et al.* [7,8,36], which formalized[1] a general "confined guessing" technique applicable to a variety of (not only lattice) settings. Here we describe their results, limited to the case of lattice signatures, and specialized to algebraic/ring lattices. Among other things, [7] gives a standard model lattice signature with public keys consisting of a single matrix, and therefore requiring only $O(m) = \tilde{\Omega}(n)$ storage when using algebraic/ring lattices. However, this comes at a substantial cost in terms of signature size: the digital signatures of [7] consist of $O(\log n)$ vectors. While a $O(\log n)$ increase may not seem much, it is quite a high cost when it comes to signature size, both in theory and in practice. In fact, a similar trade-off was already known since the very first direct construction of lattice signatures [24], which, as alredy discussed, produced general signatures consisting of $O(\log n)$ vectors (as well as short public keys). In other words, just like [24], the lattice signatures of [7] are not "short". (The main contribution of [7] over the classic scheme of [24], is that the results of [7] also apply to general lattices.)

Our results. We present the first standard model construction of short signatures based on (algebraic/ring) lattices with relatively small public keys: Similarly to [9,29], we achieve signatures consisting of a single vector without resorting to random oracles. At the same time, we substantially reduce the public key size from the $\Omega(n)$ vectors[2] of previously best short lattice signatures [9,29] to just $O(\log n)$ vectors. Our scheme is stateless, i.e., all signatures can be produced independently by running the signing algorithm on input the secret key and message to be signed. We also give an even more efficient scheme that further improves the public key size from $O(\log n)$ to just $O(\log \log n)$ vectors (and at the same time also improves the tightness of the reduction,) almost matching the *asymptotic* performance of schemes in the random oracle model [13,21,22,14,12,6]. This last improvement comes at the cost of statefulness: the signer has to keep some state information between signatures. However the state information is extremely simple: all that the signer has to do is to maintain a counter keeping track of how many signatures have already been produced.

We remark that it is always possible to reduce the public key size by increasing the size of the signatures, simply by compressing the public key using a collision resistant hash function (which is easily built from lattices [26,5,23,31]), and including the original public key in each signature. So, our first scheme (with

---

[1] The technique first appeared in the work of Hohenberger and Waters [18,17] and was also used in [10].

[2] Remember we are in the ring setting, so only one vector is required to represent each matrix.

Scheme	Pub. Key $\mathcal{R}_q^{1\times k}$ mat.	Secret Key $\mathcal{R}_q^{k\times k}$ mat.	Signature $\mathcal{R}_q^k$ vec.	Reduction loss	SIS parameter β
[13](ROM)	1	1	1	1	$\tilde{\Omega}(n)$
[24](Trees)	1	1	$\log n$	Q	$\tilde{\Omega}(n^2)$
[11]	n	n	n	Q	$\tilde{\Omega}(n^{3/2})$
[9,29]	n	n	1	Q	$\tilde{\Omega}(n^{7/2}), \tilde{\Omega}(n^{5/2})$
[7]	1	1	$\log_c n$	$O(Q^2/\epsilon)^c$	$\tilde{\Omega}(n^{5/2})$
Stateless (Sec. 3)	$\log_c n$	$\log_c n$	1	$O(Q^2/\epsilon)^c$	$\tilde{\Omega}(n^{7/2})$
Stateful[3]	$2\log_c(\log n)$	$2\log_c(\log n)$	1	$2Q^c$	$\tilde{\Omega}(n^{3/2})$

$\mathcal{R}_q = \mathbb{Z}_q[X]/f(X)$ for some (cyclotomic) polynomial f of degree n, $q = n^{O(1)}$, and $k = O(\log q)$. Q denotes the number of signature queries made by the attacker and ϵ is its success probability. The value $c > 1$ is an arbitrary constant that governs the security/efficiency trade off. The reduction loss is the ratio ϵ'/ϵ between the success probability ϵ' of the reduction and the success probability ϵ of the attacker.

Fig. 1. Comparison to previous work on lattice signatures in the ring setting

$O(\log n)$ vectors in the public key and short signatures) subsumes the results of [7] in the algebraic/ring lattice setting with $O(\log n)$ vectors per signatures.

The efficiency of our lattice constructions, compared to previous schemes (all adapted to the ring setting), is detailed in Figure 1.The trick leading to our stateful signature scheme can also be applied to improve the generic construction of [7]. The description of our generic results is deferred to the full version of our paper.

Techniques. Our results are obtained by combining several techniques previously used in the construction of lattice-based signatures. Most notably, we use the "vanishing trapdoor" technique from [9], and the more recent "confined guessing" method of [7,18,17]. In fact, the key generation, signing and verification algorithms bear strong similarities with previously proposed schemes. However, the combination appears to be novel and nontrivial. In particular, while both the results in [9] and those in [7] are presented for general lattices, the way they are combined in our work makes essential use of the commutativity properties of ring/algebraic lattices. More specifically, our proof of security exploits a key homomorphic property of lattice trapdoors (see Lemma 6) which requires certain matrix products to commute. This is trivially verified in the ring setting, where one of the matrices corresponds to a ring scalar, but glamorously fails when the construction is adapted to arbitrary lattices.

Open problems. Interestingly, the methods employed in this paper to obtain short lattice signatures with small public key seem specific to the ring/algebraic lattice setting. Only our generic result (see the full version of this article) with signatures of $\log \log n$ many vectors applies to arbitrary lattices. We remark that the question of reducing the public key size is mostly important in the ring

---

[3] See full version of this article.

setting: when using general lattices, even a single matrix takes quadratic storage, so there is little hope to reduce the public key size to linear or quasilinear in the security parameter. Still, it would be nice to achieve results similar to those in our paper, but for general lattices: is there a standard model signature scheme based on general lattices with short signatures (consisting of a single vector) and small public keys (consisting of $O(\log n)$ matrices)?

Another important open problem is to further improve the efficiency of our scheme, and obtain short signatures where the public key is just $O(1)$ matrices (or vectors, in the ring setting). Indeed, schemes offering both short public key and short signatures[4] in the standard model have been constructed based on the Computational Diffie-Hellman (CDH) and RSA problems [18,17].

2   Preliminaries

2.1   Signatures

Definition 1. *A signature scheme* SS *is a triple* (KeyGen, Sign, Verif) *of PPT (probabilistic polynomial time) algorithms, together with message spaces* $\mathcal{M}_n$. *It is correct if, for all messages* $\mu \in \mathcal{M}_n$, Verif$(pk, \mu, \sigma) = 1$ *holds true, except with negligible probability (in* n) *over the choice of* $(sk, pk) \leftarrow$ KeyGen(1^n) *and* $\sigma \leftarrow$ Sign(sk, μ).

The standard definitions of security for digital signature schemes (under adaptive and non-adaptive attacks) is given in Figure 2.

EUF-naCMA$_{\text{SS}}(n, \mathcal{A})$	**EUF-CMA**$_{\text{SS}}(n, \mathcal{A})$
$\mathcal{A}$ chooses q messages $(\mu^{(j)}) \in \mathcal{M}_n$	$(sk, pk) \leftarrow$ KeyGen(1^n), $\mathcal{A}$ receives pk
$(sk, pk) \leftarrow$ KeyGen(1^n)	For $j = 0 \ldots Q - 1$:
For all $j = 0 \ldots Q - 1$:	$\quad \mathcal{A}$ chooses $\mu^{(j)}$
$\quad \sigma^{(j)} \leftarrow$ Sign$(sk, \mu^{(j)})$.	$\quad \mathcal{A}$ receives $\sigma^{(j)} \leftarrow$ Sign$(sk, \mu^{(j)})$
$\mathcal{A}$ receives $pk, \sigma^{(0)} \ldots \sigma^{(Q-1)}$.	$\mathcal{A}$ sends an attempted forgery $(\mu^{\diamond}, \sigma^{\diamond})$
$\mathcal{A}$ sends an attempted forgery $(\mu^{\diamond}, \sigma^{\diamond})$	$\mathcal{A}$ wins if Verif$(pk, \mu^{\diamond}, \sigma^{\diamond}) = 1$ and
$\mathcal{A}$ wins if Verif$(pk, \mu^{\diamond}, \sigma^{\diamond}) = 1$ and	$\mu^{\diamond} \notin \{\mu^{(j)}\}$.
$\mu^{\diamond} \notin \{\mu^{(j)}\}$.	

A signature scheme SS = (KeyGen, Sign, Verif) is **EUF-naCMA**-secure (or Existentially Unforgeable under non-adaptive Chosen Message Attacks) if no PPT adversary $\mathcal{A}$ wins the **EUF-naCMA**$_{\text{SS}}$ game (left) with non-negligible probability $n^{-O(1)}$. The scheme is **EUF-CMA**-secure (or Existentially Unforgeable under adaptive Chosen Message Attacks) if no PPT adversary $\mathcal{A}$ wins the **EUF-CMA**$_{\text{SS}}$ game (right) with non-negligible probability $n^{-O(1)}$.

Fig. 2. Definition of security for digital signature schemes

---

[4] Here by "short" we mean consisting of $O(1)$ group elements.

From Non-Adaptive to Full Security There are two standard techniques to transform non adaptively-secure signature schemes to fully secure ones: Chameleon Hashing and One Time Signatures both of which can be implemented using lattices [25,13]. For a description of the solution based on Chameleon Hashing see the full version of this article.

2.2   Lattices and Gaussian Distributions

A (full rank) n-dimensional *lattice* is the set $\Lambda = \mathcal{L}(\mathbf{B}) = \{\mathbf{B}\mathbf{z}\colon \mathbf{z} \in \mathbb{Z}^n\}$ of all integer linear combinations of n basis vectors $\mathbf{B} = [\mathbf{b}_1, \ldots, \mathbf{b}_n] \in \mathbb{R}^{n \times n}$. We use notation $(x_1, \ldots, x_n)$ for *column* vectors, and similarly write $(\mathbf{A}, \mathbf{B})$ for the result of vertically stacking two matrices. The dual lattice Λ^* is the set of all $\mathbf{v} \in \mathbb{R}^n$ such that $\langle \mathbf{v}, \mathbf{x} \rangle \in \mathbb{Z}$ for every $\mathbf{x} \in \Lambda$. If $\mathbf{B}$ is a basis of Λ, then $\mathbf{B}^* = \mathbf{B}^{-t}$ is a basis of Λ^*. Many cryptographic applications use a particular family of so-called q-*ary* integer lattices, which contain $q\mathbb{Z}^m$ as a sublattice for some (typically small) integer q. For positive integers n, and q, let $\mathbf{A} \in \mathbb{Z}_q^{n \times m}$ be arbitrary and define the following full-rank m-dimensional q-ary lattices:

$$\Lambda^{\perp}(\mathbf{A}) = \{\mathbf{z} \in \mathbb{Z}^m : \mathbf{A}\mathbf{z} = \mathbf{0} \bmod q\}$$
$$\Lambda(\mathbf{A}) = \{\mathbf{z} \in \mathbb{Z}^m : \exists\, \mathbf{s} \in \mathbb{Z}_q^n \text{ s.t. } \mathbf{z} = \mathbf{A}^t\mathbf{s} \bmod q\}.$$

It is easy to check that $\Lambda^{\perp}(\mathbf{A})$ and $\Lambda(\mathbf{A})$ are dual lattices, up to a q scaling factor: $q \cdot \Lambda^{\perp}(\mathbf{A})^* = \Lambda(\mathbf{A})$, and vice-versa. For any $\mathbf{u} \in \mathbb{Z}_q^n$ admitting an integral solution to $\mathbf{A}\mathbf{x} = \mathbf{u} \bmod q$, define the coset (or "shifted" lattice) $\Lambda_{\mathbf{u}}^{\perp}(\mathbf{A}) = \{\mathbf{z} \in \mathbb{Z}^m : \mathbf{A}\mathbf{z} = \mathbf{u} \bmod q\} = \Lambda^{\perp}(\mathbf{A}) + \mathbf{x}$. In the Small Integer Solution problem $(\text{SIS}_{p,n,m,\beta})$, one is given a matrix $\mathbf{A} \in \mathbb{Z}_q^{n \times m}$ and is asked to find a nonzero vector $\mathbf{s} \in \Lambda^{\perp}(\mathbf{A})$ such that $\|\mathbf{s}\| \leq \beta$ where $\|\mathbf{s}\| = \sqrt{\sum_i s_i^2}$ is the euclidean norm. The geometric quality of a matrix $\mathbf{A} \in \mathbb{R}^{m \times n}$ is measured by its spectral norm $s_1(\mathbf{A}) = \sup_{\mathbf{x}} \|\mathbf{A}\mathbf{x}\|/\|\mathbf{x}\|$.

The n-dimensional Gaussian function $\rho_s \colon \mathbb{R}^n \to (0, 1]$ is defined as $\rho_s(\mathbf{x}) = \exp(-\pi \cdot \|\mathbf{x}/s\|^2)$. For any (countable) set $X \subseteq \mathbb{R}^n$, let $\rho_s(X) = \sum_{\mathbf{x} \in X} \rho_s(\mathbf{x})$. The *smoothing parameter* of a lattice $\eta_\epsilon(\Lambda)$ [30] is the smallest s such that $\rho_{1/s}(\Lambda^*) \leq 1 + \epsilon$. The discrete gaussian distribution $D_{\Lambda,s}$ over a lattice Λ is defined as $D_{\Lambda,s}(\mathbf{x}) = \rho_s(\mathbf{x})/\rho_s(\Lambda)$ for all $\mathbf{x} \in \Lambda$.

We say that a random variable X over $\mathbb{R}$ is *subgaussian* with parameter $s > 0$ if for all $t \in \mathbb{R}$, the (scaled) moment-generating function satisfies $\mathrm{E}[\exp(2\pi t X)] \leq \exp(\pi s^2 t^2)$. If X is subgaussian, then its tails are dominated by a Gaussian of parameter s, i.e., $\Pr[|X| \geq t] \leq 2\exp(-\pi t^2/s^2)$ for all $t \geq 0$. More generally, we say that a random matrix $\mathbf{X}$ is subgaussian (of parameter s) if all its one-dimensional marginals $\mathbf{u}^t \mathbf{X} \mathbf{v}$ for unit vectors $\mathbf{u}, \mathbf{v}$ are subgaussian (of parameter s). It follows immediately from the definition that the concatenation of independent subgaussian vectors with common parameter s, interpreted either as a vector or as a matrix, is subgaussian with parameter s. For any lattice $\Lambda \subset \mathbb{R}^n$ and $s > 0$, the distribution $D_{\Lambda,s}$ is subgaussian with parameter s.

We will need the following standard result from the non-asymptotic theory of random matrices; for further details, see [37].

Lemma 1. *Let* $\mathbf{X} \in \mathbb{R}^{n \times m}$ *be a subgaussian random matrix with parameter* s. *There exists a universal constant* $C \approx 1/\sqrt{2\pi}$ *such that for any* $t \geq 0$, *we have* $s_1(\mathbf{X}) \leq C \cdot s \cdot (\sqrt{m} + \sqrt{n} + t)$ *except with probability at most* $2\exp(-\pi t^2)$.

2.3   Rings and Ideal Lattices

We consider lattice problems restricted to ideal lattices [28,23,32]. Most of our results apply to ideal/module lattices over arbitrary cyclotomic rings, but for simplicity we focus our presentation on so-called "SWIFFT" rings [26,5]. These are rings of the form $\mathcal{R} = \mathbb{Z}[X]/(\Phi_{2n}(X))$ or $\mathcal{R}_q = (\mathcal{R}/q\mathcal{R})$, where n is a power of 2, q is an integer, and $\Phi_{2n}(X) = X^n + 1$ is the cyclotomic polynomial of degree n. For our construction we will require that $\Phi_{2n}(X)$ does not split into low degree polynomials modulo the prime factors of q. More concretely we choose $q = 3^k$ and rely on the following.

Fact 1 (Irreducible factors of $\Phi_{2^k}(X)$ modulo 3. Corollary of [20, Theorem 2.47]). *For any* $k \geq 3$ *and* $2n = 2^k$ *we have* $\Phi_{2n}(X) \equiv (X^{n/2} + X^{n/4} - 1) \cdot (X^{n/2} - X^{n/4} - 1) \bmod 3$ *and both factors are irreducible in* $\mathbb{F}_3[X]$.

Lemma 2 (Hensel Lemma). *Let* $\mathcal{R} = \mathbb{Z}[X]/(F(X))$ *for some monic polynomial* $F \in \mathbb{Z}[X]$. *For any prime* p, *if* $u \in \mathcal{R}_{p^e}$ *is invertible* $\bmod p$ *(i.e. it is invertible in* $\mathcal{R}_p$*) then* u *is also invertible in* $\mathcal{R}_{p^e}$.

Corollary 1. *let* $n \geq 4$ *be a power of 2,* $q \geq 3$ *a power of 3, and set* $\mathcal{R}_q = \mathbb{Z}[X]/(\Phi_{2n}(X), q)$. *Then, any nonzero polynomial* $t \in \mathcal{R}_q$ *of degree* $d < n/2$ *and coefficients in* $\{0, \pm 1\}$ *is invertible in* $\mathcal{R}_q$.

Elements in $\mathcal{R}$ have a natural representation as polynomials of degree $n-1$ with coefficients in $\mathbb{Z}$, and $\mathcal{R}$ can be identified (as an additive group) with the integer lattice $\mathbb{Z}^n$, where each ring element $\mathbf{a} = a_0 + a_1 x + \ldots + a_{n-1} x^{n-1} \in \mathcal{R}$ is associated with the coefficient vector $(a_0, \ldots, a_{n-1}) \in \mathbb{Z}^n$. We use the identification $\mathcal{R} = \mathbb{Z}^n$ to define standard lattice quantities like the euclidean length of a ring element $\|\mathbf{a}\| = \sqrt{\sum_i |a_i|^2}$, or the spectral norm of a ring element $s_1(r) = \sup_x \|r \cdot x\|/\|x\|$. The ring $\mathcal{R}$ is also identified with the sub-ring of anti-circulant square matrices of dimension n by regarding each ring element $r \in \mathcal{R}$ as a linear transformation $x \mapsto r \cdot x$ over (the coefficient embedding) of $\mathcal{R}$. Notice that the definition of spectral norm of a ring element is consistent with the definition of spectral norm of the corresponding anticirculant matrix. The following lemma provides a useful bound on the spectral norm of ring elements.

Lemma 3. *For any ring element* $r \in \mathcal{R}$, *we have* $s_1(r) \leq \|r\|_1 = \sum_i |r_i|$.

Proof. Let $\omega_k = e^{\pi i(2k-1)/n}$ (for $k = 1, \ldots, n$) be the complex roots of the cyclotomic polynomial Φ_{2n}. Consider the image of r under the canonical embedding $\sigma \colon \mathcal{R} \to \mathbb{C}^n$, which is defined as $\sigma(r) = (r(\omega_1), \ldots, r(\omega_n))$. Using the fact that $\sigma \colon \mathcal{R} \to \mathbb{C}^n$ is a ring homomorphism (with the product $\odot$ in $\mathbb{C}^n$ defined componentwise) and a scaled isometry (satisfying $\|\sigma(r)\| = \sqrt{n} \cdot \|r\|$) we get

$$s_1(r) = \sup_x \frac{\|r \cdot x\|}{\|x\|} = \sup_x \frac{\|\sigma(r \cdot x)\|}{\|\sigma(x)\|} = \sup_x \frac{\|\sigma(r) \odot \sigma(x)\|}{\|\sigma(x)\|} \leq \|\sigma(r)\|_\infty.$$

Since for any i, $|\omega_i| = 1$, we have $|r(\omega_i)| = \left|\sum_j r_j \omega_i^j\right| \leq \sum |r_j| = \|r\|_1$. It follows that $s_1(r) \leq \|\sigma(r)\|_\infty = \max_i |\sigma(r)_i| \leq \|r\|_1$.   □

The discrete Gaussian distribution over the ring $D_{\mathcal{R},s} \equiv D_{\mathbb{Z},s}^n$ is defined as usual by identifying the ring $\mathcal{R}$ with $\mathbb{Z}^n$ under the coefficient embedding. It follows that the discrete gaussian distribution over the ring $x \leftarrow D_{\mathcal{R},s}$ is sub-gaussian of parameter s when x is regarded as a vector. For the anti-circulant matrix representation, we have the following fact, (proof in App. A).

Fact 2. *If* $\mathbf{R} \leftarrow D_{\mathcal{R},s}^{w \times k}$, *then with overwhelming probability we have* $s_1(\mathbf{R}) \leq s\sqrt{n} \cdot O(\sqrt{w} + \sqrt{k} + \omega(\sqrt{\log n}))$.

The euclidean length of vectors in $\mathcal{R}_q^k$ is defined similarly by identifying $\mathbb{Z}_q$ with the set of representatives $\{-(q-1)/2, \ldots, +(q-1)/2\}$. Similarly, we define the q-ary lattices $\Lambda(\mathbf{A})$ and $\Lambda^\perp(\mathbf{A})$ when $\mathbf{A} \in \mathcal{R}_q^{n \times m}$ is a matrix over the ring $\mathcal{R}_q$ using the standard isomorphism of $\mathcal{R}_q$ and the sub-ring of anticirculant matrices in $\mathbb{Z}_q^{n \times n}$.

Definition 2. *In the Small Integer Solution over Rings problem* (RingSIS$_{q,n,m,\beta}$), *one is given a row vector* $\mathbf{A} \in \mathcal{R}_q^{1 \times m}$, *and is asked to find a nonzero vector* $\mathbf{x} \in \Lambda_q^\perp(\mathbf{A})$ *such that* $\|\mathbf{x}\| \leq \beta$.

Let $\mathcal{U}_m$ be the uniform distribution over m-dimensional row vectors of ring elements $\mathbf{A} = [\mathbf{a}_1, \mathbf{a}_2, \ldots, \mathbf{a}_m] \in \mathcal{R}_q^{1 \times m}$. The smoothness proof from [13] can be adapted to this specific ring case (proof in App. A). A more general ring regularity result can be found [27, Theorem 7.4], but unfortunately it gives a larger bound (by a factor n) on required standard deviation s than our specialized lemma.

Lemma 4 (Smoothness Lemma). *Let* $\mathcal{R}_q = \mathbb{Z}[X]/(\Phi_{2n}(X), q)$ *for* $n \geq 4$ *a power of 2 and* $q = 3^k$ *a power of 3. Let* $w \geq 2\lceil \log_2 q \rceil + 2$ *and* $s \geq \omega(\sqrt{\ln nw})$. *With overwhelming probability over the choice of* $\mathbf{A} \leftarrow \mathcal{U}_w$, *if* $\mathbf{x}_i \leftarrow D_{\mathcal{R},s}$ *(for* $i = 1, \ldots, w$*) are chosen independently at random, then the sum* $\sum_i \mathbf{a}_i \cdot \mathbf{x}_i$ *is within negligible statistical distance from the uniform distribution over* $\mathcal{R}$.

A handy corollary used several time in our proof is the following.

Corollary 2 (Min-entropy bound). *Set* $\mathcal{R}_q$ *as above, and let* $w \geq 2\lceil \log_2 q \rceil + 3$, $s \geq \omega(\sqrt{\ln nw})$. *With overwhelming probability over the choice of* $\mathbf{A} \leftarrow \mathcal{U}_w$, *if* $\mathbf{x}_i \leftarrow D_{\mathcal{R},s}$ *(for* $i = 1, \ldots, w$*) are chosen independently at random, then for any nonzero vector* $\mathbf{V} \in \mathcal{R}^w \setminus \{\mathbf{0}\}$ *the conditional min-entropy of* $\sum_i \mathbf{v}_i \cdot \mathbf{x}_i$ *given* $\sum_i \mathbf{a}_i \cdot \mathbf{x}_i$ *is at least* $\Omega(n)$.

2.4 Lattice Trapdoors

We use the strong lattice trapdoor construction and algorithms of [29]. For a modulus $q = 3^k$ and integer dimension n, define the gadget matrix $\mathbf{G} = \left[\mathbf{I}_n \mid 3 \cdot \mathbf{I}_n \mid \ldots \mid 3^{k-1} \cdot \mathbf{I}_n\right] \in \mathbb{Z}_q^{n \times kn}$.

Definition 3. *For any* $\mathbf{A} \in \mathbb{Z}_q^{n \times (m+kn)}$, *and (invertible)* $\mathbf{H} \in \mathbb{Z}_q^{n \times n}$, *a* $\mathbf{G}$-*trapdoor for* $\mathbf{A}$ *with tag* $\mathbf{H}$ *is a matrix* $\mathbf{R} \in \mathbb{Z}_q^{m \times kn}$ *such that* $\mathbf{A}(\mathbf{R}, \mathbf{I}) = \mathbf{HG}$. *The definition is extended to trapdoors* $\mathbf{R} \in \mathbb{Z}_q^{m' \times kn}$ *with* $m' \leq m$ *by padding them with zero columns so that* $[\mathbf{R}, \mathbf{O}] \in \mathbb{Z}_q^{m \times kn}$.

The quality of a trapdoor $\mathbf{R}$ is measured by the spectral norm $s_1(\mathbf{R})$, and [29] gives efficient algorithms to generate uniformly random matrices $\mathbf{A}$ together with high quality trapdoors, and to sample cosets $\Lambda_{\mathbf{u}}^{\perp}(\mathbf{A})$ with Gaussian distribution D_s for sufficiently large s. Notice that the tag $\mathbf{H}$ can immediately be recovered from $\mathbf{A}$ and $\mathbf{R}$ as the first block of $\mathbf{HG}$, and does not need to be specified explicitly. But when one says that $\mathbf{R}$ is a trapdoor, it is usually assumed that the associated tag $\mathbf{H}$ is an invertible matrix.

Theorem 3 ([29]). *There is an efficient algorithm* SampleD$(\mathbf{A}, \mathbf{u}, \mathbf{R}, s)$ *that on input a matrix* $\mathbf{A} \in \mathbb{Z}_q^{n \times (m+kn)}$, *a syndrome* $\mathbf{u} \in \mathbb{Z}_q^n$, *a* $\mathbf{G}$-*trapdoor* $\mathbf{R} \in \mathbb{Z}_q^{m \times kn}$ *for* $\mathbf{A}$, *and parameter* $s > \omega(\sqrt{\log n}) \cdot s_1(\mathbf{R})$, *produces a sample from the distribution* $D_{\Lambda_{\mathbf{u}}^{\perp}(\mathbf{A}), s}$.

The efficient trapdoor generation algorithm of [29] follows immediately from the definition of $\mathbf{G}$-trapdoor: one simply chooses $\mathbf{A}' \in \mathbb{Z}_q^{n \times m}$ uniformly at random, samples a trapdoor matrix $\mathbf{R} \in \mathbb{Z}_q^{m \times nk}$ with small entries, and outputs $\mathbf{A} = [\mathbf{A}', \mathbf{HG} - \mathbf{A}'\mathbf{R}]$. As pointed out in [29], the algorithm is immediately adapted to ideal lattices, using the observation that the identity matrix $\mathbf{I}_n$ is precisely the matrix corresponding to the ring element $1 \in \mathcal{R}$, so the gadget matrix $\mathbf{G}$ can be regarded as a row vector of ring elements $[1, 3, 9, \ldots, 3^{k-1}] \in \mathcal{R}^{1 \times k}$. The trapdoor generation algorithm is then analyzed using Theorem 4, and the trapdoor quality is bounded applying Fact 2 to the concatenation of subgaussian random variables $\mathbf{r}_i \leftarrow D_{\mathcal{R},s} \equiv D_{\mathbb{Z},s}^n$. The formal result is stated below.

Theorem 4. *There is a polynomial time algorithm* GenTrap$(\mathbf{A}', \mathbf{H}, s)$ *that on input a matrix* $\mathbf{A}' \in \mathcal{R}_q^{1 \times w}$, *tag* $\mathbf{H} \in \mathcal{R}_q$, *and parameter* $s > \omega(\sqrt{\ln nw})$, *outputs a matrix* $\mathbf{A}'' \in \mathcal{R}_q^{1 \times k}$ *and a* $\mathbf{G}$-*trapdoor* $\mathbf{R} \in \mathcal{R}_q^{w \times k}$ *for* $\mathbf{A} = [\mathbf{A}', \mathbf{A}'']$ *with tag* $\mathbf{H}$ *such that* $s_1(\mathbf{R}) = s \cdot O(\sqrt{w} + \sqrt{k} + \omega(\sqrt{\log n}))$. *Moreover, if* $w \geq 2(\lceil \log_2 q \rceil + 1)$ *then with overwhelming probability over the choice of* $\mathbf{A}' \leftarrow \mathcal{U}_w$, *the distribution of* $\mathbf{A}''$ *is statistically close to uniform.*

In order to allow for the generation of trapdoors for multiple matrices that share the same $\mathbf{A}'$, we made $\mathbf{A}'$ an explicit input to the trapdoor generation algorithm. When $\mathbf{A}' \leftarrow \mathcal{U}_w$ is chosen freshly at random, we simply write GenTrap$(w, \mathbf{H}, s)$ and let GenTrap output the whole $\mathbf{A} = [\mathbf{A}', \mathbf{A}'']$.

Notice that $\mathbf{G}$-trapdoors generated in the ring setting also satisfy the definition of $\mathbf{G}$-trapdoor for general lattices. So, Theorem 3 can be used as it is, simply by viewing ring trapdoors $\mathbf{R} \in \mathcal{R}_q^{w \times k}$ as matrices $\mathbf{R} \in \mathbb{Z}^{wn \times kn}$ under the standard embedding from $\mathcal{R}$ to the subring of anticirculant matrices. For convenience, we reformulate Theorem 3 as a corollary specialized to the ring setting.

Corollary 3. *There is an efficient algorithm* SampleD($\mathbf{A}, \mathbf{u}, \mathbf{R}, s$) *that on input a matrix* $\mathbf{A} \in \mathcal{R}_q^{1 \times (w+k)}$, *a syndrome* $\mathbf{u} \in \mathcal{R}_q$, *a* $\mathbf{G}$-*trapdoor* $\mathbf{R} \in \mathcal{R}_q^{w \times k}$ *for* $\mathbf{A}$ *with invertible tag* $\mathbf{H} \in \mathcal{R}$, *and parameter* $s > \omega(\sqrt{\log n}) \cdot s_1(\mathbf{R})$, *produces a sample statistically close to the distribution* $D_{\Lambda_{\mathbf{u}}^{\perp}(\mathbf{A}), s}$.

We remark that GenTrap can be called with arbitrary (not necessarily invertible) tags $\mathbf{H}$. The algorithm still outputs a uniformly random $\mathbf{A}$ and small $s_1(\mathbf{R})$, but the inversion algorithm of Corollary 3 cannot be used with such invalid trapdoors.

3   Our Scheme

The scheme is parametrized by an integer n which we assume is a power of 2, and a modulus $q = 3^k$ which we assume to be a power of 3. (Other parameter settings are possible, but we consider these specific values for concreteness.) These parameters define the ring $\mathcal{R}_q = \mathbb{Z}[X]/(\Phi_{2n}(X), q)$, where (for n a power of 2) $\Phi_{2n}(X) = X^n + 1$ is the cyclotomic polynomial of degree n. The scheme also uses the parameters $w = 2\lceil \log_2 q \rceil + 2$, $m = w + k$, $s = n^{3/2} \cdot \omega(\log n)^{3/2}$, and a collection of tags defined below. We recall that the polynomial $\Phi_{2n}(X)$ is irreducible in $\mathbb{Z}[X]$, but it can be factored in $\mathbb{F}_p[X]$ for some primes p. Our choice of $q = 3^k$ ensures that, in $\mathbb{F}_3[X]$, the polynomial $\Phi_{2n}(X)$ factors into the product of just 2 irreducible polynomials of degree $n/2$. (See Fact 1.) In particular, by Corollary 1, any nonzero polynomial of degree less than $n/2$ with coefficients in $\{0, \pm 1\}$ is invertible in $\mathcal{R}_q$.

Tags For any real constants $c > 1$ and $\alpha \geq \frac{1}{c-1}$ (fixed throughout the rest of this section) define the sets of *tag prefixes* $\mathcal{T}_i = \{0, 1\}^{c_i}$ of (strictly increasing) lengths $c_0 = 0$, $c_i = \lfloor \alpha c^i \rfloor$ for $i \in \{1, \ldots, d\}$ where $d = \lfloor \log_c(n/(2\alpha)) \rfloor = O(\log n)$. We identify each tag prefix $t = [t_0, \ldots, t_{c_i-1}] \in \mathcal{T}_i$ with a corresponding ring element $t(X) = \sum_{j < c_i} t_j X^j \in \mathcal{R}_q$ with binary coefficients $t_j \in \{0, 1\}$ and degree less than $c_i \leq c_d \leq n/2$. It follows that for any two distinct tag prefixes $t, t' \in \mathcal{T}_i$, the difference $(t(X) - t'(X))$ is invertible in $\mathcal{R}_q$. For any *full tag* $t \in \mathcal{T} = \mathcal{T}_d$ and $i \leq d$, we write $t_{\leq i} \in \mathcal{T}_i$ for its prefix of length c_i, and $t_{[i]}$ for the (ring) difference $t_{\leq i}(X) - t_{\leq i-1}(X) \in \mathcal{R}_q$.

Unlike previous work using tags [29,11,9], our construction relies not only on the algebraic (invertibility) properties of tags, but also on their geometric properties, described in the following lemma.

Lemma 5. *For any* $i \leq d$ *and tags* $t, t' \in \mathcal{T}$, *one has* $s_1((t - t')_{[i]}) \leq c_i - c_{i-1}$.

Proof. Since the difference $(t - t')_{[i]}$ is a trinary polynomial with at most $c_i - c_{i-1}$ nonzero coefficients, we have $\|(t - t')_{[i]}\|_1 \leq c_i - c_{i-1}$. It follows from Lemma 3 that $s_1((t - t')_{[i]}) \leq \|(t - t')_{[i]}\|_1 \leq c_i - c_{i-1}$. $\qquad \square$

3.1   Our Scheme

Key Generation naSS.KeyGen(n): The key generation algorithm runs $(\mathbf{A}, \mathbf{R}) \leftarrow$ GenTrap$(w, \mathbf{I}, \sigma)$ with $\sigma = \omega(\sqrt{\log n})$, and chooses $\mathbf{A}_{[0]}, \mathbf{A}_{[1]}, \ldots \mathbf{A}_{[d]}, \mathbf{U} \in \mathcal{R}_q^{1 \times k}$ and $\mathbf{v} \in \mathcal{R}_q$ uniformly at random. It then outputs the secret key $sk = \mathbf{R}$, and public key $pk = (\mathbf{A}, \mathbf{A}_{[0]}, \mathbf{A}_{[1]}, \ldots \mathbf{A}_{[d]}, \mathbf{U}, \mathbf{v})$. The public key implicitly defines a collection of matrices $\mathbf{A}_t = [\mathbf{A}|\mathbf{A}_{[0]} + \sum_{i=1}^{d} t_{[i]} \cdot \mathbf{A}_{[i]}]$ indexed by the tags $t \in \mathcal{T}$.

Since $\sigma = \omega(\sqrt{\log n})$, by Theorem 4 and Lemma 2, the distribution of $\mathbf{A} \in \mathcal{R}_q^{1 \times m}$ is statistically close to $\mathcal{U}_m$, and $\mathbf{R}$ is a $\mathbf{G}$-trapdoor for $\mathbf{A}$ (and therefore also for all $\mathbf{A}_t$) with invertible tag $\mathbf{I}$ and quality $s_1(\mathbf{R}) \leq \sqrt{n} \cdot \omega(\log n)$.

Signature naSS.Sign($sk = \mathbf{R}, \boldsymbol{\mu} \in \{0,1\}^{nk} \subset \mathcal{R}_q^k$): Parse $\boldsymbol{\mu}$ as a vector of $\mathcal{R}_q^k$ splitting the nk bits into k binary polynomials. Choose a uniformly random tag $t \in \mathcal{T}$, and compute the matrix $\mathbf{A}_t$ and ring element $\mathbf{u} = \mathbf{U} \cdot \boldsymbol{\mu} + \mathbf{v}$. Then, use the $\mathbf{G}$-trapdoor $\mathbf{R}$ to sample a vector $\mathbf{s} \leftarrow$ SampleD$(\mathbf{A}, \mathbf{u}, \mathbf{R}, s)$. Output the pair $\sigma = (t, \mathbf{s})$ as the signature.

Verification naSS.Verif($pk, \boldsymbol{\mu} \in \{0,1\}^{nk} \subset \mathcal{R}_q^k, \sigma = (t, \mathbf{s})$): Compute $\mathbf{A}_t$ and $\mathbf{u} = \mathbf{U} \cdot \boldsymbol{\mu} + \mathbf{v}$ as in the signing algorithm. Then, check that $\|\mathbf{s}\| \leq s\sqrt{nm}$ and that $\mathbf{A}_t \cdot \mathbf{s} = \mathbf{u}$.

Correctness The correctness of the scheme is easily verified: Since $s > \omega(\sqrt{\log n}) \cdot s_1(\mathbf{R})$, by Corollary 3 the vector $\mathbf{s}$ produced during the signature generation process follows the distribution $D_{\Lambda_{\mathbf{u}}^{\perp}(\mathbf{A}_t), s}$ and has length at most $s\sqrt{nm} = O(s\sqrt{nk})$ with overwhelming probability. So, the signature $(t, \mathbf{s})$ is accepted by the verification algorithm.

3.2   Security Proof

The security of the scheme is based on the following homomorphic property of $\mathbf{G}$-trapdoors over rings. We remark that the property makes essential use of the commutativity of matrices corresponding to ring elements in $\mathcal{R}_q$, so it does not trivially adapts to general lattices, unless one restricts the set of tags to scalar matrices.

Lemma 6. *For $i = 0, \ldots, d$, let $\mathbf{R}_{[i]} \in \mathcal{R}^{w \times k}$ be a $\mathbf{G}$-trapdoor for $[\mathbf{A}, \mathbf{A}_{[i]}]$ with tag $\mathbf{H}_{[i]} \in \mathcal{R}_q$, where $\mathbf{A}_{[i]} \in \mathcal{R}_q^{1 \times k}$. Then, any linear combination $\mathbf{R} = \sum_i c_i \cdot \mathbf{R}_{[i]}$ with $c_i \in \mathcal{R}_q$ is a $\mathbf{G}$-trapdoor for $[\mathbf{A}, \sum_i c_i \mathbf{A}_{[i]}]$ with tag $\mathbf{H} = \sum_i c_i \mathbf{H}_{[i]}$.*

Proof. By definition of $\mathbf{G}$-trapdoor, we know that $[\mathbf{A}, \mathbf{A}_{[i]}](\mathbf{R}_{[i]}, \mathbf{I}) = \mathbf{H}_{[i]}\mathbf{G}$ for all i. Therefore

$$\left[\mathbf{A}, \sum_i c_i \mathbf{A}_{[i]}\right](\mathbf{R}, \mathbf{I}) = \mathbf{AR} + \sum_i c_i \mathbf{A}_{[i]} = \sum_i c_i (\mathbf{AR}_{[i]} + \mathbf{A}_{[i]})$$

$$= \sum_i c_i [\mathbf{A}, \mathbf{A}_{[i]}](\mathbf{R}_{[i]}, \mathbf{I}) = \sum_i c_i \mathbf{H}_{[i]}\mathbf{G} = \mathbf{HG}.$$

Therefore $\mathbf{R}$ is a $\mathbf{G}$-trapdoor with tag $\mathbf{H}$.   □

Theorem 5 (EUF-naCMASecurity). *Under the* $\text{RingSIS}_{n,m,q,\beta}$ *assumption for* $\beta = \tilde{O}(n^{7/2})$, *the above scheme* naSS *is* **EUF-naCMA** *secure. More precisely, if there exists an attacker* $\mathcal{A}$ *against* **EUF-naCMA**$_{\text{naSS}}$ *that runs in time* T, *makes at most* Q *queries where* $1 \leq Q \leq 2^{o(n)}$ *and succeeds with probability* $\epsilon \geq 2^{-o(n)}$, *then, there exists an algorithm* $\mathcal{S}^{\mathcal{A}}$ *that runs in time* $T' = T + \text{poly}(n)$, *and solves* $\text{SIS}(n, w, q, \beta)$ *with probability* $\epsilon' \geq \Omega\left(\frac{\epsilon^{1+c}}{Q^{2c}}\right)$.

The rest of the section is devoted to the proof of the theorem.

Confined Guessing Stage We assume we have an attacker $\mathcal{A}$ against the EUF-naCMA security of naSS that makes at most $Q = 2^{o(n)}$ signature queries, and succeeds with probability $\epsilon \geq 2^{-o(n)}$. Let $i^\star$ the smallest index such that $2Q^2/\epsilon \leq |\mathcal{T}_{i^\star}|$. (Notice that such index exists because $2Q^2/\epsilon = 2^{o(n)} \leq 2^{\lfloor \frac{n}{2c} \rfloor} \leq |\mathcal{T}|$.) This guarantees that, if one chooses Q tags at random in $\mathcal{T}_{i^\star}$, then they will be all distinct except with probability at most $\epsilon/2$.

The simulator $\mathcal{S}$ receives Q non-adaptive signature queries $\boldsymbol{\mu}^{(0)} \dots \boldsymbol{\mu}^{(Q-1)}$ from $\mathcal{A}$. For each message $\boldsymbol{\mu}^{(j)}$, the simulator $\mathcal{S}$ chooses a uniformly random tag $t^{(j)} \in \mathcal{T}$. If a collision of prefixes happens (i.e., if $t_{\leq i^\star}^{(j)} = t_{\leq i^\star}^{(k)}$ for some $j \neq k$) the simulator aborts. (This happens with probability at most $\epsilon/2$.) Otherwise, $\mathcal{S}$ chooses a prefix $t_{\leq i^\star}^\star \in \mathcal{T}_{i^\star}$ uniformly at random. (The rest of the tag $t^\star$ will be specified later on.) The hope is that the adversary will output a forgery $(t^\diamond, \mathbf{s}^\diamond)$ such that $t_{\leq i^\star}^\diamond = t_{\leq i^\star}^\star$. We will make the adversary's view statistically independent from the choice of $t_{\leq i^\star}^\star \in \mathcal{T}_{i^\star}$, so that $t_{\leq i^\star}^\diamond = t_{\leq i^\star}^\star$ will hold true with probability $1/|\mathcal{T}_{i^\star}|$.

Simulating Key Generation and Signatures The simulator also receives a RingSIS challenge, the row vector $\mathbf{A} \leftarrow \mathcal{U}_m$, from which it will build the public key. This is done by running $(\mathbf{A}_{[i]}, \mathbf{R}_{[i]}) \leftarrow \text{GenTrap}(\mathbf{A}, \mathbf{H}_{[i]}, \sigma')$ with $\sigma' = \omega(\sqrt{\log n})$ for $i = 0, \dots, d$ and

$$\mathbf{H}_{[i]} = \begin{cases} 0 \in \mathcal{R}_q & \text{if } i > i^\star \\ 1 \in \mathcal{R}_q & \text{if } 1 \leq i \leq i^\star \\ -t_{\leq i^\star}^\star & \text{if } i = 0. \end{cases}$$

Since $\omega(\sqrt{\log n}) \leq \sigma'$, by Theorem 4 the matrices $\mathbf{A}_{[i]}$ are statistically close to uniform. Moreover, by Fact 2, each $\mathbf{R}_{[i]} \in \mathcal{R}^{m \times k}$ is a $\mathbf{G}$-trapdoor for $[\mathbf{A}, \mathbf{A}_{[i]}]$ with $s_1(\mathbf{R}_{[i]}) \leq \sqrt{n} \cdot \omega(\log n)$. Therefore, by Lemma 6, $\mathbf{R}_t = \mathbf{R}_{[0]} + \sum_{i=1}^d t_{[i]} \cdot \mathbf{R}_{[i]}$ is a $\mathbf{G}$-trapdoor for $\mathbf{A}_t = [\mathbf{A}, \mathbf{A}_{[0]} + \sum_i t_{[i]} \cdot \mathbf{A}_{[i]}]$ with tag $\mathbf{H}_t = t_{\leq i^\star} - t_{\leq i^\star}^\star$. The quality of this trapdoor is

$$s_1(\mathbf{R}_t) \leq s_1(\mathbf{R}_{[0]}) + \sum_i s_1(t_{[i]} \cdot \mathbf{R}_{[i]}) \leq \left(1 + \sum_i s_1(t_{[i]})\right) \max_i s_1(\mathbf{R}_{[i]})$$

$$\leq \left(1 + \sum_i (c_i - c_{i-1})\right) \sqrt{n} \cdot \omega(\log n) = n^{3/2} \cdot \omega(\log n)$$

where we have used the geometric bound $s_1(t_{[i]}) \leq c_i - c_{i-1}$ from Fact 5. So, the simulator can use $\mathbf{R}_t$ as a trapdoor to sign messages with tag t as long as $\mathbf{H}_t$ is invertible. We observe that $\mathbf{H}_t = \mathbf{0}$ whenever $t_{\leq i^\star}^\star = t_{\leq i^\star}$ (i.e., when $t_{\leq i^\star}^\star$ is a prefix of t), and it is invertible otherwise. So, the simulator can efficiently answer all signature queries except at most for one index j such that $t_{\leq i^*}^{(j)} = t_{\leq i^*}^\star$. If such index exists, set $\boldsymbol{\mu}^\star = \boldsymbol{\mu}^{(j)}$ and $t^\star = t^{(j)}$ (recall that we've only chosen the prefix $t_{\leq i^*}^\star$ of $t^\star$ at the confined guessing stage), otherwise $\mathcal{S}$ chooses a random $\boldsymbol{\mu}^\star$ and a random $t^\star$ extension of $t_{\leq i^*}^\star$. We will use our last degree of freedom $\mathbf{v}$ to "program" a signature for this only message $\boldsymbol{\mu}^\star$: choose a signature $\mathbf{s}^\star \leftarrow D_{\mathcal{R},s}^m$, and set $\mathbf{v} = \mathbf{A}_{t^\star}\mathbf{s}^\star - \mathbf{U}\boldsymbol{\mu}^\star$. Applying Lemma 4, we check that $\mathbf{v}$ is close to uniform and independent of $\mathbf{A}_{t^\star}$, $\mathbf{U}$ and $\boldsymbol{\mu}$. This shows how to efficiently simulate public key and signatures that are indistinguishable from a real attack.

Notice that we have not specified how to choose $\mathbf{U}$ yet. In order to for the simulator to exploit the forgery, we want $\mathbf{U} = \mathbf{A}\mathbf{R}_\mathbf{U}$ for some $\mathbf{R}_\mathbf{U}$ with small entries. We can set $\mathbf{R}_\mathbf{U} \leftarrow D_{\mathcal{R},\sigma'}$ so that, by Lemma 4, $\mathbf{U} = \mathbf{A}\mathbf{R}_\mathbf{U}$ is statistically close to uniform, and $s_1(\mathbf{R}_\mathbf{U}) = \sqrt{n} \cdot \omega(\log n)$.

Exploiting the forgery After all those shenanigans from the simulators $\mathcal{S}$, with probability at least $\epsilon/2$, the adversary outputs a forgery $(t^\diamond, \mathbf{s}^\diamond)$ for some message $\boldsymbol{\mu}^\diamond$ of his choice. The simulator's secret hope that $t_{\leq i^*}^\diamond = t_{\leq i^*}^\star$ is fulfilled with probability $1/|\mathcal{T}_{i^*}|$; if not, $\mathcal{S}$ aborts. Otherwise we have

$$\mathbf{A}_{t^\star} \cdot \mathbf{s}^\star = \mathbf{U} \cdot \boldsymbol{\mu}^\star + \mathbf{v} \qquad \text{and} \qquad \mathbf{A}_{t^\diamond} \cdot \mathbf{s}^\diamond = \mathbf{U} \cdot \boldsymbol{\mu}^\diamond + \mathbf{v}$$

Recall that for any tag $t \in \mathcal{T}$ we have $\mathbf{A}_t = [\mathbf{A}|\mathbf{H}_t\mathbf{G} - \mathbf{A}\mathbf{R}_t]$ ($\mathbf{R}_t$ is a $\mathbf{G}$-trapdoor of $\mathbf{A}_t$ with tag $\mathbf{H}_t$); additionally the condition $t_{\leq i^*}^\diamond = t_{\leq i^*}^\star$ ensures $\mathbf{H}_{t^\star} = \mathbf{H}_{t^\diamond} = 0$. We derive

$$[\mathbf{A}|-\mathbf{A}\mathbf{R}_{t^\star}|-\mathbf{A}\mathbf{R}_\mathbf{U}] \cdot \begin{bmatrix} \mathbf{s}_1^\star \\ \mathbf{s}_2^\star \\ \boldsymbol{\mu}^\star \end{bmatrix} = \mathbf{v} = [\mathbf{A}|-\mathbf{A}\mathbf{R}_{t^\diamond}|-\mathbf{A}\mathbf{R}_\mathbf{U}] \cdot \begin{bmatrix} \mathbf{s}_1^\diamond \\ \mathbf{s}_2^\diamond \\ \boldsymbol{\mu}^\diamond \end{bmatrix} .$$

In particular we obtain $\mathbf{A}\mathbf{w} = \mathbf{0}$ for

$$\mathbf{w} = (\mathbf{s}_1^\star - \mathbf{s}_1^\diamond - (\mathbf{R}_{t^\star} \cdot \mathbf{s}_2^\star - \mathbf{R}_{t^\diamond} \cdot \mathbf{s}_2^\diamond) - \mathbf{R}_\mathbf{U}(\boldsymbol{\mu}^\star - \boldsymbol{\mu}^\diamond)) .$$

Quite obviously, $\mathbf{w}$ is small (we will quantify below). Less obviously, it is nonzero, except with negligible probability. We split our analysis into 4 different cases, corresponding to different types of forgeries $(\boldsymbol{\mu}^\star, t^\star, \mathbf{s}^\star) \neq (\boldsymbol{\mu}^\diamond, t^\diamond, \mathbf{s}^\diamond)$:

case 1 $\mathbf{s}_2^\star \neq \mathbf{s}_2^\diamond$. Even revealing $\mathbf{R}_\mathbf{U}$ and all $\mathbf{R}_{[i]}$ for $i > 0$, one has that $\mathbf{R}_{[0]} \cdot (\mathbf{s}_1^\star - \mathbf{s}_1^\diamond)$ conditioned on the knowledge of $\bar{\mathbf{A}}$ and $\mathbf{A}_{[0]} = \mathbf{A}\mathbf{R}_{[0]}$ contains at least $\Omega(n)$ bits of min-entropy, using Corollary 2. In particular the probability that $\mathbf{w} = \mathbf{0}$ is less than $2^{-\Omega(n)}$.

case 2 $\boldsymbol{\mu}^\star \neq \boldsymbol{\mu}^\diamond$. Even revealing all $\mathbf{R}_{[i]}$ for $i \geq 0$, one has that $\mathbf{R}_\mathbf{U} \cdot (\mathbf{s}_1^\star - \mathbf{s}_1^\diamond)$ conditioned on the knowledge of $\bar{\mathbf{A}}$ and $\mathbf{U} = \mathbf{A}\mathbf{R}_\mathbf{U}$ contains at least $\Omega(n)$ bits of min-entropy, using Corollary 2. In particular the probability that $\mathbf{w} = \mathbf{0}$ is less than $2^{-\Omega(n)}$.

case 3 $\mathbf{s}_1^\star = \mathbf{s}_1^\diamond$, $t^\star \neq t^\diamond$. Choose some i such that $t_{[i]}^\star \neq t_{[i]}^\diamond$. Even revealing $\mathbf{R}_\mathbf{U}$ and all $\mathbf{R}_{[j]}$ for $j \neq i$, one has that $\mathbf{R}_{[i]} \cdot \mathbf{s}_1^\star$ conditioned on the knowledge of $\bar{\mathbf{A}}$ and $\mathbf{A}_{[i]} = \mathbf{A}\mathbf{R}_{[i]}$ contains at least $\Omega(n)$ bits of min-entropy, using Corollary 2. So does $(t^\star_{[i]} - t^\diamond_{[i]}) \cdot \mathbf{R}_{[i]} \cdot \mathbf{s}^\star_1$ since $t^\star_{[i]} - t^\diamond_{[i]}$ is an invertible element of $\mathcal{R}_q$ (Corollary 1). In particular the probability that $\mathbf{w} = \mathbf{0}$ is less than $2^{-\Omega(n)}$.

case 4 $\mathbf{s}^\star_2 = \mathbf{s}^\diamond_2, \boldsymbol{\mu}^\star = \boldsymbol{\mu}^\diamond, t^\star = t^\diamond, \mathbf{s}^\star_1 \neq \mathbf{s}^\diamond_1$. In this case one notices that $\mathbf{w} = \mathbf{s}^\star_1 - \mathbf{s}^\diamond_1 \neq \mathbf{0}$ and concludes.

Size of the extracted SIS *solution* Because $\mathbf{s}^\star, \mathbf{s}^\diamond$ are valid signatures, $\|\mathbf{s}^\star\|, \|\mathbf{s}^\diamond\| \leq s\sqrt{m} \leq n^2 w \cdot \omega(\log n)^{3/2}$. Additionally $s_1(\mathbf{R}_t) \leq n^{3/2} \cdot \omega(\log n)$ for any tag $t \in \mathcal{T}$, as proved above, and $\|\boldsymbol{\mu}^\star\|, \|\boldsymbol{\mu}^\diamond\| \leq \sqrt{m} = O(\sqrt{nk})$ and $\mathbf{R}_\mathbf{U} \leq \sqrt{n} \cdot \omega(\log n)$. Combining all those bounds we obtain

$$\|\mathbf{w}\| \leq n^{7/2} \cdot \log n \cdot \omega(\log n)^{5/2}.$$

Success probability of the simulation The success probability ϵ' of the simulator is at least $(\epsilon - \epsilon/2)/|\mathcal{T}_{i^\star}| - 2^{-\Omega(n)}$ where

- ϵ is the success probability of the attacker,
- $\epsilon/2$ bounds the probability of a collision of tags,
- $1/|\mathcal{T}_{i^\star}|$ is the probability that the confined guess is correct, i.e., $t^\diamond_{\leq i^\star} = t^\star_{\leq i^\star}$, and
- $2^{-\Omega(n)}$ bounds the probability that the extracted SIS solution is zero.

Our choice of $i^\star$ (see confined guessing stage) guarantees that $2^{c_{i^\star}-1} < \frac{2Q^2}{\epsilon} \leq 2^{c_{i^\star}} = |\mathcal{T}_{i^\star}|$. We also have $c_{i^\star} \leq \alpha c^{i^\star} = c(\alpha c^{i^\star-1}) < c(c_{i^\star-1} + 1)$. Therefore $|\mathcal{T}_{i^\star}| = 2^{c_{i^\star}} \leq 2^{c \cdot (c_{i^\star-1}+1)} \leq \left(\frac{4Q^2}{\epsilon}\right)^c$. Overall the success probability of solving the SIS instance is at least

$$\epsilon' \geq \frac{\epsilon}{2}\left(\frac{\epsilon}{4Q^2}\right)^c - 2^{-\Omega(n)} = \Omega\left(\frac{\epsilon^{1+c}}{Q^{2c}}\right).$$

□

Acknowledgments. The authors wish to thank Sorina Ionica for helpful conversations, as well as the anonymous CRYPTO'14 reviewers for pointing out several issues in a preliminary version of this paper. This research was supported in part by the DARPA PROCEED program and NSF grant CNS-1117936. Opinions, findings and conclusions or recommendations expressed in this material are those of the author(s) and do not necessarily reflect the views of DARPA or NSF.

References

1. Agrawal, S., Boneh, D., Boyen, X.: Efficient lattice (H)IBE in the standard model. In: Gilbert, H. (ed.) EUROCRYPT 2010. LNCS, vol. 6110, pp. 553–572. Springer, Heidelberg (2010)
2. Agrawal, S., Boneh, D., Boyen, X.: Lattice basis delegation in fixed dimension and shorter-ciphertext hierarchical IBE. In: Rabin, T. (ed.) CRYPTO 2010. LNCS, vol. 6223, pp. 98–115. Springer, Heidelberg (2010)

3. Ajtai, M.: Generating hard instances of lattice problems. In: Complexity of Computations and Proofs, Quaderni di Matematica, vol. 13, pp. 1–32 (2004); Preliminary version in STOC 1996
4. Ajtai, M., Dwork, C.: A public-key cryptosystem with worst-case/average-case equivalence. In: Proceedings of STOC 1997, pp. 284–293. ACM (May 1997)
5. Arbitman, Y., Dogon, G., Lyubashevsky, V., Micciancio, D., Peikert, C., Rosen, A.: Swifftx: A proposal for the sha-3 standard. Submission to NIST (2008), http://csrc.nist.gov/groups/ST/hash/sha-3/Round1/documents/SWIFFTX.zip
6. Bai, S., Galbraith, S.D.: An improved compression technique for signatures based on learning with errors. In: CT-RSA, pp. 28–47 (2014)
7. Böhl, F., Hofheinz, D., Jager, T., Koch, J., Seo, J.H., Striecks, C.: Practical signatures from standard assumptions. In: Johansson, T., Nguyen, P.Q. (eds.) EUROCRYPT 2013. LNCS, vol. 7881, pp. 461–485. Springer, Heidelberg (2013)
8. Böhl, F., Hofheinz, D., Jager, T., Koch, J., Striecks, C.: Confined guessing: New signatures from standard assumptions. Cryptology ePrint Archive, Report 2013/171 (2013), http://eprint.iacr.org/2013/171
9. Boyen, X.: Lattice mixing and vanishing trapdoors: A framework for fully secure short signatures and more. In: Nguyen, P.Q., Pointcheval, D. (eds.) PKC 2010. LNCS, vol. 6056, pp. 499–517. Springer, Heidelberg (2010)
10. Brakerski, Z., Kalai, Y.T.: A framework for efficient signatures, ring signatures and identity based encryption in the standard model. Cryptology ePrint Archive, Report 2010/086 (2010), http://eprint.iacr.org/2010/086
11. Cash, D., Hofheinz, D., Kiltz, E., Peikert, C.: Bonsai trees, or how to delegate a lattice basis. Journal of Cryptology 25(4), 601–639 (2012)
12. Ducas, L., Durmus, A., Lepoint, T., Lyubashevsky, V.: Lattice signatures and bimodal gaussians. In: Canetti, R., Garay, J.A. (eds.) CRYPTO 2013, Part I. LNCS, vol. 8042, pp. 40–56. Springer, Heidelberg (2013)
13. Gentry, C., Peikert, C., Vaikuntanathan, V.: Trapdoors for hard lattices and new cryptographic constructions. In: Ladner, R.E., Dwork, C. (eds.) 40th ACM STOC, Victoria, British Columbia, Canada, May 17–20, pp. 197–206. ACM Press (2008)
14. Güneysu, T., Lyubashevsky, V., Pöppelmann, T.: Practical lattice-based cryptography: A signature scheme for embedded systems. In: Prouff, E., Schaumont, P. (eds.) CHES 2012. LNCS, vol. 7428, pp. 530–547. Springer, Heidelberg (2012)
15. Hoffstein, J., Pipher, J., Schanck, J., Silverman, J.H., Whyte, W.: Practical signatures from the partial fourier recovery problem. Cryptology ePrint Archive, Report 2013/757 (2013), http://eprint.iacr.org/2013/757
16. Hoffstein, J., Pipher, J., Silverman, J.H.: NTRU: A ring-based public key cryptosystem. In: Buhler, J.P. (ed.) ANTS 1998. LNCS, vol. 1423, pp. 267–288. Springer, Heidelberg (1998)
17. Hohenberger, S., Waters, B.: Realizing hash-and-sign signatures under standard assumptions. In: Joux, A. (ed.) EUROCRYPT 2009. LNCS, vol. 5479, pp. 333–350. Springer, Heidelberg (2009)
18. Hohenberger, S., Waters, B.: Short and stateless signatures from the RSA assumption. In: Halevi, S. (ed.) CRYPTO 2009. LNCS, vol. 5677, pp. 654–670. Springer, Heidelberg (2009)
19. Lamport, L.: Constructing digital signatures from a one-way function. Technical Report SRI-CSL-98, SRI International Computer Science Laboratory (October 1979)
20. Lidl, R., Niederreiter, H.: Finite Fields. In: Encyclopedia of Mathematics and its Applications, vol. 20, Addison-Wesley, Reading (1983)

21. Lyubashevsky, V.: Fiat-shamir with aborts: Applications to lattice and factoring-based signatures. In: Matsui, M. (ed.) ASIACRYPT 2009. LNCS, vol. 5912, pp. 598–616. Springer, Heidelberg (2009)

22. Lyubashevsky, V.: Lattice signatures without trapdoors. In: Pointcheval, D., Johansson, T. (eds.) EUROCRYPT 2012. LNCS, vol. 7237, pp. 738–755. Springer, Heidelberg (2012)

23. Lyubashevsky, V., Micciancio, D.: Generalized compact knapsacks are collision resistant. In: Bugliesi, M., Preneel, B., Sassone, V., Wegener, I. (eds.) ICALP 2006. LNCS, vol. 4052, pp. 144–155. Springer, Heidelberg (2006)

24. Lyubashevsky, V., Micciancio, D.: Asymptotically efficient lattice-based digital signatures. In: Canetti, R. (ed.) TCC 2008. LNCS, vol. 4948, pp. 37–54. Springer, Heidelberg (2008)

25. Lyubashevsky, V., Micciancio, D.: Asymptotically efficient lattice-based digital signatures. In: Canetti, R. (ed.) TCC 2008. LNCS, vol. 4948, pp. 37–54. Springer, Heidelberg (2008)

26. Lyubashevsky, V., Micciancio, D., Peikert, C., Rosen, A.: SWIFFT: A modest proposal for FFT hashing. In: Nyberg, K. (ed.) FSE 2008. LNCS, vol. 5086, pp. 54–72. Springer, Heidelberg (2008)

27. Lyubashevsky, V., Peikert, C., Regev, O.: A toolkit for ring-LWE cryptography. In: Johansson, T., Nguyen, P.Q. (eds.) EUROCRYPT 2013. LNCS, vol. 7881, pp. 35–54. Springer, Heidelberg (2013)

28. Micciancio, D.: Generalized compact knapsacks, cyclic lattices, and efficient one-way functions. Computational Complexity 16(4), 365–411 (2007); Preliminary version in FOCS 2002

29. Micciancio, D., Peikert, C.: Trapdoors for lattices: Simpler, tighter, faster, smaller. In: Pointcheval, D., Johansson, T. (eds.) EUROCRYPT 2012. LNCS, vol. 7237, pp. 700–718. Springer, Heidelberg (2012)

30. Micciancio, D., Regev, O.: Worst-case to average-case reductions based on Gaussian measure. SIAM Journal on Computing 37(1), 267–302 (2007); Preliminary version in FOCS 2004

31. Peikert, C., Rosen, A.: Efficient collision-resistant hashing from worst-case assumptions on cyclic lattices. In: Halevi, S., Rabin, T. (eds.) TCC 2006. LNCS, vol. 3876, pp. 145–166. Springer, Heidelberg (2006)

32. Peikert, C., Rosen, A.: Lattices that admit logarithmic worst-case to average-case connection factors. In: Proceedings of STOC, pp. 478–487. ACM (June 2007)

33. Regev, O.: On lattices, learning with errors, random linear codes, and cryptography. Journal of ACM 56(6), 34 (2009); Preliminary version in STOC 2005

34. Rompel, J.: One-way functions are necessary and sufficient for secure signatures. In: 22nd ACM STOC, Baltimore, Maryland, USA, May 14-16, pp. 387–394. ACM Press (1990)

35. Rückert, M.: Strongly unforgeable signatures and hierarchical identity-based signatures from lattices without random oracles. In: Sendrier, N. (ed.) PQCrypto 2010. LNCS, vol. 6061, pp. 182–200. Springer, Heidelberg (2010)

36. Seo, J.H.: Short signatures from Diffie-Hellman: Realizing short public key. Cryptology ePrint Archive, Report 2012/480 (2012), http://eprint.iacr.org/2012/480

37. Vershynin, R.: Introduction to the non-asymptotic analysis of random matrices. CoRR, abs/1011.3027 (2010), http://www-personal.umich.edu/~romanv/papers/non-asymptotic-rmt-plain.pdf

A   Missing Proofs

Fact 2. For a vector $\mathbf{v} \in R^n$ over ring R, let $\mathrm{Diag}(\mathbf{v})$ denotes the diagonal matrice with entries $v_1 \ldots v_n$. Notice that the component wise product of two vectors $\mathbf{f} \odot \mathbf{g}$ can be written as the matrix-vector product $\mathrm{Diag}(\mathbf{f}) \cdot \mathbf{g}$. This gives the identity $\sigma(f \cdot g) = \mathrm{Diag}(\sigma(f)) \cdot \sigma(g)$ for $f, g \in \mathcal{R}$ with $\sigma : \mathcal{R} \to \mathbb{C}^n$ denoting the canonical embedding:

$$\sigma : f \in \mathcal{R} \mapsto (f(\omega_1), \ldots f(\omega_\ell)) \in \mathbb{C}^n \text{ where } \omega_\ell = e^{(2\ell-1)\imath\pi/n}$$

. Let $\mathbf{R} = (r_{i,j}) \leftarrow D_{\mathcal{R},s}^{w \times k}$; and set

$$\mathbf{D} = \begin{bmatrix} \mathbf{D}_{1,1} & \cdots & \mathbf{D}_{1,k} \\ \vdots & & \vdots \\ \mathbf{D}_{w,1} & \cdots & \mathbf{D}_{w,k} \end{bmatrix} \in \mathbb{C}^{nw \times nk} \quad \text{and} \quad \mathbf{D}_{i,j} = \mathrm{Diag}(\sigma(r_{i,j})) \in \mathbb{C}^{n \times n}.$$

We extend the canonical embedding $\sigma : \mathcal{R} \to \mathbb{C}^n$ to vectors in $\mathcal{R}^d$ as its componentwise application; $\sigma(\mathbf{v}) = (\sigma(v_1), \ldots \sigma(v_k)) \in \mathbb{C}^{nk}$. With this notation, we have $\sigma(\mathbf{R} \cdot \mathbf{v}) = \mathbf{D} \cdot \sigma(\mathbf{v})$; and because the canonical embedding σ is a scaled isometry, we have $s_1(\mathbf{R}) = s_1(\mathbf{D})$.

Permuting rows and column, $\mathbf{D}$ can be rewritten as the block-diagonal matrix $\mathbf{B} = \mathrm{Diag}(\mathbf{B}_1, \ldots \mathbf{B}_n) \in \mathbb{C}^{nw \times nk}$, $\mathbf{B}_\ell \in \mathbb{C}^{w \times k}$ where the coefficients of $\mathbf{B}_\ell$ are all the embeddings $\sigma_\ell(r_{i,j}) = r_{i,j}(\omega_\ell)$ for $(i,j) \in \{1 \ldots w\} \times \{1 \ldots k\}$. The coefficients of $\mathrm{Re}(\mathbf{B}_\ell)$ (the real part of $\mathbf{B}_\ell$) are independent and sub-gaussian of parameter $s\sqrt{n}$. Indeed

$$\mathrm{Re}(\mathbf{B}_\ell) = \sum_{k=0}^{n-1} \mathrm{Re}(\omega_\ell^k) \cdot (r_{i,j})_k$$

where the $(r_{i,j})_k$ are independent and sub-gaussian of parameter s while $|\mathrm{Re}(\omega_\ell^k)| \leq 1$. Therefore by Lemma 1

$$s_1(\mathrm{Re}(\mathbf{B}_\ell)) \leq s\sqrt{n} \cdot O(\sqrt{w} + \sqrt{k} + \omega(\sqrt{\log n}))$$

with overwhelming probability. The same results hold for the imaginary part $\mathrm{Im}(\mathbf{B}_\ell)$ of $\mathbf{B}_\ell$. We conclude

$$s_1(\mathbf{D}) \leq s_1(\mathbf{B}) \leq \max_\ell s_1(\mathbf{B}_\ell) \leq \max_\ell \sqrt{s_1(\mathrm{Re}(\mathbf{B}_\ell))^2 + s_1(\mathrm{Im}(\mathbf{B}_\ell))^2}$$

$$\leq s\sqrt{n} \cdot O(\sqrt{w} + \sqrt{k} + \omega(\sqrt{\log n})).$$

□

Lemma 4. The proof is adapted from [13, Lemma 5.3]. Consider the lattice $\Lambda(\mathbf{A}^\top)$ spanned by the columns of $\mathbf{A}^\top$ and the vectors of $q\mathbb{Z}^{nw}$; it is the (scaled) dual of $\Lambda^\perp(\mathbf{A})$. We will first show that the minimal distance $\lambda_1^\infty(\Lambda(\mathbf{A}^\top))$ is at least $q/12$ with overwhelming probability, and conclude using [13, Lemma 2.6] that $\eta_\epsilon(\Lambda^\perp(\mathbf{A})) \leq \omega(\sqrt{\ln nw})$ for some negligible function $\epsilon(n)$.

Recall that the irreducible factors of $\Phi_{2n}(X) \bmod 3$ are $P_1(X) = X^{n/2} + X^{n/4} - 1$ and $P_2(X) = X^{n/2} - X^{n/4} - 1$. Setting $\mathfrak{p}_1 = (P_1(X))$, $\mathfrak{p}_2 = (P_2(X))$ the nonzero ideals of $\mathcal{R}_q$ are exactly $\mathfrak{p}_1, 3\mathfrak{p}_1 \ldots 3^{k-1}\mathfrak{p}_1$; $\mathfrak{p}_2, 3\mathfrak{p}_2 \ldots 3^{k-1}\mathfrak{p}_2$ and $(1), (3), (3^2), \ldots (3^{k-1})$.

Now, fix some $x \in \mathcal{R} \setminus \{0\}$, et set $\mathfrak{I} = (x)$, it is one of the nonzero ideal listed above. Let $r \geq n/2$ denotes its rank. Our goal is to prove, that over the randomness of $\mathbf{A} \in \mathcal{R}^{1 \times w}$, the probability that $\mathbf{A}x$ falls in in the hypercube $\mathcal{C}^w = \{\mathbf{v} \in \mathcal{R}^w | \|\mathbf{v}\|_\infty < q/12\}$ is less than $2^{-O(wr)}$. Because x is a generator of $\mathfrak{I}$ the distribution of $\mathbf{A}x$ is uniform over $\mathfrak{I}^w$. We proceed by bounding the ratio $|\mathcal{C} \cap \mathfrak{I}|/|\mathfrak{I}|$.

Case 1: ($\mathfrak{I} = (3^h)$ for $h \in \{0 \ldots k-1\}$). Observe that $|\mathcal{C} \cap \mathfrak{I}| \leq |\{3^h \mathbb{Z} \cap (-q/12, q/12)\}|^n \leq \lceil 3^{k-h}/6 \rceil^n$; which leads to

$$|\mathcal{C} \cap \mathfrak{I}|/|\mathfrak{I}| \leq \left(\frac{3^{k-h}/6 + 1}{3^{k-h}}\right)^n \leq \left(\frac{1}{6} + \frac{1}{3^{k-h}}\right)^n \leq 2^{-n}.$$

Case 2: ($\mathfrak{I} = 3^h \mathfrak{p}_i$ for $h \in \{0 \ldots k-1\}$). Start by noting that any element e of $\mathfrak{I}$ can be uniquely written $e = P_i(X) \cdot s$ where $s = \sum_{i=0}^{n/2-1} s_i X^i$ is a polynomial and of degree strictly less than $n/2$ in the ideal (3^h) of $\mathcal{R}$. Also note that $\|e\|_\infty \leq q/12$ implies $\|s\|_\infty \leq q/12$, indeed for $i \in \{0 \ldots n/4-1\}$ we have $e_i = -s_i$ and for $i \in \{n/4 \ldots n/2-1\}$ we have $e_{i+n/2} = s_i$. Using a similar counting argument on valid values of s we derive

$$|\mathcal{C} \cap \mathfrak{I}|/|\mathfrak{I}| \leq \left(\frac{3^{k-h}/6 + 1}{3^{k-h}}\right)^{n/2} \leq \left(\frac{1}{6} + \frac{1}{3^{k-h}}\right)^{n/2} \leq 2^{-n/2}.$$

Taking the union bound over all nonzero x we conclude that $\lambda_1^\infty(\Lambda(\mathbf{A}^\top)) \geq q/12$ except with probability $q^n \cdot 2^{-nw/2} \leq 2^{-\Omega(n)}$. □

Corollary 2. Without loss of generality assume that $\mathbf{v}_1 \neq 0$. Applying the previous Lemma 4 on $\sum_{i>2} \mathbf{a}_i \cdot \mathbf{x}_i$, the knowledge of $\sum_i \mathbf{a}_i \cdot \mathbf{x}_i = \mathbf{a}_1 \cdot \mathbf{x}_1 + \sum_{i \geq 2} \mathbf{a}_i \cdot \mathbf{x}_i$ reveals only negligible any information about $\mathbf{x}_1$. Also note that $\mathbf{x}_1 \bmod 3$ is negligibly close to uniform ($\eta_\epsilon(3\mathbb{Z}) \leq \omega(\sqrt{\ln n})$ for some negligible function $\epsilon(n)$).

Setting $\mathfrak{I} = (\mathbf{v}_1) \neq (0)$ we deduce that $\mathbf{v}_1 \cdot \mathbf{x}_1 \bmod 3\mathfrak{I}$ is almost uniform in $\mathfrak{I}/3\mathfrak{I}$. Recall from the previous proof that the only nonzero ideals of $\mathcal{R}_q$ are exactly $\mathfrak{p}_1, 3\mathfrak{p}_1 \ldots 3^{k-1}\mathfrak{p}_1$; $\mathfrak{p}_2, 3\mathfrak{p}_2 \ldots 3^{k-1}\mathfrak{p}_2$ and $(1), (3), (3^2), \ldots (3^{k-1})$ where both $\mathfrak{p}_1$ and $\mathfrak{p}_2$ are ideals of rank $n/2$. This implies that $|\mathfrak{I}/3\mathfrak{I}| = 3^{n/2}$ or 3^n. We conclude that $\mathbf{v}_1 \cdot \mathbf{x}_1$ has at least $\Omega(n)$ bits of entropy and so has $\sum_i \mathbf{v}_i \cdot \mathbf{x}_i$. □

New and Improved
Key-Homomorphic Pseudorandom Functions

Abhishek Banerjee[*] and Chris Peikert[**]

School of Computer Science, Georgia Institute of Technology, Atlanta, GA, USA

Abstract. A *key-homomorphic* pseudorandom function (PRF) family $\{F_s \colon D \to R\}$ allows one to efficiently compute the value $F_{s+t}(x)$ given $F_s(x)$ and $F_t(x)$. Such functions have many applications, such as distributing the operation of a key-distribution center and updatable symmetric encryption. The only known construction of key-homomorphic PRFs without random oracles, due to Boneh *et al.* (CRYPTO 2013), is based on the learning with errors (LWE) problem and hence on worst-case lattice problems. However, the security proof relies on a very strong LWE assumption (i.e., very large approximation factors), and hence has quite inefficient parameter sizes and runtimes.

In this work we give new constructions of key-homomorphic PRFs that are based on much weaker LWE assumptions, are much more efficient in time and space, and are still highly parallel. More specifically, we improve the LWE approximation factor from exponential in the input length to exponential in its *logarithm* (or less). For input length λ and 2^λ security against known lattice algorithms, we improve the key size from λ^3 to λ bits, the public parameters from λ^6 to λ^2 bits, and the runtime from λ^7 to $\lambda^{\omega+1}$ bit operations (ignoring polylogarithmic factors in λ), where $\omega \in [2, 2.373]$ is the exponent of matrix multiplication. In addition, we give even more efficient ring-LWE-based constructions whose key sizes, public parameters, and *incremental* runtimes on consecutive inputs are all *quasi-linear* $\tilde{O}(\lambda)$, which is optimal up to polylogarithmic factors. To our knowledge, these are the first *low-depth* PRFs (whether key homomorphic or not) enjoying any of these efficiency measures together with nontrivial proofs of 2^λ security under any conventional assumption.

1 Introduction

A *pseudorandom function (PRF)* family [GGM84] $\mathcal{F} = \{F_s \colon D \to R\}$ is a finite set of (deterministic) functions with common domain D and range R (both

---

[*] Research supported by the second author's grants.
[**] This material is based upon work supported by the National Science Foundation under CAREER Award CCF-1054495, by the US-Israel Binational Science Foundation Grant 2010296, by the Alfred P. Sloan Foundation, and by the Defense Advanced Research Projects Agency (DARPA) and the Air Force Research Laboratory (AFRL) under Contract No. FA8750-11-C-0098. The views expressed are those of the authors and do not necessarily reflect the official policy or position of the National Science Foundation, the BSF, the Sloan Foundation, DARPA or the U.S. Government.

J.A. Garay and R. Gennaro (Eds.): CRYPTO 2014, Part I, LNCS 8616, pp. 353–370, 2014.

finite), for which a randomly chosen $F_s \leftarrow \mathcal{F}$ cannot be efficiently distinguished from a uniformly random function $U \colon D \to R$, given adaptive oracle access. The index s of function F_s is often called its *(secret) key* or *seed*. The family $\mathcal{F}$ is *key homomorphic* if the set of keys has a group structure and if there is an efficient algorithm that, given $F_s(x)$ and $F_t(x)$ (but not s or t), outputs $F_{s+t}(x)$.

Naor, Pinkas, and Reingold [NPR99] constructed, in the random oracle model, a very simple key-homomorphic PRF family based on the decisional Diffie-Hellman problem, and gave applications like distributing the operation of a Key Distribution Center. Recently, Boneh *et al.* [BLMR13] constructed the first key-homomorphic PRFs *without* random oracles, and described many more applications (all of which are very efficient in their use of the PRF), including symmetric-key proxy re-encryption, updatable encryption, and PRFs secure against related-key attacks (cf. [BC10, LMR14]). The construction of Boneh *et al.* is based on the (appropriately parameterized) *learning with errors* (LWE) problem [Reg05], and builds upon ideas used in the *non*-key-homomorphic LWE-based PRFs of Banerjee, Peikert, and Rosen [BPR12].

One drawback of the construction and proof from [BLMR13] is its rather strong LWE assumption and, by consequence, large parameters and runtimes. For example, to obtain a PRF of input length λ with exponential 2^λ provable security against known lattice attacks, the secret keys and public parameters respectively need to be at least λ^3 and λ^6 bits, and the runtime to evaluate the function is at least λ^7 bit operations (to produce λ^2 output bits), not counting some polylogarithmic $\log^{O(1)} \lambda$ factors. It is worth noting that among the several LWE-based PRFs given in [BPR12], the most highly parallelizable "direct" construction (which can be implemented in $TC^0 \subseteq NC^1$) relies on roughly the same strong assumptions and so has similarly low efficiency as the one from [BLMR13]. However, the synthesizer-based construction (in $TC^1 \subseteq NC^2$) and sequential GGM-based one from [BPR12] can be proved secure under much weaker LWE assumptions, and hence can have much better parameters and runtimes. A natural question, therefore, is whether there exist *key-homomorphic* PRFs with similar security and efficiency characteristics.

Our results. In this work we answer the above question in the affirmative, by giving new constructions of key-homomorphic PRFs that have substantially better efficiency, and still enjoy very high parallelism. As compared with [BLMR13], we improve the key size from λ^3 to λ bits, the public parameters from λ^6 to λ^2 bits, and the runtime from λ^7 to $\lambda^{\omega+1}$ bit operations (always omitting $\log^{O(1)} \lambda$ factors), where $\omega \in [2, 2.373]$ is the exponent of matrix multiplication. Functions having these parameters can be implemented in $TC^1 \subseteq NC^2$, though seemingly not in TC^0 or NC^1.

We also give even more efficient key-homomorphic PRFs based on the *ring-LWE* problem [LPR10, LPR13]. Compared with the ring-based analogue of [BLMR13], and again ignoring $\log^{O(1)} \lambda$ factors, here our keys and public parameters are only λ bits (improving upon λ^3 and λ^4, respectively), and the runtime is only λ^2 bit operations to produce λ output bits (from λ^5 to produce λ^2). In addition, the *incremental* computation of our PRF on successive inputs (e.g., in a counter-like mode) has runtime only λ. See Figure 1 for a full comparison with [BPR12, BLMR13].

To our knowledge, ours are the first *low-depth* PRFs (whether key homomorphic or not) having nontrivial proofs of exponential 2^λ security under any conventional assumption along with *quasi-optimal* $\tilde{O}(\lambda)$ key sizes or incremental runtimes, or quasilinear $\tilde{O}(\lambda)$ nonincremental runtime per output bit. For example, the GGM construction [GGM84] can have small keys and quasilinear nonincremental runtime per output bit (using a quasi-optimal PRG), but it is highly sequential. The Naor-Reingold constructions [NR95, NR97], which are highly parallel, have at least quadratic λ^2 key sizes and runtime per output bit, even assuming exponential security of the underlying hard problems. And factoring-based constructions [NRR00] fare much worse due to subexponential-time factoring algorithms.

In their parallelism and underlying LWE assumptions, our functions are qualitatively very similar to the synthesizer- and GGM-based ones from [BPR12] (see Figure 1); however, the constructions and proofs are completely different. Instead, our construction can be seen as a substantial generalization of the one of Boneh *et al.* [BLMR13], in that theirs is an instantiation of ours with a linear-depth "left spine" tree. By contrast, our construction can be securely instantiated with *any* binary tree, thanks to a new proof technique that may be of use elsewhere. The shape of the tree determines the final parameters and parallelism of the resulting function: roughly speaking, its "left depth" determines the strength of the LWE assumption in the proof, while its "right depth" determines its parallelism. Interestingly, a complete binary tree turns out to be *very far from optimal* for the parameters we care about. Optimal trees can be found efficiently using dynamic programming, and provide input lengths that are roughly the *square* of those yielded by complete binary trees. This is all discussed in detail in the next section, where we present and analyze our construction.

Other related work. Our construction is reminiscent of those from several recent works on fully homomorphic encryption, attribute-based encryption, and garbled circuits, e.g., [GSW13, BV14, BGG+14]. In particular, these works obtain relatively good LWE assumptions and parameters by appropriately scheduling "bit decomposition" operations to ensure small noise growth, usually at the expense of increased sequentiality. Our work also falls within this theme, though our proof techniques are completely different.

Organization. In Section 2 we give our construction and a detailed analysis of its security and efficiency. In Section 3 we give the proof of the security theorem, first providing an overview of the key ideas in Section 3.1, and giving the formal proof in Section 3.3 (after recalling some necessary technical background in Section 3.2).

Reference	KH?	Expan	Sequen	Key		Params		Time/Out		Out	
this work	Y	1	$\lambda - 1$	λ	$[\lambda]$	λ^2	$[\lambda]$	λ^ω	$[\lambda]$	λ	$[\lambda]$
this work	Y	$\log_4 \lambda$	$\log_4 \lambda$	λ	$[\lambda]$	λ^2	$[\lambda]$	λ^ω	$[\lambda]$	λ	$[\lambda]$
[BLMR13]	Y	$\lambda - 1$	1	λ^3	$[\lambda^3]$	λ^6	$[\lambda^4]$	λ^5	$[\lambda^3]$	λ^2	$[\lambda^2]$
[BPR12, GGM]	N	1	λ	λ	$[\lambda]$	λ^2	$[\lambda]$	λ^2	$[\lambda]$	λ	$[\lambda]$
[BPR12, synth]	N	$\log_2 \lambda$	$\log_2 \lambda$	λ^3	$[\lambda^2]$	0	$[0]$	$\lambda^{\omega-1}$	$[\lambda]$	λ^2	$[\lambda]$
[BPR12, direct]	N	λ	1	λ^5	$[\lambda^3]$	0	$[0]$	λ^4	$[\lambda^2]$	λ^2	$[\lambda^2]$

Fig. 1. Example instantiations of our key-homomorphic PRF (for input length λ and provable 2^λ security against the best known lattice algorithms) as compared with prior lattice-based PRFs. "KH" denotes whether the construction is key homomorphic, while "Expan" and "Sequen" are respectively the expansion and sequentiality (as defined in Equations (2.4), (2.7)) of the tree T used in the instantiation (or, for prior constructions, their close analogues). Omitting polylogarithmic $\log^{O(1)} \lambda$ factors, "Key" and "Params" are respectively the bit lengths of the secret key and public parameters; "Time/Out" is the best known runtime (in bit operations) per output bit, where $\omega \in [2, 2.373]$ is the exponent of matrix multiplication; and "Out" is the output length in bits. The quantities in brackets refer to the ring-based construction given in Section 2.4.

2   Construction and Analysis

In this section we define and analyze our key-homomorphic PRF, and compare it with prior LWE-based constructions. The construction involves various parameters (e.g., matrix dimension n, modulus q, tree T) which are all chosen so that the algorithms are polynomial-time in the security parameter λ. As in [BLMR13], we work in a model where the PRF family is defined with respect to some *random public parameters* that are known to all parties, including the adversary. These parameters may be generated by a trusted party, or by the user along with the secret key.

We first recall some standard background. For an integer modulus $q \geq 1$, let $\mathbb{Z}_q = \mathbb{Z}/q\mathbb{Z}$ denote the quotient ring of integers modulo q. For an integer $p \leq q$, define the modular "rounding" function $\lfloor \cdot \rceil_p \colon \mathbb{Z}_q \to \mathbb{Z}_p$ as $\lfloor x \rceil_p = \lfloor \frac{p}{q} \cdot x \rceil$, and extend it coordinate-wise to vectors and matrices over $\mathbb{Z}_q$. Let $\ell = \lceil \log q \rceil$ and define the "gadget" (column) vector

$$\mathbf{g} = (1, 2, 4, \ldots, 2^{\ell-1}) \in \mathbb{Z}_q^\ell,$$

and the (deterministic) "binary decomposition" function $\mathbf{g}^{-1} \colon \mathbb{Z}_q \to \{0,1\}^\ell$ as follows: identifying each $a \in \mathbb{Z}_q$ with its integer residue in $\{0, \ldots, q-1\}$, let $\mathbf{g}^{-1}(a) = (x_0, x_1, \ldots, x_{\ell-1}) \in \{0,1\}^\ell$ where $a = \sum_{i=0}^{\ell-1} x_i 2^i$ is the binary representation of a. Note that by definition, $\langle \mathbf{g}, \mathbf{g}^{-1}(a)\rangle = a$ for all $a \in \mathbb{Z}_q$, which explains our choice of notation.[1]

Similarly, for vectors and matrices over $\mathbb{Z}_q$ we define the function $\mathbf{G}^{-1}\colon \mathbb{Z}_q^{n\times m} \to \{0,1\}^{n\ell \times m}$ by applying $\mathbf{g}^{-1}$ entry-wise. Notice that for all $\mathbf{A} \in \mathbb{Z}_q^{n\times m}$ we have

$$\mathbf{G}\cdot\mathbf{G}^{-1}(\mathbf{A}) = \mathbf{A}, \quad \text{where} \quad \mathbf{G} = \mathbf{g}^t \otimes \mathbf{I}_n = \operatorname{diag}(\mathbf{g}^t,\dots,\mathbf{g}^t) \in \mathbb{Z}_q^{n\times n\ell} \quad (2.1)$$

is the block matrix with n copies of $\mathbf{g}^t$ as diagonal blocks, and zeros elsewhere.

For a full (but not necessarily complete) binary tree T—i.e., one in which every non-leaf node has two children—let $|T|$ denote the number of its leaves. If $|T| \geq 1$ (i.e., T is not the empty tree), let $T.l, T.r$ respectively denote the left and right subtrees of T (which may be empty trees).

We now define our function families.

Definition 2.1. *Given matrices* $\mathbf{A}_0, \mathbf{A}_1 \in \mathbb{Z}_q^{n\times n\ell}$ *and a full binary tree* T *of at least one node, define the function* $\mathbf{A}_T\colon \{0,1\}^{|T|} \to \mathbb{Z}_q^{n\times n\ell}$ *recursively as*

$$\mathbf{A}_T(x) = \begin{cases} \mathbf{A}_x & \text{if } |T| = 1 \\ \mathbf{A}_{T.l}(x_l)\cdot\mathbf{G}^{-1}(\mathbf{A}_{T.r}(x_r)) & \text{otherwise,} \end{cases} \quad (2.2)$$

where in the second case we parse $x = x_l \| x_r$ *for* $x_l \in \{0,1\}^{|T.l|}, x_r \in \{0,1\}^{|T.r|}$.

Construction 2.1 (Key-Homomorphic PRF). The function family

$$\mathcal{F}_{\mathbf{A}_0,\mathbf{A}_1,T,p} = \left\{ F_{\mathbf{s}}\colon \{0,1\}^{|T|} \to \mathbb{Z}_p^{n\ell} \right\}$$

is parameterized by matrices $\mathbf{A}_0, \mathbf{A}_1 \in \mathbb{Z}_q^{n\times n\ell}$, a binary tree T, and a modulus $p \leq q$, which may all be considered public parameters. A member of the family is indexed by some $\mathbf{s} \in \mathbb{Z}_q^n$, and is defined as

$$F_{\mathbf{s}}(x) := \left\lfloor \mathbf{s}^t \cdot \mathbf{A}_T(x) \right\rceil_p. \quad (2.3)$$

For security based on LWE, we take $\mathbf{A}_0, \mathbf{A}_1$ and the secret key $\mathbf{s}$ to be uniformly random over $\mathbb{Z}_q$; see Theorem 2.1 below for a formal security statement. Similarly to LWE, it may also be possible to prove security when the entries of $\mathbf{s}$ are drawn from the LWE error distribution (see [ACPS09]). However, most applications of key-homomorphic PRFs need to use uniformly random secret keys anyway, so we do not pursue this question further.

Because rounding is nearly linear, i.e., $\lfloor a + b \rceil_p = \lfloor a \rceil_p + \lfloor b \rceil_p + e$ for some $e \in \{0, \pm 1\}$, it is easy to see that the family $\mathcal{F}_{\mathbf{A}_0,\mathbf{A}_1,T,p}$ defined above is "almost"

---

[1] These are just particular definitions of $\mathbf{g}, \mathbf{g}^{-1}$ that we fix for convenience. Our constructions and proofs only require that $\mathbf{g}^{-1}$ be deterministic, and that $\mathbf{g}^{-1}(a)$ be a "short" integer vector such that $\langle \mathbf{g}, \mathbf{g}^{-1}(a)\rangle = a$ for all $a \in \mathbb{Z}_q$. Alternatives include using a signed ternary decomposition, or a larger (or mixed-radix) base; the bounds in the security theorem are easily adapted to such choices.

additively key homomorphic, as defined in [BLMR13]. That is, for any keys $F_\mathbf{s}, F_\mathbf{t}$ in the family, we have

$$F_{\mathbf{s}+\mathbf{t}}(x) = F_\mathbf{s}(x) + F_\mathbf{t}(x) + \mathbf{e}^t,$$

where $\|\mathbf{e}\|_\infty \leq 1$. As long as the entries of the error term $\mathbf{e}$ are sufficiently smaller than the output modulus p, this near-homomorphism is sufficient for all the applications described in [BLMR13], and for obtaining security against related-key attacks [LMR14].

Notice that the vast majority of the cost of computing $F_\mathbf{s}(x)$ is in computing $\mathbf{A}_T(x)$, which can done "publicly" without any knowledge of the secret key $\mathbf{s}$.[2] This property can be very important for the efficiency of certain applications, such as the *homomorphic* evaluation of $F_\mathbf{s}$ given an encryption of $\mathbf{s}$. In addition, notice that if $\mathbf{A}_T(x)$ has been computed and all the intermediate matrices saved, then $\mathbf{A}_T(x')$ can be *incrementally* computed much more efficiently for an x' that differs from x in just a single bit. Specifically, one only needs to recompute the matrices for the internal nodes of T on the path from the leaf corresponding to the changed bit to the root. As in [BPR12], this can significantly speed up successive evaluations of $F_\mathbf{s}$ on related inputs, e.g., in a counter-like mode using a Gray code.

Relation to [BLMR13]. Our key-homomorphic PRF can be viewed as a substantial generalization of the one of Boneh *et al.* [BLMR13]. Specifically, their construction can be obtained from ours by instantiating it with a tree T that consists of a "left spine" with leaves for all its right children. Because all the right subtrees are just leaves, the only matrices ever decomposed with $\mathbf{G}^{-1}$ are $\mathbf{A}_0$ and $\mathbf{A}_1$. Therefore, we can replace them in the public parameters by the binary matrices $\mathbf{B}_b = \mathbf{G}^{-1}(\mathbf{A}_b)$, yielding the construction $F_\mathbf{r}(x) = \lfloor \mathbf{r}^t \cdot \prod_{i=1}^{|x|} \mathbf{B}_{x_i} \rceil_p$ from [BLMR13].[3]

The use of a "left-spine" tree T (as in [BLMR13]) yields an instantiation which is *maximally parallel*—in our language (defined below), it has *sequentiality* $s(T) = 1$. The major drawback is that it also has maximal *expansion* $e(T) = |T| - 1$. In our security theorem (Theorem 2.1 below), the LWE approximation factor and modulus q grow *exponentially* with $e(T)$, so using a tree with large expansion leads to a very strong hardness assumption, and therefore large secret keys and public parameters. By contrast, using trees T with better expansion-sequentiality tradeoffs allows us obtain much better key sizes and efficiency. See the discussion in the following subsections and Figure 1 for further details.

---

[2] For a few choices of the tree T, it can be faster to compute $\mathbf{s}^t \cdot \mathbf{A}_T(x)$ left-to-right without explicitly computing $\mathbf{A}_T(x)$, but such trees are rare and yield bad parameters.

[3] Here we have ignored the small detail that in our construction, the matrix $\mathbf{A}_{x_1}$ corresponding to the leftmost leaf in the tree is not decomposed, so our instantiation is actually $F_\mathbf{s}(x) = \lfloor \mathbf{s}^t \cdot \mathbf{A}_{x_1} \cdot \prod_{i=2}^{|x|} \mathbf{B}_{x_i} \rceil$. However, it is easy to verify that in the construction of [BLMR13], the secret key may be of the form $\mathbf{r}^t = \mathbf{s}^t \mathbf{G}$ for some $\mathbf{s} \in \mathbb{Z}_q^n$. Then $\mathbf{r}^t \mathbf{B}_{x_1} = \mathbf{s}^t \mathbf{A}_{x_1}$, which corresponds to our construction.

2.1 Security

In our security proof, the modulus q and underlying LWE error rate, and hence also the dimension n needed to obtain a desired level of provable security, are largely determined by a certain parameter of the tree T which we call the *expansion* $e(T)$. Essentially, the expansion is the maximum number of terms of the form $\mathbf{G}^{-1}(\cdot)$ that are ever consecutively multiplied together when we unwind the recursive definition of $\mathbf{A}_T$, or $\mathbf{A}_{T'}$ for related trees T' considered in the security proof. Formally, the expansion of T is defined by the recurrence

$$e(T) = \begin{cases} 0 & \text{if } |T| = 1 \\ \max\{e(T.l) + 1\,,\ e(T.r)\} & \text{otherwise.} \end{cases} \tag{2.4}$$

This is simply the "left depth" of the tree, i.e., the maximum length of a root-to-leaf path, counting only edges from parents to their left children.

We can now state our main security theorem.

Theorem 2.1. *Let T be any full binary tree, χ be some distribution over $\mathbb{Z}$ that is subgaussian with parameter $r > 0$ (e.g., a bounded or discrete Gaussian distribution with expectation zero), and*

$$q \geq p \cdot r\sqrt{|T|} \cdot (n\ell)^{e(T)} \cdot \lambda^{\omega(1)}. \tag{2.5}$$

Then over the uniformly random and independent choice of $\mathbf{A}_0, \mathbf{A}_1 \in \mathbb{Z}_q^{n \times n\ell}$, the family $\mathcal{F}_{\mathbf{A}_0, \mathbf{A}_1, T, p}$ with secret key chosen uniformly from $\mathbb{Z}_q^n$ is a secure PRF family, under the decision-$\mathsf{LWE}_{n,q,\chi}$ assumption.

An outline of the proof, which contains all the main and new ideas, is given in Section 3.1; the formal proof appears in in Section 3.3.

Notice that the modulus-to-noise ratio for the underlying LWE problem is $q/r \approx (n \log q)^{e(T)}$, i.e., exponential in the expansion $e(T)$. Known reductions [Reg05, Pei09, BLP+13] (for $r \geq 3\sqrt{n}$) guarantee that such an LWE instantiation is at least as hard as (quantumly) approximating various lattice problems in the worst case to within $\approx q/r$ factors on n-dimensional lattices. Known algorithms for achieving such factors take time exponential in $n/\log(q/r) = \tilde{\Omega}(n/e(T))$, so in order to obtain provable 2^λ security against the best known lattice algorithms, the best parameters we can use are

$$n = e(T) \cdot \tilde{\Theta}(\lambda) \quad \text{and} \quad \log q = e(T) \cdot \tilde{\Theta}(1). \tag{2.6}$$

These parameters determine the runtimes and key sizes of the construction, as analyzed below.

We conclude this discussion of security by remarking that, as in [BPR12, BLMR13], and in contrast with essentially all lattice-based *encryption* schemes, it is possible that our PRF is actually secure for *much smaller* parameters than our proof requires. For example, taking $q = \text{poly}(n)$ even for large $e(T)$, with $p|q$ to ensure that rounding produces "unbiased" output, may actually be secure—but we do not know how to prove it. (We also do not know of any effective attacks against such parameters.) The reason for this possibility is that the function itself does not actually expose any low-error-rate LWE samples to the attacker; they are used only in the proof as part of a thought experiment. Whether any of the constructions from this work or [BPR12, BLMR13] can be proved secure for smaller parameters under a standard assumption is a fascinating open question. For the remainder of the paper, we deal only with parameters for which we can *prove* security under (ring-)LWE.

2.2   Size, Time, and Depth

Here we briefly analyze the secret key and public parameter sizes, runtime, and circuit depth of our PRFs, always normalizing to 2^λ provable security under standard lattice assumptions. In some cases these quantities are not very practical (or even asymptotically good), especially when the tree T has large expansion. In Section 2.4 we give a much more efficient construction using ring-LWE, which can be quasi-optimal in key size, public parameters, and depth (simultaneously).

The secret key, which is a uniformly random element of $\mathbb{Z}_q^n$, has size $\Theta(n \log q)$, which is $e(T)^2 \cdot \tilde{\Theta}(\lambda)$ by Equation (2.6). The public parameters, being two $n \times n\ell$ matrices over $\mathbb{Z}_q$, are $\Theta(n^2 \log^2 q) = e(T)^4 \cdot \tilde{\Theta}(\lambda^2)$ bits.

For runtime, computing $\mathbf{A}_T(x)$ from scratch takes one decomposition with $\mathbf{G}^{-1}$ and one $(n \times n\ell)$-by-$(n\ell \times n\ell)$ matrix multiplication over $\mathbb{Z}_q$ per internal node of T. (As mentioned above, incremental computation of $\mathbf{A}_T(x)$ on related inputs can be much faster.) Using naïve matrix multiplication, this is a total of $\Theta(|T| \cdot n^3 \log^2 q)$ ring operations in $\mathbb{Z}_q$, which translates to $e(T)^6 \cdot \tilde{\Theta}(\lambda^4)$ bit operations by Equation (2.6) (even using quasi-linear-time multiplication in $\mathbb{Z}_q$, which is needed only when $\log q \neq \tilde{O}(1)$). This can be improved somewhat using asymptotically faster matrix multiplication, but still remains a rather large $\Omega(|T| \cdot n^\omega \log^2 q)$, where $\omega \geq 2$ is the exponent of matrix multiplication.

For certain trees T our construction is highly parallelizable, i.e., it can be computed by a low-depth circuit. First, notice that each $\mathbb{Z}_q$-entry of $\mathbf{s}^t \cdot \mathbf{A}_T(x)$ (and hence each $\mathbb{Z}_p$-entry of the PRF output) can be computed independently. This is because each column of $\mathbf{A}_T(x)$ can be computed independently, by induction and the fact that $\mathbf{G}^{-1}$ works independently on the columns of $\mathbf{A}_{T.r}(x_r)$. Next, since linear operations over $\mathbb{Z}_q$ can be computed by depth-one arithmetic circuits (with unbounded fan-in), the circuit depth of our construction is proportional to the maximum nesting depth of $\mathbf{G}^{-1}(\cdot)$ expressions when we fully unwind the definition of $\mathbf{A}_T$. We call this the *sequentiality* $s(T)$ of the tree T, which is formally defined by the recurrence

$$s(T) = \begin{cases} 0 & \text{if } |T| = 1 \\ \max\{e(T.l),\ e(T.r) + 1\} & \text{otherwise.} \end{cases} \qquad (2.7)$$

This is simply the "right depth" of the tree, i.e., the maximum length of a root-to-leaf path, counting only edges from parents to their right children.

2.3   Instantiations

Here we discuss some interesting instantiations of the tree T and the efficiency properties of the resulting functions; see Figure 1 for a summary. Generally speaking, for a given tree size $|T|$ (the PRF input length) there is a tradeoff between expansion $e(T)$ and sequentiality $s(T)$. Flipping this around, given bounds e, s we are interested in obtaining a largest possible tree T such that $e(T) \leq e$ and $s(T) \leq s$; let $t(e, s)$ denote the size of such a tree. At first blush, it may be surprising that under the simplifying restriction $e = s$, a complete binary tree of depth s is *very far* from optimal! To see this, notice that

$$t(e, s) = \begin{cases} 1 & \text{if } e = 0 \text{ or } s = 0 \\ t(e - 1, s) + t(e, s - 1) & \text{otherwise.} \end{cases} \tag{2.8}$$

The base cases follow from the fact that only a single leaf satisfies the bounds, and in the recursive case, the first and second terms respectively denote the sizes of the optimal left and right subtrees. It is easy to verify that this recurrence is simply the one that defines the *binomial coefficients*:

$$t(e, s) = \binom{e + s}{e} = \binom{e + s}{s}.$$

One can also efficiently construct an optimal tree for given e, s using dynamic programming.

For example, if we restrict to $e = s$, then by Stirling's approximation we get that $t(e, s) = \binom{2s}{s} \approx 4^s / \sqrt{s\pi}$. Said another way, we can get a PRF with input length $|T|$ where the expansion and sequentiality are both $\approx \log_4(|T|)$. By contrast, a complete binary tree with these parameters has size only $2^s \approx \sqrt{|T|}$. By Theorem 2.1 and Equation (2.6), this means we can get a PRF with input length λ and security 2^λ having sequentiality $O(\log \lambda)$ and secret keys of quasi-optimal bit length $\tilde{O}(\lambda)$.

By ignoring parallelism, one can reduce the expansion even further by letting T be a "right spine" with leaves for all its left children. Then $e(T) = 1$ and $s(T) = |T| - 1$, yielding even better parameters: the underlying LWE assumption has a nearly polynomial $n^{\omega(1)}$ approximation factor, and for security level 2^λ we still obtain secret keys of quasi-optimal bit length $\tilde{O}(\lambda)$; moreover, here the hidden factors are at least a $\log \lambda$ factor smaller than in the case above.

2.4   Ring Variant

Due to the several matrix multiplications (of dimension at least n) involved in computing $\mathbf{A}_T(x)$, our LWE-based construction is not very practically efficient. Fortunately, we can obtain a much more efficient analogue based on the ring-LWE problem [LPR10]. Here we just describe the construction and analyze its efficiency. The proof of security based on ring-LWE proceeds in essentially the same way as the one for our main construction, and is therefore omitted.

For concreteness, let $R \cong \mathbb{Z}[X]/(X^n + 1)$ where n is a power of two, which is known as the $2n$th cyclotomic ring. (The construction and analysis may be generalized to arbitrary cyclotomic rings using the tools developed in [LPR13].) For a modulus q, let $R_q = R/qR \cong \mathbb{Z}_q[X]/(X^n + 1)$, and define a suitable "gadget" vector $\mathbf{g} \in R_q^\ell$ and deterministic function $\mathbf{g}^{-1} \colon R_q \to R^\ell$, so that $\mathbf{g}^{-1}(a)$ is "short" and $\langle \mathbf{g}, \mathbf{g}^{-1}(a) \rangle = a$ for all $a \in R_q$. (E.g., we may let $\mathbf{g} = (1, 2, 4, \ldots, 2^{\ell-1}) \in R_q^\ell$ and define $\mathbf{g}^{-1}(a)$ so that each of its R-entries has $\{0, 1\}$-coefficients with respect to an appropriate "short" $\mathbb{Z}$-basis of R.) Extend $\mathbf{g}^{-1}$ to row vectors over R_q by applying $\mathbf{g}^{-1}$ entry-wise.

Construction 2.2. Fix some *row* vectors $\mathbf{a}_0, \mathbf{a}_1 \in R_q^\ell$, and for a binary tree T, define $\mathbf{a}_T \colon \{0, 1\}^{|T|} \to R_q^\ell$ recursively as

$$\mathbf{a}_T(x) = \begin{cases} \mathbf{a}_x & \text{if } |T| = 1 \\ \mathbf{a}_{T.l}(x_l) \cdot \mathbf{g}^{-1}(\mathbf{a}_{T.r}(x_r)) & \text{otherwise,} \end{cases} \tag{2.9}$$

where in the second case we parse $x = x_l \| x_r$ for $x_l \in \{0, 1\}^{|T.l|}, x_r \in \{0, 1\}^{|T.r|}$.

We define the function family

$$\mathcal{F}_{\mathbf{a}_0, \mathbf{a}_1, T, p} = \left\{ F_s \colon \{0, 1\}^{|T|} \to R_p^\ell \right\},$$

which is parameterized by row vectors $\mathbf{a}_0, \mathbf{a}_1 \in R_q^\ell$, a binary tree T, and a modulus $p \le q$. A member of the family is indexed by some $s \in R$ (or R_q), and is defined as

$$F_s(x) := \lfloor s \cdot \mathbf{a}_T(x) \rceil_p. \tag{2.10}$$

Analysis. Evaluating $\mathbf{a}_T(x)$ from scratch takes one decomposition with $\mathbf{g}^{-1}$ and one vector-matrix multiplication of dimension $\ell = \log q$ over R_q per internal node of T, for a total of $O(|T| \cdot \ell^2)$ ring operations in R_q. Ring operations in R_q can be performed in $O(n \log n)$ scalar operations over $\mathbb{Z}_q$, and $\mathbf{g}^{-1}$ can be computed in $O(n \log q)$ time. Using a tree T with expansion $e(T) = \tilde{O}(1)$, by Equation (2.6) we can get a PRF with input length λ and 2^λ security (under conventional assumptions) running in $\tilde{O}(\lambda^2)$ bit operations to output at least λ bits. When T has polylogarithmic depth, the incremental cost per invocation is reduced to $\tilde{O}(\lambda)$ bit operations, which is quasi-optimal.

As an optimization, and analogously to the LWE-based construction, each R_q-entry of $\mathbf{a}_T(x) \in R_q^\ell$ can be computed *independently* in $O(|T| \cdot \ell)$ ring operations each. Therefore, we can compute each R_p-entry of the output (yielding at least n output bits) in just $O(|T| \cdot \ell)$ ring operations. This may be useful in applications that do not need the entire large output length.

3   Security Proof

In this section we prove the security theorem, Theorem 2.1, which says that $F_\mathbf{s}(x) = \lfloor s^t \cdot \mathbf{A}_T(x) \rceil_p$ from Construction 2.1 is a PRF under the LWE assumption, for appropriate parameters.

3.1   Proof Outline

We start with an overview of the proof, which highlights the central (new) ideas. (For technical reasons, the formal proof proceeds a bit differently than this outline, but the main ideas are the same.) The basic strategy, first used in [BPR12], is to define a sequence of hybrid games where the function inside the rounding operation $\lfloor \cdot \rfloor_p$ changes in ways that are indistinguishable to the adversary, either statistically or computationally. As in prior works [BPR12, BLMR13], these changes include introducing small additive terms that are "rounded away" and hence preserve the input-output behavior (with high probability), and replacing LWE instances with uniformly random ones. In addition, we introduce a new proof technique described within.

Let T be any full binary tree, and suppose its leftmost leaf v is at depth $d > 1$. (If $d = 1$, then $|T| = 1$ and the function is trivially a PRF based on the "learning with rounding" problem, which is as hard as LWE for our choice of parameters, or even slightly better ones [BPR12, AKPW13].) In the real attack game, the adversary has oracle access to $F_{\mathbf{s}}(\cdot)$, which, by unwinding the definition of $\mathbf{A}_T$, is of the form

$$F_{\mathbf{s}}(x) = \left\lfloor \mathbf{s}^t \cdot \mathbf{A}_T(x) \right\rceil_p = \left\lfloor \mathbf{s}^t \cdot \mathbf{A}_{x_0} \cdot \underbrace{\prod_{i=1}^{d} \mathbf{G}^{-1}(\mathbf{A}_{T_i}(x'_i))}_{\mathbf{S}_T(x')} \right\rceil_p ,$$

where subtree T_i is the right child of v's ith ancestor, and $x = x_0 \| x' = x_0 \| x'_1 \| \cdots \| x'_d$ where $|x_0| = 1$ and $|x'_i| = |T_i|$ for all i.

We next consider a hybrid game in which $\mathbf{s}^t \cdot \mathbf{A}_b$ for $b \in \{0, 1\}$ is replaced by an LWE vector $\mathbf{s}^t \cdot \mathbf{A}_b + \mathbf{e}^t_b$, for some short error vectors $\mathbf{e}_0, \mathbf{e}_1$. That is, the adversary instead has oracle access to the function

$$F'_{\mathbf{s}, \mathbf{e}_0, \mathbf{e}_1}(x) := \left\lfloor (\mathbf{s}^t \cdot \mathbf{A}_{x_0} + \mathbf{e}^t_{x_0}) \cdot \mathbf{S}_T(x') \right\rceil_p = \left\lfloor \mathbf{s}^t \cdot \mathbf{A}_T(x) + \mathbf{e}^t_{x_0} \cdot \mathbf{S}_T(x') \right\rceil_p.$$

Because $\mathbf{e}_{x_0}$ and any matrix of the form $\mathbf{G}^{-1}(\cdot)$ are short, $\mathbf{e}^t_{x_0} \cdot \mathbf{S}_T(x')$ is short. More precisely, its entries are of magnitude bounded by $\approx (n \log q)^d$, which is much less than q/p because $d \le e(T)$ and by assumption on q. Therefore, the additive term $\mathbf{e}^t_{x_0} \cdot \mathbf{S}_T(x')$ is very unlikely the change the final rounded value, i.e., with high probability $F'_{\mathbf{s}, \mathbf{e}_0, \mathbf{e}_1}(x) = F_{\mathbf{s}}(x)$ for all the adversary's queries x. Therefore, this hybrid game is statistically indistinguishable from the real attack.

In the next hybrid game, we replace each $\mathbf{s}^t \cdot \mathbf{A}_b + \mathbf{e}^t_b$ for $b \in \{0, 1\}$ by uniformly random and independent $\mathbf{u}^t_b$, i.e., the adversary has access to the function

$$F''_{\mathbf{u}_0, \mathbf{u}_1}(x) := \left\lfloor \mathbf{u}^t_{x_0} \cdot \mathbf{S}_T(x') \right\rceil_p = \left\lfloor \mathbf{u}^t_{x_0} \cdot \mathbf{G}^{-1}(\mathbf{A}_{T_1}(x'_1)) \cdot \underbrace{\prod_{i=2}^{d} \mathbf{G}^{-1}(\mathbf{A}_{T_i}(x'_i))}_{\mathbf{S}'_T(x')} \right\rceil_p . \quad (3.1)$$

Since $\mathbf{S}_T(x')$ can be efficiently computed from the public parameters $\mathbf{A}_b$ and the adversary's queries x, this game is computationally indistinguishable from the previous one, under the LWE assumption.

At this point, we would like to be able to proceed by replacing the terms $\mathbf{u}_{x_0}^t \cdot \mathbf{G}^{-1}(\mathbf{A}_{T_1}(x_1'))$ with some "noisy" variants, then replace those with uniform and independent vectors for all values of $x_0 \| x_1'$, etc. Indeed, this is possible if x_1' consists of a *single bit* (i.e., if $|T_1| = 1$ and hence $\mathbf{A}_{T_1}(x_1') = \mathbf{A}_{x_1'}$), using "non-uniform LWE" exactly as is done in [BLMR13]. Unfortunately, non-uniform LWE does not appear to be sufficient when x_1' is more than one bit (i.e., when $|T_1| > 1$), because the matrices $\mathbf{A}_{T_1}(x_1')$ are not independent for different values of x_1'. And requiring $|T_i| = 1$ for all i makes T have maximal expansion $e(T) = |T| - 1$.

Our main new proof technique is a way to deal with the above issue. Going back to Equation (3.1), as "wishful thinking" suppose that each $\mathbf{u}_b$ was of the form $\mathbf{u}_b^t = \mathbf{s}_b^t \cdot \mathbf{G}$ for some (uniform, say) $\mathbf{s}_b \in \mathbb{Z}_q^n$. Then the $\mathbf{G}$ factor would undo the decomposition $\mathbf{G}^{-1}(\cdot)$, and the adversary would have access to the function

$$F_{\mathbf{s}_0,\mathbf{s}_1}'''(x) := \left\lfloor \mathbf{s}_{x_0}^t \cdot \mathbf{A}_{T_1}(x_1') \cdot \mathbf{S}_T'(x') \right\rceil_p = \left\lfloor \mathbf{s}_{x_0}^t \cdot \mathbf{A}_{T'}(x') \right\rceil_p,$$

where T' is the full binary tree obtained from T by removing its leftmost leaf v and promoting v's sibling subtree T_1 to replace their parent. Notice that the above function is just two independent members of our function family instantiated with tree T'. Moreover, T' has expansion $e(T') \le e(T)$, because expansion is just "left depth." Therefore, the above function would be a PRF simply by induction on $|T|$.

Unfortunately, our "wishful thinking" fails in a very strong sense: a uniformly random $\mathbf{u}^t$ is highly likely to be *very far* from any vector of the form $\mathbf{s}^t \cdot \mathbf{G}$. However, because $\mathbf{G}^t \cdot \mathbb{Z}_q^n$ is a *subgroup* of $\mathbb{Z}_q^{n\ell}$, a uniformly random vector $\mathbf{u} \in \mathbb{Z}_q^{n\ell}$ *can* be decomposed as $\mathbf{u}^t = \mathbf{s}^t \cdot \mathbf{G} + \mathbf{v}^t$ where $\mathbf{s} \in \mathbb{Z}_q^n$ is uniform, and $\mathbf{v}$ is uniform in (some canonical set of representatives of) the quotient group $\mathbb{Z}_q^{n\ell} / (\mathbf{G}^t \cdot \mathbb{Z}_q^n)$ and *independent* of $\mathbf{s}$. Therefore, the function in Equation (3.1) is equivalent to the function

$$F_{\mathbf{s}_0,\mathbf{s}_1,\mathbf{v}_0,\mathbf{v}_1}'''(x) := \left\lfloor \mathbf{s}_{x_0}^t \cdot \mathbf{A}_{T'}(x') \quad + \quad \mathbf{v}_{x_0}^t \cdot \mathbf{S}_T(x') \right\rceil_p,$$

where T' and x' are exactly as in the previous paragraph. Note that $\mathbf{v}_b$ is *not* short, so the extra additive term above does not simply "round away"—but we do not need it to. The main point is that $\mathbf{v}_b$ may be chosen *independently* of (and hence without knowledge of) $\mathbf{s}_b$ by the simulator, and then the additive term may be efficiently computed from it and other public information. Essentially, this allows us to complete the proof by induction on $|T|$. (Again, the actual proof is structured a bit differently, to allow us to simulate the independent additive terms *inside* the rounding operation.)

3.2 Additional Background

Games and indistinguishability. In our security proof, we model interaction with the adversary through a series of probabilistic experiments called *games*. For an adversary $\mathcal{A}$ interacting with two games H_0 and H_1, the *distinguishing advantage* of $\mathcal{A}$, which is implicitly a funtion of the security parameter λ, is defined as $\mathbf{Adv}_{H_0,H_1}(\mathcal{A}) = |\Pr[\mathcal{A} \text{ accepts in } H_0] - \Pr[\mathcal{A} \text{ accepts in } H_1]|$. Two games H_0 and H_1 are *computationally distinguishable*, denoted $H_0 \overset{c}{\approx} H_1$, if $\mathbf{Adv}_{H_0,H_1}(\mathcal{A}) = \mathrm{negl}(\lambda)$ for any efficient adversary $\mathcal{A}$.

Learning with errors. We use the following form of the learning with errors (LWE) problem, due to Regev [Reg05]. For a positive integer dimension n, a modulus $q \geq 2$, and a probability distribution χ over $\mathbb{Z}$, the decision-$\mathsf{LWE}_{n,q,\chi}$ assumption is that for for any polynomially bounded m, w,

$$(\mathbf{A} \leftarrow \mathbb{Z}_q^{n \times m}, \mathbf{B}^t = \mathbf{S}^t \cdot \mathbf{A} + \mathbf{E}^t \in \mathbb{Z}_q^{w \times m}) \overset{c}{\approx} (\mathbf{A} \leftarrow \mathbb{Z}_q^{n \times m}, \mathbf{B}^t \leftarrow \mathbb{Z}_q^{w \times m}),$$

where on the left $\mathbf{S}^t \leftarrow \mathbb{Z}_q^{w \times n}$ and $\mathbf{E}^t \leftarrow \chi^{w \times m}$. (The assumption for $w = 1$ implies the assumption for larger w, by a routine hybrid argument.)

For $\chi = D_{\mathbb{Z},r}$ where $r \geq 3\sqrt{n}$, and under mild conditions on the form of the modulus q, the decision-$\mathsf{LWE}_{n,q,\chi}$ assumption holds true assuming that various problems on n-dimensional lattices are hard for quantum algorithms to approximate to within $\tilde{O}(n \cdot q/r)$ factors in the worst case [Reg05]; see also [Pei09, BLP+13] and references therein for similar statements assuming only classical (non-quantum) hardness.

3.3 Proof of Security Theorem

In this section we give the formal proof of Theorem 2.1.

To aid the proof we first define a couple of auxiliary function families. The first family simply consists of the "pre-rounded" counterparts of the functions $F_{\mathbf{s}} \in \mathcal{F} = \mathcal{F}_{\mathbf{A}_0,\mathbf{A}_1,T,p}$.

Definition 3.1. *For $\mathbf{A}_0, \mathbf{A}_1 \in \mathbb{Z}_q^{n \times n\ell}$ and a full binary tree T, the family $\mathcal{G} = \mathcal{G}_{\mathbf{A}_0,\mathbf{A}_1,T}$ is the set of functions $G_{\mathbf{s}} \colon \{0,1\}^{|T|} \to \mathbb{Z}_q^{n\ell}$ indexed by some $\mathbf{s} \in \mathbb{Z}_q^n$, and defined as $G_{\mathbf{s}}(x) := \mathbf{s}^t \cdot \mathbf{A}_T(x)$ (where we define $\mathbf{A}_T(\varepsilon) := \mathbf{G}$ for the empty tree T). We endow $\mathcal{G}$ with the distribution where $\mathbf{s} \leftarrow \mathbb{Z}_q^n$ is chosen uniformly at random.*

Note that $F_{\mathbf{s}}(x) = \lfloor G_{\mathbf{s}}(x) \rceil_p$.

The next family $\tilde{\mathcal{G}}$ consists of functions that are certain "noisy" versions of the functions in $\mathcal{G}$. The family $\mathcal{E}$ of "error functions" used in the definition is a family of functions from $\{0,1\}^{|T|}$ to $\mathbb{Z}^{n\ell}$, and is formally defined in Definition 3.5 below. An important point is that the functions in $E \in \mathcal{E}$ have *exponentially* large keys, but they may be efficiently sampled "lazily," as values $E(x)$ are needed. See the discussion following Definition 3.5 for details.

Definition 3.2. *For $\mathbf{A}_0, \mathbf{A}_1 \in \mathbb{Z}_q^{n \times n\ell}$ and a full binary tree T, the family $\tilde{\mathcal{G}} = \tilde{\mathcal{G}}_{\mathbf{A}_0,\mathbf{A}_1,T}$ is the set of functions $\tilde{G}_{\mathbf{s},E} \colon \{0,1\}^{|T|} \to \mathbb{Z}_q^{n\ell}$ indexed by some $G_{\mathbf{s}} \in \mathcal{G}$ and $E \in \mathcal{E} = \mathcal{E}_{\mathbf{A}_0,\mathbf{A}_1,T}$, and defined as $\tilde{G}_{\mathbf{s},E}(x) := G_{\mathbf{s}}(x) + E(x)$. We endow $\tilde{\mathcal{G}}$ with the distribution where $G_{\mathbf{s}} \leftarrow \mathcal{G}$ and $E \leftarrow \mathcal{E}$ are chosen independently.*

The proof of Theorem 2.1 consists of showing that with overwhelming probability, the rounding of $G_{\mathbf{s}} \in \mathcal{G}$ agrees with the rounding of essentially any corresponding $\tilde{G}_{\mathbf{s},E} \in \tilde{\mathcal{G}}$ on all the attacker's queries, because the outputs of the error functions $E \in \mathcal{E}$ are small. This proof follows very similar to the style of the proof of the "degree-k" PRF of [BPR12], and thus we relegate the details to the full version. The main crux of the theorem, which we show in Theorem 3.1 below, is in proving that $\tilde{\mathcal{G}}$ is a PRF family *without* any rounding, and hence with rounding as well. It follows that the rounding of $G_\mathbf{s} \leftarrow \mathcal{G}$ (i.e., $F_\mathbf{s} \leftarrow \mathcal{F}$) cannot be distinguished from a uniformly random function, as desired.

We now formally define the "error function" family $\mathcal{E} = \mathcal{E}_{\mathbf{A}_0,\mathbf{A}_1,T}$. To define the error functions we first need a couple of simple definitions.

Definition 3.3 (Pruning). *For a full binary tree T of at least one node, define its pruning $T' = pr(T)$ inductively as follows: if $|T.l| \leq 1$ then $T' := T.r$; otherwise, $T'.l := pr(T.l)$ and $T'.r := T.r$. We let $T^{(i)}$ denote the ith successive pruning of T, i.e., $T^{(0)} = T$ and $T^{(i)} = pr(T^{(i-1)})$.*

In other words, pruning a tree node removes its leftmost leaf v and replaces the subtree rooted at v's parent (if it exists) with the subtree rooted at v's sibling. Notice that pruning cannot increase the tree's expansion (i.e., left depth; see Equation (2.4)): $e(T') \leq e(T)$.

Definition 3.4. *Given $\mathbf{A}_0, \mathbf{A}_1 \in \mathbb{Z}_q^{n \times n\ell}$ and a full binary tree T of at least one node, define the function $\mathbf{S}_T : \{0,1\}^{|T|-1} \rightarrow \mathbb{Z}^{n\ell \times n\ell}$ recursively as follows:*

$$\mathbf{S}_T(x) = \begin{cases} \mathbf{I} \text{ (the identity matrix)} & \text{if } |T| = 1 \\ \mathbf{S}_{T.l}(x_l) \cdot \mathbf{G}^{-1}(\mathbf{A}_{T.r}(x_r)) & \text{otherwise,} \end{cases} \tag{3.2}$$

where $x = x_l \| x_r$ for $|x_l| = |T.l| - 1$, $|x_r| = |T.r|$.

Notice that if $T' = pr(T)$ and $x = x_1 \| x' \in \{0,1\}^{|T|}$ for $|x_1| = 1$, then it follows directly from the definitions (recalling that $\mathbf{A}_\varepsilon(\varepsilon) = \mathbf{G}$) and by induction that

$$\mathbf{A}_T(x) = \mathbf{A}_{x_1} \cdot \mathbf{S}_T(x'), \tag{3.3}$$
$$\mathbf{G} \cdot \mathbf{S}_T(x') = \mathbf{A}_{T'}(x'). \tag{3.4}$$

Definition 3.5 (Error Functions). *For public matrices $\mathbf{A}_0, \mathbf{A}_1 \in \mathbb{Z}_q^{n \times n\ell}$ and a full binary tree T, the family $\mathcal{E} = \mathcal{E}_{\mathbf{A}_0,\mathbf{A}_1,T}$ consists of functions from $\{0,1\}^{|T|}$ to $\mathbb{Z}^{n\ell}$, defined inductively as follows.*

- *For $|T| = 0$, the sole function in $\mathcal{E}$ is defined simply as $E(\varepsilon) := \mathbf{0}$.*
- *For $|T| \geq 1$, a function in $\mathcal{E}$ is indexed by some $\mathbf{e}_0, \mathbf{e}_1 \in \mathbb{Z}^{n\ell}$ and $E'_0, E'_1 \in \mathcal{E}' = \mathcal{E}_{\mathbf{A}_0,\mathbf{A}_1,T'}$, where T' is the pruning of T. For $x = x_1 \| x' \in \{0,1\}^{|T|}$, the function is defined as*

$$E_{\mathbf{e}_0,\mathbf{e}_1,E'_0,E'_1}(x) := \mathbf{e}_{x_1}^t \cdot \mathbf{S}_T(x') + E'_{x_1}(x').$$

For a given error function distribution χ over $\mathbb{Z}$, we endow $\mathcal{E}$ with the distribution where $\mathbf{e}_0, \mathbf{e}_1 \leftarrow \chi^{n\ell}$ and $E'_0, E'_1 \leftarrow \mathcal{E}'$ are all chosen independently.

Note that a function $E \in \mathcal{E}$ is fully specified by *exponentially* (in $|T|$) many error vectors (namely, one $\mathbf{e}_w$ for each $w \in \{0,1\}^{\leq |T|}$), and the value $E(x)$ is fully determined by those $\mathbf{e}_w$ where w is a prefix of x (and $\mathbf{A}_0, \mathbf{A}_1$). This large number of error vectors is what prevents $\tilde{\mathcal{G}}$ itself from being usable as a PRF family. However, as needed in the proof of Theorem 2.1, a function $E \leftarrow \mathcal{E}$ can be sampled "lazily" as values $E(x)$ are needed, since each value of $E(x)$ depends on only a small number of the error vectors. The fact that the output of the error function is "small" with very high probability is also used in the proof of the theorem. Proving this fact is a standard technical exercise, and it is furnished in the full version.

We now prove that the function family $\tilde{\mathcal{G}}$ from Definition 3.2 is pseudorandom.

Theorem 3.1. *For any $n, q \geq 1$ and error distribution χ over $\mathbb{Z}$, any full binary tree T, and over the uniformly random and independent choice of $\mathbf{A}_0, \mathbf{A}_1 \in \mathbb{Z}_q^{n \times n\ell}$, the family $\tilde{\mathcal{G}} = \tilde{\mathcal{G}}_{\mathbf{A}_0, \mathbf{A}_1, T}$ is pseudorandom, assuming the hardness of* decision-$\mathsf{LWE}_{n,q,\chi}$.

Proof. We proceed through a series of games, one for each bit of the input. In each successive game, we modify the function family $\tilde{\mathcal{G}}$ a little, until we are left with the family of all functions from $\{0,1\}^{|T|}$ to $\mathbb{Z}_q^{n\ell}$ (with uniform distribution), and we show that each successive game is computationally indistinguishable under the LWE assumption from the theorem statement.

To define the games formally, we first need some notation. For a bit string x of length at least i, let $x_{(i)} = x_1 x_2 \cdots x_i$ denote the string of its first i bits, and let $x^{(i)}$ denote the remainder of the string. Where $\mathbf{A}_0, \mathbf{A}_1$ and T are clear from context, let $\mathcal{G}^{(i)} = \mathcal{G}_{\mathbf{A}_0, \mathbf{A}_1, T^{(i)}}$ and similarly for $\mathcal{E}^{(i)}$. Let $\mathcal{P} \subset \mathbb{Z}_q^{n\ell}$ denote an arbitrary set of representatives of the quotient group $\mathbb{Z}_q^{n\ell} / \mathbf{G}^t \cdot \mathbb{Z}_q^n$, and define a family of auxiliary functions $\mathcal{V}^{(i)} = \mathcal{V}^{(i)}_{\mathbf{A}_0, \mathbf{A}_1, T}$ as follows.

Definition 3.6. *For public matrices $\mathbf{A}_0, \mathbf{A}_1 \in \mathbb{Z}_q^{n \times n\ell}$, a full binary tree T, and $0 \leq i \leq |T|$, the family $\mathcal{V}^{(i)} = \mathcal{V}^{(i)}_{\mathbf{A}_0, \mathbf{A}_1, T}$ consists of functions from $\{0,1\}^{|T|}$ to $\mathbb{Z}^{n\ell}$, and is defined inductively as follows.*

- *The sole function in $\mathcal{V}^{(0)}$ is defined simply as $V(x) := \mathbf{0}$.*
- *For $i \geq 1$, a function in $\mathcal{V}^{(i)}$ is indexed by some $\mathbf{v}_w \in \mathbb{Z}^{n\ell}$ for every $w \in \{0,1\}^i$, and some $V' \in \mathcal{V}^{(i-1)}$. The function is defined as*

$$V_{\{\mathbf{v}_w\}, V'}(x) := \mathbf{v}_{x_{(i)}}^t \cdot \mathbf{S}_{T^{(i-1)}}(x^{(i)}) + V'(x).$$

We endow $\mathcal{V}^{(i)}$ with the distribution where the $\mathbf{v}_w \leftarrow \mathcal{P}$ and $V' \leftarrow \mathcal{V}^{(i-1)}$ are all chosen independently and uniformly.

Similarly to the family $\mathcal{E}$ of error functions, the description of a function in $\mathcal{V}^{(i)}$ consists of an exponential (in i) number of $\mathbf{v}_w$ vectors, and can be sampled lazily.

We now define game H_i for $0 \leq i \leq |T|$.

Game H_i. Choose $\mathbf{A}_0, \mathbf{A}_1 \leftarrow \mathbb{Z}_q^{n \times n\ell}$ independently, and lazily sample $G_{\mathbf{s}_w} \leftarrow \mathcal{G}^{(i)}$ and $E_w \leftarrow \mathcal{E}^{(i)}$ for each $w \in \{0,1\}^i$, and $V \leftarrow \mathcal{V}^{(i)}$. Give the adversary $\mathbf{A}_0, \mathbf{A}_1$ and oracle access to the function

$$H(x) := G_{\mathbf{s}_{x_{(i)}}}(x^{(i)}) + E_{x_{(i)}}(x^{(i)}) + V(x). \tag{3.5}$$

Claim. Game H_0 corresponds to the real attack game against the family $\tilde{\mathcal{G}}$, and game $H_{|T|}$ corresponds to oracle access to a uniformly random function.

The first claim follows by definition of $\tilde{\mathcal{G}} = \tilde{\mathcal{G}}_{\mathbf{A}_0,\mathbf{A}_1,T}$, and because $\mathcal{V}^{(0)}$ consists solely of the zero function. For the second claim, for $i = |T|$ we have $x_{(i)} = x$, $x^{(i)} = \varepsilon$, and $T^{(i)} = \varepsilon$ (the empty tree), so by Definitions 3.1, 3.5, and 3.6,

$$H(x) = G_{\mathbf{s}_x}(\varepsilon) + E_x(\varepsilon) + V(x) = \mathbf{s}_x^t \cdot \mathbf{G} + \mathbf{v}_x^t + V'(x).$$

Since $\mathbf{s}_x \in \mathbb{Z}_q^n, \mathbf{v}_x \in \mathcal{P}$ are uniformly random and independent for each x, and $\mathcal{P}$ is a set of representatives of the quotient group $\mathbb{Z}_q^{n\ell}/\mathbf{G}^t \cdot \mathbb{Z}_q^n$, the values $\mathbf{s}_x^t \cdot \mathbf{G} + \mathbf{v}_x^t \in \mathbb{Z}_q^{n\ell}$ are uniformly random and independent. Since V' is independent of these as well, H is a uniformly random function.

It remains to prove that successive games are computationally indistinguishable. To do so we define the following games H_i' for $1 \le i \le |T|$.

Game H_i'. Choose $\mathbf{A}_0, \mathbf{A}_1 \leftarrow \mathbb{Z}_q^{n\times n\ell}$ independently, and lazily sample $\mathbf{u}_w \leftarrow \mathbb{Z}_q^{n\ell}$ and $E_w \leftarrow \mathcal{E}^{(i)}$ for each $w \in \{0,1\}^i$, and $V' \leftarrow \mathcal{V}^{(i-1)}$. Give the adversary $\mathbf{A}_0, \mathbf{A}_1$ and oracle access to the function

$$H'(x) = \mathbf{u}_{x_{(i)}}^t \cdot \mathbf{S}_{T^{(i-1)}}(x^{(i)}) + E_{x_{(i)}}(x^{(i)}) + V(x). \tag{3.6}$$

Claim. For $1 \le i \le |T|$, games H_i and H_i' are equivalent.

We can write each uniformly random $\mathbf{u}_w \in \mathbb{Z}_q^{n\ell}$ for $w \in \{0,1\}^i$ as $\mathbf{u}_w^t = \mathbf{s}_w^t \cdot \mathbf{G}^t + \mathbf{v}_w^t$, where $\mathbf{s}_w \in \mathbb{Z}_q^n$ and $\mathbf{v}_w \in \mathcal{P}$ are uniformly random and independent. Therefore, we can rewrite the function $H'(\cdot)$ from Equation (3.6) as

$$\begin{aligned}
H'(x) &= \left(\mathbf{s}_{x_{(i)}}^t \cdot \mathbf{G} + \mathbf{v}_{x_{(i)}}^t\right) \cdot \mathbf{S}_{T^{(i-1)}}(x^{(i)}) + E_{x_{(i)}}(x^{(i)}) + V'(x) \\
&= \mathbf{s}_{x_{(i)}}^t \cdot \mathbf{G} \cdot \mathbf{S}_{T^{(i-1)}}(x^{(i)}) + E_{x_{(i)}}(x^{(i)}) + \left(\mathbf{v}_{x_{(i)}}^t \cdot \mathbf{S}_{T^{(i-1)}}(x^{(i)}) + V'(x)\right) \\
&= G_{\mathbf{s}_{x_{(i)}}}(x^{(i)}) + E_{x_{(i)}}(x^{(i)}) + V(x),
\end{aligned}$$

where in the final equality we have used Equation (3.4), and we have defined $V(x)$ to be the second parenthesized component of the previous expression. Notice that all the functions $G_{\mathbf{s}_{x_{(i)}}}$, $E_{x_{(i)}}$, and V are drawn independently from $\mathcal{G}^{(i)}$, $\mathcal{E}^{(i)}$, and $\mathcal{V}^{(i)}$ (respectively), and this proves the claim.

Claim. For $0 \le i \le |T| - 1$, games H_i and H_{i+1}' are computationally indistinguishable under the LWE assumption from the theorem statement.

To prove the claim, we design an efficient simulator $\mathcal{S}$ which receives as input a pair of matrices $(\mathbf{A}, \mathbf{B}^t) \in \mathbb{Z}_q^{n\times 2n\ell} \times \mathbb{Z}_q^{Q\times 2n\ell}$, where $Q = \text{poly}(\lambda)$ is the minimum of 2^i and the number of queries that the adversary makes. The simulator parses $\mathbf{A} = [\mathbf{A}_0 \mid \mathbf{A}_1]$ where $\mathbf{A}_0, \mathbf{A}_1 \in \mathbb{Z}_q^{n\times n\ell}$ and gives them to the adversary. It lazily samples a $V \leftarrow \mathcal{V}^{(i)}$ and an $E_w \leftarrow \mathcal{E}^{(i+1)}$ for every $w \in \{0,1\}^{i+1}$. Then for each query x from the adversary, if a vector $\mathbf{b}^t_{x_{(i)}}$ has not already been defined, it lets $\mathbf{b}^t_{x_{(i)}}$ be a previously unused row of $\mathbf{B}^t$. It parses $\mathbf{b}^t_{x_{(i)}} = (\mathbf{b}^t_{x_{(i)}\|0} \mid \mathbf{b}^t_{x_{(i)}\|1})$, where $\mathbf{b}_{x_{(i)}\|b} \in \mathbb{Z}_q^{n\ell}$ for each $b \in \{0,1\}$. It then answers the query with the value

$$J(x) := \mathbf{b}^t_{x_{(i+1)}} \cdot \mathbf{S}_{T^{(i)}}(x^{(i+1)}) + E_{x_{(i+1)}}(x^{(i+1)}) + V(x).$$

We now analyze the behavior of $\mathcal{S}$ for the two distributions of $(\mathbf{A}, \mathbf{B}^t)$ from the decision-LWE problem. In both cases, $\mathbf{A}$ is uniformly random and so the public parameters are properly distributed. When $\mathbf{B}$ is uniformly random, it can be seen by inspection that the function J is drawn from the same distribution as the function H' in game H'_{i+1} described in Equation (3.6), so the simulator exactly emulates game H'_{i+1}.

We now analyze the other case, namely, $\mathbf{B}^t = \mathbf{S}^t \cdot \mathbf{A} + \mathbf{E}^t$ for independent $\mathbf{S}^t \leftarrow \mathbb{Z}_q^{Q \times n}$ and $\mathbf{E}^t \leftarrow \chi^{Q \times 2n\ell}$. Then letting $\mathbf{s}^t_{x_{(i)}}, (\mathbf{e}^t_{x_{(i)}\|0} \mid \mathbf{e}^t_{x_{(i)}\|1})$ respectively be the rows of $\mathbf{S}^t, \mathbf{E}^t$ corresponding to the row of $\mathbf{B}^t$ used as $\mathbf{b}^t_{x_{(i)}}$, we have

$$J(x) = \left(\mathbf{s}^t_{x_{(i)}} \cdot \mathbf{A}_{x_{i+1}} + \mathbf{e}^t_{x_{(i)}\|x_{i+1}}\right) \cdot \mathbf{S}_{T^{(i)}}(x^{(i+1)}) + E_{x_{(i+1)}}(x^{(i+1)}) + V(x)$$

$$= \mathbf{s}^t_{x_{(i)}} \cdot \mathbf{A}_{T^{(i)}}(x^{(i)}) + \left(\mathbf{e}^t_{x_{(i)}\|x_{i+1}} \cdot \mathbf{S}_{T^{(i)}}(x^{(i+1)}) + E_{x_{(i)}\|x_{i+1}}(x^{(i+1)})\right) + V(x)$$

$$= G_{\mathbf{s}_{x_{(i)}}}(x^{(i)}) + E_{x_{(i)}}(x^{(i)}) + V(x),$$

where in the second equality we have used Equation (3.3), and in the last expression we have defined $E_{x_{(i)}}(x^{(i)})$ to be the parenthesized component from the previous expression. Notice that by the distributions of all the variables, the functions $G_{\mathbf{s}_w}$, E_w (for each queried prefix $w \in \{0,1\}^i$) and V are all drawn independently from $\mathcal{G}^{(i)}$, $\mathcal{E}^{(i)}$, and $\mathcal{V}^{(i)}$, so in this case the simulator exactly emulates game H_i.

Because the two LWE input distributions are computationally indistinguishable by assumption and $\mathcal{S}$ is efficient, it follows that H_i and H'_{i+1} are computationally indistinguishable, and the claim is proved.

By repeated application of the claims above, we have that $H_0 \overset{c}{\approx} H'_1 \equiv H_1 \overset{c}{\approx} H'_2 \equiv \cdots \equiv H_{|T|-1} \overset{c}{\approx} H'_{|T|} \equiv H_{|T|}$, and so $H_0 \overset{c}{\approx} H_{|T|}$ by the triangle inequality. This completes the proof of Theorem 3.1.

References

[ACPS09]   Applebaum, B., Cash, D., Peikert, C., Sahai, A.: Fast cryptographic primitives and circular-secure encryption based on hard learning problems. In: Halevi, S. (ed.) CRYPTO 2009. LNCS, vol. 5677, pp. 595–618. Springer, Heidelberg (2009)

[AKPW13]   Alwen, J., Krenn, S., Pietrzak, K., Wichs, D.: Learning with rounding, revisited - new reduction, properties and applications. In: Canetti, R., Garay, J.A. (eds.) CRYPTO 2013, Part I. LNCS, vol. 8042, pp. 57–74. Springer, Heidelberg (2013)

[BC10]   Bellare, M., Cash, D.: Pseudorandom functions and permutations provably secure against related-key attacks. In: Rabin, T. (ed.) CRYPTO 2010. LNCS, vol. 6223, pp. 666–684. Springer, Heidelberg (2010)

370   A. Banerjee and C. Peikert

[BGG⁺14] Boneh, D., Gentry, C., Gorbunov, S., Halevi, S., Nikolaenko, V., Segev,
G., Vaikuntanathan, V., Vinayagamurthy, D.: Fully key-homomorphic
encryption, arithmetic circuit ABE and compact garbled circuits. In:
Nguyen, P.Q., Oswald, E. (eds.) EUROCRYPT 2014. LNCS, vol. 8441,
pp. 533–556. Springer, Heidelberg (2014)
[BLMR13] Boneh, D., Lewi, K., Montgomery, H., Raghunathan, A.: Key homomor-
phic PRFs and their applications. In: Canetti, R., Garay, J.A. (eds.)
CRYPTO 2013, Part I. LNCS, vol. 8042, pp. 410–428. Springer, Hei-
delberg (2013)
[BLP⁺13] Brakerski, Z., Langlois, A., Peikert, C., Regev, O., Stehlé, D.: Classical
hardness of learning with errors. In: STOC, pp. 575–584 (2013)
[BPR12] Banerjee, A., Peikert, C., Rosen, A.: Pseudorandom functions and lattices.
In: Pointcheval, D., Johansson, T. (eds.) EUROCRYPT 2012. LNCS,
vol. 7237, pp. 719–737. Springer, Heidelberg (2012)
[BV14] Brakerski, Z., Vaikuntanathan, V.: Lattice-based FHE as secure as PKE.
In: ITCS, p. 1 (2014)
[GGM84] Goldreich, O., Goldwasser, S., Micali, S.: How to construct random func-
tions. J. ACM 33(4), 792–807 (1984); Preliminary version in FOCS 1984
[GSW13] Gentry, C., Sahai, A., Waters, B.: Homomorphic encryption from learn-
ing with errors: Conceptually-simpler, asymptotically-faster, attribute-
based. In: Canetti, R., Garay, J.A. (eds.) CRYPTO 2013, Part I. LNCS,
vol. 8042, pp. 75–92. Springer, Heidelberg (2013)
[LMR14] Lewi, K., Montgomery, H., Raghunathan, A.: Improved constructions of
PRFs secure against related-key attacks. In: Boureanu, I., Owesarski, P.,
Vaudenay, S. (eds.) ACNS 2014. LNCS, vol. 8479, pp. 44–61. Springer,
Heidelberg (2014)
[LPR10] Lyubashevsky, V., Peikert, C., Regev, O.: On ideal lattices and learn-
ing with errors over rings. Journal of the ACM 60(6), 43:1–43:35 (2013);
Gilbert, H. (ed.) EUROCRYPT 2010. LNCS, vol. 6110, pp. 445–465.
Springer, Heidelberg (2010)
[LPR13] Lyubashevsky, V., Peikert, C., Regev, O.: A toolkit for ring-LWE cryp-
tography. In: Johansson, T., Nguyen, P.Q. (eds.) EUROCRYPT 2013.
LNCS, vol. 7881, pp. 35–54. Springer, Heidelberg (2013)
[NPR99] Naor, M., Pinkas, B., Reingold, O.: Distributed pseudo-random functions
and KDCs. In: Stern, J. (ed.) EUROCRYPT 1999. LNCS, vol. 1592, pp.
327–346. Springer, Heidelberg (1999)
[NR95] Naor, M., Reingold, O.: Synthesizers and their application to the parallel
construction of pseudo-random functions. J. Comput. Syst. Sci. 58(2),
336–375 (1995); Preliminary version in FOCS 1995
[NR97] Naor, M., Reingold, O.: Number-theoretic constructions of efficient
pseudo-random functions. J. ACM 51(2), 231–262 (1997); Preliminary
version in FOCS 1997
[NRR00] Naor, M., Reingold, O., Rosen, A.: Pseudorandom functions and factoring.
SIAM J. Comput. 31(5), 1383–1404 (2000); Preliminary version in STOC
2000
[Pei09] Peikert, C.: Public-key cryptosystems from the worst-case shortest vector
problem. In: STOC, pp. 333–342 (2009)
[Reg05] Regev, O.: On lattices, learning with errors, random linear codes, and
cryptography. J. ACM 56(6), 1–40 (2005); Preliminary version in STOC
2005

Homomorphic Signatures with Efficient Verification for Polynomial Functions

Dario Catalano[1], Dario Fiore[2], and Bogdan Warinschi[3]

[1] Università di Catania, Italy
catalano@dmi.unict.it
[2] IMDEA Software Institute, Spain
dario.fiore@imdea.org
[3] University of Bristol, UK
bogdan@compsci.bristol.ac.uk

Abstract. A homomorphic signature scheme for a class of functions $\mathcal{C}$ allows a client to sign and upload elements of some data set D on a server. At any later point, the server can derive a (publicly verifiable) signature that certifies that some y is the result computing some $f \in \mathcal{C}$ on the basic data set D. This primitive has been formalized by Boneh and Freeman (Eurocrypt 2011) who also proposed the only known construction for the class of multivariate polynomials of fixed degree $d \geq 1$. In this paper we construct new homomorphic signature schemes for such functions. Our schemes provide the first alternatives to the one of Boneh-Freeman, and improve over their solution in three main aspects. First, our schemes do not rely on random oracles. Second, we obtain security in a stronger fully-adaptive model: while the solution of Boneh-Freeman requires the adversary to query messages in a given data set all at once, our schemes can tolerate adversaries that query one message at a time, in a fully-adaptive way. Third, signature verification is more efficient (in an amortized sense) than computing the function from scratch. The latter property opens the way to using homomorphic signatures for publicly-verifiable computation on outsourced data. Our schemes rely on a new assumption on leveled graded encodings which we show to hold in a generic model.

1 Introduction

Cryptographic mechanisms for building trust are essential for the shift towards a world where weak clients leverage access to all-powerful servers to remotely store and compute on data. Trust issues include availability of storage, privacy of data, authenticity of delegated computation, etc. which in turn take a multitude of forms. For example, privacy concerns range from simply ensuring the secrecy of stored data, to additionally allowing for search over outsourced data and/or optimizing storage space. This paper contributes to the area of *verifiable computation*, and specifically to the setting in which a client delegates the computation of one or more functions $F_1, F_2, \ldots, F_n$ on one or more of its data sets $D_1, D_2, \ldots, D_m$. The crucial requirement here is that the answer y returned

J.A. Garay and R. Gennaro (Eds.): CRYPTO 2014, Part I, LNCS 8616, pp. 371–389, 2014.

by the server, purportedly the result of $F_i(D_j)$, can be efficiently verified. Efficiency has multiple dimensions, but two are needed to avoid trivial solutions: the client should not have to store all data D_j on which the server computes and/or verification should be faster than simply computing F_i on D_j.

In addition to the different forms of efficiency one may require, the problem of verifiable computation also comes in several different scenarios. For example, the function computed by the server may be fixed or changing, the data stored may be fixed or incrementally updated, the client may have access to multiple (non-communicating) servers, the verification of the result may be interactive, etc. In this work we focus on the scenario where the client has access to a single server, he can incrementally add data on the server, the functions to be computed are not known in advance, and the verification of the result is non-interactive and can be done publicly.

To place our contribution in the landscape of solutions for verifiable computation and to facilitate the comparison with existent solutions, we note that previously proposed protocols for verifiable computation use one of two techniques. The first type of solutions (which for brevity we call proof-based) build on foundations going back to Micali's computationally sound proofs [26]. The idea is for the server to provide (or to prove knowledge of) a certificate for the NP statement: $y = F(D)$. The earlier work used probabilistically checkable proofs (PCPs) [26], whereas recent results rely on succinct arguments (SNARGs) or succinct arguments of knowledge (SNARKs) [7,22] where the dependency between the length of the statement and the proof is greatly reduced. Other protocols where proofs are not explicitly mentioned can be thought of as instantiations where the proofs are encrypted information-theoretic secure MAC [21,27].

The second type of solutions use homomorphic authenticators; we refer to these constructions as authenticator-based. In these constructions, one attaches to every input data an unforgeable *authenticator*. The main property is that any operation (gate) used in the computation which takes as input correctly authenticated data, produces a result together with a valid authenticator. Solutions exist in both the symmetric and the public-key setting. Depending on how the authenticator is verified, we distinguish between homomorphic message authentication codes [24,10,4] and homomorphic signatures [9]. Clearly, the difficulty of the problem increases with the class of functions one considers. For example, there are numerous signature schemes homomorphic with respect to linear functions over vector spaces [1,8,23,2,13,14,19,3,12]. In contrast, there has been little progress on signature schemes homomorphic with respect to non-linear polynomials. The only known construction is provided by Boneh and Freeman [9] who construct a homomorphic signature scheme for multivariate polynomials of constant degree.

Summary of our contribution and relation to previous work. In this paper we provide the first alternative to the homomorphic signature scheme of Boneh and Freeman (henceforth BF), which is the work closest to ours. Our result improves over the BF solution in three main aspects. First, we solve a problem left open in [9], as unlike the BF scheme, our construction does *not* rely on the random oracle assumption. Second, our scheme is proven secure in a stronger adaptive model: in the BF scheme the adversary is restricted to query signatures on messages belonging to a given data set all at once; in contrast, our construction is proven secure against adversaries that can query *one message at a time* in a fully adaptive way. Finally, our construction enjoys *efficient verification* in that verifying a signature against a function f can be done faster than computing f (and in particular does not require storing the input data). More accurately, this property holds in an amortized sense: after a single (local) pre-computation of f, one can verify the evaluation of f on any dataset more efficiently. This property has been recently identified, defined and realized for homomorphic MACs in [4]. Our construction is the *first* to achieve efficient verification for homomorphic signatures, and therefore it opens the way to using homomorphic signatures for verifiable computation.

We remark that other constructions of homomorphic authenticators are either in the symmetric key setting [24,10,4], or are for the restricted class of linear functions [1,8,23,2,13,14,19,3,12]. Below we discuss the benefits that our solution brings to the broader field of verifiable computation. We start with general remarks on the benefits that authenticator-based solutions hold over proof-based ones.

INCREMENTAL, COMPOSITIONAL VERIFIABLE COMPUTATION. Homomorphic authenticators naturally give rise to incremental/composable verifiable computation: the output of some computation on authenticated data is already authenticated so it can be fed as input for follow-up computation. This property is of particular interest to parallelize computations (e.g., MapReduce). Emulating this composition within the proof-based frameworks is possible [7] but it leads to complex statements and less natural realizations. For an extensive discussion of this issue see [24].

FLEXIBLE SCENARIOS. Furthermore, homomorphic authenticators are applicable to a broader range of scenarios as neither the data to be computed on, nor the function to be applied need to be known in advance. For example the data can be incrementally updated (by authenticating and uploading new pieces of data), and the function to be applied can be selected at any point by the server (without having to wait for some parameters generated by the client). In contrast, in (most) proof-based solutions the function needs to be known at the moment when data is uploaded, or a copy of the data needs to be kept locally by the client [21,15,6,18,27,7,22]. Perhaps the biggest advantage of verifiable computation based on authenticators is that verification does not need the input data; indeed we only need to check that the result comes with a valid authenticator. Just like for incremental computation, an analogous result can be obtained with proof-based constructions through theoretically beautiful but practically cumbersome solutions. For example, one can fix the computation performed by the server to be some universal circuit and then see the actual function to be computed as part of the data that is uploaded. While the dependency between data and functions is broken, verification would still need the whole data (and function description) as input.

Improving flexibility is also addressed by the notions of memory delegation and streaming delegation [16] in which a client can outsource a large memory to a server, keeps a small local state, and can later delegate and verify computations on the outsourced memory. This setting is very general and is close to the one achieved by using homomorphic authenticators. As mentioned in [24], a difference between memory delegation and homomorphic authenticators is that the former considers a single user who outsources the data all at once and keeps a state associated with the data. In contrast, by using homomorphic authenticators various users may independently upload several data items without sharing any state (beyond the fixed signing key).

COMPLEXITY ASSUMPTIONS. In terms of the usual trade-off between efficiency and the underlying assumptions our scheme fares well. Most proof-based constructions rely on proofs (SNARGs, SNARKs) for which instantiations either rely on the random oracle model [26] or employ non-falsifiable assumptions. Our scheme is in the standard model and is based on problems in the groups underlying a multi-linear map. Our scheme can be instantiated with any of the existing graded encoding schemes [20,17] and hence it will increase in efficiency with any progress on the implementation of the latter primitive [25].

High level idea of our construction. Our scheme signs messages in $\mathbb{Z}_p$ and is homomorphic with respect to polynomial functions on $\mathbb{Z}_p^n$ (where n is the size of the data set); the degree of the polynomial is d (which is bounded).

To realize our construction we proceed in three main stages. First we construct an homomorphic scheme (with the same domain, and homomorphic with respect to the same class of functions) secure in a weaker sense: in an attack, the adversary asks all of the messages to be signed non-adaptively before the scheme is initialized. This is the technically most difficult part of the paper. Then we provide a generic transformation that strengthens any weakly-secure homomorphic signature for degree-d polynomials to an adaptive-secure one, i.e. one that withstands adaptive chosen-message attacks. The third step is to optimize the resulting construction when instantiated with the weakly-secure scheme that we develop. Below we provide an overview of these steps, starting with the generic transformation. Then we describe the main ideas that go into the construction of our weakly-secure scheme. To conclude, we discuss the efficiency of the scheme that we obtain from our weakly-secure scheme via both the generic and the optimized transformation.

In both schemes we encode a message in $\mathbb{Z}_p$ as the free term of a polynomial of degree at most d. Messages in the data set are encoded in polynomials of degree one, whereas the results of computations will be encoded by higher degree polynomials. Start with a weakly secure homomorphic signature scheme Π. The signing key for the scheme we construct consists of $d+1$ different signing keys for Π, say $sk_1, sk_2, \ldots, sk_{d+1}$. If message m is encoded by some polynomial t, then a signature on m is of the form $(\sigma_1, \sigma_2, \ldots, \sigma_{d+1})$, where σ_i is a signature using the weakly secure scheme on $t(i)$ using sk_i. Since we only work with polynomials of degree at most d, the $d+1$ points that are signed uniquely determine the polynomial t, hence the message m. Homomorphicity of the scheme that we construct follows from that of the underlying scheme. Given signatures $(\sigma_1^1, \sigma_2^1, \ldots, \sigma_{d+1}^1)$ and $(\sigma_1^2, \sigma_2^2, \ldots, \sigma_{d+1}^2)$ for messages m_1 and m_2, a signature on $m_1 \circ m_2$ (where $\circ$ is one of the operations in $\mathbb{Z}_p$) is $(\sigma_1^1 \circ \sigma_1^2, \sigma_2^1 \circ \sigma_2^2, \ldots, \sigma_{d+1}^1 \circ \sigma_{d+1}^2)$. Without going into the details, a key idea of using the encoding of messages into polynomials is that a simulator can adaptively sign arbitrary messages, while having access only to signatures (of Π) on a set of random messages.

Our construction of the weakly-secure signature scheme is based on graded encodings [20]. The overview here uses the (more idealized) leveled multilinear maps setting. The basic idea is that a signature on a data set message m_i is a level-1 element of the form $\Lambda = g^{(r_i - m_i x)b}$, where g^{r_i} is some public information, b is the secret key and g, g^x, g^b are in the public key[1]. Given signatures on messages m_1 and m_2 one obtains a signature on the sum by simply computing $\Lambda_1 \cdot \Lambda_2$. To obtain a signature on the multiplication $m_1 \cdot m_2$, we apply the graded map to $g^{(r_1 - m_1 x)b}$ and $g^{(r_2 - m_2 x)b}$ and obtain something of the form $g_2^{[r_1 r_2 - (r_1 m_2 + r_2 m_1)x + m_1 m_2 x^2]b^2}$ where g_2 is a generator of $\mathbb{G}_2$. The main issue with the resulting signature is verification: here, one should either know the original messages m_1, m_2 (which is what we want to avoid) or keep track in the signature of the middle term in the exponent (which we also want to avoid since this term grows with successive multiplications). We solve this problem with two main ideas: (1) we publish a randomized version g^{abx} of the secret value g^{bx}, and (2) we create a twin version of every signature which has the form $\Gamma = g^{(r - mx)ab}$. This way, a signature on the multiplication of $m_1 \cdot m_2$ is obtained by applying the graded map to $\Lambda_1 = g^{(r_1 - m_1 x)b}$ and $\Gamma_2 = g^{(r_2 - m_2 x)ab}$, which produces something of the form $g_2^{[r_1 r_2 - (r_1 m_2 + r_2 m_1)x + m_1 m_2 x^2]ab^2}$. Then, by using g^{abx}, the latter value can now be "cleaned up" (by multiplying appropriately computed values) to obtain $g_2^{[r_1 r_2 - m_1 m_2 x^2]ab^2}$. More generally, we show how to clean the multiplication of arbitrary signatures to always produce a signature of the form $g_i^{[f(r) - f(m)x^i]a^{i-1}b^i}$, where i is the degree of polynomial f, g_i is the generator in $\mathbb{G}_i$, m is the vector of original messages, and r is the vector of r_i' in the publicly known $g^{r_1}, g^{r_2}, \ldots$. Related issues that we solve include enabling verification of these signatures, and ensuring that the cleaning information does not enable the creation of forgeries. Also, while the simplified description above works for signatures in a single dataset, our full realization provides a way to deal with multiple datasets. The construction sketched above is homomorphic for polynomials of degree-d, if instantiated with $2d$-linear maps, and is proven weakly-secure under a new, constant-size, assumption that we prove hard in the generic multilinear group model. As a final note, we observe that this construction enjoys efficient verification, which (intuitively) follows from that one can precompute $f(r)$ and reuse it to verify *all* signatures for the same f. In terms of efficiency, in this weakly-secure construction every signature consists of the message m and two group elements—Λ, Γ—and is in principle of constant size. When instantiated with currently known graded encoding schemes, each of these group elements

---

[1] We emphasize that our signatures are quite different, and we only use these to explain the intuition.

(aka encodings) is of size $O(d^2 + d \log n)$ (ignoring the security parameter), if we want to support n-variate polynomials of degree d.

By applying our generic transformation we obtain an adaptive-secure homomorphic signature in which signatures have size $O(d)$, which turns into $O(d^3 + d^2 \log n)$ when instantiated with known graded encoding schemes [20,17,25]. We also show a more optimized transformation tailored to our weakly-secure scheme, which yields a more efficient adaptive-secure homomorphic signature where, for instance, the size of the public and the secret key does not grow by a factor of d. Furthermore, we show that our weakly-secure scheme can be also proven adaptive-secure, though by assuming a stronger, interactive, assumption.

2   Preliminaries

2.1   Leveled Multilinear Maps and Graded Encodings

In this section we recall the definition of leveled multilinear maps and the computational assumptions used in our scheme. Candidate implementations of this abstraction have been recently proposed [20,17,25] in the form of *graded encodings*, a concept similar to generic, leveled multilinear maps.

In generic, symmetric, leveled multilinear maps we assume the existence of an algorithm $\mathcal{G}(1^\lambda, k)$ that, on input the security parameter and an integer k indicating the number of levels (i.e., the number of allowed pairing operations), generates the description pp of leveled multilinear groups $(\mathbb{G}_1, \ldots, \mathbb{G}_k)$, each of large prime order $p > 2^\lambda$. We let g_i be a canonical generator of $\mathbb{G}_i$; we assume that pp includes $g_1 \in \mathbb{G}_1$. The groups are such that there exists a set of bilinear maps $\{e_{i,j} : \mathbb{G}_i \times \mathbb{G}_j \to \mathbb{G}_{i+j}\}_{i,j \geq 1, i+j \leq k}$ such that $\forall a, b \in \mathbb{Z}_p : e_{i,j}(g_i^a, g_j^b) = g_{i+j}^{ab}$. When obvious from the context we drop the indices i, j from $e_{i,j}$. We work with symmetric bilinear maps and we let the canonical generators $g_i \in \mathbb{G}_i$ be obtained by repeatedly applying the map to g_1, i.e. we let $g_i = e(g_1, g_{i-1})$.

HARDNESS ASSUMPTION.   Below we define the computational assumption that underlies the security of our scheme. In the full version we justify the assumption by proving it holds in a generic model for level multilinear maps. The assumption can also be tested using recently proposed automated techniques [5]. Informally, the assumption says that given the level-1 encodings $g_1^a, g_1^b, g_1^{ab}, g_1^x, g_1^{xa}, g_1^{abx}$ with $a, b, x \in \mathbb{Z}_p$ random, it must be hard to compute a level-k encoding of $a^{k-1}(bx)^k$ (i.e., $g_k^{a^{k-1}(bx)^k}$). More formally:

Definition 1 (k-Augmented-Power Multilinear Diffie-Hellman). *Let* pp *be the description of a set of multilinear groups and* $g_1 \in \mathbb{G}_1$ *be a random generator. Let* $a, b, x \xleftarrow{\$} \mathbb{Z}_p$ *be chosen at random. We define the advantage of an adversary* $\mathcal{A}$ *in solving the k-APMDH problem as* $\mathbf{Adv}_{\mathcal{A}}^{APMDH}(\lambda) = \Pr[\mathcal{A}(g_1, g_1^a, g_1^b, g_1^{ab}, g_1^x, g_1^{ax}, g_1^{abx}) = g_k^{a^{k-1}(bx)^k}]$, *and we say that the k-APMDH assumption holds for* $\mathcal{G}$ *if for every PPT* $\mathcal{A}$, $\mathbf{Adv}_{\mathcal{A}}^{APMDH}(\lambda)$ *is negligible in* λ.

Graded Encodings.   Informally speaking, a k-graded encoding system for a ring R includes a system of sets $\{S_i^{(\alpha)} \subset \{0,1\}^* : i \in [0, k], \alpha \in R\}$ such that for every fixed $i \in [0, k]$ the sets $\{S_i^{(\alpha)} : \alpha \in R\}$ are disjoint. The set $S_i^{(\alpha)}$ contains the level-i encodings of $\alpha \in R$. As a first requirement, the system needs an algorithm to obtain an encoding $a_i \in S_i^{(\alpha)}$ of some ring element α (notice that such encoding can be randomized). Additionally, the encoding system is homomorphic in a graded sense. Namely, let us abuse notation and assume that every set $S_i^{(\alpha)}$ is a ring where $+, \cdot$ are the usual addition/multiplication operations. Then, for any $a_i \in S_i^{(\alpha)}$ and $b_i \in S_i^{(\beta)}$ we have $a_i + b_i \in S_i^{(\alpha+\beta)}$. Furthermore, for $a_i \in S_i^{(\alpha)}$ and $b_j \in S_j^{(\beta)}$ we have $a_i \cdot b_j \in S_{i+j}^{(\alpha \cdot \beta)}$, if $i + j \leq k$. Finally, the encoding system has an algorithm to test if a given a is an encoding of 0 in the last level k, i.e., if $a \in S_k^{(0)}$. We refer to [20] or the full version of our work for a more precise description of graded encodings.

2.2 Homomorphic Signatures for Multi-labeled Programs

In this section we provide the definition of homomorphic signatures. Our definition is essentially the same as the one proposed by Freeman in [19] except that we adapt it to work in the model of *multi-labeled programs* introduced in [4] as an extension to labeled programs [24,10].

Multi-labeled Programs. A *labeled program* $\mathcal{P}$ consists of a tuple $(f, \tau_1, \ldots, \tau_n)$ such that $f : \mathcal{M}^n \to \mathcal{M}$ is a function on n variables (e.g., a circuit), and $\tau_i \in \{0, 1\}^*$ is the label of the i-th variable input of f. Labeled programs can be composed in the following way. Given $\mathcal{P}_1, \ldots, \mathcal{P}_t$ and a function $g : \mathcal{M}^t \to \mathcal{M}$, the *composed program* $\mathcal{P}^*$ is the one obtained by evaluating g on the outputs of $\mathcal{P}_1, \ldots, \mathcal{P}_t$, and is compactly denoted as $\mathcal{P}^* = g(\mathcal{P}_1, \ldots, \mathcal{P}_t)$. The labeled inputs of $\mathcal{P}^*$ are all distinct labeled inputs of $\mathcal{P}_1, \ldots, \mathcal{P}_t$, i.e., all inputs with the same label are grouped together in a single input of the new program. Let $f_{id} : \mathcal{M} \to \mathcal{M}$ be the canonical identity function and $\tau \in \{0, 1\}^*$ be a label. Then $\mathcal{I}_\tau = (f_{id}, \tau)$ is the *identity program* for input label τ. Using this notation, observe that any program $\mathcal{P} = (f, \tau_1, \ldots, \tau_n)$ can be expressed as the composition of n identity programs $\mathcal{P} = f(\mathcal{I}_{\tau_1}, \ldots, \mathcal{I}_{\tau_n})$.

A *multi-labeled program* $\mathcal{P}_\Delta$ is a pair $(\mathcal{P}, \Delta)$ in which $\mathcal{P} = (f, \tau_1, \ldots, \tau_n)$ is a labeled program and $\Delta \in \{0, 1\}^*$ is a binary string called the *data set identifier*. Multi-labeled programs allow for composition within the same data set in the most natural way, i.e., given multi-labeled programs $(\mathcal{P}_1, \Delta), \ldots, (\mathcal{P}_t, \Delta)$ sharing the same data set identifier Δ, and given a function $g : \mathcal{M}^t \to \mathcal{M}$, the *composed multi-labeled program* $\mathcal{P}_\Delta^*$ is the pair $(\mathcal{P}^*, \Delta)$ where $\mathcal{P}^*$ is the composed program $g(\mathcal{P}_1, \ldots, \mathcal{P}_t)$, and Δ is the data set identifier shared by all the $\mathcal{P}_i$. Similarly to the labeled case, we define a multi-labeled identity program as $\mathcal{I}_{(\Delta, \tau)} = ((f_{id}, \tau,), \Delta)$.

Definition 2 (Homomorphic Signatures). *A homomorphic signature scheme* HomSig *is a tuple of probabilistic, polynomial-time algorithms* (KeyGen, Sign, Ver, Eval) *satisfying four properties:* authentication correctness, evaluation correctness, succinctness, *and* security. *More precisely:*

KeyGen$(1^\lambda, \mathcal{L})$ *takes a security parameter* λ, *the description of the label space* $\mathcal{L}$ *(possibly fixing a maximum data set size* N*), and outputs a public key* vk

and a secret key sk. The public key vk defines implicitly a message space $\mathcal{M}$
and a set $\mathcal{F}$ of admissible functions.

Sign(sk, Δ, τ, m) takes a secret key sk, a data set identifier Δ, a label $\tau \in \mathcal{L}$, a
message $m \in \mathcal{M}$, and it outputs a signature σ.

Ver(vk, $\mathcal{P}_\Delta, m, \sigma$) takes a public key vk, a multi-labeled program $\mathcal{P}_\Delta = ((f, \tau_1,$
$\ldots, \tau_n), \Delta)$ with $f \in \mathcal{F}$, a message $m \in \mathcal{M}$, and a signature σ. It outputs
either 0 (reject) or 1 (accept).

Eval(vk, $f, \boldsymbol{\sigma}$) takes a public key vk, a function $f \in \mathcal{F}$ and a tuple of signatures
$\{\sigma_i\}_{i=1}^n$ (assuming that f takes n inputs). It outputs a new signature σ.

AUTHENTICATION CORRECTNESS. Intuitively, a homomorphic signature satisfies authentication correctness if the signatures generated by Sign(sk, Δ, τ, m)
verify correctly for m as the output of the identity program $\mathcal{I}_{(\Delta, \tau)}$. Formally,
HomSig has authentication correctness if for a given label space $\mathcal{L}$, all key pairs
(sk, vk) $\xleftarrow{\$}$ KeyGen($1^\lambda, \mathcal{L}$), any label $\tau \in \mathcal{L}$, data set identifier $\Delta \in \{0,1\}^*$, and
any signature $\sigma \xleftarrow{\$}$ Sign(sk, Δ, τ, m), Ver(vk, $\mathcal{I}_{(\Delta, \tau)}, m, \sigma$) outputs 1 with all but
negligible probability.

EVALUATION CORRECTNESS. Informally, this property says that running the
evaluation algorithm on signatures $(\sigma_1, \ldots, \sigma_n)$ such that σ_i verifies for m_i as the
output of a multi-labeled program $(\mathcal{P}_i, \Delta)$, produces a signature σ which verifies
for $f(m_1, \ldots, m_n)$ as the output of the composed program $(f(\mathcal{P}_1, \ldots, \mathcal{P}_n), \Delta)$.
More formally, fix a key pair (sk, vk) $\xleftarrow{\$}$ KeyGen($1^\lambda, \mathcal{L}$), a function $g : \mathcal{M}^t \to \mathcal{M}$
and any set of program/message/signature triples $\{(\mathcal{P}_i, m_i, \sigma_i)\}_{i=1}^t$ such that
Ver(vk, $\mathcal{P}_i, m_i, \sigma_i$) = 1. If $m^* = g(m_1, \ldots, m_t)$, $\mathcal{P}^* = g(\mathcal{P}_1, \ldots, \mathcal{P}_t)$, and $\sigma^* =$
Eval(vk, $g, (\sigma_1, \ldots, \sigma_t)$), then Ver(vk, $\mathcal{P}^*, m^*, \sigma^*$) = 1 holds with all but negligible
probability.

SUCCINCTNESS. A homomorphic signature scheme is succinct if, for a fixed
security parameter λ, the size of the signatures depends at most logarithmically
on the data set size N.

SECURITY. We say that a homomorphic signature scheme HomSig is secure if
for every PPT adversary $\mathcal{A}$ we have $\Pr[\text{HomUF-CMA}_{\mathcal{A}, \text{HomSig}}(\lambda) = 1] \leq \epsilon(\lambda)$
where $\epsilon(\lambda)$ is a negligible function, and the experiment $\text{HomUF-CMA}_{\mathcal{A}, \text{HomSig}}(\lambda)$
is defined as follows.

Key generation The challenger runs (vk, sk) $\xleftarrow{\$}$ KeyGen($1^\lambda, \mathcal{L}$) and gives vk to
the adversary.

Signing Queries The adversary can adaptively submit queries of the form
(Δ, τ, m), where Δ is a dataset identifier, $\tau \in \mathcal{L}$, and $m \in \mathcal{M}$. The challenger
proceeds as follows: If (Δ, τ, m) is the first query with data set identifier Δ,
then the challenger initializes an empty list $T_\Delta = \emptyset$ for Δ. If T_Δ does not
already contain a tuple $(\tau, \cdot)$ (i.e., the adversary never asked for a query
$(\Delta, \tau, \cdot)$), the challenger computes $\sigma \xleftarrow{\$}$ Sign(sk, Δ, τ, m), returns σ to $\mathcal{A}$ and
updates the list $T_\Delta \leftarrow T_\Delta \cup (\tau, m)$. If $(\tau, m) \in T_\Delta$ (i.e., the adversary had already queried the tuple (Δ, τ, m)), then the challenger replies with the same
signature generated before. If T_Δ contains a tuple (τ, m') for some message
$m' \neq m$, then the challenger ignores the query.

Forgery The previous stage is repeated a polynomial number of times until the adversary outputs a tuple $(\mathcal{P}^*_{\Delta^*}, m^*, \sigma^*)$.

Finally, the experiment outputs 1 if the tuple returned by the adversary is a forgery, and 0 otherwise. However, to do this we need to provide a way for characterizing forgeries in this model. To this end, we recall the notion of well-defined program w.r.t. a list T_Δ [19]. A labeled program $\mathcal{P}^* = (f^*, \tau_1^*, \ldots, \tau_n^*)$ is *well-defined with respect to* T_{Δ^*} if one of the following two cases holds:

- there exist messages $m_1, \ldots, m_n$ such that the list T_{Δ^*} contains all tuples $(\tau_1^*, m_1), \ldots, (\tau_n^*, m_n)$. Intuitively, this means that the challenger has generated signatures for the entire input space of f for data set Δ^*.
- there exist indices $i \in \{1, \ldots, n\}$ such that $(\tau_i^*, \cdot) \notin T_{\Delta^*}$ (i.e., $\mathcal{A}$ never asked signing queries of the form $(\Delta^*, \tau_i^*, \cdot)$), and the function $f^*(\{m_j\}_{(\tau_j, m_j) \in T_{\Delta^*}} \cup \{\tilde{m}_j\}_{(\tau_j, \cdot) \notin T_{\Delta^*}})$ outputs the same value for all possible choices of $\tilde{m}_j \in \mathcal{M}$. Intuitively, this case means that the inputs that were not signed in the experiment never contribute to the computation of f.

The experiment HomUF-CMA outputs 1 if and only if $\mathsf{Ver}(\mathsf{vk}, \mathcal{P}^*_{\Delta^*}, m^*, \sigma^*) = 1$ and one of the following conditions holds:

- *Type 1 Forgery:* no list T_{Δ^*} was created during the game, i.e., during the experiment no message m has ever been signed with respect to a data set identifier Δ^*.
- *Type 2 Forgery:* $\mathcal{P}^*$ is well-defined w.r.t. T_{Δ^*} and $m^* \neq f^*(\{m_j\}_{(\tau_j, m_j) \in T_{\Delta^*}})$, i.e., m^* is not the correct output of the labeled program $\mathcal{P}^*$ when executed on previously signed messages $(m_1, \ldots, m_n)$.
- *Type 3 Forgery:* $\mathcal{P}^*$ is *not* well-defined w.r.t. T_{Δ^*}.

As pointed out by Freeman [19], for a general class of functions it may not be possible for the challenger to efficiently decide whether a given program is well-defined or not. Freeman shows that for the case of linearly-homomorphic signatures this is not an issue. More precisely he shows that any adversary who outputs a Type-3 forgery can be converted into one that outputs a Type-2 forgery. Below, we show two simple propositions that allow to overcome this issue for the case of homomorphic signatures whose class of supported functions are arithmetic circuits of degree d, over a finite field of order p such that $d/p < 1/2$. The first proposition is taken from [11] and provides a way to probabilistically test whether a program is well-defined.

Proposition 1 ([11]). *Let $\lambda, n \in \mathbb{N}$ and let $\mathcal{F}$ be the class of arithmetic circuits $f : \mathbb{F}^n \to \mathbb{F}$ over a finite field $\mathbb{F}$ of order p and such that the degree of f is at most d, for $\frac{d}{p} < \frac{1}{2}$. Then, there exists a probabilistic polynomial-time algorithm that for any given $f \in \mathcal{F}$, decides if there exists $y \in \mathbb{F}$ such that $f(\boldsymbol{u}) = y, \forall \boldsymbol{u} \in \mathbb{F}^n$ (i.e., if f is constant) and is correct with probability at least $1 - 2^{-\lambda}$.*

The second proposition below is the analogue of the one proven by Freeman, which shows that any adversary who outputs a Type-3 forgery can be converted into one that outputs a Type-2 forgery. This result has been proven for homomorphic MACs in [11]. Here we extend it to homomorphic signatures. For lack of space, its proof appears in the full version.

Proposition 2. *Let $\lambda \in \mathbb{N}$ be the security parameter, and let $\mathcal{F}$ be the class of arithmetic circuits $f : \mathbb{F}^n \to \mathbb{F}$ over a finite field $\mathbb{F}$ of order p and such that the degree of f is at most d, for $\frac{d}{p} < \frac{1}{2}$. Let HomSig be a signature scheme with message space $\mathbb{F}$, and let $\mathcal{E}_b$ be the event that the adversary returns a Type-b forgery (for $b = 1, 2, 3$) in experiment HomUF-CMA. Then, if for any adversary $\mathcal{B}$ we have that $\Pr[\mathsf{HomUF\text{-}CMA}_{\mathcal{B},\mathsf{HomSig}}(\lambda) = 1 \wedge \mathcal{E}_2] \leq \epsilon$, then for any adversary $\mathcal{A}$ producing a Type-3 forgery it holds $\Pr[\mathsf{HomUF\text{-}CMA}_{\mathcal{A},\mathsf{HomSig}}(\lambda) = 1 \wedge \mathcal{E}_3] \leq \epsilon + 2^{-\lambda}$.*

Weakly-Secure Homomorphic Signatures. In our work we also consider a weaker notion of unforgeability for homomorphic signatures. We define experiment Weak-HomUF-CMA$_{\mathcal{A},\mathsf{HomSig}}$ which is a variant of HomUF-CMA$_{\mathcal{A},\mathsf{HomSig}}$. The difference is that before key generation $\mathcal{A}$ declares *all* the signing queries that it will make (i.e., messages), but without necessarily specifying the data set names, i.e., $\mathcal{A}$ outputs $\{m_{\tau,j}\}_{\tau \in \mathcal{L}}$, for $j = 1$ to Q, where Q is the number of different queried datasets. Once applying the above change, in the signing query phase, $\mathcal{A}$ will only specify a data set Δ_j and will receive signatures on $\{(\Delta_j, \tau, m_{\tau,j})\}_{\tau \in \mathcal{L}}$. Also, notice that with this change, there are no Type-3 forgeries as the data sets are always full.

While this security notion may look rather weak, in Section 3 we show a generic way to convert any weakly-secure homomorphic signature for arithmetic circuits of degree d to an adaptively secure one (for the same class of functions.)

2.3   Homomorphic Signatures with Efficient Verification

We propose the notion of homomorphic signatures with efficient verification, which naturally extends to the public-key setting the analogous notion introduced for homomorphic MACs in [4]. Roughly speaking, this property says that the verification algorithm can be split in two phases. In an offline phase, given the verification key vk and a labeled program $\mathcal{P}$, one precomputes a concise key vk$_{\mathcal{P}}$. The latter key can then be used to verify signatures (in the online phase) w.r.t. $\mathcal{P}$ and *any* dataset Δ. Crucially, vk$_{\mathcal{P}}$ can be reused an unbounded number of times, and the verification cost of the online phase is much less than running $\mathcal{P}$. As in [4], this efficiency property is defined in an amortized sense, so that verification is more efficient when the same program $\mathcal{P}$ is executed on different data sets. This property enables the use of homomorphic signatures for publicly-verifiable delegation of computation on outsourced data.

The formal definition follows.

Definition 3. *Let HomSig $=$ (KeyGen, Sign, Ver, Eval) be a homomorphic signature scheme for multi-labeled programs. HomSig satisfies efficient verification if there exist two additional algorithms (VerPrep, EffVer) such that:*

VerPrep(vk, $\mathcal{P}$): *on input the verification key* vk *and a labeled program* $\mathcal{P} = (f, \tau_1, \ldots, \tau_n)$, *this algorithm generates a concise verification key* vk$_\mathcal{P}$. *We stress that this verification key does* not *depend on any data set identifier* Δ.

EffVer(vk$_\mathcal{P}$, Δ, m, σ): *given a verification key* vk$_\mathcal{P}$, *a data set identifier* Δ, *a message* $m \in \mathcal{M}$ *and a signature* σ, *the efficient verification algorithm outputs 0 (reject) or 1 (accept).*

The above algorithms are required to satisfy the following two properties:

CORRECTNESS. *Let* $(\mathsf{sk}, \mathsf{vk}) \xleftarrow{\$} \mathsf{KeyGen}(1^\lambda)$ *be honestly generated keys, and* $(\mathcal{P}_\Delta, m, \sigma)$ *be any program/message/signature tuple with* $\mathcal{P}_\Delta = (\mathcal{P}, \Delta)$ *such that* $\mathsf{Ver}(\mathsf{vk}, \mathcal{P}_\Delta, m, \sigma) = 1$. *Then, for every* $\mathsf{vk}_\mathcal{P} \xleftarrow{\$} \mathsf{VerPrep}(\mathsf{vk}, \mathcal{P})$, $\mathsf{EffVer}(\mathsf{vk}_\mathcal{P}, \Delta, m, \sigma) = 1$ *holds with all but negligible probability.*

AMORTIZED EFFICIENCY. *Let* $\mathcal{P}_\Delta = (\mathcal{P}, \Delta)$ *be a program, let* $(m_1, \ldots, m_n) \in \mathcal{M}^n$ *be any vector of inputs, and let* $t(n)$ *be the time required to compute* $\mathcal{P}(m_1, \ldots, m_n)$. *If* $\mathsf{vk}_\mathcal{P} \leftarrow \mathsf{VerPrep}(\mathsf{vk}, \mathcal{P})$, *then the time required for* $\mathsf{EffVer}(\mathsf{vk}_\mathcal{P}, \Delta, m, \tau)$ *is* $t' = o(t(n))$.

Notice that in our efficiency requirement, we do not include the time needed to compute vk$_\mathcal{P}$. This is justified by the fact that, being vk$_\mathcal{P}$ independent of Δ, the same vk$_\mathcal{P}$ can be re-used in many verifications involving the same labeled program $\mathcal{P}$ but many different Δ. Namely, the cost of computing vk$_\mathcal{P}$ is *amortized* over many verifications of the same function on different data sets.

3 From Weakly-Secure to Adaptive-Secure Homomorphic Signatures

In this section we show how to convert a weakly-secure homomorphic signature that works for arithmetic circuits of degree k, into an adaptive-secure one supporting the same class of functionalities. The only restriction is that the message space is expected to be some finite field, e.g., $\mathbb{Z}_p$ for a prime p, that does not depend on the secret key. In the full version we show how to extend these ideas to the case where the messages and the polynomials supported by the homomorphic signature scheme are defined over the integers.

The basic idea behind the conversion is to interpret the message one wants to sign as the free term of a random degree-1 (univariate) polynomial $t(z)$ defined over a finite field. Next, rather than signing m, one signs $(k + 1)$ points of this polynomial, e.g., $t(1), \ldots, t(k + 1)$, by using $(k + 1)$ different secret keys. To homomorphically evaluate a function over such signatures, one executes the Eval algorithm in a point-wise fashion. Interestingly, the homomorphic properties of the underlying signature scheme remains preserved because of analogous properties of polynomials. The formal description of the scheme follows.

Let HomSig $=$ (KeyGen, Sign, Eval, Ver) be a weakly-secure scheme with message space $\mathbb{Z}_p$, our (adaptive-secure) homomorphic signature HomSig* $=$ (KeyGen*, Sign*, Eval*, Ver*) works as follows.

KeyGen$^*(1^\lambda, k, \mathcal{L})$. Let λ be the security parameter, $k \in \mathbb{N}^+$ be a constant denoting the bound on the degree of the supported polynomials, and $\mathcal{L} \subset \{0,1\}^*$ be a set of admissible labels $\mathcal{L} = \{\tau_1, \ldots, \tau_N\}$, for some $N = \mathsf{poly}(\lambda)$. The algorithms runs $(k+1)$ times KeyGen$(1^\lambda, k, \mathcal{L})$. Denoting by $(\mathsf{vk}_i, \mathsf{sk}_i)$ the public key/secret key pair obtained from the i-th execution of KeyGen, the algorithm outputs $\mathsf{sk} = (\mathsf{sk}_1, \ldots, \mathsf{sk}_{k+1})$, $\mathsf{vk} = (\mathsf{vk}_1, \ldots, \mathsf{vk}_{k+1})$. The message space $\mathcal{M}$ is $\mathbb{Z}_p$

Sign$^*(\mathsf{sk}, \Delta, \tau, m)$. The signing algorithm takes as input the secret key $\mathsf{sk} = (\mathsf{sk}_1, \ldots, \mathsf{sk}_{k+1})$, a data set identifier $\Delta \in \{0,1\}^*$, a label $\tau \in \mathcal{L}$ and a message $m \in \mathbb{Z}_p$. The signing procedure consists of two main steps. First it generates a random degree-1 (univariate) polynomial $t(z)$ such that $t(0) = m \in \mathbb{Z}_p$. Second, for $i = 1, \ldots, k+1$, it signs $t(i)$ using $\sigma_i \xleftarrow{\$} \mathsf{Sign}(\mathsf{sk}_i, \Delta, \tau, t(i))$. In other words, each $t(i)$ is signed with respect to a *different* signing key sk_i. The signing algorithm returns $\sigma = ((\sigma_1, t(1)), \ldots, (\sigma_{k+1}, t(k+1)))$.

Eval$^*(\mathsf{vk}, f, \boldsymbol{\sigma})$. The public evaluation algorithm takes as input the public key vk, an arithmetic circuit $f : \mathbb{Z}_p^n \to \mathbb{Z}_p$ and a vector $\boldsymbol{\sigma}$ of n signatures $\sigma^{(1)}, \ldots, \sigma^{(n)}$ such that $\sigma^{(i)}$ is a $(k+1)$-tuple $((\sigma_1^{(i)}, t^{(i)}(1)), \ldots, (\sigma_{k+1}^{(i)}, t^{(i)}(k+1)))$. Eval* computes a signature $\sigma = ((\sigma_1, t(1)), \ldots, (\sigma_{k+1}, t(k+1)))$, by computing $\sigma_i \leftarrow \mathsf{Eval}(\mathsf{vk}_i, f, (\sigma_i^{(1)}, \ldots, \sigma_i^{(n)}))$ and $t(i) \leftarrow f(t^{(1)}(i), \ldots, t^{(n)}(i))$.

Ver$^*(\mathsf{vk}, \mathcal{P}_\Delta, m, \sigma)$. Let $\mathcal{P}_\Delta = ((f, \tau_1, \ldots, \tau_n), \Delta)$ be a multi-labeled program such that $f : \mathbb{Z}_p^n \to \mathbb{Z}_p$ is an arithmetic circuit of degree $d \le k$. Let $m \in \mathbb{Z}_p$ and $\sigma = ((\sigma_1, t(1)), \ldots, (\sigma_{k+1}, t(k+1)))$.

First of all, Ver* checks that the signatures on all the values $t(i)$ are correct. To do so, it runs $b_i \leftarrow \mathsf{Ver}(\mathsf{vk}_i, \mathcal{P}_\Delta, t(i), \sigma_i)$, $\forall i = 1, \ldots, k+1$. If $b_1 = \ldots = b_{k+1} = 1$ Ver* proceeds to the next step, otherwise it stops and returns 0.

So, if the values $t(i)$ in the signature are valid, Ver* uses these values to interpolate a polynomial $t(z)$ of degree (at most) k. More precisely, this is done as follows: if the degree of the arithmetic circuit f is k, $t(z)$ is interpolated using all the $t(i)$'s; if, on the other hand, f is of degree $d < k$, the algorithm first interpolates $t(z)$ using $t(1), \ldots, t(d+1)$ and then checks that $t(z)$ is correct with respect to $t(d+2), \ldots, t(k+1)$.[2] Finally, Ver* checks whether $t(0) = m$ or not. Again, if any of the above tests fail the algorithm outputs 0, otherwise it outputs 1.

To complete the description of HomSig* we give the algorithms for efficient verification:

VerPrep$^*(\mathsf{vk}, \mathcal{P})$. Let $\mathcal{P} = (f, \tau)$ be a labeled program for an arithmetic circuit $f \in \mathbb{Z}_p^n \to \mathbb{Z}_p$ with labels $\tau = (\tau_1, \ldots, \tau_n)$. For $i = 1$ to $(k+1)$ the algorithm runs $\mathsf{vk}_\mathcal{P}^{(i)} = \mathsf{VerPrep}(\mathsf{vk}_i, \mathcal{P})$ and returns the efficient verification key $\mathsf{vk}_\mathcal{P} = (\mathsf{vk}_\mathcal{P}^{(1)}, \ldots, \mathsf{vk}_\mathcal{P}^{(k+1)})$.

EffVer$^*(\mathsf{vk}_\mathcal{P}, \Delta, m, \sigma)$. Let $(\sigma = ((\sigma_1, t(1)), \ldots, (\sigma_{(k+1)}, t(k+1)))$. For $i = 1$ to $(k+1)$, the online verification algorithm runs $b_i \leftarrow \mathsf{EffVer}(\mathsf{vk}_\mathcal{P}^{(i)}, \Delta, t(i), \sigma_i)$. If

---

[2] This is done by simply recomputing the interpolated polynomial on points $(d+2), \ldots, (k+1)$.

the $t(i)$'s correctly interpolate to m and $\wedge_{i=1}^{k+1} b_i = 1$, output 1. Otherwise output 0. Notice that if the EffVer provides efficient verification, then EffVer* has efficient verification as well.

In the following theorem (its proof is in the full version), we show that if HomSig is a weakly-secure scheme, our transformation yields an adaptive-secure homomorphic signature.

Theorem 1. *If* HomSig *is a weakly-secure homomorphic signature scheme for arithmetic circuits of degree* $d \leq k$ *then* HomSig* *is an adaptive-secure homomorphic signature scheme for the same class of circuits.*

4   A Weakly-Secure Homomorphic Signature

In this section we describe our construction of homomorphic signatures with efficient verification from leveled multilinear maps. When working with $2k$-linear maps, our scheme can support the evaluation of arithmetic circuits of degree k. The scheme presented in this section is proven weakly-secure under the AP-MDH assumption (Definition 1). This construction can then be turned into an adaptive-secure scheme by either applying our generic transformation of Section 3, or by tailoring our generic technique to this scheme.

Here we describe the scheme using the abstraction of leveled multilinear maps. A discussion about implementing the scheme with graded encodings is given later in this section and more details appear in the full version.

Without loss of generality our scheme works with arithmetic circuits in which addition gates take inputs of the same degree. Notice that any arithmetic circuit $f : \mathbb{F}^n \to \mathbb{F}$ of degree d can be converted into another circuit $\tilde{f} : \mathbb{F}^{n+1} \to \mathbb{F}$ of the same degree d such that $\tilde{f}$ can compute the same function of f. The idea of the transformation is very simple: one first adds to $\tilde{f}$ (say at the end) one additional input wire, labeled by u; then, whenever there is an addition gate taking inputs x_1, x_2 such that $deg(x_1) < deg(x_2)$, one multiplies x_1 by u as many times as needed to obtain a wire x_1' such that $deg(x_1') = deg(x_2)$. Finally, by assigning 1 to the input labeled by u, it is easy to see that $\tilde{f}(m_1, \ldots, m_n, 1) = f(m_1, \ldots, m_n)$. From now on we assume that the circuits used in our scheme have this form.

In what follows we provide a full-detailed description of our construction, which is rather intricate. We refer the reader to the introduction for a more intuitive explanation of our ideas.

To build our scheme we use a regular signature scheme $\Sigma' = (\mathsf{KeyGen'}, \mathsf{Sign'}, \mathsf{Ver'})$, a pseudorandom function $F : \mathcal{K} \times \{0,1\}^* \to \mathbb{Z}_p^2$ with key space $\mathcal{K}$, and an implementation of leveled multilinear groups whose description is generated by $\mathcal{G}$. Our homomorphic signature scheme HomSig = (KeyGen, Sign, Eval, Ver) works as follows.

KeyGen$(1^\lambda, k, \mathcal{L})$. Let λ be the security parameter, $k \in \mathbb{N}^+$ be a constant denoting the bound on the degree of the supported polynomials, and $\mathcal{L} \subset \{0,1\}^*$ be a set of admissible labels $\mathcal{L} = \{u\} \cup \{\tau_1, \ldots, \tau_N\}$, for some $N = \mathrm{poly}(\lambda)$. Here "u" (which stands for "unity") is a special additional label that is used for the modified arithmetic circuits in which addition gates always take in homogenous monomials. The set of labels is implicitly defining the maximum data set size N supported by the scheme. The key generation algorithm works as follows.

- Generate a key pair $(\mathsf{sk}', \mathsf{vk}') \xleftarrow{\$} \mathsf{KeyGen}'(1^\lambda)$ for the regular signature scheme.
- Choose a random seed $K \xleftarrow{\$} \mathcal{K}$ for the PRF $F_K : \{0,1\}^* \to \mathbb{Z}_p^2$.
- Run $\mathcal{G}(1^\lambda, 2k)$ to generate the description of $(2k)$-linear groups $\mathbb{G}_1, \ldots, \mathbb{G}_{2k}$ of order p, where p is a prime number of roughly λ bits. In the scheme, we use group elements with subscripts to denote the group they live in. Also, for an $h_1 \in \mathbb{G}_1$, we denote by $h_i \in \mathbb{G}_1$ the i-fold graded multiplication of h_1. Analogous notation is used for other group elements.
- Choose random elements $g_1, h_1 \xleftarrow{\$} \mathbb{G}_1$ as well as $N+1$ random values $R_\tau \xleftarrow{\$} \mathbb{G}_1, \forall \tau \in \mathcal{L}$.

Finally, output $\mathsf{sk} = (\mathsf{sk}', K)$, $\mathsf{vk} = (\mathsf{vk}', g_1, h_1, \{R_\tau\}_{\tau \in \mathcal{L}})$, and let the message space $\mathcal{M}$ be $\mathbb{Z}_p$.

$\mathsf{Sign}(\mathsf{sk}, \Delta, \tau, m)$. The signing algorithm takes as input the secret key $\mathsf{sk} = (\mathsf{sk}', K)$, a data set identifier $\Delta \in \{0,1\}^*$, a label $\tau \in \mathcal{L}$ and a message $m \in \mathbb{Z}_p$. The signing procedure consists of two main steps. First, it uses the pseudorandom function to (re-)derive some common parameters for the dataset Δ and signs the public part of these parameters using the regular signature scheme. Second, it uses the secret part of the parameters for Δ to create the homomorphic component of the signature which is the one strictly bound to (Δ, τ, m). The latter procedure is the core of our technique. We describe it below as a separate subroutine.

- $\mathsf{HomSign}(\mathsf{vk}, a, b, \tau, m)$: this algorithm simply computes $\Lambda_1 = (R_\tau h_1^{-m})^b$, $\Gamma_1 = \Lambda_1^a$, and returns $\nu = (m, \Lambda_1, \Gamma_1)$.

The full signing algorithm proceeds as follows.

1. Derive two integers $(a, b) \leftarrow F_K(\Delta)$ using the pseudorandom function, and compute $A_1 = g_1^a, B_1 = g_1^b, C_1 = g_1^{ab}, T_1 = h_1^a, U_1 = h_1^{ab}$.
2. Run the routine $\mathsf{HomSign}(\mathsf{vk}, a, b, u, 1)$ described above, to compute a triple $\nu_{\Delta,u} = (1, \Lambda_u, \Gamma_u) \in \mathbb{Z}_p \times \mathbb{G}_1^2$. The tuple $\nu_{\Delta,u}$ is essentially the homomorphic component of a signature of "1" with respect to the special label "u" and for the dataset Δ. This signature $\nu_{\Delta,u}$ is needed to perform the homomorphic evaluations on the modified circuits.
3. Let $\mathsf{pp}_\Delta = (\Delta, A_1, B_1, C_1, T_1, U_1, \nu_{\Delta,u})$ be the public parameters of dataset Δ. Then sign pp_Δ using the regular signature scheme, i.e., compute $\sigma_\Delta \leftarrow \mathsf{Sign}'(\mathsf{sk}', \mathsf{pp}_\Delta)$.
4. Run $\mathsf{HomSign}(\mathsf{vk}, a, b, \tau, m)$ to generate a tuple $\nu = (m, \Lambda_1, \Gamma_1) \in \mathbb{Z}_p \times \mathbb{G}_1^2$.

Finally, the signing algorithm returns the signature $\sigma = (\mathsf{pp}_\Delta, \sigma_\Delta, \nu)$. Observe that when generating many signatures for the same dataset Δ the steps 1–3 can be executed only once.

$\mathsf{Eval}(\mathsf{vk}, f, \boldsymbol{\sigma})$. The public evaluation algorithm takes as input the public key vk, an arithmetic circuit $f : \mathbb{Z}_p^n \to \mathbb{Z}_p$ and a vector $\boldsymbol{\sigma}$ of n signatures

$\sigma^{(1)}, \ldots, \sigma^{(n)}$ such that $\sigma^{(i)} = (\mathsf{pp}_\Delta^{(i)}, \sigma_\Delta^{(i)}, \nu_i)$ for $i = 1, \ldots, n$. Eval computes a signature $\sigma = (\mathsf{pp}_\Delta, \sigma_\Delta, \nu)$ as follows.

First, set $\mathsf{pp}_\Delta = \mathsf{pp}_\Delta^{(1)}$ and $\sigma_\Delta = \sigma_\Delta^{(1)}$. Namely, we take the common parameters of the first signature in the vector. Observe that our notion of evaluation correctness works for signatures in the same data set, i.e., all these signatures are supposed to share the same parameters.

In the second stage, Eval computes the homomorphic component ν by homomorphically evaluating the circuit f over the values $\{\nu_i\}_{i=1}^n$. To do so, it proceeds over f gate by gate.

At every gate f_g, given two values ν_1, ν_2 (or a value ν_1 and a constant $c \in \mathbb{Z}_p$), Eval runs the algorithm $\nu \leftarrow \mathsf{GateEval}(\mathsf{vk}, \mathsf{pp}_\Delta, f_g, \nu_1, \nu_2)$ described below that returns a new value ν, which is in turn passed on as input to the next gate in the circuit. When the computation reaches the last gate of the circuit f, Eval outputs the value ν obtained by running GateEval on such last gate. On input $\nu_1 = (m_1, \Lambda_i^{(1)}, \Gamma_i^{(1)}) \in \mathbb{Z}_p \times \mathbb{G}_i^2$ and $\nu_2 = (m_2, \Lambda_j^{(2)}, \Gamma_j^{(2)}) \in \mathbb{Z}_p \times \mathbb{G}_j^2$, $\mathsf{GateEval}(\mathsf{vk}, \mathsf{pp}_\Delta, f_g, \nu_1, \nu_2)$ proceeds as follows. For an *addition* gate f_+, it computes $m = m_1 + m_2$, $\Lambda_i = \Lambda_i^{(1)} \cdot \Lambda_i^{(2)}$, and $\Gamma_i = \Gamma_i^{(1)} \cdot \Gamma_i^{(2)}$. For a *multiplication-by-constant* gate $f_\times$ and constant $c \in \mathbb{Z}_p$, it computes $m = c \cdot m_1$, $\Lambda_i = (\Lambda_i^{(1)})^c$, and $\Gamma_i = (\Gamma_i^{(1)})^c$. For a *multiplication* gate $f_\times$, it computes $m = m_1 \cdot m_2$, $\Lambda_d = e(\Lambda_i^{(1)}, \Gamma_j^{(2)}) \cdot e(\Lambda_i^{(1)}, U_j^{m_2}) \cdot e(U_i^{m_1}, \Lambda_j^{(2)})$, and $\Gamma_d = e(\Gamma_i^{(1)}, \Gamma_j^{(2)}) \cdot e(\Gamma_i^{(1)}, U_j^{m_2}) \cdot e(U_i^{m_1}, \Gamma_j^{(2)})$.

Ver$(\mathsf{vk}, \mathcal{P}_\Delta, m, \sigma)$. Let $\mathcal{P}_\Delta = ((f, \tau_1, \ldots, \tau_n), \Delta)$ be a multi-labeled program such that $f : \mathbb{Z}_p^n \to \mathbb{Z}_p$ is an arithmetic circuit of degree $d \leq k$. Let $m \in \mathbb{Z}_p$ and $\sigma = (\mathsf{pp}_\Delta, \sigma_\Delta, \nu)$ be a signature with $\nu = (m, \Lambda_d, \Gamma_d) \in \mathbb{Z}_p \times \mathbb{G}_d^2$. First, run Ver$'(\mathsf{vk}', \mathsf{pp}_\Delta, \sigma_\Delta)$ to check that σ_Δ is a valid signature on pp_Δ for the same Δ taken as input by the verification algorithm. If σ_Δ is valid, then proceed as follows. Otherwise, stop and return 0 (reject).

Use the graded maps to evaluate the circuit f on the values $(R_{\tau_1}, \ldots, R_{\tau_n})$. Namely, replace additions in f (for inputs of degree i) by the group operation in $\mathbb{G}_i$, whereas a multiplication in f, with inputs of degree i and j respectively, is replaced by evaluating the graded map $e_{i,j}$. We compactly denote this operation as $R = f(R_{\tau_1}, \ldots, R_{\tau_n}) \in \mathbb{G}_d$. Next, output 1 only if the following two equations are satisfied:

$$e(R \cdot h_d^{-m}, g_d^{a^{d-1}b^d}) = e(\Lambda_d, g_d) \tag{1}$$

$$e(\Lambda_d, A_1) = e(\Gamma_d, g_1) \tag{2}$$

Finally, to complete the description of HomSig we give the algorithms for efficient verification:

VerPrep$(\mathsf{vk}, \mathcal{P})$. Let $\mathcal{P} = (f, \tau)$ be a labeled program for an arithmetic circuit $f \in \mathbb{Z}_p^n \to \mathbb{Z}_p$ with labels $\tau = (\tau_1, \ldots, \tau_n)$. The algorithm computes $R = f(R_{\tau_1}, \ldots, R_{\tau_n}) \in \mathbb{G}_d$, h_d, $g_d^{a^{d-1}b^d} = e(C_{d-1}, B_1)$, and returns the concise verification key $\mathsf{vk}_\mathcal{P} = (\mathsf{vk}', g_1, h_d, g_d^{a^{d-1}b^d}, R)$.

EffVer($\mathsf{vk}_{\mathcal{P}}, \Delta, m, \sigma$). The online verification is basically the same as Ver except that the values $R, h_d, g_d^{a^{d-1}b^d}$ have been already computed in the off-line phase and are now part of the online algorithm's input. Notice that the computational complexity of the online verification depends only on the complexity of computing the group operations and the bilinear maps in equations (1), (2). Using current graded encoding schemes, the cost essentially becomes $\mathsf{poly}(k, \log N)$ which is much less than the cost of evaluating an N-variate polynomial of degree k.

It is easy to see that running the combination of VerPrep and EffVer produces the same result as running Ver.

Very intuitively, the correctness of the scheme follows by that, for any operation $+, \times$, GateEval preserves the form of the signatures, i.e., $\Lambda_i = (Rh_i^{-m})^{a^{i-1}b^i}$ and $\Gamma_i = \Lambda_i^a$.

In the following theorem we prove that HomSig is a weakly-secure homomorphic signature scheme. For lack of space, the proof of security and a formal proof of correctness appear in the full version.

Theorem 2. *If Σ' is an unforgeable signature scheme, F is a pseudorandom function, and $\mathcal{G}$ is the generator of $2k$-linear groups such that the $2k$-APMDH assumption holds for $\mathcal{G}$, then HomSig is a weakly-secure homomorphic signature scheme for arithmetic circuits of degree k.*

Achieving Adaptive Security. In order to achieve adaptive security for the scheme described above, we discuss three different approaches. The first one is to apply our generic transformation of Section 3. In the transformed scheme, both public/secret keys and the signatures are longer by a factor of d, that for the class of functions considered in this work is assumed to be independent of n. As a second possibility, we exploit the specific structure of our weakly-secure scheme, and show a more optimized transformation which avoids increasing the size of public and secret keys, i.e., they remain of the same size as in HomSig. Finally, as a third possibility, we show that, under a stronger, interactive variant of the APMDH assumption, the scheme HomSig is by itself adaptive-secure. The optimized transformation and the adaptive security of HomSig appear in the full version of our work.

Instantiating the Scheme with Graded Encodings. In the full version of our paper we show how to translate the scheme presented above to the setting of graded encodings [20,17]. Here we discuss the changes incurred by our scheme to accommodate the differences between multilinear maps and (known) graded encoding schemes. Recall that graded encodings can be randomized. In addition: (1) the ring R in which the encoded values live is not public, i.e., the order p of the encoding sets S_i may not be publicly known (although a lower bound on p is public); (2) one cannot (publicly) encode arbitrary elements "in the exponent"; (3) in order for the zero-test to work properly, one can support only a bounded number of operations over the encodings. To address the first difference, our scheme signs messages that are integers within a certain bound B, and as the class of admissible functions we consider N-variate polynomials of constant degree k over the integers. We can then bound the size of all reachable outputs (obtained by applying an admissible f on integers in $\mathbb{Z}_B$) – say it is B^* – and finally we instantiate the parameters of the graded encoding scheme accordingly so that the order p of the ring is such that $p > B^*$. For the second difference, we note that graded encodings allow one to encode arbitrary elements with the knowledge of a trapdoor which, in our case, can be made available to the signer. In the key generation we let the signer use this procedure to publish level-1 encodings of the $\log B^*$ powers of 2 (i.e., the equivalent of $h_1^{2^j}$). This way, upon verification, an encoding of m (i.e., h_d^m) can be obtained by adding up the encodings of the appropriate powers of 2, according to the bit-decomposition of m (i.e., $h_d^m = e(\prod_{j:m_j=1} h_1^{2^j}, h_{i-1})$). This operation can be done by "consuming the noise" of at most $\log B^*$ additions. To address the third difference, we note that the solutions to (1) and (2) already provide a bound on the maximum number of operations (additions and multiplications) that will be performed over the encodings when running the homomorphic evaluation algorithm. Using such bounds it is then possible to take appropriately large parameters of the graded encodings that can accommodate this number of operations.

Acknowledgements. The research of Dario Fiore has been partially supported by the European Commission's Seventh Framework Programme Marie Curie Cofund Action AMAROUT II (grant no. 291803), and by the Madrid Regional Government under project PROMETIDOS-CM (ref. S2009/TIC1465). The work of Bogdan Warinschi has been supported in part by ERC Advanced Grant ERC-2010-AdG-267188-CRIPTO, by EPSRC via grant EP/H043454/1, and has received funding from the European Union Seventh Framework Programme (FP7/2007-2013) under grant agreement 609611 (PRACTICE).

References

1. Agrawal, S., Boneh, D.: Homomorphic mACs: MAC-based integrity for network coding. In: Abdalla, M., Pointcheval, D., Fouque, P.-A., Vergnaud, D. (eds.) ACNS 2009. LNCS, vol. 5536, pp. 292–305. Springer, Heidelberg (2009)
2. Attrapadung, N., Libert, B.: Homomorphic network coding signatures in the standard model. In: Catalano, D., Fazio, N., Gennaro, R., Nicolosi, A. (eds.) PKC 2011. LNCS, vol. 6571, pp. 17–34. Springer, Heidelberg (2011)
3. Attrapadung, N., Libert, B., Peters, T.: Computing on authenticated data: New privacy definitions and constructions. In: Wang, X., Sako, K. (eds.) ASIACRYPT 2012. LNCS, vol. 7658, pp. 367–385. Springer, Heidelberg (2012)
4. Backes, M., Fiore, D., Reischuk, R.M.: Verifiable delegation of computation on outsourced data. In: Sadeghi, A.-R., Gligor, V.D., Yung, M. (eds.) ACM CCS 2013, pp. 863–874. ACM Press (November 2013)
5. Barthe, G., Fagerholm, E., Fiore, D., Mitchell, J., Scedrov, A., Schmidt, B.: Automated analysis of cryptographic assumptions in generic group models. In: Garay, J.A., Gennaro, R. (eds.) CRYPTO 2014, Part I. LNCS, vol. 8616, pp. 95–112. Springer, Heidelberg (2014)

6. Benabbas, S., Gennaro, R., Vahlis, Y.: Verifiable delegation of computation over large datasets. In: Rogaway, P. (ed.) CRYPTO 2011. LNCS, vol. 6841, pp. 111–131. Springer, Heidelberg (2011)

7. Bitansky, N., Canetti, R., Chiesa, A., Tromer, E.: Recursive composition and bootstrapping for SNARKS and proof-carrying data. In: Boneh, D., Roughgarden, T., Feigenbaum, J. (eds.) 45th ACM STOC, pp. 111–120. ACM Press (June 2013)

8. Boneh, D., Freeman, D., Katz, J., Waters, B.: Signing a linear subspace: Signature schemes for network coding. In: Jarecki, S., Tsudik, G. (eds.) PKC 2009. LNCS, vol. 5443, pp. 68–87. Springer, Heidelberg (2009)

9. Boneh, D., Freeman, D.M.: Homomorphic signatures for polynomial functions. In: Paterson, K.G. (ed.) EUROCRYPT 2011. LNCS, vol. 6632, pp. 149–168. Springer, Heidelberg (2011)

10. Catalano, D., Fiore, D.: Practical homomorphic mACs for arithmetic circuits. In: Johansson, T., Nguyen, P.Q. (eds.) EUROCRYPT 2013. LNCS, vol. 7881, pp. 336–352. Springer, Heidelberg (2013)

11. Catalano, D., Fiore, D., Gennaro, R., Nizzardo, L.: Generalizing homomorphic mACs for arithmetic circuits. In: Krawczyk, H. (ed.) PKC 2014. LNCS, vol. 8383, pp. 538–555. Springer, Heidelberg (2014)

12. Catalano, D., Fiore, D., Gennaro, R., Vamvourellis, K.: Algebraic (trapdoor) one-way functions and their applications. In: Sahai, A. (ed.) TCC 2013. LNCS, vol. 7785, pp. 680–699. Springer, Heidelberg (2013)

13. Catalano, D., Fiore, D., Warinschi, B.: Adaptive pseudo-free groups and applications. In: Paterson, K.G. (ed.) EUROCRYPT 2011. LNCS, vol. 6632, pp. 207–223. Springer, Heidelberg (2011)

14. Catalano, D., Fiore, D., Warinschi, B.: Efficient network coding signatures in the standard model. In: Fischlin, M., Buchmann, J., Manulis, M. (eds.) PKC 2012. LNCS, vol. 7293, pp. 680–696. Springer, Heidelberg (2012)

15. Chung, K.-M., Kalai, Y., Vadhan, S.P.: Improved delegation of computation using fully homomorphic encryption. In: Rabin, T. (ed.) CRYPTO 2010. LNCS, vol. 6223, pp. 483–501. Springer, Heidelberg (2010)

16. Chung, K.-M., Kalai, Y.T., Liu, F.-H., Raz, R.: Memory delegation. In: Rogaway, P. (ed.) CRYPTO 2011. LNCS, vol. 6841, pp. 151–168. Springer, Heidelberg (2011)

17. Coron, J.-S., Lepoint, T., Tibouchi, M.: Practical multilinear maps over the integers. In: Canetti, R., Garay, J.A. (eds.) CRYPTO 2013, Part I. LNCS, vol. 8042, pp. 476–493. Springer, Heidelberg (2013)

18. Fiore, D., Gennaro, R.: Publicly verifiable delegation of large polynomials and matrix computations, with applications. In: Yu, T., Danezis, G., Gligor, V.D. (eds.) ACM CCS 2012, pp. 501–512. ACM Press (October 2012)

19. Freeman, D.M.: Improved security for linearly homomorphic signatures: A generic framework. In: Fischlin, M., Buchmann, J., Manulis, M. (eds.) PKC 2012. LNCS, vol. 7293, pp. 697–714. Springer, Heidelberg (2012)

20. Garg, S., Gentry, C., Halevi, S.: Candidate multilinear maps from ideal lattices. In: Johansson, T., Nguyen, P.Q. (eds.) EUROCRYPT 2013. LNCS, vol. 7881, pp. 1–17. Springer, Heidelberg (2013)

21. Gennaro, R., Gentry, C., Parno, B.: Non-interactive verifiable computing: Outsourcing computation to untrusted workers. In: Rabin, T. (ed.) CRYPTO 2010. LNCS, vol. 6223, pp. 465–482. Springer, Heidelberg (2010)

22. Gennaro, R., Gentry, C., Parno, B., Raykova, M.: Quadratic span programs and succinct nIZKs without pCPs. In: Johansson, T., Nguyen, P.Q. (eds.) EUROCRYPT 2013. LNCS, vol. 7881, pp. 626–645. Springer, Heidelberg (2013)

23. Gennaro, R., Katz, J., Krawczyk, H., Rabin, T.: Secure network coding over the integers. In: Nguyen, P.Q., Pointcheval, D. (eds.) PKC 2010. LNCS, vol. 6056, pp. 142–160. Springer, Heidelberg (2010)
24. Gennaro, R., Wichs, D.: Fully homomorphic message authenticators. In: Sako, K., Sarkar, P. (eds.) ASIACRYPT 2013, Part II. LNCS, vol. 8270, pp. 301–320. Springer, Heidelberg (2013)
25. Langlois, A., Stehle, D., Steinfeld, R.: GGHLite: More efficient multilinear maps from ideal lattices. In: Advances in Cryptology – Eurocrypt 2014, Springer, Heidelberg (2014)
26. Micali, S.: Computationally sound proofs. SIAM J. Comput. 30(4), 1253–1298 (2000)
27. Parno, B., Raykova, M., Vaikuntanathan, V.: How to delegate and verify in public: Verifiable computation from attribute-based encryption. In: Cramer, R. (ed.) TCC 2012. LNCS, vol. 7194, pp. 422–439. Springer, Heidelberg (2012)

Structure-Preserving Signatures from Type II Pairings

Masayuki Abe[1], Jens Groth[2*], Miyako Ohkubo[3], and Mehdi Tibouchi[1]

[1] NTT Secure Platform Laboratories, Japan
[2] University College London, UK
[3] Security Fundamentals Lab, NSRI, NICT, Japan

Abstract. We investigate structure-preserving signatures in asymmetric bilinear groups with an efficiently computable homomorphism from one source group to the other, i.e., the Type II setting. It has been shown that in the Type I and Type III settings, structure-preserving signatures need at least 2 verification equations and 3 group elements. It is therefore natural to conjecture that this would also be required in the intermediate Type II setting, but surprisingly this turns out not to be the case. We construct structure-preserving signatures in the Type II setting that only require a single verification equation and consist of only 2 group elements. This shows that the Type II setting with partial asymmetry is different from the other two settings in a way that permits the construction of cryptographic schemes with unique properties.

We also investigate lower bounds on the size of the public verification key in the Type II setting. Previous work on structure-preserving signatures has explored lower bounds on the number of verification equations and the number of group elements in a signature but the size of the verification key has not been investigated before. We show that in the Type II setting it is necessary to have at least 2 group elements in the public verification key in a signature scheme with a single verification equation.

Our constructions match the lower bounds so they are optimal with respect to verification complexity, signature sizes and verification key sizes. In fact, in terms of verification complexity, they are the most efficient structure preserving signature schemes to date.

We give two structure-preserving signature schemes with a single verification equation where both the signatures and the public verification keys consist of two group elements each. One signature scheme is strongly existentially unforgeable, the other is fully randomizable. Having such simple and elegant structure-preserving signatures may make the Type II setting the easiest to use when designing new structure-preserving cryptographic schemes, and lead to schemes with the greatest conceptual simplicity.

Keywords: Structure-preserving signatures, Type II pairings, strong existential unforgeability, randomizability, lower bounds.

---

* The research leading to these results has received funding from the Engineering and Physical Sciences Research Council grant EP/J009520/1 and the European Research Council under the European Union's Seventh Framework Programme (FP/2007-2013) / ERC Grant Agreement n. 307937.

J.A. Garay and R. Gennaro (Eds.): CRYPTO 2014, Part I, LNCS 8616, pp. 390–407, 2014.

1   Introduction

Structure-preserving signatures [3] are pairing-based signatures that consist of group elements and are verified by testing equality of products of pairings of group elements. They are useful building blocks in modular design of cryptographic protocols, in particular in combination with non-interactive zero-knowledge (NIZK) proofs of knowledge about group elements [22]. There are numerous applications of structure-preserving signatures, such as blind signatures [3,17], group signatures [3,17,26], homomorphic signatures [25,9], delegatable anonymous credentials [16], compact verifiable shuffles [14], network encoding [8], oblivious transfer [20,12], tightly secure encryption [23,2], anonymous e-cash [28], etc.

Galbraith, Paterson and Smart [18] classify pairings $e : \mathbb{G}_1 \times \mathbb{G}_2 \to \mathbb{G}_T$ into three types depending on whether $\mathbb{G}_1 = \mathbb{G}_2$ (Type I), or there is an efficiently computable homomorphism $\psi : \mathbb{G}_2 \to \mathbb{G}_1$ (Type II), or there is no efficiently computable homomorphism in either direction between $\mathbb{G}_1$ and $\mathbb{G}_2$ (Type III). Structure-preserving signatures have been analyzed in the symmetric Type I setting [4] and in the fully asymmetric Type III setting [7], and in both cases it has been shown that a structure-preserving signatures requires at least 2 verification equations and 3 group elements in the signatures.

It is thus natural to conjecture that 2 verification equations and 3 group elements would be needed in the intermediate Type II setting as well; and indeed this is the case if the messages belong to $\mathbb{G}_1$. However, when the messages belong to $\mathbb{G}_2$ we find the conjecture to be false, and give constructions of structure-preserving signatures with only one verification equation and 2 group elements in the signatures. This is significant from a high level pairing-based cryptography perspective, as it provides a concrete example of a property that can be obtained in the Type II setting but not in the other settings. Therefore, contrary to expectations, we settle Chatterjee and Menezes' open question of whether schemes based on Type II pairings can always be converted to Type III pairings at no efficiency loss [15] in the negative.

Having a single verification equation make the structure-preserving signature schemes quite efficient. As we discuss in Sect. 2.1 even though Type III pairings are more efficient in some respects with current techniques (certain group elements have a more compact representation), Type II pairings are competitive, especially in terms of speed. Our proposed scheme is the most efficient construction to date in terms of verification complexity. Furthermore, with only one verification equation, structure-preserving signatures become conceptually simpler and easier to use for the designer of cryptographic schemes. Groth-Sahai proofs for the Type II setting [22,19] also incur a smaller overhead when there is only one verification equation.

We give two constructions of structure-preserving signatures. One is randomizable, which means that a signature on a message can be randomized to look like a new fresh signature on the message. This randomization is useful because it ensures that one of the group elements in the signature is uniformly random, which is a convenient feature when building anonymization protocols: this random group element can be revealed in the clear without showing what the

Table 1. Most efficient structure-preserving signatures schemes for all three types of pairings, in terms of signature size, verification key size and number of verification equations. Boldface values are known to be optimal for their respective pairing types. Verification key size is inclusive of group elements that can be shared in a common reference string used by all signers.

Setting	Signature	Verification key	Equations
Type III [4]	**3**	2	**2**
Type I [7]	3	3	**2**
Type II (this work)	**2**	**2**	**1**

original signature was. In other contexts, it is desirable that the signature cannot be tampered with, and our second construction satisfies this property: it is strongly unforgeable.

Prior work has explored lower bounds in the Type I and Type III settings, and established that 2 verification equations are required, and that signatures must consist of at least 3 group elements in both of those cases [4,7]. A third dimension of efficiency is the size of the verification key of the signature scheme. In this paper, we obtain the first lower bounds on verification key size in the literature on structure-preserving signatures: in the Type II setting, a verification key for a single verification equation signature scheme must have at least two group elements. A summary of the best known constructions and efficiency bounds for all three types of pairings is provided in Table 1.

Related Work. The term "structure-preserving signatures" was first introduced by Abe et al. [3], but the notion appears in earlier works as well. Groth [21] proposed the first structure-preserving signature scheme, but the construction involves hundreds of group elements and is not practical. Green and Hohenberger [20] constructed a structure-preserving signature scheme secure against random message attacks but which is not known to be secure against adaptive chosen message attacks. Cathalo, Libert and Yung [13] constructed a signature scheme that is structure-preserving in a relaxed sense that permits the verification key to include target group elements. Hofheinz and Jager [23] and Abe et al. [1,2] investigated the possibility of basing structure-preserving signatures on standard assumptions. They proposed structure-preserving signatures based on the decision linear (DLIN) assumption. The use of a nice security assumption, however, comes at the price of reduced efficiency.

Abe et al. [4] showed that structure-preserving signatures in Type III bilinear groups require at least 3 group elements and 2 verification equations. They also gave structure-preserving signatures matching those bounds that are secure in the generic bilinear group model.

Abe et al. [5] later showed that 3-element signatures cannot be proved secure under a non-interactive assumption using black-box reductions, so strong assumptions are needed to get optimal efficiency in the Type III setting. It is an open question whether a similar impossibility of basing optimal structure-preserving signatures on non-interactive assumptions also holds in the Type II setting (we conjecture it does). However, to get a more conservative non-interactive assumption we modify our first structure-preserving signature scheme in the full version of this paper [6] to base it on a standard non-interactive hardness assumption. The modification requires adding an extra group element to the verification key and the signature but the scheme still has only a single verification equation.

Recently Abe et al. [7] investigated the symmetric setting (Type I) and found that the same lower bound of 3 group elements and 2 verification equations applies. They also presented a unified structure-preserving signature scheme working in all three types of settings and meeting this bound, which means a structure-preserving signature scheme with 3 group elements and 2 verification equations exists (and is the best construction published so far) in the Type II setting we investigate. They also considered the question of verification key size and their scheme requires 3 additional elements in addition to the description of the bilinear group. However, two of these group elements can be fixed in a common reference string together with the description of the bilinear group and may therefore be reused by structure-preserving signature schemes leaving only one variable group element in the verification key. It is an open question whether such a technique applies in the Type II setting.

2   Preliminaries

2.1   Bilinear Groups

Let $\mathcal{G}$ be a bilinear group generator, which given the security parameter k returns a bilinear group description $(p, \mathbb{G}_1, \mathbb{G}_2, \mathbb{G}_T, e, \psi, G, H) \leftarrow \mathcal{G}(1^k)$ such that

- $\mathbb{G}_1, \mathbb{G}_2, \mathbb{G}_T$ are cyclic groups of order p, which is a k-bit prime
- $\psi : \mathbb{G}_2 \to \mathbb{G}_1$ is a homomorphism such that $\psi(H) = G$, hence $\psi(H^a) = G^a$ for all $a \in \mathbb{Z}$
- G generates $\mathbb{G}_1$, H generates $\mathbb{G}_2$ and $e(G, H)$ generates $\mathbb{G}_T$
- $e : \mathbb{G}_1 \times \mathbb{G}_2 \to \mathbb{G}_T$ is a bilinear map, i.e., $e(G^a, H^b) = e(G, H)^{ab}$ for all $a, b \in \mathbb{Z}$
- There are efficient algorithms for computing group operations, evaluating the homomorphism ψ and the bilinear map e, comparing group elements and deciding membership of the groups

Generic Algorithms. In a bilinear group $(p, \mathbb{G}_1, \mathbb{G}_2, \mathbb{G}_T, e, \psi, G, H)$ generated by $\mathcal{G}$ we refer to deciding group membership, computing group operations in $\mathbb{G}_1$, $\mathbb{G}_2$ or $\mathbb{G}_T$, comparing group elements and evaluating the homomorphism or the bilinear map as the generic bilinear group operations. The signature schemes we construct only use generic bilinear group operations.

As a matter of notation, we will use capital letters $G, H, M, R, S, T, U, V, W$ for group elements in $\mathbb{G}_1$ and $\mathbb{G}_2$. We will use small letters $1, m, r, s, t, u, v, w$ for the corresponding discrete logarithms of group elements with respect to base G or H.

Type II Pairings. Galbraith, Paterson and Smart [18] classify bilinear groups into three types according to the efficient morphisms that exist between the source groups $\mathbb{G}_1$ and $\mathbb{G}_2$. Type I pairings have $\mathbb{G}_1 = \mathbb{G}_2$ and $G = H$, i.e., ψ is the identity function (or equivalently, it is an efficiently computable and efficiently invertible isomorphism). Type II pairings have an efficiently computable isomorphism ψ from one source group to the other but none in the reverse direction. Type III pairings have no efficiently computable isomorphism from either source group to the other, i.e., in the definition given above ψ would not be efficiently computable. We will throughout this paper work in the Type II setting.

Type II pairings are usually constructed from the same type of pairing-friendly ordinary elliptic curves as Type III pairings. In contrast with Type III pairings, however, $\mathbb{G}_2$ is then chosen as some subgroup of order p in the p-torsion of the curve other than the trace-zero subgroup (and the homomorphism ψ is then the trace map). As a result, there is no efficient way to hash to $\mathbb{G}_2$ in the Type II setting, but this is of course an irrelevant feature for structure-preserving cryptographic schemes since they only rely on the structure-preserving generic operations and avoid structure-destroying primitives such as cryptographic hash functions.

In terms of efficiency, Type II pairings compare quite favorably to Type I pairings (especially at higher security levels, and particularly now that low-characteristic pairings are known to be broken [24,10]), and are close to Type III pairings: in fact, a Type II pairing computation can be reduced to a Type III one at the cost of one multiplication in $\mathbb{G}_1$ [18, Note 10]. The size of the representation of elements in $\mathbb{G}_1$ is also the same in the Type II and Type III settings, and usually much smaller than in Type I pairings. However, Type II pairings do not support compression using twists for elements in $\mathbb{G}_2$, and hence their representation tends to be larger than in the Type III setting (by a factor of 1 to 6 depending on the embedding degree), and arithmetic in $\mathbb{G}_2$ is accordingly slower.

This has prompted suggestions, for example by Chatterjee and Menezes [15], that Type II pairings were "merely less efficient implementations of Type III pairings", and that cryptographic schemes designed in the Type II setting should adapt to the Type III setting at the cost of slightly different security proofs or assumptions. The present paper shows that this belief is incorrect, in the sense that certain Type II primitives (viz. structure-preserving signatures with a single verification equation) have no secure counterpart in the Type III setting.

2.2   Secure Signature Schemes

A digital signature scheme (with setup algorithm $\mathcal{P}$) is a quadruple of efficient algorithms $(\mathcal{P}, \mathcal{K}, \mathcal{S}, \mathcal{V})$. The setup algorithm $\mathcal{P}$ takes the security parameter and outputs a public parameter PP. The key generation algorithm $\mathcal{K}$ takes PP as input and returns a public verification key VK and a secret signing key SK. We will always assume that VK includes PP and that SK includes VK. The signing algorithm $\mathcal{S}$ takes a signing key SK and a message M in the message space $\mathcal{M}$ defined by PP and VK as input and returns a signature Σ. The verification algorithm $\mathcal{V}$ takes the verification key VK, a message M and the signature Σ and returns either 1 (accept) or 0 (reject).

Definition 1 (Correctness). *We say the signature scheme $(\mathcal{P}, \mathcal{K}, \mathcal{S}, \mathcal{V})$ is correct if for all probabilistic polynomial time adversaries $\mathcal{A}$*

$$\Pr\left[\begin{array}{l} PP \leftarrow \mathcal{P}(1^k) \\ (VK, SK) \leftarrow \mathcal{K}(PP) \\ M \leftarrow \mathcal{A}(SK) \\ \Sigma \leftarrow \mathcal{S}_{SK}(M) \end{array} : M \in \mathcal{M} \ \wedge \ \mathcal{V}_{VK}(M, \Sigma) = 1\right] = 1 - \mathrm{negl}(k).$$

We say the signature scheme is perfectly correct if the probability is exactly 1.

All the signature schemes we construct will have perfect correctness. The lower bounds on the other hand will hold even for signature schemes that are only computationally correct as defined above.

A signature scheme is said to be existentially unforgeable if it is hard to forge a signature on a new message that has not been signed before. The adversary may see signatures on other messages before making the forgery. We distinguish between a random message attack (RMA), where the adversary gets pairs of random messages and corresponding signatures, and an adaptive chosen message attack (CMA) where the adversary can choose arbitrary messages and receive signatures on them. Our signature schemes will be existentially unforgeable against the strong adaptive chosen message attack, but our lower bounds on the complexity of signature schemes will hold even for the weaker random message attacks.

Definition 2 (EUF-CMA). *A signature scheme $(\mathcal{P}, \mathcal{K}, \mathcal{S}, \mathcal{V})$ is existentially unforgeable under adaptive chosen message attack if for all non-uniform polynomial time $\mathcal{A}$*

$$\Pr\left[\begin{array}{l} PP \leftarrow \mathcal{P}(1^k) \\ (VK, SK) \leftarrow \mathcal{K}(PP) \\ (M, \Sigma) \leftarrow \mathcal{A}^{\mathcal{S}_{SK}(\cdot)}(VK) \end{array} : M \notin Q \ \wedge \ \mathcal{V}_{VK}(M, \Sigma) = 1\right] = \mathrm{negl}(k),$$

where Q is the set of queries made by $\mathcal{A}$ to the signing oracle.

Sometimes it is also useful to prevent the adversary from issuing a new signature for a message that has already been signed. A signature scheme is strongly existentially unforgeable if it is hard to find a signature on a message that has not been signed before and also hard to find a new signature for a message that has already been signed. This notion, denoted by sEUF-CMA, is formally captured in the same way as the definition of EUF-CMA except for additionally requiring $(M, \Sigma) \notin Q$ where Q is the set of message-signature pairs from $\mathcal{A}$'s queries to the signing oracle.

We get the definition for existential unforgeability against random message attack (EUF-RMA) by modifying the signing oracle to picking $M \leftarrow \mathcal{M}$ at random, computing $\Sigma \leftarrow \mathcal{S}_{SK}(M)$ and returning (M, Σ) to the adversary whenever the signing oracle is queried.

Corresponding security notions for one-time signature schemes can be obtained by restricting the adversary to only calling the signing oracle once in the above definitions.

Randomizable Signatures. In some applications it is desirable to have randomizable signatures, i.e., given a signature it is possible to randomize it such that it looks like a fresh signature on the message. The randomization is carried out by a randomization algorithm $\mathcal{R}$ that takes as input a verification key VK, a message M and a signature Σ and returns a randomized signature Σ.

Definition 3 (Randomizability). *A signature scheme $(\mathcal{P}, \mathcal{K}, \mathcal{S}, \mathcal{V})$ is said to be (perfectly) randomizable if there exists a randomization algorithm $\mathcal{R}$ such that for all $k \in \mathbb{N}$ and all interactive adversaries $\mathcal{A}$*

$$\Pr\left[\begin{array}{l} PP \leftarrow \mathcal{P}(1^k) \\ (VK, SK) \leftarrow \mathcal{K}(PP) \\ (M, \Sigma) \leftarrow \mathcal{A}(SK) \\ \Sigma_0 \leftarrow \mathcal{S}_{SK}(M) \\ \Sigma_1 \leftarrow \mathcal{R}_{VK}(M, \Sigma) \\ b \leftarrow \{0, 1\} \end{array} : \mathcal{V}_{VK}(M, \Sigma) = 1 \ \wedge \ \mathcal{A}(\Sigma_b) = b\right] \leq \frac{1}{2}.$$

2.3   Structure-Preserving Signature Schemes

We study structure-preserving signature schemes [3] on bilinear groups generated by group generator $\mathcal{G}$. In a structure preserving signature scheme the verification key, the messages and the signatures consist only of group elements from $\mathbb{G}_1$ and $\mathbb{G}_2$ and the verification algorithm evaluates the signature by deciding group membership of elements in the signature, using the homomorphism ψ and by evaluating pairing product equations, which are equations of the form

$$\prod_i \prod_j e(X_i, Y_j)^{a_{ij}} = 1,$$

where $X_1, X_2, \ldots \in \mathbb{G}_1$, $Y_1, Y_2, \ldots \in \mathbb{G}_2$ are group elements appearing in PP, VK, M and Σ and $a_{11}, a_{12}, \ldots \in \mathbb{Z}_p$ are constants stored in PP. Structure-preserving signatures are extremely versatile because they mix well with other pairing-based protocols. Groth-Sahai proofs [22] are for instance designed with pairing product equations in mind and can therefore easily be applied to structure-preserving signatures.

Definition 4 (Structure-preserving signatures). *A signature scheme $(\mathcal{P}, \mathcal{K}, \mathcal{S}, \mathcal{V})$ is said to be structure preserving over bilinear group generator $\mathcal{G}$ if*

- *PP includes a bilinear group $(p, \mathbb{G}_1, \mathbb{G}_2, \mathbb{G}_T, e, \psi, G, H)$ generated by $\mathcal{G}$, group elements in $\mathbb{G}_1$ and $\mathbb{G}_2$, and constants in $\mathbb{Z}_p$,*
- *the verification key consists of PP and group elements in $\mathbb{G}_1$ and $\mathbb{G}_2$,*
- *the messages consist of group elements in $\mathbb{G}_1$ and $\mathbb{G}_2$,*
- *the signatures consist of group elements in $\mathbb{G}_1$ and $\mathbb{G}_2$, and*
- *the verification algorithm only needs to decide membership in $\mathbb{G}_1$ and $\mathbb{G}_2$, use the homomorphism ψ, and evaluate pairing product equations.*

Generic signer. Abe et al. [3] did not explicitly require the signing algorithm to only use generic group operations when they defined structure-preserving signatures. However, all existing structure-preserving signatures in the literature have generic signing algorithms and we believe it would be a surprising result in itself to construct a structure-preserving signature with a non-generic signer. Our constructions have generic signer algorithms and some of our lower bounds will assume the signer is generic.

3   Randomizable Structure-Preserving Signatures

We will now show that in the Type II setting it is possible to construct an EUF-CMA secure structure-preserving signature scheme with a single verification equation. This is surprising since both in the symmetric Type I setting and the fully asymmetric Type III setting structure-preserving signature schemes require at least two verification equations [4,7].

The signature scheme is given in Fig. 1. It has a single verification equation and both signatures and verification keys consist of two group elements. This is optimal with respect to both verification complexity, signature size and verification key size as we demonstrate in Sect. 5.

As an additional benefit, the signature scheme is perfectly randomizable. We show a simple randomization algorithm that converts a signature into a new randomized signature that looks exactly like a fresh signature on the message. It is worth observing that while the natural formalization of randomizability gives both the message and the signature to the randomization algorithm our randomization algorithm does not need the message and simply ignores it and randomizes the signature directly. There may be applications where this is a feature.

The signature scheme is designed with Groth-Sahai proofs in mind. If we randomize a signature, we may reveal the random group element R without this leaking any information about the message or the original signature from which the randomized signature was derived. When R is public the verification equation become linear, which makes Groth-Sahai proofs very efficient.

It is easy to see that the signature scheme is perfectly correct. Randomized signatures are perfectly indistinguishable from real signatures since both types of signatures are uniquely determined by the uniformly random non-trivial group element R. We will now prove that the signature scheme is existentially unforgeable under adaptive chosen message attack.

Theorem 1. *The signature scheme in Fig. 1 is* EUF-CMA *secure in the generic bilinear group model.*

Proof. A generic adversary only uses generic group operations. This means that in $\mathbb{G}_1$ and $\mathbb{G}_2$ it can only compute linear combinations of group elements from the verification key and the signatures it has seen and use the map ψ to map elements from $\mathbb{G}_2$ to $\mathbb{G}_1$. Linear combinations on verification key elements and signature elements correspond to formal polynomials (of degree ranging from

---

Setup $\mathcal{P}(1^k)$:   Return $PP = (p, \mathbb{G}_1, \mathbb{G}_2, \mathbb{G}_T, e, \psi, G, H) \leftarrow \mathcal{G}(1^k)$.

Key generation $\mathcal{K}(PP)$:   Choose $v, w \leftarrow \mathbb{Z}_p$ and compute the keys $VK = (PP, V, W)$ and $SK = (PP, v, w)$ as

$$V \leftarrow G^v \qquad\qquad W \leftarrow G^w.$$

Signing $\mathcal{S}_{SK}(M)$:   On $M \in \mathbb{G}_2$ choose $r \leftarrow \mathbb{Z}_p$ and compute signature $\Sigma = (R, S)$ as

$$R \leftarrow H^r \qquad\qquad S \leftarrow M^v H^{r^2 + w}.$$

Randomization $\mathcal{R}_{VK}(M, (R, S))$:   Pick $\alpha \leftarrow \mathbb{Z}_p^*$ and compute the randomized signature $\Sigma' = (R', S')$ as

$$R' \leftarrow R H^\alpha \qquad\qquad S' \leftarrow S R^{2\alpha} H^{\alpha^2}.$$

Verification $\mathcal{V}_{VK}(M, (R, S))$:   Accept if and only if $M, R, S \in \mathbb{G}_2$ and

$$e(G, S) = e(V, M) e(\psi(R), R) e(W, H).$$

Fig. 1. Randomizable structure-preserving signature scheme for messages in $\mathbb{G}_2$

0 to $q + 1$ after q signature queries) in the discrete logarithms of the group elements. We will show that no linear combinations produce formal polynomials corresponding to a forgery. By the master theorem in [11] this means that the signature scheme is secure in the generic bilinear group model.

The group elements in VK are $G, V, W \in \mathbb{G}_1$ and $H \in \mathbb{G}_2$ with corresponding discrete logarithms $1, v, w$ and 1. On a query M_i with discrete logarithm m_i from the adversary, the signature oracle responds with a signature (R_i, S_i) with discrete logarithms

$$r_i \leftarrow \mathbb{Z}_p^* \qquad s_i = v m_i + r_i^2 + w.$$

Suppose the adversary after q queries constructs $(M, (R, S))$ in $\mathbb{G}_2$. Since the adversary is generic it can only construct them in $\mathbb{G}_2$ such that the discrete logarithms m, r, s are linear combinations of $1, r_1, s_1, \ldots, r_q, s_q$, i.e.,

$$m = \mu + \sum_{i=1}^{q} \mu_{r_i} r_i + \sum_{i=1}^{q} \mu_{s_i} (v m_i + r_i^2 + w),$$

$$r = \rho + \sum_{i=1}^{q} \rho_{r_i} r_i + \sum_{i=1}^{q} \rho_{s_i} (v m_i + r_i^2 + w),$$

$$s = \sigma + \sum_{i=1}^{q} \sigma_{r_i} r_i + \sum_{i=1}^{q} \sigma_{s_i} (v m_i + r_i^2 + w).$$

Similarly, the discrete logarithm m_i of a signing query is a linear combination of $1, r_1, s_1, \ldots, r_{i-1}, s_{i-1}$.

We will show that the signature scheme is EUF-CMA secure, i.e., an adversary cannot construct a valid signature (R, S) on M where the discrete logarithms

m, r, s satisfy the verification equation

$$s = vm + r^2 + w$$

unless it reuses $M = M_j$ from a previous query.

We can write $s = vm + r^2 + w$ as

$$\sigma + \sum_{i=1}^{q} \left(\sigma_{r_i} r_i + \sigma_{s_i} (vm_i + r_i^2 + w) \right) - v \left(\mu + \sum_{i=1}^{q} \mu_{r_i} r_i + \mu_{s_i} (vm_i + r_i^2 + w) \right)$$

$$= \left(\rho + \sum_{i=1}^{q} \rho_{r_i} r_i + \sum_{i=1}^{q} \rho_{s_i} (vm_i + r_i^2 + w) \right)^2 + w.$$

We first look at the terms r_i^4. Observe that all elements $m, r, s, m_1, \ldots, m_q$ constructed using generic bilinear group operations in $\mathbb{G}_2$, i.e., linear combinations of the discrete logarithms, can only have degree $0, 1$ or 2 in r_i. This shows that each term r_i^4 has coefficient 0 in $s - vm$. On the other side of the verification equation each term r_i^4 has coefficient $\rho_{s_i}^2$. Therefore $\rho_{s_i} = 0$ for all $i = 1, \ldots, q$.

In $s - vm$ the coefficients of all combinations $r_i r_j$ are 0 for $i \neq j$. On the other side of the verification equation in the product r^2 they have coefficients $\rho_{r_i} \rho_{r_j}$. This means for all $i \neq j$ we have $\rho_{r_i} \rho_{r_j} = 0$ and therefore there can be at most one $\rho_{r_j} \neq 0$. We now have $r = \rho + \rho_{r_j} r_j$ giving us that $s = vm + r^2 + w$ can be written as

$$\sigma + \sum_{i=1}^{q} \sigma_{r_i} r_i + \sum_{i=1}^{q} \sigma_{s_i} (vm_i + r_i^2 + w)$$

$$= v \left(\mu + \sum_{i=1}^{q} \mu_{r_i} r_i + \sum_{i=1}^{q} \mu_{s_i} (vm_i + r_i^2 + w) \right) + \left(\rho + \rho_{r_j} r_j \right)^2 + w$$

for some $j \in \{1, \ldots, q\}$.

Comparing the coefficients of r_i^2 from the two sides of the verification equation we get $\sigma_{s_j} = \rho_{r_j}^2$ and $\sigma_{s_i} = 0$ for $i \neq j$. The coefficients of w on the two sides of the verification equation gives us $\sigma_{s_j} = 1$. Then the verification equation is described as

$$v\, m_j + r_j^2 + \sigma + \sum_{i=1}^{q} \sigma_{r_i} r_i + w$$

$$= v \left(\mu + \sum_{i=1}^{q} \mu_{r_i} r_i + \sum_{i=1}^{q} \mu_{s_i} (vm_i + r_i^2 + w) \right) + \left(\rho + \rho_{r_j} r_j \right)^2 + w$$

for some $j \in \{1, \ldots, q\}$.

Looking at coefficient of terms that involve v we then get $vm_j = vm$, which shows us $m = m_j$ and therefore $M = M_j$. $\qquad \square$

In some cases it is desirable to sign many group elements at once. The signature scheme we presented can easily be modified to sign n group elements at once by changing the verification equation to:

$$e(G, S) = \prod_{i=1}^{n} e(V_i, M_i) e(\psi(R), R) e(W, H)$$

and modifying the key generation and signing processes accordingly. The security proof for the generalized scheme is virtually the same as the proof for Theorem 1.

4 Strongly Unforgeable Structure-Preserving Signatures

For some applications it is desirable to use a strongly existentially unforgeable signature scheme. It is in general harder to get strong unforgeability because now also the signatures have to be immutable, but we will present a construction that preserves optimality with respect to verification complexity, signature size and verification key size.

Fig. 2 gives a structure-preserving signature scheme with a single verification equation, 2 element verification keys and 2 element signatures. It is easy to see that it is perfectly correct. Signature verification requires only two pairing evaluations (not counting the constant factor $e(G, H)$), which makes this scheme the most efficient structure preserving signature so far in terms of verification complexity (faster than all previous Type I and Type III constructions by a significant margin). In the full version of this paper [6], we prove that it is strongly existentially unforgeable under adaptive chosen message attack.

Theorem 2. *The signature scheme in Fig. 2 is* sEUF-CMA *secure in the generic bilinear group model.*

5 Lower Bounds in the Type II Setting

We will now establish lower bounds for the complexity of structure-preserving signature schemes in the Type II setting. Unlike the Type I and the Type III

Setup $\mathcal{P}(1^k)$: Return $PP = (p, \mathbb{G}_1, \mathbb{G}_2, \mathbb{G}_T, e, \psi, G, H) \leftarrow \mathcal{G}(1^k)$.
Key generation $\mathcal{K}(PP)$: Choose $v, w \leftarrow \mathbb{Z}_p$ and compute $VK = (PP, V, W)$ and $SK = (PP, v, w)$ using

$$V \leftarrow G^v \qquad W \leftarrow G^w.$$

Signing $\mathcal{S}_{SK}(M)$: On $M \in \mathbb{G}_2$ choose $t \leftarrow \mathbb{Z}_p^*$ and compute signature $\Sigma = (R, S)$ as

$$R \leftarrow H^{t-w} \qquad S \leftarrow M^{\frac{v}{t}} H^{\frac{1}{t}}.$$

Verification $\mathcal{V}_{VK}(M, (R, S))$: Accept if and only if $M, R, S \in \mathbb{G}_2$ and

$$e(W\psi(R), S) = e(V, M) e(G, H).$$

Fig. 2. Strong structure-preserving signature scheme for messages in $\mathbb{G}_2$

settings where two verification equation are needed we have already seen in Sections 3 and 4 that it is possible to use only one verification equation in the Type II setting. However, these signature schemes only work for messages in $\mathbb{G}_2$. We start by showing this is necessarily so, a structure-preserving signature scheme for messages in $\mathbb{G}_1$ cannot have a single verification equation.

Theorem 3. *A structure-preserving signature scheme for messages in $\mathbb{G}_1$ must have at least two verification equations. This holds even for one-time signatures with security against random message attack.*

Proof. Suppose we have a structure-preserving signature scheme with a single verification equation for messages in $\mathbb{G}_1$. We will construct a one-time random message attack on the scheme. The attacker queries the signing oracle and get a signature on a random message $M \in \mathbb{G}_1$. Let S be a group element in the signature that appears non-trivially in the verification equation.

If $S \in \mathbb{G}_1$ we can write the verification equation as $e(M, X) = e(S, Y)Z$ where X, Y, Z are expression that do not include any M or S terms. We now have $e(M\psi(Y), X) = e(S\psi(X), Y)Z$, which means replacing the group element S with $S^* = S\psi(X)$ in the signature gives us a forgery on $M^* = M\psi(Y)$.

If $S \in \mathbb{G}_2$ we can write the verification equation as $e(M, S^a X) \cdot e(\psi(S)^b Y, S) = Z$ for some $a, b \in \mathbb{Z}_p$ and expressions X, Y, Z that do not have any M or S terms. Pick $r \leftarrow \mathbb{Z}_p^*$ and define $\Delta = (S^a X)^{\frac{1}{r}}$. Replace S with $S^* = S\Delta$ to get a signature on $M^* = M(M^a \psi(\Delta)^b Y \psi(S)^{2b})^{-\frac{1}{a+r}}$. For the signature to be non-trivial in M we must have $S^a X \neq 1$ with overwhelming probability, giving us that Δ is uniformly random in $\mathbb{G}_1^*$ and, therefore, that $M^* \neq M$ with high probability, so we do obtain a forgery.   □

With Theorem 3 in mind, we will in the rest of this section only consider structure-preserving signatures on $M \in \mathbb{G}_2$. We will now show our main result in this section, which is that the verification key must have at least two group elements. The following theorem follows as a corollary to Lemmata 1, 2 and 3.

Theorem 4. *A structure-preserving signature scheme with a single verification equation and a generic signer must have at least two group elements in the verification key. This holds even for one-time signatures secure under random message attack.*

Lemma 1. *A structure-preserving signature for $M \in \mathbb{G}_2$ with a single verification equation cannot have a non-redundant signature element $S \in \mathbb{G}_1$. This holds even for one-time signatures with security under random message attack.*

Proof. We will construct an one-time random message attack similar to the one for the proof of Theorem 3 with the roles of M and S reversed. The attacker queries the signing oracle and gets a signature on a random message $M \in \mathbb{G}_2$. Let $S \in \mathbb{G}_1$ be an element in the signature that appears non-trivially in the verification equation, i.e., it does not have negligible probability of being paired with 1.

We can write the verification equation as $e(S, M^aX) \cdot e(\psi(M)^bY, M) = Z$ for some $a, b \in \mathbb{Z}_p$ and expressions X, Y, Z that do not have any M or S terms. Pick $r \leftarrow \mathbb{Z}_p^*$ and define $\Delta = (M^aX)^{\frac{1}{r}}$. Replace S in the signature with $S^* = S(S^a\psi(\Delta)^bY\psi(M)^{2b})^{-\frac{1}{a+r}}$ to get a signature on $M^* = M\Delta$. For the signature scheme to be non-redundant in S the probability of $M^aX \neq 1$ has to be non-negligible and in that case Δ is uniformly random in $\mathbb{G}_1^*$, giving us $M^* \neq M$ so that we do obtain a forgery.   □

Lemma 2. *There is no structure-preserving signature for $M \in \mathbb{G}_2$ with a single verification equation and a key consisting of a single group element $V \in \mathbb{G}_2$. This holds even for one-time signatures with security under random message attack.*

Proof. From Lemma 1 we can without loss of generality consider only signature schemes where all signature elements belong to $\mathbb{G}_2$. A group element $S \in \mathbb{G}_2$ from the signature appears as $e(\psi(S), S^aX)$ in the verification equation, where X is an expression that does not contain an S-term. If $a \neq 0$ we can substitute S with $S' = SX^{\frac{1}{2a}}$ to get the simpler term $e(\psi(S'), S')^a$ in the verification equation. Moreover, if $a = 0$ but X involves another signature element T^b for $b \neq 0$ we can by substituting T with $T' = TS^{-1}$ get a term $e(\psi(S), S^{a'})$ with $a' = b \neq 0$. Using these two diagonalization techniques we can without loss of generality write the single verification equation for the structure-preserving signature $(S_1, \ldots, S_n) \in \mathbb{G}_2^n$ on $M \in \mathbb{G}_2$ as

$$e(\psi(M), M^aX) \prod_{i \in I} e(\psi(S_i), M^{b_i}Y_i) \cdot \prod_{j \in J} e(\psi(S_j), S_j)^{c_j} = Z,$$

where I, J are disjoint subsets of $\{1, \ldots, n\}$, $b_i \neq 0, Y_i \neq 1$ or both for each $i \in I$, $c_j \neq 0$ for each $j \in J$, and X, Y_i and Z are expressions that only involve constant terms and the verification key.

The adversary starts by getting a signature $(S_1, \ldots, S_n)$ on a random message M. If $I \neq \emptyset$ we can use the method from the proof of Lemma 1 to modify S_i into S_i^* giving a forgery on $M^* \neq M$. If $I = \emptyset$ and $a = 0$ we can use the method from the proof of Theorem 3 to obtain a forgery on a message $M^* \neq M$.

The remaining case is when $I = \emptyset$ and $a \neq 0$, i.e., the verification equation is

$$e(\psi(M), M^aX) \cdot \prod_{j \in J} e(\psi(S_j), S_j)^{c_j} = Z,$$

with $a \neq 0$ and each $c_j \neq 0$. But in this case the equation can be seen as a quadratic equation in M with two solutions. The signature on M is also a signature on $M^* = M^{-1}X^{-\frac{1}{a}}$. This gives us an existential forgery unless $M^* = M$, which only happens in the unlikely event that $X = M^{-2a}$.   □

Lemma 3. *There is no structure-preserving signature for $M \in \mathbb{G}_2$ with a single verification equation, a generic signer and a verification key consisting of a single group element $V \in \mathbb{G}_1$. This holds even for one-time signatures with security under random message attack.*

Proof. As in the proof of Theorem 2 we can rewrite the verification equation as

$$e(\psi(M), M^a H^x) \cdot \prod_{i \in I} e(\psi(S_i), M^{b_i} H^{y_i})$$

$$\cdot \prod_{j \in J} e(\psi(S_j), S_j)^{c_j} \cdot e(V, M^d \prod_{k=1}^{n} S_k^{e_k} H^f) = e(G, H)^z,$$

where I, J are disjoint subsets of $\{1, \ldots, n\}$, $x, y_i, z \in \mathbb{Z}_p$ are constant terms, $b_i \neq 0, Y_i \neq 1$ or $e_i \neq 0$ for each $i \in I$, and $c_j \neq 0$ for each $j \in J$. We will consider three cases: all $e_k = 0$ and $d = 0$ but $f \neq 0$, all $e_k = 0$ but $d \neq 0$, and without loss of generality $e_1 \neq 0$.

In the first case $d = 0$ and all $e_k = 0$ but without loss of generality $f = 1$. The adversary makes a one-time random message attack to get a signature $(S_1, \ldots, S_n)$ on a random message M. We can now make an analysis similar to the proof of Lemma 2 to create an existential forgery on a message $M^* \neq M$.

In the second case all $e_k = 0$ but $d \neq 0$. The adversary picks $M = H^{-\frac{f}{d}}$ such that the $e(V, *)$ part cancels out. If there is an $i \in I$ such that $M^{b_i} H^{y_i} = H^{-b_i \frac{f}{d} + y_i} \neq 1$, we can pick all other signature elements $S_k = 1$ for $k \neq i$ and since we know all the discrete logarithm solve for the discrete logarithm s_i of S_i to get a signature on $M = H^{\frac{f}{d}}$. Else if there is no such $i \in I$, then we have an equation in the discrete logarithms of the signature such that $m(am + x) + \sum_{j \in J} c_j s_j^2 = z$ with $m = -\frac{d}{f}$. By the completeness of the signature scheme, this equation is solvable in the unknowns s_j and can be efficiently solved [27], which gives us a signature on M.

Finally, in the third case without loss of generality $e_1 \neq 0$. We can substitute S_1 with $(M^d \prod_{k=1}^{n} S_k^{e_k} H^f)^{-\frac{1}{e_1}}$ to get a structure-preserving signature scheme with a verification equation of the form

$$e(\psi(M), M^a H^x) \cdot \prod_{i \in I \setminus \{1\}} e(\psi(S_i), M^{b_i} H^{y_i}) \cdot \prod_{j \in J \setminus \{1\}} e(\psi(S_j), S_j)^{c_j}$$

$$= e(VG^\gamma \psi(M)^\mu \prod_{k \in I \cup J} \psi(S_k)^{\sigma_k}, S_1) \cdot e(G, H)^z,$$

for some $\mu, \sigma_k \in \mathbb{Z}_p$ and with suitable modifications of a, x, b_i, y_i and z.

Our strategy now is to pick $S_1 = 1$ to eliminate the effect of the verification key V. If there is a $b_i \neq 0$ or $y_i \neq 0$, we can pick $m \leftarrow \mathbb{Z}_p$ at random and set $S_k = 1$ for $k \neq i$ and $S_i = H^{\frac{z - m(am + x)}{b_i m + y}}$ to get a signature on $M = H^m$.

If all $b_i = y_i = 0$ but there is some $j \in J \setminus \{1\}$ where $c_j \neq 0$ we instead set $S_k = 1$ for $k \neq j$ and solve the bivariate quadratic $m(am + x) + c_j s_j^2 = z$ in $\mathbb{Z}_p[m, s_j]$, which can be done efficiently [27] unless $a = x = 0$ and c_j and $z \neq 0$ have different quadratic residuosity. However, if $a = x = 0$ the adversary can use a one-time random message to get a signature on M. The adversary picks $r \leftarrow \mathbb{Z}_p^*$ and replaces S_j with $S_j^* = S_j S_k^r$ to get a signature on $M^* = M S_j^{-2c_j r} S_k^{-\sigma_j - c_j r^2}$. For the verification equation to be non-trivial in M, with overwhelming probability $S_k \neq 1$ and therefore $M^* \neq M$ so we have an existential forgery.

The remaining case is when both all $b_i = y_i = 0$ and all $c_j = 0$, i.e., the verification equation is

$$e(\psi(M), M^a H^x) = e(V G^\gamma \psi(M)^\mu \prod_{k \in I \cup J} \psi(S_k)^{\sigma_k}, S_1) \cdot e(G, H)^z.$$

If $z = 0$ we immediately get a signature $(1, \ldots, 1)$ on the message $M = 1$, so let us from now on consider the case where $z \neq 0$.

If there is a $\sigma_k \neq 0$ for $k \neq 1$, we can substitute S_k with $G^\gamma M^\mu \prod_{\ell \in (I \cup J)} S_\ell^{\sigma_\ell}$ to get verification equation $e(\psi(M), M^a H^x) = e(V S_k, S_1) \cdot e(G, H)^z$. The adversary gets a signature on a random message M and replaces S_k with $S_k^* = S_k M^{2a} S_1^a H^x$ to get a signature on $M^* = M S_1$. With overwhelming probability $S_1 \neq 1$, since otherwise the signature would not affect any part of the equation, giving us $M^* \neq M$.

Finally, let us consider the case where we only have a single signature element S_1 that is used in a non-trivial way, i.e., the verification equation is $e(\psi(M), M^a H^x S_1^{-\mu}) = e(V G^\gamma \psi(S_1)^{\sigma_1}, S_1) \cdot e(G, H)^z$. If $a \neq 0$ the attacker can use a one-time random message attack to get a signature on a random message M, which is also a signature on $M^* = M^{-1}(H^x S_1^{-\mu})^{-\frac{1}{a}}$. We have $M^* \neq M$ unless $S_1^\mu = M^{2a} H^x$ but since a and x are known to the adversary this would mean the adversary could forge signatures on arbitrary messages. If $a = 0$ and $x \neq 0$ we can pick $S_1 = 1$ to give us a signature on $M = H^{\frac{z}{x}}$. Finally, if $a = 0$ and $x = 0$ we cannot sign the message using a generic signer. A generic signer computes $S_1 = M^\alpha H^\beta$ using known $\alpha, \beta \in \mathbb{Z}_p$ and is unlikely over the choice of the unknown discrete logarithm of M to solve the equality $-\mu(\alpha m + \beta) = (v + \gamma + \alpha m + \beta)(\alpha m + \beta) + z$ unless it is possible to use $\alpha = 0$, in which case the signature is independent of M and therefore either invalid or valid for every message. □

We have now established a lower bound of 2 group elements in the verification key for structure-preserving signatures with a single verification equation and that this lower bound even holds for one-time random message attacks. In the full version of this paper [6] we give a structure-preserving one-time signature where the signature is a single group element, so such a lower bound does not hold for the signature size. However, if the adversary is allowed to obtain multiple signatures on random messages we can establish a lower bound of 2 group elements for the signatures.

Theorem 5. *A structure-preserving signature scheme with a generic signer that is existentially unforgeable against random message attacks must have at least 2 group elements for messages in $\mathbb{G}_2$ and at least 3 group elements for messages in $\mathbb{G}_1$.*

Proof. Suppose that we have a structure-preserving signature scheme with just one group element in the signature and a single verification equation. The verification equation can for any given message be seen as a quadratic or linear equation in the discrete logarithm of the signature, so there are at most two potential signatures on the message. We conclude from Lemma 4 below that a structure-preserving signature with a single verification equation must consist of at least two group elements.

If there is more than one verification equation we now have two linear or quadratic equations in the signature elements. For messages in $\mathbb{G}_1$ we know by Theorem 3 that at least two verification equations are needed. Both equations must place non-trivial constraints on the signature or else we could reduce to a single verification equation. Again by Lemma 4, we therefore get that for messages in $\mathbb{G}_1$ at least 3 signature elements are needed, since with one or two group elements in the signature there would be at most 4 possible signatures satisfying the verification equations. $\square$

Lemma 4. *A structure-preserving signature scheme with a generic signer that is existentially unforgeable against random message attacks must for each message have a superpolynomial number of potential signatures.*

Proof. Suppose that for a message $M \in \mathbb{G}_2$ there are only a polynomial number of signatures $\Sigma = (R_1, \ldots, R_m, S_1, \ldots, S_n) \in \mathbb{G}_1^m \times \mathbb{G}_2^n$. Since the signer is generic this means there is a set $\{(\vec{\alpha}, \vec{\beta}, \vec{\gamma}, \vec{\delta})\}_{i=1}^{\text{poly}(k)}$ of vectors in $(\mathbb{Z}_p^m)^2 \times (\mathbb{Z}_p^n)^2$ creating signature vectors $\Sigma = (\psi(M)^{\vec{\alpha}} G^{\vec{\beta}}, M^{\vec{\gamma}} H^{\vec{\delta}})$ by entry-wise exponentiation. Given signatures Σ_0 and Σ_1 on random messages M_0 and M_1 we have $\frac{1}{\text{poly}(k)^2}$ probability that they are constructed with the same $(\vec{\alpha}, \vec{\beta}, \vec{\gamma}, \vec{\delta})$ pair. In that case

$$\Sigma^* = \Sigma_0^r \Sigma_1^{1-r} = \left(\psi(M_0^r M_1^{1-r})^{\vec{\alpha}} G^{\vec{\beta}}, (M_0^r M_1^{1-r})^{\vec{\gamma}} H^{\vec{\delta}} \right)$$

is a signature on $M^* = M_0^r M_1^{1-r}$ for all $r \in \mathbb{Z}_p$. A similar proof applies to the case where $M \in \mathbb{G}_1$. $\square$

References

1. Abe, M., Chase, M., David, B., Kohlweiss, M., Nishimaki, R., Ohkubo, M.: Constant-size structure-preserving signatures: Generic constructions and simple assumptions. In: Wang, X., Sako, K. (eds.) ASIACRYPT 2012. LNCS, vol. 7658, pp. 4–24. Springer, Heidelberg (2012)
2. Abe, M., David, B., Kohlweiss, M., Nishimaki, R., Ohkubo, M.: Tagged one-time signatures: Tight security and optimal tag size. In: Kurosawa, K., Hanaoka, G. (eds.) PKC 2013. LNCS, vol. 7778, pp. 312–331. Springer, Heidelberg (2013)
3. Abe, M., Fuchsbauer, G., Groth, J., Haralambiev, K., Ohkubo, M.: Structure-preserving signatures and commitments to group elements. In: Rabin, T. (ed.) CRYPTO 2010. LNCS, vol. 6223, pp. 209–236. Springer, Heidelberg (2010)
4. Abe, M., Groth, J., Haralambiev, K., Ohkubo, M.: Optimal structure-preserving signatures in asymmetric bilinear groups. In: Rogaway, P. (ed.) CRYPTO 2011. LNCS, vol. 6841, pp. 649–666. Springer, Heidelberg (2011)

5. Abe, M., Groth, J., Ohkubo, M.: Separating short structure-preserving signatures from non-interactive assumptions. In: Lee, D.H., Wang, X. (eds.) ASIACRYPT 2011. LNCS, vol. 7073, pp. 628–646. Springer, Heidelberg (2011)
6. Abe, M., Groth, J., Ohkubo, M., Tibouchi, M.: Structure-preserving signatures from type II pairings. Cryptology ePrint Archive, Report 2014/312 (2014), http://eprint.iacr.org/
7. Abe, M., Groth, J., Ohkubo, M., Tibouchi, M.: Unified, minimal and selectively randomizable structure-preserving signatures. In: Lindell, Y. (ed.) TCC 2014. LNCS, vol. 8349, pp. 688–712. Springer, Heidelberg (2014)
8. Attrapadung, N., Libert, B., Peters, T.: Computing on authenticated data: New privacy definitions and constructions. In: Wang, X., Sako, K. (eds.) ASIACRYPT 2012. LNCS, vol. 7658, pp. 367–385. Springer, Heidelberg (2012)
9. Attrapadung, N., Libert, B., Peters, T.: Efficient completely context-hiding quotable and linearly homomorphic signatures. In: Kurosawa, K., Hanaoka, G. (eds.) PKC 2013. LNCS, vol. 7778, pp. 386–404. Springer, Heidelberg (2013)
10. Barbulescu, R., Gaudry, P., Joux, A., Thomé, E.: A heuristic quasi-polynomial algorithm for discrete logarithm in finite fields of small characteristic. In: Nguyen, P.Q., Oswald, E. (eds.) EUROCRYPT 2014. LNCS, vol. 8441, pp. 1–16. Springer, Heidelberg (2014)
11. Boneh, D., Boyen, X.: Short signatures without random oracles and the SDH assumption in bilinear groups. J. Cryptology 21(2), 149–177 (2008)
12. Camenisch, J., Dubovitskaya, M., Enderlein, R.R., Neven, G.: Oblivious transfer with hidden access control from attribute-based encryption. In: Visconti, I., De Prisco, R. (eds.) SCN 2012. LNCS, vol. 7485, pp. 559–579. Springer, Heidelberg (2012)
13. Cathalo, J., Libert, B., Yung, M.: Group encryption: Non-interactive realization in the standard model. In: Matsui, M. (ed.) ASIACRYPT 2009. LNCS, vol. 5912, pp. 179–196. Springer, Heidelberg (2009)
14. Chase, M., Kohlweiss, M., Lysyanskaya, A., Meiklejohn, S.: Malleable proof systems and applications. In: Pointcheval, D., Johansson, T. (eds.) EUROCRYPT 2012. LNCS, vol. 7237, pp. 281–300. Springer, Heidelberg (2012)
15. Chatterjee, S., Menezes, A.: On cryptographic protocols employing asymmetric pairings — The role of Ψ revisited. Discrete Applied Mathematics 159(13), 1311–1322 (2011)
16. Fuchsbauer, G.: Commuting signatures and verifiable encryption. In: Paterson, K.G. (ed.) EUROCRYPT 2011. LNCS, vol. 6632, pp. 224–245. Springer, Heidelberg (2011)
17. Fuchsbauer, G., Vergnaud, D.: Fair blind signatures without random oracles. In: Bernstein, D.J., Lange, T. (eds.) AFRICACRYPT 2010. LNCS, vol. 6055, pp. 16–33. Springer, Heidelberg (2010)
18. Galbraith, S.D., Paterson, K.G., Smart, N.P.: Pairings for cryptographers. Discrete Applied Mathematics 156(16), 3113–3121 (2008)
19. Ghadafi, E., Smart, N.P., Warinschi, B.: Groth–Sahai proofs revisited. In: Nguyen, P.Q., Pointcheval, D. (eds.) PKC 2010. LNCS, vol. 6056, pp. 177–192. Springer, Heidelberg (2010)
20. Green, M., Hohenberger, S.: Universally composable adaptive oblivious transfer. In: Pieprzyk, J. (ed.) ASIACRYPT 2008. LNCS, vol. 5350, pp. 179–197. Springer, Heidelberg (2008)
21. Groth, J.: Simulation-sound NIZK proofs for a practical language and constant size group signatures. In: Lai, X., Chen, K. (eds.) ASIACRYPT 2006. LNCS, vol. 4284, pp. 444–459. Springer, Heidelberg (2006)

22. Groth, J., Sahai, A.: Efficient noninteractive proof systems for bilinear groups. SIAM J. Comput. 41(5), 1193–1232 (2012)
23. Hofheinz, D., Jager, T.: Tightly secure signatures and public-key encryption. In: Safavi-Naini, R., Canetti, R. (eds.) CRYPTO 2012. LNCS, vol. 7417, pp. 590–607. Springer, Heidelberg (2012)
24. Joux, A.: A new index calculus algorithm with complexity $L(1/4 + o(1))$ in very small characteristic. IACR ePrint Archive, Report 2013/095 (2013), http://eprint.iacr.org/
25. Libert, B., Peters, T., Joye, M., Yung, M.: Linearly homomorphic structure-preserving signatures and their applications. In: Canetti, R., Garay, J.A. (eds.) CRYPTO 2013, Part II. LNCS, vol. 8043, pp. 289–307. Springer, Heidelberg (2013)
26. Libert, B., Peters, T., Yung, M.: Group signatures with almost-for-free revocation. In: Safavi-Naini, R., Canetti, R. (eds.) CRYPTO 2012. LNCS, vol. 7417, pp. 571–589. Springer, Heidelberg (2012)
27. Van de Woestijne, C.E.: Deterministic equation solving over finite fields. PhD thesis, Leiden University (2006)
28. Zhang, J., Li, Z., Guo, H.: Anonymous transferable conditional E-cash. In: Keromytis, A.D., Di Pietro, R. (eds.) SecureComm 2012. LNICST, vol. 106, pp. 45–60. Springer, Heidelberg (2013)

(Hierarchical) Identity-Based Encryption from Affine Message Authentication

Olivier Blazy, Eike Kiltz, and Jiaxin Pan

Faculty of Mathematics
Horst Görtz Institute for IT-Security
Ruhr University Bochum, Germany
{olivier.blazy,eike.kiltz,jiaxin.pan}@rub.de

Abstract. We provide a generic transformation from any *affine* message authentication code (MAC) to an identity-based encryption (IBE) scheme over pairing groups of prime order. If the MAC satisfies a security notion related to unforgeability against chosen-message attacks and, for example, the k-Linear assumption holds, then the resulting IBE scheme is adaptively secure. Our security reduction is tightness preserving, i.e., if the MAC has a tight security reduction so has the IBE scheme. Furthermore, the transformation also extends to hierarchical identity-based encryption (HIBE). We also show how to construct affine MACs with a tight security reduction to standard assumptions. This, among other things, provides the first tightly secure HIBE in the standard model.

Keywords: IBE, HIBE, standard model, tight reduction.

1 Introduction

Identity-based encryption (IBE) [24] enables a user to encrypt to a recipient's identity id (e.g., an email or phone number); decryption can be done using a user secret key for id, obtained from a trusted authority. The first instantiations of an IBE scheme were given in 2001 [7,4,23]. Whereas earlier constructions relied on the random oracle model, the first adaptively secure construction in the standard model was proposed in [26]. Here adaptive security means that an adversary may select the challenge identity id^* after seeing the public key and arbitrarily many user secret keys for identities of his choice. The concept of IBE generalizes naturally to hierarchical IBE (HIBE). In an L-level HIBE, hierarchical identities are vectors of identities of maximal length L and user secret keys for a hierarchical identity can be delegated. An IBE is simply a L-level HIBE with $L = 1$.

In this work we focus on adaptively secure (H)IBE schemes in the standard model. The construction from [26] has the disadvantage of a non-tight security reduction, i.e., the security reduction reducing security of the L-level HIBE to the hardness of the underlying assumption loses at least a factor of Q^L, where Q is the maximal number of user secret key queries. Modern HIBE schemes

J.A. Garay and R. Gennaro (Eds.): CRYPTO 2014, Part I, LNCS 8616, pp. 408–425, 2014.

[25,6] only lose a factor Q, independent of L. The first tightly secure IBE was recently proposed by Chen and Wee [6] but designing a L-level HIBE for $L > 1$ and a tight (i.e., independent of Q) security reduction to a standard assumption remains an open problem.

Until now, all known constructions of (H)IBE schemes are specific, i.e., they are custom-made to a specific hardness assumption. This is in contrast to other basic cryptographic primitives such as signatures and public-key encryption, for which efficient generic transformations have been known for a long time. We would like to highlight the concept of smooth projective hash proof systems for chosen-ciphertext secure encryption [9] and an old construction by Bellare and Goldwasser [1] that transforms any pseudorandom function (PRF) plus a non-interactive zero-knowledge (NIZK) proof into a signature scheme. Until today no generic construction of a (H)IBE from any "simple" low-level cryptographic primitive is known. However, the recent IBE scheme by Chen and Wee [6] uses a specific randomized PRF at the core of their construction, but its usage is non-modular.

1.1   This Work

AFFINE MACS. In this work we put forward the notion of *affine message authentication codes* (affine MACs). An affine MAC over $\mathbb{Z}_q^n$ is a randomized MAC with a special algebraic structure over some group $\mathbb{G} = \langle g \rangle$ of prime-order q. For a vector $\mathbf{a} \in \mathbb{Z}_q^n$, define $[\mathbf{a}] := g^{\mathbf{a}} = (g^{\mathbf{a}_1}, \ldots, g^{\mathbf{a}_n})^\top \in \mathbb{G}^n$ as the implicit representation of $\mathbf{a}$ over $\mathbb{G}$. Roughly speaking, the MAC tag $\tau_{\mathsf{m}} = ([\mathbf{t}], [u])$ of an affine MAC over $\mathbb{Z}_q^n$ on message $\mathsf{m} \in \mathcal{M}$ is split into a random message-independent part $[\mathbf{t}] \in \mathbb{G}^n$ plus a message-depending affine part $[u] \in \mathbb{G}$ satisfying

$$u = \sum f_i(\mathsf{m}) \mathbf{x}_i^\top \cdot \mathbf{t} + \sum f_i'(\mathsf{m}) x_i' \in \mathbb{Z}_q, \tag{1}$$

where $f_i, f_i' : \mathcal{M} \to \mathbb{Z}_q$ are public functions and $\mathbf{x}_i \in \mathbb{Z}_q^n$, $x_i' \in \mathbb{Z}_q$ are from the secret key $\mathsf{sk}_{\mathsf{MAC}}$. Almost all group-based MACs recently considered in [10], as well as the MAC derived from the randomized Naor-Reingold PRF [21] implicitly given in [6] are affine.

FROM AFFINE MACS TO IBE. Let us fix (possibly symmetric) pairing groups $\mathbb{G}_1, \mathbb{G}_2, \mathbb{G}_T$ equipped with a bilinear map $e : \mathbb{G}_1 \times \mathbb{G}_2 \to \mathbb{G}_T$. Let $\mathcal{D}_k$-MDDH be any Matrix Diffie-Hellman Assumption [11][1] that holds in $\mathbb{G}_1$, e.g., k-Linear or DDH.

Our main result is a *generic transformation* $\mathsf{IBE}[\mathsf{MAC}_n, \mathcal{D}_k]$ from any affine message authentication code MAC_n over $\mathbb{Z}_q^n$ into an IBE scheme. If MAC_n (defined over $\mathbb{G}_2$) is PR-CMA-secure (pseudorandom against chosen message attacks,

---

[1] The $\mathcal{D}_k$-MDDH assumption over $\mathbb{G}_1$ captures naturally all subspace decisional assumptions over prime order groups. Concretely, it states that given $[\mathbf{A}]_1 \in \mathbb{G}^{(k+1)\times k}$, the value $[\mathbf{A} \cdot \mathbf{w}]_1 \in \mathbb{G}_1^{k+1}$ is pseudorandom, where $\mathbf{A} \in \mathbb{Z}_q^{(k+1)\times k}$ gets chosen according to distribution $\mathcal{D}_k$ and $\mathbf{w} \in \mathbb{Z}_q^k$. Examples include k-Linear and DDH ($k = 1$).

a decisional variant of the standard UF-CMA security for MACs) and the $\mathcal{D}_k$-MDDH assumption holds in $\mathbb{G}_1$, then $\mathsf{IBE}[\mathsf{MAC}_n, \mathcal{D}_k]$ is an adaptively secure (and anonymous) IBE scheme. Furthermore, the security reduction of $\mathsf{IBE}[\mathsf{MAC}_n, \mathcal{D}_k]$ is as tight as the one of MAC_n. The size of the public IBE parameters depends on the size of the MAC secret key $\mathsf{sk_{MAC}}$, whereas the IBE ciphertexts and user secret keys always contain $n + k + 1$ group elements. We stress that our transformation works with any $k \geq 1$ and any $\mathcal{D}_k$-MDDH Assumption, hence $\mathcal{D}_k$ can be chosen to match the security assumption of MAC_n.

We also extend our generic transformation to HIBE schemes. In particular, we have two generic HIBE constructions depending on different properties of the underlying affine MACs. If the affine MAC is *delegatable* (to be defined in Section 5.1), we obtain an adaptively secure L-level HIBE $\mathsf{HIBE}[\mathsf{MAC}_n, \mathcal{D}_k]$. If the affine MAC is furthermore *anonymity-preserving*, we obtain an *anonymous* and adaptively secure L-level HIBE $\mathsf{AHIBE}[\mathsf{MAC}_n, \mathcal{D}_k]$. Both of the constructions have the same tightness properties as the MAC, and their ciphertexts sizes are the same as in the IBE case. Due to different delegation methods, $\mathsf{AHIBE}[\mathsf{MAC}_n, \mathcal{D}_k]$ has slightly shorter public parameters, but larger user secret keys than $\mathsf{HIBE}[\mathsf{MAC}_n, \mathcal{D}_k]$. Due to space restrictions, the anonymity-preserving transformation $\mathsf{AHIBE}[\mathsf{MAC}_n, \mathcal{D}_k]$ is only given in the full version [3].

Let us highlight again the fact that the underlying object is a *symmetric* primitive (a MAC) that we transform to an *asymmetric* primitive (an IBE scheme). Furthermore, as a MAC is a very simple and well-understood object, we hope that our transformation can contribute to understanding the more complex object of an IBE scheme.

Two Delegatable Affine MACs. To instantiate our transformations, we consider two specific delegatable affine MACs. Our first construction, $\mathsf{MAC_{NR}}[\mathcal{D}_k]$, is a generalization of the MAC derived from the randomized Naor-Reingold PRF [6] to any $\mathcal{D}_k$-MDDH Assumption. (Unfortunately, the MAC based on the original deterministic Naor-Reingold PRF is not affine.) We show that it is affine over $\mathbb{Z}_q^n$ with $n = k$ and delegatable. We prove PR-CMA-security with an (almost) tight security reduction to $\mathcal{D}_k$-MDDH. (Almost tight, as the security reduction loses a factor $O(m)$, where m is the length of the message space.) This leads to the first HIBE with a tight security reduction to a standard assumption. Ciphertexts and user secret keys of $\mathsf{HIBE}[\mathsf{MAC_{NR}}[\mathcal{D}_k], \mathcal{D}_k]$ only contain $2k+1$ group elements which is 3 in case we use $k = 1$ and the SXDH Assumption (i.e., DDH in $\mathbb{G}_1$ and $\mathbb{G}_2$). Interestingly, our SXDH-based IBE scheme can be seen as a "two-copy version" of Waters' IBE [26] which does not have a tight security reduction. The disadvantage of $\mathsf{MAC_{NR}}[\mathcal{D}_k]$ is that the public parameters of $\mathsf{IBE}[\mathsf{MAC_{NR}}[\mathcal{D}_k], \mathcal{D}_k]$ are linear in the bit-size of the identity space.

Our second construction, $\mathsf{MAC_{HPS}}[\mathcal{D}_k]$, is based on a hash proof system given in [11] for any $\mathcal{D}_k$-MDDH problem. A hash proof system is known to imply a UF-CMA-secure MAC [10]. We extend this result to PR-CMA-security, where the reduction loses a factor of Q, the number of MAC queries. Furthermore, $\mathsf{MAC_{HPS}}[\mathcal{D}_k]$ is affine over $\mathbb{Z}_q^{k+1}$ (i.e., $n = k + 1$) and delegatable. Whereas public parameters of the L-level HIBE $\mathsf{HIBE}[\mathsf{MAC_{HPS}}[\mathcal{D}_k], \mathcal{D}_k]$ only depend on

L, ciphertexts and user secret keys contain $2k + 2$ group elements which is 4 in case of the SXDH assumption ($k = 1$). We remark that the efficiency of HIBE[MAC$_{\mathsf{HPS}}[\mathcal{D}_k], \mathcal{D}_k]$ is roughly the same as a HIBE proposed in [6]. Additionally, we show MAC$_{\mathsf{HPS}}[\mathcal{D}_k]$ is also anonymity-preserving, which implies an anonymous (but non-tight) HIBE, AHIBE[MAC$_{\mathsf{HPS}}[\mathcal{D}_k], \mathcal{D}_k]$, while the delegatable MAC$_{\mathsf{NR}}[\mathcal{D}_k]$ is unlikely to be anonymity-preserving.

Table 1 summarizes all known (H)IBE scheme and their parameters.

EXTENSIONS. In fact, our generic transformation even gives (hierarchical) ID-based hash proof system from any (delegatable) affine MAC and the $\mathcal{D}_k$-MDDH assumption. From an (H)ID-based hash proof system one readily obtains a chosen-ciphertext secure (H)IBE [16]. Furthermore, any (H)IBE directly implies a (Hierarchical ID-based) signature scheme [12]. The signature obtained from IBE[MAC$_{\mathsf{NR}}[\mathcal{D}_k], \mathcal{D}_k]$ has a tight security reduction. Even though it is not entirely structure preserving, it can still be used to obtain a constant-size IND-CCA-secure public-key encryption scheme with a tight security reduction in the multi-user and multi-challenge setting [14,2].

1.2 Technical Details

OUR TRANSFORMATION. The high level idea behind our generic transformation IBE[MAC, $\mathcal{D}_k]$ from any affine MAC over $\mathbb{Z}_q^n$ to an IBE scheme is the transfor-

Table 1. Comparison between known adaptively secure IBEs with identity-space $\mathcal{ID} = \{0,1\}^\lambda$ and L-level HIBEs with identity-space $\mathcal{ID} = (\{0,1\}^\lambda)^L$ in prime order groups from standard assumptions. For $|\mathsf{pk}|$ (public-key size), $|\mathsf{usk}|$ (user secret-key size), and $|\mathsf{C}|$ (ciphertext size), we count the sum of all elements in $\mathbb{G}_1, \mathbb{G}_2, \mathbb{G}_T$, and $\mathbb{Z}_q$. Q is the number of user secret key queries by the adversary. Schemes from this paper are: IBE$_{\mathsf{HPS}}$:= IBE[MAC$_{\mathsf{HPS}}[\mathcal{D}_k], k$-LIN], IBE$_{\mathsf{NR}}$:= IBE[MAC$_{\mathsf{NR}}[\mathcal{D}_k], k$-LIN], HIBE$_{\mathsf{HPS}}$:= HIBE[MAC$_{\mathsf{HPS}}[\mathcal{D}_k], k$-LIN], HIBE$_{\mathsf{NR}}$:= HIBE[MAC$_{\mathsf{NR}}[\mathcal{D}_k], k$-LIN] and AHIBE$_{\mathsf{HPS}}$:= AHIBE[MAC$_{\mathsf{HPS}}[\mathcal{D}_k], k$-LIN].

	Scheme	$|\mathsf{pk}|$	$|\mathsf{usk}|$	$|\mathsf{C}|$	Anonymity?	Loss	Assumption
IBE	Wat05 [26]	$4 + \lambda$	2	2	$-$	$O(\lambda Q)$	DBDH
	Wat09 [25]	13	9	10	$-$	$O(Q)$	2-LIN
	Lew12 [17]	25	6	6	$\checkmark$	$O(Q)$	2-LIN
	CLL$^+$12 [5]	9	4	4	$-$	$O(Q)$	SXDH
	JR13 [15]	7	5	4	$-$	$O(Q)$	SXDH
	CW13 [6]	$2k^2(2\lambda+1)+k$	$4k$	$4k$	$-$	$O(\lambda)$	k-LIN
	IBE$_{\mathsf{HPS}}$	$3k^2 + 4k$	$2k+2$	$2k+2$	$\checkmark$	$O(Q)$	k-LIN
	IBE$_{\mathsf{NR}}$	$2\lambda k^2 + 2k$	$2k+1$	$2k+1$	$\checkmark$	$O(\lambda)$	k-LIN
HIBE	Wat05 [26]	$O(\lambda L)$	$O(\lambda L)$	$1 + L$	$-$	$O(\lambda Q)^L$	DBDH
	Wat09 [25]	$O(L)$	$O(L)$	$O(L)$	$-$	$O(Q)$	2-LIN
	CW13 [6]	$O(Lk^2)$	$O(Lk)$	$2k+2$	$-$	$O(Q)$	k-LIN
	HIBE$_{\mathsf{HPS}}$	$O(Lk^2)$	$O(Lk)$	$2k+2$	$-$	$O(Q)$	k-LIN
	HIBE$_{\mathsf{NR}}$	$O(L\lambda k^2)$	$O(L\lambda k)$	$2k+1$	$-$	$O(L\lambda)$	k-LIN
	AHIBE$_{\mathsf{HPS}}$	$O(Lk^2)$	$O(Lk^2)$	$2k+2$	$\checkmark$	$O(Q)$	k-LIN

mation from Bellare and Goldwasser [1] from a MAC (originally, a PRF) and a NIZK to a signature scheme. We use the same approach but define the user secret keys to be Bellare-Goldwasser signatures. The (H)IBE encryption functionality makes use of the special properties of the algebraic MAC and (tuned) Groth-Sahai proofs.

Concretely, the public key pk of the IBE scheme contains special perfectly hiding commitments $[\mathbf{Z}]_1$ to the MAC secret keys $\mathsf{sk_{MAC}}$, which also depend on the $\mathcal{D}_k$-MDDH assumption. The user secret key $\mathsf{usk[id]}$ of an identity id contains the MAC tag $\tau_{\mathsf{id}} = ([\mathbf{t}]_2, [u]_2) \in \mathbb{G}_2^{n+1}$ on id, plus a tuned Groth-Sahai [13] non-interactive zero-knowledge (NIZK) proof π that τ_{id} was computed correctly with respect to the commitments $[\mathbf{Z}]_1$ containing $\mathsf{sk_{MAC}}$. Since the MAC is affine, the NIZK proof $\pi \in \mathbb{G}^k$ is very compact. The next observation is that the NIZK verification equation for π is a linear equation in the (committed) MAC secret keys and hence a randomized version of it gives rise to the IBE ciphertext and a decryption algorithm.

SECURITY PROOF. The security proof can also be sketched easily at a high level. We first apply a Cramer-Shoup argument [8], where we decrypt the IBE challenge ciphertext using the MAC secret key $\mathsf{sk_{MAC}}$. Next, we make the challenge ciphertext inconsistent which involves one application of the $\mathcal{D}_k$-MDDH assumption. Now we can use the NIZK simulation routine to simulate the NIZK proof π from the user secret key $\mathsf{usk[id]} = (\tau_{\mathsf{id}}, \pi)$. At this point, as the commitments perfectly hide the MAC secret keys $\mathsf{sk_{MAC}}$, the only part of the security experiment still depending on $\mathsf{sk_{MAC}}$ is τ_{id} from $\mathsf{usk[id]}$ plus the computation of the challenge ciphertext. Now we are in the position to make the reduction to the symmetric primitive. We can use the PR-CMA symmetric security of MAC to argue directly about the pseudorandomness of the IBE challenge ciphertext. An IBE with pseudorandom ciphertexts is both IND-CPA secure and anonymous.

1.3   Other Related Work

Recently, Wee [27] proposed an information-theoretic primitive called *predicate encodings* that characterize the underlying algebraic structure of a number of predicate encoding schemes, including known IBE [19] and attribute-based encryption (ABE) [18] schemes. The main conceptual difference to affine MACs is that predicate encodings is a purely information-theoretic object. Furthermore, the framework by Wee is inherently limited to composite order groups.

Waters introduced the dual system framework [25] in order to facility tighter proofs for (H)IBE systems and beyond. The basic idea is that there exists functional and semi-functional ciphertexts and user secret keys, that are computationally indistinguishable. Decrypting a ciphertext with a user secret key is successful unless both are semi-functional. The $\mathcal{D}_k$-MDDH assumptions are specifically tailored to the dual system framework as they provide natural subspace assumptions over $\mathbb{G}^{k+1}$. Previous dual system constructions [25,19,6] usually first construct a scheme over composite-order groups and then transform it into prime-order groups. As the transformation uses a subspace assumption over $\mathbb{G}^{k+1}$

for *each component* of the composite-order group, ciphertexts and user secret keys contain at least $2(k + 1)$ group elements. An exception is a recent *direct construction* in prime-order groups by Jutla and Roy [15]. Their scheme is based on the SXDH assumption (*i.e.*, $k = 1$) and achieves slightly better ciphertext size of 3 group elements plus one element from $\mathbb{Z}_q$. Even though our construction and proof strategy is inspired by the Bellare-Goldwasser NIZK approach and Cramer-Shoup's hash proof systems, we still roughly follow the dual system framework. However, as we give a direct construction in prime-order groups, our IBE scheme $\mathsf{IBE}[\mathsf{MAC_{NR}}[\mathcal{D}_k], \mathcal{D}_k]$ has ciphertexts and user secret keys of size $2k + 1$, breaking the "$2(k + 1)$ barrier".

Lewko and Waters [20] consider the difficulty of a security proof for L-level HIBEs that does not proving exponentially in L. Essentially, they prove that any scheme with rerandomizable user secret keys (over the space of all "functional" user secret keys) will suffer an exponential degradation in security. While some of our tightly-secure HIBEs are rerandomizable, they are only rerandomizable over the space of all user secret keys generated by the user secret key generation algorithm. Hence, our tightly-secure HIBE does not contradict the negative results of [20].

1.4 Open Problems

We leave finding a PR-CMA-secure algebraic MAC with a tight security reduction and constant-size secret keys as an open problem. Given our main result this would directly imply a tightly-secure (H)IBE with constant-size public parameters. Furthermore, we leave finding a tightly-secure and anonymity-preserving delegatable affine MAC as an open problem, which would imply a tightly-secure anonymous HIBE.

Finally, we think that the concept of algebraic MACs can be extended such that our transformation also covers more general predicate encoding schemes, including attribute-based encryption.

2 Definitions

2.1 Notation

If $\mathbf{x} \in \mathcal{B}^n$, then $|\mathbf{x}|$ denotes the length n of the vector. Further, $x \leftarrow_\$ \mathcal{B}$ denotes the process of sampling an element x from set $\mathcal{B}$ uniformly at random.

GAMES. We use games for our security reductions. A game G is defined by procedures INITIALIZE and FINALIZE, plus some optional procedures $P_1, \ldots, P_n$. All procedures are given using pseudo-code, where initially all variables are undefined. An adversary $\mathcal{A}$ is executed in game G if it first calls INITIALIZE, obtaining its output. Next, it may make arbitrary queries to P_i (according to their specification), again obtaining their output. Finally, it makes one single call to FINALIZE($\cdot$) and stops. We define $\mathsf{G}^{\mathcal{A}}$ as the output of $\mathcal{A}$'s call to FINALIZE.

2.2 Pairing Groups and Matrix Diffie-Hellman Assumption

Let GGen be a probabilistic polynomial time (PPT) algorithm that on input 1^λ returns a description $\mathcal{G} = (\mathbb{G}_1, \mathbb{G}_2, \mathbb{G}_T, q, g_1, g_2, e)$ of asymmetric pairing groups where $\mathbb{G}_1$, $\mathbb{G}_2$, $\mathbb{G}_T$ are cyclic groups of order q for a λ-bit prime q, g_1 and g_2 are generators of $\mathbb{G}_1$ and $\mathbb{G}_2$, respectively, and $e : \mathbb{G}_1 \times \mathbb{G}_2$ is an efficiently computable (non-degenerated) bilinear map. Define $g_T := e(g_1, g_2)$, which is a generator in $\mathbb{G}_T$.

We use implicit representation of group elements as introduced in [11]. For $s \in \{1, 2, T\}$ and $a \in \mathbb{Z}_q$ define $[a]_s = g_s^a \in \mathbb{G}_s$ as the *implicit representation* of a in $\mathbb{G}_s$. More generally, for a matrix $\mathbf{A} = (a_{ij}) \in \mathbb{Z}_q^{n \times m}$ we define $[\mathbf{A}]_s$ as the implicit representation of $\mathbf{A}$ in $\mathbb{G}_s$:

$$[\mathbf{A}]_s := \begin{pmatrix} g_s^{a_{11}} & \cdots & g_s^{a_{1m}} \\ & & \\ g_s^{a_{n1}} & \cdots & g_s^{a_{nm}} \end{pmatrix} \in \mathbb{G}_s^{n \times m}$$

We will always use this implicit notation of elements in $\mathbb{G}_s$, i.e., we let $[a]_s \in \mathbb{G}_s$ be an element in $\mathbb{G}_s$. Note that from $[a]_s \in \mathbb{G}_s$ it is generally hard to compute the value a (discrete logarithm problem in $\mathbb{G}_s$). Further, from $[b]_T \in \mathbb{G}_T$ it is hard to compute the value $[b]_1 \in \mathbb{G}_1$ and $[b]_2 \in \mathbb{G}_2$ (pairing inversion problem). Obviously, given $[a]_s \in \mathbb{G}_s$ and a scalar $x \in \mathbb{Z}_q$, one can efficiently compute $[ax]_s \in \mathbb{G}_s$. Further, given $[a]_1, [a]_2$ one can efficiently compute $[ab]_T$ using the pairing e. For $\mathbf{a}, \mathbf{b} \in \mathbb{Z}_q^k$ define $e([\mathbf{a}]_1, [\mathbf{b}]_2) := [\mathbf{a}^\top \mathbf{b}]_T \in \mathbb{G}_T$.

We recall the definition of the matrix Diffie-Hellman (MDDH) assumption [11].

Definition 1 (Matrix Distribution). *Let $k \in \mathbb{N}$. We call $\mathcal{D}_k$ a matrix distribution if it outputs matrices in $\mathbb{Z}_q^{(k+1) \times k}$ of full rank k in polynomial time.*

Without loss of generality, we assume the first k rows of $\mathbf{A} \leftarrow_\$ \mathcal{D}_k$ form an invertible matrix. The $\mathcal{D}_k$-Matrix Diffie-Hellman problem is to distinguish the two distributions $([\mathbf{A}], [\mathbf{Aw}])$ and $([\mathbf{A}], [\mathbf{u}])$ where $\mathbf{A} \leftarrow_\$ \mathcal{D}_k$, $\mathbf{w} \leftarrow_\$ \mathbb{Z}_q^k$ and $\mathbf{u} \leftarrow_\$ \mathbb{Z}_q^{k+1}$.

Definition 2 ($\mathcal{D}_k$-Matrix Diffie-Hellman Assumption $\mathcal{D}_k$-MDDH). *Let $\mathcal{D}_k$ be a matrix distribution and $s \in \{1, 2, T\}$. We say that the $\mathcal{D}_k$-Matrix Diffie-Hellman ($\mathcal{D}_k$-MDDH) Assumption holds relative to GGen in group $\mathbb{G}_s$ if for all PPT adversaries $\mathcal{D}$,*

$$\mathbf{Adv}_{\mathcal{D}_k, \mathsf{GGen}}(\mathcal{D}) := |\Pr[\mathcal{D}(\mathcal{G}, [\mathbf{A}]_s, [\mathbf{Aw}]_s) = 1] - \Pr[\mathcal{D}(\mathcal{G}, [\mathbf{A}]_s, [\mathbf{u}]_s) = 1]| = negl(\lambda),$$

where $\mathcal{G} \leftarrow \mathsf{GGen}(1^\lambda)$, $\mathbf{A} \leftarrow_\$ \mathcal{D}_k$, $\mathbf{w} \leftarrow_\$ \mathbb{Z}_q^k$, $\mathbf{u} \leftarrow_\$ \mathbb{Z}_q^{k+1}$.

For each $k \geq 1$, [11] specifies distributions $\mathcal{L}_k$, $\mathcal{C}_k$, $\mathcal{SC}_k$, $\mathcal{IL}_k$ such that the corresponding $\mathcal{D}_k$-MDDH assumption is the k-Linear assumption, the k-Cascade, the k-Symmetric Cascade, and the Incremental k-Linear Assumption, respectively. All assumptions are generically secure in bilinear groups and form a hierarchy of increasingly weaker assumptions. The distributions of $\mathbf{A}$ are exemplified for $k = 2$, where $a_1, \ldots, a_6 \leftarrow_\$ \mathbb{Z}_q$.

$$\mathcal{C}_2 : \begin{pmatrix} a_1 & 0 \\ 1 & a_2 \\ 0 & 1 \end{pmatrix}, \quad \mathcal{SC}_2 : \begin{pmatrix} a_1 & 0 \\ 1 & a_1 \\ 0 & 1 \end{pmatrix}, \quad \mathcal{L}_2 : \begin{pmatrix} a_1 & 0 \\ 0 & a_2 \\ 1 & 1 \end{pmatrix}, \quad \mathcal{U}_2 : \begin{pmatrix} a_1 & a_2 \\ a_3 & a_4 \\ a_5 & a_6 \end{pmatrix}.$$

It was also shown in [11] that $\mathcal{U}_k$-MDDH is implied by all other $\mathcal{D}_k$-MDDH assumptions. If $\mathbf{A}$ is chosen from $\mathcal{SC}_k$, then $[\mathbf{A}]_s$ can be represented with 1 group element; if $\mathbf{A}$ is chosen from $\mathcal{L}_k$ or $\mathcal{C}_k$, then $[\mathbf{A}]_s$ can be represented with k group elements; If $\mathbf{A}$ is chosen from $\mathcal{U}_k$, then $[\mathbf{A}]_s$ can be represented with $(k + 1)k$ group elements. Hence, $\mathcal{SC}_k$-MDDH offers the same security guarantees as k-Linear, while having the advantage of a more compact representation.

3 Message Authentication Codes

We use the standard definition of a (randomized) message authentication code MAC = (Gen$_{\mathsf{MAC}}$, Tag, Ver), where $\mathsf{sk}_{\mathsf{MAC}} \leftarrow_\$ \mathsf{Gen}_{\mathsf{MAC}}(\mathsf{par})$ returns a secret key, $\tau \leftarrow_\$ \mathsf{Tag}(\mathsf{sk}_{\mathsf{MAC}}, \mathsf{m})$ returns a tag τ on message m from some message space $\mathcal{M}$, and $\mathsf{Ver}(\mathsf{sk}_{\mathsf{MAC}}, \mathsf{m}, \tau) \in \{0, 1\}$ returns a verification bit.

3.1 Affine MACs

Affine MACs over $\mathbb{Z}_q^n$ are group-based MACs with a specific algebraic structure.

Definition 3. *Let* par *be system parameters containing a group* $\mathcal{G} = (\mathbb{G}_2, q, g_2)$ *of prime-order* q *and let* $n \in \mathbb{N}$. *We say that* MAC = (Gen$_{\mathsf{MAC}}$, Tag, Ver) *is affine over* $\mathbb{Z}_q^n$ *if the following conditions hold:*

1. Gen$_{\mathsf{MAC}}$(par) *returns* $\mathsf{sk}_{\mathsf{MAC}}$ *containing* $(\mathbf{B}, \mathbf{x}_0, \ldots, \mathbf{x}_\ell, x'_0, \ldots, x'_{\ell'})$, *where* $\mathbf{B} \in \mathbb{Z}_q^{n \times n'}$, $\mathbf{x}_i \in \mathbb{Z}_q^n$, $x'_j \in \mathbb{Z}_q$, *for some* $n', \ell, \ell' \in \mathbb{N}$.
2. Tag($\mathsf{sk}_{\mathsf{MAC}}, \mathsf{m} \in \mathcal{B}^\ell$) *returns a tag* $\tau = ([\mathbf{t}]_2, [u]_2) \in \mathbb{G}_2^n \times \mathbb{G}_2$, *computed as*

$$\mathbf{t} = \mathbf{Bs} \in \mathbb{Z}_q^n \qquad \text{for } \mathbf{s} \leftarrow_\$ \mathbb{Z}_q^{n'} \tag{2}$$

$$u = \sum_{i=0}^{\ell} f_i(\mathsf{m})\mathbf{x}_i^\top \mathbf{t} + \sum_{i=0}^{\ell'} f'_i(\mathsf{m})x'_i \in \mathbb{Z}_q \tag{3}$$

   for some public defining functions $f_i : \mathcal{M} \to \mathbb{Z}_q$ *and* $f'_i : \mathcal{M} \to \mathbb{Z}_q$. *Vector* $\mathbf{t}$ *is the randomness and* u *is the (deterministic) message-depending part.*
3. Ver($\mathsf{sk}_{\mathsf{MAC}}, \mathsf{m}, \tau = ([\mathbf{t}]_2, [u]_2)$) *verifies if (3) holds.*

The standard security notion for probabilistic MACs is unforgeability against chosen-message attacks UF-CMA [10]. In this work we require *pseudorandom against chosen-message attacks* (PR-CMA), which is slightly stronger than UF-CMA. Essentially, we require that the values used for one single verification equation (3) on message m^* are pseudorandom over $\mathbb{G}_1$ and $\mathbb{G}_T$.

INITIALIZE:	CHAL(m^*): //one query
$sk_{MAC} \leftarrow_\$ Gen_{MAC}(par)$	$h \leftarrow_\$ \mathbb{Z}_q^*$
Return ε	$\mathbf{h}_0 = \sum f_i(m^*)\mathbf{x}_i \cdot h \in \mathbb{Z}_q^n;$
	$h_1 = \sum f_i'(m^*)x_i' \cdot h \in \mathbb{Z}_q$
EVAL(m):	$\boxed{\mathbf{h}_0 \leftarrow_\$ \mathbb{Z}_q^n;\ h_1 \leftarrow_\$ \mathbb{Z}_q}$
$\mathcal{Q}_\mathcal{M} = \mathcal{Q}_\mathcal{M} \cup \{m\}$	Return $([h]_1, [\mathbf{h}_0]_1, [h_1]_T)$
Return $([\mathbf{t}]_2, [u]_2) \leftarrow_\$ Tag(sk_{MAC}, m)$	
	FINALIZE($d \in \{0,1\}$):
	Return $d \wedge (m^* \notin \mathcal{Q}_\mathcal{M})$

Fig. 1. Games PR-CMA$_{real}$ and $\boxed{\text{PR-CMA}_{rand}}$ for defining PR-CMA security. In all procedures, the boxed statements redefining $(\mathbf{h}_0, h_1)$ are only executed in game PR-CMA$_{rand}$.

Let $\mathcal{G} = (\mathbb{G}_1, \mathbb{G}_2, \mathbb{G}_T, q, g_1, g_2, e)$ be an asymmetric pairing group such that $(\mathbb{G}_2, g_2, q)$ is contained in par. We define the PR-CMA security via games PR-CMA$_{real}$ and PR-CMA$_{rand}$ from Figure 1. Note that the output $([h]_1, [\mathbf{h}_0]_1, [h_1]_T)$ of CHAL(m^*) in game PR-CMA$_{real}$ can be viewed as a "token" for message m^* to check verification equation (3) for arbitrary tags $([\mathbf{t}]_2, [u]_2)$ via equation $e([h]_1, [u]_2) \stackrel{?}{=} e([\mathbf{t}]_1, [\mathbf{h}_0]_1) \cdot [h_1]_T$. Intuitively, the pseudorandomness of $[h_1]_T$ is responsible for indistinguishablity and of $[\mathbf{h}_0]_1$ to prove anonymity of the IBE scheme.

Definition 4. *An affine* MAC *over* $\mathbb{Z}_q^n$ *is* PR-CMA-*secure if for all PPT* $\mathcal{A}$,
$Adv_{MAC}^{pr-cma}(\mathcal{A}) := \Pr[\text{PR-CMA}_{real}^\mathcal{A} \Rightarrow 1] - \Pr[\text{PR-CMA}_{rand}^\mathcal{A} \Rightarrow 1]$ *is negligible, where the experiments are defined in Figure 1.*

3.2 An Affine MAC from the Naor-Reingold PRF

Unfortunately, the (deterministic) Naor-Reingold pseudorandom function is not affine. We use the following randomized version $MAC_{NR}[\mathcal{D}_k] = (Gen_{MAC}, Tag, Ver)$ of it based on any matrix assumption $\mathcal{D}_k$. For the special case $\mathcal{D}_k = \mathcal{L}_k$, it was implicitly given in [6]. For a matrix $\mathbf{A} \in \mathbb{Z}_q^{(k+1) \times k}$ we denote the upper k rows by $\overline{\mathbf{A}} \in \mathbb{Z}_q^{k \times k}$ and the last row by $\underline{\mathbf{A}} \in \mathbb{Z}_q^{1 \times k}$.

Gen$_{MAC}$(par):	Tag(sk_{MAC}, m):	Ver(sk_{MAC}, τ, m):
$\mathbf{A} \leftarrow_\$ \mathcal{D}_k;\ \mathbf{B} := \overline{\mathbf{A}} \in \mathbb{Z}_q^{k \times k}$	$\mathbf{s} \leftarrow_\$ \mathbb{Z}_q^k;\ \mathbf{t} = \mathbf{Bs}$	If $u = (\sum_{i=1}^{|m|} \mathbf{x}_{i,m_i}^\top)\mathbf{t} + x_0'$
$\mathbf{x}_{1,0}, \ldots, \mathbf{x}_{m,1} \leftarrow_\$ \mathbb{Z}_q^k$	$u = (\sum_{i=1}^{|m|} \mathbf{x}_{i,m_i}^\top)\mathbf{t} + x_0'$	then ret 1
$x_0' \leftarrow_\$ \mathbb{Z}_q$	Ret $\tau = ([\mathbf{t}]_2, [u]_2)$	Else ret 0
Ret $\qquad sk_{MAC}$	$\in \mathbb{G}_2^k \times \mathbb{G}_2$	
$(\mathbf{B}, \mathbf{x}_{1,0}, \ldots, \mathbf{x}_{m,1}, x_0')$	$=$	

Note that $MAC_{NR}[\mathcal{D}_k]$ is n-affine over $\mathbb{Z}_q^n$ with message space $\mathcal{M} = \{0,1\}^m$. Writing $\mathbf{x}_{i,b} = \mathbf{x}_{2i+b}$ we have $n = n' = k$, $\ell' = 0$, $\ell = 2m+1$ and functions

$f_0(\mathsf{m}) = f_1(\mathsf{m}) = 0$, $f_0'(\mathsf{m}) = 1$, and $f_{2i+b}(\mathsf{m}) = (\mathsf{m}_i = b)$ for $1 \leq i \leq m$, where m_i is the i-th bit of m. (To perfectly fit our definition, $\mathbf{x}_{i,b}$ should be renamed to $\mathbf{x}_{2i+b}$, but we conserve the other notations for better readability.)

Theorem 1. $\mathsf{MAC}_{\mathsf{NR}}[\mathcal{D}_k]$ *is tightly* PR-CMA-*secure under the* $\mathcal{D}_k$-MDDH *assumption. In particular, for all adversaries* $\mathcal{A}$ *there exists an adversary* $\mathcal{D}$ *with* $\mathbf{T}(\mathcal{A}) \approx \mathbf{T}(\mathcal{D})$ *and* $\mathsf{Adv}^{\mathsf{pr\text{-}cma}}_{\mathsf{MAC}_{\mathsf{NR}}[\mathcal{D}_k]}(\mathcal{A}) \leq 4m(\mathbf{Adv}_{\mathcal{D}_k,\mathsf{GGen}}(\mathcal{D}) - 1/(q-1))$.

Note that the security bound is (almost) tight, as m is the bit-length of message space $\mathcal{M}$. The proof follows the ideas from [6,22]. We use m hybrids, where in hybrid i *all* the (maximal Q) values $\mathbf{x}_{i,1-\mathsf{m}_i^*}^\top \cdot \mathbf{t}$ in the response to an EVAL query are replaced by uniform randomness. Here m^* is the message from the challenge query. We use the Q-fold $\mathcal{D}_k$-MDDH assumption [11] (which gives Q-many real-or-random $\mathcal{D}_k$-MDDH tuples) to interpolate between the hybrids, where the reductions guesses m_i^* correctly with probability $1/2$. As the Q-fold $\mathcal{D}_k$-MDDH assumption is tightly implied by the standard $\mathcal{D}_k$-MDDH assumption [11], the proof follows. A formal proof can be found in the full version [3].

We remark, that one can define an alternative version of $\mathsf{MAC}_{\mathsf{NR}}[\mathcal{D}_k]$ by setting $\mathbf{x}_0 := \sum \mathbf{x}_{i,0}$, $\mathbf{x}_i := \mathbf{x}_{i,1} - \mathbf{x}_{i,0}$ and $u = (\mathbf{x}_0^\top + \sum_{i=1}^{|\mathsf{m}|} \mathsf{m}_i \mathbf{x}_i^\top)\mathbf{t} + x_0'$. This MAC has a shorter secret key and can also be shown to be PR-CMA. (However, it does not satisfy the stronger security notion of HPR-CMA needed in Sect. 5.)

3.3   An Affine MAC from Hash Proof System

Let $\mathcal{D}_k$ be a matrix distribution. We now combine the hash proof system for the subset membership problem induced by the $\mathcal{D}_k$-MDDH assumption from [11] with the generic MAC construction from [10] and obtain the following $\mathsf{MAC}_{\mathsf{HPS}}[\mathcal{D}_k]$ for $\mathcal{M} = \mathbb{Z}_q^\ell$. Algorithm $\mathsf{Gen}_{\mathsf{MAC}}(\mathsf{par})$ picks $\mathbf{B} \leftarrow_\$ \mathcal{D}_k$, $\mathbf{x}_0, \dots, \mathbf{x}_\ell \leftarrow_\$ \mathbb{Z}_q^{k+1}$, and $x_0' \leftarrow_\$ \mathbb{Z}_q$. The MAC secret-key is $\mathsf{sk}_{\mathsf{MAC}} = (\mathbf{B}, \mathbf{x}_0, \dots, \mathbf{x}_\ell, x_0')$.

$\mathsf{Tag}(\mathsf{sk}_{\mathsf{MAC}}, \mathsf{m})$:	$\mathsf{Ver}(\mathsf{sk}_{\mathsf{MAC}}, \tau, \mathsf{m})$:
Parse $\mathsf{sk}_{\mathsf{MAC}} = (\mathbf{B}, \mathbf{x}_0, \dots, \mathbf{x}_\ell, x_0')$	Parse $\mathsf{sk}_{\mathsf{MAC}} = (\mathbf{B}, \mathbf{x}_0, \dots, \mathbf{x}_\ell, x_0')$
$\mathbf{s} \leftarrow_\$ \mathbb{Z}_q^k$; $\mathbf{t} = \mathbf{Bs} \in \mathbb{Z}_q^{k+1}$	If $u = (\mathbf{x}_0^\top + \sum_{i=1}^{|\mathsf{m}|} \mathsf{m}_i \cdot \mathbf{x}_i^\top)\mathbf{t} + x_0'$
$u = (\mathbf{x}_0^\top + \sum_{i=1}^{|\mathsf{m}|} \mathsf{m}_i \cdot \mathbf{x}_i^\top)\mathbf{t} + x_0'$	then return 1
Return $\tau = ([\mathbf{t}]_2, [u]_2) \in \mathbb{G}_2^{k+1} \times \mathbb{G}_2$	Else return 0

Note that $\mathsf{MAC}_{\mathsf{HPS}}[\mathcal{D}_k]$ is n-affine over $\mathbb{Z}_q^n$ with $n = k + 1$, $n' = k$, $\ell' = 0$, and defining functions $f_0(\mathsf{m}) = 1$, $f_i(\mathsf{m}) = \mathsf{m}_i$, and $f_0'(\mathsf{m}) = 1$, where m_i is the i-th component of m. For the moment we use $\ell = 1$ which already gives a MAC with exponential message space $\mathcal{M} = \mathbb{Z}_q$.

Combining [11,10] we obtain that $\mathsf{MAC}_{\mathsf{HPS}}[\mathcal{D}_k]$ is UF-CMA under the $\mathcal{D}_k$-MDDH assumption. The proof extends to show even PR-CMA security. Compared to $\mathsf{MAC}_{\mathsf{NR}}[\mathcal{D}_k]$, we lose the tight reduction, but gain much shorter public parameters. A formal proof can be found in the full version [3].

Theorem 2. $\mathsf{MAC}_{\mathsf{HPS}}[\mathcal{D}_k]$ *is* PR-CMA-*secure under the* $\mathcal{D}_k$-MDDH *assumption. In particular, for all adversaries* $\mathcal{A}$ *there exists an adversary* $\mathcal{D}$ *with* $\mathbf{T}(\mathcal{A}) \approx$

$\mathbf{T}(\mathcal{D})$ and $\mathsf{Adv}^{\text{pr-cma}}_{\mathsf{MAC}_{\mathsf{HPS}}[\mathcal{D}_k]}(\mathcal{A}) \leq 2Q(\mathbf{Adv}_{\mathcal{D}_k,\mathsf{GGen}}(\mathcal{D}) + 1/q)$, where Q is the maximal number of queries to $\mathrm{EVAL}(\cdot)$.

4   Identity-Based Encryption from Affine MACs

In this section, we will present our transformation $\mathsf{IBE}[\mathsf{MAC}, \mathcal{D}_k]$ from affine MACs to IBE based on the $\mathcal{D}_k$-MDDH assumption.

4.1   Identity-Based Key Encapsulation

We now recall syntax and security of IBE in terms of an ID-based key encapsulation mechanism IBKEM. Every IBKEM can be transformed into an ID-based encryption scheme IBE using a (one-time secure) symmetric cipher.

Definition 5 (Identity-Based Key Encapsulation Scheme). *An identity-based key encapsulation (IBKEM) scheme* IBKEM *consists of four PPT algorithms* IBKEM = (Gen, USKGen, Enc, Dec) *with the following properties.*

- *The probabilistic key generation algorithm* Gen(1^λ) *returns the (master) public/secret key* (pk, sk). *We assume that* pk *implicitly defines a message space* $\mathcal{M}$, *an identity space* $\mathcal{ID}$, *a key space* $\mathcal{K}$, *and ciphertext space* $\mathcal{C}$.
- *The probabilistic user secret key generation algorithm* USKGen(sk, id) *returns the user secret-key* usk[id] *for identity* id $\in \mathcal{ID}$.
- *The probabilistic encapsulation algorithm* Enc(pk, id) *returns the symmetric key* K $\in \mathcal{K}$ *together with a ciphertext* C $\in \mathcal{C}$ *with respect to identity* id.
- *The deterministic decapsulation algorithm* Dec(usk[id], id, C) *returns the decapsulated key* K $\in \mathcal{K}$ *or the reject symbol* $\bot$.

For perfect correctness we require that for all $\lambda \in \mathbb{N}$, *all pairs* (pk, sk) *generated by* Gen(1^λ), *all identities* id $\in \mathcal{ID}$, *all* usk[id] *generated by* USKGen(sk, id) *and all* (K, C) *output by* Enc(pk, id):

$$\Pr[\mathsf{Dec}(\mathsf{usk}[\mathsf{id}], \mathsf{id}, \mathsf{C}) = \mathsf{K}] = 1.$$

The security requirements for an IBKEM we consider here are indistinguishability and anonymity against chosen plaintext and identity attacks (IND-ID-CPA and ANON-ID-CPA). Instead of defining both security notions separately, we define pseudorandom ciphertexts against chosen plaintext and identity attacks (PR-ID-CPA) which means that challenge key and ciphertext are both pseudorandom. Note that PR-ID-CPA trivially implies IND-ID-CPA and ANON-ID-CPA.

We define PR-ID-CPA-security of IBKEM formally via the games given in Figure 2.

Definition 6 (PR-ID-CPA Security). *An identity-based key encapsulation scheme* IBKEM *is* PR-ID-CPA-*secure if for all PPT* $\mathcal{A}$, $\mathsf{Adv}^{\text{pr-id-cpa}}_{\mathsf{IBKEM}}(\mathcal{A}) := |\Pr[\mathrm{PR\text{-}ID\text{-}CPA}^{\mathcal{A}}_{\text{real}} \Rightarrow 1] - \Pr[\mathrm{PR\text{-}ID\text{-}CPA}^{\mathcal{A}}_{\text{rand}} \Rightarrow 1]|$ *is negligible.*

Procedure INITIALIZE:	Procedure ENC(id^*): //one query
$(pk, sk) \leftarrow_\$ Gen(1^\lambda)$	$(K^*, C^*) \leftarrow_\$ Enc(pk, id^*)$
Return pk	$\boxed{K^* \leftarrow_\$ \mathcal{K}; C^* \leftarrow_\$ \mathcal{C}}$
	Return (K^*, C^*)
Procedure USKGEN(id):	
$\mathcal{Q}_{ID} \leftarrow \mathcal{Q}_{ID} \cup \{id\}$	Procedure FINALIZE(β):
Return usk[id] $\leftarrow_\$ USKGen(sk, id)$	Return $(id^* \notin \mathcal{Q}_{ID}) \wedge \beta$

Fig. 2. Security Games PR-ID-CPA$_{real}$ and $\boxed{\text{PR-ID-CPA}_{rand}}$ for defining PR-ID-CPA-security

4.2 The Transformation

Let $\mathcal{D}_k$ be a matrix distribution that outputs matrices $\mathbf{A} \in \mathbb{Z}_q^{(k+1) \times k}$. Let MAC be an affine MAC over $\mathbb{Z}_q^n$ with message space $\mathcal{ID}$. Our IBKEM IBKEM[MAC, $\mathcal{D}_k$] = (Gen, USKGen, Enc, Dec) for key-space $\mathcal{K} = \mathbb{G}_T$ and identity space $\mathcal{ID}$ is defined in Figure 3.

Gen(par):	Enc(pk, id):
$\mathbf{A} \leftarrow_\$ \mathcal{D}_k$	$\mathbf{r} \leftarrow_\$ \mathbb{Z}_q^k$
skMAC $\leftarrow_\$$ GenMAC(par)	$\mathbf{c}_0 = \mathbf{Ar} \in \mathbb{Z}_q^{k+1}$
Parse skMAC $= (\mathbf{B}, \mathbf{x}_0, \ldots, \mathbf{x}_\ell, x'_0, \ldots, x'_{\ell'})$	$\mathbf{c}_1 = (\sum_{i=0}^\ell f_i(id)\mathbf{Z}_i) \cdot \mathbf{r} \in \mathbb{Z}_q^n$
For $i = 0, \ldots, \ell$:	$C = ([\mathbf{c}_0]_1, [\mathbf{c}_1]_1)$
$\quad \mathbf{Y}_i \leftarrow_\$ \mathbb{Z}_q^{k \times n}; \mathbf{Z}_i = (\mathbf{Y}_i^\top \mid \mathbf{x}_i) \cdot \mathbf{A} \in \mathbb{Z}_q^{n \times k}$	$K = (\sum_{i=0}^{\ell'} f'_i(id)z'_i) \cdot \mathbf{r} \in \mathbb{Z}_q$
For $i = 0, \ldots, \ell'$:	Return $(K = [K]_T, C)$
$\quad \mathbf{y}'_i \leftarrow_\$ \mathbb{Z}_q^k; z'_i = (\mathbf{y}_i'^\top \mid x'_i) \cdot \mathbf{A} \in \mathbb{Z}_q^{1 \times k}$	
pk $:= (\mathcal{G}, [\mathbf{A}]_1, ([\mathbf{Z}_i]_1)_{0 \le i \le \ell}, ([\mathbf{z}'_i]_1)_{0 \le i \le \ell'})$	Dec(usk[id], id, C):
sk $:= (skMAC, (\mathbf{Y}_i)_{0 \le i \le \ell}, (\mathbf{y}'_i)_{0 \le i \le \ell'})$	Parse usk[id] $= ([\mathbf{t}]_2, [u]_2, [\mathbf{v}]_2)$
Return (pk, sk).	Parse $C = ([\mathbf{c}_0]_1, [\mathbf{c}_1]_1)$
	$K = e([\mathbf{c}_0]_1, \begin{bmatrix} \mathbf{v} \\ u \end{bmatrix}_2) \cdot e([\mathbf{c}_1]_1, [\mathbf{t}]_2)^{-1}$
USKGen(sk, id):	Return $K \in \mathbb{G}_T$.
$([\mathbf{t}]_2, [u]_2) \leftarrow_\$ Tag(skMAC, id)$	
$\mathbf{v} = \sum_{i=0}^\ell f_i(id)\mathbf{Y}_i\mathbf{t} + \sum_{i=0}^{\ell'} f'_i(id)\mathbf{y}'_i \in \mathbb{Z}_q^k$	
Return usk[id] $:= ([\mathbf{t}]_2, [u]_2, [\mathbf{v}]_2) \in \mathbb{G}_2^{n+1+k}$	

Fig. 3. Definition of the transformation IBKEM[MAC, $\mathcal{D}_k$]

The intuition behind our construction is that the values $[\mathbf{Z}_i]_1, [\mathbf{z}'_i]_1$ from pk can be viewed as perfectly hiding commitments to the secrets keys skMAC = $(\mathbf{x}_1, \ldots, \mathbf{x}_\ell, x'_1, \ldots, x'_{\ell'})$ of MAC. User secret key generation computes the MAC tag $\tau = ([\mathbf{t}]_2, [u]_2) \leftarrow_\$ Tag(skMAC)$ plus a "non-interactive zero-knowledge proof" $[\mathbf{v}]_2$ proving that τ was computed correctly with respect to the commitments.

As the MAC is affine, the NIZK proof has a very simple structure. The encryption algorithm is derived from a randomized version of the NIZK verification equation. Here we again make use of the affine structure of MAC.

To show correctness of $\mathsf{IBKEM}[\mathsf{MAC}, \mathcal{D}_k]$, let (K, C) be the output of $\mathsf{Enc}(\mathsf{pk}, \mathsf{id})$ and let $\mathsf{usk}[\mathsf{id}]$ be the output of $\mathsf{USKGen}(\mathsf{sk}, \mathsf{id})$. By Equation (3) in Section 3, we have

$$e([\mathbf{c}_0]_1, \begin{bmatrix} \mathbf{v} \\ u \end{bmatrix}_2) = \left[(\mathbf{Ar})^\top \cdot \begin{pmatrix} \sum_{i=0}^{\ell} f_i(\mathsf{id})\mathbf{Y}_i\mathbf{t} + \sum_{i=0}^{\ell'} f_i'(\mathsf{id})\mathbf{y}_i' \\ \sum_{i=0}^{\ell} f_i(\mathsf{id})\mathbf{x}_i^\top\mathbf{t} + \sum_{i=0}^{\ell'} f_i'(\mathsf{id})x_i' \end{pmatrix} \right]_T,$$

$$e([\mathbf{c}_1]_1, [\mathbf{t}]_2) = \left[(\mathbf{Ar})^\top \left(\sum f_i(\mathsf{id})\mathbf{Y}_i \\ \sum f_i(\mathsf{id})\mathbf{x}_i^\top \right) \cdot \mathbf{t} \right]_T,$$

and the quotient of the two elements yields $\mathsf{K} = [(\sum_{i=0}^{\ell'} f_i'(\mathsf{id})\mathbf{z}_i') \cdot \mathbf{r}]_T$.

Theorem 3. *Under the $\mathcal{D}_k$-MDDH assumption relative to GGen in $\mathbb{G}_1$ and the PR-CMA-security of MAC, $\mathsf{IBKEM}[\mathsf{MAC}, \mathcal{D}_k]$ is a PR-ID-CPA-secure IBKEM. Particularly, for all adversaries $\mathcal{A}$ there exist adversaries $\mathcal{B}_1$ and $\mathcal{B}_2$ with $\mathbf{T}(\mathcal{B}_1) \approx \mathbf{T}(\mathcal{A}) \approx \mathbf{T}(\mathcal{B}_2)$ and $\mathsf{Adv}_{\mathsf{IBKEM}[\mathsf{MAC},\mathcal{D}_k]}^{\mathsf{pr-id-cpa}}(\mathcal{A}) \leq \mathbf{Adv}_{\mathcal{D}_k,\mathsf{GGen}}(\mathcal{B}_1) + \mathsf{Adv}_{\mathsf{MAC}}^{\mathsf{pr-cma}}(\mathcal{B}_2)$.*

The proof can be found in the full version [3].

5   Hierarchical Identity-Based Encryption from Delegatable Affine MACs

In this section, we will define syntax and security requirements of *delegatable affine MACs* and describe our transformation $\mathsf{HIBE}[\mathsf{MAC}, \mathcal{D}_k]$ from delegatable affine MACs to HIBE based on any $\mathcal{D}_k$-MDDH assumption. In the full version [3] we recall syntax and IND-HID-CPA security of a hierarchical ID-based key encapsulation mechanism (HIBKEM).

5.1   Delegatable Affine MACs

Definition 7. *An affine MAC over $\mathbb{Z}_q^n$ (Definition 3) is delegatable, if the message space is $\mathcal{M} = \mathcal{B}^{\leq m}$ for some finite base set $\mathcal{B}$, $\ell' = 0$ with $f_0'(\mathsf{m}) = 1$, and there exists a public function $l : \mathcal{M} \to \{0, \ldots, \ell\}$ such that for all $\mathsf{m}' \in \mathcal{M}$ with $\mathsf{m}' = (\mathsf{m}_1, \ldots, \mathsf{m}_{p+1}) \in \mathcal{B}^{p+1}$ and length p prefix $\mathsf{m} = (\mathsf{m}_1, \ldots, \mathsf{m}_p)$ of m, we have $l(\mathsf{m}) \leq l(\mathsf{m}')$ and*

$$f_i(\mathsf{m}') = \begin{cases} f_i(\mathsf{m}) & 0 \leq i \leq l(\mathsf{m}) \\ 0 & l(\mathsf{m}') < i \leq \ell \end{cases}.$$

Note that for a delegatable MAC, equation (3) simplifies to

$$u = \left(\sum_{i=0}^{l(\mathsf{m})} f_i(\mathsf{m})\mathbf{x}_i^\top + \sum_{i=l(\mathsf{m})+1}^{l(\mathsf{m}')} f_i(\mathsf{m}')\mathbf{x}_i^\top \right) \mathbf{t} + f_0'(\mathsf{m})x_0'.$$

Intuitively, this property will be used for HIBE user secret key delegation.

SECURITY REQUIREMENTS. Let MAC be a delegatable affine MAC over $\mathbb{Z}_q^n$ with message space $\mathcal{M} = \mathcal{B}^{\leq m} := \bigcup_{i=1}^m \mathcal{B}^i$. To build a HIBE, we require a new notion denoted as $\mathsf{HPR_0\text{-}CMA}$ security. It differs from $\mathsf{PR\text{-}CMA}$ security in two ways. Firstly, additional values needed for HIBE delegation are provided to the adversary through the call to INITIALIZE and EVAL. Secondly, CHAL always returns a real $\mathbf{h}_0$ which is the reason why our HIBE is not anonymous. (In fact, the additional values actually allow the adversary to distinguish real from random $\mathbf{h}_0$.)

Let $\mathcal{G} = (\mathbb{G}_1, \mathbb{G}_2, \mathbb{G}_T, q, g_1, g_2, e)$ be an asymmetric pairing group such that $(\mathbb{G}_2, g_2, q)$ is contained in par. Consider the games from Figure 4.

<u>INITIALIZE:</u>	<u>CHAL(m^*):</u> // one query
$\mathsf{sk_{MAC}} = (\mathbf{B}, (\mathbf{x}_i)_{0 \leq i \leq \ell}, x_0') \leftarrow_\$ \mathsf{Gen_{MAC}(par)}$	$h \leftarrow_\$ \mathbb{Z}_q$
Return $([\mathbf{B}]_2, ([\mathbf{x}_i^\top \mathbf{B}]_2)_{0 \leq i \leq \ell})$	$\mathbf{h}_0 = \sum f_i(m_i^*)\mathbf{x}_i \cdot h \in \mathbb{Z}_q^n$
	$h_1 = x_0' \cdot h \in \mathbb{Z}_q$
<u>EVAL(m):</u>	$\boxed{h_1 \leftarrow_\$ \mathbb{Z}_q}$
$\mathcal{Q_M} = \mathcal{Q_M} \cup \{m\}$	Return $([h]_1, [\mathbf{h}_0]_1, [h_1]_T)$
$([t]_2, [u]_2) \leftarrow_\$ \mathsf{Tag(sk_{MAC}}, m)$	
For $i = (m) + 1, \ldots, \ell$:	<u>FINALIZE($\beta \in \{0,1\}$):</u>
$d_i = \mathbf{x}_i^\top \mathbf{t} \in \mathbb{Z}_q; d_i' = \mathbf{x}_i^\top \mathbf{t}' \in \mathbb{Z}_q$	Return $\beta \wedge (\mathsf{Prefix}(m^*) \cap \mathcal{Q_M} = \emptyset)$
Return $([\mathbf{t}]_2, [u]_2, [\mathbf{t}']_2, [u']_2, ([d_i]_2)_{(m)+1 \leq i \leq \ell})$	

Fig. 4. Games $\mathsf{HPR\text{-}CMA_{real}}$, and $\boxed{\mathsf{HPR_0\text{-}CMA_{rand}}}$ for defining $\mathsf{HPR_0\text{-}CMA}$ security

Definition 8. *A delegatable affine* MAC *over* $\mathbb{Z}_q^n$ *is* $\mathsf{HPR_0\text{-}CMA}$-*secure if for all PPT* $\mathcal{A}$, $\mathsf{Adv}_{\mathsf{MAC}}^{\mathsf{hpr_0\text{-}cma}}(\mathcal{A}) := \Pr[\mathsf{HPR\text{-}CMA}_{\mathsf{real}}^{\mathcal{A}} \Rightarrow 1] - \Pr[\mathsf{HPR_0\text{-}CMA}_{\mathsf{rand}}^{\mathcal{A}} \Rightarrow 1]$ *is negligible.*

5.2 Examples of Delegatable Affine MACs

We first note that $\mathsf{MAC_{NR}}[\mathcal{D}_k]$ from Section 3 with message space $\mathcal{M} = \{0,1\}^{\leq m}$ is delegatable.

Theorem 4. *Under the* $\mathcal{D}_k$-*MDDH assumption,* $\mathsf{MAC_{NR}}[\mathcal{D}_k]$ *is tightly* $\mathsf{HPR_0\text{-}CMA}$ *secure. In particular, for all adversaries* $\mathcal{A}$ *there exists an adversary* $\mathcal{D}$ *with* $\mathbf{T}(\mathcal{A}) \approx \mathbf{T}(\mathcal{D})$ *and* $\mathsf{Adv}_{\mathsf{MAC_{NR}}[\mathcal{D}_k]}^{\mathsf{hpr_0\text{-}cma}}(\mathcal{A}) \leq 6m(\mathbf{Adv}_{\mathcal{D}_k,\mathsf{GGen}}(\mathcal{D}) - 1/(q-1))$.

The proof is similar to that of Theorem 1, with the difference that the reduction between games G_i and G_{i-1} now has to guess $m_i^* \in \{0, 1, \perp\}$, where $\perp$ means that $|m^*| < i$. Furthermore, $\mathbf{h}_0$ from CHAL(m^*) is not pseudorandom in the delegatable case, since $([\mathbf{B}]_2, ([\mathbf{x}_i^\top \mathbf{B}]_2)_{0 \leq i \leq m})$ are disclosed from INITIALIZE and then it is easy to check if $\mathbf{h}_0$ is well-formed under m^* by using the pairing. A formal proof of Theorem 4 is given in [3].

We now turn to $\mathsf{MAC}_{\mathsf{HPS}}[\mathcal{D}_k]$ from Section 3 with message space $\mathcal{M} = \mathcal{B}^{\leq m} = (\mathbb{Z}_q^*)^{\leq m}$. Again, it can be verified to be delegatable. One should remark the change on $\mathcal{B}$, where we now define $\mathcal{B} = \mathbb{Z}_q^*$ to avoid having a collision between the MAC of m and the MAC of $\mathsf{m}\|0$.

Theorem 5. *Under the $\mathcal{D}_k$-MDDH assumption, $\mathsf{MAC}_{\mathsf{HPS}}[\mathcal{D}_k]$ is HPR_0-CMA-secure. In particular, for all adversaries $\mathcal{A}$ there exists an adversary $\mathcal{D}$ with $\mathbf{T}(\mathcal{A}) \approx \mathbf{T}(\mathcal{D})$ and $\mathsf{Adv}_{\mathsf{MAC}_{\mathsf{HPS}}[\mathcal{D}_k]}^{\mathsf{hpr}_0\text{-}\mathsf{cma}}(\mathcal{A}) \leq 2Q(\mathbf{Adv}_{\mathcal{D}_k,\mathsf{GGen}}(\mathcal{D}) + 1/q)$, where Q is the maximal number of queries to $\mathrm{EVAL}(\cdot)$.*

A formal proof can be found in the full version [3].

5.3   The Transformation

Let $\mathcal{D}_k$ be a matrix distribution that outputs matrices $\mathbf{A} \in \mathbb{Z}_q^{(k+1)\times k}$. Let MAC be a delegatable affine MAC over $\mathbb{Z}_q^n$ with message space $\mathcal{M} = \mathcal{B}^{\leq m}$. Our $\mathsf{HIBKEM}[\mathsf{MAC}, \mathcal{D}_k] = (\mathsf{Gen}, \mathsf{USKGen}, \mathsf{USKDel}, \mathsf{Enc}, \mathsf{Dec})$ for key-space $\mathcal{K} = \mathbb{G}_T$ and hierarchical identity space $\mathcal{ID} = \mathcal{M} = \mathcal{B}^{\leq m}$ is defined as in Figure 5. Compared to the IBE construction from Section 4, the main difference is that Gen also returns a delegation key dk which allows re-randomization of every $\mathsf{usk}[\mathsf{id}]$. Further, USKGen also outputs user delegation keys $\mathsf{udk}[\mathsf{id}]$ allowing USKDel to delegate.

To show correctness of $\mathsf{HIBKEM}[\mathsf{MAC}, \mathcal{D}_k]$, first note that $(\hat{u}, \hat{\mathbf{v}})$ computed in USKDel is a correct user secret key for id', $\hat{u} = \sum_{i=0}^{l(\mathsf{id}')} f_i(\mathsf{id}')\mathbf{x}_i^\top \mathbf{t} + x_0'$ and $\hat{\mathbf{v}} = \sum_{i=0}^{l(\mathsf{id}')} f_i(\mathsf{id}')\mathbf{Y}_i\mathbf{t} + \mathbf{y}_0'$. In the next step they get rerandmozied as $u' = \sum_{i=0}^{l(\mathsf{id}')} f_i(\mathsf{id}')\mathbf{x}_i^\top(\mathbf{t} + \mathbf{B}\mathbf{s}')$ and $\mathbf{v}' = \sum_{i=0}^{l(\mathsf{id}')} f_i(\mathsf{id}')\mathbf{Y}_i(\mathbf{t} + \mathbf{B}\mathbf{s}') + \mathbf{y}_0'$. Consequently, $\mathsf{usk}[\mathsf{id}']$ from USKDel has the same distribution as the one output by USKGen. By applying the similar correctness argument from $\mathsf{HIBKEM}[\mathsf{MAC}, \mathcal{D}_k]$, we can show that a correctly generated ciphertext can be correctly decapsulated by using a correct user secret key.

The next theorem shows our construction is a IND-HID-CPA-secure HIBKEM. Its proof can be found in [3]. We remark that $\mathsf{HIBKEM}[\mathsf{MAC}, \mathcal{D}_k]$ can never be anonymous as one can always check whether $\mathbf{c}_0 \cdot \sum f_i(\mathsf{id})(\mathbf{E}_i^\top\|\mathbf{d}_i) = c_1 \cdot \mathbf{B}$ using the pairing.

Theorem 6. *If MAC is HPR_0-CMA-secure and the $\mathcal{D}_k$-MDDH assumption holds in $\mathbb{G}_1$ then $\mathsf{HIBKEM}[\mathsf{MAC}, \mathcal{D}_k]$ is IND-HID-CPA secure. For all adversaries $\mathcal{A}$ there exist adversaries $\mathcal{B}_1$ and $\mathcal{B}_2$ with $\mathbf{T}(\mathcal{B}_1) \approx \mathbf{T}(\mathcal{A}) \approx \mathbf{T}(\mathcal{B}_2)$ and*

$$\mathsf{Adv}_{\mathsf{HIBKEM}[\mathsf{MAC},\mathcal{D}_k]}^{\mathsf{ind}\text{-}\mathsf{hid}\text{-}\mathsf{cpa}}(\mathcal{A}) \leq \mathbf{Adv}_{\mathcal{D}_k,\mathsf{GGen}}(\mathcal{B}_1) + \mathsf{Adv}_{\mathsf{MAC}}^{\mathsf{hpr}_0\text{-}\mathsf{cma}}(\mathcal{B}_2).$$

Gen(par):
$\mathbf{A} \leftarrow_\$ \mathcal{D}_k$; $\mathsf{sk}_{\mathsf{MAC}} \leftarrow_\$ \mathsf{Gen}_{\mathsf{MAC}}(\mathsf{par})$
Parse $\mathsf{sk}_{\mathsf{MAC}} = (\mathbf{B}, \mathbf{x}_0, \ldots, \mathbf{x}_\ell, x'_0, \ldots, x'_{\ell'})$
For $i = 0, \ldots, \ell$:
$\quad \mathbf{Y}_i \leftarrow_\$ \mathbb{Z}_q^{k \times n}; \mathbf{Z}_i = (\mathbf{Y}_i^\top \mid \mathbf{x}_i) \cdot \mathbf{A} \in \mathbb{Z}_q^{n \times k}$
$\quad \mathbf{d}_i = \mathbf{x}_i^\top \cdot \mathbf{B} \in \mathbb{Z}_q^{n'}; \mathbf{E}_i = \mathbf{Y}_i \cdot \mathbf{B} \in \mathbb{Z}_q^{k \times n'}$
$\mathbf{y}'_0 \leftarrow_\$ \mathbb{Z}_q^k; z'_0 = (\mathbf{y}'_0{}^\top \mid x'_0) \cdot \mathbf{A} \in \mathbb{Z}_q^{1 \times k}$
$\mathsf{pk} := (\mathcal{G}, [\mathbf{A}]_1, ([\mathbf{Z}_i]_1)_{0 \leq i \leq \ell}, [z'_0]_1)$
$\mathsf{dk} := ([\mathbf{B}]_2, ([\mathbf{d}_i]_2, [\mathbf{E}_i]_2)_{0 \leq i \leq \ell})$
$\mathsf{sk} := (\mathsf{sk}_{\mathsf{MAC}}, (\mathbf{Y}_i)_{0 \leq i \leq \ell}, \mathbf{y}'_0)$
Return $(\mathsf{pk}, \mathsf{dk}, \mathsf{sk})$

USKGen($\mathsf{sk}, \mathsf{id} \in \mathcal{ID}$):
$\overline{([\mathbf{t}]_2, [u]_2) \leftarrow_\$ \mathsf{Tag}(\mathsf{sk}_{\mathsf{MAC}}, \mathsf{id})}$
$// \ \mathbf{t} \in \mathbb{Z}_q^n; u = \sum f_i(\mathsf{id})\mathbf{x}_i^\top \mathbf{t} + x'_0 \in \mathbb{Z}_q$
$\mathbf{v} = \sum_{i=0}^{l(\mathsf{id})} f_i(\mathsf{id}) \mathbf{Y}_i \mathbf{t} + \mathbf{y}'_0 \in \mathbb{Z}_q^k$
For $i = l(\mathsf{id}) + 1, \ldots, \ell$:
$\quad d_i = \mathbf{x}_i^\top \mathbf{t} \in \mathbb{Z}_q; \mathbf{e}_i = \mathbf{Y}_i \mathbf{t} \in \mathbb{Z}_q^k$
$\mathsf{usk}[\mathsf{id}] := ([\mathbf{t}]_2, [u]_2, [\mathbf{v}]_2)$
$\mathsf{udk}[\mathsf{id}] := ([d_i]_2, [\mathbf{e}_i]_2)_{l(\mathsf{id}) < i \leq \ell}$
Return $(\mathsf{usk}[\mathsf{id}], \mathsf{udk}[\mathsf{id}])$

Enc(pk, id):
$\mathbf{r} \leftarrow_\$ \mathbb{Z}_q^k$; $\mathbf{c}_0 = \mathbf{A}\mathbf{r} \in \mathbb{Z}_q^{k+1}$
$\mathbf{c}_1 = (\sum_{i=0}^{l(\mathsf{id})} f_i(\mathsf{id})\mathbf{Z}_i) \cdot \mathbf{r} \in \mathbb{Z}_q^n$
$K = \mathbf{z}'_0 \cdot \mathbf{r} \in \mathbb{Z}_q$.
Return $K = [K]_T$ and $\mathsf{C} = ([\mathbf{c}_0]_1, [\mathbf{c}_1]_1)$

USKDel($\mathsf{usk}[\mathsf{id}], \mathsf{udk}[\mathsf{id}], \mathsf{id}, \mathsf{id}_{p+1}$):
$\overline{\text{Parse } \mathsf{id} \in \mathcal{B}^p, \mathsf{id}_{p+1} \in \mathcal{B}}$
$\mathsf{id}' := (\mathsf{id}_1, \ldots, \mathsf{id}_p, \mathsf{id}_{p+1}) \in \mathcal{B}^{p+1}$
If $p \geq m$, then return $\bot$
//Delegation of u and $\mathbf{v}$:
$\hat{u} = u + \sum_{i=l(\mathsf{id})+1}^{l(\mathsf{id}')} f_i(\mathsf{id}')d_i \in \mathbb{Z}_q$
$\hat{\mathbf{v}} = \mathbf{v} + \sum_{i=l(\mathsf{id})+1}^{l(\mathsf{id}')} f_i(\mathsf{id}')\mathbf{e}_i \in \mathbb{Z}_q^k$
//Rerandomization of $\hat{u}$ and $\hat{\mathbf{v}}$:
$\mathbf{s}' \leftarrow_\$ \mathbb{Z}_q^{n'}$
$\mathbf{t}' = \mathbf{t} + \mathbf{B}\mathbf{s}' \in \mathbb{Z}_q^n$
$u' = \hat{u} + \sum_{i=0}^{l(\mathsf{id}')} f_i(\mathsf{id}')\mathbf{d}_i \mathbf{s}' \in \mathbb{Z}_q$
$\mathbf{v}' = \hat{\mathbf{v}} + \sum_{i=0}^{l(\mathsf{id}')} f_i(\mathsf{id}')\mathbf{E}_i \mathbf{s}' \in \mathbb{Z}_q^k$
//Rerandomization of d'_i and $\mathbf{e}_i$:
For $i = l(\mathsf{id}') + 1, \ldots, \ell$:
$\quad d'_i = d_i + \mathbf{d}_i \mathbf{s}' \in \mathbb{Z}_q$
$\quad \mathbf{e}'_i = \mathbf{e}_i + \mathbf{E}_i \mathbf{s}' \in \mathbb{Z}_q^k$
$\mathsf{usk}[\mathsf{id}'] := ([\mathbf{t}']_2, [u']_2, [\mathbf{v}']_2)$
$\mathsf{udk}[\mathsf{id}'] := ([d'_i]_2, [\mathbf{e}'_i]_2)_{l(\mathsf{id}') < i \leq \ell}$
Return $(\mathsf{usk}[\mathsf{id}'], \mathsf{udk}[\mathsf{id}'])$

Dec($\mathsf{usk}[\mathsf{id}], \mathsf{id}, \mathsf{C}$):
$\overline{\text{Parse } \mathsf{usk}[\mathsf{id}] = ([\mathbf{t}]_2, [u]_2, [\mathbf{v}]_2)}$
Parse $\mathsf{C} = ([\mathbf{c}_0]_1, [\mathbf{c}_1]_1)$
$K = e([\mathbf{c}_0]_1, \begin{bmatrix} \mathbf{v} \\ u \end{bmatrix}_2) \cdot e([\mathbf{c}_1]_1, [\mathbf{t}]_2)^{-1}$
Return $K \in \mathbb{G}_T$

Fig. 5. Definition of the transformation $\mathsf{HIBKEM}[\mathsf{MAC}, \mathcal{D}_k]$

5.4 Anonymity-Preserving Transformation

In this section, we sketch an alternative (but less efficient) transformation, which is anonymity-preserving. Due to space limitations, we only give the idea behind our construction and refer to the full version for details.

Our transformation is based on the notion of APR-CMA-security (anonymity-preserving pseudorandomness against chosen-message attacks) for a delegatable affine MAC MAC over $\mathbb{Z}_q^n$ with message space $\mathcal{M} = \mathcal{B}^{\leq m} := \bigcup_{i=1}^m \mathcal{B}^i$. It differs from HPR-CMA-security (Section 5.1) in the sense that $\mathrm{EVAL}(\mathsf{m})$ will output the terms for usk rerandomization, not $\mathrm{INITIALIZE}$ and that in the random game, CHAL returns uniform $(\mathbf{h}_0, h_1)$. Unfortunately, $\mathsf{MAC}_{\mathsf{NR}}[\mathcal{D}_k]$ is unlikely to be APR-CMA-secure, but $\mathsf{MAC}_{\mathsf{HPS}}[\mathcal{D}_k]$ with message space $\mathcal{M} = \mathcal{B}^{\leq m} = (\mathbb{Z}_q^*)^{\leq m}$ is provably APR-CMA-secure.

Compared to the HIBE construction from Section 5.3, the new transformation $\mathsf{AHIBKEM}[\mathsf{MAC}, \mathcal{D}_k]$ uses a different rerandomization method for $\mathsf{usk} :=$

$([\mathbf{t}]_2, [u]_2, [\mathbf{v}]_2)$: USKGen outputs a random basis $\mathbf{T}$ which allows rerandomization of $\mathbf{t}$; similarly, $\mathbf{u}$ and $\mathbf{V}$ are generated for rerandomizing u and $\mathbf{v}$. In the full version we prove that if MAC is an APR-CMA-secure and the $\mathcal{D}_k$-MDDH assumption holds in $\mathbb{G}_1$, then AHIBKEM[MAC, $\mathcal{D}_k$] is PR-HID-CPA-secure, i.e., IND-HID-CPA-secure and anonymous.

Acknowledgements. All authors were (partially) supported by the Sofja Kovalevskaja Award of the Alexander von Humboldt Foundation and the German Federal Ministry for Education and Research. Jiaxin Pan was also partially supported by the German Israel Foundation.
We thank Hoeteck Wee for various comments and helpful discussions.

References

1. Bellare, M., Goldwasser, S.: New paradigms for digital signatures and message authentication based on non-interactive zero knowledge proofs. In: Brassard, G. (ed.) CRYPTO 1989. LNCS, vol. 435, pp. 194–211. Springer, Heidelberg (1990)
2. Blazy, O., Kakvi, S., Kiltz, E., Pan, J.: Tightly-secure signatures from chameleon hash functions. unpublished (2013)
3. Blazy, O., Kiltz, E., Pan, J.: (Hierarchical) identity-based encryption from affine message authentication. Cryptology ePrint Archive, Full version of this paper (2014)
4. Boneh, D., Franklin, M.: Identity-based encryption from the Weil pairing. In: Kilian, J. (ed.) CRYPTO 2001. LNCS, vol. 2139, pp. 213–229. Springer, Heidelberg (2001)
5. Chen, J., Lim, H.W., Ling, S., Wang, H., Wee, H.: Shorter IBE and signatures via asymmetric pairings. In: Abdalla, M., Lange, T. (eds.) Pairing 2012. LNCS, vol. 7708, pp. 122–140. Springer, Heidelberg (2013)
6. Chen, J., Wee, H.: Fully (almost) tightly secure IBE and dual system groups. In: Canetti, R., Garay, J.A. (eds.) CRYPTO 2013, Part II. LNCS, vol. 8043, pp. 435–460. Springer, Heidelberg (2013)
7. Cocks, C.: An identity based encryption scheme based on quadratic residues. In: Honary, B. (ed.) Cryptography and Coding 2001. LNCS, vol. 2260, pp. 360–363. Springer, Heidelberg (2001)
8. Cramer, R., Shoup, V.: A practical public key cryptosystem provably secure against adaptive chosen ciphertext attack. In: Krawczyk, H. (ed.) CRYPTO 1998. LNCS, vol. 1462, pp. 13–25. Springer, Heidelberg (1998)
9. Cramer, R., Shoup, V.: Universal hash proofs and a paradigm for adaptive chosen ciphertext secure public-key encryption. In: Knudsen, L.R. (ed.) EUROCRYPT 2002. LNCS, vol. 2332, pp. 45–64. Springer, Heidelberg (2002)
10. Dodis, Y., Kiltz, E., Pietrzak, K., Wichs, D.: Message authentication, revisited. In: Pointcheval, D., Johansson, T. (eds.) EUROCRYPT 2012. LNCS, vol. 7237, pp. 355–374. Springer, Heidelberg (2012)
11. Escala, A., Herold, G., Kiltz, E., Ràfols, C., Villar, J.: An algebraic framework for Diffie-Hellman assumptions. In: Canetti, R., Garay, J.A. (eds.) CRYPTO 2013, Part II. LNCS, vol. 8043, pp. 129–147. Springer, Heidelberg (2013)
12. Gentry, C., Silverberg, A.: Hierarchical ID-based cryptography. In: Zheng, Y. (ed.) ASIACRYPT 2002. LNCS, vol. 2501, pp. 548–566. Springer, Heidelberg (2002)

13. Groth, J., Sahai, A.: Efficient non-interactive proof systems for bilinear groups. In: Smart, N.P. (ed.) EUROCRYPT 2008. LNCS, vol. 4965, pp. 415–432. Springer, Heidelberg (2008)

14. Hofheinz, D., Jager, T.: Tightly secure signatures and public-key encryption. In: Safavi-Naini, R., Canetti, R. (eds.) CRYPTO 2012. LNCS, vol. 7417, pp. 590–607. Springer, Heidelberg (2012)

15. Jutla, C.S., Roy, A.: Shorter quasi-adaptive NIZK proofs for linear subspaces. In: Sako, K., Sarkar, P. (eds.) ASIACRYPT 2013, Part I. LNCS, vol. 8269, pp. 1–20. Springer, Heidelberg (2013)

16. Kiltz, E., Pietrzak, K., Stam, M., Yung, M.: A new randomness extraction paradigm for hybrid encryption. In: Joux, A. (ed.) EUROCRYPT 2009. LNCS, vol. 5479, pp. 590–609. Springer, Heidelberg (2009)

17. Lewko, A.: Tools for simulating features of composite order bilinear groups in the prime order setting. In: Pointcheval, D., Johansson, T. (eds.) EUROCRYPT 2012. LNCS, vol. 7237, pp. 318–335. Springer, Heidelberg (2012)

18. Lewko, A., Okamoto, T., Sahai, A., Takashima, K., Waters, B.: Fully secure functional encryption: Attribute-based encryption and (Hierarchical) inner product encryption. In: Gilbert, H. (ed.) EUROCRYPT 2010. LNCS, vol. 6110, pp. 62–91. Springer, Heidelberg (2010)

19. Lewko, A., Waters, B.: New techniques for dual system encryption and fully secure HIBE with short ciphertexts. In: Micciancio, D. (ed.) TCC 2010. LNCS, vol. 5978, pp. 455–479. Springer, Heidelberg (2010)

20. Lewko, A., Waters, B.: Why proving HIBE systems secure is difficult. In: Nguyen, P.Q., Oswald, E. (eds.) EUROCRYPT 2014. LNCS, vol. 8441, pp. 58–76. Springer, Heidelberg (2014)

21. Naor, M., Reingold, O.: Number-theoretic constructions of efficient pseudo-random functions. In: 38th FOCS, pp. 458–467. IEEE Computer Society Press (October 1997)

22. Naor, M., Reingold, O.: On the construction of pseudo-random permutations: Luby-Rackoff revisited (extended abstract). In: 29th ACM STOC, pp. 189–199. ACM Press (May 1997)

23. Sakai, R., Ohgishi, K., Kasahara, M.: Cryptosystems based on pairing. In: SCIS 2000, Okinawa, Japan (January 2000)

24. Shamir, A.: Identity-based cryptosystems and signature schemes. In: Blakely, G.R., Chaum, D. (eds.) CRYPTO 1984. LNCS, vol. 196, pp. 47–53. Springer, Heidelberg (1985)

25. Waters, B.: Dual system encryption: Realizing fully secure IBE and HIBE under simple assumptions. In: Halevi, S. (ed.) CRYPTO 2009. LNCS, vol. 5677, pp. 619–636. Springer, Heidelberg (2009)

26. Waters, B.: Efficient identity-based encryption without random oracles. In: Cramer, R. (ed.) EUROCRYPT 2005. LNCS, vol. 3494, pp. 114–127. Springer, Heidelberg (2005)

27. Wee, H.: Dual system encryption via predicate encodings. In: Lindell, Y. (ed.) TCC 2014. LNCS, vol. 8349, pp. 616–637. Springer, Heidelberg (2014)

Witness Encryption
from Instance Independent Assumptions

Craig Gentry[1], Allison Lewko[2], and Brent Waters[3,*]

[1] IBM Research, T.J. Watson, Yorktown Heights, NY, USA
cbgentry@us.ibm.com
[2] Columbia University, New York, NY, USA
alewko@cs.columbia.edu
[3] University of Texas at Austin, TX, USA
bwaters@cs.utexas.edu

Abstract. Witness encryption was proposed by Garg, Gentry, Sahai, and Waters as a means to encrypt to an instance, x, of an NP language and produce a ciphertext. In such a system, any decryptor that knows of a witness w that x is in the language can decrypt the ciphertext and learn the message. In addition to proposing the concept, their work provided a candidate for a witness encryption scheme built using multilinear encodings. However, one significant limitation of the work is that the candidate had no proof of security (other than essentially assuming the scheme secure).

In this work we provide a proof framework for proving witness encryption schemes secure under instance independent assumptions. At the highest level we introduce the abstraction of *positional witness encryption* which allows a proof reduction of a witness encryption scheme via a sequence of 2^n hybrid experiments where n is the witness length of the NP-statement. Each hybrid step proceeds by looking at *a single witness candidate* and using the fact that it does not satisfy the NP-relation to move the proof forward. We show that this "isolation strategy" enables one to create a witness encryption system that is provably secure from assumptions that are (maximally) independent of any particular encryption instance. We demonstrate the viability of our approach by implementing this strategy using level n-linear encodings where n is the witness length. Our complexity assumption has $\approx n$ group elements, but does not otherwise depend on the NP-instance x.

1 Introduction

Witness encryption, as introduced by Garg, Gentry, Sahai, and Waters [14], is a primitive that allows one to encrypt to an instance of an **NP** language L.

* Supported by NSF CNS-0915361 and CNS-0952692, CNS-1228599 DARPA through the U.S. Office of Naval Research under Contract N00014-11-1-0382, DARPA N11AP20006, Google Faculty Research award, the Alfred P. Sloan Fellowship, Microsoft Faculty Fellowship, and Packard Foundation Fellowship.

J.A. Garay and R. Gennaro (Eds.): CRYPTO 2014, Part I, LNCS 8616, pp. 426–443, 2014.

An encryptor will take in an instance x along with a message m and run the encryption algorithm to produce a ciphertext CT. Later a user will be able to decrypt the ciphertext and recover m if they know of a witness w showing that x is in the language L according to some witness relation $R(\cdot, \cdot)$. The security of witness encryption states that, for any ciphertext created for an instance x that *is not in the language* L, it must be hard to distinguish whether the ciphertext encrypts m_0 or m_1. Concepts related to witness encryption include: (in the computational setting) point-filter functions [16], and (in the statistical setting for languages in SZK) non-interactive instance-dependent commitments [23], including efficiently-extractable ones [15].

The primitive of encrypting to an instance is intriguing in its own right, and Garg et. al. show that it has many compelling applications, including public key encryption with very fast key generation, identity-based encryption [22,3,8], attribute-based encryption [21] (ABE) for arbitrary circuits, and ABE for Turing Machines [17]. The work of [17] goes on to develop even further applications, such as reusable garbling schemes for Turing machines.

These powerful applications motivate the quest for constructions of witness encryption with strong provable security guarantees. In [14], they gave a witness encryption construction for the **NP**-complete Exact Cover problem [19] using multilinear encodings (first suggested in [5] and first constructed by Garg, Gentry, and Halevi [11], with an alternative construction later provided by Coron, Lepoint and Tibouchi [9]).

While the GGSW construction candidate demonstrates the plausibility of realizing secure witness encryption, they were unable to reduce the security of their system to anything simpler than directly assuming the security of their construction. Instead they applied what we will call an *instance dependent* family of assumptions that they called the "Decision Graded Encoding No-Exact-Cover Problem." The assumption is that for each instance x not in the language, no PPT attacker can distinguish between two particular distributions of multilinear encodings. The distributions directly embed the Exact Cover instance x and are almost identical to the structure of the ciphertexts from the construction.

The Importance and Difficulty of Using Simple Assumptions While a generic group argument might give some confidence that it will be difficult to find an attack on a scheme, a reduction to an assumption simpler than the scheme itself is much more desirable. First, such a reduction will often provide critical insight and understanding into why the scheme is secure. Second, the ideas behind proof reductions often transcend their original settings and will be of use elsewhere. Having a single, concrete assumption also provides a clearer focus for cryptanalysis efforts to stress-test a candidate scheme.

Prior to this work, no known schemes could be reduced to instance-independent assumptions. This is also the case for all known indistinguishability obfuscation schemes. For example, [12] explicitly reduces to a instance-dependent family of assumption, while [20] implicitly does this through a meta-assumption.

Our goal is to create techniques for building witness encryption systems that are provably secure under radically simpler assumptions. To achieve this, we must first confront an intuitive barrier that is formalized as an impossibility result in [14] (with some restrictions). The idea is that any black-box security reduction to an instance-independent assumption for a witness encryption scheme must (in some sense) verify that a statement is false. Otherwise, we could use the reduction to break the assumption by "fooling it" on a true statement for which we know a witness, and hence can simulate an attack. Since the best known methods for solving **NP-hard** problems take exponential time, this implies an instance independent reduction will have an exponential security loss.

Our Strategy To address the barrier above, we devise a proof technique that employs a reduction which gradually "learns" that the instance x is not in the language. Consider a instance $x \notin L$ with witness candidates of n bits. Our strategy is to allow a reduction to build a hybrid argument by isolating and examining each witness candidate, w, in sequence and utilizing the fact that $R(x, w) = 0$ (i.e. the witness is not valid) to progress the hybrid to the next step. (Since there are 2^n witness candidates, the proof strategy will inherently use complexity leveraging, as will any reduction strategy that falls within the confines of the [14] impossibility result.) In this way, we obtain a "true reduction" that represents a new understanding of the security of witness encryption.

To implement this hybrid approach, we will need a technique that somehow allows a proof to compactly "save" its work for all of the witnesses it has examined. Our starting point will be a broadcast encryption (BE) [10] system proposed by Boneh and Waters [6] in 2006, which was the first collusion resistant system to be proved adaptively secure. Instead of proving security all at once, they employed a method of altering the challenge ciphertext over a sequence of N hybrid experiments for a BE system of N users. At the center of their approach was a new abstraction that they called augmented broadcast encryption. An augmented BE system has an encryption algorithm $Encrypt_{\mathrm{AugBE}}(\mathrm{PK}, S, t, m)$ that takes as input a public key PK, a set of user indices $S \subseteq [0, N-1]$, an index $t \in [0, N]$, and a message m. This produces a ciphertext CT. The semantics of the system are that a user with index $u \in [0, N-1]$ [1] can decrypt the ciphertext and learn the message only if $u \in S$ *and* $u \geq t$. These are like the semantics of standard broadcast encryption, but with the added constraint of the index t. Augmented broadcast encryption has two security properties. The first is that no poly-time attack can distinguish between $Encrypt_{\mathrm{AugBE}}(\mathrm{PK}, S, t, m)$ and $Encrypt_{\mathrm{AugBE}}(\mathrm{PK}, S, t+1, m)$ if the attacker does not have the key for index t or if $t \notin S$. The second property is that the scheme is semantically secure if we encrypt to index $t = N$, thus cutting off all the user keys whether or not they are in S.

It is straightforward to make a standard broadcast encryption using an augmented one, as we can create a broadcast ciphertext to the set S by simply calling $Encrypt_{\mathrm{AugBE}}(\mathrm{PK}, S, t = 0, m)$. By setting $t = 0$, the range condition is never invoked. The advantage of using this condition comes into the proof where

---

[1] The Boneh-Waters paper uses indices $1, \dots, N$ for the users. We shift this to $0, \dots, N-1$ to better match our exposition.

we want to prove that no attack algorithm can distinguish an encryption to an adaptively chosen set S^* (meaning it is chosen after seeing the public key) if it is only given keys for $u \notin S^*$. The proof proceeds by a sequence of indistinguishable hybrid experiments where at the i-th hybrid the challenge ciphertext is generated for index $t = i$. Finally, we move to $t = N$ and the second property then implies security of the scheme. Even though there were N indices, the abstraction and hybrid sequence allowed for a proof to isolate one user at a time. The BW construction melded a broadcast system with the Boneh-Sahai-Waters [4] traitor tracing [7] system to enforce the range condition.

Positional Witness Encryption With these concepts in mind, we can turn back to the problem of devising a proof strategy for witness encryption for an **NP**-complete language L. The first step we take is the introduction of a primitive that we call positional witness encryption. A positional witness encryption system has an encryption algorithm $Encrypt_{\mathrm{PWE}}(1^\lambda, x, t, m)$ that takes as input a security parameter 1^λ, a string x, a position index $t \in [0, 2^n]$, and a message m and outputs a ciphertext CT. Here we let n be the witness length of x and let $N = 2^n$. One can decrypt a ciphertext by producing a witness w such that $R(x, w) = 1$ *and* $w \geq t$ where w is interpreted as an integer. Essentially, this has the same correctness semantics as standard witness encryption, but with the range condition added. The security properties are as follows:

- Positional Indistinguishability: If $R(x, t) = 0$ then no poly-time attacker can distinguish between $Encrypt_{\mathrm{PWE}}(1^\lambda, x, t, m)$ and $Encrypt_{\mathrm{PWE}}(1^\lambda, x, t+1, m)$.
- Message Indistinguishability: No poly-time attacker can distinguish between $Encrypt_{\mathrm{PWE}}(1^\lambda, x, t = 2^n, m_0)$ and $Encrypt_{\mathrm{PWE}}(1^\lambda, x, t = 2^n, m_1)$ for all equal length messages m_0, m_1.

We point out that the security definition of positional witness encryption is not explicitly constrained to $x \notin L$ in any place. However, if some $x \notin L$, then for all witnesses $w \in [0, 2^n - 1]$ (interpreting the bitstring w as an integer) we have that $R(x, w) = 0$. This leads to a natural construction and proof strategy for witness encryption. To witness encrypt a message m to an instance x, we call $Encrypt_{\mathrm{PWE}}(1^\lambda, x, t = 0, m)$. To prove security, we design a sequence of indistinguishable hybrids where we increase the value of t at each step until we get to $t = N = 2^n$ where we can invoke message indistinguishability. Each step can be made since $x \notin L$ implies $R(x, w) = 0$. The hybrids cause a 2^n loss of security relative to the security of the positional witness encryption and this should be compensated for in setting the security parameter. [2]

The potential advantage of positional witness encryption is that it offers a hybrid strategy where the core security property is focused on whether a *single* witness satisfies a relation. However, there is still a very large gap between

---

[2] We note that complexity leveraging is used elsewhere in "computing on encrypted data". For example, current solutions of Attribute-Based Encryption for circuits [13,18] are naturally selectively secure and require complexity leveraging to achieve adaptive security.

imagining this primitive and realizing it. First, we need a data structure that can both securely hide t and compactly store it (e.g. ciphertexts cannot grow proportional to the number of witnesses $N = 2^n$). Next, we need to be able to somehow embed an instance x of an **NP**-complete problem. This must be done in such a way that the security proof can isolate a property that depends on whether $R(x, w) = 0$ for each witness candidate w and use this to increment the positional data in a manner oblivious to an attacker.

Tribes Schemes and Their Uses We begin our realization by introducing a data structure that we call a tribes matrix, which will be flexible enough to encode both a position and a CNF formula. A tribes matrix will induce a boolean function from n-bit inputs to a single output bit. We then introduce a cryptographic primitive called a tribes scheme that will hide some properties of the matrix while still enabling evaluation of the corresponding boolean function. The benefit of this middle layer of abstraction is that it portions the work into a manageable hybrid security proof at the abstract level and creates a rather slim and concise target for lower level instantiations. This naturally increases the potential for instantiating our framework with a variety of different assumptions.

The name "tribes" was chosen because of the structural similarities between the induced boolean function and the tribes function commonly considered in boolean function analysis (e.g. [2]). In the tribes function, n inputs are thought of as people that are partitioned into ℓ tribes, and the function outputs 1 if and only if at least one tribe takes value 1 unanimously. In our case, we define 3-dimensional $n \times \ell \times 2$ matrix, where we think of it as having n rows, ℓ columns, and 2 "slots" for each row and column pair. The slots take values from a 2-symbol alphabet, notated as $\{U, B\}$ and are $\{0, 1\}$-indexed. The B stands for "blocked" and the U stands for "unblocked." To evaluate the boolean function on a n bit input $x_1, \ldots, x_n$, we consider each of the ℓ columns as a tribe, but in each row i we take the value in the slot indexed by x_i (this means that the input bits specify the composition of the tribe from a pre-existing set of values, rather than providing the values themselves). If some tribe is unanimously "blocked," the function outputs 1, otherwise it outputs 0. For a tribes matrix denoted by M, we will denote the associated boolean function by f_M.

When we embed a tribes matrix into a tribes scheme, we seek to allow access to evaluating the function without revealing full information about the matrix entries. Of course, some properties of the matrix entries can be inferred from black-box access to the function, and this is fine; we only seek to hide a very specific kind of structure that does not affect the function evaluation on any input. For this, we define the notions of "inter-column" and "intra-column" security, and the combination essentially requires that the tribes schemes for two matrices that differ in a single slot value are indistinguishable if there a simple reason why this slot value does not affect the boolean function.

More precisely, suppose we wish to hide the value of a slot in row i^*, column k. If there is a column j such that the corresponding slot has value B, and furthermore in all rows $i \neq i^*$, occurrences of B in column k are always matched by occurrences of B in column j, then we can change the value of this slot in row i^* without affecting the boolean function. To see this, observe that regardless of the value at this slot in column k, for any input where tribe k is unanimously blocked, tribe j is also unanimously blocked. Inter-column security requires that we can furthermore hide this change in the sense of computational indistinguishability. The second property of intra-column security ensures that if both slots of a particular row in a single column have the value U, then any other slot in the column can be changed without an attacker noticing.

We next consider how one might encode positions and CNF formulas into a tribes matrix. Our approach is to encode these two objects separately, and then simply concatenate the matrices. To encode a position t, we wish to produce a tribes matrix M where the boolean function f_M will output 1 for every witness $y < t$, and will output 0 otherwise. The key observation is that every potential witness $y < t$ will have some bit j where it first departs from t (starting from the most significant bit), and in this bit y will be 0 and t will be 1. We leverage this by designing the jth column of M to be blocked precisely for such y.

To encode a CNF formula in a tribes matrix, we build a column corresponding to each clause, where the rows are indexed by the variables, $x_1, \ldots, x_n$. To fill in the slots of row i in column j, we see if the literal x_i or its negation $\overline{x_i}$ appear in the jth clause. If x_i appears, we put a U in the 1-indexed slot. If $\overline{x_i}$ appears, we put a U in the 0-indexed slot. For any remaining slots, we put B. This yields a column that is blocked precisely for inputs that do not satisfy the clause. We therefore get a matrix whose associated boolean function outputs 1 if and only if the CNF formula is unsatisfied.

From this, we can construct a positional witness encryption scheme. To encrypt a message bit to a particular position and formula, the encryptor forms tribes matrices as above, concatenates them, and concatenates one extra column to encode the message (it will contain all U's if the bit is 0 and all B's if the bit is one). It finally embeds this matrix in a tribes scheme, which serves as the ciphertext. A decryptor can then evaluate the boolean function to recover the message. (If $R(x, w) = 1$ and $w \geq t$, then the output of the tribes evaluation will reflect the message; otherwise, it outputs 1 regardless of the message.)

To prove positional indistinguishability, we proceed through a hybrid argument that relies upon the inter-column security of the cryptographic tribes scheme to incrementally change the matrix entries to an encoding of the next position. During this process, we will need to leverage the fact that the current position represents a witness that *does not satisfy* the CNF formula. The key observation here is that there will be at least one clause that is not satisfied, and the column for that clause can be used to make changes to entries in another column through the inter-column security game.

Instantiating a Tribes Scheme Finally, what is left is to instantiate a tribes scheme and reduce the inter-column and intra-column security requirements to a computational assumption. We give three related constructions each from multilinear algebraic groups with an n linear map, which required for a tribes instance of n rows.

Our first instantiation uses *composite* order symmetric multilinear groups. The group's order is a product $n + \ell$ primes for an n by ℓ tribes matrix. Our next instantiation (which can be found in the full version) utilizes asymmetric groups to reduce the order of the group to the product of ℓ primes. Also in the full version of this paper, we modify the instantiation to be in prime order using a translation based on eigenvectors. Each instance is based on a pair of multilinear map assumptions that depend on n (or n and ℓ), but are independent of the contents of the tribes matrix. The assumptions we use in the composite order symmetric context for example, are given in Section 5, and we call them the multilinear subgroup decision and multilinear subgroup elimination assumptions, as they are rather natural variants of subgroup decision assumptions typically used in bilinear groups. In fact, in the full version we explain how to use only the multilinear subgroup elimination assumption. We also justify the prime order variants of our assumptions in the multilinear generic group model and show how to translate from algebraic groups into the multilinear encodings of Coron, Lepoint and Tibouchi (CLT) [9].

2   Positional Witness Encryption

We will first give our definition of a positional witness encryption system and then show how it implies standard witness encryption by a hybrid argument.

We define a *positional witness encryption* scheme for an **NP** language L. Let $R(\cdot, \cdot)$ be the corresponding witness relation and let $n = n(|x|)$ be the witness length for a particular witness x. The system consists of two algorithms:

> **Encryption.** The algorithm $Encrypt_{\mathrm{PWE}}(1^\lambda, x, t, m)$ takes as input a security parameter 1^λ, an unbounded-length string x, a position index $t \in [0, 2^n]$ (we let $n = n(x)$) and a message $m \in \mathcal{M}$ for some (fixed and finite) message space $\mathcal{M}$, and outputs a ciphertext CT.
>
> **Decryption.** The algorithm $Decrypt_{\mathrm{PWE}}(\mathrm{CT}, w)$ takes as input a ciphertext CT and a length n string w, and outputs a message m or the symbol $\perp$. (We assume the ciphertext specifies the instance x and therefore $n = n(|x|)$ is known.)

Given a string $w \in \{0, 1\}^n$ we will sometimes slightly abuse notation and also refer to w as an integer in $[0, 2^n - 1]$ where the most significant bit is the leftmost bit. In other words, we consider the integer $\Sigma_{i=1, \cdots, n} w_i \cdot 2^{n-i}$, where w_i is the i-th bit of the string w.

Definition 1 ((Perfect) Correctness of Positional Witness Encryption). *For any security parameter λ, for any $m \in \mathcal{M}$, and for any $x \in L$ such that $R(x, w)$ holds for $w \geq t$, we have that*

$$Decrypt_{\mathrm{PWE}}\big(Encrypt_{\mathrm{PWE}}(1^\lambda, x, t, m), w\big) = m.$$

2.1 Security of Positional Witness Encryption

Message Indistinguishability The security of a positional witness encryption for language L is given as two security properties. The first is message indistinguishability, which is parameterized by an instance x and two equal length messages m_0, m_1. Intuitively, the security property states that if one encrypts to the "final" position $t = 2^n$ (where n is the witness length of x) then no attacker can distinguish whether a ciphertext is an encryption of m_0 or m_1. *We emphasize that this security property is entirely independent of whether $x \in L$.* We define the (parameterized) advantage of an attacker as

$$\mathsf{MsgPWE\,Adv}_{\mathcal{A},x,m_0,m_1}(\lambda) =$$

$$\left| \Pr[\mathcal{A}(Encrypt_{\mathrm{PWE}}(1^\lambda, x, t=2^n, m_1))=1] - \Pr[\mathcal{A}(Encrypt_{\mathrm{PWE}}(1^\lambda, x, t=2^n, m_0))=1] \right|.$$

Definition 2 (Message Indistinguishability Security of Positional Witness Encryption).

We say that a positional witness encryption scheme for a language L with witness relation $R(\cdot, \cdot)$ is Message Indistinguishability secure if for any probabilistic poly-time attack algorithm $\mathcal{A}$ there exists a negligible function in the security parameter $negl(\cdot)$ such that for all instances x and equal length messages m_0, m_1 we have $\mathsf{MsgPWE\,Adv}_{\mathcal{A},x,m_0,m_1}(\lambda) \le negl(\lambda)$.

We let $\mathsf{MsgPWE\,Adv}_{\mathcal{A},x}(\lambda)$ be the maximum value of $\mathsf{MsgPWE\,Adv}_{\mathcal{A},x,m_0,m_1}(\lambda)$ over the pairs $m_0, m_1 \in \mathcal{M}$ for each λ.

Position Indistinguishability The second security game is positional indistinguishability. Informally, this security game states that it is hard to distinguish between an encryption to position t from an encryption to $t + 1$ when t is not a valid witness – that is, $R(x, t) = 0$. (Here we slightly abuse notation in the other direction by interpreting the integer t as a bit string.) Positional indistinguishability security is parameterized by an instance x, a message m, and a position $t \in [0, 2^n - 1]$ where n is the witness length of x. We define the (parameterized) advantage of an attacker as

$$\mathsf{PosPWE\,Adv}_{\mathcal{A},x,m,t}(\lambda) =$$

$$\left| \Pr[\mathcal{A}(Encrypt_{\mathrm{PWE}}(1^\lambda, x, t+1, m)) = 1] - \Pr[\mathcal{A}(Encrypt_{\mathrm{PWE}}(1^\lambda, x, t, m)) = 1] \right|.$$

Definition 3 (Position Indistinguishability Security of Positional Witness Encryption).

We say that a positional witness encryption scheme for a language L with witness relation $R(\cdot, \cdot)$ is Position Indistinguishability secure if for any probabilistic poly-time attack algorithm $\mathcal{A}$ there exists a negligible function in the security parameter $negl(\cdot)$ such that for all instances x, all message m, and any $t \in [0, 2^n - 1]$ where $R(x, t) = 0$ we have $\mathsf{PosPWE\,Adv}_{\mathcal{A},x,m,t}(\lambda) \le negl(\lambda)$.

We let $\mathsf{PosPWE\,Adv}_{\mathcal{A},x}(\lambda)$ be the maximum value of $\mathsf{PosPWE\,Adv}_{\mathcal{A},x,m,t}(\lambda)$ over $m \in \mathcal{M}$ and $t \in [0, 2^n]$ where $R(x, t) = 0$ for each λ.

We further require that both the message length and the problem statement length must be bounded by some polynomial of the security parameter.

A quick note on the witness encryption definition. We provide the definition of witness encryption in the appendix of the full version. The definition of our appendix follows the original of Garg, Gentry, Sahai, and Waters [14], but with two modifications. First, we restrict ourselves to perfect correctness for simplicity. Second, in defining soundness security we use a notation that the scheme is secure if for all PPT attackers there exists a negligible function negl(·) such that for any $x \notin L$ the attacker must only be able to distinguish encryption with probability at most negl(λ). The GGSW definition had a different ordering of quantifiers which allowed the bounding negligible function for a particular attacker to depend on the instance x. Bellare and Tung Hoang [1] showed that this formulation was problematic for multiple applications of witness encryption. Our positional witness encryption definition follows a similar corrected ordering of quantifiers.

Building Witness Encryption from Positional Witness Encryption Building standard witness encryption from Positional WE is straightforward. The proofs follows from the hybrid outlined in the introduction. We describe this formally in the full version of this paper.

3   Tribes Schemes

A tribes matrix M is an $n \times \ell \times 2$ 3-dimensional matrix, with entries belonging to the two symbol alphabet $\{B, U\}$, which stand for "blocked" and "unblocked". We consider $[n] = \{1, 2, \ldots, n\}$ as indexing the rows, $[\ell]$ as indexing the columns, and $\{0, 1\}$ as indexing the "slots" (i.e. we think of M as an $n \times \ell$ matrix whose entries are pairs of slots, each containing a symbol from $\{B, U\}$).

Such a matrix M defines a boolean function f_M from $\{0, 1\}^n$ to $\{0, 1\}$ as follows. Given an input $x = (x_1, x_2, \ldots, x_n) \in \{0, 1\}^n$, we examine each column of M. Suppose, for example, that we are considering column j. We cycle through the n rows of this column, and while considering row i, we take the value of the slot whose index matches x_i. If the column contains *at least one* value U in these slots, then we define the value of the column to be 0. Otherwise, we define it to be 1. Finally, if there exists a column with value 1, we define the output of the function to be 1, otherwise it is 0.

More formally, we define f_M as:

$$f_M(x) := \begin{cases} 1, \exists j \text{ s.t. } M_{i,j,x_i} = B \ \forall i \in [n]; \\ 0, \text{ otherwise.} \end{cases}$$

The name for these matrices is inspired by the tribes function, an interesting object in boolean function analysis. In that domain, one considers the input boolean vector as specifying the "votes" of a population that is organized into tribes, and the output is 1 if and only if there exists a tribe that unanimously voted 1. This is not a perfect analogy to our setting, since we view the input not as specifying these votes directly but rather as selecting each vote from a predetermined set of two possible values. Nonetheless, we adopt the "tribes" terminology as a helpful device for reinforcing the key feature here that the output of our function is 1 if and only some column takes on unanimous values of B when the input is used for indexing the slots.

3.1 Tribes Schemes

We next use the notion of tribes matrices to define a cryptographic primitive that we will call a tribes scheme. A tribes scheme will have two algorithms. The first algorithm, Create, will take in tribes matrices and generate objects that we will call cryptographic tribes. This algorithm is randomized. The second algorithm, Eval, will take in a cryptographic tribe and an input and compute the boolean function described above for the tribes matrix that is incorporated in the cryptographic tribe. This algorithm is deterministic.

$Create(\lambda, M) \to T$ The creation algorithm takes in a security parameter λ and a tribes matrix M and outputs a cryptographic tribe T.

$Eval(T, x \in \{0,1\}^n) \to \{0,1\}$ The evaluation algorithm takes in a cryptographic tribe T and a boolean vector x and outputs a value $\{0,1\}$.

Correctness We require perfect correctness, meaning that for every tribes matrix $M \in \{B, U\}^{n \times \ell \times 2}$, for any security parameter λ, and for any input vector $x \in \{0,1\}^n$, we have that

$$Eval(Create(\lambda, M), x) = f_M(x).$$

3.2 Tribes Security Properties

We will define two security properties for a tribes scheme. Both will be defined as typical distinguishing games between a challenger and an attacker. We call the first of these the intra-column game, as it only relies on a condition within a single column of the underlying tribes matrix. We call the other the inter-column game, as it involves a relationship between two columns that allows us to change a "U" symbol to a "B".

Intra-column Game This game is parameterized by a security parameter λ, a tribes matrix M, an index j of a column in M such that there is some row i^* where both slots take the value U[3], and an alternate column $C \in \{B, U\}^{n \times 2}$ such that the row i^* also has both slots equal to U. All of these parameters are given both to the challenger and to the attacker.

The challenger samples a uniformly random bit $b \in \{0,1\}$. If $b = 0$, it runs $Create(\lambda, M)$ to produce a cryptographic tribe T. If $b = 1$, it forms M' by replacing the jth column of M with C, and then runs $Create(\lambda, M')$ to produce T. It gives T to the attacker, who must then guess the value of the bit b.

---

[3] Of course, for an arbitrary tribes matrix, such a column may not exist. This is an extra condition we are imposing on M, and this property is only defined for such M.

Definition 4. *We say a tribes scheme has* intra-column security *if for every polynomial time attacker $\mathcal{A}$, there exists a negligible function $negl(\lambda)$ such that the attacker's advantage in the Intra-column Game is $\leq negl(\lambda)$, for any valid settings of M, j, C. Note that the negligible function depends only on $\mathcal{A}$ and λ, and is independent of the dimensions of M, for example.*

Inter-column Game This game is parameterized by a security parameter λ, a tribes matrix M, two indices j and k of columns of M, an index i^* of a row of M, and a slot index β such that $M_{i^*,j,\beta} = B$. We require the following condition on the jth and k^{th} columns of M. For every row i and slot γ *except for $i = i^*$ and $\gamma = 1 - \beta$*, if $M_{i,k,\gamma} = B$, then $M_{i,j,\gamma} = B$ as well (i.e. when there is only one U among these values, it is always in column k)[4]. All of these parameters are given both to the challenger and to the attacker.

The challenger samples a uniformly random bit $b \in \{0, 1\}$. If $b = 0$, it runs $Create(\lambda, M)$ to produce a cryptographic tribe T. If $b = 1$, it forms M' by copying M except for flipping just one entry: $M'_{i^*,k,\beta} = B$ if $M_{i^*,k,\beta} = U$, and $M'_{i^*,k,\beta} = U$ if $M_{i^*,k,\beta} = B$. It then runs $Create(\lambda, M')$ to produce T. The challenger gives T to the attacker, who finally must guess the value of the bit b.

Definition 5. *We say a tribes scheme has* inter-column security *if for every polynomial attacker $\mathcal{A}$, there exists a negligible function $negl(\lambda)$ such that the attacker's advantage in the Inter-Column Game is $\leq negl(\lambda)$, for any valid settings of M, j, k, i, β. Note that the negligible function depends only on $\mathcal{A}$ and λ, and is independent of the dimensions of M, for example.*

In the full version, we define a tribes-lite scheme as a relaxed notion of a tribes scheme, where only inter-column security is required. We then demonstrate how to be build a tribes scheme from a tribes-lite scheme.

Required Security To be useful for ultimately building witness encryption, the required security of all of our security games is that they must be $negl(\lambda) \cdot 2^{-n}$ where $negl(\lambda)$ is some negligible function. The demand for the 2^{-n} term is passed down from the positional hybrid of the previous Section 2. In the next section we show how to build positional WE from a Tribes scheme. Since that reduction involves only a polynomial number of hybrids in n (and thus λ) these are absorbed in the negligible function.

4   Constructing a Positional Witness Encryption Scheme from a Tribes Scheme

We will now show how to build a positional witness encryption scheme from a tribes scheme.

---

[4] Again, these are extra conditions we are imposing on M, j, k, β in order for this game to be applicable.

4.1   Encoding a CNF in a Tribal Matrix

Suppose we have a CNF formula ϕ with n variables and ℓ clauses. In other words, we can write $\phi = \phi_1 \wedge \phi_2 \wedge \ldots \wedge \phi_\ell$, where each ϕ_i is a clause over the variables $X_1, \ldots, X_n$ and their negations, denoted $\overline{X_1}, \ldots, \overline{X_n}$. We will define an $n \times \ell \times 2$ tribes matrix M^ϕ.

In order to set the entries of the jth column of M^ϕ, we consider the jth clause ϕ_j. For each row i, we do the following: **(A)** If X_i appears in ϕ_j, we set $M^\phi_{i,j,1} = U$. **(B)** If $\overline{X_i}$ appears in ϕ_j, we set $M^\phi_{i,j,0} = U$. **(C)** For any entries $M^\phi_{i,j,\beta}$ not yet defined, set $M^\phi_{i,j,\beta} = B$. We note the following property of M^ϕ:

Lemma 1. *If we consider a boolean string $x \in \{0,1\}^n$ as an assignment of truth values to the variables $X_1, \ldots, X_n$ of ϕ, then if clause ϕ_j is unsatisfied by x, column j of M^ϕ will evaluate to value 1, and hence $f_{M^\phi}(x) = 1$. If x satisfies ϕ, then $f_{M^\phi}(x) = 0$.*

Proof. Suppose clause ϕ_j is unsatisfied by the assignment x. For each $i \in [n]$, we consider M^ϕ_{i,j,x_i}. If $x_i = 0$, then ϕ_j unsatisfied implies that $\overline{X_i}$ does not appear in ϕ_j, and so $M^\phi_{i,j,0} = B$. Similarly, if $x_i = 1$, then ϕ_j unsatisfied implies X_i does not appear in ϕ_j, so $M^\phi_{i,j,1} = B$. Thus, $f_{M^\phi}(x) = 1$. Conversely, if x satisfies ϕ, then for each column j, there exists some row i such that either $x_i = 0$ and $\overline{X_i}$ appears in ϕ_j or $x_i = 1$ and X_i appears in ϕ_j. Either way, $M^\phi_{i,j,x_i} = U$. Hence, $f_{M^\phi}(x) = 0$.

4.2   Encoding a Position in a Tribal Matrix

Suppose we have a position t considered as a binary string $t = (t_1, t_2, \ldots, t_n) \in \{0,1\}^n$. We will define an $n \times n \times 2$ tribes matrix M^t. We describe M^t by specifying how to fill in the jth column of M^t. To Set Column j:

$$\text{For } i < j, \quad M^t_{i,j,0} = B, \quad M^t_{i,j,1} = \begin{cases} U, & \text{if } t_i = 0; \\ B, & \text{if } t_i = 1. \end{cases}$$

$$\text{For } i = j, \quad M^t_{i,j,0} = \begin{cases} U, & \text{if } t_i = 0; \\ B, & \text{if } t_i = 1. \end{cases} \quad M^t_{i,j,1} = U$$

$$\text{For } i > j, \quad M^t_{i,j,0} = B = M^t_{i,j,1}.$$

We now establish some relevant properties of M^t. First, we observe that the associated boolean function f_{M^t} evaluates to 1 for every boolean string $y < t$ and evaluates to 0 for every $y \geq t$. Here, we use "$<$" and "$\geq$" to denote the order induced by the usual ordering of integers, when we think of t, y as binary expansions with t_1, y_1 being the most significant bits.

Lemma 2. *If $y < t$, then $f_{M^t}(y) = 1$.*

Proof. Since $y < t$, there must be some index $k \in [n]$ such that $t_i = y_i$ for all $i < k$ and $t_k = 1$ while $y_k = 0$. We consider the kth column of M^t. We claim that for all i, $M^t_{i,k,y_i} = B$. To see this, we can consult our description of the kth column of M^t above, noting that for $i < k$, whenever $y_i = 1$, then $t_i = 1$ as well (by definition of k). Thus, $f_{M^t}(y) = 1$.

Lemma 3. *If $y \geq t$, then $f_{M^t}(y) = 0$.*

Proof. We let $k \in [n]$ denote an index such that $y_i = t_i$ for all $i \leq k$, and $y_{k+1} = 1$, $t_{k+1} = 0$, if $k + 1 \leq n$. For a column j where $j \leq k$, we observe that $M_{j,j,y_j} = U$, since $y_j = t_j$. For any column j where $j > k$, we observe that $M_{k+1,j,y_{k+1}} = U$. This is because $t_{k+1} = 0$ and $y_{k+1} = 1$. Hence, $f_{M^t}(y) = 0$.

This defines an effective encoding of positions t from 0 to $2^n - 1$ (considering t as an integer). We also require an encoding of 2^n. We define M^{2^n} to be the same as M^{2^n-1}, except that the first diagonal entry has both slots equal to B. We observe that $f_{M^{2^n}}(y) = 1$ for all n-bit values y, since the first column is all filled with B values.

4.3 Our Positional Witness Encryption Scheme

We let our message space be $\{0, 1\}$.

Encrypt$_{\text{PWE}}(1^\lambda, \phi, t, m)$ The encryptor constructs M^ϕ and M^t as above. For $m \in \{0, 1\}$, it constructs an additional column C^m (which is $n \times 2$) as follows. If $m = 1$, $C^m_{i,0} = C^m_{i,1} = B$ for all i, and if $m = 0$, $C^m_{i,0} = C^m_{i,1} = U$ for all i. It also constructs a completely unblocked column S, meaning that $S_{i,0} = S_{i,1} = U$ for all i. Note that appending such a column to a tribes matrix will not affect the evaluation function. (This "scratch column" S will be useful in the proof of security.)

It then forms an $n \times (\ell + n + 2) \times 2$ tribes matrix M as $M := M^\phi || M^t || C^m || S$, meaning that the first ℓ columns are taken to be M^ϕ, the next n columns are taken to be M^t, and the final two columns are taken to be C^m and S. The encryptor then calls $Create(\lambda, M)$ to produce a tribes scheme T, and sets $\text{CT} := T$.

Decrypt$_{\text{PWE}}(\text{CT}, w)$ The decryptor runs $Eval(\text{CT}, w)$ and outputs the result.

4.4 Security of our Positional Witness Encryption Scheme

We now prove security of the positional witness encryption based on the two tribes properties on inter-column and intra-column security. The most complex part is the proof of position hiding, which is given over a sequence of hybrid steps. At a very high level the proof (for indistinguishability of position t from $t+1$) proceeds in two stages. In the first stage the reduction algorithm identifies a clause, j, in the CNF formula that the witness candidate $w = t$ does not satisfy. Such a clause must exist for this to be a valid instance of the positional game. The proof then uses the j-th column in the CNF portion of the matrix to (undetectably) change the scratch column S from having U's in each of its $2 \cdot n$ slots to having a B in row i slot t_i for each i. (The other n slots remain U.) The security properties are used to argue that such a change is indistinguishable to an attacker. This copy action into the scratch column will cause the column to evaluate as "blocked" on input t and remain evaluating to unblocked on all other inputs. Intuitively, this will have no impact on the overall evaluation since the j-th column caused a block on input t anyway — providing a conceptual sanity check for our claim. Intuitively, this stage reflects the fact that t is not a valid witness and represents this fact in the scratch column.

The next stage of our proof solely involves the scratch column and position matrix. A series of hybrid steps will use the scratch column to update the positional part of the matrix from position t to position $t + 1$ by "assimilating" the scratch column from the previous stage. At the end of these steps that scratch column will again become unblocked in all slots and thus matching the end goal of our argument. Our proof can be found in the full version.

5   An Instantiation in a Symmetric Model of Composite Order Multilinear Groups

We provide a description of our first instance of a tribes schemes by instantiating it in symmetric composite order multilinear groups. Its proof of security and intuition on the assumptions appear in the full version.

5.1   An Abstract Model of Composite Order Multilinear Groups

We let G denote a (cyclic) group of order $N = p_1 p_2 \cdots p_r$, where p_1, ..., p_r are distinct primes. We let G_T also denote a cyclic group of order N. We suppose that we have a k-linear map $E : G^k \to G_T$. We assume this is non-degenerate, meaning that if g generates G, then $E(g, g, \ldots, g)$ generates G_T. We write the group operations multiplicatively, and we let $1_G, 1_{G_T}$ denote the identity elements in G and G_T respectively.

For each prime p_i dividing the group order of N, we have a subgroup G_{p_i} of order p_i inside G. We let g_{p_i} denote a generator for G_{p_i}. We let $G_{p_1 p_2}$, for example, denote the subgroup of order $p_1 p_2$ that is generated by $g_{p_1} g_{p_2}$.

These subgroups are "orthogonal" under G, meaning (for example) that if $h \in G_{p_1 p_2 \cdots p_{i-1} p_{i+1} p_r}$, then for any $g_2, \ldots, g_{k-1} \in G$,

$$E(h, g_2, \ldots, g_{k-1}, g_{p_i}) = 1_{G_T}.$$

More generally, each element of G can be expressed as $g_{p_1}^{\alpha_1} g_{p_2}^{\alpha_2} \cdots g_{p_r}^{\alpha_r}$. Thus if we have k elements of G that are input to E, we can write them as $g_{p_1}^{\alpha_{1,1}} g_{p_2}^{\alpha_{2,1}} \cdots g_{p_r}^{\alpha_{r,1}}$,

$\ldots, g_{p_1}^{\alpha_{1,k}} g_{p_2}^{\alpha_{2,k}} \cdots g_{p_r}^{\alpha_{r,k}}$, and by multi-linearity of E and orthogonality we then have:

$$E(g_{p_1}^{\alpha_{1,1}} g_{p_2}^{\alpha_{2,1}} \cdots g_{p_r}^{\alpha_{r,1}}, \ldots, g_{p_1}^{\alpha_{1,k}} g_{p_2}^{\alpha_{2,k}} \cdots g_{p_r}^{\alpha_{r,k}}) = E(g_{p_1}^{\alpha_{1,1}}, \ldots, g_{p_1}^{\alpha_{1,k}}) \cdots E(g_{p_r}^{\alpha_{r,1}}, \ldots, g_{p_r}^{\alpha_{r,k}}).$$

We let $\mathcal{G}(\lambda, r, k)$ denote a group generation algorithm that takes in a security parameter λ, a desired number of prime factors r, and a desired level of multilinearity k and outputs a description of a group G as above. We assume the description includes a generator $g \in G$, the group order N, the primes $p_1, \ldots, p_r$ comprising N, and efficient algorithms for the group operation in G, the group operation in G_T, and the multilinear map E. Note that with a generator g for the full group plus knowledge of the prime factors, one can efficiently produce a generator for any subgroup of order dividing N.

Computational Assumption 1_S Our first computational assumption in the symmetric setting will be parameterized by positive integers n and ν. It will concern a group of order $N = a_1 \ldots a_n b_1 \ldots b_\nu c$, where $a_1, \ldots, a_n, b_1, \ldots, b_\nu, c$ are $n+\nu+1$ distinct primes. We give out generators $g_{a_1}, \ldots, g_{a_n}, g_{b_1}, \ldots, g_{b_\nu}$ for each prime order subgroup *except for the subgroup of order c*. For each $i \in [n]$, we also give out a group element h_i sampled uniformly at random from the subgroup of order $c a_1 \cdots a_{i-1} a_{i+1} \cdots a_n$. The challenge term is a group element $T \in G$ that is either sampled uniformly at random from the subgroup or order $c a_1 \cdots a_n$ or uniformly at random from the subgroup of order $a_1 \cdots a_n$. The task is to distinguish between these two distributions of T.

We name this assumption the (n, ν)-*multilinear subgroup elimination assumption*.

Computational Assumption 2_S Our second computational assumption will be parameterized by positive integers n and ν. It will again concern a group of order $N = a_1 \ldots a_n b_1 \ldots b_\nu c$, where $a_1, \ldots, a_n, b_1, \ldots, b_\nu, c$ are $n + \nu + 1$ distinct primes. As in Assumption 1, we give out generators $g_{a_1}, \ldots, g_{a_n}, g_{b_1}, \ldots, g_{b_\nu}$ for each prime order subgroup *except for the subgroup of order c*. The challenge term is a group element T that is either sampled uniformly at random the subgroup of order $c a_n$ or the subgroup of order a_n. The task is to distinguish between these two distributions of T.

We name this assumption the (n, ν)-*multilinear subgroup decision assumption*.

5.2   Instantiating a Tribes Scheme

Suppose we wish to build a tribes scheme for $n \times \ell \times 2$ tribes matrices, and we have a generation algorithm $\mathcal{G}$ for producing composite order multilinear groups. We construct a tribes scheme as follows:

Create(λ, M): The creation algorithm takes in a security parameter λ and an $n \times \ell \times 2$ tribes matrix M (entries in $\{U, B\}$). It then calls $\mathcal{G}(\lambda, r = n + \ell, n)$ to produce a group G of order $N = p_1 \cdots p_n q_1 \cdots q_\ell$ equipped with an n-linear map E. It will produce $2n$ group elements, each indexed by a row $i \in [n]$ and a slot $\beta \in \{0, 1\}$. We let $g_{i,\beta}$ be sampled as follows. First, for each $i' \neq i$, a uniformly random element $s_{i'}$ of the subgroup of order $p_{i'}$ is sampled. Next, for each column index $j \in [\ell]$, if $M_{i,j,\beta} = B$, then z_j is sampled as a uniformly random element of the subgroup of order q_j. If $M_{i,j,\beta} = U$, then $z_j := 1_G$. (All of these values are freshly resampled for each i, β.) We set:

$$g_{i,\beta} := \prod_{i' \neq i} s_{i'} \prod_{j=1}^{\ell} z_j.$$

The tribes scheme T consists of these $2n$ elements $\{g_{i,\beta}\}$ (we assume this implicitly includes a description of G that enables efficient computation of the group operation and E, and the full group order N, but *not* the individual primes comprising N).

Eval(T, x): The evaluation algorithm takes in a tribes scheme T and a boolean vector $x = (x_1, \ldots, x_n) \in \{0, 1\}^n$. It computes $E(g_{1,x_1}, g_{2,x_2}, \ldots, g_{n,x_n})$ and checks whether this is equal to 1_{G_T} or not. If is it the identity, it outputs 0. Otherwise, it outputs 1.

Correctness We first establish that $Eval(Create(\lambda, M), x) = f_M(x)$. We first observe that (by orthogonality of distinct prime order subgroups) $E(g_{1,x_1}, g_{2,x_2}, \ldots, g_{n,x_n})$ can be considered as a product of contributions within each prime order subgroup. Consider a prime p_i. No component in the subgroup of order p_i appears in g_{i,x_i} (regardless of the value of x_i), so this contribution is trivial (just the identity element). For analyzing the contribution of the q_j primes, we consider two cases. Suppose $\exists$ a column j such that $M_{i,j,x_i} = B$ for all $i \in [n]$. This is equivalent to supposing that $f_M(x) = 1$. In this case, there is a random component in the subgroup of order q_j incorporated in *every* g_{i,x_i}, so the contribution will be (with high probably) a non-identity element in the q_j order subgroup of G_T. This cannot be "canceled out" by a contribution in any other prime order subgroup, so the result will be $\neq 1_{G_T}$ in this case, resulting in an output that matches f_M. In the other case, no such column j exists. This means that for every column j, there is some g_{i,x_i} which lacks a component in the order q_j subgroup, hence causing a result of 1_{G_T}, and the output again matches f_M.

In the full version, we show that inter-column security for this tribes scheme is implied by the multilinear subgroup elimination assumption, and intra-column security is implied by the multilinear subgroup decision assumption. One can alternatively rely solely on the multilinear subgroup elimination assumption to build a tribes-lite scheme first and then derive a tribes scheme from it.

Acknowledgements. We thank Mihir Bellare and Amit Sahai for helpful discussions and comments.

References

1. Bellare, M., Hoang, V.T.: Adaptive witness encryption and asymmetric password-based cryptography. Cryptology ePrint Archive, Report 2013/704 (2013), http://eprint.iacr.org/
2. Ben-Or, M., Linial, N.: Collective coin flipping, robust voting schemes and minima of banzhaf values. In: FOCS, pp. 408–416 (1985)
3. Boneh, D., Franklin, M.K.: Identity-based encryption from the weil pairing. SIAM J. Comput. 32(3), 586–615 (2003); Extended abstract in Kilian, J. (ed.) CRYPTO 2001. LNCS, vol. 2139, pp. 213–615. Springer, Heidelberg (2001)
4. Boneh, D., Sahai, A., Waters, B.: Fully collusion resistant traitor tracing with short ciphertexts and private keys. In: Vaudenay, S. (ed.) EUROCRYPT 2006. LNCS, vol. 4004, pp. 573–592. Springer, Heidelberg (2006)
5. Boneh, D., Silverberg, A.: Applications of multilinear forms to cryptography. Contemporary Mathematics 324, 71–90 (2003)
6. Boneh, D., Waters, B.: A fully collusion resistant broadcast, trace, and revoke system. In: ACM Conference on Computer and Communications Security, pp. 211–220 (2006)
7. Chor, B., Fiat, A., Naor, M.: Tracing traitors. In: Desmedt, Y.G. (ed.) CRYPTO 1994. LNCS, vol. 839, pp. 257–270. Springer, Heidelberg (1994)
8. Cocks, C.: An identity based encryption scheme based on quadratic residues. In: IMA Int. Conf., pp. 360–363 (2001)
9. Coron, J.-S., Lepoint, T., Tibouchi, M.: Practical multilinear maps over the integers. In: Canetti, R., Garay, J.A. (eds.) CRYPTO 2013, Part I. LNCS, vol. 8042, pp. 476–493. Springer, Heidelberg (2013)
10. Fiat, A., Naor, M.: Broadcast encryption. In: Stinson, D.R. (ed.) CRYPTO 1993. LNCS, vol. 773, pp. 480–491. Springer, Heidelberg (1994)
11. Garg, S., Gentry, C., Halevi, S.: Candidate multilinear maps from ideal lattices. In: Johansson, T., Nguyen, P.Q. (eds.) EUROCRYPT 2013. LNCS, vol. 7881, pp. 1–17. Springer, Heidelberg (2013)
12. Garg, S., Gentry, C., Halevi, S., Raykova, M., Sahai, A., Waters, B.: Candidate indistinguishability obfuscation and functional encryption for all circuits. In: FOCS, pp. 40–49. IEEE Computer Society (2013)
13. Garg, S., Gentry, C., Halevi, S., Sahai, A., Waters, B.: Attribute-based encryption for circuits from multilinear maps. Cryptology ePrint Archive, Report 2013/128 (2013), http://eprint.iacr.org/
14. Garg, S., Gentry, C., Sahai, A., Waters, B.: Witness encryption and its applications. In: STOC, pp. 467–476 (2013)
15. Garg, S., Ostrovsky, R., Visconti, I., Wadia, A.: Resettable statistical zero knowledge. In: Cramer, R. (ed.) TCC 2012. LNCS, vol. 7194, pp. 494–511. Springer, Heidelberg (2012)
16. Goldwasser, S., Kalai, Y.T.: On the impossibility of obfuscation with auxiliary input. In: 46th Annual IEEE Symposium on Foundations of Computer Science, FOCS 2005, pp. 553–562. IEEE (2005)
17. Goldwasser, S., Kalai, Y.T., Popa, R.A., Vaikuntanathan, V., Zeldovich, N.: How to run turing machines on encrypted data. In: Canetti, R., Garay, J.A. (eds.) CRYPTO 2013, Part II. LNCS, vol. 8043, pp. 536–553. Springer, Heidelberg (2013)
18. Gorbunov, S., Vaikuntanathan, V., Wee, H.: Predicate encryption for circuits. In: STOC (2013)

19. Karp, R.M.: Reducibility among combinatorial problems. In: Complexity of Computer Computations, pp. 85–103 (1972)
20. Pass, R., Seth, K., Telang, S.: Indistinguishability obfuscation from semantically-secure multilinear encodings. Cryptology ePrint Archive, Report 2013/781 (2013), http://eprint.iacr.org/
21. Sahai, A., Waters, B.: Fuzzy identity-based encryption. In: Cramer, R. (ed.) EUROCRYPT 2005. LNCS, vol. 3494, pp. 457–473. Springer, Heidelberg (2005)
22. Shamir, A.: Identity-based cryptosystems and signature schemes. In: Blakely, G.R., Chaum, D. (eds.) CRYPTO 1984. LNCS, vol. 196, pp. 47–53. Springer, Heidelberg (1985)
23. Tompa, M., Woll, H.: Random self-reducibility and zero knowledge interactive proofs of possession of information. In: FOCS, pp. 472–482 (1987)

RSA Key Extraction via Low-Bandwidth Acoustic Cryptanalysis*

Daniel Genkin[1], Adi Shamir[2], and Eran Tromer[3]

[1] Technion and Tel Aviv University, Israel
danielg3@cs.technion.ac.il
[2] Weizmann Institute of Science, Israel
adi.shamir@weizmann.ac.il
[3] Tel Aviv University, Israel
tromer@cs.tau.ac.il

Abstract. Many computers emit a high-pitched noise during operation, due to vibration in some of their electronic components. These acoustic emanations are more than a nuisance: as we show in this paper, they can leak the key used in cryptographic operations. This is surprising, since the acoustic information has very low bandwidth (under 20 kHz using common microphones, and a few hundred kHz using ultrasound microphones), which is many orders of magnitude below the GHz-scale clock rates of the attacked computers. We describe a new *acoustic cryptanalysis* attack which can extract full 4096-bit RSA keys from the popular GnuPG software, within an hour, using the sound generated by the computer during the decryption of some chosen ciphertexts. We experimentally demonstrate such attacks, using a plain mobile phone placed next to the computer, or a more sensitive microphone placed 10 meters away.

1 Introduction

1.1 Overview

Cryptanalytic side-channel attacks target implementations of cryptographic algorithms which, while perhaps secure at the mathematical level, inadvertently leak secret information through indirect channels: variations in power consumption, electromagnetic emanations, timing variations, contention for CPU resources such as caches, and so forth (see [And08] for a survey). Acoustic emanations are another potential channel, but so far it was used only in order to eavesdrop to slow electromechanical components such as keyboards and printers [AA04, ZZT05, BDG+10].

In this paper, we focus on a different source of computer noise: vibration of electronic components in the computer, sometimes heard as a faint high-pitched tone or hiss (commonly called "coil whine", though often generated by capacitors). These acoustic emanations, typically caused by voltage regulation circuits, are correlated with system activity since CPUs drastically change their power

---

* The authors thank Lev Pachmanov for programming and experiments support.

J.A. Garay and R. Gennaro (Eds.): CRYPTO 2014, Part I, LNCS 8616, pp. 444–461, 2014.
© International Association for Cryptologic Research 2014

draw according to the operations they execute, but in a very coarse way due to the low bandwidth, which does not enable the attacker to "hear" individual instructions executed on a multi-GHz computer.[1]

The first indication that acoustic emanation from electronic computers is of cryptanalytic interest was by Shamir and Tromer [ST04], observing that different RSA keys have distinguishable acoustic fingerprints. However, no approach has been proposed to extract actual key bits from the faint, noisy and low-bandwidth acoustic information. In fact, a recent survey stated that while "acoustic effects have been suggested as possible side channel, the quality of the resulting measurements is likely to be low" [KJJR11].

Acoustic Cryptanalysis. In this paper we show that despite this skepticism, full key recovery via pure *acoustic cryptanalysis* is feasible on common software and hardware. As a typical case study, we focused on GnuPG (GNU Privacy Guard) [Gpg], which is a popular cross-platform open-source implementation of the OpenPGP standard. We first verified that different secret keys can be *distinguished* by the spectrum of the sound made when they are used. We then developed a new *key extraction* attack which can find the full 4096-bit RSA secret keys used by GnuPG running on a laptop computer, within an hour, by analyzing only the sound picked up by either a plain cellular phone placed next to the computer, or by a sensitive microphone from a distance of 10 meters. In a nutshell, our attack relies on crafting special chosen RSA ciphertexts that cause numerical cancellations deep inside GnuPG's modular exponentiation algorithm. This causes the special value zero to appear frequently in the innermost loop of the algorithm, where it affects control flow. A single iteration of that loop is much too fast for direct acoustic observation. However, in our attack the effect is repeated and amplified over many thousands of iterations, resulting in a gross leakage effect that is discernible for hundreds of milliseconds and distinguishable in the acoustic spectrum. Thus, our attack not only causes key-dependent side-channel leakage in GnuPG's RSA implementation, but moreover utilizes the GnuPG's *own code* in order to amplify the aforementioned leakage.

Chosen Ciphertexts by e-mail. Our key extraction technique requires the decryption of multiple ciphertexts which are adaptively chosen by the attacker. Prior works which used chosen plaintexts or ciphertexts required direct access to the input of the protected device, or attacked network protocols such as SSL/TLS or WEP. To break GnuPG, we used a new attack vector based on Enigmail [Eni], which is a popular plugin to the Thunderbird e-mail client that enables transparent signing and encryption of e-mail messages using GnuPG, following the OpenPGP and PGP/MIME standards. For "new e-mail" notifications, Enigmail automatically decrypts each e-mail as soon as it is received, provided that the GnuPG passphrase is cached or empty. In this case, an attacker can e-mail suitably-crafted messages to the victims (backdated, so they go unnoticed), and

---

[1] Above a few hundred kHz, sound propagation in the air has a very short range, due to non-linear attenuation and distortion effects (viscosity, relaxation and diffusion at the molecular level). Most microphones are limited to about 20 kHz.

observe the acoustic signature of their automatic decryption, thereby closing the adaptive attack loop without manual intervention by the recipient.

Applicability. Our observations apply to many laptop computers made by various vendors and running various operating systems. Signal quality and effective attack distance vary greatly, and seem to be correlated with the computer's age (i.e., older computers tend to emit stronger and more informative sounds). Our acoustic attacks can be applied in a large variety of situations. For example, any electronic device which has an internal microphone can be used to spy on itself, using an unprivileged application which has access to the microphone listen to the sounds made by a privileged security application (even when the two applications run in two different virtual machines). On a cellular phone, the whole attack can be packaged into a simple software "app" which can close the adaptive chosen-ciphertext loop in real time, using the phone's signal processing capabilities and wireless data connectivity. The phone can then be used to spy on a nearby laptop computer, for example, during an hour-long face-to-face business meeting between two persons who place their gadgets on the same table. In another scenario, the attacker can place in advance a hidden acoustic bug near the likely location of the attacked laptop, e.g., in a lecture podium used by a visiting speaker, or in a hotel desk.

Physical Countermeasures. Many of the physical side-channel countermeasures used in highly sensitive applications, such as air gaps, Faraday cages, and power supply filters, provide no protection against acoustic leakage. In particular, Faraday cages containing computers require ventilation, which is typically provided by means of vents covered with perforated sheet metal or metal honeycomb. These are very effective at attenuating compromising electromagnetic radiation ("TEMPEST"), but — as we empirically verified — are nearly transparent to acoustic emanations. This can make the acoustic attack one of the few remaining options when the target is heavily protected by expensive shielding that blocks all the standard sources of electronic emanations.

Current Status. Our attacks can be applied to all recent versions of the GnuPG 1.x series (up to the latest, 1.4.15, released on 15 Oct. 2013), including the side-channel mitigation introduced in GnuPG 1.4.14 (which ironically helps our attack by amplifying the aforementioned effect of the zero value in the innermost loop). After disclosing our detailed attack to the GnuPG developers and main distributors, we suggested several suitable countermeasures, and verified that the new versions of GnuPG 1.x and of libgcrypt (which underlies GnuPG 2.x), released concurrently with this paper's first public posting, correctly implement our countermeasures and resist the current key-extraction attack.

1.2   Related Work

Auditory eavesdropping on human conversations is a common practice, first published several millenia ago [Gen]. Analysis of sound emanations from mechanical devices is a newer affair, with precedents in military context such as identification of vehicles (e.g., submarines) via the sound signature of their engine or propeller. Wright [Wri87, pp. 103–107] provides an account of MI5 and GCHQ using a phone tap to eavesdrop on an electromechanical Hagelin cipher machine, by counting the clicks during the rotors' secret-setting procedure. Keystroke timing patterns (which can be acquired acoustically) are known to be a way to identify users, and more recently, also to leak information about the typed text (see [SWT01] and the references therein). Later work by Asonov and Agrawal [AA04], improved by Zhuang et al. [ZZT05], Berger et al. [BWY06], and by Halevi and Saxena [HS10], shows that keys can also be distinguished individually by their sound, due to minute differences in mechanical properties such as their position on a slightly vibrating printed circuit board. Backes et al. [BDG⁺10] show that the sound produced by dot matrix printers can be used to recover printed English text.

NSA's partially-declassified "TEMPEST Fundamentals" [Nat82] mentions acoustic emanations, but defines them narrowly as "emanations in the form of free-space acoustical energy produced by the operation of a *[sic]* purely mechanical or electromechanical device equipment". Other official publications, such as the latest FIPS 140-3 draft [Nat09], describe a large variety of side channel attacks, but do not mention acoustic emanations. One is thus led to conclude that this attack vector was not believed to pose a threat to nonmechanical systems.

2   Observing Acoustic Leakage

In this section we show that it is possible, using acoustic emanations, to glean information about the CPU operations of various laptop computers. We show that it is possible for an attacker to learn the instructions executed by the target computer, solely by observing its acoustic emanations using a microphone. Moreover, we show a rudimentary cryptographic side channel, namely distinguishing GnuPG RSA keys (which will be further developed in the subsequent sections).

Lab-grade Experimental Setup. For the experiments in this section, meant to best characterize the emitted signal, we used carefully-optimized lab-grade equipment with very high sensitivity and frequency range. Specifically, we measured the acoustic emanations using a Brüel&Kjær 4190 and 4939 microphone capsules, connected to a Brüel&Kjær 2669 pre-amplifier, powered by a Brüel&Kjær 2804 microphone power supply. The signal is low-pass filtered at 1.9 MHz, amplified using a (customized) Mini-Circuits ZPUL-30P amplifier, and then high-pass filtered at 10 kHz. The result is digitized using an National Instruments PXIe 6356. For further details, see the extended version of this paper [GST13].

Culprit Components. The exploited acoustic emanations are clearly not caused by fan rotation, hard disk activity, or audio speakers — as readily verified by disabling these. Rather, they are caused by vibrations of electrical components in the power supply circuitry, familiar to many as the faint high-pitched whine produced by some devices and power adapters (commonly called "coil whine", though not always emanating from coils). The physical source of the

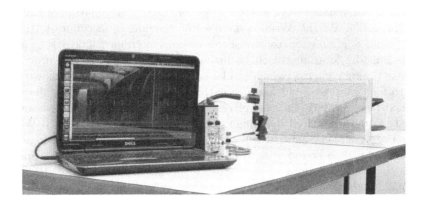

Fig. 1. Portable measurement setup recording a laptop through an EMI-shielded vent panel (1/2"-thick cross-cell double-honeycomb mesh, Holland Shielding Systems Honeycomb Ventilation 9500).

relevant emanations is difficult to characterize precisely, since it varies between target machines, and is typically located in the hard-to-reach innards. Still, experimentation with the microphone placement invariably located the strongest useful signals in the vicinity of the on-board voltage regulation circuit supporting the CPU. Indeed, modern CPUs change their power consumption dramatically depending on software load, and we conjecture that this affects, and modulates, the dynamics of the pulse-width-modulation-based voltage regulator. More remote stages of the power supply (e.g., laptops' external AC-DC "power brick" adapter) sometimes exhibit software-dependent acoustic leakage as well.

Microphone Placement. The placement of the microphone relative to the laptop body has a great influence on the obtained signal. Ideally, we would like to measure acoustic emanations as close as possible to the CPU's on-board power supply located on the laptop's motherboard, but without intrusion or disassembly. Luckily, laptop computers have a substantial cooling system for heat dissipation, with a fan that requires large exhaust holes. In addition, there are numerous holes and gaps for ports such as USB, Express Card slot, and Ethernet port. Each of the above ports has proven useful on some computers.

Acoustic or EM? To ascertain that the obtained signal is truly acoustic rather than electromagnetic interference picked up by the microphone, we placed a sheet of non-conductive sound-absorbing material (e.g., cork or thick cloth) in front of the microphone. This always resulted in a severe attenuation of the recorded signals. Thus, we conclude that the measured signals are indeed acoustic.

EM Shielding. As discussed in Section 1.1, standard TEMPEST electromagnetic shielding, such as metal meshes and perforated metal, can be nearly-transparent to acoustic emanations, as we verified on a professionally-produced EMI mesh shield (see Figure 1, and note that as before, covering the mesh with cardboard severely attenuated the signal).

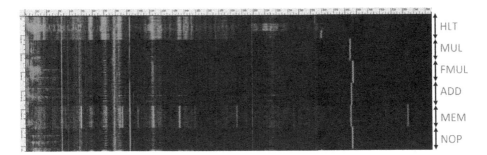

Fig. 2. Acoustic measurement frequency spectrogram of a recording of different CPU operations using the Brüel&Kjær 4939 microphone capsule. The horizontal axis is frequency (0–350 kHz), the vertical axis is time (1.7 sec), and intensity is proportional to the instantaneous energy in that frequency band.

2.1 Distinguishing Various CPU Operations

We begin our analysis of acoustic leakage by attempting to distinguish various operations performed by the CPU of the target computer. For this purpose we wrote a simple program that executes (partially unrolled) loops containing one of the following x86 instructions: HLT (CPU sleep), MUL (integer multiplication), FMUL (floating-point multiplication), main memory access (forcing L1 and L2 cache misses), and REP NOP (short-term idle). While it was possible to distinguish between some CPU operations on almost all the machines we tested, some machines have a particularly rich leakage spectrum. Figure 2 shows a recording of the Evo N200 laptop while executing our program using the Brüel&Kjær 4939 high frequency microphone capsule. As can be seen in Figure 2, the leakage of the Evo N200 is present all over the 0–350 kHz spectrum. Moreover, different types of operations can be easily distinguished. Similar types of leakage (although less prominent) was detected on numerous other machines as well.

2.2 GnuPG Key Distinguishability

The results in Section 2.1 demonstrate that it is possible, even when using a very low-bandwidth measurement of 35 kHz, to obtain information about the code executed by the target machine. While this is certainly some form of leakage, it is not clear how to use this information to form a real key extraction attack on the target machine. In this section, we show that some useful information about a secret key used by the target machine can be obtained from the acoustic information, even though it is still unclear how to use it to derive the full key. In particular, we demonstrate that the acoustic information obtained during a single RSA secret operation (such as ciphertext decryption or signature generation) suffices in order to determine which of several randomly generated keys was used by the target machine during that operation. Throughout this paper, we target a standard and commonly used RSA implementation, GnuPG (GNU Privacy

Guard) [Gpg], a popular open source implementation of the OpenPGP standard available on all major operating systems.[2]

The Sound of a Single RSA Secret Key. Figure 3 depicts the spectrogram of five RSA signing operations in sequence, using the same message and five randomly generated 4096-bit keys. Each signing operation is preceded by a short delay, during which the CPU is in a sleep state. Figure 3 contains several interesting effects. The delays are manifested as bright horizontal strips. Between these strips, the five signing operations can be clearly distinguished. Halfway through each signing operation there is a transition at several frequency bands (marked by yellow arrows), corresponding to the transition between exponentiation modulo the secret p to exponentiation modulo the secret q, in the RSA decryption implementation of GnuPG, which uses the Chinese Remainder Theorem.

Distinguishing between RSA Secret Keys. Having observed that the acoustic signature of modular integer exponentiation depends on the modulus involved. Thus, one may expect different keys to cause different sounds. This is indeed the case, as demonstrated in Figure 3. It is readily observed that each signature (and in fact, each exponentiation using modulus p or q) has a unique spectral signature. This ability to distinguish keys is of interest in traffic-analysis applications.[3] It is likewise possible to distinguish between algorithms, between different implementations of an algorithm, and between different computers (even of the same model) running the same algorithm. Again, this effect is consistent and reproducible (in various frequency ranges) on various machines and manufacturers. Finally, for the case of ElGamal decryption, various secret keys can also be acoustically distinguished.

3 Overview of GnuPG RSA Key Extraction

In this section, we present our acoustic RSA key extraction attack, and discuss its performance (e.g., extracting a whole RSA key within about one hour using just the acoustic emanations from the target machine). For concreteness, in the following we consider GnuPG 1.4.14 and key size of 4096 bit (i.e., 2048 bit primes p, q), which should be secure beyond the year 2031 [BBB+12].

3.1 GnuPG's Modular Exponentiation Routine

GnuPG's Mathematical Library. Algebraic operations on large integers (which are much larger than the machine's word size) are implemented using

---

[2] We focus on the GnuPG 1.x series and its internal cryptographic library (including the side-channel countermeasures recently added in GnuPG 1.4.14 following the work of [YF13]). The effects presented in the paper were observed on a variety of operating systems (Windows 2000 through 7, Fedora Core 2 through 19), GnuPG versions 1.2.4 through 1.4.15, and many target machines.

[3] For example, observing that an embassy has now decrypted a message using a rarely-used key, heard before only in specific diplomatic circumstances, can be valuable.

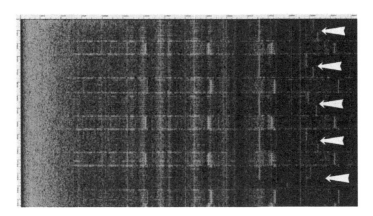

Fig. 3. Acoustic signature (1.4 sec, 0–40 kHz) of five GnuPG RSA signatures executed on a Lenovo ThinkPad T61. Recorded using the lab-grade setup and the Brüel&Kjær 4190 microphone capsule. The transitions between p and q are marked by yellow arrows.

software routines. GnuPG uses an internal mathematical library called MPI, which is based on the GMP library [Gmp], to store and perform mathematical operations on large integers. MPI stores each large integer in an array of *limbs*, each consisting of a 32-bit word (on the x86 architecture used in our tests).

We now review the modular exponentiation used in GnuPG's implementation of RSA (as introduced in GnuPG v1.4.14). GnuPG uses a side-channel protected variant of the square-and-multiply modular exponentiation algorithm, processing the bits of the exponent d from most significant bit to the least significant one. Algorithm 1 is a pseudocode of the modular exponentiation algorithm used in GnuPG. The operation SIZE_IN_LIMBS(x) returns the number of limbs in the t-bit number x, namely $\lceil t/32 \rceil$. Understanding this top-level exponentiation routine suffices for the high-level description of our attack. For details about GnuPG's underlying multiplication routines, necessary for understanding the attack's success, see Section 4.

Since GnuPG represents large numbers in arrays of 32 bit limbs, GnuPG optimizes the number of modulo reductions by always checking (at the end of every multiplication and squaring) whether the number of limbs in the partially computed result exceeds the number of limbs in the modulus. If so, a modular reduction operation is performed. If not, reduction will not decrease the limb count, and thus is not performed. This measurement of size in terms of limbs, as opposed to bits, slightly complicates our attack. Note that due to a recently introduced side-channel mitigation technique (following the work of [YF13]), this code always performs the multiplications, regardless of the bits of d.

3.2 The Attack Algorithm

Our attack is an adaptive chosen-ciphertext attack, which exposes the secret factor q one bit at a time, from MSB to LSB (similarly to Boneh and Brumley's

---

Algorithm 1. GnuPG's modular exponentiation (see function mpi_powm in mpi/mpi-pow.c).

Input: Three integers c, d and q in binary representation such that $d = d_n \cdots d_1$.
Output: $m = c^d \mod q$.

1: **procedure** MODULAR_EXPONENTIATION(c, d, q)
2:     **if** SIZE_IN_LIMBS(c) > SIZE_IN_LIMBS(q) **then**
3:         $c \leftarrow c \mod q$
4:     $m \leftarrow 1$
5:     **for** $i \leftarrow n$ **downto** 1 **do**
6:         $m \leftarrow m^2$ ▷ Karatsuba or grade-school squaring
7:         **if** SIZE_IN_LIMBS(m) > SIZE_IN_LIMBS(q) **then**
8:             $m \leftarrow m \mod q$
9:         $t \leftarrow m \cdot c$ ▷ Karatsuba or grade-school multiplication
10:        **if** SIZE_IN_LIMBS(t) > SIZE_IN_LIMBS(q) **then**
11:            $t \leftarrow t \mod q$
12:            **if** $d_i = 1$ **then**
13:                $m \leftarrow t$
14:        **return** m
15:    **end procedure**

---

timing attack [BB05]). For each bit q_i of q, starting from the most significant bit position ($i = 2048$), we assume that key bits $q_{2048} \cdots q_{i+1}$ were correctly recovered, and check the two hypotheses about q_i. Eventually, we learn all of q and thus recover the factorization of n. Note that after recovering the top half the bits of q, it is possible to use Coppersmith's attack [Cop97] (following Rivest and Shamir [RS85]) to recover the remaining bits, or to continue extracting them using the side channel.[4]

Ciphertext Choice for Modified GnuPG. Let us first consider a modified version of GnuPG's modular exponentiation routine (Algorithm 1), where the size comparisons done in line 2 are removed and line 3 is always executed.

GnuPG always generates RSA keys such that the most significant bit of q is set, i.e., $q_{2048} = 1$. Assume that we have already recovered the topmost $i - 1$ bits of q, and let $g^{i,1}$ be the 2048 bit ciphertext whose topmost $i - 1$ bits are the same as those recovered from q, whose i-th bit is 0, and whose remaining (low) bits are 1. Consider the RSA decryption of $g^{i,1}$. Two cases are possible, depending on q_i.

- $q_i = 1$. Then $g^{i,1} < q$. The ciphertext $g^{i,1}$ is passed as the variable c to Algorithm 1, in which (with the modification introduced earlier) the modular reduction of c in line 3 returns c (since $c = g^{i,1} < q$). Thus, the structure of c (a 2048 bit number whose $i - 1$ lowest bits are set to 1) is preserved, and it is passed to the multiplication routine in line 9.

---

[4] The same technique applies to p. However, on many machines we noticed that the second modular exponentiation (modulo q) exhibits a better signal-to-noise ratio, possibly because the target's power circuitry has by then stabilized.

$-\ q_i\ =\ 0$. Then $q \leq g^{i,1}$. Thus, when $g^{i,1}$ is passed as the variable c to Algorithm 1, the modular reduction of c in line 3 changes the value of c. Since c and q share the same topmost $2048 - i$ bits, the reduction amounts to computing $c \leftarrow c - q$, which is a random-looking number of size $i - 1$ bits. This is then passed to the multiplication routine in line 9.

Thus, depending on q_i, the second operand to the multiplication routine will be either full-size and repetitive or shorter and random-looking. We may hope that the multiplication routine's implementation will behave differently in these two cases, and thus result in key-dependent side-channel leakage. Note that the multiplication is performed 2048 times with that same second operand, which will hopefully amplify the difference and create a distinct leakage pattern that persists for the duration of the exponentiation. As we shall see, there is indeed a difference, which lasts for hundreds of milliseconds and can thus be detected even by very low bandwidth leakage channels such as our acoustic technique.[5]

Ciphertext Choice for Unmodified GnuPG. Unfortunately, line 2 in Algorithm 1 makes its reduction decision on the basis of the limb count of c. This poses a problem for the above attack, since even if $g^{i,1} \geq q$, both $g^{i,1}$ and q have the same number of limbs (64 limbs each). Thus, the branch in line 2 of Algorithm 1 is *never* taken, so c is never reduced modulo q, and the multiplication routine always gets a long and repetitive second operand.

This can be solved in either of two ways. First, GnuPG's binary format parsing algorithm is willing to allocate space for leading zeros. Thus, one may just ask for a decryption of $g^{i,1}$ with additional limbs of leading zeros. GnuPG will pass on this suboptimal representation to the modular exponentiation routine, causing the branch in line 2 of Algorithm 1 to be *always* taken and the reduction to always take place, allowing us to perform the attack.

While the above observation can be easily remedied by changing the parsing algorithm to not allocate leading zero limbs, there is another way (which is harder to fix) to ensure that the branch in line 2 of Algorithm 1 is always taken. Note that the attacker has access to the public RSA modulus $n = pq$, which is 128 limbs (4096 bits) long. Moreover, by definition it holds that $n = 0 \mod q$. Thus, by requesting the decryption of the 128 limb number $g^{i,1} + n$, the attacker can still ensure that the branch in line 2 of Algorithm 1 will be *always* taken and proceed with the attack.

3.3   Acoustic Leakage of the Bits of q

In this section we present empirical results on acoustic leakage of the bits of q using our attack. As argued in Section 3.2, we expect that during the entire modular exponentiation operation using the prime q, the acoustic leakage will depend on the value of the single bit being attacked.

---

[5] Ironically, the latest GnuPG implementations use the side-channel mitigation technique of always multiplying the intermediate results by the input, but this only helps our attack, since it doubles the number of multiplications and replaces their random timing with a repetitive pattern that is easier to record and analyze.

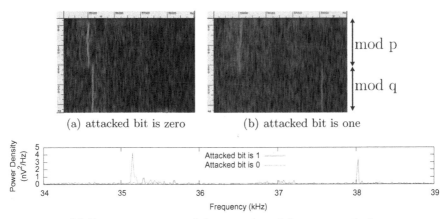

(c) Frequency spectra of the second modular exponentiation

Fig. 4. Acoustic emanations of RSA decryption for various values of the attacked bit ($q_{2039} = 1$ and $q_{2038} = 0$)

Figure 4(a) shows a typical recording of RSA decryption when the value of the attacked bit of q is 0 and Figure 4(b) shows a recording of RSA decryption when the value of the attacked bit of q is 1. Several effects are shown in the figures. Recall that GnuPG first performs modular exponentiation using the secret prime p and then performs another modular exponentiation using the secret prime q. As in figure 3, the transition between p and q is clearly visible in Figures 4(a) and 4(b). Note, then, that the acoustic signatures of the modular exponentiation using the prime q (the second exponentiation) are quite different in Figures 4(a) and 4(b). This is the effect utilized to extract the bits of q.

The spectral signatures in Figure 4(c) were computed from the acoustic signatures of the second modular exponentiation in Figures 4(a) and 4(b). For each signature, we computed the median frequency spectrum (i.e., the median value in each frequency bin over the sliding-window FFT spectra). Again, the differences in the frequency spectra between a 0-valued bit and a 1-valued bit are clearly visible and can be used to extract the bits of q.

Unfortunately, the differences in acoustic leakage between 0-valued bits and 1-valued bits as presented in this section become less prominent as the attack progresses. Thus, in order to extract the entire 2048 bit prime q, additional analysis and improvements to the basic attack algorithm are needed (see the extended version [GST13] for details).

3.4   Overall Attack Performance

We conducted our attack in a variety of measurement setups, on various target machines and software configurations. The attack's success, and its running time (due to repeated measurements and backtracking), depend on many

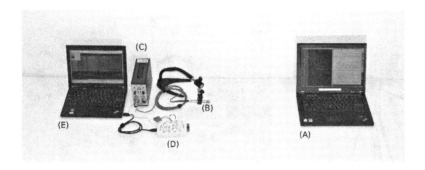

Fig. 5. Photograph of our portable setup. In this photograph (A) is a Lenovo ThinkPad T61 target, (B) is a Brüel&Kjær 4190 microphone capsule mounted on a Brüel&Kjær 2669 preamplifier held by a flexible arm, (C) is a Brüel&Kjær 5935 microphone power supply and amplifier, (D) is a National Instruments MyDAQ device with a 10 kHz RC high-pass filter cascaded with a 150 kHz RC low-pass filter on its A2D input, and (E) is a laptop computer performing the attack. Full key extraction is possible in this configuration, from a distance of 1 meter (see Section 3.4).

physical parameters. These include the machine model and age, the signal acquisition hardware, the microphone positioning, ambient noise, room acoustics, and temperature (affecting fan activity). The following are examples of successful key-extraction experiments.

Ultrasound-Frequency Attack. Extracting the topmost 1024 bits of q (and thereby, the whole key) from GnuPG 1.4.14, running on a Lenovo ThinkPad T61 laptop, in a typical office environment, takes approximately 1 hour. Since this laptop's useful signal is at approximately 35 kHz, there is no need to use the full capabilities of the lab-grade setup (see Section 2), and instead we used a portable (briefcase-sized) setup consisting of a Brüel&Kjær 4190 microphone capsule connected (via a Brüel&Kjær 2669 pre-amplifier) to a Brüel&Kjær 5935 amplifier and microphone power supply, filtered, and digitized using a National Instruments MyDAQ device. Figure 5 depicts this portable setup.

Audible-Frequency Attack. Low frequency sound propagates in the air, and through obstacles, better than high frequency sound. Moreover, larger capsule diaphragms allow better sensitivity but reduce the frequency range. Indeed, some machines, such as the Lenovo ThinkPad X300 and ThinkPad T23, exhibit useful leakage at lower leakage frequency, 15–22 kHz (i.e., within audible range). This allows us to use the very sensitive Brüel&Kjær 4145 microphone capsule and extract the key from some machines at the range of around 1 meter. Moreover, by placing the Brüel&Kjær 4190 microphone in a parabolic reflector, we were able to extract all the bits automatically from the range of 4 meters. (See Figure 6 for a similar setup.) With human-assisted signal processing, we extended the range up to 10 meters.

(a) Parabolic microphone   (b) Distant attack

Fig. 6. (a) Brüel&Kjær 4145 microphone capsule and 2669 preamplifier, attached to a transparent parabolic reflector (56 cm diameter), on a tripod. (b) same, connected to the portable measurement setup, attacking a target laptop from a distance of 4 meters.

Fig. 7. Physical setup of a key recovery attack. A mobile phone (Samsung Note II) is placed 30 cm from a target laptop. The phone's internal microphone points towards the laptop's fan vents. Full key extraction is possible with this configuration and distance.

Mobile-Phone Attack. Lowering the leakage frequency also allows us to use lower quality microphones such as the ones in smartphones. We used several Android phones, with similar results: HTC Sensation, Samsung Galaxy S II and Samsung Galaxy Note II. Recorded using a custom Android app, accessing the internal microphone. Due to the lower signal-to-noise ratio and frequency response of the phone's internal microphone, our attack is limited in frequency (about 24 kHz) and in range (about 30 cm). However, it is still possible to perform it on certain target computers, simply by placing the phone's microphone near to and directed towards the fan exhaust vent of the target while running the attack (see Figure 7). Unlike previous setups, all that is required from the attacker in order to actually mount the attack is to download a suitable application to the phone, and place it appropriately near the target computer.

4   Analyzing the Code of GnuPG RSA

In this section we analyze how our attack affects the code of GnuPG's multiplication routine and causes the differences presented in Section 3.3. We begin

---

Algorithm 2. GnuPG's basic multiplication code (see functions mul_n_basecase and mpihelp_mul in mpi/mpih−mul.c).

Input: Two numbers $a = a_k \cdots a_1$ and $b = b_n \cdots b_1$ of size k and n limbs respectively.
Output: $a \cdot b$.

1: **procedure** MUL_BASECASE(a, b)
2:     **if** $b_1 \leq 1$ **then**
3:         **if** $b_1 = 1$ **then**
4:             $p \leftarrow a$
5:         **else**
6:             $p \leftarrow 0$
7:     **else**
8:         $p \leftarrow$ MUL_BY_SINGLE_LIMB(a, b_1) $\triangleright\ p \leftarrow a \cdot b_1$
9:     **for** $i \leftarrow 2$ **to** n **do**
10:         **if** $b_i \leq 1$ **then**
11:             **if** $b_i = 1$ **then** $\triangleright$ (and if $b_i = 0$ do nothing)
12:                 $p \leftarrow$ ADD_WITH_OFFSET(p, a, i) $\triangleright\ p \leftarrow p + a \cdot 2^{32 \cdot i}$
13:             **else**
14:                 $p \leftarrow$ MUL_AND_ADD_WITH_OFFSET(p, a, b_i, i) $\triangleright\ p \leftarrow p + a \cdot b_i \cdot 2^{32 \cdot i}$
15:     **return** p
16: **end procedure**

---

by describing the multiplication algorithms used by GnuPG (Section 4.1) and then proceed (Sections 4.2 and 4.3) to describe the effects of our attack on the internal values computed during the execution of these algorithms.

4.1 GnuPG's Multiplication Routine

GnuPG's large-integer multiplication routine combines two multiplication algorithms: a basic grade-school multiplication routine, and a variant of a recursive Karatsuba multiplication algorithm [KO62]. The chosen combination of algorithms is based on the size of the operands, measured in whole limbs. Our attack (usually) utilizes the specific implementation of the Karatsuba multiplication routine in order to make an easily-observable connection between the control flow inside the grade-school multiplication routine and the current bit of q. This lets us leak q one bit at a time.

GnuPG's Basic Multiplication Routine. The side-channel weakness we exploit in GnuPG's code resides inside the basic multiplication routines. Both of the basic multiplication routines used by GnuPG are almost identical implementations of the simple, quadratic-complexity grade-school multiplication algorithm, with optimizations for multiplication by limbs equal to 0 or 1 (see Algorithm 2).

Note how MUL_BASECASE handles zero limbs of b. In particular, when a zero limb of b is encountered, none of the operations MUL_BY_SINGLE_LIMB, ADD_WITH_OFFSET and MUL_AND_ADD_WITH_OFFSET are performed and the loop in line 9 continues to the next limb of b. This particular optimization is critical for our attack. Specifically, our chosen ciphertext will cause the private key bit q_i to affect the number of zero limbs of b given to MUL_BASECASE, thus affecting the control flow in lines 3 and 11, and thereby the side-channel emanations.

GnuPG's Karatsuba Multiplication Routine. The basic multiplication routine described above is invoked by both the modular exponentiation routine described in Section 3.1 and by the Karatsuba multiplication routine implementing a variant of the Karatsuba multiplication algorithm with some optimizations. GnuPG's variant of the Karatsuba multiplication algorithm relies on

$$uv = (2^{2n} + 2^n)u_{\mathsf{H}}v_{\mathsf{H}} + 2^n(u_{\mathsf{H}} - u_{\mathsf{L}})(v_{\mathsf{L}} - v_{\mathsf{H}}) + (2^n + 1)v_{\mathsf{L}}u_{\mathsf{L}} \ , \tag{1}$$

where $u_{\mathsf{H}}, v_{\mathsf{H}}$ are the most significant halves of u and v respectively and $u_{\mathsf{L}}, v_{\mathsf{L}}$ are the least significant halves of u and v respectively. Note the subtraction $v_{\mathsf{L}} - v_{\mathsf{H}}$ in Equation 1. Recall that in Section 3 we created a connection between the bits of q and specific values of c. Concretely, for the case where $q_i = 1$, then c is a 2048 bit number such that its first $2048 - i$ bits are the same as q, its i-th bit is zero, and the rest of its bits are ones. Conversely, for the case where the $q_i = 0$, we have that c consists of $i - 1$ random-looking bits.

The code of GnuPG passes c (with some whole-limb truncations) directly to the Karatsuba multiplication routine as the second operand v. Thus, this structure of c has the property that the result of computing $v_{\mathsf{L}} - v_{\mathsf{H}}$ will have almost all of its limbs equal to zero when the current bit of q is 1 and have all of its limbs be random-looking when the current bit of q is 0. Thus, when the recursion eventually reaches its base case, MUL_BASECASE, it will be the case that if the current bit of q is 1, the values of the second operand b supplied to MUL_BASECASE (in some branches of the recursion) will have almost all of its limbs equal to zero, and when the current bit of q is 0, the values of the second operand b supplied to MUL_BASECASE in all branches of the recursion will be random-looking. Next, by (indirectly) measuring the number of operations performed by MUL_BASECASE, we shall be able to deduce the number of zero limbs in c and thus whether the correct bit of q is 0 or 1.

4.2   Attacking the Most Significant Limb of q

In this section we analyze the effects of our attack on MUL_BASECASE (Algorithm 2). Note that in this case, the cipher text c used in the main loop of the modular exponentiation routine (Algorithm 1) always contains at least 2017 bits (64 limbs), meaning that MUL is used for multiplication.

Since in this case both c and m are of the same length and since the constant KARATSUBA_THRESHOLD is defined to be 16, the Karatsuba multiplication routine generates a depth-4 recursion tree where each node has 3 children before using the basic multiplication code (MUL_BASECASE) on 8-limb (256 bit) numbers located on the leaves of the tree. Combining this observation with the case analysis in Section 3.2, we see that for each bit i of q:

- If $q_i = 1$, then the second operand b of MUL mainly consists of limbs having all their bits set to 1.
  Thus, during the first call to the Karatsuba multiplication routine, the result of $v_L - v_H$ contains mostly zero limbs, causing the second operand of all the calls to MUL_BASECASE resulting from the recursive call for computing $(u_H - u_L) \cdot (v_L - v_H)$ to contain mostly zero limbs.
- If $q_i = 0$, then the second operand b of MUL consists of random-looking limbs.
  Thus, during the first call to the Karatsuba multiplication routine, result of $v_L - v_H$ contains very few (if any) zero limbs, causing the second operand of *all* the calls to MUL_BASECASE to consist of mostly non-zero limbs.

Next, recall that the control flow in MUL_BASECASE depends on the number of non-zero limbs present in its second operand. The drastic change in the number of zero limbs in the second is detectable by our side-channel measurements. Thus, we are able to leak the bits of q, one bit at a time, by creating the connection between the current bit of q and the number of zero limbs in the second operand of MUL_BASECASE using our carefully chosen cipher texts.

Finally, note that the Karatsuba multiplication algorithm is (indirectly) called during the main loop of the modular exponentiation routine (Algorithm 1) once per bit of d_q as computed by the RSA decryption operation. Since d_q is a 2048 bit number, we get that the leakage generated by line 11 in MUL_BASECASE (Algorithm 2) is repeated once for every zero limb of b for a total of 7 times during an execution of MUL_BASECASE, which is in turn repeated once for every leaf resulting from the first computation of $(u_H - u_L) \cdot (v_L - v_H)$ for a total of 9 times in each of the 2048 multiplications in that loop. Thus, we get that leakage generated by line 11 in MUL_BASECASE is repeated $2048 \cdot 9 \cdot 7 = 129024$ times. This repetition is what allows the leakage generated by line 11 in MUL_BASECASE to be detected using only low bandwidth measurements.

4.3   The Remaining Bits of q

Unfortunately, the analysis in Section 4.2 does not precisely hold for the remaining bits of q. Recall that by our choice of ciphertexts, at the beginning of MODULAR_EXPONENTIATION both $c - n$ and q always agree on some prefix of their bits and this prefix becomes longer as the attack progresses and more bits of q are extracted. Thus, since $c - n < 2q$, the reduction of c modulo q always returns $c - q - n$ for the case of $c - n \geq q$ and $c - n$ otherwise.

In particular, after the first limb of q has been successfully extracted, for the case where $q_i = 0$, the value of c after the modular reduction in line 2 of MODULAR_EXPONENTIATION (Algorithm 1) is shorter than 64 limbs. Since in line 9, while m remains a 64 limb number, the part of the multiplication routine responsible for handling operands of different sizes is used. Thus, instead of a single call to the Karatsuba multiplication routine, might make several recursive calls to itself as well as several calls to the Karatsuba multiplication routine.

Nonetheless, there is still a connection between the secret bits of q and the structure of the value of c passed to multiplication routine by the modular exponentiation routine (Algorithm 1), as follows. For any $1 \leq i \leq 2048$, one of the following two cases holds. If $q_i = 1$, then c is a 2048 bit number such that the first $2048 - i$ bits are the same as q, the i-th bit is zero, and the rest of the bits are ones. If $q_i = 0$, then c consists of $i - 1$ random-looking bits. While the analysis in this case is not as precise as in Section 4.2, the number of zero limbs in the second operand of MUL_BASECASE still allows us to extract the bits of q.

The acoustic distinguishability of the two cases does vary with bit index, and in particular is harder for the range $q_{1850}, \ldots, q_{1750}$. Using additional techniques (discussed in the extended version of this paper [GST13]), we recover these problematic bits and continue our attack.

5   Conclusion

In this paper we developed a new side channel attack, exploiting low-bandwidth computation-dependent acoustic emanations which easily escape computers' chassis (and even expensive Faraday cages). We demonstrated extraction of full RSA keys within a reasonable amount of time using commonly available and easily concealed components. Some algorithmic countermeasures, such as ciphertext normalization and randomization, are effective against our key-extraction attack (and thus implemented in GnuPG consequentially to our results' disclosure), though not against acoustic key-distinguishing, and are discussed in the extended version of this paper [GST13].

Acknowledgments.   Lev Pachmanov wrote much of the software setup used in our experiments, including custom signal acquisition programs. Avi Shtibel, Ezra Shaked and Oded Smikt assisted in constructing and configuring the experimental setup. Assa Naveh assisted in various experiments, and offered valuable suggestions. Sharon Kessler provided copious editorial advice. We thank Werner Koch, lead developer of GnuPG, for the prompt response to our disclosure and the productive collaboration in adding suitable countermeasures. We are indebted to Pankaj Rohatgi for inspiring the origin of this research; to Nir Yaniv for audio recording advice and use of the Nir Space Station studio; and to National Instruments Israel for donating PCI-6052E and MyDAQ hardware.

This work was sponsored by the Check Point Institute for Information Security; by European Union's Tenth Framework Programme (FP10/2010-2016) under grant agreement 259426 ERC-CaC, by the the Leona M. & Harry B. Helmsley Charitable Trust; by the Israeli Ministry of Science and Technology; by the Israeli Centers of Research Excellence I-CORE program (center 4/11); and by NATO's Public Diplomacy Division in the Framework of "Science for Peace".

References

[AA04]   Asonov, D., Agrawal, R.: Keyboard acoustic emanations. In: IEEE Symposium on Security and Privacy, pp. 3–11 (2004)

[And08]   Anderson, R.J.: Security engineering — a guide to building dependable distributed systems, 2nd edn. Wiley (2008)

[BB05]   Brumley, D., Boneh, D.: Remote timing attacks are practical. Computer
            Networks 48(5), 701–716 (2005)
[BBB+12]   Barker, E., Barker, W., Burr, W., Polk, W., Smid, M.: NIST SP 800-57:
            Recommendation for key management — part 1: General (2012)
[BDG+10]   Backes, M., Dürmuth, M., Gerling, S., Pinkal, M., Sporleder, C.: Acous-
            tic side-channel attacks on printers. In: USENIX Security Symposium,
            pp. 307–322 (2010)
[BWY06]   Berger, Y., Wool, A., Yeredor, A.: Dictionary attacks using keyboard acous-
            tic emanations. In: ACM Conference on Computer and Communications
            Security, pp. 245–254 (2006)
[Cop97]   Coppersmith, D.: Small solutions to polynomial equations, and low expo-
            nent RSA vulnerabilities. J. Cryptology 10(4), 233–260 (1997)
[Eni]   The Enigmail Project. Enigmail: A simple interface for OpenPGP email
            security
[Gen]   Genesis 27:5
[Gmp]   GNU multiple precision arithmetic library
[Gpg]   The GNU Privacy Guard
[GST13]   Genkin, D., Shamir, A., Tromer, E.: RSA key extraction via low-
            bandwidth acoustic cryptanalysis (extended version). IACR Cryptology
            ePrint Archive, 2013:857 (2013)
[HS10]   Halevi, T., Saxena, N.: On pairing constrained wireless devices based on
            secrecy of auxiliary channels: the case of acoustic eavesdropping. In: ACM
            Conference on Computer and Communications Security, pp. 97–108 (2010)
[KJJR11]   Kocher, P., Jaffe, J., Jun, B., Rohatgi, P.: Introduction to differential power
            analysis. Journal of Cryptographic Engineering 1(1), 5–27 (2011)
[KO62]   Karatsuba, A., Ofman, Y.: Multiplication of Many-Digital Numbers by Au-
            tomatic Computers. Proceedings of the USSR Academy of Sciences 145,
            293–294 (1962)
[Nat82]   National Security Agency. NACSIM 5000: TEMPEST fundamentals
            (February 1982)
[Nat09]   National Institute of Standards and Technology. FIPS 140-3: Draft security
            requirements for cryptographic modules, revised draft (2009)
[RS85]   Rivest, R.L., Shamir, A.: Efficient factoring based on partial informa-
            tion. In: Pichler, F. (ed.) EUROCRYPT 1985. LNCS, vol. 219, pp. 31–34.
            Springer, Heidelberg (1986)
[ST04]   Shamir, A., Tromer, E.: Acoustic cryptanalysis: on nosy people and noisy
            machines. Eurocrypt rump session (2004)
[SWT01]   Song, D.X., Wagner, D., Tian, X.: Timing analysis of keystrokes and timing
            attacks on SSH. In: USENIX Security Symposium, vol. (2001)
[Wri87]   Wright, P.: Spycatcher. Viking Penguin (1987)
[YF13]   Yarom, Y., Falkner, K.E.: Flush+reload: a high resolution, low noise,
            L3 cache side-channel attack. IACR Cryptology ePrint Archive, 2013:448
            (2013)
[ZZT05]   Zhuang, L., Zhou, F., Tygar, J.D.: Keyboard acoustic emanations revis-
            ited. In: ACM Conference on Computer and Communications Security,
            pp. 373–382 (2005)

On the Impossibility of Cryptography
with Tamperable Randomness

Per Austrin[1,*], Kai-Min Chung[2,**], Mohammad Mahmoody[3,***],
Rafael Pass[4,†], and Karn Seth[4]

[1] KTH Royal Institute of Technology, Stockholm, Sweden
austrin@kth.se
[2] Academica Sinica, Taipei, Taiwan
kmchung@iis.sinica.edu.tw
[3] University of Virginia, Charlottesville, VA, USA
mohammad@cs.virginia.edu
[4] Cornell University, New York, NY, USA
{rafael,karn}@cs.cornell.edu

Abstract. We initiate a study of the security of cryptographic primitives in the presence of efficient tampering attacks to the randomness of honest parties. More precisely, we consider p-tampering attackers that may *efficiently* tamper with each bit of the honest parties' random tape with probability p, but have to do so in an "online" fashion. Our main result is a strong negative result: We show that any secure encryption scheme, bit commitment scheme, or zero-knowledge protocol can be "broken" with probability p by a p-tampering attacker. The core of this result is a new Fourier analytic technique for biasing the output of bounded-value functions, which may be of independent interest.

We also show that this result cannot be extended to primitives such as signature schemes and identification protocols: assuming the existence of one-way functions, such primitives can be made resilient to $(1/\mathrm{poly}(n))$-tampering attacks where n is the security parameter.[1]

Keywords: Tampering, Randomness, Encryption.

---

[*] Work done while at University of Toronto, funded by NSERC
[**] Supported in part by NSF Award CNS-1217821.
[***] Research done while in part supported by NSF Award CNS-1217821.
[†] Pass is supported in part by a Alfred P. Sloan Fellowship, Microsoft New Faculty Fellowship, NSF Award CNS-1217821, NSF CAREER Award CCF-0746990, NSF Award CCF-1214844, AFOSR YIP Award FA9550-10-1-0093, and DARPA and AFRL under contract FA8750-11-2- 0211. The views and conclusions contained in this document are those of the authors and should not be interpreted as representing the official policies, either expressed or implied, of the Defense Advanced Research Projects Agency or the US Government.
[1] The full version of the paper is available at [4].

1   Introduction

A traditional assumption in cryptography is that the only way for an attacker to gather or control information is by receiving and sending messages to honest parties. In particular, it is assumed that the attacker may not access the *internal states* of honest parties. However, such assumptions on the attacker—which we refer to as *physical assumptions*—are quite strong (and even unrealistic). In real-life, an attacker may through a "physical attack" learn some "leakage" about the honest parties' internal states and may even tamper with their internal states. For instance, a computer virus may (e.g., using a, so-called, buffer overflow attack [2, 18, 32]) be able to bias the randomness of an infected computer. Understanding to what extents the traditional physical assumptions can be relaxed, to capture such realistic attacks, is of fundamental importance.

Indeed, in recent years *leakage-resilient cryptography*—that is, the design of cryptographic schemes and protocols that remain secure even when the attacker may receive (partial) leakage about the internal state of the honest parties—has received significant attention (see e.g., [1, 7, 10, 12–14, 21, 25, 30, 31, 34]).

In this work, we focus on understanding the power of *tampering attacks*—that is, attacks where the adversary may partially modify (i.e., tamper with) the internal state of honest parties. Early results in the 1990's already demonstrate that tampering attacks may be very powerful: by just slightly tampering with the computation of specific implementations of some cryptographic schemes (e.g., natural implementations of RSA encryption [33]), Boneh, DeMillo and Lipton [6] demonstrated that the security of these schemes can be broken completely.

Previous works on tamper-resilient cryptography consider tampering of computation [3,5,6,16,24,30] and tampering with the memory of an honest party who holds a secret (e.g., a signing or a decryption algorithm) [9,15,19,26,29,30]. This line of research culminated in strong *compilers* turning any polynomial-size circuit C into a new "tamper-resilient" polynomial-size circuit C'; tamper-resilience here means that having "grey-box" access to C' (i.e., having black-box access while tampering with the computation of C') yields no more "knowledge" than simply having black-box access to C. These works, thus, show how to keep a secret hidden from a tampering attacker. Our focus here is somewhat different. In analogy with recent work of leakage-resilient security, we aim to investigate to what extent a tampering attacker may violate the *security* of a cryptographic protocol by tampering with the internal state of honest parities.

For concreteness, let us focus on the security of public-key encryption schemes (but as we shall see shortly, our investigation applies to many more cryptographic tasks such as zero-knowledge proofs and secure computation). Roughly speaking, we require a tamper-resilient encryption scheme to guarantee that ciphertexts conceal the encrypted messages, even if the internal computation of the sender (of the ciphertext) has been tampered with.[2] As first observation note that if the attacker may completely change the computation of the sender, he could simply make the sender send the message in the clear. Thus, to hope for any reasonable notion of tamper-resilient security we need to restrict the attacker's ability to tamper with the computation.

Tampering with Randomness. Among various computational resources, randomness might be one of the hardest to protect against tampering. This is due to the fact that randomness is usually *generated* (perhaps based on some "physical" resources available to the system) and any malicious attacker who is able to change the bits along their generation can mount a tampering attack against the randomness. Indeed given the breakthrough results of [22, 27, 28] it is becoming even more clear that randomness is one of the most vulnerable aspects of a cryptographic system. Thus, a very basic question is to what extent we can protect our systems against tampering with randomness. In this work we initiate a formal study of this question by considering tampering attacks to the randomness of the honest players; namely we study the following basic question:

Can security of cryptographic primitives be preserved under tampering attacks to the randomness of honest parties?

Note that we need to restrict the tampering ability of the attacker, for otherwise the adversary can effectively make the scheme deterministic by always fixing the randomness of the honest parties to all zeros. But it is well-known that deterministic encryption schemes cannot be semantically secure. Therefore, here we initiate study of the power of *weak* types of tampering attacks to the randomness of the honest parties.

General Model: The Tampering Virus. We envision the adversary as consisting of two separate entities: (1) a *classical attacker* who interacts with the honest parties only by sending/receiving messages to/from them (without any side-channels), and (2) a *tampering circuit* (a.k.a. the "virus") who observes the internal state of the honest parties and may *only* tamper with their randomness (but may not communicate with the outside world, and in particular with the classical attacker). The tampering circuit only gets to tamper with a *small fraction* of the random bits, and in an *efficient* manner. Note that this model

---

[2] Let us remark that the simulation property of tamper-resilient compilers do not necessarily guarantee that if the sender algorithm is compiled into a "tamper-resilient" version, then the encryption scheme is tamper-resilient. This is due to the fact that the simulation property of those compilers only guarantee that an attacker cannot learn more from tampering with the sender strategy than it could have with black-box access to it. But in the case of encryption schemes, it is actually the *input* to the algorithm (i.e., the message to be encrypted) that we wish to hide (as opposed to some secret held by the algorithm). See the full version for a more detailed comparison with previous work.

excludes a scenario in which the virus (even efficiently) samples the *whole* randomness, regardless of the original randomness sampled by the system, because in this cases all of the sampled tampered bits might be different from the system's original random seed. However, here we study weak attackers who only tamper with a *small fraction* of the random bits. In fact, previous works on *resettable cryptography* [8] can be interpreted as achieving tamper resilience against adversaries who tamper with *all* of the randomness of the honest parties by *resampling* the randomness of the honest parties and executing them again and again. This is incomparable to our model, since our adversary does not have control over the honest parties' execution (to run them again), but is more powerful in that it could change the value of some of the random bits.

Online Tampering. Let $0 < p < 1$ be the parameter describing the "power" of adversary (which defines the fraction of tampered bits). It still remains to clarify how the tampering is done over these bits. The first naive model would allow the adversary to tamper with a p fraction of the bits *after* all the bits are sampled by the system (and, thus, are known to the virus as well). However, this is not realistic since the sequence of random bits used by the system could always be sampled in an *online* manner; namely, the system could sample the i-th random bit whenever it needs it and might use it "on the fly". Therefore, a tampering adversary also needs to tamper with them one-by-one, efficiently, and in an on-line fashion.

More precisely, we consider a so-called p-*tampering attack*, where the adversary gets to tamper with the random tape of the honest players as follows. The randomness of honest parties is generated bit-by-bit, and for each generated bit x_i the efficient tampering circuit gets to tamper with it with independent probability p having only knowledge of previously generated random bits $x_1, x_2, \ldots, x_{i-1}$ (but not the value of the random bits tossed in the future). Roughly speaking, requiring security with respect to p-tampering attacks amounts to requiring that security holds even if the honest players' randomness comes from a *computationally efficient* analog of a Santha-Vazirani (SV) source [35]. Recall that a random variable $X = (X_1, \ldots, X_n)$ over bit strings is an SV source with parameter δ if for every $i \in [n]$ and every $(x_1, \ldots, x_i) \in \{0,1\}^i$, it holds that $\delta \leq \mathsf{P}[X_i = x_i | X_1 = x_1, \ldots, X_{i-1} = x_{i-1}] \leq 1 - \delta$. It is easy to see that the random variable resulting from performing a p-tampering attack on a uniform n-bit string is an SV source with parameter $(1 - p)/2$; in fact, any SV source is equivalent to performing a *computationally unbounded* p-tampering attack on a uniform n-bit string.

The main focus of this work is on the following question:

Can security be achieved under p-tampering attacks?

2   Our Results and Techniques

Our main result is a strong negative answer to the question above for a variety of basic cryptographic primitives. A p-tampering attacker can break all of the following with advantage $\Omega(p)$: (**1**) the security of any CPA-secure (public-key or private-key) encryption scheme, (**2**) the zero-knowledge property of any efficient-prover proof (or argument) system for nontrivial languages, (**3**) the hiding property of any commitment scheme, and (**4**) the security of any protocol for computing a "nontrivial" finite function. More formally, we prove the following theorems.

Theorem 1 (Impossibility of Encryption). *Let Π be any CPA-secure public-key encryption scheme. Then a p-tampering attacker can break the security of Π with advantage $\Omega(p)$. Moreover, the attacker only tampers with the random bits of the* encryption *(not the key-generation) without* knowing the message.

A similar impossibility result holds for private-key encryption schemes in which the tampering adversary can also tamper with the randomness of the key-generation phase.[3]

Theorem 2 (Impossibility of Zero-Knowledge). *Let (P, V) be an efficient prover proof/argument system for a language $L \in$ NP such that the view of any p-tampering verifier can be simulated by an efficient simulator with indistinguishability gap $o(p)$, then the language L is in BPP.*

Theorem 3 (Impossibility of Commitments). *Let (S, R) be a bit-commitment scheme. Then, either an efficient malicious sender can break the biding with advantage $\Omega(p)$ (without tampering), or an efficient malicious p-tampering receiver can break the hiding with advantage $\Omega(p)$.*

Following [19] we consider two-party functions $f \colon \mathcal{D}_1 \times \mathcal{D}_2 \mapsto \mathcal{R}$ where only one player gets the output. A function f is called trivial in this context, if there is a deterministic single-message (i.e., only one player speaks) protocol for computing f that is information theoretically secure.

Theorem 4 (Impossibility of Secure Computation). *The security of any protocol for computing a two-party non-trivial function can be broken with advantage $\Omega(p)$ through a p-tampering attack.*

Relation to Subliminal Channels. Cryptographic research on "subliminal channels" [36] and the related field of "kleptography" [37] study whether a cryptographic scheme can be "misused" for a purpose other than the original purpose it was designed for (e.g., by putting an undetectable trapdoor in the systems). The existence of subliminal channels between "outside" and "inside" adversaries could be a huge security concern in certain scenarios such as voting schemes [17]. Our Theorem 1 (and its private-key variant) show that *any* efficient encryption scheme always has a subliminal channel between an outsider adversary and an insider virus who is (only) able to tamper with the randomness of the encryption device no matter how the encryption algorithm tries to "detect" a virus who is signaling a bit of information to the adversary.

---

[3] Note that this is necessary, because the one-time pad encryption is deterministic during its encryption phase.

Tampering with Randomness vs. Imperfect Randomness. Our negative results are closely related to the impossibility result of Dodis et al. [11] on the "impossibility of cryptography with imperfect randomness", where the security of cryptographic primitives are analyzed assuming that the honest parties only have access to randomness coming from an SV source (as opposed to the randomness being perfectly uniform). [11] present several strong impossibility results for secure realizability of cryptography primitives in a setting where players only have access to such imperfect randomness. The SV sources they consider for their impossibility results, however, may not be efficiently computable.

The key-difference between tamper-resilient security in our setting and security with imperfect randomness is that we restrict to randomness sources that are efficiently sampled through an (online) p-tampering attack; thus achieving tamper-resilient security becomes easier than resilience to imperfect randomness. Note that even if one can *efficiently* sample from the sources employed by [11], that still does not solve our main question, because by sampling fresh randomness for the system the adversary is indeed tampering with *all* of the random seed. As we discussed above, however, in such scenario the adversary can always fix the randomness to zero and so we are essentially down to the deterministic case. Another, perhaps less important difference, is that for primitives with simulation-based security, we allow the simulator to depend on the p-tampering attacker, whereas in [11] the simulator must work for any randomness source; this further makes achieving tamper-resilient security easier than resilience to imperfect randomness.

Positive Results. We complement the above negative results by demonstrating some initial positive results: Assuming the existence of one-way functions, for any $p = n^{-\alpha}$, where $\alpha > 0$ is a constant and n is the security parameter, every implementation of signature schemes, identification protocols, and witness hiding protocols can be made resilient against p-tampering attackers. We also present a relaxed notion of semantic security for encryption schemes that can be achieved under $n^{-\alpha}$-tampering attacks. We show that for these primitives, security holds even if the randomness source "min-entropy loss" of at most $O(\log n)$. We next show how to use PRGs to ensure that a tampering attacker will only be able to decrease the overall (pseudo) min-entropy by $O(\log n)$. The above mentioned primitives already imply the existence of one-way functions [23], thus preventing against $n^{-\alpha}$-tampering attacks can be achieved for these primitives unconditionally. Finally, we present positive results for tamper-resilient key-agreement and secure multi-party computation in the presence of (at least) two honest players. For further details see the full version [4] of the paper.

2.1   Our Techniques

Our main technical contribution is to develop new methods for biasing Boolean, and more generally, bounded-value functions, using a p-tampering attack.

Biasing Bounded-Value Functions. Our first (negative) result uses elementary Fourier analysis to prove an efficient version of the Santha-Vazirani theorem: Any balanced (or almost balanced) efficiently computable Boolean function f can be biased by $\Omega(p)$ through an efficient p-tampering attack.

Specifically, let U_n denote the uniform distribution over $\{0,1\}^n$ and let $U_n^{\mathsf{Tam},p}$ denote the distribution obtained after performing a p-tampering attack on U_n using a tampering algorithm Tam; more precisely, let $U_n^{\mathsf{Tam},p} = (X_1, \ldots, X_n)$ where with probability $1 - p$, X_i is a uniform random bit, and with probability p, $X_i = \mathsf{Tam}(1^n, X_1, \ldots, X_{i-1})$.

Theorem 5 (Biasing Boolean Functions: Warm-up). *There exists an oracle machine* Tam *with input parameters n and $\varepsilon < 1$ that runs in time* $\mathrm{poly}(n/\varepsilon)$ *and for every $n \in N$ and $\varepsilon \in (0,1)$, every Boolean function $f : \{0,1\}^n \to \{-1,1\}$, and every $p < 1$, for $\mu = \mathsf{E}[f(U_n)]$ it holds that*

$$\mathsf{E}[f(U_n^{\mathsf{Tam}^f,p})] \geq \mu + p \cdot (1 - |\mu| - \varepsilon).$$

The tampering algorithm Tam is extremely simple and natural; it just greedily picks the bit that maximizes the bias at every step. More precisely, $\mathsf{Tam}^f(x_1, \ldots, x_{i-1})$ estimates the value of

$$\mathsf{E}_{U_{n-i}}[f(x_1, \ldots, x_{i-1}, b, U_{n-i})]$$

for both of $b = 0$ and $b = 1$ by sampling, and sets x_i to the bit b with larger estimated expectation.

Theorem 5 suffices for our impossibility result for tamper-resilient zero-knowledge. For all our remaining impossibility results, however, we need a more general version that also deals with bounded value functions $f : \{0,1\}^n \to [-1,1]$. Our main technical theorem provides such a result.

Theorem 6 (Main Technical Theorem: Biasing Bounded-Value Functions). *There exists an efficient oracle machine* Tam *such that for every $n \in N$, every bounded-value function $f : \{0,1\}^n \to [-1,1]$, and every $p < 1$,*

$$\mathsf{E}[f(U_n^{\mathsf{Tam}^f,p})] \geq \mathsf{E}[f(U_n)] + \frac{p \cdot \mathrm{Var}[f(U_n)]}{5}.$$

Note that in Theorem 6 the dependence on the variance of f is necessary because f may be the constant function $f(x) = 0$, whereas for the case of balanced Boolean functions this clearly cannot happen. Let us also point out that we have not tried to optimize the constant $1/5$, and indeed it seems that a more careful analysis could be used to bring it down since for small p the constant gets close to 1.

The greedy algorithm does not work in the non-Boolean case anymore. The problem, roughly speaking, is that a greedy strategy will locally try to increase the expectation, but that might lead to choosing a wrong path. As a "counterexample" consider a function f such that: conditioned on $x_1 = 0$ f is a constant function ε, but conditioned on $x_1 = 1$, f is a Boolean function with average

$-\varepsilon$. For such f, the greedy algorithm will set $x_1 = 0$ and achieves bias at most ε, while by choosing $x_1 = 1$ more bias could be achieved. To circumvent this problem we use a "mildly greedy" strategy: we take only one sample of $f(\cdot)$ by choosing $x'_i, x'_{i+1}, \ldots, x'_n$ at random ($x_1, \ldots, x_{i-1}$ are already fixed). Then, we keep the sampled x'_i with probability proportional to how much the output of f is close to our "desired value", and flip the value of x'_i otherwise.

More precisely, $\mathsf{Tam}(1^n, x_1, \ldots, x_{i-1})$ proceeds as follows:

- Samples $(x'_i, x'_{i+1}, \ldots, x'_n) \leftarrow U_{n-i+1}$ and compute
  $y = f(x_1, \ldots, x_{i-1}, x'_i, \ldots, x'_n)$.
- Sample $\mathsf{Tam}(1^n, x_1, \ldots, x_{i-1})$ from a Boolean random variable with average $y \cdot x'_i$ (i.e. output x'_i with probability $\frac{1+y}{2}$, and $-x'_i$ with probability $\frac{1-y}{2}$).

Note that our mildly greedy strategy is even easier to implement than the greedy one: to tamper with each bit, it queries f *only once*.

Impossibility Results for Tamper-Resilient Cryptography. We employ the biasing algorithms of Theorems 5 and 6 to obtain our negative results using the following blue-print: We first prove a *computational version* of the "splitting lemma" of [11] (Lemma 7 below which follows from Corollary 3.2 in [11]). Then we will use the same arguments as those of [11] to derive our impossibility results.

Lemma 7 ([11]). *Let f_0 and f_1 be two efficient functions from $\{0,1\}^m$ to $\{0,1\}^{\mathrm{poly}(m)}$ such that $\mathrm{Pr}_{x \leftarrow U_m}[f_0(x) \neq f_1(x)] \geq 1/\mathrm{poly}(n)$. Then there an Santha-Vazirani source of randomness X with parameter $1/2 - 1/\mathrm{poly}(n)$ such that $f_0(X)$ is computationally distinguishable from $f_1(X)$.*

We use our Theorem 6 to prove the following computational version of Lemma 7 which allows one to distinguish the functions f_0, f_1 by tampering with their random input.

Lemma 8 (Computational Splitting Lemma). *Let f_0 and f_1 be two efficient functions from $\{0,1\}^m$ to $\{0,1\}^{\mathrm{poly}(m)}$ and $\mathrm{Pr}_{x \leftarrow U_m}[f_0(x) \neq f_1(x)] \geq \varepsilon > 1/\mathrm{poly}(m)$. Then one can efficiently find a $\mathrm{poly}(m)$-size function f and a $\mathrm{poly}(m)$-size tampering circuit Tam such that*

$$\mathrm{Pr}[f(f_1(U_n^{\mathsf{Tam},p})) = 1] \geq \mathrm{Pr}[f(f_0(U_n^{\mathsf{Tam},p})) = 1] + \Omega(\varepsilon \cdot p).$$

Proof Outline. We derive Lemma 8 from Theorem 6 as follows. We use Theorem 6 to bias the *difference* function $g_f(x) = f(f_1(x)) - f(f_0(x))$ (with domain $\{-1, 0, +1\}$) towards 1 by a tampering circuit Tam. It is easy to see that if f is Boolean, doing this is equivalent to the goal of Lemma 8. We show that if one samples f from a family of pairwise independent Boolean functions, then the resulting function $g_f(\cdot)$ has sufficient variance as needed by Theorem 6.

We use our Lemma 8 similar to the way Lemma 7 is employed in [11] to derive our impossibility results for tamper resilient: encryption schemes, commitments, and two-party secure function evaluation protocols. For all these primitives an adversary uses Lemma 8 to generate a tampering circuit Tam that later on lets him distinguish the corresponding challenges (generated using the tampered randomness) and break the security.

Zero-Knowledge. Zero-knowledge proofs in the setting of [11] require a *universal* simulator that simultaneously handles a large class of imperfect randomness sources. We can also use our Lemma 8 to rule out such tamper-resilient zero-knowledge proofs. In the computational setting, however, it is the malicious verifier who generates the bad source of randomness, and so we shall allow the simulator to depend on the tampering circuit as well. Thus, the simulator in our setting has more power. This prevents us from applying Lemma 8 directly.

We proceed in using the following high level outline. In the first step, we generalize a result by Goldreich and Oren [20] showing that non-trivial zero-knowledge protocols cannot have deterministic provers. Our generalization to [20] shows that non-trivial zero-knowledge protocols require having prover messages with min-entropy $\omega(\log n)$. This means that the verifier can apply a (seeded) randomness extractor to the transcript and obtain one almost unbiased bit. In a second step, we show how to use (the proof of) Theorem 5 to tamper with the prover's randomness so as to signal bits of the witness to the verifier.

This outline, however, oversimplifies: is it not the case that every non-trivial zero-knowledge protocol requires the prover messages to have min-entropy $\omega(\log n)$; in fact, for some "easy" instances, the prover may not communicate at all. Rather, we demonstrate that an "instance-based" version of the min-entropy extension of the Goldreich and Oren [20] theorem holds, and using it we can prove that *either* the prover's messages have high min-entropy (and thus the witness can be leaked to the verifier), or the instance can be decided "trivially". It follows that in either case, we can correctly decide the instance and thus the language must be trivial.

3   Biasing Functions via Online Tampering

In this section we study how much the output of a bounded function can be biased through a tampering attack, and we will formally prove Theorem 5 and Theorem 6. For the full proofs of the applications of these two theorems (sketched in previous section) we refer the reader to the full version of the paper [4].

First we formally define an online tampering process and a tampering source of randomness (as a result of an online tampering attack performed on a uniform source of randomness).

Definition 9. *A distribution $X = (X_1, \ldots, X_n)$ over $\{-1, 1\}^n$ is an (efficient) p-tampering source if there exists an (efficient) tampering algorithm Tam such that X can be generated in an online fashion as follows: For $i = 1, \ldots, n$,*

$$X_i = \begin{cases} \mathsf{Tam}(1^n, X_1, \ldots, X_{i-1}) & \text{with probability } p, \\ U_1^i & \text{with probability } 1 - p, \end{cases}$$

where U_1^i denotes a uniformly random bit over $\{-1, 1\}$. In other words, with probability p, Tam gets to tamper the next bit with the knowledge of the previous bits (after the tampering)[4]. The tampering algorithm Tam might also receive an auxiliary input and use it in its tampering strategy.[5] We use $U_n^{\mathsf{Tam},p}$ to denote the p-tampered source obtained by the above tampering process with tampering algorithm Tam.

Note that in the definition above, the tampering algorithm Tam might be completely oblivious to the parameter p. By referring to Tam as a p-tampering algorithm, we emphasize on the fact that Tam's algorithm might depend on p.

Remark 10. Every p-tampering source is also a Santha-Vazirani source [35] with parameter $\delta = (1-p)/2$. In fact, it is not hard to see that without the efficiency consideration, the two notions are equivalent.

3.1 Preliminaries: Calculating the Effect of a Single Variable

Recall that the Fourier coefficients of any function $f : \{-1,1\}^n \to [-1,1]$ are indexed by the subsets S of $[n]$ and are defined as $\hat{f}(S) := \mathsf{E}_x[f(x)\chi_S(x)]$, where $\chi_S(x) := \prod_{i \in S} x_i$. Note that the Fourier coefficient of the empty set $\hat{f}(\emptyset)$ is simply the expectation $\mathsf{E}[f(U_n)]$.

For every prefix $x_{\leq i} = (x_1, \ldots, x_i)$, let $f_{x_{\leq i}} : \{-1,1\}^{n-i} \to [-1,1]$ be the restriction of f on $x_{\leq i}$, i.e., $f_{x_{\leq i}}(x_{i+1}, \ldots, x_n) := f(x_1, \ldots, x_n)$. We note that the variables of $f_{x_{\leq i}}$ are named $(x_{i+1}, \ldots, x_n)$ and thus the Fourier coefficients of $f_{x_{\leq i}}$ are $\hat{f}_{x_{\leq i}}(S)$'s with $S \subseteq \{i+1, \ldots, n\}$. The following basic identity can be proved by straightforward calculation.

$$\hat{f}_{x_1}(\emptyset) = \hat{f}(\emptyset) + \hat{f}(\{1\}) \cdot x_1. \tag{1}$$

Recall that $\hat{f}(\emptyset)$ and $\hat{f}_{x_1}(\emptyset)$ are simply expectations. One interpretation of the above identity is that $\pm\hat{f}(\{1\})$ is the change of expectation when we set $x_1 = \pm 1$. This is thus useful for analyzing the bias introduced as the result of a tampering attack.

Using the above identity with a simple induction, we can express $f(x)$ as a sum of Fourier coefficients of restrictions of f. Namely, $f(x)$ equals to the expectation $\hat{f}(\emptyset)$ plus the changes in expectation when we set x_i bit by bit.

Lemma 11. $f(x) = \hat{f}(\emptyset) + \sum_{i=1}^{n} \hat{f}_{x_{\leq i-1}}(\{i\}) \cdot x_i$ for every $x \in \{-1,1\}^n$.

---

[4] In a stronger variant of tampering attacks, the attacker might be completely stateful and memorize the original values of the previous bits before and after tampering and also the places where the tampering took place, and use this extra information in its future tampering. Using the weaker stateless attacker of Definition 9 only makes our negative results stronger. Our *positive* results hold even against stateful attackers.

[5] The auxiliary input could, e.g., be the information that the tampering algorithm receives about the secret state of the tampered party; this information might not be available at the time the tampering circuit is generated by the adversary.

Proof. By expanding $\hat{f}_{x_{\leq j}}(\emptyset) = \hat{f}_{x_{\leq j-1}}(\emptyset) + \hat{f}_{x_{\leq j-1}}(\{j\}) \cdot x_j$, (implied by Equation (1)) and a simple induction on j it follows that:

$$f(x) = \hat{f}_{x_{\leq j}}(\emptyset) + \sum_{i=j+1}^{n} \hat{f}_{x_{\leq i-1}}(\{i\}) \cdot x_i,$$

which proves the lemma.

As a corollary, the above lemma implies that the sum of Fourier coefficients (of restrictions of f) in absolute value is at least $|f(x)| - |\hat{f}(\emptyset)|$.

Corollary 12. *For every $x \in \{-1,1\}^n$, it holds that $\sum_{i=1}^{n} \left| \hat{f}_{x_{\leq i-1}}(\{i\}) \right| \geq |f(x)| - |\hat{f}(\emptyset)|$.*

Proof. By triangle inequality we have

$$\sum_{i=1}^{n} \left| \hat{f}_{x_{\leq i-1}}(\{i\}) \right| = \sum_{i=1}^{n} \left| \hat{f}_{x_{\leq i-1}}(\{i\}) \cdot x_i \right| \geq \left| \sum_{i=1}^{n} \hat{f}_{x_{\leq i-1}}(\{i\}) \cdot x_i \right|$$
$$= |f(x) - \hat{f}(\emptyset)| \geq |f(x)| - |\hat{f}(\emptyset)|$$

where the second equality uses Lemma 11.

3.2   The Boolean Case

A seminal result by Santha and Vazirani [35] shows that for every balanced Boolean function f (e.g., a candidate "extractor"), there exists a p-tampering source X that biases the output of f by at least p. We now present a strengthening of this result that additionally shows that if the function f is efficiently computable, then the source X could be an efficient p-tampering one (and only needs to use f as a black box). In the language of extractors, our result thus proves a strong impossibility result for deterministic randomness extraction from "efficient" Santha-Vazirani sources. Our proof of the generalized result is quite different (and in our eyes simpler) than classic proofs of the Santha-Vazirani theorem and may be of independent interest.

In fact, we present two different proofs. The first one achieves optimal bias p for balanced f, whereas the second uses an extremely simple "*lazy greedy*" tampering algorithm that makes only a *single* query to f and achieves bias $p/3$ for balanced f.

Theorem 13 (Theorem 5 restated). *There exists an oracle machine* Tam *with input parameters n and $\varepsilon < 1$ that runs in time $\mathrm{poly}(n/\varepsilon)$ and for every $n \in N$ and $\varepsilon \in (0,1)$, every Boolean function $f : \{0,1\}^n \to \{-1,1\}$, and every $p < 1$, for $\mu = \mathsf{E}[f(U_n)]$ it holds that*

$$\mathsf{E}[f(U_n^{\mathsf{Tam}^f,p})] \geq \mu + p \cdot (1 - |\mu| - \varepsilon).$$

Proof (Proof of Theorem 5). Let us first present a proof with an inefficient tampering algorithm achieving bias $p \cdot (1 - |\mu|)$; next, we show how to make it efficient while not loosing much in bias. On input $x_{\leq i-1} = (x_1, \ldots, x_{i-1})$, Tam sets $x_i = \mathrm{sgn}(\hat{f}_{x_{\leq i-1}}(\{i\}))$. By Equation (1), $\hat{f}_{x_{\leq i-1}}(\{i\})$ corresponds to the change in expectation of $f_{x_{\leq i-1}}$ when setting the value of x_i. This amounts to greedily choosing the x_i that increases the expectation. Let $X = U_n^{\mathsf{Tam},p}$. By applying Lemma 11 and the linearity of expectations, we have

$$\mathsf{E}[f(X)] = \hat{f}(\emptyset) + \sum_{i=1}^{n} \mathsf{E}_X \left[\hat{f}_{X_{\leq i-1}}(\{i\}) \cdot X_i \right]$$

$$= \hat{f}(\emptyset) + \sum_{i=1}^{n} \mathsf{E}_{X_{\leq i-1}} \left[\hat{f}_{X_{\leq i-1}}(\{i\}) \cdot \mathsf{E}[X_i | X_{\leq i-1}] \right].$$

Since Tam tampers with the i'th bit with independent probability p, therefore

$$\mathsf{E}[X_i | X_{\leq i-1}] = p \cdot \mathrm{sgn}(\hat{f}_{X_{\leq i-1}}(\{i\}))$$

and so it holds that

$$\mathsf{E}[f(X)] = \hat{f}(\emptyset) + p \cdot \sum_{i=1}^{n} \mathsf{E}_X \left[\left| \hat{f}_{X_{\leq i-1}}(\{i\}) \right| \right] = \hat{f}(\emptyset) + p \cdot \mathsf{E}_X \left[\sum_{i=1}^{n} \left| \hat{f}_{X_{\leq i-1}}(\{i\}) \right| \right]$$

$$\geq \hat{f}(\emptyset) + p \cdot (1 - \hat{f}(\emptyset))$$

where the last inequality follows by Corollary 12.

Note that the above tampering algorithm Tam in general is not efficient since computing $\hat{f}_{x_{\leq i-1}}(\{i\})$ exactly may be hard. However, we show that Tam may approximate $\hat{f}_{x_{\leq i-1}}(\{i\})$ using $M = \Theta(\frac{n^2}{\varepsilon^2} \cdot \log \frac{n}{\varepsilon})$ samples, and set x_i according to the sign of the approximation of $\hat{f}_{x_{\leq i-1}}(\{i\})$, while still inducing essentially the same bias. This clearly can be done efficiently given oracle access to f. As before, let $X = U_n^{\mathsf{Tam}^f, p}$ denote the corresponding p-tampering source. To lower bound $\mathsf{E}[f(X)]$, we note that the only difference from the previous case is that $\mathsf{Tam}(1^n, x_{\leq i-1})$ is no longer always outputting $\mathrm{sgn}(\hat{f}_{x_{\leq i-1}}(\{i\}))$. Nevertheless, we claim that for every $x_{<i}$ it holds that

$$\hat{f}_{x_{\leq i-1}}(\{i\}) \cdot \mathsf{E}[X_i | X_{\leq i-1} = x_{\leq i-1}] \geq p \cdot \left(|\hat{f}_{x_{\leq i-1}}(\{i\})| - \varepsilon/n \right)$$

since either (i) $|\hat{f}_{x_{\leq i-1}}(\{i\})| \geq \varepsilon/2n$ in which case (by a Chernoff bound) Tam outputs $\mathrm{sgn}(\hat{f}_{x_{\leq i-1}}(\{i\}))$ with probability at least $1 - \varepsilon/2n$, or (ii) $|\hat{f}_{x_{\leq i-1}}(\{i\})| < \varepsilon/2n$ in which case the inequality holds no matter what Tam outputs since $|\mathsf{E}[X_i | X_{\leq i-1} = x_{\leq i-1}]| \leq p$. A lower bound on $\mathsf{E}[f(X)]$ then follows by the same analysis as before:

$$\mathsf{E}[f(X)] \geq \hat{f}(\emptyset) + p \cdot \sum_{i=1}^{n} \mathsf{E}_X \left[\left| \hat{f}_{X_{\leq i-1}}(\{i\}) \right| - \varepsilon/n \right] \geq \mu + p \cdot (1 - |\hat{f}(\emptyset)| - \varepsilon).$$

Before presenting the second proof, we state the following lemma, which follows similarly to lemma 11, but instead it relies on a squared version of Equation (1). See the full version for a proof.

Lemma 14. *For every* $x \in \{-1,1\}^n$,

$$f(x)^2 = \hat{f}(\emptyset)^2 + \sum_{i=1}^n \left(\hat{f}_{x_{\leq i-1}}(\{i\})^2 + 2\hat{f}_{x_{\leq i-1}}(\emptyset) \cdot \hat{f}_{x_{\leq i-1}}(\{i\}) \cdot x_i \right).$$

We continue to present the second proof using a *"lazy greedy"* tampering algorithm that makes a *single* query to f and achieves bias $p/3$ for balanced f.

Theorem 15. *There exists an oracle machine* Tam *that makes a single query to its oracle such that for every* $n \in N$, *every Boolean function* $f : \{0,1\}^n \to \{-1,1\}$, *and every* $p < 1$, *for* $\mu = E[f(U_n)]$ *it holds that*

$$E[f(U_n^{\mathsf{Tam}^f,p})] \geq \mu + p \cdot (1 - \mu^2)/3.$$

Proof. We consider a *lazy greedy tampering algorithm* LTam that on input $x_{\leq i-1} = (x_1, \ldots, x_{i-1})$, samples uniformly random $(x_i', \ldots, x_n') \leftarrow U_{n-i+1}$, queries $y = f(x_1, x_{i-1}, x_i', \ldots, x_n')$, and outputs $X_i = x_i'$ if $y = 1$ and $X_i = -x_i'$ if $y = -1$. Namely, LTam samples a random completion of $x_{\leq i-1}$ and output the sampled bit x_i' iff the sample evaluates to 1.

Interestingly, this simple lazy greedy LTam implicitly "plays the first Fourier coefficient" in expectation in the sense that $E[\mathsf{Tam}(x_{\leq i-1})] = \hat{f}_{x_{\leq i-1}}(\{i\})$.

Claim. For every $x_{\leq i-1}$, $E[\mathsf{LTam}(x_{\leq i-1})] = \hat{f}_{x_{\leq i-1}}(\{i\})$.

Proof. Let $X_i = \mathsf{LTam}(x_{\leq i-1})$. We have $E[X_i] = E_{x \geq i \leftarrow U_{n-i+1}}[f(x) \cdot x_i] = \hat{f}_{x_{\leq i-1}}(\{i\})$.

To analyze LTam, we derive two equalities analogous to that in the proof of Theorem 5, and the theorem follows by combining the two equalities. First, since LTam gets to tamper with bit i with independent probability p and $E[\mathsf{LTam}(x_{\leq i-1})] = \hat{f}_{x_{\leq i-1}}(\{i\})$, by Lemma 3.3, we have that $E[X_i|X_{\leq i-1}] = p \cdot \hat{f}_{x_{\leq i-1}}(\{i\})$. Thus,

$$E[f(X)] = \hat{f}(\emptyset) + p \cdot \sum_{i=1}^n E_X\left[\hat{f}_{X_{\leq i-1}}(\{i\})^2\right]. \qquad (2)$$

Similarly, by applying Lemma 14 and the linearity of expectations, we have

$$E[f(X)^2] = \hat{f}(\emptyset)^2 + \sum_{i=1}^n \left(E_X[\hat{f}_{X_{\leq i-1}}(\{i\})^2] \right)$$
$$+ \sum_{i=1}^n \left(2E_{X_{\leq i-1}}[\hat{f}_{X_{\leq i-1}}(\emptyset) \cdot \hat{f}_{X_{\leq i-1}}(\{i\}) \cdot E[X_i|X_{\leq i-1}]] \right).$$

Simplifying using the fact that f is Boolean, the trivial bound $|\hat{f}_{X_{\leq i-1}}(\emptyset)| \leq 1$, and $\mathsf{E}[X_i | X_{\leq i-1}] = p \cdot \hat{f}_{X_{\leq i-1}}(\{i\})$ gives

$$1 \leq \hat{f}(\emptyset)^2 + (1 + 2p) \cdot \sum_{i=1}^{n} \mathsf{E}_X[\hat{f}_{X_{\leq i-1}}(\{i\})^2]. \tag{3}$$

Plugging Equation (3) in Equation (2) yields

$$\mathsf{E}[f(X)] \geq \hat{f}(\emptyset) + \frac{p}{1 + 2p}\left(1 - \hat{f}(\emptyset)^2\right) \geq \mu + p \cdot \left(1 - \mu^2\right)/3,$$

which completes the proof of Theorem 15.

3.3   Tampering with Bounded-Value Functions—The General Case

We further consider the more general case of tampering non-Boolean, bounded-value functions. We present an efficient tampering algorithm that biases the expectation of the function by an amount linear in the variance of the function.

Theorem 16 (Theorem 6 restated). *There exists an efficient oracle machine* Tam *such that for every* $n \in N$, *every bounded-value function* $f : \{0,1\}^n \to [-1,1]$, *and every* $p < 1$,

$$\mathsf{E}[f(U_n^{\mathsf{Tam}^f, p})] \geq \mathsf{E}[f(U_n)] + \frac{p \cdot \mathrm{Var}[f(U_n)]}{5}.$$

We prove Theorem 5 using *lazy greedy* tampering algorithm again. As before, we let LTam take a single sample, and make decision based on the outcome of the sample, but since f is not Boolean, we make randomized decision based on the function value on the sample. Specifically, on input $x_{\leq i-1} = (x_1, \ldots, x_{i-1})$:

- LTam samples uniformly random $(x_i', \ldots, x_n') \leftarrow U_{n-i+1}$, and computes $y = f(x_1, x_{i-1}, x_i', \ldots, x_n')$.
- LTam outputs $X_i = x_i'$ with probability $(1 + y)/2$, and $X_i = -x_i'$ with probability $(1 - y)/2$. Note that X_i has expectation $\mathsf{E}[X_i] = y \cdot x_i'$.

The following claim says that LTam "implicitly plays the first Fourier coefficient" in expectation.

Claim. For every $x_{\leq i-1} \in \{-1,1\}^{i-1}$, $\mathsf{E}[\mathsf{LTam}(x_{\leq i-1})] = \hat{f}_{x_{\leq i-1}}(\{i\})$.

Proof. Let $X_i = \mathsf{LTam}(x_{\leq i-1})$. We have:

$$\mathsf{E}[X_i] = \mathsf{E}_{x \geq i \leftarrow U_{n-i+1}}[f(x) \cdot x_i] = \hat{f}_{x_{\leq i-1}}(\{i\}).$$

Let $X = U_n^{\mathsf{LTam}^f, p}$. Also, let the mean $\mathsf{E}[f(U_n)] = \mu$, the second moment $\mathsf{E}[f(U_n)^2] = \nu$, and the variance $\mathrm{Var}[f] = \sigma^2$ be denoted so. The analyze of the lazy greedy algorithm LTam for the non-Boolean case is significantly more involved. We first follow an analogous step in the analysis of Boolean case to derive an inequality between $\mathsf{E}[f(X)]$ and $\mathsf{E}[f(X)^2]$, then rely on a potential function analysis to derive a second inequality relation between $\mathsf{E}[f(X)]$ and $\mathsf{E}[f(X)^2]$, and then derive a lower bound on $\mathsf{E}[f(X)]$ by combining the two. The two inequalities are stated in the following lemmas.

Lemma 17. $\mathsf{E}[f(X)] - \mu \geq \frac{p}{1+2p} \cdot \left(\mathsf{E}[f(X)^2] - \nu + \sigma^2 \right).$

Lemma 18. $\mathsf{E}[f(X)] + \frac{\mathsf{E}[f(X)^2]}{2} + \frac{\mathsf{E}[f(X)^2]^2}{4} \geq \mu + \frac{\nu}{2} + \frac{\nu^2}{4}.$

We first use the above two lemmas to show that $\mathsf{E}[f(X)] - \mu \geq (p\sigma^2)/5$, which implies that $\mathsf{E}[f(X)] \geq \mu + p \cdot \mathrm{Var}[f]/5$, as desired. If $\mathsf{E}[f(X)^2] \geq \nu$, then Lemma 17 implies

$$\mathsf{E}[f(X)] - \mu \geq \frac{p}{1+2p} \cdot \sigma^2 \geq \frac{1}{5} \cdot p\sigma^2.$$

For the case that $\mathsf{E}[f(X)^2] < \nu$, let $\alpha \triangleq \nu - \mathsf{E}[f(X)^2] \geq 0$. Lemma 18 implies

$$\mathsf{E}[f(X)] - \mu \geq \frac{1}{2}(\nu - \mathsf{E}[f(X)^2]) + \frac{1}{4}(\nu^2 - \mathsf{E}[f(X)^2]^2) \geq \frac{\alpha}{2}$$

which together with Lemma 17 implies that

$$[f(X)] - \mu \geq \max \left\{ \frac{p}{1+2p} \cdot (\sigma^2 - \alpha), \frac{\alpha}{2} \right\} \geq \frac{p}{1+4p} \geq \frac{p\sigma^2}{5}.$$

Now we prove Lemmas 17 and 18. The proof of Lemma 17 is a generalization of the analysis for biasing Boolean functions.

Proof (Proof of Lemma 17). By applying Lemma 11 and the linearity of expectations, we have

$$\mathsf{E}[f(X)] = \hat{f}(\emptyset) + \sum_{i=1}^{n} \mathsf{E}_X \left[\hat{f}_{X_{\leq i-1}}(\{i\}) \cdot X_i \right]$$

$$= \hat{f}(\emptyset) + \sum_{i=1}^{n} \mathsf{E}_{X_{\leq i-1}} \left[\hat{f}_{X_{\leq i-1}}(\{i\}) \cdot \mathsf{E}[X_i | X_{\leq i-1}] \right].$$

Since LTam gets to tamper with bit i with independent probability p, by Lemma 3.3 we have that $\mathsf{E}[X_i | X_{\leq i-1}] = p \cdot \hat{f}_{X_{\leq i-1}}(\{i\})$. Thus,

$$\mathsf{E}[f(X)] = \hat{f}(\emptyset) + p \cdot \sum_{i=1}^{n} \mathsf{E}_X \left[\hat{f}_{X_{\leq i-1}}(\{i\})^2 \right]. \tag{4}$$

Similarly, by applying Lemma 14 and the linearity of expectations, we have

$$\mathsf{E}[f(X)^2] = \hat{f}(\emptyset)^2 + \sum_{i=1}^{n} \left(\mathsf{E}_X[\hat{f}_{X_{\leq i-1}}(\{i\})^2] \right)$$

$$+ \sum_{i=1}^{n} \left(2\mathsf{E}_{X_{\leq i-1}}[\hat{f}_{X_{\leq i-1}}(\emptyset) \cdot \hat{f}_{X_{\leq i-1}}(\{i\}) \cdot \mathsf{E}[X_i | X_{\leq i-1}]] \right).$$

Simplifying using the trivial bound $|\hat{f}_{X_{\leq i-1}}(\emptyset)| \leq 1$ and $\mathsf{E}[X_i|X_{\leq i-1}] = p \cdot \hat{f}_{X_{\leq i-1}}(\{i\})$ gives

$$\mathsf{E}[f(X)^2] \leq \hat{f}(\emptyset)^2 + (1 + 2p) \cdot \sum_{i=1}^{n} \mathsf{E}_X[\hat{f}_{X_{\leq i-1}}(\{i\})^2]. \tag{5}$$

The lemma follows by combining Equations (4) and (5):

$$\mathsf{E}[f(X)] \geq \hat{f}(\emptyset) + \frac{p}{1 + 2p}\left(\mathsf{E}[f(X)^2] - \hat{f}(\emptyset)^2\right) = \mu + \frac{p}{1 + 2p}\left(\mathsf{E}[f(X)^2] - \nu + \sigma^2\right)$$

where the last equality uses the fact that $\hat{f}(\emptyset)^2 = \mu^2 = \nu - \sigma^2$.

The proof of Lemma 18 is less trivial. Our key observation is the following useful property of the lazy greedy tampering algorithm LTam: consider the function f together with an arbitrary function $g : \{-1, 1\}^n \to [-1, 1]$ (ultimately, we shall set $g(x) = f(x)^2$, but in the discussion that follows, g can be completely unrelated to f). While intuitively we expect the expectation of f to be increasing after tampering, it is clearly possible that the tampering causes the expectation of g to decrease. Nevertheless, we show that for a properly defined potential function combining the expectations of f and g, the potential is guaranteed to be non-decreasing after tampering. Namely, we prove the following lemma whose proof can be found in the full version of the paper.

Lemma 19. *Let* $g : \{-1, 1\}^n \to [-1, 1]$ *be an arbitrary function. For every prefix* $x_{\leq i} \in \{-1, 1\}^i$, *define a potential*

$$\Phi(x_{\leq i}) := \hat{f}_{x_{\leq i}}(\emptyset) + \frac{\hat{g}_{x_{\leq i}}(\emptyset)}{2} + \frac{\hat{g}_{x_{\leq i}}(\emptyset)^2}{4},$$

and let $\Phi := \Phi(x_{\leq 0})$. *Then it holds that* $\mathsf{E}[\Phi(X)] \geq \Phi$.

Lemma 18 now follows easily.

Proof (Proof of Lemma 18). By applying Lemma 19 with $g = f^2$ and noting that $\hat{g}(\emptyset) = \nu$, we have

$$\mathsf{E}[f(X)] + \frac{\mathsf{E}[f(X)^2]}{2} + \frac{\mathsf{E}[f(X)^2]^2}{4} \geq \mu + \frac{\nu}{2} + \frac{\nu^2}{4}.$$

References

1. Akavia, A., Goldwasser, S., Vaikuntanathan, V.: Simultaneous hardcore bits and cryptography against memory attacks. In: Reingold, O. (ed.) TCC 2009. LNCS, vol. 5444, pp. 474–495. Springer, Heidelberg (2009)
2. One, A.: Smashing the stack for fun and profit. Phrack Magazine 7(49), File 14 (1996)

3. Anderson, R., Kuhn, M.: Tamper resistance – a cautionary note. In: Proceedings of the Second USENIX Workshop on Electronic Commerce, pp. 1–11 (November 1996)
4. Austrin, P., Chung, K.-M., Mahmoody, M., Pass, R., Seth, K.: On the impossibility of cryptography with tamperable randomness. Cryptology ePrint Archive, Report 2013/194 (2013), http://eprint.iacr.org/
5. Biham, E., Shamir, A.: Differential fault analysis of secret key cryptosystems. In: Kaliski Jr., B.S. (ed.) CRYPTO 1997. LNCS, vol. 1294, pp. 513–525. Springer, Heidelberg (1997)
6. Boneh, D., DeMillo, R.A., Lipton, R.J.: On the importance of checking cryptographic protocols for faults. In: Fumy, W. (ed.) EUROCRYPT 1997. LNCS, vol. 1233, pp. 37–51. Springer, Heidelberg (1997)
7. Brakerski, Z., Kalai, Y.T., Katz, J., Vaikuntanathan, V.: Overcoming the hole in the bucket: Public-key cryptography resilient to continual memory leakage. In: FOCS, pp. 501–510. IEEE Computer Society (2010)
8. Canetti, R., Goldreich, O., Goldwasser, S., Micali, S.: Resettable zero-knowledge (extended abstract). In: STOC, pp. 235–244 (2000)
9. Choi, S.G., Kiayias, A., Malkin, T.: BiTR: Built-in tamper resilience. In: Lee, D.H., Wang, X. (eds.) ASIACRYPT 2011. LNCS, vol. 7073, pp. 740–758. Springer, Heidelberg (2011)
10. Dachman-Soled, D., Kalai, Y.T.: Securing circuits against constant-rate tampering. IACR Cryptology ePrint Archive, 2012:366 (2012); Informal publication
11. Dodis, Ong, Prabhakaran, Sahai: On the (im)possibility of cryptography with imperfect randomness. In: FOCS: IEEE Symposium on Foundations of Computer Science, FOCS (2004)
12. Dodis, Y., Goldwasser, S., Tauman Kalai, Y., Peikert, C., Vaikuntanathan, V.: Public-key encryption schemes with auxiliary inputs. In: Micciancio, D. (ed.) TCC 2010. LNCS, vol. 5978, pp. 361–381. Springer, Heidelberg (2010)
13. Dodis, Y., Haralambiev, K., López-Alt, A., Wichs, D.: Cryptography against continuous memory attacks. In: FOCS, pp. 511–520. IEEE Computer Society (2010)
14. Dziembowski, S., Pietrzak, K.: Leakage-resilient cryptography. In: FOCS, pp. 293–302. IEEE Computer Society (2008)
15. Dziembowski, S., Pietrzak, K., Wichs, D.: Non-malleable codes. In: Yao, A.C.-C. (ed.) ICS, pp. 434–452. Tsinghua University Press (2010)
16. Faust, S., Pietrzak, K., Venturi, D.: Tamper-proof circuits: How to trade leakage for tamper-resilience. In: Aceto, L., Henzinger, M., Sgall, J. (eds.) ICALP 2011, Part I. LNCS, vol. 6755, pp. 391–402. Springer, Heidelberg (2011)
17. Ariel, J.: Feldman and Josh Benaloh. On subliminal channels in encrypt-on-cast voting systems. In: Proceedings of the 2009 Conference on Electronic Voting Technology/Workshop on Trustworthy Elections, EVT/WOTE 2009, p. 12. USENIX Association, Berkeley (2009)
18. Frykholm, N.: Countermeasures against buffer overflow attacks. Technical report, RSA Data Security, Inc., pub-RSA:adr (November 2000)
19. Gennaro, R., Lysyanskaya, A., Malkin, T., Micali, S., Rabin, T.: Algorithmic tamper-proof (atp) security: Theoretical foundations for security against hardware tampering. In: Naor, M. (ed.) TCC 2004. LNCS, vol. 2951, pp. 258–277. Springer, Heidelberg (2004)
20. Goldreich, O., Oren, Y.: Definitions and properties of zero-knowledge proof systems. Journal of Cryptology 7(1), 1–32 (1994)
21. Goldwasser, S., Rothblum, G.: How to compute in the presence of leakage (2012)

22. Heninger, N., Durumeric, Z., Wustrow, E., Halderman, J.A.: Mining your Ps and Qs: Detection of widespread weak keys in network devices. In: Proceedings of the 21st USENIX Security Symposium (August 2012)
23. Impagliazzo, R., Luby, M.: One-way functions are essential for complexity based cryptography. In: Proceedings of the 30th Annual Symposium on Foundations of Computer Science (FOCS), pp. 230–235 (1989)
24. Ishai, Y., Prabhakaran, M., Sahai, A., Wagner, D.: Private circuits II: Keeping secrets in tamperable circuits. In: Vaudenay, S. (ed.) EUROCRYPT 2006. LNCS, vol. 4004, pp. 308–327. Springer, Heidelberg (2006)
25. Kalai, Y., Lewko, A., Rao, A.: Formulas resilient to short-circuit errors (2012)
26. Kalai, Y.T., Kanukurthi, B., Sahai, A.: Cryptography with tamperable and leaky memory. In: Rogaway, P. (ed.) CRYPTO 2011. LNCS, vol. 6841, pp. 373–390. Springer, Heidelberg (2011)
27. Lenstra, A.K., Hughes, J.P., Augier, M., Bos, J.W., Kleinjung, T., Wachter, C.: Public keys. In: Safavi-Naini, R., Canetti, R. (eds.) CRYPTO 2012. LNCS, vol. 7417, pp. 626–642. Springer, Heidelberg (2012)
28. Lenstra, A.K., Hughes, J.P., Augier, M., Bos, J.W., Kleinjung, T., Wachter, C.: Ron was wrong, whit is right. Cryptology ePrint Archive, Report 2012/064 (2012), http://eprint.iacr.org/
29. Liu, F.-H., Lysyanskaya, A.: Algorithmic tamper-proof security under probing attacks. In: Garay, J.A., De Prisco, R. (eds.) SCN 2010. LNCS, vol. 6280, pp. 106–120. Springer, Heidelberg (2010)
30. Liu, F.-H., Lysyanskaya, A.: Tamper and leakage resilience in the split-state model. In: Safavi-Naini, R., Canetti, R. (eds.) CRYPTO 2012. LNCS, vol. 7417, pp. 517–532. Springer, Heidelberg (2012)
31. Micali, S., Reyzin, L.: Physically observable cryptography (extended abstract). In: Naor, M. (ed.) TCC 2004. LNCS, vol. 2951, pp. 278–296. Springer, Heidelberg (2004)
32. Pincus, J.D., Baker, B.: Beyond stack smashing: Recent advances in exploiting buffer overruns. IEEE Security & Privacy 2(4), 20–27 (2004)
33. Rivest, R.L., Shamir, A., Adleman, L.M.: A method for obtaining digital signatures and public-key cryptosystems. Communications of the ACM 21(2), 120–126 (1978)
34. Rothblum, G.N.: How to compute under $\mathcal{AC}^0$ leakage without secure hardware. In: Safavi-Naini, R., Canetti, R. (eds.) CRYPTO 2012. LNCS, vol. 7417, pp. 552–569. Springer, Heidelberg (2012)
35. Santha, M., Vazirani, U.V.: Generating quasi-random sequences from semi-random sources. J. Comput. Syst. Sci. 33(1), 75–87 (1986)
36. Simmons, G.J.: Subliminal channels; past and present. ETT 5(4), 15 (1994)
37. Young, A., Yung, M.: The dark side of "Black-box" cryptography, or: Should we trust capstone? In: Koblitz, N. (ed.) CRYPTO 1996. LNCS, vol. 1109, pp. 89–103. Springer, Heidelberg (1996)

Multiparty Key Exchange, Efficient Traitor Tracing, and More from Indistinguishability Obfuscation

Dan Boneh and Mark Zhandry

Stanford University, CA, USA
{dabo,zhandry}@cs.stanford.edu

Abstract. In this work, we show how to use indistinguishability obfuscation (iO) to build multiparty key exchange, efficient broadcast encryption, and efficient traitor tracing. Our schemes enjoy several interesting properties that have not been achievable before:

- Our multiparty non-interactive key exchange protocol does not require a trusted setup. Moreover, the size of the published value from each user is independent of the total number of users.
- Our broadcast encryption schemes support *distributed* setup, where users choose their own secret keys rather than be given secret keys by a trusted entity. The broadcast ciphertext size is *independent* of the number of users.
- Our traitor tracing system is fully collusion resistant with short ciphertexts, secret keys, and public key. Ciphertext size is logarithmic in the number of users and secret key size is independent of the number of users. Our public key size is polylogarithmic in the number of users. The recent functional encryption system of Garg, Gentry, Halevi, Raykova, Sahai, and Waters also leads to a traitor tracing scheme with similar ciphertext and secret key size, but the construction in this paper is simpler and more direct. These constructions resolve an open problem relating to differential privacy.
- Generalizing our traitor tracing system gives a private broadcast encryption scheme (where broadcast ciphertexts reveal minimal information about the recipient set) with optimal size ciphertext.

Several of our proofs of security introduce new tools for proving security using indistinguishability obfuscation.

1 Introduction

An obfuscator is a machine that takes as input a program, and produces a second program with identical functionality that in some sense hides how the original program works. An important notion of obfuscation called *indistinguishability obfuscation* (iO) was proposed by Barak et al. [BGI⁺01] and further studied by Goldwasser and Rothblum [GR07]. Indistinguishability obfuscation asks that obfuscations of any two (equal-size) programs that compute the same function are computationally indistinguishable. The reason iO has become so important

J.A. Garay and R. Gennaro (Eds.): CRYPTO 2014, Part I, LNCS 8616, pp. 480–499, 2014.

is a recent breakthrough result of Garg, Gentry, Halevi, Raykova, Sahai, and Waters [GGH+13b] that put forward the first candidate construction for an efficient iO obfuscator for general boolean circuits. The construction builds upon the multilinear map candidates of Garg, Gentry, and Halevi [GGH13a] and Coron, Lepoint, and Tibouchi [CLT13].

In subsequent work, Sahai and Waters [SW13] showed that indistinguishability obfuscation is a powerful cryptographic primitive: it can be used to build public-key encryption from pseudorandom functions, selectively-secure short signatures, deniable encryption, and much more. Hohenberger, Sahai, and Waters [HSW13] showed that iO can be used to securely instantiate the random oracle in several random-oracle cryptographic systems.

Our results. In this paper, we show further powerful applications for indistinguishability obfuscation. While the recent iO constructions make use of multilinear maps, the converse does not seem to hold: we do not yet know how to build multilinear maps from iO. Nevertheless, we show that iO *can* be used to construct many of the powerful applications that follow from multilinear maps. The resulting iO-based constructions have surprising features that could not be previously achieved, not even using the current candidate multilinear maps. All of our constructions employ the punctured PRF technique introduced by Sahai and Waters [SW13].

1.1   Multiparty Non-Interactive Key Exchange

Our first construction uses iO to construct a multiparty non-interactive key exchange protocol (NIKE) from a pseudorandom generator. Recall that in a NIKE protocol, N parties each post a single message to a public bulletin board. All parties then read the board and agree on a shared key k that is secret from any eavesdropper who only sees the bulletin board. The classic Diffie-Hellman protocol solves the two-party case $N = 2$. The first three-party protocol was proposed by Joux [Jou04] using bilinear maps. Boneh and Silverberg [BS03] gave a protocol for general N using multilinear maps. The candidate multilinear map constructions by Garg, Gentry, and Halevi [GGH13a] using ideal lattices, and by Coron, Lepoint, and Tibouchi [CLT13] over the integers, provide the first implementations for N parties, but require a trusted setup phase. Prior to this work, these were the only known constructions for NIKE.

We construct new NIKE protocols from a general indistinguishability obfuscator. Our basic protocol is easy to describe: each user generates a random seed s for a pseudorandom generator G whose output is at least twice the size of the seed. The user posts $G(s)$ to the bulletin board. When N users wish to generate a shared group key, they each collect all the public values from the bulletin board and run a certain *public* obfuscated program P_{KE} (shown in Figure 1) on the public values along with their secret seed. The program outputs the group key.

Inputs: public values $x_1, \ldots x_N \in \mathcal{X}^N$, an index $i \in [N]$, and a secret seed $s \in \mathcal{S}$
Embedded constant: pseudorandom function PRF with an embedded random key

1. If $x_i \neq G(s)$, output $\perp$
2. Otherwise, output $\mathsf{PRF}(x_1, x_2, \ldots, x_N)$

Fig. 1. The program P_{KE}

We show that this protocol is secure in a semi-static model [FHKP13]: an adversary that is allowed to (non-adaptively) corrupt participants of its choice cannot learn the shared group key of a group of uncorrupt users of its choice. The proof uses the punctured PRF technique of Sahai and Waters, but interestingly requires the full power of the constrained PRFs of Boneh and Waters [BW13] for arbitrary circuit constraints. In addition, we show that the point-wise punctured PRFs used by Sahai and Waters are sufficient to prove security, but only in a weaker static security model where the adversary cannot corrupt users. We leave the construction of a fully adaptively secure NIKE (in the sense of [FHKP13]) from iO as a fascinating open problem.

In the full version [BZ14], we observe that our iO-based NIKE can be easily extended to an identity-based multiparty key exchange. Existing ID-NIKE protocols are based on multilinear maps [FHPS13].

Comparison to existing constructions. While NIKE can be built directly from multilinear maps, our iO-based protocol has a number of advantages:

- No trusted setup. Existing constructions [GGH13a, CLT13] require a trusted setup to publish public parameters: whoever generates the parameters can expose the secret keys for all groups just from the public values posted by members of the group. A variant of our iO-based construction requires no trusted setup, and in fact requires no setup at all. We simply have user number 1 generate the obfuscated program P_{KE} and publish it along with her public values. The resulting scheme is the first statically secure NIKE protocol with no setup requirements. In the full version [BZ14] we enhance the construction and present a NIKE protocol with no setup that is secure in the stronger semi-static model. This requires changing the scheme to defend against a potentially malicious program P_{KE} published by a corrupt user. To do so we replace the secret seed s by a digital signature generated by each user. Proving security from iO requires the signature scheme to have a special property we call *constrained public-keys*, which may be of indpendent interest. We construct such signatures from iO.
- Short public values. In current multilinear-based NIKE protocols, the size of the values published to the bulletin board is at least linear in the number of users N. In our basic iO-based construction, the size of published values is independent of N.

– Since the published values are independent of any public parameters, the same published values can be used in multiple NIKE environments setup by different organizations.

It is also worth noting that since our NIKE is built from a generic iO mechanism, it may eventually depend on a weaker complexity assumption than those needed for secure multilinear maps.

1.2 Broadcast Encryption

Broadcast encryption [FN94] lets an encryptor broadcast a message to a subset of recipients. The system is said to be collusion resistant if no set of non-recipients can learn information about the plaintext. The efficiency of a broadcast system is measured in the ciphertext overhead: the number of bits in the ciphertext beyond what is needed to describe the recipient set and encrypt the payload message using a symmetric cipher. The shorter the overhead, the better (an overhead of zero is optimal). We survey some existing constructions in related work below.

Using a generic conversion from NIKE to broadcast encryption described in the full version [BZ14], we obtain two collusion-resistant broadcast systems. The first is a secret-key broadcast system with *optimal* broadcast size. The second is a public-key broadcast system with constant overhead, namely *independent* of the number of recipients. In both systems, decryption keys are constant size (i.e. independent of the number of users). The encryption key, however, is linear in the number of users as in several other broadcast systems [BGW05, GW09, DPP07, BW13].

By starting from our semi-static secure NIKE, we obtain a semi-static secure broadcast encryption (as defined by Gentry and Waters [GW09]). Then applying a generic conversion due to Gentry and Waters [GW09], we obtain a fully adaptively secure public-key broadcast encryption system with the shortest known ciphertext overhead.

Our public-key broadcast encryption has a remarkable property that has so far not been possible, not even using the candidate multilinear maps. The system is a public-key *distributed* broadcast system: users generate secret keys on their own and simply append their corresponding public values to the broadcast public key. In contrast, in existing low-overhead public-key broadcast systems surveyed below, users are assigned their secret key by a trusted authority who has the power to decrypt all broadcasts. In our iO-based public-key system, there is no trusted authority.

Another interesting aspect of the construction is that the PRG used in the scheme (as in the program P_{KE}) can be replaced by the RSA public key encryption system where the RSA secret key plays the role of the PRG seed and the corresponding RSA public key plays the role of the PRG output. Then, our broadcast system shows that iO makes it possible to use *existing* certified RSA keys in a short-ciphertext broadcast encryption system and in a NIKE protocol. To prove security using iO we need the following property of RSA: there is a distribution of invalid RSA public-keys (e.g. products of three random large primes) that is computationally indistinguishable from a distribution of real RSA public keys (i.e. products of two random large primes). This property also holds for other public-key systems such as Regev's lattice encryption scheme, but does not hold for systems like basic ElGamal encryption.

1.3   Recipient-Private Broadcast Encryption

A broadcast encryption system is said to be recipient-private if broadcast ciphertexts reveal nothing about the intended set of recipients [BBW06, LPQ12, FP12]. Valid recipients will learn that they are members of the recipient set (by successfully decrypting the ciphertext), but should learn nothing else about the set. Until very recently, the best recipient-private broadcast systems had a broadcast size of $O(\lambda \cdot N)$, proportional to the product of the security parameter λ and the number of users N.

Using iO, we construct a recipient-private broadcast system with a broadcast size of $O(\lambda + N)$, proportional to the *sum* of the security parameter and the number of users. This is the best possible broadcast size. If one is allowed to leak the size k of the recipient set (and nothing else) then we construct a system where the broadcast size is proportional to $O(\lambda + k \log N)$, which is again the best possible. Building such systems has been open for some time [BBW06] and is now resolved using iO.

Our approach to building a recipient-private broadcast system is to embed an encryption of the intended recipient set in the broadcast header. We publish an obfuscated program in the public key that begins by decrypting the encrypted recipient set in the broadcast header. It then decrypts the message body only if the recipient can provide a proof that it is one of the intended recipients. Interestingly, encrypting the recipient set in a way that lets us prove security using iO is non-trivial. The problem is that using a generic CPA-secure scheme is insecure due to potential malleability attacks on the encrypted recipient set that can break recipient privacy. Using an authenticated encryption scheme to prevent the malleability attack is problematic because forged valid ciphertexts exist (even though they may be difficult to construct), and this prevents us from proving security using iO. The difficulty arises because iO can only be applied to two programs that agree on *all* inputs, including hard-to-compute ones.

Instead of using authenticated encryption, we encrypt the recipient set using a certain malleable encryption scheme that lets us translate an encryption of a recipient set S to an encryption of some other recipient set S'. We use indistinguishability of obfuscations to argue that an attacker cannot detect this change, thereby proving recipient privacy.

The recent succinct functional encryption scheme of Garg et al. [GGH+13b] can also be used to build recipient-private broadcast encryption from iO. However, our construction is quite different and is simpler and more direct. For example, it does not use non-interactive zero-knowledge proofs. Moreover, our scheme has shorter secret keys: $O(1)$ as a function of N compared to $N^{O(1)}$. The main drawback of our scheme is the larger public key: $N^{O(1)}$ compared to $O(1)$.

1.4 Traitor Tracing with Short Ciphertexts, Secret Keys, and Public Keys

Private broadcast-encryption is further motivated by its application to traitor tracing systems [CFN94]. Recall that traitor tracing systems, introduced by Chor, Fiat, and Naor, help content distributors identify the origin of pirate decryption boxes, such as pirate cable-TV set top decoders. Boneh, Sahai, and Waters [BSW06] showed that a private broadcast encryption system that can broadcast privately to any of the $N + 1$ sets $\emptyset, \{1\}, \{1, 2\}, \dots, \{1, \dots, N\}$ is sufficient for building an N-user traitor tracing system. The ciphertext used in the traitor tracing system under normal operation is simply a broadcast to the full set $\{1, \dots, N\}$, allowing all decoders to decrypt. Therefore, the goal is, as before, to build a private broadcast system for this specific set system where ciphertext overhead is minimized. Such systems are called *private linear broadcast encryption* (PLBE).

Adapting our iO-based private broadcast system to the linear set system above, we obtain a collusion resistant traitor tracing system where ciphertext size is $O(\lambda + \log N)$ where λ is the security parameter and N is the total number of users in the system. Moreover, secret keys are short: their length is λ, independent of N. However, this scheme has large public keys, polynomial in N and λ. The main reason public keys are large is that the malleable encryption scheme we need requires polynomial size circuits for encryption and decryption.

Fortunately we can reduce the public-key size to only $\mathsf{poly}(\log N, \lambda)$ without affecting secret-key or ciphertext size. We do so by adapting the authenticated encryption approach discussed in the previous section: when embedding the encrypted recipient set in the broadcast ciphertext we also embed a MAC of the encrypted set. The decryption program will reject a broadcast ciphertext with an invalid MAC. To prove security we need to puncture the MAC algorithm at all possible recipient sets. Naively, since in a PLBE there are $N + 1$ recipient sets, the resulting program size would be linear in N thereby resulting in large secret keys. Instead, we step through a sequence of hybrids where at each hybrid we puncture the MAC at exactly one point. This sequence of hybrids ensures that the obfuscated decryption program remains small. Once this sequential puncturing process completes, security follows from security of an embedded PRF. We emphasize that this proof technique works for proving security of a PLBE because of the small number of possible recipient sets.

The functional encryption scheme of Garg et al. [GGH+13b] can also be used to obtain collusion resistant traitor tracing, however as for private broadcast encryption, our construction is conceptually simpler and has shorter secret keys.

Connection to Differential Privacy Dwork et al. [DNR+09] show that efficient traitor tracing schemes imply the impossibility of any differentially private data release mechanism. A data release mechanism is a procedure that outputs a data structure that supports approximations to queries of the form "what fraction of records have property P?" Informally, a data release mechanism is differentially private if it does not reveal whether any individual record is in the database.

Applying the counter-example of [DNR+09] to our traitor tracing scheme, we obtain a database of N records of size λ and $N2^{O(\lambda)}$ queries. Moreover, the records are just independent uniform bit strings. Even with these small and simple records and relatively few queries, no polynomial time (in λ and N) differentially private data release mechanism is possible, so long as our construction is secure. The first scheme this counter example was applied to is the traitor tracing scheme of Boneh, Sahai, and Waters [BSW06], giving records of size $O(\lambda)$, but with a query set of size $2^{\tilde{O}(\sqrt{N})}$, exponential in N.

Ullman [Ull13] shows that, assuming one-way functions exist, there is no algorithm that takes a database of N records of size λ and an arbitrary set of approximately $O(N^2)$ queries, and approximately answers each query in time $\mathsf{poly}(N, \lambda)$ while preserving differential privacy. This result also uses the connection between traitor tracing and differential privacy, but is qualitatively different from ours. Their result applies to algorithms answering any *arbitrary* set of $O(N^2)$ queries while maintaining differential privacy, whereas we demonstrate a *fixed* set of $O(N2^{\lambda})$ queries that are impossible to answer efficiently.

Constrained PRFs. Recall that constrained PRFs, needed in iO proofs of security, are PRFs for which there are constrained keys than enable the evaluation of the PRF at a subset of the PRF domain and nowhere else [BW13, KPTZ13, BGI13]. The next section gives a precise definition. Our last construction shows that iO, together with a one-way function, are sufficient to build a constrained PRF for arbitrary circuit constraints. Consequently, all our constructions that utilize circuit constrained PRFs can be directly built from iO and a one-way function without additional assumptions. In fact, Moran and Rosen [MR13] show, under the assumption that NP is not solvable in probabilistic polynomial time in the worst case, that indistinguishability obfuscation implies one-way functions. Previously, constrained PRFs for arbitrary circuit constraints were built using multilinear maps [BW13].

1.5   Related Work

While some works have shown how to obfuscate simple functionalities such as point functions [Can97, CMR98, LPS04, Wee05], inner products [CRV10], and d-CNFs [BR13a], it is only recently that obfuscation for poly-size circuits became possible [GGH+13b, BR13b, BGK+13] and was applied to building higher level cryptographic primitives [SW13, HSW13].

Broadcast encryption. Fully collusion resistant broadcast encryption has been widely studied. Revocation systems [NNL01, HS02, GST04, DF02, LSW10] can encrypt to $N - r$ users with ciphertext size of $O(r)$. Further combinatorial solutions [NP00, DF03] achieve similar parameters. Algebraic constructions [BGW05, GW09, DPP07] using bilinear maps achieve constant (but non-zero) ciphertext overhead and some are even identity-based [GW09, Del07, SF07]. Multilinear maps give secret-key broadcast systems with optimal ciphertext size and short private keys [BS03, FHPS13, BW13]. They also give public-key broadcast systems with short ciphertexts and short public keys (using an $O(\log N)$-linear map) [BWZ14], but using the existing multilinear candidates, those systems are not distributed: users must be given their private keys by a central authority. The difficulty with using existing N-linear maps for distributed public-key broadcast encryption is that the encoding of a single element requires $\Omega(N)$ bits, and therefore a short ciphertext cannot include even a single element.

Recipient-private broadcast encryption. The first constructions for private broadcast encryption [BBW06, LPQ12] required a ciphertext header whose size is proportional to the product of the security parameter and the number of recipients. More recently, Fazio and Perera [FP12] presented a system with a weaker privacy guarantee called *outsider anonymity*, but where the header size is proportional to the number of revoked users. Kiayias and Samari [KS13] even provide lower bounds showing that certain types of natural constructions cannot improve on these bounds.

The functional encryption scheme of Garg et al. [GGH+13b] can also be used to build recipient-private broadcast encryption from iO. Our scheme is conceptually simpler, and avoids the need for non-interactive zero-knowledge proofs. Moreover, our scheme has shorter secret keys: $O(1)$ in N compared to $N^{O(1)}$ — though for private *linear* broadcast, their secret keys are $\mathsf{polylog}(N)$. The main drawback of our scheme is the large public key size: $N^{O(1)}$ compared to $O(\log N)$.

Traitor tracing. The literature on traitor tracing is vast and here we only discuss results on fully collusion resistant systems. Since the trivial fully-collusion resistant system has ciphertext size that is linear in the number of users, we are only interested in fully collusion resistant systems that achieve sub-linear size ciphertext. The first such system [BSW06, BW06], using bilinear maps, achieved $\sqrt{n}$ size ciphertexts with constant size keys. Other schemes based on different assumptions achieve similar parameters [GKSW10, Fre10]. Combinatorial constructions can achieve constant size ciphertexts [BN08, Sir07], but require secret keys whose size is quadratic (or worse) in the number of users. In most traitor tracing systems, the tracing key must be kept secret. Some systems, including ours, allow anyone to run the tracing algorithm [Pfi96, PW97, WHI01, KY02, CPP05, BW06].

Recently, Koppula, Ramchen, and Waters [KRW13] provide counter-examples to the conjecture that all bit encryption schemes are circularly secure. Concurrently and independent of our work, they use a valid/invalid key strategy that is similar to our strategy of replacing correctly generated public parameters with incorrect parameters, but in a very different context.

2 Preliminaries: Definitions and Notation

In this section, we briefly discuss notation and the building blocks for our constructions: indistinguishability obfuscation and constrained pseudorandom functions. A more complete description of these primitives appears in the full version [BZ14].

Notation We let $[N] = \{1, \cdots, N\}$ denote the positive integers from 1 to N. For a set S we denote by $x \leftarrow S$ the uniform random variable on S. For a randomized algorithm $\mathcal{A}$, we denote by $x \leftarrow \mathcal{A}(y)$ the random variable defined by the output of $\mathcal{A}$ on input y.

Indistinguishability Obfuscation. An indistinguishability obfuscation iO is a probabilistic polynomial time algorithm that takes a circuit C and produces an obfuscated circuit $C' = \text{iO}(C)$. We require that $C'(x) = C(x)$ for all inputs x. For security, we require that, for any two circuits C_1 and C_2 that agree on all inputs, no probabilistic polynomial time adversary can distinguish the obfuscation $C_1' = \text{iO}(C_1)$ from $C_2' = \text{iO}(C_2)$. The first candidate construction of such obfuscators is due to Garg et al. [GGH+13b].

Constrained Pseudorandom Functions. A constrained pseudorandom function (PRF) [BW13, KPTZ13, BGI13] PRF for a class of subsets $\mathcal{S}$ is a pseudorandom function for which there is an efficient algorithm that takes the secret key k for PRF and a set $S \in \mathcal{S}$, and outputs a circuit PRF_k^S which satisfies

$$\text{PRF}_k^S(x) = \begin{cases} \text{PRF}_k(x) & \text{if } x \in S \\ \bot & \text{if } x \notin S \end{cases}.$$

For security, we require that the circuit PRF_k^S reveals no information about $\text{PRF}_k(x)$ for points $x \notin S$. From this point forward, we will omit reference to the secret key k. We are interested in several classes of subsets $\mathcal{S}$. A *punctured* PRF is a constrained PRF for all sets whose complements are polynomial in size. The PRF construction of Goldreich, Goldwasser, and Micali [GGM86] satisfies this notion. An *inverval* constrained PRF allows sets of the form $\{1, \ldots, \ell\}$ and $\{\ell', \ldots, D\}$ where the domain is $\{1, \ldots, D\}$. The construction of [GGM86] also satisfies this stronger notion. We also consider constrained PRFs for circuit predicates, where $\mathcal{S}$ consists of all sets accepted by polynomial-sized circuits. Boneh and Waters [BW13] show how to build a constrained PRF for circuit predicates using multilinear maps. In the full version [BZ14], we show that such PRFs can also be built from indistinguishability obfuscation and any punctured PRF.

3   Key Exchange from Indistinguishability Obfuscation

In this section, we show how to realize multiparty non-interactive key exchange (NIKE) from general indistinguishability obfuscation. Intuitively, a NIKE protocol allows a group of users to simultaneously publish a single message, and all will derive the same shared group key. The first such protocols [BS03, GGH13a, CLT13] are based on multilinear maps. Our construction, based on a generic iO obfuscator, has the following properties:

- Using a punctured pseudorandom function, our protocol achieves a *static* notion of security, similar to existing protocols.
- Using a constrained pseudorandom function for circuit predicates, our protocol achieves a stronger notion of security called *semi-static* security. We show in the full version [BZ14] how to use iO to construct constrained pseudorandom functions for circuit predicates from any secure puncturable PRF.

- While our base protocol requires a trusted setup phase, our setup phase can be run *independently* of the messages sent by users. In the full version [BZ14] we use this property to remove the setup phase altogether, arriving at the first NIKE protocol *without trusted setup*. We provide protocols for both static and semi-static security.

We begin by first defining NIKE protocols and their security. To setup the NIKE protocol for N users, run a procedure Setup(λ, N), which outputs public parameters params. Then each party $i \in [N]$ runs a publish algorithm Publish(params, i), which generates two values: a user secret key sk_i and a user public value pv_i. User i keeps sk_i as his secret, and publishes pv_i to the other users. Finally, each user runs a key generation algorithm KeyGen(params, i, sk_i, $\{\mathsf{pv}_j\}_{j\in[N]}$) using their secret and all other user's public values, which outputs a shared key k.

For correctness, we require that each user derives the same secret key. That is, for all $i, i' \in [N]$,

$$\mathsf{KeyGen}(\mathsf{params}, i, \mathsf{sk}_i, \{\mathsf{pv}_j\}_{j\in[N]}) = \mathsf{KeyGen}(\mathsf{params}, i', \mathsf{sk}_{i'}, \{\mathsf{pv}_j\}_{j\in[N]})$$

For security, here we only consider a static notion of security. In the full version [BZ14], we also consider a stronger semi-static security notion. Fix a bit b, and consider the following experiment. The challenger runs params $\leftarrow$ Setup(λ, N). For $i \in [N]$, the challenger also runs $(\mathsf{sk}_i, \mathsf{pv}_i) \leftarrow$ Publish(params, i). Set $k_0 = \mathsf{KeyGen}(\mathsf{params}, 1, \mathsf{sk}_i, \{\mathsf{pv}_j\}_{j\in[N]})$ and $k_1 \leftarrow \mathcal{K}$. Give the adversary $\{\mathsf{pv}_j\}_{j\in[N]}, k_b$. For $b = 0, 1$ let W_b be the event that $b' = 1$ and define $\mathtt{AdvKE}(\lambda) = |\Pr[W_0] - \Pr[W_1]|$.

Definition 1. *A multiparty key exchange protocol* (Setup, Publish, KeyGen) *is statically secure if, for any polynomials N, and any PPT adversary $\mathcal{A}$, the function* $\mathtt{AdvKE}(\lambda)$ *is negligible.*

3.1   Our Construction

We now build a multiparty non-interactive key exchange (NIKE) from indistinguishability obfuscation and pseudorandom generators. The idea is the following: each party generates a seed s_i as their secret key, and publishes $x_i = \mathsf{PRG}(s_i)$ as their public value, where PRG is a pseudorandom generator. In the setup-phase, a key k is chosen for a punctured pseudorandom function PRF. The shared secret key will be the function PRF evaluated at the concatenation of the samples x_i. To allow the parties to compute the key, the setup will publish an obfuscated program for PRF which requires knowledge of a seed to operate. In this way, each of the parties can compute the key, but anyone else will not know any of the seeds, and will therefore be unable to compute the key.

The construction is as follows:

Construction 1. *Let* PRF *be a constrained pseudorandom function, and let* PRG : $\{0,1\}^{\lambda} \to \{0,1\}^{2\lambda}$ *be a pseudorandom generator. Let* iO *be a program indistinguishability obfuscator.*

- Setup(λ, G, N): *Sets up the key exchange protocol supporting at most N users and allowing any group of at most G users to compute a shared secret key. Choose a random key to obtain an instance of a pseudorandom function* PRF. *Build the program P_{KE} in Figure 2, padded to the appropriate length[1]. Also choose a random $x_0 \in \{0,1\}^{2\lambda}$. Output $P_{iO} = iO(P_{KE})$ and x_0 as the public parameters.*
- Publish(λ): *Party i chooses a random seed $s_i \in \{0,1\}^\lambda$ as a secret key, and publish $x_i = $ PRG(s_i)*
- KeyGen($P_{iO}, x_0, i, s_i, S, \{x_i\}_{i \in S}$): *Abort if $|S| > G$ or $i \notin S$. Let $S(j)$ denote the jth index in S, and $S^{-1}(k)$ for $k \in S$ be the inverse. Let*

$$\hat{x}_j = \begin{cases} x_{S(j)} & \text{if } j \leq |S| \\ x_0 & \text{if } j > |S| \end{cases}$$

Run P_{iO} on $(\hat{x}_1, ..., \hat{x}_G, S^{-1}(i), s_i)$ to obtain $k = $ PRF$(\hat{x}_1, ..., \hat{x}_G)$ or $\perp$.

Inputs: $\hat{x}_1, \ldots \hat{x}_G \in \mathcal{X}^G$, $i \in [G]$, $s \in \mathcal{S}$
Constants: PRF

1. If $\hat{x}_i \neq $ PRG(s), output $\perp$
2. Otherwise, output PRF$(\hat{x}_1, \hat{x}_2, \ldots, \hat{x}_G)$

Fig. 2. The program P_{KE} (same as Figure 1)

Correctness is trivial by inspection. For security, we consider two cases. If PRF is a punctured PRF, then we get static security. If PRF is a constrained PRF for circuit predicates, then our construction actually achieves a semi-static notion of security (as defined in the full version [BZ14]). Security is summarized by the following theorem:

Theorem 2. *If* PRG *is a secure pseudorandom generator,* PRF *a secure punctured PRF, and* iO *a secure indistinguishability obfuscator, then Construction 1 is a statically secure NIKE. If, in addition,* PRF *is a secure constrained PRF for circuit predicates, then Construction 1 is semi-statically secure.*

Removing trusted setup. Before proving Theorem 2, notice that if the adversary is able to learn the random coins used by Setup, he will be able to break the scheme. All prior key exchange protocols [GGH13a, CLT13] also suffer from this weakness. However, note that, unlike previous protocols, Publish does not depend on params. This allows us to remove trusted setup as follows: each user

---

[1] To prove security, we will replace P_{KE} with the obfuscation of another program P'_{KE}, which may be larger than P_{KE}. In order for the obfuscations to be indistinguishable, both programs must have the same size.

runs both Setup and Publish, and publishes their own public parameters params_i along with x_i. Then in KeyGen, choose some params_i in a canonical way (say, corresponding smallest x_i when treated as an integer), and run the original key exchange protocol using $\mathsf{params} = \mathsf{params}_i$. Now there is no setup, and static security follows from the static security of the original scheme. However, in the full version [BZ14] we show that semi-static does not follow. Instead, in the full version we give a modified construction that achieves semi-static security.

The proof of Theorem 2 is given in the full version [BZ14]. Here we sketch the main idea:

Proof sketch. For simplicity, assume $G = N$, though it is easy to generalize to $N > G$. In the static security game, the challenger draws N random seeds s_i^*, and lets $x_i^* = \mathsf{PRG}(s_i^*)$. It also constructs an obfuscation of the program P_{KE} in Figure 2. Then it gives this obfuscation, all of the x_i^*, and a challenge key k^* to the adversary $\mathcal{A}$. $\mathcal{A}$ then outputs its guess for whether $k^* = \mathsf{PRF}(x_1^*, \ldots, x_N^*)$ or not. We first slightly change the game by choosing the x_i^* uniformly at random in $\{0, 1\}^{2\lambda}$. The security of PRG shows that this modification at most negligibly changes $\mathcal{A}$'s advantage. Because the image of PRG is so much bigger than its domain, with high probability, none of the x_i^* are in the image of PRG. Thus, we can modify P_{KE} to obtain a new program P'_{KE} that aborts whenever an input x_i equals x_i^* for some i and this does not change the functionality of P_{KE}. Now P'_{KE} never evaluates PRF on the point $(x_1^*, \ldots, x_N^*)$, so we can puncture PRF at that point, and include only the punctured program in P'_{KE}. The indistinguishability of obfuscations shows that these modifications are undetectable by $\mathcal{A}$. We can simulate the view of $\mathcal{A}$ using only the punctured PRF, and $\mathcal{A}$ still succeeds with non-negligible probability. However, $\mathcal{A}$ now distinguishes the correct value of PRF at the puncture point from a truly random value, violating the security of PRF. For semi-static security, we need to puncture PRF at all points corresponding to the various subsets the adversary may challenge on, which we do using a PRF for circuit predicates.   □

4 Traitor Tracing with Small Parameters

In this section, we present a private linear broadcast encryption (PLBE) scheme, which has short ciphertexts, secret keys, and public keys. Boneh, Sahai, and Waters [BSW06] show that this implies a fully collusion resistant traitor tracing system with the same parameters.

Our approach gives a more general primitive called a recipient private broadcast system. Informally, a recipient private broadcast system allows the broadcaster to broadcast a message to a subset of N users. For security, we require that any user outside of the recipient set cannot learn the message, and each user only learns one bit of information about the recipient set: whether or not they are in it. We give a formal definition in the full version [BZ14]. Private linear broadcast encryption is recipient broadcast encryption where the only recipient sets allowed are $\emptyset = [0], [1], \ldots, [N]$.

To setup a PLBE scheme for N users, run a setup procedure $\mathsf{Setup}(\lambda, N)$, which outputs public parameters params and user secret keys $\{\mathsf{sk}_i\}_{i \in [N]}$ for each user. Distribute sk_i to user i. To encrypt to a set $[j]$, run an encryption algorithm $\mathsf{Enc}(\mathsf{params}, j)$ to obtain a header Hdr and message encryption key k. Use k to encrypt the message, and broadcast Hdr along with the resulting ciphertext. A user $i \leq j$ decrypts by running $\mathsf{Dec}(\mathsf{params}, \mathsf{sk}_i, \mathsf{Hdr})$, which outputs the message encryption key that can be used to actually decrypt the ciphertext.

For correctness, we require that any encryption to a set $[i]$ can be decrypted by any user in $[j]$. In other words, if $(\mathsf{Hdr}, k) \leftarrow \mathsf{Enc}(\mathsf{params}, j)$, then $\mathsf{Dec}(\mathsf{params}, \mathsf{sk}_i, \mathsf{Hdr}) = k$ for $i \leq j$.

For security, we have two experiments: semantic security and recipient privacy. For semantic security, fix a bit b and consider the following experiment. The adversary commits to a set $[j]$. The challenger then runs $\mathsf{params}, \{\mathsf{sk}_i\}_{i \in [N]} \leftarrow \mathsf{G}(\lambda, N)$, and then gives params as well as the secret keys $\{\mathsf{sk}_i\}_{i > j}$ for users not in $[j]$ to the adversary. The challenger also runs $(\mathsf{Hdr}, k_0) \leftarrow \mathsf{Enc}(\mathsf{params}, j)$, and generates $k_1 \leftarrow \mathcal{K}$, and gives Hdr, k_b. The adversary outputs a guess b' for b. For $b = 0, 1$ let W_b be the event that $b' = 1$ and $\mathsf{PLBE}_{\mathsf{SS}}^{(\mathrm{adv})}(\lambda) = |\Pr[W_0] - \Pr[W_1]|$.

For recipient privacy, also fix a bit b. The adversary commits to a user i^*. The challenger runs $\mathsf{params}, \{\mathsf{sk}_i\}_{i \in [N]} \leftarrow \mathsf{G}(\lambda, N)$, and gives params as well as the secret keys $\{\mathsf{sk}_i\}_{i \neq i^*}$ for all users except i^* to the adversary. The challenger also runs $(\mathsf{Hdr}, k) \leftarrow \mathsf{Enc}(\mathsf{params}, i^* - b)$ and gives Hdr to the adversary. The adversary outputs a guess b' for b. For $b = 0, 1$ let W_b be the event that $b' = 1$ and $\mathsf{PLBE}_{\mathsf{RP}}^{(\mathrm{adv})}(\lambda) = |\Pr[W_0] - \Pr[W_1]|$.

Definition 2. *A private linear broadcast (PLBE) scheme* ($\mathsf{Setup}, \mathsf{Publish}, \mathsf{KeyGen}$) *is secure if, for any polynomials N, and any PPT adversary $\mathcal{A}$, the functions $\mathsf{PLBE}_{\mathsf{SS}}^{(\mathrm{adv})}(\lambda)$ and $\mathsf{PLBE}_{\mathsf{RP}}^{(\mathrm{adv})}(\lambda)$ are negligible.*

4.1 Private Broadcast Encryption: First Construction

Construction overview. Since a broadcast ciphertext should reveal as little as possible about the recipient set S our plan is to embed an encryption of the set S in the broadcast ciphertext. The public-key will contain an obfuscated program that decrypts the encrypted recipient set S and then outputs a message decryption key only if the recipient can prove it is a member of S. However, encrypting the set S so that we can prove security using iO is non-trivial, and requires a certain type of encryption system.

In more detail, each user's private key will be a random seed s_i, and we let $x_i = \mathsf{PRG}(s_i)$ as in the previous section. We need to allow user i to learn the message decryption key for all sets S containing i. To that end, we include in the public key an obfuscated program that takes three inputs: an encrypted recipient set, an index i, and a seed s_i. The program decrypts the encrypted set, checks that the index i is in the set, and that the seed s_i is correct for that index (i.e. $x_i = \mathsf{PRG}(s_i)$). If all the checks pass, the program evaluates some pseudorandom function on the ciphertext to obtain the message decryption key and outputs that key.

We immediately see a problem with the description above: the obfuscated program must, at a minimum, have each of the x_i embedded in it, making the program and hence the public key linear in size. To keep the public key short, we instead generate the seeds s_i using a pseudorandom function PRF_{sk}: $s_i = \mathsf{PRF}_{sk}(i)$. We then have the program compute the x_i on the fly as $x_i = \mathsf{PRG}(\mathsf{PRF}_{sk}(i))$.

Another problem with the above description is that encrypting the recipient set S using a generic CPA-secure encryption scheme is insufficient for providing recipient privacy. The problem is that ciphertexts may be malleable: an attacker may be able to transform an encryption of a set S containing user i into an encryption of a set S' containing user j instead (that is, j is in S' if and only if i is in S). Now the attacker can use user j's secret key to decrypt the broadcast ciphertext. If decryption succeeds the attacker learns that user i is in the original ciphertext's recipient set, despite not having user i's secret key. This violates recipient privacy.

To solve this problem, we authenticate the encrypted recipient set using a message authentication code (MAC). However, proving security is a bit challenging because the decryption program must include the secret MAC key, and we need to ensure that this key does not leak to the attacker. We do so by implementing the MAC using a constrained PRF that supports interval constraints. We then prove that this is sufficient to thwart the aforementioned malleability attacks and allows us to prove security of the scheme.

We now present our private linear broadcast construction (i.e. the case where $\mathcal{S} = \mathsf{Lin}_N$). We first present a private-key variant, where a secret broadcast key is required to encrypt. In the full version [BZ14], we show how to make the system public-key. We discuss extending this construction to other set systems at the end of the section.

Construction 3. *Our traitor tracing scheme consists of three algorithms* (Setup, Enc, Dec) *defined as follows:*

- Setup(λ, N): *Let* $\mathsf{PRF}_{enc} : \{0,1\}^{2\lambda} \to [N]$ *and* $\mathsf{PRF}_{key} : \{0,1\}^{2\lambda} \times \{0,\ldots,N\}$ $\to \{0,1\}^{\lambda}$ *be punctured PRFs and* $\mathsf{PRF}_{mac} : \{0,1\}^{2\lambda} \times \{0,\ldots,N\} \to \{0,1\}^{\lambda}$ *and* $\mathsf{PRF}_{sk} : [N] \to \{0,1\}^{\lambda}$ *be interval constrained PRFs. Let* $s_i \leftarrow \mathsf{PRF}_{sk}(i)$ *for each* $i \in [N]$. *Let* P_{TT-Dec} *be the program in Figure 3, padded to the appropriate length. User* i*'s secret key is* s_i, *and the public parameters are* params $= P_{\mathsf{Dec}} = \mathsf{iO}(P_{TT-Dec})$.
- Enc$((\mathsf{PRF}_{enc}, \mathsf{PRF}_{mac}, \mathsf{PRF}_{key}), [j])$: *Pick a random* $r \in \{0,1\}^{2\lambda}$, *and let* $c_1 \leftarrow \mathsf{PRF}_{enc}(r) + j \mod (N+1)$. *Let* $c_2 \leftarrow \mathsf{PRF}_{mac}(r, c_1)$. *Finally, let* $k \leftarrow \mathsf{PRF}_{key}(r, c_1)$. *Output* $(\mathsf{Hdr} = (r, c_1, c_2), k)$.
- Dec$(\mathsf{params}, s_i, i, r, c)$: *Run* $k \leftarrow P_{\mathsf{Dec}}(r, c, s_i, i)$.

A public-key system. As described, our scheme requires a secret broadcast key in order to encrypt. However, using the trick of Sahai and Waters [SW13], we show in the full version [BZ14] how to include in the public parameters an obfuscated program that allows anyone to encrypt.

Inputs: r, c_1, c_2, s, i
Constants: $\mathsf{PRF}_{enc}, \mathsf{PRF}_{mac}, \mathsf{PRF}_{key}, \mathsf{PRF}_{sk}$

1. Let $j \leftarrow c_1 - \mathsf{PRF}_1(r) \mod (N+1)$
2. Let $x \leftarrow \mathsf{PRG}(\mathsf{PRF}_{sk}(i))$
3. Let $y \leftarrow \mathsf{PRG}(\mathsf{PRF}_{mac}(r, c_1))$
4. Check that $\mathsf{PRG}(s) = x$, $\mathsf{PRG}(c_2) = y$, and $i \leq j$.   If check fails, output $\perp$ and stop
5. Otherwise, output $\mathsf{PRF}_{key}(r, c_1)$

Fig. 3. The program $P_{TT-\mathsf{Dec}}$

In our public key scheme, secret keys have length λ, and ciphertexts have size $3\lambda + \log(N+1)$. The public key consists of two obfuscated programs. The size of these programs is only dependent polylogarithmically on the number of users, so the obfuscated programs will have size $\mathsf{poly}(\log N, \lambda)$. Therefore, we simultaneously achieve small ciphertexts, secret keys, and public keys. Security is given by the following theorem:

Theorem 4. *If PRF_{enc} and PRF_{key} are secure punctured PRFs, PRF_{mac} and PRF_{sk} are secure interval constrained PRFs, and PRG is a secure pseudorandom generator, then $(\mathsf{Setup}, \mathsf{Enc}, \mathsf{Dec})$ in Construction 3 is an adaptively secure private linear broadcast encryption scheme.*

The proof is given in the full version [BZ14]. Here we sketch the main ideas:

Proof sketch. We must prove that our scheme is both semantically secure and has recipient privacy. For semantic security, the adversary $\mathcal{A}$ commits to a set $[j^*]$, and receives the secret keys s_i for all $i > j^*$. $\mathcal{A}$ also receives the obfuscation of $P_{TT-\mathsf{Dec}}$, as well as a challenge (r^*, c_1^*, c_2^*) that is an encryption to the set $[j^*]$, and a key k^*. $\mathcal{A}$ must distinguish the correct k^* from random. Our first step is to modify $P_{TT-\mathsf{Dec}}$ by puncturing PRF_{sk} at each key that $\mathcal{A}$ does not receive, and hard-code the values $x_i = \mathsf{PRG}(\mathsf{PRF}_{sk}(i))$ into $P_{TT-\mathsf{Dec}}$ so that is correctly decrypts all valid ciphertexts. This does not change the functionality of $P_{TT-\mathsf{Dec}}$. We then replace these x_i with random values in $\{0,1\}^{2\lambda}$, and the security of PRF_{sk} and PRG shows that this change is undetectable by $\mathcal{A}$. Now, with overwhelming probability, none of the x_i for $i \leq j^*$ are in the image of PRG, so we can modify $P_{TT-\mathsf{Dec}}$ to abort if $i \leq j^*$. On the challenge ciphertext, $j = j^*$, the $P_{TT-\mathsf{Dec}}$ will also abort if $i > j^*$, meaning $P_{TT-\mathsf{Dec}}$ will always abort. Therefore, we can puncture PRF_{key} at (r^*, c_1^*) without modifying the functionality of $P_{TT-\mathsf{Dec}}$. The indistinguishability of obfuscations shows that these changes are undetectable. However, $\mathcal{A}$ now distinguishes the correct value of PRF_{key} ad (r^*, c_1^*) from random, violating the security of PRF_{key}. One problem with the above proof is that hard-coding all x_i into $P_{TT-\mathsf{Dec}}$ expands its size considerably. We show in the full version [BZ14] how to puncture PRF_{sk}, one user at a time, using a sequence of hybrids while keeping $P_{TT-\mathsf{Dec}}$ small.

For recipient privacy, the proof is more complicated but similar. Here, $\mathcal{A}$ commits to a j^*, and receives all secret keys except those for user j^* and and encryption to the set $[j^* - b]$ for some $b \in \{0,1\}$. $\mathcal{A}$ must determine b. Similar to the semantic security case, we puncture PRF_{sk} at j^* and replace $x_{j^*} = \mathsf{PRG}(\mathsf{PRF}_{sk}(j^*))$ with a truly random value in $\{0,1\}^{2\lambda}$, and then modify $P_{TT-\mathsf{Dec}}$ so that it aborts if $i = j^*$. Now we puncture PRF_{mac} at all points (r^*, c_1), and hard-code $y_{c_1} = \mathsf{PRG}(\mathsf{PRF}_{mac}(r^*, c_1))$ into $P_{TT-\mathsf{Dec}}$. For each $c_1 \neq c_1^*$, we replace y_{c_1} with a truly random value in $\{0,1\}^{2\lambda}$. Similar to the semantic security proof, we have to puncture iteratively in order to keep the program size small. At this point, the only (r^*, c_1, c_2) that authenticates is the challenge ciphertext itself. This means we can puncture PRF_{enc} at r^*, and hard-code the necessary values to decrypt the challenge ciphertext, including $z^* = \mathsf{PRF}_{enc}(r^*)$. The security of PRF_{enc} shows that we can replace z^* with a truly random value. At this point, $c_1^* = j^* - b + z^*$. As we show in the full version [BZ14], in the $b = 0$ case, we can replace z^* with $z^* - 1$ without changing the functionality of $P_{TT-\mathsf{Dec}}$. However, moving to $z^* - 1$ also moves us to the $b = 1$ case, meaning that $\mathcal{A}$ actually breaks the indistinguishability of obfuscations. $\qquad\square$

4.2 Extension to Other Set Systems

Construction 3 easily extends the other classes of recipient sets — for example, the set of all subsets of $[N]$, or the subsets of size exactly r. Ciphertexts will simply be an encryption of (the description of) the recipient set, and the obfuscated program will output the PRF applied to the ciphertext only if the user can supply a valid seed for one of the users in the set. However, now the number of possible recipient sets is exponential, and consequently our security reduction becomes non-polynomial. In the full version [BZ14], we give a different construction that has a polynomial security proof for these classes of recipient sets. However, the public key size becomes $N^{O(1)}$.

Acknowledgments. This work is supported by NSF, DARPA, IARPA, and others, as listed in the full version.

References

[BBW06]   Barth, A., Boneh, D., Waters, B.: Privacy in encrypted content distribution using private broadcast encryption. In: Di Crescenzo, G., Rubin, A. (eds.) FC 2006. LNCS, vol. 4107, pp. 52–64. Springer, Heidelberg (2006)

[BGI+01]   Barak, B., Goldreich, O., Impagliazzo, R., Rudich, S., Sahai, A., Vadhan, S.P., Yang, K.: On the (Im)possibility of obfuscating programs. In: Kilian, J. (ed.) CRYPTO 2001. LNCS, vol. 2139, pp. 1–18. Springer, Heidelberg (2001)

[BGI13]   Boyle, E., Goldwasser, S., Ivan, I.: Functional signatures and pseudorandom functions. Cryptology ePrint Archive, Report 2013/401 (2013)

[BGK+13]   Barak, B., Garg, S., Kalai, Y.T., Paneth, O., Sahai, A.: Protecting ob-
           fuscation against algebraic attacks. Cryptology ePrint Archive, Report
           2013/631 (2013), http://eprint.iacr.org/

[BGW05]   Boneh, D., Gentry, C., Waters, B.: Collusion resistant broadcast en-
           cryption with short ciphertexts and private keys. In: Shoup, V. (ed.)
           CRYPTO 2005. LNCS, vol. 3621, pp. 258–275. Springer, Heidelberg
           (2005)

[BN08]   Boneh, D., Naor, M.: Traitor tracing with constant size ciphertext.
           In: ACM Conference on Computer and Communications Security,
           pp. 501–510 (2008)

[BR13a]   Brakerski, Z., Rothblum, G.N.: Black-box obfuscation for d-cnfs. Cryptol-
           ogy ePrint Archive, Report 2013/557 (2013), http://eprint.iacr.org/

[BR13b]   Brakerski, Z., Rothblum, G.N.: Virtual black-box obfuscation for all cir-
           cuits via generic graded encoding. Cryptology ePrint Archive, Report
           2013/563 (2013), http://eprint.iacr.org/

[BS03]   Boneh, D., Silverberg, A.: Applications of multilinear forms to cryptog-
           raphy. Contemporary Mathematics 324, 71–90 (2003)

[BSW06]   Boneh, D., Sahai, A., Waters, B.: Fully Collusion Resistant Traitor Trac-
           ing with Short Ciphertexts and Private Keys. In: Vaudenay, S. (ed.)
           EUROCRYPT 2006. LNCS, vol. 4004, pp. 573–592. Springer, Heidelberg
           (2006)

[BW06]   Boneh, D., Waters, B.: A fully collusion resistant broadcast trace and
           revoke system with public traceability. In: ACM Conference on Computer
           and Communication Security, CCS (2006)

[BW13]   Boneh, D., Waters, B.: Constrained Pseudorandom Functions and Their
           Applications. In: Sako, K., Sarkar, P. (eds.) ASIACRYPT 2013, Part II.
           LNCS, vol. 8270, pp. 280–300. Springer, Heidelberg (2013)

[BWZ14]   Boneh, D., Waters, B., Zhandry, M.: Low overhead broadcast encryption
           from multilinear maps. In: Garay, J.A., Gennaro, R. (eds.) CRYPTO
           2014, Part I. LNCS, vol. 8616, pp. 206–223. Springer, Heidelberg (2014)

[BZ14]   Boneh, D., Zhandry, M.: Multiparty key exchange, efficient traitor trac-
           ing, and more from indistinguishability obfuscation. Full version available
           at the Cryptology ePrint Archives http://eprint.iacr.org/2013/642.

[Can97]   Canetti, R.: Towards realizing random oracles: Hash functions that hide
           all partial information. In: Kaliski Jr., B.S. (ed.) CRYPTO 1997. LNCS,
           vol. 1294, pp. 455–469. Springer, Heidelberg (1997)

[CFN94]   Chor, B., Fiat, A., Naor, M.: Tracing traitors. In: Desmedt, Y.G. (ed.)
           CRYPTO 1994. LNCS, vol. 839, pp. 257–270. Springer, Heidelberg (1994)

[CLT13]   Coron, J.-S., Lepoint, T., Tibouchi, M.: Practical Multilinear Maps over
           the Integers. In: Canetti, R., Garay, J.A. (eds.) CRYPTO 2013, Part I.
           LNCS, vol. 8042, pp. 476–493. Springer, Heidelberg (2013)

[CMR98]   Canetti, R., Micciancio, D., Reingold, O.: Perfectly One-Way Probabilis-
           tic Hash Functions. In: Proc. of STOC 1998, pp. 131–140 (1998)

[CPP05]   Chabanne, H., Phan, D.H., Pointcheval, D.: Public traceability in traitor
           tracing schemes. In: Cramer, R. (ed.) EUROCRYPT 2005. LNCS,
           vol. 3494, pp. 542–558. Springer, Heidelberg (2005)

[CRV10]   Canetti, R., Rothblum, G.N., Varia, M.: Obfuscation of hyperplane mem-
           bership. In: Micciancio, D. (ed.) TCC 2010. LNCS, vol. 5978, pp. 72–89.
           Springer, Heidelberg (2010)

[Del07]   Delerablée, C.: Identity-Based Broadcast Encryption with Constant Size Ciphertexts and Private Keys. In: Kurosawa, K. (ed.) ASIACRYPT 2007. LNCS, vol. 4833, pp. 200–215. Springer, Heidelberg (2007)

[DF02]   Dodis, Y., Fazio, N.: Public key broadcast encryption for stateless receivers. In: Feigenbaum, J. (ed.) DRM 2002. LNCS, vol. 2696, pp. 61–80. Springer, Heidelberg (2003)

[DF03]   Dodis, Y., Fazio, N.: Public key broadcast encryption secure against adaptive chosen ciphertext attack. In: Desmedt, Y.G. (ed.) PKC 2003. LNCS, vol. 2567, pp. 100–115. Springer, Heidelberg (2002)

[DNR+09]   Dwork, C., Naor, M., Reingold, O., Rothblum, G.N., Vadhan, S.: On the complexity of differentially private data release: efficient algorithms and hardness results. In: Proceedings of STOC 2009 (2009)

[DPP07]   Delerablée, C., Paillier, P., Pointcheval, D.: Fully collusion secure dynamic broadcast encryption with constant-size ciphertexts or decryption keys. In: Takagi, T., Okamoto, T., Okamoto, E., Okamoto, T. (eds.) Pairing 2007. LNCS, vol. 4575, pp. 39–59. Springer, Heidelberg (2007)

[FHKP13]   Freire, E.S.V., Hofheinz, D., Kiltz, E., Paterson, K.G.: Non-interactive key exchange. In: Kurosawa, K., Hanaoka, G. (eds.) PKC 2013. LNCS, vol. 7778, pp. 254–271. Springer, Heidelberg (2013)

[FHPS13]   Freire, E.S.V., Hofheinz, D., Paterson, K.G., Striecks, C.: Programmable Hash Functions in the Multilinear Setting. In: Canetti, R., Garay, J.A. (eds.) CRYPTO 2013, Part I. LNCS, vol. 8042, pp. 513–530. Springer, Heidelberg (2013)

[FN94]   Fiat, A., Naor, M.: Broadcast encryption. In: Stinson, D.R. (ed.) CRYPTO 1993. LNCS, vol. 773, pp. 480–491. Springer, Heidelberg (1994)

[FP12]   Fazio, N., Perera, I.M.: Outsider-anonymous broadcast encryption with sublinear ciphertexts. In: Fischlin, M., Buchmann, J., Manulis, M. (eds.) PKC 2012. LNCS, vol. 7293, pp. 225–242. Springer, Heidelberg (2012)

[Fre10]   Freeman, D.M.: Converting pairing-based cryptosystems from composite-order groups to prime-order groups. In: Gilbert, H. (ed.) EUROCRYPT 2010. LNCS, vol. 6110, pp. 44–61. Springer, Heidelberg (2010)

[GGH13a]   Garg, S., Gentry, C., Halevi, S.: Candidate multilinear maps from ideal lattices. In: Johansson, T., Nguyen, P.Q. (eds.) EUROCRYPT 2013. LNCS, vol. 7881, pp. 1–17. Springer, Heidelberg (2013)

[GGH+13b]   Garg, S., Gentry, C., Halevi, S., Raykova, M., Sahai, A., Waters, B.: Candidate indistinguishability obfuscation and functional encryption for all circuits. In: Proc. of FOCS 2013 (2013)

[GGM86]   Goldreich, O., Goldwasser, S., Micali, S.: How to Construct Random Functions. Journal of the ACM (JACM) 33(4), 792–807 (1986)

[GKSW10]   Garg, S., Kumarasubramanian, A., Sahai, A., Waters, B.: Building efficient fully collusion-resilient traitor tracing and revocation schemes. In: ACM Conference on Computer and Communications Security, pp. 121–130 (2010)

[GR07]   Goldwasser, S., Rothblum, G.N.: On best-possible obfuscation. In: Vadhan, S.P. (ed.) TCC 2007. LNCS, vol. 4392, pp. 194–213. Springer, Heidelberg (2007)

[GST04]   Goodrich, M.T., Sun, J.Z., Tamassia, R.: Efficient tree-based revocation in groups of low-state devices. In: Franklin, M. (ed.) CRYPTO 2004. LNCS, vol. 3152, pp. 511–527. Springer, Heidelberg (2004)

[GW09]   Gentry, C., Waters, B.: Adaptive security in broadcast encryption systems (with short ciphertexts). In: Joux, A. (ed.) EUROCRYPT 2009. LNCS, vol. 5479, pp. 171–188. Springer, Heidelberg (2009)

[HS02]   Halevy, D., Shamir, A.: The LSD broadcast encryption scheme. In: Yung, M. (ed.) CRYPTO 2002. LNCS, vol. 2442, pp. 47–60. Springer, Heidelberg (2002)

[HSW13]   Hohenberger, S., Sahai, A., Waters, B.: Replacing a random oracle: Full domain hash from indistinguishability obfuscation. Cryptology ePrint Archive, Report 2013/509 (2013)

[Jou04]   Joux, A.: A One Round Protocol for Tripartite Diffie-Hellman. Journal of Cryptology 17(4), 263–276 (2004)

[KPTZ13]   Kiayias, A., Papadopoulos, S., Triandopoulos, N., Zacharias, T.: Delegatable pseudorandom functions and applications. In: Proceedings ACM CCS (2013)

[KRW13]   Koppula, V., Ramchen, K., Waters, B.: Separations in circular security for arbitrary length key cycles. Cryptology ePrint Archive, Report 2013/683 (2013), http://eprint.iacr.org/

[KS13]   Kiayias, A., Samari, K.: Lower bounds for private broadcast encryption. In: Kirchner, M., Ghosal, D. (eds.) IH 2012. LNCS, vol. 7692, pp. 176–190. Springer, Heidelberg (2013)

[KY02]   Kiayias, A., Yung, M.: Breaking and repairing asymmetric public-key traitor tracing. In: Feigenbaum, J. (ed.) DRM 2002. LNCS, vol. 2696, pp. 32–50. Springer, Heidelberg (2003)

[LPQ12]   Libert, B., Paterson, K.G., Quaglia, E.A.: Anonymous broadcast encryption: Adaptive security and efficient constructions in the standard model. In: Fischlin, M., Buchmann, J., Manulis, M. (eds.) PKC 2012. LNCS, vol. 7293, pp. 206–224. Springer, Heidelberg (2012)

[LPS04]   Lynn, B., Prabhakaran, M., Sahai, A.: Positive results and techniques for obfuscation. In: Cachin, C., Camenisch, J.L. (eds.) EUROCRYPT 2004. LNCS, vol. 3027, pp. 20–39. Springer, Heidelberg (2004)

[LSW10]   Lewko, A.B., Sahai, A., Waters, B.: Revocation systems with very small private keys. In: IEEE Symposium on Security and Privacy, pp. 273–285 (2010)

[MR13]   Moran, T., Rosen, A.: There is no indistinguishability obfuscation in pessiland. Cryptology ePrint Archive, Report 2013/643 (2013), http://eprint.iacr.org/

[NNL01]   Naor, D., Naor, M., Lotspiech, J.: Revocation and tracing schemes for stateless receivers. In: Kilian, J. (ed.) CRYPTO 2001. LNCS, vol. 2139, pp. 41–62. Springer, Heidelberg (2001)

[NP00]   Naor, M., Pinkas, B.: Efficient trace and revoke schemes. In: Frankel, Y. (ed.) FC 2000. LNCS, vol. 1962, pp. 1–20. Springer, Heidelberg (2001)

[Pfi96]   Pfitzmann, B.: Trials of traced traitors. In: Anderson, R. (ed.) IH 1996. LNCS, vol. 1174, pp. 49–64. Springer, Heidelberg (1996)

[PW97]   Pfitzmann, B., Waidner, M.: Asymmetric fingerprinting for larger collusions. In: Proceedings of the ACM Conference on Computer and Communication Security, pp. 151–160 (1997)

[SF07]   Sakai, R., Furukawa, J.: Identity-Based Broadcast Encryption. IACR Cryptology ePrint Archive (2007)

[Sir07]   Sirvent, T.: Traitor tracing scheme with constant ciphertext rate against powerful pirates. In: Workshop on Coding and Cryptography (2007)

[SW13]   Sahai, A., Waters, B.: How to Use Indistinguishability Obfuscation: Deniable Encryption, and More. Cryptology ePrint Archive, Report 2013/454 (2013), `http://eprint.iacr.org/`

[Ull13]   Ullman, J.: Answering $n^{\{2+o(1)\}}$ counting queries with differential privacy is hard. In: STOC, pp. 361–370 (2013)

[Wee05]   Wee, H.: On obfuscating point functions. In: Proc. of STOC 2005, p. 523 (2005)

[WHI01]   Watanabe, Y., Hanaoka, G., Imai, H.: Efficient asymmetric public-key traitor tracing without trusted agents. In: Naccache, D. (ed.) CT-RSA 2001. LNCS, vol. 2020, pp. 392–407. Springer, Heidelberg (2001)

Indistinguishability Obfuscation
from Semantically-Secure Multilinear Encodings

Rafael Pass*, Karn Seth, and Sidharth Telang

Cornell University, New York, NY, USA
{rafael,karn,sidtelang}@cs.cornell.edu

Abstract. We define a notion of semantic security of multilinear (a.k.a. graded) encoding schemes, which stipulates security of a class of algebraic "decisional" assumptions: roughly speaking, we require that for every nuPPT distribution D over two *constant-length* sequences m_0, m_1 and auxiliary elements z such that all arithmetic circuits (respecting the multilinear restrictions and ending with a zero-test) are *constant* with overwhelming probability over (m_b, z), $b \in \{0, 1\}$, we have that encodings of m_0, z are computationally indistinguishable from encodings of m_1, z. Assuming the existence of semantically secure multilinear encodings and the LWE assumption, we demonstrate the existence of indistinguishability obfuscators for all polynomial-size circuits.

1 Introduction

The goal of *program obfuscation* is to "scramble" a computer program, hiding its implementation details (making it hard to "reverse-engineer"), while preserving the functionality (i.e, input/output behavior) of the program. Precisely defining what it means to "scramble" a program is non-trivial: on the one hand, we want a definition that can be plausibly satisfied, on the other hand, we want a definition that is useful for applications.

Hada [Had00] and Barak, Goldreich, Impagliazzo, Rudich, Sahai, Vadhan, and Yang [BGI+01] show that simulation-based notion such as *virtual black-box obfuscation (VBB)* [BGI+01]—which, roughly speaking, require that everything that can be learn from the code of the obfuscated program can be simulated using just black-box access to the functionality—run into strong impossibility results.

We here focus on the notion of *indistinguishability obfuscation*, first defined by Barak *et al.* [BGI+01] and explored by Garg, Gentry, Halevi, Raykova, Sahai, and Waters [GGH+13b]. Roughly speaking, this notion requires that obfuscations $\mathcal{O}(C_1)$ and $\mathcal{O}(C_2)$ of any two *equivalent* circuits C_1 and C_2 (i.e., whose

* Work supported in part by a Microsoft Faculty Fellowship, NSF Award CNS-1217821, NSF CAREER Award CCF-0746990, NSF Award CCF-1214844, AFOSR YIP Award FA9550-10-1-0093, and DARPA and AFRL under contract FA8750-11-2-0211. The views and conclusions contained in this document are those of the authors and should not be interpreted as representing the official policies, either expressed or implied, of the Defense Advanced Research Projects Agency or the US Government.

J.A. Garay and R. Gennaro (Eds.): CRYPTO 2014, Part I, LNCS 8616, pp. 500–517, 2014.
© International Association for Cryptologic Research 2014

outputs agree on all inputs) from some class $\mathcal{C}$ are computationally indistinguishable. In a very recent breakthrough result, Garg, Gentry, Halevi, Raykova, Sahai, and Waters [GGH+13b] provided the first candidate constructions of indistinguishability obfuscators for all polynomial-size circuits, based on so-called *multilinear (a.k.a. graded) encodings* [BS03, Rot13, GGH13a]—for which candidate constructions were recently discovered in the seminal work of Garg, Gentry and Halevi [GGH13a], and more recently, alternative constructions were provided by Coron, Lepoint and Tibouchi [CLT13].

The obfuscator construction of Garg et al proceeds in two steps. They first provide a candidate construction of an indistinguishability obfuscator for NC^1 (this construction is essentially assumed to be secure); next, they demonstrate a "bootstrapping" theorem showing how to use fully homomorphic encryption (FHE) schemes [Gen09] and indistinguishability obfuscators for NC^1 to obtain indistinguishability obfuscators for all polynomial-size circuits. Further constructions of obfuscators for NC^1 were subsequently provided by Brakerski and Rothblum [BR14] and Barak, Garg, Kalai, Paneth and Sahai [BGK+13]—in fact, these constructions achieve the even stronger notion of virtual-black-box obfuscation in idealized "generic" multilinear encoding models.

In parallel with the development of candidate obfuscation constructions, several surprising applications of indistinguishability have emerged (see e.g., [[GGH+13b, SW14, HSW14, BZ14, GGHR14, BCP14, BCPR14, GGG+14], [KNY14,KMN+14]]). Furthermore, as shown by Goldwasser and Rothblum [GR07], indistinguishability obfuscators provide a very nice "best-possible" obfuscation guarantee: if a functionality can be VBB obfuscated (even non-efficiently!), then any indistinguishability obfuscator for this functionality is VBB secure.

1.1 Towards "Provably-Secure" Obfuscation

But despite these amazing developments, the following question remains open:

> *Can the security of general-purpose indistinguishability obfuscators be reduced to some "natural" intractability assumption?*

The principal goal of the current paper is to make progress toward addressing this question.

Note that while the construction of indistinguishability obfuscation of Garg et al is based on *some* intractability assumption, the assumption is very tightly tied to their scheme—in essence, the assumption stipulates that their scheme is a secure indistinguishability obfuscator.

The VBB constructions of Brakerski and Rothblum [BR14] and Barak et al [BGK+13] give us more confidence in the plausible security of their obfuscators, in that they show that at least "generic" attacks—that treat multilinear encoding as if they were "physical envelopes" on which multilinear operations can be performed—cannot be used to break security of the obfuscators. But at the same time, non-generic attacks against their scheme are known—since general-purpose VBB obfuscation is impossible. Thus, it is not clear to what extent security arguments in the generic multilinear encoding model should make us

more confident that these constructions satisfy e.g., a notion of indistinguishability obfuscation.[1] In particular, the question of to what extent one can capture
"real-world" security properties from security proofs in the generic model through
a "meta-assumption" (regarding multilinear encoding) was raised (but not investigated) in [BGK+13]; see Remark 1 there. In this work, we initiate a study of
this question.

1.2   Security of Multilinear (Graded) Encodings

Towards explaining the assumptions we consider, let us start by briefly recalling
multilinear (a.k.a. graded) encoding schemes [GGH13a, GGH+13b]. Roughly
speaking, such schemes enable anyone that has access to a *public parameter* pp
and *encodings* $E_S^x = \mathsf{Enc}(x, S)$, $E_S^y = \mathsf{Enc}(y, S')$ of ring elements x, y under the
sets $S, S' \subset [k]$ to *efficiently*:[2]

  – compute an encoding $E_{S \cup S'}^{x \cdot y}$ of $x \cdot y$ under the set $S \cup S'$, as long as $S \cap S' = \emptyset$;
  – compute an encoding E_S^{x+y} of $x + y$ under the set S as long as $S = S'$;
  – compute an encoding E_S^{x-y} of $x - y$ under the set S as long as $S = S'$.

(Given just access to the public-parameter pp, generating an encoding to a particular element x may not be efficient; however, it can be efficiently done given
access to the *secret parameter* sp.) Additionally, given an encoding E_S^x where
the set S is the whole universe $[k]$—called the "target set"—we can efficiently
check whether $x = 0$ (i.e., we can "zero-test" encodings under the target set $[k]$.)
In essence, multilinear encodings enable computations of certain restricted set
of arithmetic circuits (determined by the sets S under which the elements are
encoded) and finally determine whether the output of the circuit is 0; we refer
to these as the *legal arithmetic circuits*.

Semantical Security of Multilinear (Graded) Encodings. The above description only explains the *functionality* of multlinear encodings, but does not
discuss *security*. As far as we are aware, there have been two approaches to defining security of multilinear encodings. The first approach, initiated in [GGH13a],
stipulates specific hardness assumptions closely related to the DDH assumption.
The second approach instead focuses on *generic attackers* and assumes that the

---

[1] In fact, mirroring ideas from [GGSW13], assuming the existence of indistinguishability obfuscation and one-way functions it is easy to come up with a method to
sample C_1, C_2, z such that with high probability $C_1(z) \neq C_2(z)$ (and thus, given z,
we can easily distinguish obfuscations of them), yet the pair of circuits (C_1, C_2) are
indistinguishable from a pair of functionally equivalent circuits. Thus, there are "fake
attacks" on indistinguishability obfuscation that cannot be efficiently distinguished
from a real attack.

[2] Just as [BR14, BGK+13], we here rely on "set-based" graded encoding; these
were originally called "generalized" graded encodings in [GGH13a]. Following
[GGH+13b, BGK+13] (and in particular the notion of a "multilinear jigsaw puzzles" in [GGH+13b]), we additionally enable anyone with the secret parameter to
encode *any* elements (as opposed to just *random* elements as in [GGH13a]).

attacker does not get to see the actual encodings but instead can only access them through legal arithmetic circuits.

In this work, we consider the first approach, but attempt to capture a general *class* of algebraic "decisional" assumptions (such as the the graded DDH assumption of [GGH13a]) which holds against generic attackers (and as such, it can be viewed as a merge of the two approaches). In essence, our notion of (single-message) *semantical security* attempts to capture the intuition that encodings of elements m_0 and m_1 (under the set S) are indistinguishable in the presence of encodings of "auxiliary" elements z (under sets T), as long as m_0, m_1, z are sampled from *any* "nice" distribution D; in the context of a graded DDH assumption, think of z as a vector of independent uniform elements, m_0 as the product of the elements in z and m_1 as an independent uniform element. We analogously consider stronger notions of *constant-message* and *multi-message* semantical security, where m_0, m_1 (and S) are replaced by either constant-length or arbitrary polynomial-length vectors $\boldsymbol{m_0}, \boldsymbol{m_1}$ of elements (and sets $\boldsymbol{S}$).

Defining what makes a distribution D "nice" turns out to be quite non-trivial: A first (and minimal) approach—similar to e.g., the uber assumption of [BBG05] in the context of bilinear maps—would be to simply require that D samples elements $\boldsymbol{m_0}, \boldsymbol{m_1}, z$ such that no generic attacker can distinguish $\boldsymbol{m_0}, z$ and $\boldsymbol{m_0}, z$. As we discuss in Section 1.3, the most natural formalization of this approach can be attacked assuming standard cryptographic hardness assumptions. The distribution D considered in the attack, however, is "unnatural" in the sense that encodings of $\boldsymbol{m_b}, z$ actually leak information about $\boldsymbol{m_b}$ even to generic attackers (in fact, this information fully determines the bit b, it is just that it cannot be computed in polynomial time).

Our notion of a *valid* message distribution disallows such information leakage w.r.t. generic attacks. More precisely, we require that every (even unbounded-size) legal arithmetic circuit C is *constant* over $(\boldsymbol{m_b}, z)$, $b \in \{0,1\}$ with overwhelming probability; that is, there exists some bit c such that with overwhelming probability over $m_0, m_1, z \leftarrow D$, $C(\boldsymbol{m_b}, z) = c$ for $b \in \{0,1\}$ (recall that a legal arithmetic circuit needs to end with a zero-test and thus the output of the circuit will be either 0 or 1). We refer to any distribution D satisfying this property as being *valid*, and our formal definition of semantic security now only quantifies over such valid message distributions.

Obfuscation from Semantically-Secure Multilinear Encodings. As a starting point, we observe that slight variants of the constructions of [BR14, BGK+13] can be shown to satisfy indistinguishability obfuscation for NC^1 assuming *multi-message* semantically-secure multilinear encodings. In fact, any VBB secure obfuscation in the generic model where the construction only releases encodings of elements (as the constructions of [BR14, BGK+13] do) satisfies indistinguishability obfuscation assuming a slight strengthening of multi-message semantic security where validity only consider *polynomial-size* (as opposed to arbitrary-size) legal arithmetic circuits:[3] let $\boldsymbol{m_0}$ denote the elements corresponding to an obfuscation of some program Π_0, and $\boldsymbol{m_1}$ the elements corresponding

---

[3] We thank Sanjam Garg for this observation.

to an obfuscation of some functionally equivalent program Π_1. VBB security implies that all polynomial-size legal arithmetic circuits are constant with overwhelming probability over both m_0 and m_1 (as any such query can be simulated given black-box access to the functionality of the program), and thus encodings of m_0 and m_1 (i.e., obfuscations of Π_0 and Π_1) are indistinguishable. By slightly tweaking the construction of [BGK+13] and the analysis[4], we can extend this to hold against *all* (arbitrary-size) legal arithmetic circuits, and thus indistinguishability of the encodings (which implies indistinguishability of the obfuscations) follows as a direct consequence of the multi-message security assumption.

While this observation does takes us a step closer towards basing the security of obfuscation on a simple, natural, assumption, it is unappealing in that the assumption itself directly implies the security of the scheme (without any security reduction).

Our central result shows how to construct indistinguishability obfuscators for NC^1 based on the existence of *constant-message* semantically-secure multilinear encodings; in the sequel, we simply refer to such schemes as being semantically secure (dropping "constant-message" from the notation). Note that the constant-message restriction not only simplifes (and reduces the complexity) of the assumption, it also takes us a step closer to the more standard GDDH assumption. (As far as we know, essentially all "DDH-type" assumptions in "standard"/bilinear or multilinear settings consider a constant-message setting, stipulating indistinguishability of either a *single* or a *constant* number of elements in the presence of polynomially many auxiliary elements. It is thus safe to say that such constant-message indistinguishability assumptions are significantly better understood than their multi-message counterpart.)

Theorem 1 (Informally stated). *Assume the existence of semantically secure multilinear encodings. Then there exists an indistinguishability obfuscator for NC^1.*

As far as we know, this is the first result presenting indistinguishability obfuscators for NC^1 based on any type of assumption with a "non-trivial" security reduction w.r.t. arbitrary nuPPT attackers.

The core of our result is a general technique for transforming any obfuscator for matrix branching programs that satisfies a weak notion of *neighboring-matrix* indistinguishability obfuscation—which roughly speaking only requires indistinguishability of obfuscations of branching programs that differ only in a constant number of matrices—into a "full-fledged" indistinguishability obfuscator. We next show how to adapt the construction of [BGK+13] and its analysis to satisfy neighboring-matrix indistinguishability obfuscation based on semantical secure multilinear encodings; on a high-level, the security analysis in the generic model is useful for proving that the particular message distribution we consider is "valid".[5]

---

[4] Briefly, we need to tweak the construction to ensure a "perfect" simulation property.

[5] As we explain in more details later, to use our transformation, we need to deal with branching programs that satisfy a slightly more liberal definition of a branching program than what is used in earlier works. This is key reason why we need to modify the construction and analysis from [BGK+13].

If additionally assuming the existence of a leveled FHE [RAD78, Gen09] with decryption in NC^1—implied, for instance, by the LWE assumption [BV11, BGV12]—our construction can be bootstrapped up to obtain indistinguishability obfuscators for all polynomial-size circuits by relying on the technique from [GGH+13b].

Theorem 2 (Informally stated). *Assume the existence of semantically secure multilinear encodings and a leveled FHE with decryption in NC^1. Then there exists indistinguishability obfuscators for P/poly.*

Semantical Security w.r.t. Restricted Classes of Distributions. Our most basic notion of semantical security requires indistinguishability to hold w.r.t. to *any* "valid" message distribution. This may seem like a strong assumption. Firstly, such a notion can clearly not be satisfied by a *deterministic* encoding schemes (as envisioned in the original work of [BS03])—we can never expect encodings of 0 and 1 (under a non target set, and without any auxiliary inputs) to be indistinguishable. Secondly, even if we have a randomized encoding scheme in mind (such as the candidates of [GGH13a, CLT13]), giving the attacker access to encodings of *arbitrary* elements may be dangerous: As mentioned in [GGH13a], attacks (referred to as "weak discrete logarithm attacks") on their scheme are known in settings where the attacker can get access to "non-trivial" encodings of 0 under any *non-target* set $S \subset [k]$. (We mention that, as far as we know, no such attacks are currently known on the candidate construction of [CLT13].)

For the purposes of the results in our paper, however, it suffices to consider a notion of semantical security w.r.t. *restricted classes of distributions* D. In particular, to deal with both of the above issues, we consider "high-entropy" distributions D that sample elements m_0, m_1, z such that 1) each individual element has high-entropy, and 2) any element, associated with a *non-target* set $S \subset [k]$, that can be obtained by applying "legal" algebraic operations to (m_b, z) (for $b \in \{0, 1\}$) has high-entropy (and thus is non-zero with overwhelming probability).[6] We refer to such message distributions as being *entropically valid*.

Basing Security on "Single-Distribution" Semantical Security. The assumption that a scheme satisfies semantical security may be viewed as an (exponential-size) *class* of algebraic "decisional" assumptions (or as a "meta"-assumption, just like the "uber assumption" of [BBG05]): we have one assumption for each valid message distributions D. Indeed, to prove indistinguishability of obfuscations of two circuits C_0, C_1, we rely on an instance in this class which is a function of the circuits C_0, C_1—in the language of [GGSW13, GLW14], security is thus based on an "instance-dependent" assumption.

This view-point also clarifies that semantical security is not an *efficiently falsifiable* assumption [Nao03]: the problem is that there may not exist an efficient way of checking whether a distribution D is valid (as this requires checking that

---

[6] Technically, by high-entropy, we here mean that the min-entropy is at least $\log|R| - O(\log\log|R|)$ where R is the ring associated with the encodings; that is, the min-entropy is "almost" optimal (i.e., $\log|R|$).

all legal arithmetic circuits are constant with overwhelming probability, which in our particular case would require checking whether C_0 and C_1 are functionally equivalent).

We finally observe that both of these issues can be overcome if we make subexponential hardness assumptions: there exists a single (uniform PPT samplable) distribution Sam over (nuPPT message distributions D) that are *provably* entropically valid such that it suffices to assume the existence of an encoding scheme that is entropic semantically secure w.r.t., this particular distribution with *subexponentially small indistinguishability gap.*[7] Note that this is a single, non-interactive and efficiently falsifiable, decisional assumption.

1.3   Alternative Notions of Semantical Security

We finally investigate various ways of defining a "super" (or uber) assumption for multilinear encodings. As mentioned above, a natural way of defining security of multilinear encodings would be to require that for specific classes of problems, generic attacks cannot be beaten (this is the approach alluded to in [BGK+13]). Perhaps the most natural instantiation of this in the context of a multilinear DDH assumption would be to require that for any distribution D over m_0, m_1, z (where m_0, m_1 are constant-length sequences), if encodings of m_0, z and and m_0, z are indistinguishable w.r.t. to generic attackers, then they are also indistinguishable w.r.t. arbitrary nuPPT attackers; in essence, "if an algerbraic decisional assumption holds w.r.t. to generic attacks, then it also holds with respect to nuPPT attackers". We refer to this notion of security as *extractable uber security.*[8]

Our second main result shows that, assuming the existence of a leveled FHE with decryption in NC^1, there do not exist extractable uber-secure multilinear encodings (even if we only require security to hold w.r.t high-entropy distributions D).

Theorem 3 (Informally stated). *Assume the existence of a leveled FHE with decryption in* NC^1. *Then no multilinear encodings can satisfy extractable (entropic) uber security.*

The high-level idea behind this result is to rely on the "conflict" between the feasibility of VBB obfuscation in the generic model of [BGK+13] and the impossibility of VBB obfuscation in the "standard" model [BGI+01]: we let m_b, z contain a generically-secure VBB obfuscation of a program Π_b that hides b given just black-box access to Π_b, yet b can be recovered given the code of Π_b. By generic security of the obfuscation, it follows that *efficient* generic attackers

---

[7] These results were added to our e-Print report April 25, 2014, motivated in part by [GLW14] (which bases witness encryption on an instant-independent assumption) and a conversation with Amit Sahai.

[8] We use the adjective "extractable" as this security notion implies that if an nuPPT attacker can distinguish encodings, then the arithmetic circuits needed to distinguish the elements can be efficiently extracted out.

cannot distinguish $\boldsymbol{m}_0, \boldsymbol{z}$ and $\boldsymbol{m}_1, \boldsymbol{z}$ yet, "non-generic" (i.e., standard PPT) attackers can. In our formal treatment, to rule out *constant-message* (as opposed to multi-message) security, we rely on a variant of the obfuscator presented in this paper, enhanced using techniques from [BGK+13].

We emphasize that in the above attack it is crucial that we restrict to efficient (nuPPT) generic attacks. In the full version of the paper we consider several plausible ways of defining uber security for multilinear encodings, which circumvent the above impossibility results by requiring indistinguishability of encodings only if the encodings are *statistically* close w.r.t. *unbounded* generic attackers (that are restricted to making polynomially many zero-test queries). We highlight that none of these assumptions are needed for our construction of an indistinguishability obfuscation and are stronger than semantical security, but they may find other applications.

2 Definition of Semantically Secure Graded Encodings

2.1 Graded Encoding Schemes

Graded (multilinear) encoding schemes were originally introduced in the work of Garg, Gentry and Halevi [GGH13a]. Just as [BR14, BGK+13], we here rely on "set-based" (or "asymmetric") graded encoding; these were originally called "generalized" graded encodings in [GGH13a]. Following [GGH+13b, BGK+13] and the notion of "multilinear jigsaw puzzles" from [GGH+13b], we additionally enable anyone with the secret parameter to encode *any* elements (as opposed to just *random* elements as in [GGH13a]).

Definition 1 ((k, R)-Graded Encoding Scheme). *A (k, R)-graded encoding scheme for $k \in \mathbb{N}$ and ring R is a collection of sets $\{E_S^\alpha : \alpha \in R, S \subseteq [k]\}$ with the following properties*

- *For every $S \subseteq [k]$ the sets $\{E_S^a : a \in R\}$ are disjoint.*
- *There are associative binary operations $\oplus$ and $\ominus$ such that for every $\alpha_1, \alpha_2 \in R$, $S \subseteq [k]$, $u_1 \in E_S^{\alpha_1}$ and $u_2 \in E_S^{\alpha_2}$ it holds that $u_1 \oplus u_2 \in E_S^{\alpha_1 + \alpha_2}$ and $u_1 \ominus u_2 \in E_S^{\alpha_1 - \alpha_2}$ where '+' and '−' are the addition and subtraction operations in R.*
- *There is an associative binary operation $\otimes$ such that for every $\alpha_1, \alpha_2 \in R$, $S_1, S_2 \subseteq [k]$ such that $S_1 \cap S_2 = \emptyset$, $u_1 \in E_{S_1}^{\alpha_1}$ and $u_2 \in E_{S_2}^{\alpha_2}$ it holds that $u_1 \otimes u_2 \in E_{S_1 \cup S_2}^{\alpha_1 \cdot \alpha_2}$ where '·' is multiplication in R.*

Definition 2 (Graded Encoded Scheme). *A graded encoding scheme $\mathcal{E}$ is associated with a tuple of PPT algorithms, $(\mathsf{InstGen}_\mathcal{E}, \mathsf{Enc}_\mathcal{E}, \mathsf{Add}_\mathcal{E}, \mathsf{Sub}_\mathcal{E}, \mathsf{Mult}_\mathcal{E}, \mathsf{isZero}_\mathcal{E})$ which behave as follows:*

- *Instance Generation: $\mathsf{InstGen}_\mathcal{E}$ takes as input the security parameter 1^n and multilinearity parameter 1^k, and outputs secret parameters sp and public parameters pp which describe a (k, R)-graded encoding scheme $\{E_S^\alpha : \alpha \in R, S \subseteq [k]\}$. We refer to E_S^α as the set of encodings of the pair (α, S). We restrict to graded encoding schemes where R is $\mathbb{Z}_p$ and p is a prime exponential in n and k.*

- *Encoding:* $\mathsf{Enc}_{\mathcal{E}}$ *takes as input the secret parameters* sp, *an element* $\alpha \in R$ *and set* $S \subseteq [k]$, *and outputs a random encoding of the pair* (α, S).
- *Addition:* $\mathsf{Add}_{\mathcal{E}}$ *takes as input the public parameters* pp *and encodings* $u_1 \in E_{S_1}^{\alpha_1}, u_2 \in E_{S_2}^{\alpha_2}$, *and outputs an encoding of the pair* $(\alpha_1 + \alpha_2, S)$ *if* $S_1 = S_2 = S$ *and outputs* $\perp$ *otherwise.*
- *Negation:* $\mathsf{Sub}_{\mathcal{E}}$ *takes as input the public parameters* pp *and encodings* $u_1 \in E_{S_1}^{\alpha_1}, u_2 \in E_{S_2}^{\alpha_2}$, *and outputs an encoding of the pair* $(\alpha_1 - \alpha_2, S)$ *if* $S_1 = S_2 = S$ *and outputs* $\perp$ *otherwise.*
- *Multiplication:* $\mathsf{Mult}_{\mathcal{E}}$ *takes as input the the public parameters* pp *and encodings* $u_1 \in E_{S_1}^{\alpha_1}, u_2 \in E_{S_2}^{\alpha_2}$, *and outputs an encoding of the pair* $(\alpha_1 \cdot \alpha_2, S_1 \cup S_2)$ *if* $S_1 \cap S_2 = \emptyset$ *and outputs* $\perp$ *otherwise.*
- *Zero testing:* $\mathsf{isZero}_{\mathcal{E}}$ *takes as input the public parameters* pp *and an encoding* $u \in E_S(\alpha)$, *and outputs 1 if and only if* $\alpha = 0$ *and* S *is the universe set* $[k]$.[9]

Whenever it is clear from the context, to simplify notation we drop the subscript $\mathcal{E}$ *when we refer to the above procedures (and simply call them* InstGen, Enc, . . .*).*

In known candidate constructions [GGH13a, CLT13], encodings are "noisy" and the noise level increases with each operation; the parameters, however, are set so that any $\mathrm{poly}(n, k)$ operations can be performed without running into trouble. For convenience of notation (and just like all other works in the area), we ignore this noise issue.[10]

Note that the above procedures allow algebraic operations on the encodings in a restricted way. Given the public parameters and encodings made under the sets $\boldsymbol{S}$, one can only perform algebraic operations that are allowed by the structure of the sets in $\boldsymbol{S}$. We call such operations $\boldsymbol{S}$-respecting and formalize this notion as follows:

Definition 3 (Set Respecting Arithmetic Circuits). *For any sequence* $\boldsymbol{S}$ *of subsets of* $[k]$, *we say that an arithmetic circuit* C *(i.e. gates perform only ring operations* $\{+, -, \cdot\}$*) is* $\boldsymbol{S}$-respecting *if it holds that*

- *Eevery input wire of* C *is tagged with some set in* $\boldsymbol{S}$.
- *For every* $+$ *and* $-$ *gate in* C, *if the tags of the two input wires are the same set* S *then the output wire of the gate is tagged with* S. *Otherwise the output wire is tagged with* $\perp$.

---

[9] In the candidate scheme given by [GGH13a], isZero may not have perfect correctness: the generated instances (pp, sp) can be "bad" with some negligible probability, so that there could exist an encoding u of a nonzero element where isZero(pp, u) = 1. However, these "bad" parameters can be efficiently detected during the execution of InstGen. We can thus modify the encoding scheme to simply set Enc(pp, e) = e whenever the parameters are "bad" (and appropriately modify Add, Sub, Mult and isZero so that the operate on "unencoded" elements). This change ensures that, for every pp, including "bad" ones, the zero test procedure isZero works with perfect correctness. We note that since bad parameters occur only with negligible probability, this change does not affect the security of the encodings.

[10] The above definition can be easily generalized to deal with the candidates by only requiring that the above conditions hold when u_1, u_2 have been obtained by $\mathrm{poly}(n, k)$ operations.

- *For every · gate in C, if the tags of the two input wires are sets S_1 and S_2 and $S_1 \cap S_2 = \emptyset$ then the output wire of the gate is tagged with $S_1 \cup S_2$. Otherwise the output wire is tagged with $\bot$.*
- *It holds that the output wire is tagged with the universe set $[k]$.[11]*

2.2 Semantical Security

We now turn to defining semantical security of graded encoding schemes. As outlined in the introduction, we start by defining the notion of a respecting (or valid) message sampler w.r.t. to sets $\boldsymbol{S}, \boldsymbol{T}$. Such a message sampler samples elements $\boldsymbol{m}_0, \boldsymbol{m}_1, \boldsymbol{z}$ such that for every $(\boldsymbol{S}, \boldsymbol{T})$-respecting circuit C, isZero$(C(\cdot))$ is *constant* over $(m_b, \boldsymbol{z})$, $b \in \{0, 1\}$ with overwhelming probability.

Definition 4 (Respecting Message Sampler). *Let $\mathcal{E}$ be a graded encoding scheme, and $\{(\boldsymbol{S}_n, \boldsymbol{T}_n)\}_{n \in \mathbb{N}}$ be an ensemble of pairs of sequences of sets over $[k_n]$. We say that a nuPPT M is a $\{(\boldsymbol{S}_n, \boldsymbol{T}_n)\}_{n \in \mathbb{N}}$-respecting message sampler (or valid w.r.t. $\{(\boldsymbol{S}_n, \boldsymbol{T}_n)\}_{n \in \mathbb{N}}$) if*

- *M on input 1^n and a public parameter pp computes the ring R associated with pp and next based on only 1^n and R generates and outputs a pair $(\boldsymbol{m}_0, \boldsymbol{m}_1)$ of sequences of $|S_n|$ ring elements and a sequence $\boldsymbol{z}$ of $|T_n|$ ring elements;*
- *There exists a polynomial $Q(\cdot, \cdot)$ such that for every $n \in \mathbb{N}$, every (sp, pp) in the support of InstGen$(1^n, 1^{k_n})$, every $(\boldsymbol{S}, \boldsymbol{T})$-respecting arithmetic circuit C, there exists a constant $c \in \{0, 1\}$ such that for any $b \in \{0, 1\}$,*

$$Pr[(\boldsymbol{m}_0, \boldsymbol{m}_1, \boldsymbol{z}) \leftarrow M(1^n, \mathsf{pp}) : \mathsf{isZero}(C(\boldsymbol{m}_b, \boldsymbol{z})) = c] \geq 1 - Q(n, k_n)/|R|.$$

Let us comment that Definition 4 allows the message sampler M to select $\boldsymbol{m}_0, \boldsymbol{m}_1, \boldsymbol{z}$ based on the ring $R = \mathbb{Z}_p$ (or else we could not pick a uniform element in the ring). On the other hand, to make the notion of valid message samplers as restrictive as possible, we prevent the message selection from depending on pp in any other way.

We can now define what it means for a graded encoding scheme to be semantically secure. Roughly speaking, we require that encodings of $(\boldsymbol{m}_0, \boldsymbol{z})$ and $(\boldsymbol{m}_1, \boldsymbol{z})$ under the sets $(\boldsymbol{S}, \boldsymbol{T})$ are indistinguishable as long as $(\boldsymbol{m}_0, \boldsymbol{m}_1, \boldsymbol{z})$ is sampled by a message sampler that is valid w.r.t. $(\boldsymbol{S}, \boldsymbol{T})$.

Definition 5 (Semantic Security). *Let $\mathcal{E}$ be a graded encoding scheme and $q(\cdot)$ and $c(\cdot)$ be polynomials. We say a graded encoding scheme $\mathcal{E}$ is (c, q)-semantically secure if for every polynomial $k(\cdot)$, every ensemble $\{(\boldsymbol{S}_n, \boldsymbol{T}_n)\}_{n \in \mathbb{N}}$ where $\boldsymbol{S}_n$ and $\boldsymbol{T}_n$ are sequences of subsets of $[k(n)]$ of length $c(k(n))$) and $q(k(n))$ respectively, for every $\{(\boldsymbol{S}_n, \boldsymbol{T}_n)\}_{n \in \mathbb{N}}$-respecting message sampler M and every nuPPT adversary A, there exists a negligible function ϵ such that for every security parameter $n \in \mathbb{N}$,*

$$|Pr[\boldsymbol{Output}_0(1^n) = 1] - Pr[\boldsymbol{Output}_1(1^n) = 1]| \leq \epsilon(n)$$

---

[11] For ease of notation, we assume that the description of a set S also contains a description of the universe set $[k]$.

where $\boldsymbol{Output}_b(1^n)$ is A's output in the following game:

- Let $(\mathsf{sp}, \mathsf{pp}) \leftarrow \mathsf{InstGen}(1^n, 1^{k(n)})$.
- Let $\boldsymbol{m_0}, \boldsymbol{m_1}, \boldsymbol{z} \leftarrow M(1^n, \mathsf{pp})$.
- Let $\boldsymbol{u_b} \leftarrow \{\mathsf{Enc}(\mathsf{sp}, \boldsymbol{m_0}[i], \boldsymbol{S}_n[i])\}_{i=1}^{c(k_n)}, \{\mathsf{Enc}(\mathsf{sp}, \boldsymbol{z}[i], \boldsymbol{T}_n[i])\}_{i=1}^{q(k(n))}$.
- Finally, run $A(1^n, \mathsf{pp}, \boldsymbol{u_b})$.

We say that $\mathcal{E}$ is (constant-message) semantically secure *if it is* $(O(1), O(k))$-*semantically secure; we say that* $\mathcal{E}$ multi-message semantically secure *if it is* $(O(k), O(k))$-*semantically secure. We additionally say that* $\mathcal{E}$ *is* subexponentially-hard semantically secure *if there exists some exists some constant* $\alpha > 0$ *such that for every nuPPT A the above indistinguishability gap is bounded by* $\varepsilon(n) = 2^{-O(n^\alpha)}$.

In analogy with the GDDH assumption [GGH13a], our notion of semantical security restricts to the case when the number of elements encoded is $O(k)$.[12] Let us end this section by remarking that (sub-exponentially hard) semantical security trivially holds against polynomial-time "generic" attackers that are restricted to "legally" operating on the encodings—in fact, it holds even against *unbounded* generic attackers that are restricted to only making polynomially (or even subexponentially) many zero-test queries: recall that each legal zero-test query is constant with overwhelming probability (whether we operate on $\boldsymbol{m_0}, \boldsymbol{z}$ or $\boldsymbol{m_1}, \boldsymbol{z}$) and thus by a Union Bound, the output of any generic attacker restricted to polynomially many zero-test queries is also constant with overwhelming probability.

Semantical Security w.r.t. Restricted Classes of Message Samplers. For our specific construction of indistinguishability obfuscators it suffices to assume the existence of *semantically secure encodings w.r.t. restricted classes of message samplers* M, where the $\{(\boldsymbol{S}_n, \boldsymbol{T}_n)\}_{n \in \mathbb{N}}$-respecting condition on M is replaced by some stronger restriction on M. It particular, it suffices to restrict to message samplers M that induce a *high-entropy*[13] distribution over $\boldsymbol{m_0}, \boldsymbol{m_1}, \boldsymbol{z}$—not only the individual elements have high min-entropy but also any element computed by applying a "non-terminal" sequence of legal arithmetic operations to $\boldsymbol{m_b}, \boldsymbol{z}$ (for $b \in \{0, 1\}$); we refer to schemes satisfying this weaker notion of semantical security *entropic semantically secure* (and refer the reader to the full version for a formal definition).

---

[12] This restriction was suggested in [BCKP14] and independently by Hoeteck Wee; our original formulation of semantical security considered an unbounded polynomial number of elements in $\boldsymbol{z}$ (but our proof of security only relied on security for $O(k)$ elements). We now refer to the unbounded notion as *unbounded semantical security*, but it will not be needed for any of our results.

[13] Technically, by high-entropy, we here mean that the min-entropy is at least $\log|R| - O(\log\log|R|)$ where R is the ring associated with the encodings; that is, the min-entropy is "almost" optimal (i.e., $\log|R|$).

3   Proof Overview

We here provide an overview of our obfuscator and its proof of security, and refer the reader to the full version [PST13] for further details.

The Basic Obfuscator. We start by providing a construction of a "basic" obfuscator; our final construction will then rely on the basic obfuscator as a black-box. The construction of this obfuscator closely follows the design principles laid out in the original work by Garg et al [GGH+13b] and follow-up constructions [BR14, BGK+13] (in fact, the basic obfuscator may be viewed as a simplified version of the obfuscator from [BGK+13]). As these works, we proceeds in three steps:

- We view the NC^1 circuit to be obfuscated as a *branching program BP* (using Barrington's Theorem [Bar86])—that is, the program is described by m pairs of matrices $(B_{i,0}, B_{i,1})$, each one labelled with an input bit $inp(i)$. The program is evaluated as follows: for each $i \in [m]$, we choose one of the two matrices $(B_{i,0}, B_{i,1})$, based on the input. Next, we compute the product of the chosen matrices, and based on the product determine the output—there is a unique "accept" (i.e., output 1) matrix, and a unique "reject" (i.e., output 0) matrix.
- The branching program BP is *randomized* using Kilian's technique [Kil88] (roughly, each pair of matrices is appropriately multiplied with the same random matrix R while ensuring that the output is the same), and then "randomized" some more—each individual matrix is multiplied by a random *scalar* α. Let us refer to this step as Rand.
- Finally the randomized matrices are encoded using multilinear encodings with the sets selected appropriately. We here rely on a (simple version) of the *straddling set* idea of [BGK+13] to determine the sets. We refer to this step as Encode.

(The original construction as well as the subsequent works also consisted of several other steps, but for our purposes these will not be needed.) The obfuscated program is now evaluated by using the multilinear operations to evaluate the branching program and finally appropriately use the zero-test to determine the output of the program.

   Roughly speaking, the idea behind the basic obfuscator is that the multilinear encodings *intuitively* ensure that any attacker getting the encoding needs to multiply matrices along paths that corresponds to some input to the branching program (the straddling sets are used to ensure that the input is used consistently in the evaluation)[14]; the scalars α, roughly speaking, ensure that a potential attacker without loss of generality can use a *single* "multiplication-path" and still succeed with roughly the same probability, and finally, Kilian's randomization steps ensures that if an attacker *only* operates on matrices along a single path that corresponds to some input x (in a consistent way), then its output can

---

[14] The encodings, however, still permit an attacker to add elements within matrices.

be perfectly simulated given just the output of the circuit on input x. (The final step relies on the fact that the output of the circuit uniquely determines product of the branching program along the path, and Kilian's randomization then ensures that the matrices along the path are random conditioned on the product being this unique value.) Thus, if an attacker can tell apart obfuscations of two programs BP_0, BP_1, there must exist some input on which they produce different outputs. The above intuitions can indeed be formalized w.r.t. *generic attackers* (that only operate on the encodings in a legal way, respecting the set restrictions), relying on arguments from [BR14, BGK+13]. This already suffices to prove that the basic obfuscator is an indistinguishability obfuscator assuming the encodings are *multi-message* semantically secure.[15]

The Merge Procedure. To base security on the weaker assumption of (constant-message) semantical security, we will add an additional program transformation steps before the Rand and Encode steps. Roughly speaking, we would like to have a method $\mathsf{Merge}(BP_0, BP_1, b)$ that "merges" BP_0 and BP_1 into a single branching program that evaluates BP_b; additionally, we require that $\mathsf{Merge}(BP_0, BP_1, 0)$ and $\mathsf{Merge}(BP_0, BP_1, 1)$ only differ in a constant number of matrices. We achieve this merge procedure by connecting together BP_0, BP_1 into a branching program of double width and adding two "switch" matrices in the beginning and the end, determining if we should go "up" or "down". Thus, to switch between $\mathsf{Merge}(BP_0, BP_1, 0)$ (which is functionally equivalent to BP_0) and $\mathsf{Merge}(BP_0, BP_1, 1)$ (which is functionally equivalent to BP_1) we just need to switch the "switch matrices". More precisely, given branching programs BP_0 and BP_1 described respectively by pairs of matrices $\{(B^0_{i,0}, B^0_{i,1}), (B^1_{i,0}, B^1_{i,1})\}_{i \in [m]}$, we construct a merged program $\mathsf{Merge}(BP_0, BP_1, b)$ described by $\{(\hat{B}^0_{i,0}, \hat{B}^0_{i,1})\}_{i \in [m+2]}$ such that

$$\hat{B}^0_{i,b} = \hat{B}^1_{i,b} = \begin{pmatrix} B^0_{(i-1),b} & 0 \\ 0 & B^1_{(i-1),b} \end{pmatrix} \quad \text{for all } 2 \leq i \leq m+1 \text{ and } b \in \{0,1\}$$

and the first and last matrices are given by:

$$\hat{B}^0_{1,b} = \hat{B}^0_{m+2,b} = I_{2w \times 2w} \qquad\qquad \text{for } b \in \{0,1\}$$

$$\hat{B}^1_{1,b} = \hat{B}^1_{m+2,b} = \begin{pmatrix} 0 & I_{w \times w} \\ I_{w \times w} & 0 \end{pmatrix} \qquad \text{for } b \in \{0,1\}$$

It directly follows from the construction that $\mathsf{Merge}(BP_0, BP_1, 0)$ and $\mathsf{Merge}(BP_0, BP_1, 1)$ differ only in the first and the last matrices (i.e., the "switch" matrices). Furthermore, it is not hard to see that $\mathsf{Merge}(BP_0, BP_1, b)$ is functionally equivalent to BP_b.

---

[15] As mentioned above, there are still some minor subtleties involved in doing this: the analyses of [BR14, BGK+13] implicitly show that all *polynomial-size* legal arithmetic circuits are constant with overwhelming probability, but by slightly tweaking the constructions and the analyses to ensure a "perfect" simulation property, we can extend these arguments to hold against *all* (arbitrary-size) legal arithmetic circuits and thus base security on multi-message semantical security.

Our candidate obfuscator is now defined as $i\mathcal{O}(B) = \mathsf{Encode}(\mathsf{Rand}(\mathsf{Merge}(BP, I, 0)))$, where I is simply a "dummy" program of the same size as BP.[16]

The idea behind the merge procedure is that to prove that obfuscations of two programs BP_0, BP_1 are indistinguishable, we can come up with a sequence of hybrid experiments that start with $i\mathcal{O}(BP_0)$ and end with $i\mathcal{O}(BP_1)$, but between any two hybrids only changes a constant number of encodings, and thus we may rely on semantic security of multilinear encodings to formalize the above intuitions. At a high level, our strategy will be to matrix-by-matrix, replace the dummy branching program in the obfuscation of BP_0 with the branching program for BP_1. Once the entire dummy branching program has been replaced by BP_1, we flip the "switch" so that the composite branching program now computes the branching program for BP_1. We then replace the branching program for BP_0 with BP_1, matrix by matrix, so that we have two copies of the branching program for BP_1. We now flip the "switch" again, and finally restore the dummy branching program, so that we end up with one copy of BP_1 and one copy of the dummy, which is now a valid obfuscation of BP_1. In this way, we transition from an obfuscation of BP_0 to an obfuscation of BP_1, while only changing a small piece of the obfuscation in each step. (On a very high-level, this approach is somewhat reminiscent of the Naor-Yung "two-key" approach in the context of CCA security [NY90] and the "two-key" bootstrapping result for indistinguishability obfuscation due to Garg et al [GGH⁺13b]—in all these approaches the length of the scheme is artificially doubled to facilitate a hybrid argument. It is perhaps even more reminiscent of the Feige-Shamir "trapdoor witness" approach for constructing zero-knowledge arguments [FS90], whereby an additional "dummy" trapdoor witness is introduced in the construction to enable the security proof.)

More precisely, consider the following sequence of hybrids.

- We start off with $i\mathcal{O}(BP_0) = \mathsf{Enc}(\mathsf{Rand}(\mathsf{Merge}(BP_0, I, 0)))$
- We consider a sequence of hybrids where we gradually change the dummy program I to become BP_1; that is, we consider $\mathsf{Encode}(\mathsf{Rand}(\mathsf{Merge}(BP_0, BP', 0)))$, where BP' is "step-wise" being populated with elements from BP_1.
- We reach $\mathsf{Encode}(\mathsf{Rand}(\mathsf{Merge}(BP_0, BP_1, 0)))$.
- We turn the "switch" : $\mathsf{Encode}(\mathsf{Rand}(\mathsf{Merge}(BP_0, BP_1, 1)))$.
- We consider a sequence of hybrids where we gradually change the BP_0 to become BP_1; that is, we consider $\mathsf{Encode}(\mathsf{Rand}(\mathsf{Merge}(BP', BP_1, 1)))$, where BP' is "step-wise" being populated with elements from BP_1.
- We reach $\mathsf{Encode}(\mathsf{Rand}(\mathsf{Merge}(BP_1, BP_1, 1)))$.
- We turn the "switch" back: $\mathsf{Encode}(\mathsf{Rand}(\mathsf{Merge}(BP_1, BP_1, 0)))$.
- We consider a sequence of hybrids where we gradually change the second BP_1 to become I; that is, we consider $\mathsf{Encode}(\mathsf{Rand}(\mathsf{Merge}(BP_1, BP', 0)))$, where BP' is "step-wise" being populated with elements from I.
- We reach $\mathsf{Encode}(\mathsf{Rand}(\mathsf{Merge}(BP_1, I, 0))) = i\mathcal{O}(BP_1)$.

---

[16] This description oversimplifies a bit. Formally, the Rand step needs to depends on the field size used in the Encode steps, and thus in our formal treatment we combine these two steps together.

By construction we have that if BP_0 and BP_1 are functionally equivalent, then so will all the hybrid programs–the key point is that we only "morph" between two branching programs on the "inactive" part of the merged branching program. Furthermore, by construction, between any two hybrids we only change a constant number of elements. Thus, if some distinguisher can tell apart $i\mathcal{O}(BP_0)$ and $i\mathcal{O}(BP_1)$, it must be able to tell apart two consecutive hybrids. But, by semantic security it then follows that some "legal" arithmetic circuit can tell apart the encodings in the two hybrids. Roughly speaking, we can now rely on simulation security of the basic obfuscator w.r.t. to just *legal* arithmetic circuits to complete the argument. A bit more precisely, based on BP_0, BP_1 and the hybrid index i, we can define a message sampler M_{i,BP_0,BP_1} that is valid (by the simulation arguments in [BGK+13]) as long as BP_0 is functionally equivalent to BP_1, yet our distinguisher manages to distinguish messages sampled from M_{i,BP_0,BP_1}, contradicting semantical security.

Dealing with Branching Programs with Non-unique Outputs. There is a catch with the final step though. Recall that to rely on Kilian's simulation argument it was crucial that there are *unique* accept and reject matrices. For our "merged" programs, this is no longer the case (the output matrix is also a function of the second "dummy" program), and thus it is no longer clear how to prove that the message distribution above is valid. We overcome this issue by noting that the *first column* of the output matrix actually is unique, and this is all we need to determine the output of the branching program; we refer to such branching programs as *fixed output-column branching programs*. Consequently it suffices to release encodings of the *just* first column (as opposed to the whole matrices) of the last matrix pair in the branching program, and we can still determine the output of the branching program. As we show, for such a modified scheme, we can also simulate the (randomized) matrices along an "input-path" given just the first column of the output matrix.

A Modular Analysis: Neighboring-Matrix $i\mathcal{O}$. In the actual proof, we provide a modular analysis of the above two steps (that may be interesting in its own right).

- We define a notion of *neighboring-matrix indistinguishability obfuscation*, which relaxes indistinguishability obfuscation by only requiring security to hold w.r.t. any two functionally equivalent branching programs that differ in at most a constant number of matrices.
- We then use the above merge procedure (and the above hybrid argument) to show that the existence of a neighboring-matrix $i\mathcal{O}$ for all "fixed output column" branching programs implies the existence of a "full-fledged" $i\mathcal{O}$.
- We finally use the "basic obfuscator" construction to show how to construct a neighboring-matrix $i\mathcal{O}$ for all fixed output column branching programs based on (constant-message) semantical security.

Basing Security on a (Single) Falsifiable Assumption. To base security on a falsifiable assumption, we rely on a different merge procedure from the work of Boyle, Chung and Pass [BCP14]: Given two NC^1 circuits C_0, C_1 taking (at most) n-bit inputs, and a string z, let $\widehat{\mathsf{Merge}}(C_0, C_1, z)$ be a circuit that on input x runs $C_0(x)$ if $x \geq z$ and $C_1(x)$ otherwise; in essence, this procedure lets us "traverse" between C_0 and C_1 while provably only changing the functionality on at most one input. ([BCP14] use this type of merged circuits to perform a binary search and prove that indistinguishability obfuscation implies differing-input obfuscation for circuits that differ in only polynomially many inputs.) We now define a notion of *neighboring-input* $i\mathcal{O}$, which relaxes $i\mathcal{O}$ by only requiring that security holds with respect to "neigboring-input" programs $\widehat{\mathsf{Merge}}(C_0, C_1, z)$, $\widehat{\mathsf{Merge}}(C_0, C_1, z+1)$ that are functionally equivalent. Note that checking whether $\widehat{\mathsf{Merge}}(C_0, C_1, z)$, $\widehat{\mathsf{Merge}}(C_0, C_1, z+1)$ are functionally equivalent is easy: they are equivalent iff $C_0(z) = C_1(z)$. (As such, the assumption that a scheme satisfies neighboring-input $i\mathcal{O}$ is already an efficiently falsfiable assumption.) Furthermore, by a simple hybrid argument over $z \in \{0,1\}^n$, *exponentially-secure* neighboring-input $i\mathcal{O}$ implies "full" $i\mathcal{O}$—exponential security is needed since we have 2^n hybrids. (We mention a very recent work by Gentry, Lewko and Waters [GLW14] in the context of *witness encryption* [GGSW13] that similarly defines a falsifiable primitive "positional witness encryption" that implies the full-fledged notion with an exponential security loss.)

Additionally, note that to show that our construction satisfies exponentially-secure neighboring-input $i\mathcal{O}$, we only need to rely on exponentially-secure semantical security w.r.t. classes of message distributions corresponding to programs of the form $\widehat{\mathsf{Merge}}(C_0, C_1, z)$, $\widehat{\mathsf{Merge}}(C_0, C_1, z+1)$. Equivalently, it suffices to rely on exponentially-secure semantical security w.r.t. a *single* distribution over sets and message samplers corresponding to uniformly selected z and programs C_0, C_1 (again, this only results in an exponential security loss). Finally, by padding the security parameter of the multilinear encodings in the construction, it actually suffices to rely on subexponential security.

Acknowledgments. We are very grateful to Benny Applebaum, Omer Paneth, Ran Canetti, Kai-Min Chung, Sanjam Garg, Craig Gentry, Shai Halevi, Amit Sahai, abhi shelat, Hoeteck Wee and Daniel Wichs for many helpful comments. We are especially gratefeul to Shai for pointing out the connection between semantical security for multilinear encodings and the "uber" assumption for bilinear maps of [BBG05], and for several very useful conversations about multilinear encodings and the security of the [GGH13a] constructions, to Amit for several helpful conversations about the presentation of our results, and Benny for suggesting we make our proof more modular (which lead to the notion of neigboring-matrix branching programs). Finally thanks to the anonymous Crypto reviewers for their useful comments. Thanks so very much!

References

[Bar86]   Barrington, D.A.M.: Bounded-width polynomial-size branching programs recognize exactly those languages in nc^1. In: STOC, pp. 1–5 (1986)

[BBG05]   Boneh, D., Boyen, X., Goh, E.-J.: Hierarchical identity based encryption
             with constant size ciphertext. In: Cramer, R. (ed.) EUROCRYPT 2005.
             LNCS, vol. 3494, pp. 440–456. Springer, Heidelberg (2005)
[BCKP14]   Bitansky, N., Canetti, R., Kalai, Y.T., Paneth, O.: On virtual grey box ob-
             fuscation for general circuits. In: Garay, J.A., Gennaro, R. (eds.) CRYPTO
             2014, Part II. LNCS, vol. 8617, pp. 108–125. Springer, Heidelberg (2014)
[BCP14]   Boyle, E., Chung, K.-M., Pass, R.: On extractability obfuscation. In: Lin-
             dell, Y. (ed.) TCC 2014. LNCS, vol. 8349, pp. 52–73. Springer, Heidelberg
             (2014)
[BCPR14]   Bitansky, N., Canetti, R., Paneth, O., Rosen, A.: On the existence of
             extractable one-way functions. In: STOC 2014 (2014)
[BGI+01]   Barak, B., Goldreich, O., Impagliazzo, R., Rudich, S., Sahai, A., Vadhan,
             S.P., Yang, K.: On the (Im)possibility of obfuscating programs. In: Kilian,
             J. (ed.) CRYPTO 2001. LNCS, vol. 2139, pp. 1–18. Springer, Heidelberg
             (2001)
[BGK+13]   Barak, B., Garg, S., Kalai, Y.T., Paneth, O., Sahai, A.: Protecting ob-
             fuscation against algebraic attacks. Cryptology ePrint Archive, Report
             2013/631 (2013)
[BGV12]   Brakerski, Z., Gentry, C., Vaikuntanathan, V.: (leveled) fully homomor-
             phic encryption without bootstrapping. In: ITCS, pp. 309–325 (2012)
[BR14]   Brakerski, Z., Rothblum, G.N.: Virtual black-box obfuscation for all cir-
             cuits via generic graded encoding. In: Lindell, Y. (ed.) TCC 2014. LNCS,
             vol. 8349, pp. 1–25. Springer, Heidelberg (2014)
[BS03]   Boneh, D., Silverberg, A.: Applications of multilinear forms to cryptogra-
             phy. Contemporary Mathematics 324(1), 71–90 (2003)
[BV11]   Brakerski, Z., Vaikuntanathan, V.: Efficient fully homomorphic encryption
             from (standard) lwe. In: FOCS, pp. 97–106 (2011)
[BZ14]   Boneh, D., Zhandry, M.: Multiparty key exchange, efficient traitor tracing,
             and more from indistinguishability obfuscation. In: Garay, J.A., Gennaro,
             R. (eds.) CRYPTO 2014, Part I. LNCS, vol. 8616, pp. 480–499. Springer,
             Heidelberg (2014)
[CLT13]   Coron, J.-S., Lepoint, T., Tibouchi, M.: Practical multilinear maps over
             the integers. In: Canetti, R., Garay, J.A. (eds.) CRYPTO 2013, Part I.
             LNCS, vol. 8042, pp. 476–493. Springer, Heidelberg (2013)
[FS90]   Feige, U., Shamir, A.: Witness indistinguishable and witness hiding pro-
             tocols. In: STOC 1990, pp. 416–426 (1990)
[Gen09]   Gentry, C.: A fully homomorphic encryption scheme. PhD thesis, Stanford
             University (2009)
[GGG+14]   Goldwasser, S., et al.: Multi-input functional encryption. In: Nguyen, P.Q.,
             Oswald, E. (eds.) EUROCRYPT 2014. LNCS, vol. 8441, pp. 578–602.
             Springer, Heidelberg (2014)
[GGH13a]   Garg, S., Gentry, C., Halevi, S.: Candidate multilinear maps from ideal lat-
             tices. In: Johansson, T., Nguyen, P.Q. (eds.) EUROCRYPT 2013. LNCS,
             vol. 7881, pp. 1–17. Springer, Heidelberg (2013)
[GGH+13b]   Garg, S., Gentry, C., Halevi, S., Raykova, M., Sahai, A., Waters, B.: Can-
             didate indistinguishability obfuscation and functional encryption for all
             circuits. In: Proc. of FOCS (2013)
[GGHR14]   Garg, S., Gentry, C., Halevi, S., Raykova, M.: Two-round secure MPC from
             indistinguishability obfuscation. In: Lindell, Y. (ed.) TCC 2014. LNCS,
             vol. 8349, pp. 74–94. Springer, Heidelberg (2014)

[GGSW13]   Garg, S., Gentry, C., Sahai, A., Waters, B.: Witness encryption and its applications. In: Proceedings of the 45th Annual ACM Symposium on Symposium on Theory of Computing, STOC 2013, pp. 467–476 (2013)

[GLW14]   Gentry, C., Lewko, A., Waters, B.: Witness encryption from instance independent assumptions. In: Garay, J.A., Gennaro, R. (eds.) CRYPTO 2014, Part I. LNCS, vol. 8616, pp. 426–443. Springer, Heidelberg (2014)

[GR07]   Goldwasser, S., Rothblum, G.N.: On best-possible obfuscation. In: Vadhan, S.P. (ed.) TCC 2007. LNCS, vol. 4392, pp. 194–213. Springer, Heidelberg (2007)

[Had00]   Hada, S.: Zero-knowledge and code obfuscation. In: Okamoto, T. (ed.) ASIACRYPT 2000. LNCS, vol. 1976, pp. 443–457. Springer, Heidelberg (2000)

[HSW14]   Hohenberger, S., Sahai, A., Waters, B.: Replacing a random oracle: Full domain hash from indistinguishability obfuscation. In: Nguyen, P.Q., Oswald, E. (eds.) EUROCRYPT 2014. LNCS, vol. 8441, pp. 201–220. Springer, Heidelberg (2014)

[Kil88]   Kilian, J.: Founding crytpography on oblivious transfer. In: Proceedings of the Twentieth annual ACM Symposium on Theory of Computing, pp. 20–31. ACM (1988)

[KMN⁺14]   Komargodski, I., Moran, T., Naor, M., Pass, R., Rosen, A., Yogev, E.: One-way functions and (im)perfect obfuscation. Cryptology ePrint Archive, Report 2014/347 (2014), http://eprint.iacr.org/

[KNY14]   Komargodski, I., Naor, M., Yogev, E.: Secret-sharing for np from indistinguishability obfuscation. CoRR, abs/1403.5698 (2014)

[Nao03]   Naor, M.: On cryptographic assumptions and challenges. In: Boneh, D. (ed.) CRYPTO 2003. LNCS, vol. 2729, pp. 96–109. Springer, Heidelberg (2003)

[NY90]   Naor, M., Yung, M.: Public-key cryptosystems provably secure against chosen ciphertext attacks. In: Proceedings of the Twenty-Second Annual ACM symposium on Theory of Computing, pp. 427–437 (1990)

[PST13]   Pass, R., Seth, K., Telang, S.: Indistinguishability obfuscation from semantically-secure multilinear encodings. Cryptology ePrint Archive, Report 2013/781 (2013)

[RAD78]   Rivest, R.L., Adleman, L., Dertouzos, M.L.: On data banks and privacy homomorphisms. In: Foundations of Secure Computation, pp. 169–179. Academia Press (1978)

[Rot13]   Rothblum, R.D.: On the circular security of bit-encryption. In: Sahai, A. (ed.) TCC 2013. LNCS, vol. 7785, pp. 579–598. Springer, Heidelberg (2013)

[SW14]   Sahai, A., Waters, B.: How to use indistinguishability obfuscation: Deniable encryption, and more. In: Proc. of STOC 2014 (2014)

On the Implausibility
of Differing-Inputs Obfuscation
and Extractable Witness Encryption
with Auxiliary Input

Sanjam Garg[1], Craig Gentry[1], Shai Halevi[1], and Daniel Wichs[2,*]

[1] IBM Research, T.J. Watson, Yorktown Heights, NY, USA
`sanjamg@cs.ucla.ed, cbgentry@us.ibm.com, shaih@alum.mit.edu`
[2] Dept. of Computer Science, Northeastern University, Boston, MA, USA
`wichs@ccs.neu.edu`

Abstract. The notion of *differing-inputs obfuscation* (diO) was introduced by Barak et al. (CRYPTO 2001). It guarantees that, for any two circuits C_0, C_1, if it is difficult to come up with an input x on which $C_0(x) \neq C_1(x)$, then it should also be difficult to distinguish the obfuscation of C_0 from that of C_1. This is a strengthening of *indistinguishability obfuscation*, where the above is only guaranteed for circuits that agree on all inputs: $C_0(x) = C_1(x)$ for all x. Two recent works of Ananth et al. (ePrint 2013) and Boyle et al. (TCC 2014) study the notion of diO in the setting where the attacker is also given some auxiliary information related to the circuits, showing that this notion leads to many interesting applications.

In this work, we show that the existence of *general-purpose* diO with *general* auxiliary input has a surprising consequence: it implies that a *specific* circuit C^* with *specific* auxiliary input aux* cannot be obfuscated in a way that hides some *specific* information. In other words, under the conjecture that such *special-purpose obfuscation* exists, we show that general-purpose diO cannot exist. We do not know if this special-purpose obfuscation assumption is implied by diO itself, and hence we do not get an unconditional impossibility result. However, the special-purpose obfuscation assumption is a falsifiable assumption which we do not know how to break for candidate obfuscation schemes. Showing the existence of general-purpose diO with general auxiliary input would necessitate showing how to break this assumption.

We also show that the special-purpose obfuscation assumption implies the impossibility of *extractable witness encryption* with auxiliary input, a notion proposed by Goldwasser et al. (CRYPTO 2013). A variant of this assumption also implies the impossibility of *"output-only dependent"* *hardcore bits* for general one-way functions, as recently constructed by Bellare and Tessaro (ePrint 2013) using diO.

---

* Research supported by NSF grants 1347350, 1314722.

J.A. Garay and R. Gennaro (Eds.): CRYPTO 2014, Part I, LNCS 8616, pp. 518–535, 2014.

1   Introduction

The formal study of program obfuscation was initiated by Hada [Had00] and Barak et al. [BGI+01, BGI+12]. Since then there have been many negative and, more recently, also positive results on obfuscation. We briefly survey both directions.

Negative Results. Hada observed that a super-strong notion of obfuscation, requiring that the obfuscated code does not leak anything beyond what can be learned given black-box oracle access to the underlying function, cannot be met unless the obfuscated function is learnable. Barak et al. define a slightly weaker (but still very strong) notion of "virtual-black-box" (VBB) obfuscation, roughly requiring that the obfuscated circuit does not leak any *predicate of the obfuscated function* beyond what can be learned given black-box oracle access to that function. The main result of Barak et al. shows the impossibility of VBB-obfuscation for general circuits. The impossibility result constructs specific albeit "contrived" functions that cannot be VBB-obfuscated, but also shows that such functions can be embedded into cryptosystems giving "contrived" constructions of cryptosystems (signature schemes, encryption, pseudo-random functions) that cannot be VBB-obfuscated. The main idea behind this result is to construct functions where obfuscated code can be "fed" into the function as an input, causing it to output extra information. Such counterexamples even exist in weak computational classes, (such as any class that simultaneously contains NC0 and a PRF [App13]), and therefore we cannot even get general VBB obfuscation for such weak classes. However, this result still leaves open the possibility that many specific functions and most natural cryptosystems can be VBB-obfuscated.

The work of Goldwasser and Kalai [GK05] considers a notion of VBB obfuscation with auxiliary input and shows that *no* pseudo-random function (even natural ones) can be VBB-obfuscated in the presence of arbitrary auxiliary input. Recent work extends this result to weaker assumptions and more restricted forms of auxiliary input [GK13, BCPR13b]. In all these works, the impossibility result constructs some contrived auxiliary input. In particular, in all these results, the auxiliary input is itself an obfuscated circuit. These negative results leave open two interesting possibilities:

- Perhaps most "natural" functions and "standard-construction" cryptosystems can be VBB-obfuscated in the presence of most "natural" auxiliary inputs, even though there are "contrived" examples of functions and auxiliary inputs that cannot be obfuscated.
- Perhaps some general form of obfuscation, weaker than VBB, is possible for general functions.

Positive Results. In the face of their impossibility result, Barak et al. proposed two weaker notions of obfuscation that may be achievable for general functions: *indistinguishability obfuscation* and *differing-inputs obfuscation*. Indistinguishability obfuscation says that for any pair of circuits C_0, C_1 that agree on all inputs $C_0(x) = C_1(x)$, it should be hard to distinguish the obfuscation of C_0 from that of C_1. The work of Garg et al. [GGH+13] gave the first candidate for general-purpose indistinguishability obfuscation based on multilinear maps, and several applications of this primitive. Many more applications have appeared since then [SW13, HSW13, GGHR13, BZ13, MO13, PPS13, GGJS13, GJKS13]. Indistinguishability obfuscation is also called the "best possible" obfuscation since anything that any obfuscator can hide, an indistinguishability obfuscator (with sufficient padding) is guaranteed to hide as well. Therefore, one can conjecture that this obfuscator satisfies stronger properties.

The works of [BR13, BGK+13] also give constructions of obfuscators satisfying even the stronger VBB property in the "generic multilinear map" model. This result is difficult to interpret since we do have "non-generic" attacks given by the prior negative results.

Reconciling the positive and negative results suggests the following interpretation: when it comes to *general* functionalities and *general* auxiliary input, one can cook-up clever "contrived" counterexamples that allow for "non-generic" attacks, and therefore one must settle for weak notions of obfuscation, like indistinguishability obfuscation. On the other hand, when it comes to *specific* functionalities with fixed auxiliary input, even strong notions of VBB obfuscation may be achievable. In particular, if we fix a specific function and auxiliary input, unless there is some "obvious" attack where the code of the function can be meaningfully used as an input (either to the function itself or to some other function given by the auxiliary input) it may be reasonable to assume that VBB obfuscation is possible in this specific case.

Differing-Inputs Obfuscation. Despite its usefulness in many recent applications, indistinguishability obfuscation is often difficult to use as a general assumption. The work of Barak et al. also proposed a stronger notion called differing-inputs obfuscation (diO). In particular, this notion says that for any distribution on circuits (C_0, C_1), if it is hard to find an input x such that $C_0(x) \neq C_1(x)$, then it should also be hard to distinguish the obfuscation of C_0 from that of C_1. The recent work of Ananth et al. [ABG+13] and Boyle et al. [BCP14] extend this notion to the setting of auxiliary input, where the attacker is given (C_0, C_1, aux) and, if it is hard to use this information to find an input x on which $C_0(x) \neq C_1(x)$, then it should also be hard to use this information to distinguish the obfuscation of C_0 and C_1. These works give several interesting applications of this notion, including the ability to obfuscate Turing Machine without the cost of converting them into a circuit.[1]

Our Result. As our main result, we show that the existence of general-purpose differing-inputs obfuscation (diO) with auxiliary-input leads to a surprising consequence: it would show the impossibility of obfuscating a *specific* circuit C^* with *specific* auxiliary input aux^* in a way that hides some *specific* information. In particular, we put forth a "counter-conjecture" that such "special purpose"

---

[1] We are not aware of any applications of diO without auxiliary input.

circuit-obfuscators exist and, under this conjecture, general-purpose diO with auxiliary input does not exist. Moreover, under the same conjecture, we also show that extractable witness encryption (with auxiliary input) does not exist. We also consider a restricted scenario of "bounded-length axillary input" where the length of the auxiliary input is bounded a-priori, and the diO obfuscator is given the length bound. We show that a variant of our 'special purpose obfuscation" conjecture (using an obfuscator for Turing Machines rather than circuits) rules this out as well. Lastly, in Appendix A, we also show that this variant of our conjecture rules out "output-only dependent" hardcore bits for general one-way functions, where the value of the hardcore bit is completely determined by the output of the function. Such hardcore bits were recently constructed using diO with bounded-length auxiliary input by Bellare and Tessaro [BT13].

What to Believe? Our "special-purpose obfuscation" conjecture is not known to be implied by differing-inputs obfuscation itself, and hence we do not get unconditional *impossibility* results. In particular, our main result leaves us with the following two opposing possibilities: (I) general-purpose diO with auxiliary input exists, (II) our special-purpose obfuscation assumption holds. We cannot objectively say which one of these is false. However, (II) is a falsifiable assumption in the formal sense of [Nao03], where an efficient challenger can check if an attack is valid. Using the obfuscator of [GGH+13] (or [BR13, BGK+13]), we currently do not know of any attacks on (II). In other words, the validity of (I) would imply the existence of an efficient algorithm whose correctness would be easy to verify, but we do not have any candidate for this algorithm. On the other hand, (I) itself is not stated as a falsifiable assumption, and there is no direct way to verify an attack against it via an efficient challenger. Indeed, we present an efficient attack that contradicts the security of (I), but there is no direct way to check if our attack is "valid" since doing so requires proving (II). Therefore, we view our result as presenting a significant challenge to the plausibility of general-purpose diO with auxiliary input. See further discussion on our conjecture in Section 4.

Consequences of Our Result. Assuming that our "special-purpose obfuscation" conjecture holds, we have ruled out the existence of general-purpose diO with auxiliary input. However, it may still be reasonable to assume that diO security and even VBB security with auxiliary input can hold in concrete cases. Many of the applications of diO in the works [ABG+13, BCP14] and follow-up works remain plausible and only rely on diO security with some concrete auxiliary input, which is unlikely to contain our "counterexample".[2] Nevertheless, to avoid our implausibility result, one would have to carefully pose a new diO assumption for the specific auxiliary input required in each new application and convincingly argue that this assumption is plausible even if the general one is

---

[2] The notable exceptions are "extractable/functional witness encryption"[BCP14] and "output-only dependent hardcore bits for any one-way function" [BT13] where the auxiliary input is external and is not fixed by the construction. Our counterexamples show that these notions are "implausible" in their general form.

not. Although this approach may be a sound, it runs counter to our goal of constructing a wide variety of cryptosystems from a few general (and plausible) assumptions.

Related Work. Our technique follows the approach of similar results [GK05] [GK13, BCPR13b, BCPR13a, BP13], all of which use one form of obfuscation to derive counterexamples for other forms of obfuscation and/or various extractability assumptions. In particular, the results of [GK05, GK13, BCPR13b] show that existence of iO implies the impossibility of VBB obfuscation of natural functionalities with (unnatural) auxiliary input, whereas [BCPR13a, BP13] show that existence of iO/diO implies impossibility of extractable functions and related extractability primitives.

Our Technique. The main idea of our technique is to create a contrived "auxiliary input" aux which is itself an obfuscated circuit. In particular, aux allows the attacker to distinguish any obfuscations of some carefully designed C_0, C_1, without gaining the ability to find an input on which they differ. The "special purpose" assumption is needed to guarantee that aux does not "leak" an input x on which $C_0(x) \neq C_1(x)$.

In more detail, the circuits C_b ($b \in \{0,1\}$) have a verification key vk of a signature scheme hard-coded in them. If they get an input $x = (m, \sigma)$ consisting of a valid message/signature pair, they output the bit b, else they just both output 0. Finding an input x on which $C_0(x) \neq C_1(x)$ requires finding a valid message/signature pair (which is hard to do even given vk). We set the auxiliary input aux to be a "special-purpose" obfuscation of a circuit C^* that has the signing key sk hard-coded and is defined as follows: given as input any circuit C with 1-bit output, it outputs $C(m, \sigma)$ where $m = H(C)$ is a hash of C and σ is a signature of m under sk. It is easy to use (an obfuscation of) C^* to distinguish C_0 and C_1 just by feeding them to C^*. However, given black-box access to C^*, we show that it is impossible to recover any message/signature pair and therefore any input x on which $C_0(x) \neq C_1(x)$. Intuitively, each call to C^* leaks one bit of information on a fresh message/signature pair, which is not enough to recover any such pair in full. We therefore put forth the conjecture that there exists a "special-purpose" method of obfuscating C^*, that does not allow the attacker to learn any message/signature pair. Under this special-purpose obfuscation assumption, the auxiliary input aux allows us to distinguish any obfuscation of C_0, C_1 but does not allow us to find any input x on which $C_0(x) \neq C_1(x)$.

2   Preliminaries and Definitions

Notation. We let λ denote the security parameter throughout the paper. We use the notation $C[\text{prm}]$ to denote a circuit that depends on a parameter prm. The parameter can be an arbitrary string, and we think of prm as being "hard wired" in the description of the corresponding circuit. The input to a circuit is specified inside parenthesis, so $C[\mathsf{prm}](x)$ describes the computation of the circuit $C[\mathsf{prm}]$ (whose definition depends on prm) on the input x.

Differing-Inputs Obfuscation. Our definition of differing-inputs obfuscation (diO) with auxiliary input follows that of Ananth et al. [ABG+13], which is also equivalent to that of Boyle et al. [BCP14]. First, we define the notion of "differing-inputs" circuits.

Definition 1. *A circuit family $\mathcal{C}$ with a sampler $(C_0, C_1, \mathsf{aux}) \leftarrow \mathsf{Sam}(1^\lambda)$ which samples $C_0, C_1 \in \mathcal{C}$ is said to be a* differing-inputs *family if for all PPT attackers $\mathcal{A}$ there is a negligible function ε such that:*

$$\Pr[C_0(x) \neq C_1(x) \ : \ (C_0, C_1, \mathsf{aux}) \leftarrow \mathsf{Sam}(1^\lambda), x \leftarrow \mathcal{A}(1^\lambda, C_0, C_1, \mathsf{aux})] \leq \varepsilon(\lambda).$$

Definition 2. *A PPT algorithm $\mathcal{O}$ is a* differing-inputs obfuscator *(diO) for a differing-inputs family $\mathcal{C}, \mathsf{Sam}$ if the following holds:*

- *Correctness: For all $\lambda \in \mathbb{N}, C \in \mathcal{C}$ and all inputs x, we have:*

$$\Pr[C'(x) = C(x) \mid C' \leftarrow \mathcal{O}(1^\lambda, C)] = 1.$$

- *Security: For all PPT distinguishers $\mathcal{D}$, there is a negligible function ε such that:*

$$|\Pr[\mathcal{D}(1^\lambda, \mathcal{O}(1^\lambda, C_0), \mathsf{aux}) = 1] - \Pr[\mathcal{D}(1^\lambda, \mathcal{O}(1^\lambda, C_1), \mathsf{aux}) = 1]| \leq \varepsilon(\lambda)$$

where $(C_0, C_1, \mathsf{aux}) \leftarrow \mathsf{Sam}(1^\lambda)$.

A PPT algorithm $\mathcal{O}$ is a general-purpose differing-inputs obfuscator *if the above holds for all differing-inputs families $\mathcal{C}, \mathsf{Sam}$.*

The works of [ABG+13, BCP14] put forth the conjecture that general-purpose diO exists, and that the obfuscator of [GGH+13] is a good candidate.

Extractable Witness Encryption. Next we define the "extractable" variant of witness encryption following Goldwasser et al. [GKP+13]. The notion of witness encryption was first defined and realized by Garg et al. [GGSW13]. Goldwasser et al. [GKP+13] conjecture that the same construction can be assumed to be extractable with auxiliary input. For simplicity, we assume that the message space is 1 bit. Next we present these definitions formally (following [GGSW13, GKP+13], but making the definitions even weaker by assuming the auxiliary input comes from an efficiently sampleable distribution and allowing the extractor to depend on this distribution).

Definition 3. *A* witness encryption *scheme for an* **NP** *language L (with corresponding witness relation R) consists of the following two polynomial-time algorithms:*

Encryption. *The algorithm* $\mathsf{Enc}(1^\lambda, x, b)$ *takes as input a security parameter* 1^λ, *an unbounded-length string* x, *and a message* $b \in \{0,1\}$ *and outputs a ciphertext* c.

Decryption. *The algorithm* $\mathsf{Dec}(c, w)$ *takes as input a ciphertext* c *and an unbounded-length string* w, *and outputs a message* b *or the symbol* $\perp$.

These algorithms satisfy the following two conditions:

– **Correctness.** *For any security parameter* λ, *for any* $b \in \{0,1\}$, *and for any* $x \in L$ *such that* $R(x, w)$ *holds, we have that*

$$\Pr\left[\mathsf{Dec}\big(\mathsf{Enc}(1^\lambda, x, b), w\big) = b\right] = 1$$

– **Extractable Security.** *For any PPT adversary* A, *polynomial-time sampler* $(x, \mathsf{aux}) \leftarrow \mathsf{Sam}(1^\lambda)$ *and for any polynomial* $q(\cdot)$, *there exists a PPT extractor* E *and a polynomial* $p(\cdot)$, *such that:*

$$\Pr\left[A(1^\lambda, x, c, \mathsf{aux}) = b \;\middle|\; \begin{array}{c} b \leftarrow \{0,1\}, (x, \mathsf{aux}) \leftarrow \mathsf{Sam}(1^\lambda), \\ c \leftarrow \mathsf{Enc}(1^\lambda, x, b) \end{array}\right] \geq \frac{1}{2} + \frac{1}{q(\lambda)}$$

$$\Rightarrow \Pr[E(1^\lambda, x, \mathsf{aux}) = w \text{ s.t. } (x, w) \in R_L \; : \; (x, \mathsf{aux}) \leftarrow \mathsf{Sam}(1^\lambda)] \geq \frac{1}{p(\lambda)}.$$

3   The Counterexample to diO and the Counter-Conjecture

We construct a family $(\mathcal{C}, \mathsf{Sam})$ which we show to be unobfuscatable with respect to differing-inputs obfuscation. However, to show that this family $(\mathcal{C}, \mathsf{Sam})$ is a differing-inputs family, we will in turn need to rely on a new "special purpose obfuscation" conjecture.

Let $\mathcal{S} = (\mathsf{KeyGen}, \mathsf{Sig}, \mathsf{Ver})$ be a signature scheme with signature size $\ell_{sig}(\lambda)$ and a deterministic signing algorithm.[3] Let $\mathcal{H} = \{\mathcal{H}_\lambda\}$ be a collision-resistant hash function (CRHF) family with output size $\ell_{hash}(\lambda)$. Define the circuit family $\mathcal{C}$ consisting of circuits $C[b, \mathsf{vk}] \in \mathcal{C}$ defined as follows:

$C[b, \mathsf{vk}](m, \sigma)$ // Hard-coded values: $b \in \{0,1\}$, vk verification key

// Input: $m \in \{0,1\}^{\ell_{hash}(\lambda)}, \sigma \in \{0,1\}^{\ell_{sig}(\lambda)}$

– Check $\mathsf{Ver}_{\mathsf{vk}}(m, \sigma) = 1$. If not output 0 else output b.

Let $\ell_{circ}(\lambda)$ be the maximal size of the circuit $C[b, \mathsf{vk}]$ when $b \in \{0,1\}$ and $(\mathsf{sk}, \mathsf{vk}) \leftarrow \mathsf{KeyGen}(1^\lambda)$.

Our counterexample to diO will consist of setting $C_0 = C[0, \mathsf{vk}]$ and $C_1 = C[1, \mathsf{vk}]$. Finding an input on which $C_0(x) \neq C_1(x)$ is equivalent to finding any valid message/signature pair $x = (m, \sigma)$ which is hard given only the description

---

[3] Any signature scheme can be converted into one with a deterministic signing algorithm by replacing the random coins with a PRF of the message.

of C_0, C_1 (which includes vk). However, we will provide an additional auxiliary input aux which makes it easy to distinguish any (bounded size) obfuscation of C_0 from that of C_1. We will need to argue that aux does not leak any valid message/signature pair, which will require a new assumption.

Let $\ell^* = \ell^*(\lambda)$ be be a length parameter (which will later be set to correspond to the size of a candidate obfuscation of the circuits $C[b, \text{vk}]$). Define the circuit family $\mathcal{C}_{\text{break}}$ consisting of circuits $C^*[H, \text{sk}] \in \mathcal{C}_{\text{break}}$ with input-length ℓ^* and 1-bit output as follows:

$\underline{C^*[H, \text{sk}](C)}$ // Hard-coded values: $H \in \mathcal{H}$, sk signing key
                          // Input: C : a circuit of size $|C| = \ell^*$ with 1-bit output.
– Compute $m = H(C)$, $\sigma = \text{Sig}_{\text{sk}}(m)$.
– Output the bit $C(m, \sigma)$.

Let $\text{sp}\mathcal{O}$ be a "special purpose" obfuscator that satisfies correctness and whose security properties we will define shortly. We define the circuit sampler $\text{Sam}_{\ell^*}(1^\lambda)$, parameterized by some polynomial $\ell^*(\cdot)$, as follows:

– Sample $(\text{sk}, \text{vk}) \leftarrow \text{KeyGen}(1^\lambda)$ and $H \leftarrow \mathcal{H}_\lambda$.
– Set $C_0 = C[0, \text{vk}], C_1 = C[1, \text{vk}] \in \mathcal{C}$.
– Set $C^* = C^*[H, \text{sk}] \in \mathcal{C}_{\text{break}}$ to be a circuit with input-length $\ell^* = \ell^*(\lambda)$ and set aux $\leftarrow \text{sp}\mathcal{O}(1^\lambda, C^*)$.
– Output C_0, C_1, aux.

It is easy to see that the the circuit family $\mathcal{C}, \text{Sam}_{\ell^*}$ is unobfuscatable since aux allows one to easily distinguish any obfuscations of C_0 and C_1 that have circuit-size at most ℓ^*. For any candidate obfuscator $\mathcal{O}$, we can choose ℓ^* sufficiently large to ensure that $\mathcal{O}$ fails.

Lemma 1. *Fix any signature/hash schemes $\mathcal{S}, \mathcal{H}$ which define the class of circuits $\mathcal{C}$, and let $\text{sp}\mathcal{O}$ be any "special-purpose obfuscator" satisfying correctness. Then for any candidate diO obfuscator $\mathcal{O}$ there is a polynomial $\ell^*(\lambda)$ such that the obfuscations of the family $(\mathcal{C}, \text{Sam}_{\ell^*})$ under $\mathcal{O}$ are easily distinguishable: there is a polynomial-time distinguisher $\mathcal{D}$ such that*

$$| \Pr[\mathcal{D}(1^\lambda, \mathcal{O}(1^\lambda, C_0), \text{aux}) = 1] - \Pr[\mathcal{D}(1^\lambda, \mathcal{O}(1^\lambda, C_1), \text{aux}) = 1]| = 1$$

where $(C_0, C_1, \text{aux}) \leftarrow \text{Sam}_{\ell^*}(1^\lambda)$.

Proof. Let $\ell_{circ}(\lambda)$ be the maximal size of the circuit $C[b, \text{vk}] \in \mathcal{C}$ when $b \in \{0, 1\}$ and $(\text{sk}, \text{vk}) \leftarrow \text{KeyGen}(1^\lambda)$. Set $\ell^*(\lambda)$ be the maximal size of $\mathcal{O}(1^\lambda, C)$ for any $C \in \mathcal{C}$ of size $|C| = \ell_{circ}(\lambda)$. The distinguisher $\mathcal{D}(1^\lambda, \widetilde{C}, \text{aux})$ simply interprets aux as a circuit and outputs $\text{aux}(\widetilde{C})$. It is easy to see that, if $\widetilde{C} = \mathcal{O}(1^\lambda, C_b)$, then $\text{aux}(\widetilde{C}) = b$ and therefore the distinguishing advantage is 1. Also the size of $\widetilde{C}$ is at most $\ell^*(\lambda)$ and hence it can be used as an input to aux.

To get a counterexample to the existence of general-purpose differing-inputs obfuscation, we need to show that, for *some* signature scheme $\mathcal{S}$, CRHF $\mathcal{H}$ and obfuscator sp$\mathcal{O}$, the family $(\mathcal{C}, \mathsf{Sam}_{\ell*})$ is a differing-inputs family for any ℓ^*. Notice that finding an input $x = (m, \sigma)$ on which $C_0(x) \neq C_1(x)$ is the same as finding a valid message/signature pair. Therefore, the above reduces to the following conjecture which says that, given the obfuscation of the "breaker" circuit C^* it is difficult to produce any valid message/signature pair.

Conjecture 1 (Special-Purpose Obfuscation). There exists a signature scheme $\mathcal{S}$, CRHF $\mathcal{H}$ and an obfuscator sp$\mathcal{O}$ such that the following hods. For any PPT attacker $\mathcal{A}$ and any polynomial $\ell^*(\cdot)$ there is a negligible $\varepsilon(\lambda)$ such that:

$$\Pr\left[\mathsf{Ver}_{\mathsf{vk}}(m, \sigma) = 1 \;\middle|\; \begin{array}{c} (\mathsf{sk}, \mathsf{vk}) \leftarrow \mathsf{KeyGen}(1^\lambda), H \leftarrow \mathcal{H}_\lambda \\ \tilde{C} \leftarrow \mathsf{spO}(1^\lambda, C^*[H, \mathsf{sk}]) \\ (m, \sigma) \leftarrow \mathcal{A}(1^\lambda, \mathsf{vk}, \tilde{C}) \end{array}\right] \leq \varepsilon(\lambda)$$

where we take the circuit $C^*[H, \mathsf{sk}] \in \mathcal{C}_{\mathsf{break}}$ with input-size $\ell^*(\lambda)$ as defined above.

If we fix some specific choice of schemes $\mathcal{S}, \mathcal{H}, \mathsf{spO}$ (e.g., a standard construction of signatures and hash functions and the obfuscation scheme of [GGH$^+$13]) then the above becomes a falsifiable assumption. We can efficiently test if an attacker $\mathcal{A}$ breaks the scheme. We now show that, under the above conjecture, the circuit family $(\mathcal{C}, \mathsf{Sam})$ defined above is a differing-inputs family.

Lemma 2. *For any signature scheme $\mathcal{S}$, CRHF $\mathcal{H}$ and an obfuscator sp$\mathcal{O}$ satisfying Conjecture 1, for any polynomial ℓ^*, the circuit family $(\mathcal{C}, \mathsf{Sam}_{\ell*})$ defined above is a differing-inputs family.*

Proof. Assume there is a PPT attacker $\mathcal{B}$ such that:

$$\Pr[C_0(x) \neq C_1(x) \;:\; (C_0, C_1, \mathsf{aux}) \leftarrow \mathsf{Sam}_{\ell*}(1^\lambda), x \leftarrow \mathcal{B}(1^\lambda, C_0, C_1, \mathsf{aux})] = \varepsilon(\lambda).$$

Since $C_0(x) \neq C_1(x)$ means that $x = (m, \sigma)$ such that $\mathsf{Ver}_{\mathsf{vk}}(m, \sigma) = 1$, we get

$$\Pr[\mathsf{Ver}_{\mathsf{vk}}(m, \sigma) = 1 \;:\; (C_0, C_1, \mathsf{aux}) \leftarrow \mathsf{Sam}_{\ell*}(1^\lambda), x \leftarrow \mathcal{B}(1^\lambda, C_0, C_1, \mathsf{aux})] = \varepsilon(\lambda).$$

Define the attacker $\mathcal{A}(1^\lambda, \mathsf{vk}, \tilde{C})$ that constructs $C_0 = C[0, \mathsf{vk}], C_1 = C[1, \mathsf{vk}]$, $\mathsf{aux} = \tilde{C}$ and calls $\mathcal{B}(1^\lambda, C_0, C_1, \mathsf{aux})$. Then

$$\Pr\left[\mathsf{Ver}_{\mathsf{vk}}(m, \sigma) = 1 \;\middle|\; \begin{array}{c} (\mathsf{sk}, \mathsf{vk}) \leftarrow \mathsf{KeyGen}(1^\lambda), H \leftarrow \mathcal{H}_\lambda, \\ (m, \sigma) \leftarrow \mathcal{A}(1^\lambda, \mathsf{vk}, \mathsf{spO}(1^\lambda, C^*[H, \mathsf{sk}])) \end{array}\right] = \varepsilon(\lambda)$$

where the input size of $C^*[H, \mathsf{sk}]$ is $\ell^*(\lambda)$. Therefore, by the conjecture, we must have $\varepsilon(\lambda)$ is negligible, which means that the $(\mathcal{C}, \mathsf{Sam}_{\ell*})$ is differing-inputs family.

Combining Lemma 2 and Lemma 1 we get the main theorem.

Theorem 1. *Under the special-purpose obfuscation conjecture (Conjecture 1), general-purpose differing-inputs obfuscators do not exist.*

4   Substantiating the Special-Purpose Obfuscation Conjecture

We now attempt to substantiate the special-purpose obfuscation conjecture (Conjecture 1). As a first step, we show that black-box access to the circuit $C^*[H, \mathsf{sk}]$ cannot be used to leak a message/signature pair. Intuitively, each query C allows the attacker to learn 1 bit of leakage $C(m, \sigma)$ on a signature of the message $m = H(C)$. Assuming the attacker cannot break collision-resistance, he cannot get get more than 1 bit of leakage on any single signature. Generically, seeing 1 bit of leakage on signatures of many different messages does not allow an attacker to come up with any valid message, signature pair. We formalize this via the following Lemma.

Lemma 3. *For any signature scheme $\mathcal{S}$ and CRHF $\mathcal{H}$, and parameter $\ell^*(\lambda)$, for any PPT attacker $\mathcal{A}$ there is a negligible $\varepsilon(\cdot)$ such that:*

$$\Pr\left[\mathsf{Ver}_{\mathsf{vk}}(m, \sigma) = 1 \;\middle|\; \begin{array}{l} (\mathsf{sk}, \mathsf{vk}) \leftarrow \mathsf{KeyGen}(1^\lambda), H \leftarrow \mathcal{H}_\lambda, \\ (m, \sigma) \leftarrow \mathcal{A}^{C^*[H, \mathsf{sk}](\cdot)}(1^\lambda, \mathsf{vk}, H) \end{array}\right] \leq \varepsilon(\lambda)$$

where $C^*[H, \mathsf{sk}] \in \mathcal{C}_{\mathsf{break}}$ is defined above and has input size $\ell^*(\lambda)$.

Proof. Fix some signature scheme $\mathcal{S}$ and CRHF $\mathcal{H}$ and PPT attacker $\mathcal{A}$. Let $q = q(\lambda)$ be an upper bound on the number of queries that $\mathcal{A}$ makes to C^* and let $\varepsilon(\lambda)$ denote the success probability of $\mathcal{A}$. We define an attacker $\mathcal{B}$ on the EU-CMA (existential unforgeability against chosen message attack) signature security of $\mathcal{S}$ as follows:

- $\mathcal{B}$ guesses an index $i \leftarrow [q]$ and a bit $b \leftarrow \{0, 1\}$ uniformly at random.
- $\mathcal{B}$ gets vk from its challenger and samples $H \leftarrow \mathcal{H}_\lambda$. It runs $\mathcal{A}(1^\lambda, \mathsf{vk}, H)$.
  - Whenever $\mathcal{A}$ makes any query other than the ith query to C^* with some input C, the attacker $\mathcal{B}$ computes $m = H(C)$ uses its signing oracle to compute $\sigma = \mathsf{Sig}_{\mathsf{sk}}(m)$. It then output $C(m, \sigma)$.
  - When $\mathcal{A}$ makes the ith query C_i to C^*, the attacker $\mathcal{B}$ simply responds with the bit b it chose randomly.
- At the end $\mathcal{B}$ outputs the value (m, σ) that $\mathcal{A}$ outputs.

Define the events:

- $\mathsf{Win}_\mathcal{B}$ is the event that $\mathcal{B}$ wins the EU-CMA signature game.
- Ver is the event that $\mathsf{Ver}_{\mathsf{vk}}(m, \sigma) = 1$.
- Col is the event that, during the course of the game, the attacker $\mathcal{A}$ submits two different circuits C, C' to its oracle such that $H(C) = H(C')$.
- Good_1 is the event that, if $\mathcal{A}$ outputs (m, σ), then no query C_j to C^* resulted in $H(C_j) = m$ other than possibly the ith query.
- Good_2 is the event that, if the ith query is C_i, and we set $m = H(C_i)$, $\sigma = \mathsf{Sig}_{\mathsf{sk}}(m)$, then $C_i(m, \sigma) = b$.

Then we have

$$\Pr[\mathsf{Win}_{\mathcal{B}}] \geq \Pr[\mathsf{Ver} \wedge \mathsf{Good}_1] \geq \Pr[\mathsf{Ver} \wedge \mathsf{Good}_1 \wedge \mathsf{Good}_2 \wedge \neg\mathsf{Col}]$$

$$\geq \Pr[\mathsf{Good}_1 \mid \mathsf{Ver} \wedge \mathsf{Good}_2 \wedge \neg\mathsf{Col}] \Pr[\mathsf{Ver} \wedge \mathsf{Good}_2 \wedge \neg\mathsf{Col}]$$

$$\geq \frac{1}{q} \Pr[\mathsf{Ver} \wedge \mathsf{Good}_2 \wedge \neg\mathsf{Col}] \tag{1}$$

$$\geq \frac{1}{q} \Pr[\mathsf{Good}_2] \Pr[\mathsf{Ver} \mid \mathsf{Good}_2] - \Pr[\mathsf{Col}]$$

$$\geq \frac{1}{2q}\varepsilon(\lambda) - \delta_{col}(\lambda) \tag{2}$$

where $\delta_{col}(\lambda) := \Pr[\mathsf{Col}]$ is negligible by the security of the CRHF. Equation (1) follows since, even if we condition on $\neg\mathsf{Col}$ and all other randomness in the game other than the choice of i, the attacker $\mathcal{A}$ made at most 1 query C_j such that $H(C_j) = m$ and therefore with probability $1/q$ over only the choice of i we have $i = j$. Equation (2) follows since the probability of Good_2 is $\frac{1}{2}$ only over the choice of b, and conditioned on Good_2, the attacker $\mathcal{B}$ perfectly simulates the obfuscation game for $\mathcal{A}$.

Since, by the security of the signature scheme, we must have $\Pr[\mathsf{Win}_{\mathcal{B}}]$ is negligible, this must also mean that $\varepsilon(\lambda)$ is negligible, which concludes the proof.

Further Informal Discussion. We stress that to rule out general-purpose diO we do *not* need the conjecture above to hold for *all* hash functions and signatures. [4] Rather, it is enough that it holds for *some* hash function and signature scheme (such as e.g., RSA PKCS #1 v1.5).

Let's consider attempts at attacking the conjecture, and give highly informal arguments for why they seem to fail. To do so, let's fix some "standard-construction" hash function and signature scheme such as RSA PKCS #1 v1.5, in which case we are also fixing the auxiliary information $\mathsf{aux} = \mathsf{vk}$. As mentioned, all of the prior obfuscation impossibility results have the same general structure which, applied to our problem, would require us to either: (i) use the obfuscated-code $\mathsf{spO}(C^*)$ to design a special input on which C^* outputs additional information [BGI+12], or (ii) interpret the auxiliary information $\mathsf{aux} = \mathsf{vk}$ as code which outputs some information when given $\mathsf{spO}(C^*)$ as an input [GK05, GK13, BCPR13b]. Since in our case vk is just the verification of a standard scheme (e.g. RSA PKCS #1 v1.5), there does not seem to be much hope in approach (ii). On the other hand, there do not seem to be any special inputs on which C^* acts in any "special way" so as to exploit approach (i). The fact that the input to C^* is itself interpreted as a circuit C and executed by C^* should give us some pause. After all, we can make C depend on $\mathsf{spO}(C^*)$. But such inputs would not be treated in any kind of special way by C^*: they

---

[4] Indeed, we suspect that one should be able to come up with some "unnatural" signature and hash function for which it does not hold (following similar counterexamples from [BGI+12, GK05, GK13, BCPR13b]).

would still only allow the attacker to leak one bit of information $C(m, \sigma)$ on an honestly generated message/signature pair.

Finally, we note that a recent result that relates iO to a limited form of diO has no bearing on our counterexample: Boyle et al. [BCP14] showed that differing-inputs obfuscation is already implied by indistinguishability obfuscation, in the special case where the two circuits C_0, C_1 only differ on polynomially many inputs. In our counterexample, the circuits C_0, C_1 differ on all valid message/signature pairs where the message-domain is super-polynomial. Therefore, we do not get any negative results for indistinguishability obfuscation.

5   Bounded-Length Auxiliary Input

Our counterexample shows that, under our special-purpose obfuscation conjecture, there is no general-purpose diO scheme that works with any auxiliary input. In particular, we constructed family (C_0, C_1, aux) where the definition of aux relies on some parameter ℓ^* such that any obfuscations of C_0 and C_1 having size at most ℓ^* are always distinguishable given aux. We can make the parameter ℓ^* arbitrary large at the expense of making the auxiliary input aux correspondingly large. This leaves open the possibility of a diO scheme that is secure for all auxiliary input of some arbitrary but a-priori bounded size. We define this as follows:

Definition 4. *We define a general-purpose diO obfuscator with bounded-length auxiliary input analogously to Definition 1 but with the following changes:*

- *The syntax of the obfuscator $\mathcal{O}(1^\lambda, 1^{\ell_{aux}(\lambda)}, C)$ now takes an additional parameter $\ell_{aux}(\lambda)$.*
- *We require that for all polynomial $\ell_{aux}(\lambda)$ security holds for differing-inputs families $(\mathcal{C}, \mathsf{Sam})$ where the size of aux in $(C_0, C_1, \mathsf{aux}) \leftarrow \mathsf{Sam}(1^\lambda)$ is bounded by $\ell_{aux}(\lambda)$.*

Our previously described counterexample does not rule out this definition. In particular, the auxiliary input aux in our counterexample is an obfuscated circuit that takes as input an obfuscation of C_b. If the obfuscation of C_b can depend on (and exceed) the size of aux, then this would not work. However, we can rule out this weaker notion of diO for bounded-length auxiliary input if we additionally assume that we have a special-purpose obfuscator $\mathsf{sp}\mathcal{O}$ which works directly on Turing Machines rather than circuits. In particular, a Turing Machine special-purpose obfuscator $\mathsf{sp}\mathcal{O}(1^\lambda, M)$ takes as input a Turing Machine M and outputs an obfuscated Turing Machine $\tilde{M}$ where $\tilde{M}$ can be evaluated on arbitrary-length inputs and produces the same output as M.

The Counterexample. Fix a signature scheme $\mathcal{S}$ and hash function family $\mathcal{H}$ as before, and define the circuit family $\mathcal{C}$ consisting of circuits $C[b, \mathsf{vk}]$ as before. We define the "breaker" Turing Machine $M^*[H, \mathsf{sk}]$ which has H and sk hardcoded in its description analogously to the way we defined the "breaker" circuit $C^*[H, \mathsf{sk}]$, as follows:

> $\underline{M^*[H, \mathsf{sk}](C)}$ // Hard-coded values: $H \in \mathcal{H}_\lambda$, sk signing key
>   // Input: C circuit with 1-bit output and arbitrary size.
> – Compute $m = H(C)$, $\sigma = \mathsf{Sig}_{\mathsf{sk}}(m)$.
> – Output $C(m, \sigma)$.

Notice that, unlike before, we no longer have any parameter ℓ^* that would fix the maximal input length of the input circuit C given to $M^*[H, \mathsf{sk}]$.

Let spO be a Turing-Machine obfuscator that satisfies correctness. We define the circuit sampler $\mathsf{Sam}_{TM}(1^\lambda)$ as follows:

- Sample $(\mathsf{sk}, \mathsf{vk}) \leftarrow \mathsf{KeyGen}(1^\lambda)$ and $H \leftarrow \mathcal{H}_\lambda$.
- Set $C_0 = C[0, \mathsf{vk}], C_1 = C[1, \mathsf{vk}] \in \mathcal{C}$.
- Set $M^* = M^*[H, \mathsf{sk}]$ and $\mathsf{aux} \leftarrow \mathsf{spO}(1^\lambda, M^*)$.
- Output C_0, C_1, aux.

Conjecture 2 (Special-Purpose TM Obfuscation). There exists a signature scheme $\mathcal{S}$, CRHF $\mathcal{H}$ and an Turing Machine obfuscator spO such that the following hods: for any PPT attacker $\mathcal{A}$ there is a negligible $\varepsilon(\lambda)$ such that:

$$\Pr\left[\mathsf{Ver}_{\mathsf{vk}}(m, \sigma) = 1 \;\middle|\; \begin{matrix}(\mathsf{sk}, \mathsf{vk}) \leftarrow \mathsf{KeyGen}(1^\lambda), H \leftarrow \mathcal{H}_\lambda, \\ \tilde{M} \leftarrow \mathsf{spO}(1^\lambda, M^*[H, \mathsf{sk}]), (m, \sigma) \leftarrow \mathcal{A}(1^\lambda, \mathsf{vk}, \tilde{M})\end{matrix}\right] \leq \varepsilon(\lambda)$$

where the Turing Machine $M^*[H, \mathsf{sk}]$ is defined above.

Theorem 2. *Under the special-purpose TM obfuscation conjecture (Conjecture 2), there is no general-purpose diO obfuscators (for circuits) that has security for bounded-length auxiliary input.*

In particular, under the conjecture, the circuit family $(\mathcal{C}, \mathsf{Sam}_{TM})$ defined above is a fixed differing-inputs family with some fixed polynomial bound on the length of the auxiliary input, yet there is no diO obfuscator for this particular family.

The proof of the above theorem is the same as that of Theorem 1.

Discussion. We note that candidate general-purpose iO and diO obfuscators for Turing Machines were constructed by [BCP14, ABG+13]. Although the security claims rely on general-purpose (circuit) diO with auxiliary input, it seems reasonable to assume that these constructions are secure in special cases, and also that they satisfies stronger security properties than merely iO and diO. In particular, using these candidate obfuscators, we do not know of any attacks on Conjecture 2. Moreover, it is still a falsifiable assumption once we fix some candidates $\mathcal{S}, \mathcal{H}, \mathsf{spO}$. On the other hand, the Turing Machine conjecture certainly seems stronger and more complex than the corresponding circuit conjecture (Conjecture 1).

6   Extending Implausibility to Extractable Witness Encryption

In previous section we showed that a "special-purpose obfuscation" conjecture (Conjecture 1) can be used to rule out existence of a general-purpose differing-inputs obfuscator. In this section we show that the same "special-purpose obfuscation" conjecture can also be used to rule out existence of extractable witness encryption. Note that this is a stronger result as general-purpose differing-inputs obfuscation is known to imply extractable witness encryption.

Theorem 3. *Under the special-purpose obfuscation conjecture (Conjecture 1), extractable witness encryption does not exist.*

Proof. We prove our theorem by giving an **NP**-relation R for which there does not exist an extractable witness encryption scheme. In order to prove this we will need to rely on our "special-purpose obfuscation" conjecture (Conjecture 1).

Let $S = (\mathsf{KeyGen}, \mathsf{Sig}, \mathsf{Ver})$ be a signature scheme with a deterministic signing algorithm. We define the **NP**-relation R_{ver} so that $(\mathsf{vk}, (m, \sigma)) \in R_{ver}$ if and only if $\mathsf{Ver}_{\mathsf{vk}}(m, \sigma) = 1$. Let $(\mathsf{Enc}, \mathsf{Dec})$ be a candidate extractable witness encryption for this relation R. Given an string vk and a ciphertext c, let $C[\mathsf{vk}, \mathsf{c}](w)$ be the circuit that takes as input a witness w and computes $\mathsf{Dec}(\mathsf{c}, w)$. Let $\ell^*(\lambda)$ be the size of $C[\mathsf{vk}, \mathsf{c}]$.

We now define the same auxiliary input as in the previous section. Let $\mathcal{H} = \{\mathcal{H}_\lambda\}$ be a collision-resistant hash function (CRHF) family with output size $\ell_{in}(\lambda)$. Define the circuit family $\mathcal{C}_{\mathsf{break}}$ consisting of circuits $C^*[H, \mathsf{sk}] \in \mathcal{C}_{\mathsf{break}}$ defined as follows:

$C^*[H, \mathsf{sk}](C)$	// Hard-coded values: $H \in \mathcal{H}_\lambda$, sk signing key
	// Input: C circuit of size $\ell^*(\lambda)$ with 1-bit output.
– Compute $m = H(C)$, $\sigma = \mathsf{Sig}_{\mathsf{sk}}(m)$.	
– Output $C(m, \sigma)$.	

Let spO be a "special purpose" obfuscator whose properties defined in Conjecture 1. We define the distribution samples $\mathsf{Sam}(1^\lambda)$ as follows:

- Sample $(\mathsf{sk}, \mathsf{vk}) \leftarrow \mathsf{KeyGen}(1^\lambda)$ and $H \leftarrow \mathcal{H}_\lambda$.
- Set $C^* = C^*[H, \mathsf{sk}] \in \mathcal{C}_{\mathsf{break}}$ and $\mathsf{aux} \leftarrow \mathsf{spO}(C^*)$.
- Output $\mathsf{vk}, \mathsf{aux}$, where vk is the **NP** statement.

Now consider an experiment where we sample $(\mathsf{vk}, \mathsf{aux}) \leftarrow \mathsf{Sam}(1^\lambda)$ and encrypt $\mathsf{c} \leftarrow \mathsf{Enc}(1^\lambda, \mathsf{vk}, b)$ where $b \leftarrow \{0, 1\}$ and vk acts as an **NP** statement. We construct an adversary A that can output b with probability 1. Our adversary $A(1^\lambda, \mathsf{vk}, \mathsf{c}, \mathsf{aux})$ simply interprets aux as a circuit and outputs $\mathsf{aux}(C[\mathsf{vk}, \mathsf{c}])$. It is easy to see that, if $\mathsf{c} = \mathsf{Enc}(1^\lambda, \mathsf{vk}, b)$, then $\mathsf{aux}(C[\mathsf{vk}, \mathsf{c}]) = b$ and therefore the adversary outputs b with probability 1.

On the other hand, we claim that no extractor E that can output valid witnesses given $(\mathsf{vk}, \mathsf{aux})$, contradicting the extractability property of the witness encryption scheme. Notice that finding a witness $w = (m, \sigma)$ for the statement consisting of a verification key vk under the relation R_{ver} is same as finding a valid message/signature pair given just the "special purpose" obfuscation aux and vk (the proof of this is similar to the proof of Lemma 2). In other words, Conjecture 1 directly implies that for any PPT candidate extractor E there is a negligible ε such that:

$$\Pr[E(1^\lambda, \mathsf{vk}, \mathsf{aux}) = (m, \sigma) \text{ s.t. } (\mathsf{vk}, (m, \sigma)) \in R_{ver} \; : \; (x, \mathsf{aux}) \leftarrow \mathsf{Sam}(1^\lambda)] \leq \varepsilon(\lambda)$$

contradicting the extractability requirement of extractable witness encryption. This completes our proof.

Bounded-Length Auxiliary Input. We could also define extractable witness encryption with bounded-length auxiliary input, where the encryption/decryption procedures can all depend on the size of the auxiliary input. This would be analogous to the definition of diO with bounded-length auxiliary input. We can rule out this notion of witness encryption with bounded-length auxiliary input under our special-purpose *Turing Machine* obfuscation assumption (Conjecture 2) analogously to our results for diO in Section 5.

7   Conclusions

We propose a seemingly reasonable "special-purpose" obfuscation conjecture under which general-purpose diO and extractable witness encryption with auxiliary input cannot exist. Furthermore a variant of this conjecture also shows the impossibility of output-only dependent hardcore bits for every one-way function. Many interesting open problems remain. Firstly, is there some inherent reason why our conjecture cannot hold? This is certainly possible, and we cannot objectively say which of the two conflicting possibilities (diO with auxiliary input vs. our conjecture) is false. However, the conjecture is a simple-to-state falsifiable assumption. Showing the possibility of general-purpose diO and extractable witness encryption would require coming up with an attack on this conjecture. On the other hand, general-purpose diO and witness encryption are not stated as falsifiable assumptions; indeed we give a candidate attack on these notions, but we cannot efficiently check if the attack is valid. In the absence of further evidence, we choose to interpret this result as giving strong evidence that general-purpose diO and extractable witness encryption are "implausible". Is there a way to convert this "implausibility" result into an "impossibility" result? On a different note, is it still reasonable to assume the existence of general-purpose diO *without* auxiliary input? We do not see any way to extend our "implausibility" result to the case without auxiliary input. Lastly, it remains as an interesting open problem to characterize the known techniques for getting obfuscation impossibility results, and come up with a strong and general obfuscation assumption that capture everything which is not directly ruled out by these techniques.

Acknowledgments. We thank Mariana Raykvoa and Amit Sahai for initial discussions relating to this work, Nir Bitansky for suggesting we look at extractable witness encryption, and Mihir Bellare for pointing us to his paper on poly-many hardcore bits and for suggesting we consider diO with bounded-length auxiliary input.

References

[ABG+13]   Ananth, P., Boneh, D., Garg, S., Sahai, A., Zhandry, M.: Differing-inputs obfuscation and applications. Cryptology ePrint Archive, Report 2013/689 (2013), http://eprint.iacr.org/

[App13]   Applebaum, B.: Bootstrapping obfuscators via fast pseudorandom functions. Cryptology ePrint Archive, Report 2013/699 (2013), http://eprint.iacr.org/

[BCP14]   Boyle, E., Chung, K.-M., Pass, R.: On extractability obfuscation. In: Lindell, Y. (ed.) TCC 2014. LNCS, vol. 8349, pp. 52–73. Springer, Heidelberg (2014)

[BCPR13a]   Bitansky, N., Canetti, R., Paneth, O., Rosen, A.: Indistinguishability obfuscation vs. auxiliary-input extractable functions: One must fall. Cryptology ePrint Archive, Report 2013/641 (2013), http://eprint.iacr.org/

[BCPR13b]   Bitansky, N., Canetti, R., Paneth, O., Rosen, A.: More on the impossibility of virtual-black-box obfuscation with auxiliary input. Cryptology ePrint Archive, Report 2013/701 (2013), http://eprint.iacr.org/

[BGI+01]   Barak, B., Goldreich, O., Impagliazzo, R., Rudich, S., Sahai, A., Vadhan, S.P., Yang, K.: On the (Im)possibility of Obfuscating Programs. In: Kilian, J. (ed.) CRYPTO 2001. LNCS, vol. 2139, pp. 1–18. Springer, Heidelberg (2001)

[BGI+12]   Barak, B., Goldreich, O., Impagliazzo, R., Rudich, S., Sahai, A., Vadhan, S.P., Yang, K.: On the (im)possibility of obfuscating programs. J. ACM 59(2), 6 (2012)

[BGK+13]   Barak, B., Garg, S., Kalai, Y.T., Paneth, O., Sahai, A.: Protecting obfuscation against algebraic attacks. Cryptology ePrint Archive, Report 2013/631 (2013), http://eprint.iacr.org/

[BP13]   Boyle, E., Pass, R.: Limits of extractability assumptions with distributional auxiliary input. Cryptology ePrint Archive, Report 2013/703 (2013), http://eprint.iacr.org/

[BR13]   Brakerski, Z., Rothblum, G.N.: Virtual black-box obfuscation for all circuits via generic graded encoding. Cryptology ePrint Archive, Report 2013/563 (2013), http://eprint.iacr.org/

[BT13]   Bellare, M., Tessaro, S.: Poly-many hardcore bits for any one-way function. Cryptology ePrint Archive, Report 2013/873 (2013), http://eprint.iacr.org/

[BZ13]   Boneh, D., Zhandry, M.: Multiparty key exchange, efficient traitor tracing, and more from indistinguishability obfuscation. Cryptology ePrint Archive, Report 2013/642 (2013), http://eprint.iacr.org/

[GGH+13]   Garg, S., Gentry, C., Halevi, S., Raykova, M., Sahai, A., Waters, B.: Candidate indistinguishability obfuscation and functional encryption for all circuits. To appear in FOCS 2013, vol. 2013, p. 451 (2013)

[GGHR13]   Garg, S., Gentry, C., Halevi, S., Raykova, M.: Two-round secure mpc from indistinguishability obfuscation. Cryptology ePrint Archive, Report 2013/601 (2013), http://eprint.iacr.org/

[GGHW13]   Garg, S., Gentry, C., Halevi, S., Wichs, D.: On the implausibility of differing-inputs obfuscation and extractable witness encryption with auxiliary input. Cryptology ePrint Archive, Report 2013/860 (2013), http://eprint.iacr.org/

[GGJS13]   Goldwasser, S., Goyal, V., Jain, A., Sahai, A.: Multi-input functional encryption. Cryptology ePrint Archive, Report 2013/727 (2013), http://eprint.iacr.org/

[GGSW13]   Garg, S., Gentry, C., Sahai, A., Waters, B.: Witness encryption and its applications. In: STOC (2013)

[GJKS13]   Goyal, V., Jain, A., Koppula, V., Sahai, A.: Functional encryption for randomized functionalities. Cryptology ePrint Archive, Report 2013/729 (2013), http://eprint.iacr.org/

[GK05]   Goldwasser, S., Kalai, Y.T.: On the impossibility of obfuscation with auxiliary input. In: FOCS, pp. 553–562 (2005)

[GK13]   Goldwasser, S., Kalai, Y.T.: A note on the impossibility of obfuscation with auxiliary input. Cryptology ePrint Archive, Report 2013/665 (2013), http://eprint.iacr.org/

[GKP+13]   Goldwasser, S., Kalai, Y.T., Popa, R.A., Vaikuntanathan, V., Zeldovich, N.: How to run turing machines on encrypted data. In: Canetti, R., Garay, J.A. (eds.) CRYPTO 2013, Part II. LNCS, vol. 8043, pp. 536–553. Springer, Heidelberg (2013)

[Had00]   Hada, S.: Zero-knowledge and code obfuscation. In: Okamoto, T. (ed.) ASIACRYPT 2000. LNCS, vol. 1976, pp. 443–457. Springer, Heidelberg (2000)

[HSW13]   Hohenberger, S., Sahai, A., Waters, B.: Replacing a random oracle: Full domain hash from indistinguishability obfuscation. Cryptology ePrint Archive, Report 2013/509 (2013), http://eprint.iacr.org/

[MO13]   Marcedone, A., Orlandi, C.: Obfuscation [implies] (ind-cpa security [does not imply] circular security). Cryptology ePrint Archive, Report 2013/690 (2013), http://eprint.iacr.org/

[Nao03]   Naor, M.: On cryptographic assumptions and challenges. In: Boneh, D. (ed.) CRYPTO 2003. LNCS, vol. 2729, pp. 96–109. Springer, Heidelberg (2003)

[PPS13]   Pandey, O., Prabhakaran, M., Sahai, A.: Obfuscation-based non-black-box simulation and four message concurrent zero knowledge for np. Cryptology ePrint Archive, Report 2013/754 (2013), http://eprint.iacr.org/

[SW13]   Sahai, A., Waters, B.: How to use indistinguishability obfuscation: Deniable encryption, and more. Cryptology ePrint Archive, Report 2013/454 (2013), http://eprint.iacr.org/

A   Output-Only Dependent Hardcore Bits

In a recent work, Bellare and Tessaro [BT13] show the existence of polynomially many hardcore bits for any one-way function. In the case of *injective* one-way functions, their construction relies on indistinguishability obfuscation. However, in the case of *arbitrary* one-way functions, it relies on diO with auxiliary input. The construction has a very interesting property which we call "output-only dependence". In particular, even if the one-way function $f(x)$ is many-to-one, the hardcore bits $h(x)$ are completely determined by $f(x)$; for any inputs x, x' such that $f(x) = f(x')$ we also get $h(x) = h(x')$. This property is interesting even in the case of a single hardcore bit, and does *not* hold for any of the known general constructions (such as for the Goldreich-Levin bit).

Unfortunately, we show that our special-purpose obfuscation assumption (for Turing Machines) also gives a counterexample to the security of the hardcore bit construction of [BT13]. More generally, we show that there is a contrived one-way function that does not have *any* output-only dependent hardcore bit. In more detail:

- Under the special-purpose obfuscation conjecture for *circuits* (Conjecture 1), we construct a one-way function that does not have any output-only dependent hardcore bits given *auxiliary input*.[5]
- Under the special-purpose obfuscation conjecture for *Turing Machines* (Conjecture 2) we get the above result even *without* auxiliary input. In particular, we construct a one-way function which does not have any output-only dependent hardcore bits.

Due to space constraints, this result appears in our full version [GGHW13].

---

[5] The result of Bellare and Tessaro [BT13] does not consider auxiliary input.

Maliciously Circuit-Private FHE

Rafail Ostrovsky[1,*], Anat Paskin-Cherniavsky[2,**],
and Beni Paskin-Cherniavsky[3]

[1] Department of Computer Science and Mathematics, UCLA, USA
rafail@cs.ucla.edu
[2] Department of Computer Science, UCLA, USA
anpc@cs.ucla.edu
[3] cben@users.sf.net

Abstract. We present a framework for transforming FHE (fully homomorphic encryption) schemes with no circuit privacy requirements into maliciously circuit-private FHE. That is, even if both maliciously formed public key and ciphertext are used, encrypted outputs only reveal the evaluation of the circuit on some well-formed input x^*. Previous literature on FHE only considered semi-honest circuit privacy. Circuit-private FHE schemes have direct applications to computing on encrypted data. In that setting, one party (a receiver) holding an input x wishes to learn the evaluation of a circuit C held by another party (a sender). The goal is to make receiver's work sublinear (and ideally independent) of $|C|$, using a 2-message protocol. The transformation technique may be of independent interest, and have various additional applications. The framework uses techniques akin to Gentry's bootstrapping and conditional disclosure of secrets (CDS [AIR01]) combining a non circuit private FHE scheme, with a homomorphic encryption (HE) scheme for a smaller class of circuits which is maliciously circuit-private. We devise the first known circuit private FHE, by instantiating our framework by various (standard) FHE schemes from the literature.

Keywords: Fully homomorphic encryption, computing on encrypted data, privacy, malicious setting.

---

* Work supported in part by NSF grants 09165174, 1065276, 1118126 and 1136174, US-Israel BSF grant 2008411, OKAWA Foundation Research Award, IBM Faculty Research Award, Xerox Faculty Research Award, B. John Garrick Foundation Award, Teradata Research Award, and Lockheed-Martin Corporation Research Award. This material is based upon work supported by the Defense Advanced Research Projects Agency through the U.S. Office of Naval Research under Contract N00014 -11 -1-0392. The views expressed are those of the author and do not reflect the official policy or position of the Department of Defense or the U.S. Government.
** Work supported in part by NSF grants 09165174, 1065276, 1118126 and 1136174. This material is based upon work supported by the Defense Advanced Research Projects Agency through the U.S. Office of Naval Research under Contract N00014 -11 -1-0392. The views expressed are those of the author and do not reflect the official policy or position of the Department of Defense or the U.S. Government.

1   Introduction

In this paper, we devise a first fully homomorphic encryption scheme (FHE) [Gen09] that satisfies (a meaningful form of) circuit privacy in the malicious setting—a setting where the public key and ciphertext input to Eval are not guaranteed to be well-formed. We present a framework for transforming FHE schemes with no circuit privacy requirements into maliciously circuit-private FHE. The transformation technique may be of independent interest, and have various additional applications. The framework uses techniques akin to Gentry's bootstrapping and conditional disclosure of secrets (CDS [AIR01]) combining a non circuit private FHE scheme, with a homomorphic encryption (HE) scheme for a smaller class of circuits which is maliciously circuit-private. We then demonstrate an instantiation of this framework using schemes from the literature.

The notion of FHE does not require circuit privacy even in the semi-honest setting (but rather standard IND-CPA security, the ability to evaluate arbitrary circuits on encrypted inputs and encrypted outputs being "compact"). In [Gen09] and [vDGHV09, appendix C], the authors show how to make their FHE schemes circuit-private in the semi-honest setting.

One natural application of (compact) FHE is the induced 2-message, 2-party protocol, where a receiver holds an input x, and a sender holds a circuit C; the receiver learns $C(x)$, while the sender learns nothing. In the first round the receiver generates a public-key pk, encrypts x to obtain c, and sends (pk, c). The sender evaluates C on (pk, c) using the schemes' homomorphism, and sends back the result. An essential requirement is that receiver's work (and overall communication) is $poly(k, n, o(|C|))$, where k is a security parameter, ideally independent of $|C|$ altogether. This application of homomorphic encryption, termed *computing on encrypted data*, was studied both in several works [IP07, BKOI07] predating Gentry's first fully homomorphic scheme, and mentioned in [Gen09].

The underlying scheme's IND-CPA security translates into the standard simulation-based notion of *privacy* against a malicious sender in the stand-alone model[1] (but not any form of correctness against a malicious sender). The circuit privacy of the scheme translates into a privacy guarantee against a malicious receiver (of the same "flavor"). While standard FHE (without extra requirements) does not imply any security guarantees against malicious receivers, the semi-honestly circuit-private schemes from (e.g.) [vDGHV09] imply standard simulation-based security against semi-honest receivers. Thus, a maliciously circuit-private scheme induces a protocol which is private against malicious corruptions in the stand-alone model.

Let us now define maliciously circuit-privacy of FHE more precisely. We say a scheme is circuit-private if it satisfies the following privacy notion ala [IP07], stating that any (pk, c) pair induces some "effective" encrypted input x^*:

---

[1] Privacy against a malicious sender comes "for free", as the protocol is 2-round, and the client speaks first.

Definition 1. *(informal). We say a C-homomorphic[2] encryption scheme* (KeyGen, Enc, Eval, Dec) *is (maliciously) circuit-private if there exists an unbounded algorithm* Sim, *such that for all security parameters k, and all pk^*, c^* there exists x^*, such that for all circuits $C \in C$ over $|x^*|$ variables* Sim$(1^k, C(x^*))$ $=^s$ Eval$(1^k, pk^*, C, c^*)$ *(statistically indistinguishable). We say the scheme is semi-honestly circuit-private if the above holds only for well-formed pk^*, c^* pairs.*

An FHE satisfying Definition 1 induces a protocol private against a malicious sender (by IND-CPA security of the FHE), but private against unbounded malicious receivers with unbounded simulation.[3]

On one hand, this privacy notion is weaker compared to full security as the simulation is not efficient; on the other hand, it is stronger in the sense that it holds against unbounded adversaries as well.

Due to impossibility results for general 2-round sender-receiver computation in the plain model (e.g. [BLV04]), this notion has become standard in the (non-interactive, plain model) setting of computing on encrypted data [NP01, AIR01, HK12, IP07] as a plausible relaxation.

It is important to note that we only consider the plain model. If preprocessing, such as CRS was allowed, the malicious case could be easily reduced to the semi-honest case. That is, given CRS, Enc could have added a NIZK proving that the key is well-formed, and that the ciphertext is a valid ciphertext under that public key. Then, Eval could explicitly check that the proof is valid, if not return $\perp$, otherwise run Eval as for the semi-honest setting (for that scheme). Some care needs to be taken even in this setting, so that the scheme for the semi-honest setting used has somewhat enhanced privacy. More specifically, it needs to hold assuming (pk, c) are in the support of valid pk and $c \in Enc_{pk}(\cdot)$ respectively, but not that the *distribution* of (pk, c) is identical to the honestly generated one. Indeed, such a semi-honestly circuit-private scheme has been put forward in [GHV10] (when applied to a perfectly correct FHE scheme). On the other hand, the semi-honestly private scheme suggested in [vDGHV09] needs that pk has the proper distribution (KeyGen(1^k)), rather than just being in the support of that distribution.

To summarize, our main "take home" theorem is as follows.

Theorem 1. *(informal) Assume an FHE scheme $\mathcal{F}$ with decryption circuits in* NC^1 *exists. Assume further there exists a maliciously circuit private HE $\mathcal{B}$ that supports bit OT exists. Then, there exists a maliciously circuit private multi-hop FHE scheme.*

There exist several instantiations of the theorem by ingredients from the literature. All known FHE schemes from the literature, such as [Gen09, vDGHV09, BV11] have efficient decryption circuits as required by the Theorem. Some candidates for $\mathcal{B}$ are [NP01, AIR01, HK12].

---

[2] In particular, for fully homomorphic schemes, C is the class of all circuits.

[3] Jumping ahead, settling for computational indistinguishability with unbounded simulation would allow for somewhat simplified constructions. However, we shoot for the best achievable privacy notion.

In terms of implications to MPC, our result can be interpreted as non-interactive 2PC protocols with asymmetric inputs as follows.

Theorem 2. *(informal) Assume the preconditions of Theorem 1 hold. Then, there exist 2-message client-server MPC protocols where the client holds x, and the server holds $n = 1^{|x|}$ a circuit C with n inputs (which may be much larger then x), and 1^k a security parameter. The client learns $C(x)$ (but nothing else, not even $|C|$), and the server learns nothing about x (but $|x|$). The privacy guarantee for the client is standard simulation-based computational privacy. The privacy guarantee for the server is based on unbounded simulation (against possibly unbounded clients). The protocols's communication is at most $poly(n, k)$ (as the client is efficient in its own input).* [4]

Multi-hop circuit-private FH. It is a desirable property of an FHE scheme that the outputs of Eval applied to any given circuit C mapping $\{0,1\}^n$ to $\{0,1\}^m$ can be fed again into Eval running on a circuit taking $\{0,1\}^m$ as input, and so on - an unbounded number of times. In the terminology of [GHV10], this property of a HE scheme is referred to as multi-hop. If only upto some i such iterations are supported, the scheme is called i-hop. The standard definition of FHE, thus corresponds to 1-hop encryption. However, there exists a simple transformation for any compact FHE into multi-hop. This is done by including an encryption of the secret key in the public key, and homomorphically decrypting the encrypted outputs received using the encrypted key bits (as in Gentry's bootstrapping theorem [Gen09, GHV10]). The maliciously circuit-private FHE resulting from our construction is also designed to be only 1-hop, but the standard transformation does not make it multi-hop, as it does not preserve malicious circuit-privacy.

In Section 3.3, we define multi-hop maliciously circuit-private HE, and sketch a modification to our 1-hop scheme, making it multi-hop. Our transformation starts with the above transformation, and adds some validation of the added key bits.

1.1   Previous Work

Circuit private FHE implicit in work on MPC. As explained above, (compact) HE naturally gives rise to non-interactive client-server protocols for computing on encrypted data (and the privacy level of the protocol depends on the notion of circuit privacy of the FHE). An essential requirement is that client's work in these protocols is sub-linear in $|C|$ (ideally independent of $|C|$).

In the other direction, standard two-message client-server protocols (inputs of similar length, only client learns output), robust against malicious receivers,

---

[4] In fact, the above result can be interpreted as general "size hiding" 2PC with asymmetric inputs. The case in Theorem 2 is a special case with $F(x, y)$ being the universal function of evaluating a circuit y on input x. In the general case, F is some polynomial-time computable function. The client learns $F(x, y)$ (but not even $|y|$), and the server learns only $|x|$. This can be implemented by letting the server set $C = F_y(x)$, and run the protocol from Theorem with input $C, |x|$ for the server and input x for the client.

induce circuit-private (not necessarily compact) HE, when applied with the universal function [CCKM00,GHV10]. Roughly the round-1 message of the client is viewed as an encryption for his input, and the senders' reply as Eval's output. This is still not an encryption scheme, as in the protocol, the client needs to "remember" the randomness generating a round-1 message in order to "decode" the reply. This is solved by setting KeyGen output a public-key, private-key pair of some public key encryption scheme as (pk, sk) respectively. Enc is augmented concatenate an encryption c_r of its randomness under pk. Eval is augmented to pass c_r as is. Intuitively, if the protocol was private against malicious clients, then so is the encryption scheme (as the randomness was known to the client anyway).

One specific construction of HE from 2PC is by combining an information-theoretic version of Yao's garbled circuits with oblivious transfer (OT) secure against malicious receivers [NP01, AIR01, HK12, IP07]. As information-theoretic Yao is only efficient for NC^1, the resulting HE scheme captures only functions in NC^1. Also, this generic construction, even in the semi-honest setting (for all known 2PC protocols from then literature) has encrypted output size at least $|C|$, while the crux of HE is having compact encrypted outputs—ideally $poly(k)$, as modern FHE schemes achieve.

Another relevant work is a recent work on 2-message 2-party evaluation of a public function $f(x, y)$, where the work of the party holding y (wlog.) is $poly(k, n, \log |f|)$, where $|f|$ is the size of f's circuit representation [DFH12]. Their protocol is maliciously UC-secure [Can01]. Instantiating f with the universal function for evaluating circuits, results in HE with good compactness properties for circuits of certain size. However, this protocol requires CRS, and thus does not translate into (circuit private) HE in the plain model.

Circuit privacy in HE literature. Without the circuit privacy requirement, FHE candidates have been proposed in a line of work following the seminal work of Gentry [Gen09, vDGHV09, BV11], to mention a few. However, these works typically do not have circuit privacy as a goal.

Circuit privacy in the semi-honest setting, for properly generated pk and c has been addressed in [Gen09, vDGHV09]. In both works, the solution method is akin to that used in additively homomorphic cryptosystems [GM84, DJ01]. The idea in the latter is to (homomorphically) add a fresh encryption of 0. In FHE, the situations is a bit more complicated, as the output of Eval typically has a different domain than "fresh" encryptions, so adding a 0 is not straightforward. However, a generalization of this technique often works (see e.g. [vDGHV09]).

Another approach suggested in [vDGHV09] for the semi-honest setting is replacing the encrypted output c with a Yao garbled circuit for decryption (with sk, c as inputs), thereby transforming any scheme into a semi-honestly circuit-private one.

The work of [GHV10] considers a generalization of circuit privacy of HE (referred there as function privacy) to a setting with multiple evaluators and single encryptor (and decryptor), where all but a single evaluator E_i can collude to learn extra information about E_i's circuit. Among other contributions, in [GHV10], the authors further abstract the Yao-based approach from [vDGHV09] as a combination of two HE scheme, one compact but not private, the other (semi-honestly) private but not compact, so that the result is both compact and (semi-honestly) private. We use this transformation as is as a first step in our transformation (along with some additional ideas formulated in [GHV10]).

As mentioned above, the malicious setting (with compact encrypted outputs) has been addressed in the context of Oblivious Transfer (OT) [NP01, AIR01, HK12] (these works can be viewed as HE for the limited class of Oblivious Transfer functions). For broader classes of functions [IP07] devise maliciously circuit-private HE for depth-bounded *branching programs* (with partial compactness).

All of the above schemes use the Conditional Disclosure of Secrets (CDS) methodology [GIKM98]. CDS is a light-weight alternative to zero-knowledge proofs, that receives a secret string, an encryption c of some x in an HE, and the corresponding pk. It discloses the secret iff. x satisfies a certain condition. CDS was originally defined for well-formed pk, c, leading to semi-honestly circuit private HE constructions [AIR01]. The CDS from [AIR01] works for additive HE with ciphertexts over groups of a prime order, and [Lip05] generalized it to groups of sufficiently "rough" composite order. For specific groups, the technique of [Lip05] turned out to generalize to situations where (pk, c) may not be well-formed. Roughly, the secret luckily remains hidden even if the CDS was obliviously performed on the (possibly malformed) encryptions and pk as if they were proper. Such CDS was used in [HK12, IP07] to obtain maliciously circuit-private HE.

1.2 Our Techniques

We devise a framework for transforming FHE schemes with no circuit privacy requirements into maliciously circuit-private FHE. We use 2 ingredients which have implementations in the literature: powerful (evaluate circuits) compact FHE without privacy $\mathcal{F}$, and weak (evaluate formulas) non-compact maliciously circuit-private HE $\mathcal{P}$ (by "compact" we refer to the strong requirement that encrypted outputs have size $poly(k)$, where k is the security parameter). Our construction proceeds in three steps:

Lemma 1 ([GHV10]). *A compact FHE without privacy can be upgraded to (compact) semi-honestly circuit private by decrypting its encrypted output under a (possibly non-compact) enhanced semi-honestly private HE, capable of evaluating the decryption circuit. The resulting scheme has enhanced semi-honest circuit privacy, assuming only that pk, c are in the support of honestly generated pairs, rather then being distributed as honestly generated pairs.*

Lemma 2 (this paper). *An enhanced semi-honestly circuit-private FHE can be upgraded to maliciously circuit-private by homomorphically validating its keys and inputs under a (possibly non-compact) maliciously circuit-private FHE capable of evaluating a circuit validating that (pk, c) are well-formed (provided a suitable witness as additional input).*

The output resulting from composing these two steps is not compact—but fortunately:

Lemma 3 (this paper,same construction as [GHV10] for semi-honest setting). *Any circuit-private FHE can be upgraded to compact by homomorphically decrypting its output under a compact (F)HE (while preserving circuit-privacy).*

Let us now elaborate on each of the steps.

Step 1. The first step transforms a "main" (compact) FHE scheme $\mathcal{M}_1$ into a semi-honestly circuit-private scheme ($\mathcal{M}_2$). An output encrypted via $\mathsf{Eval}_{\mathcal{M}_1}$ may contain extra information about the structure of C (though limited to $poly(k)$ since $\mathcal{M}_1$ is compact). An easy way to strip all information beyond the value of the function is to decrypt it already during Eval under an "auxiliary" scheme $\mathcal{A}_1$ which is semi-honestly circuit-private. To make sure the evaluator learns no secret information (sk_M, x), the decryption is done "blindly", using $\mathcal{A}_1$'s homomorphic properties. Details follow.

KeyGen generates a public key $pk = (pk_{\mathcal{M}_1}, pk_{\mathcal{A}_1}, a_{sk_{\mathcal{M}_1}} = \mathsf{Enc}_{\mathcal{A}_1}(sk_{\mathcal{M}_1}))$ and secret key $sk_{\mathcal{A}_1}$. Enc simply outputs $c_{\mathcal{M}_1} = \mathsf{Enc}_{\mathcal{M}_1}(x)$. Now, Eval first computes $out_{\mathcal{M}_1} = \mathsf{Eval}_{\mathcal{M}_1}(C, c_{\mathcal{M}_1})$, and outputs $\mathsf{Eval}_{\mathcal{A}_1}(\mathsf{Dec}_{\mathcal{M}_1}, (out_{\mathcal{M}_1}, a_{sk_{\mathcal{M}_1}}))$.[5] Dec simply applies $\mathsf{Dec}_{\mathcal{A}_1}$ to Eval's output.

Enhanced semi-honest circuit privacy of the resulting scheme follows by (semi-honest) circuit privacy of $\mathcal{A}_1$, and the correctness of $\mathcal{M}_1$. For enhanced circuit privacy, we assume perfect correctness of $\mathcal{M}_1$ rather then allowing for negligible decryption error, as is common in the FHE literature. The reason is that otherwise, the receiver could pick pairs (pk, c) for which the output of Eval is not $C(x)$ with high probability over Eval's randomness, and potentially reveal a bit about the circuit not consistent with x.

Note that since $\mathcal{M}_1$ is compact $|out_{\mathcal{A}_1}| = \mathsf{poly}(k)$ even if $\mathcal{A}_1$ is not compact. $\mathcal{M}_1, \mathcal{A}_1$ can be instantiated via almost any FHE from the literature, and (non-compact) semi-honestly circuit-private HE obtained from non-interactive 2PC protocols, such as Yao-based protocols (see above). In particular, although most FHE schemes from the literature have (negligible) decryption errors, they can be modified to have perfect correctness, while maintaining security.[6]

---

[5] The trick of "re-encrypting" under $\mathcal{A}_1$ using $\mathcal{A}_1$'s own Eval procedure to perform the decryption is similar in Gentry's bootstrapping technique in [Gen09] for transforming a "somewhat homomorphic" scheme into (unleveled) FHE. One difference is that here we can use two different schemes. Another difference is that for the purpose of reducing noise via Gentry's bootstrapping theorem, it is important to hardwire c as a string into the decryption circuit, rather then supplying an encryption of it. In step 1, we can afford introducing c into Dec in either way.

[6] Consider for example the [BV11] scheme can be modified to use Gaussian noise truncated to a value which still does not incur decryption errors. It is easy to prove that the new scheme remains secure under the same (LWE) assumption - essentailly because samples larger then some bound hae negligible probability of being sampled. A similar trick can be applied to other LWE-based FHE schemes [BV11, Bra12, GSW13].

Step 2. The above approach generally fails in the malicious setting, even with stronger ingredients. Let $\mathcal{A}_2$ be a maliciously circuit-private scheme, and $\mathcal{M}_2$ be the semi-honestly circuit-private FHE resulting from step 1. An obvious attempt is using $\mathcal{A}_2$ instead of $\mathcal{A}_1$ in the first decryption; another is repeating the construction, taking $\mathcal{M}_2$ and additionally decrypting its output under $\mathcal{A}_2$. Neither is enough.

On a high level, in a maliciously circuit-private $\mathcal{A}_2$, any $pk_{\mathcal{A}_2}, a_{sk_M}$ "induce" an encryption of *some* sk_M^* under $\mathcal{A}_2$, so, we may think of them as being well-formed. However, the following potential attack exists. Assume even that pk_M is well-formed, but c_M is arbitrarily malformed. Thus out_M is not guaranteed to be a valid encryption of some x, and may instead carry some other arbitrary (upto $poly(k)$) bits of information about C. In turn, $\mathsf{Dec}_M(c_M, sk_M^*)$ may leak some of this information (even if sk_M^* is the right key corresponding to pk_M).

To fix the above potential attack (and other similar ones) and achieve malicious circuit privacy, we validate (pk, c) of $\mathcal{M}_2$. (In particular, for $\mathcal{M}_2$ resulting from step 1, we need to also check that $a_{sk_{\mathcal{M}_1}}$ encrypts an $sk_{\mathcal{M}_1}$ that corresponds to $pk_{\mathcal{M}_1}$ as part of validating $pk_{\mathcal{M}_2}$ is well-formed.) Such validation is generally hard, so we augment $\mathsf{KeyGen}_{\mathcal{M}_2}$ and $\mathsf{Enc}_{\mathcal{M}_2}$ to supply a helpful witness (encrypted under $\mathcal{A}_2$). For starters we want the full *randomness* used by them. However known $\mathcal{A}_2$ instantiations with malicious circuit privacy against unbounded adversaries, as we require can only evaluate functions in NC^1 and $\mathsf{KeyGen}_{\mathcal{M}_2}, \mathsf{Enc}_{\mathcal{M}_2}$ need not be in NC^1.[7] But surely they are in P, and we can use the standard transform to validate polynomial work in parallel we require the witness to also include values of *all intermediate wires* of $\mathsf{KeyGen}_{\mathcal{M}_2}, \mathsf{Enc}_{\mathcal{M}_2}$, validate all gates in parallel, and have a log-depth AND tree. A similar issue arises already in step 1, where $\mathcal{A}_1$ needs to evaluate the decryption circuit of $\mathcal{M}_1$. The same trick can not by applied there, as the values to decrypt are not known to the receiver. Thus we need to assume $\mathsf{Dec}_{\mathcal{M}_1}$ is in NC^1, which is fortunately satisfied by all known schemes from the literature.

Somewhat more precisely, we transform $\mathcal{M}_2$ in the following non-blackbox way.

- $\mathsf{Enc}(x)$ outputs $c_{\mathcal{M}_2} = \mathsf{Enc}_{\mathcal{M}_2}(r', pk_{\mathcal{M}_2}, x)$ along with $a_{r_{\mathcal{M}_2}} = \mathsf{Enc}_{\mathcal{A}_2}(r)$ - a witness that $c_{\mathcal{M}_2}$ is a proper encryption (derived from r').
- $\mathsf{Eval}(C, c_{\mathcal{M}_2})$: Let $Validate(pk_{\mathcal{M}_2}, c_{\mathcal{M}_2}, out, rk_{\mathcal{M}_2}, r_{\mathcal{M}_2})$ denote a circuit where $rk_{\mathcal{M}_2}, r_{\mathcal{M}_2}$ are purported witnesses for the well-formedness of $pk_{\mathcal{M}_2}, c_{\mathcal{M}_2}$ respectively. It outputs out if $rk_{\mathcal{M}_2}, r_{\mathcal{M}_2}$ certify well-formedness of $pk_{\mathcal{M}_2}, c_{\mathcal{M}_2}$, and the all-zero vector otherwise.
    - Compute $out_{\mathcal{M}_2} = \mathsf{Eval}_{\mathcal{M}_2}(pk_{\mathcal{M}_2}, C, c_{\mathcal{M}_2})$.
    - Output $out = \mathsf{Eval}_{\mathcal{A}_2}(Validate_{pk_{\mathcal{M}_2}, c_{\mathcal{M}_2}, out_{\mathcal{M}_2}}(a_{rk_{\mathcal{M}_2}}, a_{r_{\mathcal{M}_2}}))$ (that is, fixing the suitable variables in $Validate$ to the values at the subscript).
- Dec outputs $\mathsf{Dec}_{\mathcal{M}_2}(sk_{\mathcal{M}_2}, \mathsf{Dec}_{\mathcal{A}_2, out}(sk_{\mathcal{A}_2}))$.

---

[7] Settling for unbounded simulation against bounded distinguishers, would allow to evaluate arbitrary circuits under the scheme without further complications, however, we are shooting for the strongest possible privacy guarantee.

Note that $pk_{\mathcal{M}_2}, c_{\mathcal{M}_2}, out_{\mathcal{M}_2}$ are hardwired into $Validate$, rather then encrypted via $\mathcal{A}_2$ (in Eval). The subtle reason for this, is that if $pk_{\mathcal{M}_2}$ is malformed, "encrypting" values under it can yield effective encryptions of different values, possibly making us perform the "wrong" validation.

In a nutshell, the construction works by enhanced semi-honest circuit privacy of $\mathcal{M}_2$ if $(pk_{\mathcal{M}_2}, c_{\mathcal{M}_2})$ is well-formed, and the validation procedure takes care of the fact that it is indeed well-formed (otherwise, no information whatsoever is revealed about C).

Merging steps 1 and 2 (Section 3.1). Steps 1+2 provide a clear blueprint for transforming a (compact) FHE $\mathcal{F}$ into maliciously circuit-private FHE by combining it with a maliciously circuit-private HE $\mathcal{P}$ capable of evaluating $\mathcal{F}$'s decryption and validation circuits. That is, the same maliciously circuit-private $\mathcal{P}$ may instantiate both $\mathcal{A}_1$ and $\mathcal{A}_2$. $\mathcal{M}_2$ resulting from step 1 is fed into step 2. In section 3.1 we describe the natural composition of the two steps in a single protocol, making a small shortcut that exploits the structure of $\mathcal{M}_2$ output by step 1, and the fact that $\mathcal{A}_1 = \mathcal{A}_2$. Namely, we do not have to check the well-formedness of $a_{sk_{\mathcal{M}_2}}$ as an encryption under $\mathcal{A}_2$.

Step 3. Let us look more closely at the compactness of the scheme achieved from steps 1+2. The validate circuit that needs to be evaluated has input of size $m = poly(k, n)$ (and polynomial size $|C|$). Thus, if $\mathcal{P}$ is not compact, the encrypted output size is some $poly(|C|, m, k)$ – also $poly(k, n)$.

This is acceptable for our main application of computing on encrypted data, as receiver's input is of size n. However, it would be nice to meet the current standard for FHE where encrypted output size is independent of n.

A more complicated setting is that of leveled FHE. In such schemes, $\mathsf{KeyGen}(1^k, 1^d)$ generates an additional key pk_{Eval} received by Eval (all the rest remains the same), which may grow with the bound d on depth of circuits to be evaluated. In a nutshell, such schemes are often considered in the FHE literature in order to make the underlying assumptions more plausible, avoiding so called (weak) circular security assumptions.

The encrypted output size is $poly(k, n, d)$—quite undesirable!

The idea is to combine the circuit-private (non-compact) HE $\mathcal{M}_3$ resulting from steps 1+2 with a compact FHE $\mathcal{A}_3$ with no circuit privacy "in the opposite direction" from step 1. That is, use $\mathcal{A}_3$ to homomorphically decrypt the output of $\mathcal{M}_3$ to "compress" it. Intuitively, even though the FHE $\mathcal{A}_3$ used for decrypting is not circuit-private, the resulting scheme is, because $\mathsf{Eval}_{\mathcal{A}_3}$ merely acts upon a string that we originally were willing to output "in the plain", so there is no need to protect it.

2   Preliminaries

Notation. We use $\overrightarrow{\text{arrow}}$ to denote vectors, though not always—we tend to use it to stress element-wise handling, e.g. bit-by-bit encryptions. For a function

$f(a, b, c, \ldots)$, $\overrightarrow{f}(\overrightarrow{a}, b, \overrightarrow{c}, \ldots)$ is a shorthand for $(f(a_1, b, c_1, \ldots), f(a_2, b, c_2, \ldots), \ldots)$. When considering function vectors, all inputs which are the same in all executions appear without an arrow (even if they are vectors by themselves).

For a pair of vectors u, v (u, v) denotes the vector resulting from concatenating u, v. For vectors u, v over some U^t, V^t $(u; v)$ denotes $((u_1, v_1), \ldots, (u_t, v_t))$.

For a function $f(a, b, c, \ldots)$, we denote the set of functions fixing some of its parameters (here b, c) as follows $f|_{b,c}(a, \ldots)$. $f|_{b=B,c=C}$ denotes a function fixing the parameters to particular values B, C respectively.

For randomized algorithms $A(x, r)$, we sometimes write $out \in A(x)$ as a shorthand for $out \in support(A(x, r))$. By $negl(k)$ we refer to a function that for all polynomials $p(k)$, $negl(k) < \frac{1}{p(k)}$ for all $k > K$, where K is a constant determined by p. We use the standard notions of statistical and computational indistinguishability of distribution ensembles. Usually an encryption of x under scheme $\mathcal{Y}$ will be named y_x.

Representation Models When we say a HE scheme is $\mathcal{C}$-homomorphic for a class of functions, we actually mean functions having programs C from the set $\mathcal{C}$ of programs. By a program C, we mean a string representing a function $f : \{0, 1\}^n \to \{0, 1\}$. The correspondence between programs C and the function it represents is determined by an underlying representation model U. A *representation model* $U : \{0, 1\}^* \times \{0, 1\}^* \to \{0, 1\}^*$ is an efficient algorithm taking an input (C, x), and returning $f(x)$, where f is the function represented by C. By $|C|$ we simply refer to the length of the string C (as opposed to $size(C)$, which is a related measure depending on U, such as the number of gates in a boolean circuit). For completeness, for circuits and other models we let $U(C, x) = 0$ whenever the input (C, x) is syntactically malformed.

As typical in the FHE literature, our default representation model is boolean circuits, unless stated otherwise. We use circuits over some complete set of gates, such as $\{AND, XOR, NOT\}$. Another model we will consider is boolean formulas, which are circuits with fanout 1. We assume the underlying DAGs of circuits are connected. For formulas, we assume wlog. that $depth(C) \leq c \log size(C)$ for a global constant c (that is, that they are "balanced"). For a circuit C, $size(C)$ denotes the number of wires in C's underlying graph, and $depth(C)$ the number of gates on the longest path between an input wire and the output wire of the circuit. By NC^1 we refer to the class of function (families) with uniform formulas of size $poly(n)$.

2.1   Homomorphic Encryption

Throughout the paper k denotes the security parameter taken by HE schemes. A (public-key) homomorphic encryption scheme (HE) $\mathcal{E} = (\mathsf{KeyGen}_E, \mathsf{Enc}_E, \mathsf{Eval}_E, \mathsf{Dec}_E)$ is a quadruple of PPT algorithms as follows.

$\mathsf{KeyGen}(1^k)$: Outputs a public key, secret key pair (pk, sk).

$\mathsf{Enc}(pk, b)$: Takes a public key and a bit b to encrypt, and returns an encryption c of the bit under pk.

$\mathsf{Eval}(pk, C, c = (c_1, \ldots, c_n))$: Takes a public key pk, a bit-by-bit encryption c of a bit vector $x \in \{0,1\}^n$, a function represented by a program C (encoded in a pre-fixed representation model U) and outputs an encryption out of bit $U(C, x)$. We assume wlog. that pk includes 1^k (intuitively, this is intended to handle maliciously generated public keys). We refer to outputs of Eval as "encrypted outputs".

$\mathsf{Dec}(sk, out)$: Takes a secret key sk, and a purported output out of Eval, outputs a bit.

Throughout the paper, HE is semantically secure if (KeyGen, Enc, Dec) satisfies standard IND-CPA security for public key encryption schemes as in [GM84]. An HE scheme is *weakly circular-secure* if even knowing a bit-by-bit encryption of the schemes' secret key sk, the adversary still has negligible advantage in the IND-CPA experiment. In this paper, we consider a general notion of homomorphism, under various program classes, rather then just circuits (to express weaker homomorphism properties then FHE).

Definition 2 ((U, C)-homomorphic encryption). *Let $C = \bigcup C_k$. We say a scheme $\mathcal{E}$ is (U, C)-homomorphic if for every $k > 0$ and every program $C \in C_k$ on inputs $x \in \{0,1\}^n$, the experiment*

$$(pk, sk) \xleftarrow{\$} \mathsf{KeyGen}(1^k)$$

$$out \xleftarrow{\$} \mathsf{Dec}(sk, \mathsf{Eval}(pk, C, \overrightarrow{\mathsf{Enc}}(pk, \overrightarrow{x})))$$

outputs $out = U(C, x)$ with probability 1 for all $x \in \{0,1\}$ and all random choices of the algorithms involved. We say the scheme is k-independently homomorphic if $C_k = C$ for all k.[8]

By default our schemes are k-independently homomorphic (in particular the C_k's are not explicitly defined).

If a scheme satisfies that C equals the set of all circuits, and is k-independetly homomorphic, we refer to it as "fully homomorphic" (FHE).

Definition 3. *We say a (U, C)-homomorphic scheme $\mathcal{E}$ is* compact *if there exists an output bound $B(k, n, |C|) = \mathsf{poly}(k)$ on the output of Eval on all $1^k, n$ and $C \in C_k$ on n bits.*

Another standard variant of HE we consider is *leveled* HE. In this variant, KeyGen is modified to take another parameter 1^d. KeyGen outputs keys (pk, sk), where pk includes a fixed-size part pk_{Enc}, which depends only on k; likewise, sk depends only on k. Enc is modified to accept pk_{Enc} as the public key, and only Eval receives the entire public key pk. In particular, Enc is the same for all d. The notions of compact HE is as for non-leveled schemes ($B(k, n, |C|) =$

---

[8] For instance, the notion of "somewhat homomorphic" schemes in the FHE literature corresponds to non k-independent schemes, where C_k is a set of circuits with depth bounded by some function of k.

$poly(k)$).For compact schemes, the algorithm Dec is also independent of d. We say such a leveled compact scheme is an FHE, if for all D, the (standard) HE $\mathcal{E}^D$ induced by fixing $d = D$ when calling KeyGen$(1^d, \cdot)$ induces a k-independently $\mathcal{C}$-homomorphic scheme $\mathcal{E}^D$, where C is the set of all depth-D circuits. The encrypted outputs' size is still $poly(k)$ (for a global polynomial independent of d).

Standard FHE schemes can be thought of as a special case of leveled FHE schemes where KeyGen simply ignores d. Thus, all schemes $\mathcal{E}^d$ are the same (standard) FHE scheme. We refer to this special case as *unleveled* FHE. A HE scheme is maliciously circuit private if every (pk, c) (even arbitrarily malformed) induce some "effective" input x^*.

Definition 4. *Let* $\mathcal{E} = $ (KeyGen, Enc, Eval, Dec) *denote a* $(U, \mathcal{C})$-*homomorphic scheme. We say* $\mathcal{E}$ *is (maliciously) circuit private if there exist unbounded algorithms* Sim$(1^k, pk^*, c^*, b)$, *and deterministic* Ext$(1^k, pk^*, b)$, *such that for all* k, *and all* $pk^*, c^* = (c_1^*, \ldots, c_n^*)$ *and all programs* $C : \{0, 1\}^n \to \{0, 1\} \in (U, \mathcal{C})$ *in* $\mathcal{C}_k$ *the following holds:*

- $\overrightarrow{x^*} = \overrightarrow{\mathsf{Ext}}(1^k, pk^*, \overrightarrow{c^*})$.
- Sim$(1^k, pk^*, c^*, U(C, x^*)) =^s$ Eval$(1^k, pk^*, C, c^*)$.

In particular, for circuits $C(x_1, \ldots, x_n) \in \mathcal{C}_k$ *the output distribution of* Eval *(including length) depends only on* n, k. *For leveled schemes,* Sim *and* Ext *also take a depth parameter* 1^d. *We say a scheme is semi-honestly circuit-private if the above holds, where* pk^*, c^* *belong to the set of* well-formed *public-key, ciphertext pairs.*

3   Framework

In this section we spell out the construction outlined in the introduction in more detail, including a certain simplification. All security proofs are deferred to the full version. We will need the representation model $(U_{SI}, \mathcal{C})$ (from "split-input") and the weaker notion of "input-privacy" for $\mathcal{P}$. The purpose of introducing this seemingly unnatural representation model and relaxed circuit privacy definition is to allow for simpler implementations of auxiliary HE schemes we use, and overall presentation of our result. The scenario in Theorem 3 is that a function f known to the adversary is to be homomorphically evaluated on a (partially) secret input y, together with the adversary's input x, so hiding f would be an overkill. More specifically, the implementation we use for $\mathcal{P}$ is based on Yao's garbled circuits, which works with a pair of private inputs x, y, and a public circuit C. It outputs $C(x, y)$, but nothing else about x, y. It also leaks quite a lot about C itself (there is no goal of protecting it). While we could use Yao to get circuit privacy for $f|_y$ scheme by evaluating a universal function, there is no need to. The straightforward use of Yao with $C = f$ provides exactly what we need.

Programs in the model are represented by a pair (C_p, C_s), where C_p is a circuit on some m variables, and $C_s \in \{0, 1\}^t$ for some $t \leq m$. A program (C_p, C_s) is interpreted as a function f over $n = m - |C_s|$ variables via $U_{SI}((C_p, C_s), x) = C_p(C_s, x)$. Typically, we will consider $(U_{SI}, \mathcal{C})$-homomorphic schemes where for each (C_p, C_s), all (C_p, Z) for $Z \in \{0, 1\}^*$ of length $t \leq m$ are in $\mathcal{C}$. In this case, we specify $\mathcal{C}$ as just a set of circuits.

We say a $(U_{SI}, \mathcal{C})$-homomorphic scheme is *input-private* if it satisfies Definition 4, with the only modification that Sim receives C_p as an input. (This is exactly the guarantee from converting any 2PC protocol where both parties have private input but the function is public to an HE.)

3.1 From Compact FHE to Circuit-Private (Somewhat Compact) FHE

In this section we spell out the combination of steps 1 and 2 as described in the introduction. The schemes $\mathcal{F}, \mathcal{P}$ are a compact FHE, and maliciously circuit private HE respectively. Here $\mathcal{P}$ is for the split input model, and is input-private rather then circuit-private. Although the construction would go through with standard circuit privacy of $\mathcal{P}$, this simplifies the presentation and instantiation of the framework.

Given a leveled FHE $\mathcal{F}$ we define a set of programs $\mathcal{C}_{\mathcal{F}}$ (to be interpreted via U_{SI}) as follows.

1. Let $Dec_{\mathcal{F},k}(sk_{\mathcal{F}}, out_{\mathcal{F}})$ denote the decryption circuit of $\mathcal{F}$ instantiated with security parameter k (recall Dec, Enc are independent of d).
2. Let $Validate_{k,d,n}(pk_{\mathcal{F}}, c_{\mathcal{F}}, sk'_{\mathcal{F}}, r_{FE}, out_{\mathcal{F}})$ be the circuit computing

$$Validate_{k,d,n}(\ldots) = \begin{cases} out_{\mathcal{F}} & \text{if } (pk_{\mathcal{F}}, sk_{\mathcal{F}}) \in \mathsf{KeyGen}_{\mathcal{F}}(1^k, 1^d), \text{ and} \\ & \forall i \ (c_{\mathcal{F},i} \in \mathsf{Enc}_{\mathcal{F}}(b_i, r_{FE,i}) \text{ for some bit } b_i \in \{0, 1\}) \\ \overrightarrow{0} & \text{otherwise.} \end{cases}$$

where:
  - $(pk_{\mathcal{F}}, sk_{\mathcal{F}})$ is a purported public-key private-key pair output by $\mathsf{KeyGen}_{\mathcal{F}}(1^k, 1^d)$. $sk'_{\mathcal{F}} = (sk_{\mathcal{F}}, r_{FK})$ where r_{FK} is the purported random string used in the generation of $(pk_{\mathcal{F}}, sk_{\mathcal{F}})$, along with all the intermediate outputs of gates in the circuit for KeyGen on input r_{FK}.[9]
  - $c_{\mathcal{F}} = (c_{\mathcal{F},1}, \ldots, c_{\mathcal{F},n})$ is a purported encryption under $pk_{\mathcal{F}}$ of the input bit vector and r_{FE} is a purported vector of randomness used when generating $c_{\mathcal{F}}$, along with intermediate values for the circuit (where $r_{FK,i}$ corresponds to bit x_i).
3. Let $\mathcal{C}_{\mathcal{F}}$ include all pairs of the forms $C = (Validate_{k,d,n}(\ldots), (pk_{\mathcal{F}}, c_{\mathcal{F}}, out_{\mathcal{F}}))$ and $(\mathsf{Dec}_{\mathcal{F},k}(sk_{\mathcal{F}}), out_{\mathcal{F}})$.

Construction 5. *Let $\mathcal{F}, \mathcal{P}$ be schemes as above. We construct the following scheme $\mathcal{M}_3$.*

---

[9] As explained in the intro, the goal of the intermediate values is to put the validation in NC^1, making it implementable by known circuit-private $\mathcal{P}$ with the required notion of privacy.

$\mathsf{KeyGen}_{\mathcal{M}_3}(1^k)$: let $(pk_{\mathcal{P}}, sk_{\mathcal{P}}) \xleftarrow{\$} \mathsf{KeyGen}_{\mathcal{P}}(1^k)$, $(pk_{\mathcal{F}}, sk_{\mathcal{F}}) \xleftarrow{\$}$
   $\mathsf{KeyGen}_{\mathcal{F}}(1^k, 1^d)$, and let $sk'_{\mathcal{F}} = (sk_{\mathcal{F}}, r_{KF})$, where r_{KF} is induced by
   the randomness used by $\mathsf{KeyGen}_{\mathcal{F}}$ as specified in Validate; $\overrightarrow{p_{sk'_{\mathcal{F}}}} = $
   $\overrightarrow{\mathsf{Enc}_{\mathcal{P}}}(pk_{\mathcal{P}}, sk'_{\mathcal{F}})$. Return $(pk_{\mathcal{M}_3}, sk_{\mathcal{M}_3}) = ((pk_{\mathcal{P}}, pk_{\mathcal{F}}, \overrightarrow{p_{sk'_{\mathcal{F}}}}), (sk_{\mathcal{P}}, sk_{\mathcal{F}}))$.
   Here $pk_{\mathcal{M}_3,\mathsf{Enc}} = (pk_{\mathcal{P}}, pk_{\mathcal{F},\mathsf{Enc}}, p_{sk_{\mathcal{F}}})$.

$\mathsf{Enc}_{\mathcal{M}_3}(pk_{\mathcal{M}_3} = (pk_{\mathcal{P}}, pk_{\mathcal{F},\mathsf{Enc}}, p_{sk_{\mathcal{F}}}), b \in \{0,1\})$: Return $(c, \overrightarrow{p_{r_{FE}}}) = $
   $(\mathsf{Enc}_{\mathcal{F}}(pk_{\mathcal{F}}, b), \overrightarrow{\mathsf{Enc}_{\mathcal{P}}}(pk_{\mathcal{P}}, \overrightarrow{r_{FE}})$, where r_{FE} is derived from the randomness
   used by $\mathsf{Enc}_{\mathcal{F}}$ as in Validate.

$\mathsf{Eval}_{\mathcal{M}_3}(1^k, pk_P = (pk_{\mathcal{P}}, pk_{\mathcal{F}}, p_{sk'_{\mathcal{F}}}), C, c = (c_{\mathcal{P}}; p_{r_{FE}}))$:

   1. If C is syntactically malformed, or $|x|$ does not match the number of inputs to C, replace C with the circuit returning $x_1 \wedge \overline{x_1}$.
   2. Set $out_{\mathcal{F}} = \mathsf{Eval}_{\mathcal{F}}(pk_{\mathcal{F}}, C, c_{\mathcal{F}})$, $out_{\mathcal{P}} = \mathsf{Eval}_{\mathcal{P}}(pk_{\mathcal{P}}, (\mathsf{Dec}_{\mathcal{F}}, out_{\mathcal{F}}), p_{sk_{\mathcal{F}}})$
   3. Let $(C_p, C_s) = (Validate_{k,d,n}, (out_{\mathcal{P}}, pk_{\mathcal{F}}, c_{\mathcal{F}}))$
   4. Compute and output $out = \mathsf{Eval}_{\mathcal{P}}(pk_{\mathcal{P}}, (C_p, C_s), p_{sk'_{\mathcal{F}}}, p_{r_{FE}})$.

$\mathsf{Dec}_{\mathcal{M}_3}(sk_{\mathcal{M}_3}, out_A)$: Output $y = \mathsf{Dec}_{\mathcal{F}}(sk_{\mathcal{F}}, \overrightarrow{\mathsf{Dec}_{\mathcal{P}}}(sk_{\mathcal{P}}, \overrightarrow{out}))$.

Theorem 3. *Assume a compact leveled FHE scheme $\mathcal{F} = (\mathsf{KeyGen}_{\mathcal{F}}, \mathsf{Enc}_{\mathcal{F}},$ $\mathsf{Eval}_{\mathcal{F}}, \mathsf{Dec}_{\mathcal{F}})$ and $\mathcal{P}$ a $(U_{SI}, \mathcal{C}_{\mathcal{F}})$-homomorphic, input-private scheme, exist. Consider the resulting scheme $\mathcal{M}_3$ as specified in Construction 5 above when instantiating with $\mathcal{F}, \mathcal{P}$. Then $\mathcal{M}_3$ is a circuit-private FHE. It is unleveled iff $\mathcal{M}_3$ is unleveled, and is compact iff $\mathcal{P}$ is compact. If $\mathcal{P}$ is not compact, $\mathcal{M}_3$'s output complexity is $poly(k, d, n)$ $(poly(k, n)$ if $\mathcal{F}$ is unleveled).*

3.2   Compactization of Circuit-Private FHE

When instantiated by the best known constructions from the literature, Theorem 3 only yields $poly(n, k), poly(n, d, k)$ encrypted output complexity for unleveled and leveled $\mathcal{F}$ respectively. This is so, since all circuit-private $\mathcal{C}_{\mathcal{F}}$-homomorphic $\mathcal{P}$ for some FHE $\mathcal{F}$ we know of are not compact.

In this section, we devise a simple transformation (corresponding to Lemma 3 in the introduction) for making a (leveled) scheme's output compact (only $poly(k)$), while preserving circuit privacy. This will yield leveled circuit-private FHE with optimal $(poly(k))$ compactness.

The idea is to use bootstrapping similar to that of step 1 but "in reverse" order. Namely, we take a main scheme $\mathcal{M}_3$ which is circuit-private but not compact, and "decrypt" it under a scheme $\mathcal{A}_3$ which is compact but has no circuit privacy guarantees.

Theorem 4 (Compaction theorem). *Assume a leveled $\mathcal{C}$-homomorphic circuit-private scheme $\mathcal{M}_3$ and a compact FHE scheme $\mathcal{A}_3$ exist.[10] Then the scheme $\mathcal{M}_4$ in the following construction is a compact $\mathcal{C}$-homomorphic circuit-private scheme.*

---

[10] In fact, $\mathcal{A}_3$ should only be compact and homomorphic for the circuit family it is used for. It does not need to be an FHE.

Construction 6. *Let $\mathcal{M}_3, \mathcal{A}_3$ be HE schemes as in Theorem 4.*

$\mathsf{KeyGen}_{\mathcal{M}_4}(1^k, 1^d)$: *Sample* $(pk_{\mathcal{M}_3}, sk_{\mathcal{M}_3}) \xleftarrow{\$} \mathsf{KeyGen}_{\mathcal{M}_3}(1^k, 1^d, r_k)$; *let* $sk'_{\mathcal{M}_3} = \overrightarrow{(sk_{\mathcal{M}_3}, r_k)}$; $(pk_{\mathcal{A}_3}, sk_{\mathcal{A}_3}) \xleftarrow{\$} \mathsf{KeyGen}_{\mathcal{A}_3}(1^k)$; $\overrightarrow{a_{sk'_P}} \xleftarrow{\$} \overrightarrow{\mathsf{Enc}}_{\mathcal{A}_3}(pk_{\mathcal{A}_3}, \overrightarrow{(sk_{\mathcal{M}_3}, r_k)})$. *Output* $(pk, sk) = ((pk_{\mathcal{M}_3}, pk_{\mathcal{A}_3}, a_{sk_{\mathcal{M}_3}}), sk_{\mathcal{A}_3})$. *Here* $pk_{\mathsf{Enc}} = (pk_{\mathcal{M}_3, \mathsf{Enc}}, pk_{\mathcal{A}_3}, a_{sk_{\mathcal{M}_3}})$.

$\mathsf{Enc}_{\mathcal{M}_4}(pk, b)$: *Output* $\mathsf{Enc}_{\mathcal{M}_3}(pk_{\mathcal{M}_3, \mathsf{Enc}}, b)$. [11]

$\mathsf{Eval}_{\mathcal{M}_4}(1^k, pk, C, c)$:

- $out_{\mathcal{M}_3} \xleftarrow{\$} \mathsf{Eval}_{\mathcal{M}_3}(1^k, pk_{\mathcal{M}_3}, C, c)$.
- *Let* $Dec_{\mathcal{M}_3, k}$ *denote the decryption circuit of* $\mathcal{M}_3$ *with parameter* k. *Then* $Dec_{\mathcal{M}_3, k}|_{out=out_{\mathcal{M}_3}}(sk_{\mathcal{M}_3})$ *is a circuit for decrypting (hard-wired)* $out_{\mathcal{M}_3}$ *under secret keys generated by* $\mathsf{KeyGen}_{\mathcal{M}_3}$.
- *Compute and output* $out_{\mathcal{A}_3} = \mathsf{Eval}_{\mathcal{A}_3}(1^k, pk_{\mathcal{A}_3}, Dec_{\mathcal{M}_3, k}, a_{sk_{\mathcal{M}_3}})$.

$\mathsf{Dec}_{\mathcal{M}_4}(sk = sk_{\mathcal{A}_3}, out)$: *Output* $\mathsf{Dec}_{\mathcal{A}_3}(sk_{\mathcal{A}_3}, out)$.

Combining Theorem 3 and Theorem 4, we get:

Theorem 5. *Assume a compact unleveled FHE scheme $\mathcal{F}$ and a $(U_{SI}, \mathcal{C}_{\mathcal{M}})$-homomorphic maliciously input-private scheme $\mathcal{P}$ exist. Then there exists a maliciously circuit-private compact unleveled scheme $\mathcal{M}_4$.*

Getting rid of circular security Theorem 4 still leaves open the question of obtaining compact leveled circuit-private FHE. We show that posing some mild additional efficiency requirements on $\mathcal{M}_3, \mathcal{A}_3$ in Theorem 4, we are able to modify Construction 6 to allow for a leveled A_{three}.

Theorem 6. *Assume schemes $\mathcal{F}, \mathcal{P}$ as Theorem 5 exist. Additionally, assume $\mathsf{Dec}_{\mathcal{P}, k}(out, sk)$ has depth $poly(depth(C_p), k)$, where $out = \mathsf{Eval}_{\mathcal{P}}(C_p, C_s)$ for some C_s.[12] Assume also that $\{Dec_{\mathcal{F}, k}\}$ induced by $\mathcal{F}$ is in NC^1. Then there exists a compact leveled circuit-private scheme $\mathcal{M}_4$.*

3.3   Multi-hop Circuit-Private FHE

In this chapter, we focus on unleveled FHE schemes and show how to upgrade Theorem 5 to yield a multi-hop $\mathcal{M}_4$ (under the same assumptions).

A multi-hop scheme is an FHE scheme, where Eval is modified to support (pk, C, c), where the c_i's are either outputs of Enc or of previous Evals. More formally, if c_i is a purported output of Enc we label it as a level-0 execution of Eval. We recursively define an execution of Eval to be of level l, if the highest level input c_i to it is of level $l - 1$. Formally defining (perfectly correct) multi-hop is a natural recursive extension of Definition 2. The base case of level-0 executions is correct decryption for chiphertexts generated by Enc as in Definition 2. We

---

[11] Here and elsewhere, we do not distinguish between the parts of pk used in Eval and Enc, and refer to both as pk. The distinction is implied by the context.

[12] The circuit for Dec can be efficiently computed from out.

further require that the evaluation of a level-i execution for every $i > 1$ is correct in the sense that applying Dec to its output recovers the value induced by appropriately combining the circuits involved in the graph of Evals. Similarly, a scheme is i-hop if it satisfies the above correctness requirements only for level-j executions of Eval, where $j \leq i$. Thus standard FHE correspond to 1-hop. The definitions of IND-CPA security and compactness extend for multi-hop schemes in the natural way. The definition of maliciously circuit-private multi-hop FHE is precisely Definition 4.

The construction induced by Theorem 5 is only 1-hop, but not multi-hop. There exists a straightforward transformation from compact FHE schemes into multi-hop schemes (see discussion in Section 1), but it does not necessarily preserve malicious circuit privacy. However, it can be modified to work here. We start from the scheme $\mathcal{M}_4$ resulting from our construction, instantiated with an FHE $\mathcal{F}$, and a maliciously circuit private $\mathcal{P}$, where $\mathcal{M}_1 = \mathcal{A}_3 = \mathcal{F}$, and $\mathcal{A}_1 = \mathcal{A}_2 = \mathcal{P}$, reusing keys for the same scheme. Thus, both encrypted inputs and encrypted output of $\mathcal{M}_4$ are encryptions under $\mathcal{F}$ and same key. We can thus include encryptions of the bits of $sk_{\mathcal{F}}$ in pk_{MH}. In subsequent executions of Eval using this one as input, one can first decrypt the out_i's using these key bits, and plug the decrypted out_i's into the original scheme $\mathcal{M}_4$ as the c_i's input to Eval. The main caveat is that in our construction the well-formedness of the c_i's fed to a subsequent instance of Eval needs to be certified (under $\mathcal{A}$) as part of the output of the previous Eval. The key observation is that there is no need to certify the well-formedness of the out_i's, but rather only that of the secret key bits used for decryption![13] Moreover, we only need to prove that $sk_{\mathcal{F}}$ constitute valid encryptions of some bits under $pk_{\mathcal{F}}$ specified in $pk_{\mathcal{M}_4}$ (not even correspondence to the $pk_{\mathcal{F}}$ published as part of pk!). This is ok because $sk_{\mathcal{F}}$ is short and independent of the circuit being evaluated. As these are decrypted under the specified key bits in subsequent Eval's, the result would be an enryption of some value independent of the (subsequent) C, which is what we need (as in $\mathcal{M}_4$, if validation fails, nothing is learned about C).

We defer an explicit description and full analysis of our multi-hop construction, as well as implications to MPC and comparison to [GHV10] to the full version.

4   Instantiations of the Framework

We devise instantiations of schemes $\mathcal{F}, \mathcal{P}$ as required in Theorem 6. As these requirements are strictly stronger then the requirements in Theorem 5, they immediately yield an instantiation of Theorem 5 as well. The component $\mathcal{F}$ has many instantiations from the literature.

---

[13] If we had to certify them, seemingly, we would need to give a proof on the validity of the execution of Eval, referring to its inputs. It is not clear how to make it short and protect the privacy of that Eval's circuit.

For $\mathcal{P}$ we use the following instantiation, induced by the specific construction combining an information theoretic variant of Yao's garbled circuits [IK02] with maliciously circuit-private OT-homomorphic schemes.

Lemma 4. *Assume the existence of circuit-private schemes which are homomorphic for (bit) OT. In particular, the DDH,QR,Paillier or the DCRA assumptions yield such OT schemes [AIR01, HK12]. Then there exists a circuit-private $(U_{SI}, \mathcal{C})$-homomorphic scheme $\mathcal{P}$ where $\mathcal{C}$ consists of all (balanced) formulas. Furthermore, $\mathcal{P}$ has decryption circuits $\mathsf{Dec}_{\mathcal{P},k}(sk, out)$ of depth $\mathrm{depth}(C_p)poly(k)$, where $out = \mathsf{Eval}_{\mathcal{P}}(C_p, C_s)$.*

The following corollary of Theorem 6 (5) and Lemma 4 is our working, "take-home", instantiation of the framework.

Corollary 1. *Assume a leveled FHE $\mathcal{F}$ with decryption circuits in NC^1 exists. Assume further that there exist (bit) OT-homomorphic circuit-private HE. Then there exists a circuit-private compact FHE $\mathcal{M}_4$. $\mathcal{M}_4$ is unleveled if $\mathcal{M}$ is.*

As mentioned above, $\mathcal{M}$ has many "efficient enough" instantiations. See the full version for some examples and proofs of Lemma 4 and Corollary 1 (almost immediate).

5   Future Work

The "work horse" of our bootstrapping-based transformation 1 for transforming FHE into circuit-private is a circuit-private bit-OT-homomorphic HE. The known constructions from the literature we are aware of can not base circuit-private OT on some assumption the implies FHE (such as LWE, approximate GCD etc.). We do not see good reasons, except for historical ones to why this is the case. Such a construction would give an example of compact FHE which can be made circuit-private without additional assumptions.

References

[AIR01]   Boneh, D., Kushilevitz, E., Ostrovsky, R., Skeith III, W.E.: Public key encryption that allows PIR queries. In: Menezes, A. (ed.) CRYPTO 2007. LNCS, vol. 4622, pp. 50–67. Springer, Heidelberg (2007)

[BKOI07]   Boneh, D., Kushilevitz, E., Ostrovsky, R., Skeith III, W.E.: Public Key Encryption That Allows PIR Queries. In: Menezes, A. (ed.) CRYPTO 2007. LNCS, vol. 4622, pp. 50–67. Springer, Heidelberg (2007)

[BLV04]   Barak, B., Lindell, Y., Vadhan, S.P.: Lower bounds for non-black-box zero knowledge. In: Electronic Colloquium on Computational Complexity (ECCC), vol. (83) (2004)

[Bra12]   Brakerski, Z.: Fully Homomorphic Encryption without Modulus Switching from Classical GapSVP. In: Safavi-Naini, R., Canetti, R. (eds.) CRYPTO 2012. LNCS, vol. 7417, pp. 868–886. Springer, Heidelberg (2012)

[BV11]   Brakerski, Z., Vaikuntanathan, V.: Efficient fully homomorphic encryption
            from (standard) lwe. Cryptology ePrint Archive, Report 2011/344 (2011),
            http://eprint.iacr.org/2011/344
[Can01]   Canetti, R.: Universally composable security: A new paradigm for crypto-
            graphic protocols. In: FOCS, pp. 136–145. IEEE Computer Society (2001)
[DFH12]   Damgård, I., Faust, S., Hazay, C.: Secure Two-Party Computation with
            Low Communication. In: Cramer, R. (ed.) TCC 2012. LNCS, vol. 7194,
            pp. 54–74. Springer, Heidelberg (2012)
[DJ01]   Damgård, I., Jurik, M.: A generalisation, a simplification and some appli-
            cations of paillier's probabilistic public-key system. In: Kim, K.-c. (ed.)
            PKC 2001. LNCS, vol. 1992, pp. 119–136. Springer, Heidelberg (2001)
[Gen09]   Gentry, C.: A fully homomorphic encryption scheme. PhD thesis, Stanford
            University (2009), http://crypto.stanford.edu/craig
[GHV10]   Gentry, C., Halevi, S., Vaikuntanathan, V.: i-Hop Homomorphic Encryp-
            tion and Rerandomizable Yao Circuits. In: Rabin, T. (ed.) CRYPTO 2010.
            LNCS, vol. 6223, pp. 155–172. Springer, Heidelberg (2010)
[GIKM98]   Gertner, Y., Ishai, Y., Kushilevitz, E., Malkin, T.: Protecting data pri-
            vacy in private information retrieval schemes. In: Vitter, J.S. (ed.) STOC,
            pp. 151–160. ACM (1998)
[GM84]   Goldwasser, S., Micali, S.: Probabilistic encryption. J. Comput. Syst.
            Sci. 28(2), 270–299 (1984)
[GSW13]   Gentry, C., Sahai, A., Waters, B.: Homomorphic Encryption from Learn-
            ing with Errors: Conceptually-Simpler, Asymptotically-Faster, Attribute-
            Based. In: Canetti, R., Garay, J.A. (eds.) CRYPTO 2013, Part I. LNCS,
            vol. 8042, pp. 75–92. Springer, Heidelberg (2013)
[HK12]   Halevi, S., Kalai, Y.T.: Smooth projective hashing and two-message obliv-
            ious transfer. J. Cryptology 25(1), 158–193 (2012)
[IK02]   Ishai, Y., Kushilevitz, E.: Perfect constant-round secure computation via
            perfect randomizing polynomials (2002)
[IP07]   Ishai, Y., Paskin, A.: Evaluating Branching Programs on Encrypted Data.
            In: Vadhan, S.P. (ed.) TCC 2007. LNCS, vol. 4392, pp. 575–594. Springer,
            Heidelberg (2007), Full version in,
            http://www.cs.technion.ac.il/users/wwwb/
            cgi-bin/tr-info.cgi/2012/PHD/PHD-2012-16
[Lip05]   Lipmaa, H.: An Oblivious Transfer Protocol with Log-Squared Commu-
            nication. In: Zhou, J., López, J., Deng, R.H., Bao, F. (eds.) ISC 2005.
            LNCS, vol. 3650, pp. 314–328. Springer, Heidelberg (2005)
[NP01]   Naor, M., Pinkas, B.: Efficient oblivious transfer protocols. In: Rao
            Kosaraju, S. (ed.) SODA, pp. 448–457. ACM/SIAM (2001)
[vDGHV09]   van Dijk, M., Gentry, C., Halevi, S., Vaikuntanathan, V.: Fully homo-
            morphic encryption over the integers. Cryptology ePrint Archive, Report
            2009/616 (2009), http://eprint.iacr.org/2009/616

Algorithms in HElib

Shai Halevi[1] and Victor Shoup[1,2]

[1] IBM Research, Yorktown Heights, NY, USA
[2] New York University, New York, NY, USA

Abstract. HElib is a software library that implements homomorphic encryption (HE), specifically the Brakerski-Gentry-Vaikuntanathan (BGV) scheme, focusing on effective use of the Smart-Vercauteren ciphertext packing techniques and the Gentry-Halevi-Smart optimizations. The underlying cryptosystem serves as the equivalent of a "hardware platform" for HElib, in that it defines a set of operations that can be applied homomorphically, and specifies their cost. This "platform" is a SIMD environment (somewhat similar to Intel SSE and the like), but with unique cost metrics and parameters. In this report we describe some of the algorithms and optimization techniques that are used in HElib for data movement, linear algebra, and other operations over this "platform."

1 Introduction

Homomorphic encryption (HE) [18,8] enables performing arithmetic operations on encrypted data even without knowing the secret decryption key. All HE schemes to date roughly follow the outline of Gentry's first candidate from 2009, in which fresh ciphertexts are "noisy" to ensure security and this noise grows with every operation until it becomes so large so as to cause decryption errors. This results in a "somewhat homomorphic" encryption scheme (SWHE) that can only evaluate low-depth circuits, which can then be converted to a "fully homomorphic" encryption scheme (FHE) using bootstrapping. Currently, the most asymptotically efficient SWHE schemes that we have are the RLWE-variants of Brakerski-Gentry-Vaikuntanathan scheme [6] and Brakerski's scale-invariant scheme [4], and the NTRU-based scheme [13,16]. All these schemes work in polynomial rings, and use rings of the form $R_p = \mathbb{Z}[X]/(F(X), p)$ as their native plaintext space, with F a cyclotomic polynomial and p an integer.

Smart and Vercauteren observed [19] that (for a prime p) an element in this native plaintext space can be used to encode a vector of values from a finite field $\mathbb{F}_{p^d}$, for some integer d that depends on F and p, and that homomorphic operations then induce the corresponding entry-wise operation on the encrypted vectors. Gentry, Halevi, and Smart showed [10] how to use the SV "ciphertext packing" technique to perform asymptotically efficient computation, where a (wide enough) T-gate arithmetic circuit can be evaluated homomorphically in time $T \cdot \mathrm{polylog}(k)$, with k the security parameter. Crucial to obtaining this asymptotic efficiency is the use of automorphisms as a technique to move values between the different "slots" in a given plaintext vector, following [17,6].

J.A. Garay and R. Gennaro (Eds.): CRYPTO 2014, Part I, LNCS 8616, pp. 554–571, 2014.

Turning to software implementations, HElib [12] is an open-source C++ library that implements the BGV scheme, focusing on effective use of ciphertext packing and the GHS optimizations. It includes an implementation of the BGV scheme itself with all its basic homomorphic operations, and also some higher-level procedures implementing the GHS data-movement procedures, simple linear algebra, and some other procedures. This report is focused on these higher level procedures and the various optimizations that went into implementing them.

A useful analogy to keep in mind is to think of the lower-level of HElib as implementing an "assembly language" which is executed on a "hardware platform" given by the underlying HE scheme. The "platform" defines a set of operations that can be applied homomorphically and the cost of these operations; our goal in the current work is to provide efficient implementation of simple routing and linear-algebra procedures over that "platform." Since the homomorphic operations define entry-wise operations on the vector of plaintext values, the "platform" defines for us a SIMD environment (somewhat similar to things like Intel's SSE, the Motorola/IBM AltiVec architecture, and the like). Hence the focus of this work is the design of efficient algorithms over this SIMD architecture.

A word of caution: The view of the HE "platform" as a linear array is oversimplified, and presented here for sake of readability. In reality we have something closer to a multi-dimensional array (and even this view hides some details) — see the full version [11, Section 5].

Although SIMD hardware architectures are quite common in practice (cf. [20]), we were unable to find much algorithmic literature concerning asymptotic efficiency in such environments. This is perhaps related to the fact that common hardware architectures have vectors with only a handful of entries (for example an SSE register can hold at most 32 8-bit values). On the other hand, the plaintext arrays in HElib often hold a few hundred plaintext slots (sometimes even a few thousand), making asymptotic treatment of SIMD algorithms more relevant. Another difference between the "platform" provided by HE and the common hardware SIMD platforms is their cost metrics: in HElib we need to optimize for two parameters, namely time and noise-magnitude. These correspond roughly to size and depth of the corresponding SIMD circuits, but the correspondence is not quite one-to-one, since different operations have different time and noise behavior.

Contents of this report. In Section 2 we introduce notations and describe some details of the HE "platform." In Section 3 we describe the implementation and optimizations of the GHS permutation techniques. In particular we describe a generalization of Benes networks to handle networks of arbitrary width, extending earlier work of Chang and Melham [7], and also our approach for optimizing the GHS "hypercube networks."

In Section 4 we describe our procedures for computing running- and total-sums of a vector, for replicating the entries of a vector, and for performing a matrix-vector multiplication, and in Section 5 we describe procedures for computing the norm and trace functions entry-wise on a vector

In the full version [11] we describe a procedure for evaluating a polynomial entry-wise on a vector. We also discuss how to adapt all of these procedures to work on a multi-dimensional array (which is what more naturally arises in the HE context), rather than a linear array.

2   Background and Notation

The characteristics that define the "hardware platform" for HElib are common to many contemporary HE schemes, including the ring-LWE variants of BGV [6] and Brakerski's scale-invariant scheme [4], the NTRU-based HE scheme [13,16], and maybe even some LWE-based schemes [5]. Two salient characteristics of these cryptosystems are the following:

Growing noise. All contemporary SWHE schemes use *noisy* ciphertexts, where a fresh ciphertext includes a noise component that grows with each homomorphic operation, until it is so large that it causes decryption errors. However, different operations have very different noise-growth behavior. For example, multiplication increases the noise much more than addition.

Plaintext vectors. The plaintext space of these schemes can be viewed as a vector space over some finite field (or a module over a finite ring). This means that each native plaintext of the cryptosystem corresponds to a vector of plaintext values that the application cares about. The underlying field (or ring) and the dimension of the vector are both derived from some parameters of the cryptosystems; see, e.g., [10, Appendix C.2] (in the full version) for a description. When using such cryptosystems for specific homomorphic computation, we are typically faced with a 2-parameter optimization problem, trying to minimize both the noise-growth and the running time. In a typical scenario we would first choose the system parameters, which determine the maximum allowable level of noise, and then try to minimize the running time subject to this fixed bound on the noise. Consequently most of the optimization procedures that we describe in this work has the form of optimizing the running time subject to some depth constraints.

In Table 1 we summarize the available homomorphic operations, their effect on the noise, and their running time. For each parameter (noise and time) we divide the operations into expensive, moderate, and cheap. We often think of the cheap operations as essentially for free, the expensive operations as costing one unit (of either time or noise), and the moderate operations as having a cost of 1/2 unit. We remark that the cost in Table 1 (and even the operations themselves) are merely an approximation, see the full version [11, Section 5] for some more details.

We would like to draw the reader's attention to the "moderate" noise-growth of the multiply-by-constant operation, and stress that we have to pay this "moderate" cost even if we are multiplying by a constant zero-one vector. This is different than other (additively) homomorphic schemes where multiplication by zero or one is really "for free." In our implementation we extensively use multiplications by zero-one vectors to extract from a given vector only some of the

Table 1. Homomorphic operations and their cost

Operation	Time	Noise	Comments
Addition	cheap	cheap	entry-wise addition of vectors
Constant-add	cheap	cheap	entry-wise addition of a constant vector
Multiplication	expensive	expensive	entry-wise multiplication of vectors
Constant multiply	cheap	moderate	entry-wise multiplication by constant vector
Rotation	expensive	cheap	cyclic rotation of vector by any amount
Frobenius	expensive	cheap	entry-wise Frobenius map, $X \mapsto X^{p^n}$

entries but not others. We refer to this operation as *multiplicative masking* (or *masking*, for short). We also note that using rotations and multiplicative masking we can implement shifts with zero-fill, which would be expensive in terms of running time and moderate in terms of noise.

Notations. Throughout this report we use $[n]$ for the set $\{0 .. n - 1\}$, and use zero-based indexing for vectors. For two vectors u, v, we use $u + v$ and $u \times v$ to denote entry-wise addition and multiplication.

3   Permutations and Shift-Networks

The core of the GHS homomorphic data-routing techniques [10] is the use of Benes-like networks to arbitrarily permute the slots in a ciphertext (which is needed to allow different slots to interact with each other). In this section we describe our implementation and optimizations of the GHS techniques. We begin by introducing the notion of a *shift network*, and the shift-network minimization problem.

3.1   Shift Networks

A shift network is a method to realize an arbitrary permutation in terms of rotations, multiplicative masking, and additions. We begin by describing an arbitrary permutation in terms of a single "shift column": for an arbitrary permutation $\pi : [n] \rightarrow [n]$, the shift-column corresponding to π is a vector sh_π that describes for each index i the distance that i needs to travel under π. In formula, we have $sh_\pi[i] = \pi(i) - i$ (subtraction over the integers).

We note that a shift-column gives us a simple way of applying π to an arbitrary vector v using shift operations, multiplicative masking, and additions. Namely, for every value δ that appears in sh_π we first construct a mask m_δ which is 1 in the entries where $sh_\pi[i] = \delta$ and 0 elsewhere. We then extract from v only these entries (by multiplying $m_\delta \times v$), shift the result by δ positions, and add up all the resulting vectors. Namely the permuted vector is obtained by

$$w \leftarrow \sum_{\delta \in sh_\pi} (m_\delta \times v) \gg \delta$$

where $\times$ denote entry-wise multiplication and $\gg$ denotes shift. The running-time cost of this implementation of π is proportional to the number of distinct values in sh_π. Specifically if sh_π contains t distinct non-zero values then this implementation would perform t shift operations (and some other cheap operations that we ignore). Hence we define the *cost* of sh_π as the number of distinct non-zero values in it. The cost of this operation in terms of noise is roughly a single multiply-by-constant (since adding the resulting vector has almost no effect on the noise).

If we use rotations instead of shifts, then we can apply a similar procedure but this time use a mask m'_δ which is 1 in the entries where $sh_\pi[i] = \delta \pmod{n}$ and 0 elsewhere, then set

$$w \leftarrow \sum_{\delta \in sh_\pi \bmod n} (m'_\delta \times v) \ggg \delta$$

where $\ggg$ denotes rotation. The running-time cost of the implementation would then be related to the number of distinct non-zero values in sh_π *modulo* n, and the cost in terms of noise will be a single multiply-by-constant (since rotations and additions are cheap). We thus also define the *reduced cost* of sh_π as the number of distinct non-zero values in it modulo n.

A shift network N is a sequence of shift-columns, namely, an $n \times d$ matrix (for some d), with each column representing a permutation. If the d columns represent the permutation $\pi_1, \ldots, \pi_d$ then the network as a whole represents the composed permutation $\pi = \pi_d \circ \cdots \circ \pi_1$. We say that d is the *depth* of the shift network, and the columns of N are the *levels* of the network. The (reduced) cost of the network N is just the sum of the (reduced) costs of all levels.

A shift network for π implies an algorithm for applying π to vectors, just by applying each π_i in turn using its shift vector. If the network has depth d and reduced cost c, then this implementation of π takes c multiplicative masks, c rotations, and $O(c)$ additions, and has depth of d multiplicative masking operations, d rotations, and $O(d)$ additions.

The Cheapest-shift-network (CSN) problem. Of course there are many different shift networks that implement the same permutation, and given a target permutation π we want to find the cheapest network for it. In our setting, we typically think of the depth as a constraint and the (reduced) cost as the quantity that we optimize for. Hence we get the following optimization problem:

Input: A permutation π over $[n]$ and a depth-bound B.
Output: A shift-network for π of depth $\leq B$, minimizing the (reduced) cost.

We note that the bound parameter really does matter. For example, most permutations require a cost-$\Omega(n)$ depth-1 solution, but every permutation has a cost-$O(\sqrt{n})$ depth-2 solution (and more generally cost $O(d \cdot n^{1/d})$ depth-d solution). Even the unbounded version of this problem (with $B = \infty$) seems interesting, but in our case we are typically more interested in the bounded version. We do not know of an efficient procedure for finding the least-cost network for a given permutation and depth-bound, and speculate that it is a hard problem. Below we show, however, that when restricting ourselves to a certain natural class of solutions we can efficiently find the least-cost solution in this class.

3.2 Benes Networks

A Benes network for a permutation π is a special kind of shift network, which is rather cheap and can be constructed efficiently from any permutation. We begin by reviewing basic Benes network construction for $n = 2^r$, then describe the generalization of Chang and Melham [7] to arbitrary n and our optimizations.

For $n = 2^r$, a Benes network for a permutation π on $[n]$ is a shift network of depth $2r-1$, where every level in the network has a cost at most 2. Such a network decomposes π into $2r-1$ permutations: $\pi = \sigma_{r-1} \circ \cdots \sigma_1 \circ \sigma_0 \circ \tau_1 \cdots \circ \tau_{r-1}$, where the action of each σ_k and τ_k is to move any $i \in [n]$ to either i, $i + 2^k$, or $i - 2^k$. This is a network of depth $2r - 1 = O(\log n)$ and cost at most $4r - 2 = O(\log n)$, hence it corresponds to a fairly efficient permutation algorithm.

Decomposing a permutation into a Benes network can be done via a recursive procedure. In the first step, we decompose $\pi = \sigma \circ \rho \circ \tau$, with ρ consisting of independent permutations on the top and bottom halves of the network, and then we recurse on two halves of ρ. Computing the decomposition $\pi = \sigma \circ \rho \circ \tau$ can be done using the greedy "looping algorithm." Denote $m = n/2 = 2^{r-1}$, $S_0 = \{0 .. m - 1\}$ and $S_1 = \{m .. n - 1\}$. We seek a decomposition as above such that:

(P1) σ and τ map each $i \in S_0$ to i or $i + m$, and each $i \in S_1$ to i or $i - m$;
(P2) ρ consists of a permutation on S_0 and a permutation on S_1.

We construct an undirected graph G with $2n$ nodes, L_i, R_i for $i \in [n]$ (call these "left nodes" and "right nodes"); we add an edge from L_i to $R_{\pi(i)}$ for each $i \in [n]$ (call these "permutation edges"), and also an edge from L_i to L_{i+m} and R_i to R_{i+m} for each $i < m$ (call these "conflict edges").

It is easy to see that G is 2-colorable. Indeed, a simple algorithm to 2-color the graph is to start at any node, and trace out a path that must lead back to the starting node, alternating between permutation and conflict edges. This creates an even cycle that we can color with two colors, which we then remove from G; we then repeat the procedure on the smaller graph.

Once we have a two coloring of G with each vertex ν colored by $C(\nu) \in \{0,1\}$, we define σ and τ as follows: For each left vertex L_i, we interpret a color of 0 as τ sending i to the top half and a color of 1 as τ sending i to the bottom half. So we have $\tau(i) := i$ if $i \in S_0$ and $C(L_i) = 0$, or if $i \in S_1$ and $C(L_i) = 1$, and otherwise $\tau(i) := i \pm m$.

Similarly, for each right vertex R_i, we interpret a color of 0 as σ receiving i from the bottom half and a color of 1 as σ receiving i from the top half. Hence we have $\sigma^{-1}(i) := i$ if $i \in S_0$ and $C(L_i) = 1$, or if $i \in S_1$ and $C(L_i) = 0$, and otherwise $\sigma^{-1}(i) := i \pm m$. More succinctly:

$$
\begin{aligned}
&\text{for } i \in S_0\colon \tau(i) := i + C(L_i)m, && \sigma^{-1}(i) := i + (1 - C(R_i))m; \\
&\text{for } i \in S_1\colon \tau(i) := i - (1 - C(L_i))m, && \sigma^{-1}(i) := i - C(R_i)m.
\end{aligned}
\tag{1}
$$

Setting the permutations τ and σ determines also the middle permutation ρ (which must satisfy property (P2)) and we then recurse on the two halves of ρ.

We stress that in our setting, it is crucial that the shift amounts for the permutations σ_k, τ_k are always exactly $\pm 2^k$ and 0, regardless of the permutation π. Indeed, in the above, we recurse on two different halves of ρ, and subsequent steps recurse on a large number of different permutations. Had the shift amounts depended on the actual permutations, we would have had a higher cost for the shift-columns that implement ρ.

3.3   General Benes Networks

When n is not a power of two, we could, of course, round n to the next power of two and then apply a Benes network; however, this would effectively double the cost of implementing a permutation in our setting. Chang and Melham [7] proposed a generalization of Benes networks that works for any n, not just a power of two. Below we describe this generalization and then optimize it for our setting.

Note that the procedure above for decomposing $\pi = \sigma \circ \rho \circ \tau$ work for any even n. When n is odd, we instead break the network into two "nearly equal" parts, namely one part of size $\lfloor n/2 \rfloor$, and the other of size $\lceil n/2 \rceil$. Suppose that we let the top part be the smaller of the two, so we set $m = \lfloor n/2 \rfloor$, $S_0 = \{0 .. m-1\}$ and $S_1 = \{m .. n-1\}$. Chang and Melham observed that we can adapt the procedure from above for decomposing $\pi = \sigma \circ \rho \circ \tau$ with properties (P1) and (P2) simply by insisting that the last index, $n-1$, is mapped to itself by both σ and τ, and applying the procedure from above to all the other indexes. Formally, we construct a graph G using the same rules as above, except that we add a special conflict edge between L_{n-1} and R_{n-1} (note that none of the other conflict edges are incident to either of these two nodes). The rest of the algorithm works without any change, and correctness follows from the exact same argument.

Now that we can partition both even- and odd-size networks, we can again recurse and construct a "generalized Benes network" of depth $d = 2\lceil \log n \rceil - 1$ for any permutation. However, we no longer have the property that each level of the network only has shift amounts 0 and $\pm m$ for a single shift amount m, so we can no longer bound the cost of the network by $2d$.

Trying to bound the cost of the resulting network, we observe that all the sub-permutations at a certain level of the network are almost of the same size; specifically, they have size either $\lceil n/2^k \rceil$ or $\lfloor n/2^k \rfloor$. It follows that each level has at most four non-zero shift amounts, namely $\pm \lceil n/2^{k+1} \rceil$ and $\pm \lfloor n/2^{k+1} \rfloor$, so we can bound the cost of the network by $4d$. Unfortunately, this bound still implies a factor-of-2 slowdown when n is not a power of two. Below we describe another optimization that allows us to recover the original bound of $2d$.

Further optimizations. To reduce the cost further, we observe that there are two different options for how to split the network when n is odd, and that these two options result in different shift amounts in the shift-vectors for σ and τ. Specifically, above we made the bottom part larger, which meant setting the shift amount to $m = \lfloor n/2 \rfloor$ and fixing $\sigma(n-1) = \tau(n-1) = n-1$ by adding a conflict edge between $L_{n-1} = R_{n-1}$. However, we can also make the top half larger, setting the shift amount to $m = \lceil n/2 \rceil$ and fixing $\sigma(m-1) = \tau(m-1) = m-1$ by adding a conflict edge between $L_{m-1} = R_{m-1}$. An illustration of the two bipartite graphs and the corresponding decompositions of π that we get for a size-5 permutation can be found in Figure 1.

This observation gives us the freedom to choose the shift amounts that are used in partitioning odd-size subnetworks to either $\lfloor n/2 \rfloor$ or $\lceil n/2 \rceil$, as needed to be compatible with the even-size sub-networks in that level (if any). Thus we can recursively decompose any permutation π on $[n]$ for arbitrary n as $\pi = \sigma_{r-1} \circ \cdots \sigma_1 \circ \sigma_0 \circ \tau_1 \circ \cdots \tau_{r-1}$, where $r = \lceil \log_2 n \rceil$ and the action of each σ_k and τ_k is to move any $i \in [n]$ to either i or $i \pm \Delta_k$, with the "shift amount" $\Delta_k := \lceil \lfloor n/2^{r-1-k} \rfloor /2 \rceil$. Thus, we get a shift network for π of depth $2\lceil \log_2 n \rceil - 1$ and a cost of 2 for each level, which means a $(4 \log n)$-approximation for the *unbounded* cheapest-shift-network problem;

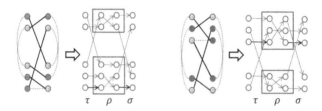

Fig. 1. Illustration of two ways to decompose a size-5 permutation as $\pi = \tau \circ \rho \circ \sigma$

3.4 Balancing Depth and Cost in Benes Networks

In our application to HE we often need to consider trade-offs between depth and cost in constructing shift networks. One natural way to enforce a depth constraint is to start from a solution to the unbounded CSN problem (such as a Benes network), and then "collapse" several consecutive levels into one, thereby reducing the depth at the price of increasing the cost.

Given a general Benes network and a bound B, we seek the "optimal way" to collapse consecutive levels so as to get a depth-B network for the same permutation. Recall that the domain size n determines the depth d of the generalized Benes network, as well as the set of possible shift amounts that may appear at each level of the network. Our approach is therefore to devise the level-collapse strategy based only on n and the bound B, rather than re-compute it for each permutation separately.

To compute the optimal level-collapse strategy for given n and B, we use a simple dynamic-programming approach. Let $d = 2\lceil \log_2 n \rceil - 1$ be the depth of a generalized Benes network for size-n permutations and let S_k be the set of shift amounts that can occur at level k in the network. (That is, $S_k = \{0, \pm\Delta_k\}$ for $k \le \lceil \log n \rceil$ and $S_k = \{0, \pm\Delta_{d-k}\}$ for $k > \lceil \log n \rceil$.)

For each pair of indexes $0 \leq j_1 \leq j_2 < d$, we let $L(j_1, j_2)$ denote the number of possible non-zero shift amounts that can occur when collapsing levels j_1 through j_2. (This number is certainly an upper bound for the cost of the corresponding shift-vector for any particular Benes network, and usually it is a fairly tight one.) Specifically, $L(j_1, j_2)$ is the number of distinct non-zero integers in the interval $(-n, n)$ that can be written as a sum $\epsilon_{j_1} + \epsilon_{j_1+1} + \cdots + \epsilon_{j_2}$, with $\epsilon_k \in S_k$ for all $k = j_1 .. j_2$. Clearly the $L(j_1, j_2)$ values can be computed efficiently (in time quasi-linear in n). Given these values, we can write a recursive formula for the optimal level-collapsing strategy for a given n, B. Specifically for each $0 \leq d' \leq d, 0 \leq B' \leq B$ let $\mathrm{Opt}(d', B')$ be the cost of the optimal way of collapsing some of the first d' columns of the depth-d network so as to get depth B'. Then we have $\mathrm{Opt}(d', B') = 0$ if $d' = 0$, $\mathrm{Opt}(d', B') = \infty$ if $d' > 0$ and $B' = 0$, and otherwise

$$\mathrm{Opt}(d', B') = \min_{\ell=1..d'} \left\{ L(d' - \ell, d' - 1) + \mathrm{Opt}(d' - \ell, B' - 1) \right\}.$$

In words, we consider collapsing the last ℓ levels into a single level of cost $L(d' - \ell, d' - 1)$, and then add to that the optimal cost for the first $d' - \ell$ levels, using the bound $B' - 1$ in place of B'.

Since there are only $O(d^2)$ values (d', B') as above, we can use standard dynamic programming techniques to compute $\mathrm{Opt}(d, B)$ and the collapsing strategy that achieves it.[1] We should note here that any $n \times d$ shift network can be collapsed to a network of depth 1 and cost at most $2n - 1$ (and reduced cost at most $n - 1$).

3.5 Hypercube Networks

A different method of constructing shift networks, which is described in [10], is via "hypercube networks": If n can be factored as $n = ab$, then we can impose on $[n]$ a two-dimensional matrix structure of a rows and b columns, using some appropriate bijective map $M : [n] \rightarrow [a] \times [b]$. Some possible choices of the map M include:

CRT order (when $\gcd(a, b) = 1$): $M(i) \mapsto (i \bmod a, i \bmod b)$;
Row-major order: M maps $0 .. b - 1$ to the first row, $b .. 2b - 1$ to the second row, etc;
column-major order: M maps $0 .. a - 1$ to the first column, $a .. 2a - 1$ to the second column, etc.

Row- and column-major orders may appear more natural, but CRT ordering (when applicable) has an advantage, because the map M is actually a ring homomorphism (viewing $[n], [a], [b]$ as the rings $\mathbb{Z}_n, \mathbb{Z}_a, \mathbb{Z}_b$, respectively). As done in [10], we will use the following decomposition lemma from [15]:

---

[1] This algorithm can be easily adapted to use reduced network costs in place of network costs, when that is the desired cost metric.

Lemma 1. *Let $S = [a] \times [b]$ be a set of ab positions, arranged as a rectangular matrix of a rows and b columns. For every permutation π over S, there exist permutations σ, ρ, τ such that $\pi = \sigma \circ \rho \circ \tau$, where σ and τ permute positions within each column, and ρ permutes positions within each row. Moreover, there is a polynomial-time algorithm that given π outputs the permutations σ, ρ, τ.*

Of course, once we decompose π as above, we can apply the same lemma recursively to each row of ρ, thus imposing an r-dimensional hypercube structure on $[n]$ and decomposing π into $2r - 1$ permutations $\pi = \pi_1 \circ \cdots \circ \pi_{2r-1}$, each of which acts along a single dimension.[2] We can then construct Benes networks for the π_i's, collapsing some of the levels within those networks so as to satisfy a bound B on the overall depth. Optimizing over this class of solutions requires finding the best splitting of n into factors, the best way to lay out the hypercube, and the best strategy for collapsing the levels of the Benes networks.

So consider $n = ab$, and a map $M : [n] \to [a] \times [b]$, which induces a correspondence between a permutation π on $[a] \times [b]$ and its representation $\bar{\pi}$ as a permutation on $[n]$. Furthermore, consider the natural generalization of the notion of a shift network to an $a \times b$ matrix: the entries in such a network are now of the form $(\Delta i, \Delta j)$, and in determining reduced costs, we consider two entries $(\Delta i, \Delta j)$ and $(\Delta i', \Delta j')$ to be equivalent if $\Delta i \equiv \Delta i' \pmod{a}$ and $\Delta j \equiv \Delta j' \pmod{b}$.

Next, consider a decomposition $\pi = \sigma \circ \rho \circ \tau$, as in Lemma 1 and let $\bar{\sigma}, \bar{\rho}, \bar{\tau}$ be the corresponding permutations on $[n]$. We can easily translate shift networks for σ, ρ, τ into shift networks for $\bar{\sigma}, \bar{\rho}, \bar{\tau}$; however, the relationship between the (reduced) costs of the shift for σ, ρ, τ and the (reduced) costs of the shift networks for $\bar{\sigma}, \bar{\rho}, \bar{\tau}$ depends on the mapping M used to impose the matrix structure on $[n]$.

CRT Order. Let λ_a, λ_b be the CRT coefficients of a, b, respectively. Then a shift amount of $(\Delta i, \Delta j)$ for a permutation on $[a] \times [b]$ translates to a shift amount that is congruent to $\lambda_a \Delta i + \lambda_b \Delta j$ modulo n for a permutation on $[n]$. Since $\lambda_a \equiv 0 \pmod{b}$ and $\lambda_b \equiv 0 \pmod{a}$, it follows that the reduced costs of the shift networks for $\bar{\sigma}, \bar{\rho}, \bar{\tau}$ are equal to the reduced costs for the networks for σ, ρ, τ. Thus, reduced costs are preserved in the translation; however, unreduced costs may not be preserved.

Row-major Order. A shift amount of $(\Delta i, \Delta j)$ for a permutation on $[a] \times [b]$ translates to a shift amount of $b\Delta i + a\Delta j$ for a permutation on $[n]$. It follows that the unreduced costs of the shift networks for $\bar{\sigma}, \bar{\rho}, \bar{\tau}$ are equal to the unreduced costs of the networks for σ, ρ, τ.

For reduced costs, the situation is a bit different. The shift networks for σ, τ have entries of the form $(\Delta i, 0)$, which translates to $b\Delta i$; it follows that the reduced costs of the shift networks for $\bar{\sigma}, \bar{\tau}$ are the same as the reduced costs of the shift networks for σ, τ. In contrast, the shift network for ρ has entries of the

---

[2] Clearly, a Benes network of width $n = 2^r$ is a special case of this construction. Unfortunately, we do not know of a generalization of Lemma 1 along the lines of the generalized Benes networks from [7].

form $(0, \Delta j)$, which translates to Δj; it follows that the reduced cost of the shift network for $\bar{\rho}$ is equal to the *unreduced cost* of the shift network for ρ.

Column-major Order. This situation is analogous to row-major order, except that reduced costs for $\bar{\sigma}, \bar{\tau}$ are equal to the *unreduced costs* for σ, τ, while for $\bar{\rho}$ we get the reduced cost of ρ.

The above observations suggest a recursive formulation to obtain a network of optimal cost for domain size n satisfying a bound B on the depth of the network. Starting from an initial domain size n, bound B, and cost metric to optimize (reduced/unreduced cost), we compare using size-n generalized Benes network to all splits $n = ab$ and all possible ways of allocating our depth budget B to the three recursive subproblems. We use row/column ordering for the $a \times b$ matrix when trying to minimize the unreduced cost, and CRT ordering when trying to minimize the reduced cost and have $\gcd(a, b) = 1$. We then recursively solve the three subproblems, trying to optimize either the reduced or unreduced cost, as needed according to the rules from above.

Let $\mathsf{SplitRcost}(n, B), \mathsf{SplitUcost}(n, B)$ denote the best reduced/unreduced cost for a size-n network with depth-bound B, and similarly let $\mathsf{BenesRcost}(n, B), \mathsf{BenesUcost}(n, B)$ be the best (reduced/unreduced) cost of a generalized Benes for these parameters. Then we have:

$$\mathsf{SplitUcost}(n, B) =$$
$$\min \left(\begin{array}{l} \mathsf{BenesUcost}(n, B), \\ \min_{\substack{ab=n \\ B_1+B_2+B_3=B}} \left(\mathsf{SplitUcost}(a, B_1) + \mathsf{SplitUcost}(b, B_2) + \mathsf{SplitUcost}(a, B_3) \right) \end{array} \right);$$

$$\mathsf{SplitRcost}(n, B) =$$
$$\min \left(\begin{array}{l} \mathsf{BenesRcost}(n, B), \\ \min_{\substack{ab=n, \gcd(a,b)=1 \\ B_1+B_2+B_3=B}} \left(\mathsf{SplitRcost}(a, B_1) + \mathsf{SplitRcost}(b, B_2) + \mathsf{SplitRcost}(a, B_3) \right), \\ \min_{\substack{ab=n, \gcd(a,b)\neq 1 \\ B_1+B_2+B_3=B}} \left(\mathsf{SplitRcost}(a, B_1) + \mathsf{SplitUcost}(b, B_2) + \mathsf{SplitRcost}(a, B_3) \right), \\ \min_{\substack{ab=n, \gcd(a,b)\neq 1 \\ B_1+B_2+B_3=B}} \left(\mathsf{SplitUcost}(a, B_1) + \mathsf{SplitRcost}(b, B_2) + \mathsf{SplitUcost}(a, B_3) \right) \end{array} \right).$$

Since there are only polynomially many (n, B) pairs, we can again use dynamic programming to solve these recurrences efficiently. We note that to count the total number of rotations required to implement a permutation on a domain of size n, the relevant quantity is the *reduced cost* of the network, i.e., $\mathsf{SplitRcost}(n, B)$. However, in calculating this reduced cost we need to know the unreduced cost of some of the subproblems that arise in the above calculation.

An illustrative timing results for some settings of the parameters are given in Table 2.

4   Replication and Linear Algebra

Since our "platform" works natively on vectors of plaintext values, it seems natural to provide support for simple vector and linear algebra operations. In this section we describe algorithmic issues in our implementation of these operations.

Table 2. Timing results for permutations in various vector sizes. The starred lines indicate that we had to choose larger parameters because of the larger depth.

Cyclotomic field	Vector size	Shift-network depth	Shift-network cost	Time
$m = 4369$	$n = 256$	3	60	4.1 sec
		7	35	2.6 sec
		10	31	2.8 sec*
$m = 8191$	$n = 630$	5	37	5.0 sec
		7	30	4.3 sec
		9	28	4.0 sec
$m = 21845$	$n = 1024$	5	66	21.2 sec
		7	45	18.3 sec*
		9	41	16.7 sec*

We begin with some basic operations for computing running sums and total sums, and then continue to replication and matrix-vector multiplication.

4.1 Running- and Total Sums

The "running sums" function $w \leftarrow \mathrm{RS}(v)$ outputs a vector w such that $w[i] = \sum_{k=0}^{i} v[k]$ for $i \in [n]$. The "total sums" function $w \leftarrow \mathrm{TS}(v)$ outputs a vector w such that $w[i] = \sum_{k=0}^{n-1} v[k]$ for $i \in [n]$. Both of these functions are implemented using a "repeated doubling" approach whose running time and depth is $O(\log n)$ additions and rotations/shifts.

Below is the code for these procedures, note that the running-sums procedure uses shifts with zero-fill (which can be implemented using rotations and multiplicative masking), while total-sums uses rotations. Here, we denote by $\mathrm{NumBits}(n)$ the number of bits in n, and $\mathrm{bit}_j(n)$ is the jth bit of n (with bit 0 being the low-order bit). The invariant throughout the total-sums procedure is that $w[i] = \sum_{k=0}^{e-1} v[i - k \bmod n]$ for $i \in [n]$; moreover, at the end of iteration j, the binary representation of e consists of bits $j \ldots \mathrm{NumBits}(n) - 1$ of n.

```
RS(v):                          TS(v):
  1  w ← v, e ← 1                 1  w ← v, e ← 1
  2  while e < n do               2  for j ← NumBits(n) − 2 down to 0 do
  3      w ← w + (w ≫ e), e ← 2·e   3      w ← w + (w ⋙ e), e ← 2·e
  4  return w                      4      if bit_j(n) = 1 then
                                   5          w ← v + (w ⋙ 1), e ← e + 1
                                   6  return w
```

We stress that although these two procedures are quite similar, the total-sum procedure uses only rotations and additions that are "cheap" in terms of noise, while the running-sums procedure uses shifts that induce "moderate" noise growth via the requisite masks.

4.2 Replication

Typical homomorphic computation has gates with large fan-out, which requires that we replicate some plaintext values many times. We have not (yet) implemented a completely generic replication method (such as the ones from [10]), but we describe procedures that we did implement for efficient replication in a few interesting special cases.

Replicating a Single Value. We begin with a procedure for replicating a single entry across the entire array. This procedure uses multiplicative masking to extract the entry, then total-sums to replicate it across the vector. It has both running time and depth of $O(\log n)$ additions and rotations and a single multiplicative masking.

Full Replication. In full replication, we take a vector v and produce vectors $\{w_i\}_{i\in[n]}$ such that each w_i has $v[i]$ in all positions. A naive solution just repeats the single-element replication n times, resulting in running time of $O(n \log n)$ additions and rotations, and n masks; and a depth of $O(\log n)$ "cheap" additions and rotations and one "moderate" masking.

We can do better than this. We begin by describing a faster, simple recursive procedure that uses just $O(n)$ additions, rotations, and masks, but has a depth of $O(\log n)$ multiplicative masking operations. We then we present a hybrid algorithm with the same linear running time, but with a masking depth of just $O(\log \log n)$.

A simple recursive procedure. Consider first the case of $n = 2^\ell$, where it is easy to apply a simple divide-and-conquer approach: in each stage we double the number of vectors while halving the number of distinct values in each vector. The following diagram illustrates this approach on a vector of size 4:

Implementing this approach, we have a recursive procedure that takes as input a vector w and an integer $h = 0 .. \ell$ (and is invoked initially with $w = v$ and $h = \ell$). The input vector w consists of $2^{\ell-h}$ repetitions of the same size-2^h vector (which we call u). The procedure computes two vectors w_L, w_R, with w_L consisting of $2^{\ell-h+1}$ repetitions of the first half of u, and w_R consisting of $2^{\ell-h+1}$ repetitions of the second half of u, and then concatenates the lists obtained by processing w_L and w_R recursively, but with h decreased by 1. The recursion stops when $h = 0$ and the singleton list $\langle w \rangle$ is returned.

RecursiveReplicate(w, h) :
```
 1   if h = 0 then return ⟨w⟩
 2   else
         set mask[i] ← bit_{h-1}(i) for i ∈ [n]   // choose half the entries
 3       w_1 ← mask × w, w_0 ← w − w_1
 4       w_L ← w_0 + (w_0 ⋙ 2^{h-1}), w_R ← w_1 + (w_1 ⋘ 2^{h-1})
 5       return RecursiveReplicate(w_L, h − 1) || RecursiveReplicate(w_R, h − 1).
```

It is not too difficult to adapt this procedure for the case where n is not a power of 2. Suppose 2^ℓ is the largest power of 2 not exceeding n. By multiplying by appropriate masks, we can construct vectors v_1 and v_2, so that v_1 equals v in the first 2^ℓ positions and is 0 everywhere else, and v_2 equals v in the last $n - 2^\ell$ positions and is 0 everywhere else. We apply RecursiveReplicate(v_1, ℓ), which gives us vectors $w_0, \ldots, w_{2^\ell - 1}$, where w_i is $v[i]$ is the first 2^ℓ positions, and 0 everywhere else. Since $2^\ell > n/2$, we can fill out the rest of each w_i as required at a cost of one mask, rotation, and addition per output vector. We apply the very same procedure to $v_2 \lll 2^\ell$, but we only need to process the first $n - 2^\ell$ vectors produced by RecursiveReplicate.

One easily verifies that the running time of this algorithm is $O(n)$ additions, rotations, and multiplicative masking; its depth is $O(\log n)$ additions, rotations, and masking.

A Shallower Full Replication Procedure. We now describe a modification of RecursiveReplicate that retains the same running time bound, while achieving a masking depth of $O(\log \log n)$, rather than $O(\log n)$. This is done by replacing the top levels of the recursive algorithm by a flatter but more time-consuming procedure (similar to the naive solution from the beginning of Section 4.2), and only switch back to the recursive procedure for the bottom few levels. We show that with a judicious choice of the number of levels to flatten, the overall running time remains $O(n)$, while the masking depth decreases to $O(\log \log n)$. Again, assume for simplicity that $n = 2^\ell$ is a power of two, and let k be a parameter, whose value we will choose to be $\log_2 \log_2 n + O(1)$.

We partition the entries in the input vector v into $n/2^k$ blocks, each of size 2^k, with block i consisting of positions $i2^k \ldots (i+1)2^k - 1$. In the first stage of the algorithm we use a "naive procedure," similar to the single-entry replication, to construct vectors v_i, $i = 0 \ldots n/2^k - 1$, where v_i consists of the entries in block i repeated $n/2^k$ times. (With our choice of the parameter $k \approx \log \log n$, this "naive part" does most of the replication work, giving us $n/\log n$ vectors with only $\log n$ distinct values in each.)

Each v_i is produced using the naive procedure, whose running time and depth are both $O(\log(n/2^k))$ additions and rotations, and a single multiplicative masking. Since we have to repeat this procedure for each v_i, the total running time of this first stage is $n/2^k \cdot \log(n/2^k)$ additions and rotations, and $O(n/2^k)$ masks. With our choice of $k \approx \log \log n$ we get running time of $n/\log n \cdot \log(n/\log n) = O(n)$.

For the second stage, we simply apply Algorithm RecursiveReplicate to (v_i, k) for $i = 0 \ldots n/2^k - 1$. The running time of the second stage is $O(n)$ additions, rotations, and masks; its depth is $k = O(\log \log n)$ additions, rotations, and masks. For example, if $n = 8$ and $k = 1$, the block size would be 2, and the first stage would produce 4 vectors. This is as illustrated in the following diagram:

4.3   Matrix/Vector Multiplication

We now proceed to describe our matrix-vector multiplication implementation, namely implementing the operation $w \leftarrow Av$ where we consider w, v as column vectors. The vectors are always encrypted, and the matrix could either be encrypted or in plaintext. The main difference between the two cases is the cost of the operations that are required to move the matrix entries around. When the matrix is encrypted, its representation (column-, row-, or diagonal-order) may have a significant impact on the cost of these data movement operations. If it is in the clear, we can use the most convenient representation.

Matrix in column-order. Assume that we are given the columns of the matrix as vectors of our underlying "platform", $A = (c_0 \mid \cdots \mid c_{n-1})$, so we have $Av = \sum_{i=0}^{n-1} v[i]c_i$. This suggests that we apply an algorithm for full replication to v, obtaining the vectors $v_0, \ldots, v_{n-1}$, and then compute $w \leftarrow \sum_{i=0}^{n-1} v_i \times c_i$. Using the HybridReplicate algorithm in §4.2, the running time of this algorithm will be $O(n)$ additions, multiplications, multiplicative masking, and rotations, and its depth is $O(n)$ additions, $O(\log n)$ rotations, $O(\log \log n)$ multiplicative masking, and a single multiplication.

Matrix in row-order. Another natural layout of A is where the rows of A are stored as vectors of the underlying "platform". In this case we could try to transpose the matrix A so as to be able use the $O(n)$ algorithm from above (or otherwise rearrange the entries of A), but this seem to require $O(n \log n)$ complexity. However, we can still get a linear-time algorithm, as follows. Suppose the rows of A are stored as vectors $r_0, \ldots, r_{n-1}$. We first compute the vectors $p_i = v \times r_i$ for $i \in [n]$. To complete the calculation, it remains to compute the entries of w by the rule $w[i] = \sum_j p_i[j]$ for $i \in [n]$. Viewing this mapping from $p_0, \ldots, p_{n-1}$ to w as a linear map, we may consider the $n \times n^2$ matrix that represents it. But observe: the transpose of this matrix represents the linear map corresponding to the replication problem; by the "transposition principle" [2,3], this immediately gives us an algorithm with the same complexity as any of our algorithms for replication: the algorithm for the transposed problem simply runs the original in reverse, with fan-out and fan-in of addition exchanging roles, and rotations having their direction reversed, and masking operations unchanged.

Matrix in diagonal order. It turns out that the most convenient representation of the matrix is diagonal order, which lets us use the parallel "systolic" multiplication algorithm, cf. [14, Figure 1-35]. Certainly, if the matrix is given to us in the clear, this is the representation of choice. As far as we know, the first usage of this method in the context of SIMD computation was in the implementations of Salsa20/ChaCha, see [1, Section 3]. We thank Daniel Bernstein for pointing out to us this method.

In detail, we represent the matrix by n vectors of the underlying "platform" $d_0, \ldots, d_{n-1}$ that contain the generalized diagonals of A, namely, $d_i = (A_{0,i}, A_{1,i+1}, \ldots, A_{n-1,n+i-1})$, so $d_i[j] = A_{j,j+i}$ (where index arithmetic is modulo n). Then the product $w = Av$ can be computed as $w \leftarrow \sum_{i=0}^{n-1} d_i \times (v \lll i)$, which takes n rotations, multiplications, and additions, and has a depth of one multiplication, one rotation, and n additions. To see that this gives the right answer, note that the j'th entry in the result is $w[j] = \sum_{i=0}^{n-1} d_i[j] \cdot (v \lll i)[j] = \sum_{i=0}^{n-1} A_{j,j+i} \cdot v[j+i] = \sum_{k=0}^{n-1} A_{j,k} \cdot v[k]$, as needed.

4.4 Performance Results

An illustrative timing results for some settings of the parameters are given in Table 3. These tests were run on a five-year-old IBM BladeCenter HS22/7870, with two Intel X5570 (4-core) processors, running at 2.93GHz. However, since HElib is (currently) not thread safe, these tests only utilized one of the eight cores available on that machine. As the operations in these procedures are "embarrassingly parallelizable" we expect that a thread-safe implementation would be about eight times faster on the same machine.

Table 3. Timing results for some operations in various vector sizes

Cyclotomic field	Vector size	Operation	Time
$m = 4369$	$n = 256$	One-Entry Replication	0.3 sec
		Full Replication	24.8 sec
		Matrix multiply	25.7 sec
$m = 8191$	$n = 630$	One-Entry Replication	0.9 sec
		Full Replication	192 sec
		Matrix multiply	84.3 sec
$m = 21845$	$n = 1024$	One-Entry Replication	3.2 sec
		Full Replication	800 sec
		Matrix multiply	473 sec

5   Computing Norms and Traces

Recall that the individual plaintext slots in a HE ciphertext can hold elements from some finite field $\mathbb{F}_{p^d}$, and that the underlying HE "platform" gives us the Frobenius operations $\sigma^i(X) = X^{p^i}$ for $i = 0 \ldots d-1$, which is applied to all the slots in a SIMD manner. These operations have the same cost as the rotation operations, namely they are "expensive" in terms of running time but "cheap" in terms of added noise.

Below we describe how to use the Frobenius operations to compute the norms and traces of the elements in the slots. Recall that the norm and trace maps are defined as follows:

Norm: $N : \mathbb{F}_{p^d} \to \mathbb{F}_p$, $N(\alpha) := \prod_{i=0}^{d-1} \sigma^i(\alpha) = \prod_{i=0}^{d-1} \alpha^{p^i} = \alpha^{(p^d-1)/(p-1)}$;

Trace: $T : \mathbb{F}_{p^d} \to \mathbb{F}_p$, $T(\alpha) := \sum_{i=0}^{d-1} \sigma^i(\alpha) = \sum_{i=0}^{d-1} \alpha^{p^i}$.

Computing traces and norms is often useful. For example, the "field switching" procedure of Gentry, Halevi, Peikert and Smart [9] relies on computing the trace. Also, computing the norm is useful in the (quite common) case where we need to compute the "not-equal-to-zero" function. That is, to map each non-zero slot to 1 while keeping the zero slots as zero, we just need to compute the function $N(X)^{p-1}$ (and in the special case $p = 2$ this is just the norm function itself). Computing the norm and trace is done directly by their definitions above, as described in the following code:

```
Norm(v):                                    Trace(v):
 1   w ← v                                    1   w ← v
 2   e ← 1                                    2   e ← 1
 3   for j ← NumBits(d) − 2 down to 0 do      3   for j ← NumBits(d) − 2 down to 0 do
 4       w ← w × σ^e(w)                       4       w ← w + σ^e(w)
 5       e ← 2 · e                            5       e ← 2 · e
 6       if bit_j(d) = 1 then                 6       if bit_j(d) = 1 then
 7           w ← v × σ(w)                      7           w ← v + σ(w)
 8           e ← e + 1                         8           e ← e + 1
 9   return w                                  9   return w
```

The running time and depth of the norm computation is $O(\log d)$ Frobenius powers and multiplications, and that of the trace computation is $O(\log d)$ Frobenius powers and additions.

Acknowledgments. Supported by the Intelligence Advanced Research Projects Activity (IARPA) via Department of Interior National Business Center (DoI/NBC) contract number D11PC20202. The U.S. Government is authorized to reproduce and distribute reprints for Governmental purposes notwithstanding any copyright annotation thereon. Disclaimer: The views and conclusions contained herein are those of the authors and should not be interpreted as necessarily representing the official policies or endorsements, either expressed or implied, of IARPA, DoI/NBC, or the U.S. Government.

References

1. Bernstein, D.J.: ChaCha, a variant of Salsa20. In: Workshop Record of SASC 2008: The State of the Art of Stream Ciphers (2008),
   http://cr.yp.to/papers.html#chacha

2. Bordewijk, J.L.: Inter-reciprocity applied to electrical networks. Applied Scientific Research B: Electrophysics, Acoustics, Optics, Mathematical Methods 6, 1–74 (1956)

3. Bostan, A., Lecerf, G., Schost, E.: Tellegen's principle into practice. In: Proceedings of the 2003 International Symposium on Symbolic and Algebraic Computation, ISSAC 2003, pp. 37–44. ACM (2003)

4. Brakerski, Z.: Fully homomorphic encryption without modulus switching from classical gapsvp. In: Safavi-Naini, R., Canetti, R. (eds.) CRYPTO 2012. LNCS, vol. 7417, pp. 868–886. Springer, Heidelberg (2012)

5. Brakerski, Z., Gentry, C., Halevi, S.: Packed ciphertexts in LWE-based homomorphic encryption. In: Kurosawa, K., Hanaoka, G. (eds.) PKC 2013. LNCS, vol. 7778, pp. 1–13. Springer, Heidelberg (2013)

6. Brakerski, Z., Gentry, C., Vaikuntanathan, V.: Fully homomorphic encryption without bootstrapping. In: Innovations in Theoretical Computer Science, ITCS 2012 (2012), http://eprint.iacr.org/2011/277

7. Chang, C., Melhem, R.: Arbitrary size benes networks. Parallel Processing Letters 07(03), 279–284 (1997)

8. Gentry, C.: Fully homomorphic encryption using ideal lattices. In: Proceedings of the 41st ACM Symposium on Theory of Computing – STOC 2009, pp. 169–178. ACM (2009)

9. Gentry, C., Halevi, S., Peikert, C., Smart, N.P.: Field switching in BGV-style homomorphic encryption. Journal of Computer Security 21(5), 663–684 (2013)

10. Gentry, C., Halevi, S., Smart, N.: Fully homomorphic encryption with polylog overhead. In: Pointcheval, D., Johansson, T. (eds.) EUROCRYPT 2012. LNCS, vol. 7237, pp. 465–482. Springer, Heidelberg (2012), Full version at http://eprint.iacr.org/2011/566

11. Halevi, S., Shoup, V.: Algorithms in HElib. Cryptology ePrint Archive, Report 2014/106 (2014), http://eprint.iacr.org/

12. Halevi, S., Shoup, V.: HElib - An Implementation of homomorphic encryption (accessed February 2014), https://github.com/shaih/HElib/

13. Hoffstein, J., Pipher, J., Silverman, J.H.: NTRU: A ring-based public key cryptosystem. In: Buhler, J.P. (ed.) ANTS 1998. LNCS, vol. 1423, pp. 267–288. Springer, Heidelberg (1998)

14. Leighton, F.T.: Introduction to Parallel Algorithms and Architectures: Arrays, Trees, Hypercubes. Morgan Kaufmann Publishers Inc., San Francisco (1992)

15. Lev, G., Pippenger, N., Valiant, L.: A fast parallel algorithm for routing in permutation networks. IEEE Transactions on Computers C-30, 93–100 (1981)

16. López-Alt, A., Tromer, E., Vaikuntanathan, V.: On-the-fly multiparty computation on the cloud via multikey fully homomorphic encryption. In: STOC, pp. 1219–1234 (2012)

17. Lyubashevsky, V., Peikert, C., Regev, O.: On ideal lattices and learning with errors over rings. In: Gilbert, H. (ed.) EUROCRYPT 2010. LNCS, vol. 6110, pp. 1–23. Springer, Heidelberg (2010)

18. Rivest, R., Adleman, L., Dertouzos, M.: On data banks and privacy homomorphisms. In: Foundations of Secure Computation, pp. 169–177. Academic Press (1978)

19. Smart, N.P., Vercauteren, F.: Fully homomorphic SIMD operations. Designs, Codes and Cryptography 71(1), 57–81 (2014)

20. SIMD. Wikipedia article (accessed February 2014), http://en.wikipedia.org/wiki/SIMD

Author Index